The Pennsylvania Anthracite Coal Industry 1860–1902

The Pennsylvania Anthracite Coal Industry 1860–1902

Economic Cycles, Business Decision-Making and Regional Dynamics

R.G. Healey

Department of Geography
University of Portsmouth

University of Scranton Press
Scranton and London

The Cover Photograph: The Brisbin Breaker, Scranton (Bunnell Collection of Glass Plate Negatives, Delaware, Lackawanna and Western Railroad Company Records, Corporate Records of the Lackawanna Division of the Erie-Lackawanna Railroad, Special Collections Research Center, Syracuse University Library – hereafter referred to in plate captions as 'Courtesy of Syracuse University Library'). Image processing and image restoration by the author. Graphic design by Grapevine-Design and Bettina Healey.

Library of Congress Cataloging-in-Publication Data

Healey, Richard G.
The Pennsylvania anthracite coal industry, 1860–1902: economic cycles, business decision-making and regional dynamics / R.G. Healey.
p. cm.
Includes bibliographical references and index.
ISBN 978-1-58966-150-9 (pbk.)
1. Anthracite coal industry—Pennsylvania—History. I. Title.
HD9547.P4H43 2007
338.2'7250974809034—dc22

2007016907

Distribution:
University of Scranton Press
Chicago Distribution Center
11030 S. Langley
Chicago, IL 60628

PRINTED IN THE UNITED STATES OF AMERICA

Contents

Contents

List of Figures

List of Plates

List of Tables

Acknowledgements

In a study of this kind, which had its origins almost thirty years ago, the accumulated debts to individuals and organizations are so numerous that it is difficult to acknowledge them all properly.

From the period when elements of this study were first conceived, thanks are due to the late Professor Richard Chorley, Professor Andrew Cliff, and Professor Graham Chapman, all at that time of the Department of Geography, University of Cambridge, for stimulating my interest in quantitative methods and the analysis of complex systems. Professor Chapman, in particular, became my long-suffering PhD. Supervisor and has been an enthusiastic supporter of this project ever since. To Derek Gregory, my thanks also, for the chance coffee room remark that led to the pages of *Pennsylvania History*, the work of Alfred Chandler, and the abiding fascination with the anthracite industry that has culminated in the present study.

The opportunity for a year of graduate study in the Department of Economics at Princeton University reinforced my enthusiasm for the study of industrial development. I was fortunate to have many of the yawning gaps in my basic knowledge of American Economic History filled by a graduate course taught by Professor Claudia Goldin, now of Harvard University. Despite the stimulating intellectual environment, as a quantitative regional geographer already tainted by the early writings of David Harvey and his colleagues, I was never quite persuaded by the prevailing paradigm of economic explanation, so ably promulgated by the Princeton Faculty members. This book is in no small measure the outcome of my struggles with these tensions, for, in the last analysis, as Richard Chorley once remarked, 'explanation is what you are prepared to accept'.

Since the late 1970s, when vacuuming the coal dust off mining company records and driving gangways into the transfer case piles of the un-catalogued Coxe Family Papers was the order of the day for visiting graduate students, a large number of people have provided every kind of library and archival assistance in tracking down and utilising the necessary record collections. In these early days, the late Revd. William Lewis and the late Hugh Gibb extended many kindnesses to me, along with Margaret Craft and Dr. Bill Siener, now Executive Director of the Buffalo and Erie County Historical Society.

More recently, the staff of a number of major libraries have provided considerable assistance. Prominent among these are the Baker Library at the Harvard Graduate School of Business, New York Public Library, Boston Public Library, the Pennsylvania Historical Society, the Library Company of Philadelphia, the Railroad Museum of Pennsylvania, and the Hagley Library. Chris Baer's unrivalled knowledge of industrial archives, and those of the Pennsylvania and Reading Railroads in particular, greatly aided the preparation of several portions of this work.

The greatest debt is due to those archives and repositories that have carried the heaviest burden of my endless requests for the company ledgers, payroll records, and faded letter copybooks, whose contents made this investigation what it has eventually become. My sincerest thanks are due to Mary Ann Moran, Executive Director, and the staff of the Lackawanna Historical Society; Jesse Teitelbaum, Executive Director, and the staff of the Luzerne County Historical Society; the staff of the George Arents Research Library, Syracuse University; Jon Stayer and his colleagues at the Pennsylvania State Archives; Lance Metz at the National Canal Museum; and Chester Kulesa, site administrator, Richard Stanislaus, Curator, and the staff of the Anthracite Heritage Museum, Scranton. Thanks are also due to the Pennsylvania Historical and Museum Commission for research funding under their Scholar-in-Residence Program 2004/5, during which substantial logistical support was again provided by Chester Kulesa. The help of Linda Shopes and Robert Dructor, Archivist, in this context, is also much appreciated. This residency has facilitated the development of an electronic reference resource on the Breakers of the Northern Coalfield, which complements material to be found in this work.

Attorney F. Charles Petrillo has generously made available both printed material and historic photographs from his collection, for which I am most grateful. John Rich, President of the Reading Anthracite Company, very kindly gave me permission to utilize the company's records at the Pottsville office, and David Christ, Company Treasurer, provided the necessary orientation. Martin Haslinsky, title searcher, enlivened many a tedious afternoon spent poring over deed book records, with his ingenious schemes for money-making in Atlantic City. Bill Best, Bill Hastie, Bob Wolensky, and John Dziak have taken a keen interest in this project and their enthusiasm to bring anthracite history to a broader audience has been an inspiration.

Over many years, successive Heads of Department and Deans have supported this investigation and the Universities of Edinburgh and Portsmouth have provided travel and research funds that greatly facilitated the necessary periods of archival fieldwork. Trem Stamp and Andy Bendall made a major contribution to the data automation aspects of this study through their work on the NE USA Historical GIS and the input of Mine Inspectors' Report data into the anthracite mines database. Latterly, considerable secretarial assistance to transform difficult drafts into publisher's text has been provided by the Department of Geography and the Faculty of Environment, University of Portsmouth.

Deryck Holdsworth of Pennsylvania State University encouraged this book project from an early stage and made a number of valuable suggestions, on the structure of the argument, that have been incorporated into the final text. Father Richard Rousseau, former Director of the University of Scranton Press supported the publication project over a number of years, and his forbearance when unexpected delays arose has been much appreciated. In the later stages, Jeff Gainey and his staff at the Press have taken on a substantial manuscript in an enthusiastic and professional manner.

As ever, the final acknowledgement belongs to the long-suffering family of the would-be author. Fortunately, my wife Bettina was aware of my commitment to 'the Anthracite' when we first met, but the gestation of the project has required considerable patience on her part, for which I am deeply grateful, not to mention her substantial and timely help with the final stages of manuscript editing. Statistically, in a work of this size, some errors are almost bound to remain, despite the utmost care in checking of statistical information from original sources (and that derived from earlier studies). The responsibility for these errors is, of course, my own.

The cover photograph is of the Brisbin Breaker of the Delaware, Lackawanna and Western Railroad Coal Department in Scranton (*courtesy of the Lackawanna Historical Society*).

Trafalgar Day, 2005
Portsmouth

Chapter 1
INTRODUCTION

OVERALL AIMS OF THE STUDY

This study has two main aims. The first is to provide a comprehensive examination of the role of economic and geographical factors in the development of the Pennsylvania anthracite coal industry during the latter part of the 19th century. Particular emphasis is placed on explaining both the regional and national contexts of the industry and the interactions between them, mediated as they were by the role of major anthracite carrying railroads. The second aim is to provide a substantial worked example of an approach to the analysis of regional economic development that has not previously been applied in the fields of historical geography or economic history. This approach derives from the pioneering work of Clark, Gertler, and Whiteman on adjustment theory, an analytical approach that will be outlined below.[1] That said, the study does not aim to be a theoretical treatise, since the author believes that engagement with the actors in the real drama will be more interesting and ultimately more convincing for the reader than endless wrestling with abstract concepts of social science or economics. Nevertheless, the underlying theoretical perspective will form a connecting thread throughout the argument as it develops.

A TALE OF TWO REGIONS

A comprehensive analysis of the anthracite or hard coal industry during the period of interest must take into consideration two distinct regional contexts. The first comprises the anthracite coalfields themselves. The second is the much larger region within which anthracite coal was distributed and marketed, both to industrial and domestic consumers.

The anthracite coalfields are of particular significance from a geographical perspective, because of their uniqueness as a 19th century industrial economy. At this time, 96 percent of the world's known anthracite reserves were concentrated in an area of less than 500 square miles of North-Eastern Pennsylvania. Since anthracite, unlike bituminous coal, required the use of expensive coal breaking and sorting equipment on the surface before it could be marketed, the growth of the industry also produced a remarkable industrial landscape of coal 'breakers'. These were large and costly structures, spread across the coalfields in their hundreds, but found nowhere else in the world. Aside from their impact on the landscape, they also had major economic implications in terms of production costs and costs of entry into the industry.

Initially, because of their proximity to eastern seaboard markets, the anthracite coalfields were of fundamental importance to the early phases of U.S. industrial expansion in the ante-bellum period. By the time of the Civil War, improving transport links in the direction of Buffalo and the Great Lakes ushered in a period of progressive expansion of markets for hard coal. This distribution region eventually encompassed large parts of the U.S. manufacturing belt between New York and Chicago. It also included portions of southern Canada, areas accessible to more remote ports on the Great Lakes, such as Duluth, and more distant coastal locations accessible by sea. The growth of the anthracite iron industry and a significant proportion of the railroad development of the North-Eastern United States were dependent on the anthracite coal industry, as were millions of urban dwellers (for domestic heating) along the eastern seaboard.

The distribution region was lent coherence by the activities of the major railroad shippers of anthracite in the East, who operated as a cartel for significant portions of the period of interest.[2] The same railroads also directly or indirectly came to control large tracts of lands in the coalfields, a trend that accelerated in the 1870s. In this way, they also performed a crucial co-ordinating role in linking the development dynamics of both the mining and the distribution regions.

ADJUSTMENT THEORY

In the course of this study, it will become clear that regional historical geography, with the particular economic emphasis developed here, is a rather different project from that major part of economic history, which takes the focus of contemporary economics on efficient resource allocation and applies it in past circumstances. Instead, this investigation aims to analyse the lived experience and decision-making strategies of economic actors in past places, and elucidate the processes by which these same actors developed, reproduced and transformed their geographical circumstances to create distinctive regions. Of necessity, the adjustment theory approach adopted here constitutes a significant challenge to the prevailing orthodoxy of mainstream economic theory, as applied in the regional domain. The reason for this is, in essence, quite simple. Neo-classical economic theory, and its mathematical formulations in particular, rely on simplifying assumptions to enable the logic of the theory to work. These assumptions exclude, *a priori*, precisely the kind of geographically specific disequilibrium processes that are shown here to be the driving forces behind the growth in the anthracite coal industry during the latter part of the 19th century. By extension therefore, the study aims to offer a stimulus to new kinds of thinking and new modes of analysis, not only in historical geography but also in economic and business history.

The significance of this challenge should not be underestimated, even by those readers who have marginal interest in demarcation battles at the boundary between subject disciplines. Busy with their rapprochement with sociology, under the watchful eyes of Anthony Giddens and David Harvey, many geographers seem to have paid little attention to the wide theoretical chasm that has opened up between their field and that of economics. A cursory reading of recent geographical literature might lead one to imagine that neo-classical economics had been abandoned as a theoretical project some time in the 1970s, shortly after the publication of Hollis and Nell's critique of concepts of rationality in economics.[3] Nothing could be further from the truth, according to a recent commentator, who stresses the central position of this theoretical framework within contemporary economics.[4] Since much of the more analytical work in economic history derives its theoretical inspiration from economics, this divergence in methodological approach has major consequences for a constructive dialogue between economic historians and historical geographers.

The development of adjustment theory by Clark and his colleagues provides the stimulus for the theoretical perspective that underlies this study, although it was originally expounded in the context of contemporary, rather than historical regional analysis. The theory is based on a trenchant critique of the fundamental assumptions of neo-classical economics concerning perfect competition and instantaneous adjustments of supply and demand in a spaceless, timeless world. They argue convincingly and with extensive empirical support that it is not equilibrium, but the disequilibria generated by responses to change in regional economic systems, that should be the focus of analysis. In the present context, the importance of their argument cannot be overstated. Quite simply, neo-classical economics denies a role in economic behavior either to temporal or spatial dynamics, leaving no analytical 'space' for regional historical geography. In contrast, when the existence of multiple disequilibria is admitted, the resulting adjustment processes over time and space suddenly become a major subject for investigation in their own right and a huge field for work in regional historical geography unfolds for the potential investigator. Clark et al.'s framework for analysing the effects of disequilibrium tendencies addresses problems of risk and uncertainty and aims to examine how the strategies and actions of individual workers, firms, and investors combine to produce the aggregate economic landscape. These actions in turn may produce unintended and unanticipated outcomes, which themselves necessitate further strategic or tactical adjustments in business activity. In consequence, decision-making, which directs action into an uncertain future, while being constrained by the results of past decisions, becomes the main focus of the investigation. This disaggregation of the market into its constituent actors, whose interactions are time- and space-dependent, opens the door for a very rich analysis of processes of region-building.

While not cast in sociological terms, and avowedly economic in emphasis, it is apparent that there is a substantial role in the theory for human agency, either in an individual or corporate form. At the same time, economic decision-making and action is confined by a range of structural constraints. Many of these will be specifically geographical, e.g., the location of investment in new plant, or the network of linkages between firms, which will in part be mediated by access to transportation networks. Over time, these structural constraints will themselves be modified as a result of future economic action. Hence adjustment theory meshes quite closely with aspects of social theory, as a more economic counterbalance to the social processes stressed by Giddens himself.[5] Also it has the important advantage that it translates readily into empirical enquiry, unlike Giddens' own conceptual framework, which Gregson has argued cannot in fact be deployed in any direct way in support of empirical investigation.[6]

On the basis of this discussion it is perhaps difficult to understand why Clark et al.'s work has not been adopted as the basis for other studies over the last two decades. One possibility is that adjustment theory has been erroneously classified with earlier simplistic quantitative analyses in economic geography, because it uses sophisticated multivariate statistics in its empirical analysis, when in fact the key contribution of the theory lies in its insight into economic processes at the regional level. A further unstated reason that may explain its lack of adoption in subsequent studies is that the data requirements for empirical investigation of adjustment processes are extremely demanding, and very few data series of the quality and duration used by Clark et al. are readily available. It should be noted however, that although their data are greatly superior to those in most other studies, the temporal resolution is quarterly or annual and the spatial resolution is usually no better than the state level. It is not difficult to argue that data at the level of the firm and on a monthly basis would greatly enhance the analysis of adjustment processes, but the data handling implications are obvious! Fortunately, in the present study, the availability for individual firms of large volumes of quantitative accounting data and qualitative commentary from correspondence, on an annual, monthly, and even weekly basis, means that in a number of cases, much more detailed identification and tracing of adjustment processes is possible, than was the case in the original investigation of present-day regional dynamics.

It is also most important, in the present context, to underline that the theoretical arguments of adjustment theory have never been countered from within either economics or economic geography and no better or contrary empirical findings have ever been produced. This is not to imply, in any sense, that the theory is flawless. Rather it is to suggest that it provides at least as good, and probably a better basis for theoretically informed empirical investigation in historically oriented regional economic analysis, than any other available approach. Most particularly, this is because it allows, and indeed encourages, disaggregation of analysis down to a level at which the key role of geography in economic processes becomes very clear.

A TALE OF MULTIPLE BUSINESS CYCLES

An immediate corollary of adopting an adjustment theory approach for the present study is that a clear path is opened up for linking broad scale economic changes at the national level to more specific regional scale adjustments and disequilibrium processes. The study therefore begins within an overview of business cycle changes in the second half of the 19th century to provide necessary context for the ensuing analysis.

At the national scale, the period of interest was characterised both by standard cyclical fluctuations involving short-term (40-month) business cycles, 10-year Juglar cycles, Kuznets (15–20-year) construction cycles and possibly the more contentious Kondratieff cycle (long wave). Superimposed on these was the impact of non-recurrent factors, such as the speculative and inflationary boom of the Civil War years, and the differential effects of labor disputes across the coalfields in the post-bellum period.

Against this background of a fluctuating economic environment, the time period was also one of extremely rapid development of population, transport, industry, and urban centers across the anthracite distribution region,[7] and there was a progressive re-orientation of industry within the manufacturing belt from serving regional to serving growing national markets.[8]

WAXING AND WANING MARKETS

While the increasing integration and extension of the Eastern railroads, which continued into the 1880s, facilitated the expansion of markets for anthracite, it also enabled bituminous coals from western Pennsylvania and Ohio[9] to enter into competition for anthracite. This was in marked contrast to the 1860s when the production of anthracite coal from a very limited geographical area, matched that of bituminous coal from the entirety of the rest of the United States.[10] From the mid 1870s, metallurgical coke from the Connellsville region[11] powered the rise of the iron and steel industries of Pittsburgh and Youngstown, Ohio. Coke progressively supplanted anthracite as a blast furnace fuel, even in iron-producing areas which had traditionally used anthracite, such as the Lehigh and Schuylkill Valleys of Eastern Pennsylvania.[12] At the same time, the Virginia Pocahontas coals[13] took an increasingly large share of the coastal steamship trade, leaving anthracite with declining industrial markets, but growing domestic demand in all accessible urban centers, as the 19th century drew to a close. It is interesting to note that these changes in specific markets across the distribution region fed back into the mining region, necessitating changes in the proportions of coal of different sizes shipped from the mines. The balance between the larger and smaller sizes swung towards the latter, as domestic demand grew to fill the gap left by declining sales of larger sizes to industrial markets. Demand for specific sizes in individual geographical markets could fluctuate quite rapidly, yet mines could not make major changes to the size proportions of their output over short time-spans and individual mines varied in their normal proportionate production of the different sizes. In consequence, a series of adjustment processes on different time scales were involved in trying to match the overall outputs of groups of mines to the varying requirements in specific markets.

CYCLES OF COAL LAND ACQUISITION

Within the Coal Regions themselves, national scale dynamics had effects that reached right down to the level of individual land tracts, mine 'patches', and new mine construction operations. Speculative cycles of coal land acquisition by railroads, mining companies, and individual wealthy investors were mirrored, after time lags of varying duration, by construction cycles of new investment in mining capacity. This in turn led to the establishment of small mining settlements, which affected the relative growth or decline of larger urban areas. At the same time, the economic experience in different parts of the coalfields varied substantially, from retrenchment and reorganization in the Southern Field, where individually owned mines came under corporate control,[14] to rapid and sustained expansion of mining activity in the Northern Field.

THE RISE OF THE CARTELS

Cartel formation by the anthracite carrying railroads, set against a background of cyclical changes in demand for coal, had a much more complex genesis than has been apparent from most previous writings on the subject.[15] The initial cartel was probably the first of its kind covering an entire industry in the United States,[16] although evidence is presented that cartel-like organizations had been developed by Northern Field companies a decade earlier to control the western anthracite trade.

Patchy success in restricting production to bolster prices during the 1870s was followed by more effective cartel operation during the 1880s. Relatively stable prices and rapid growth in shipments and revenues in this period provided the springboard for several of the railroads to transform themselves from regional coal carriers into members of the national trunk line cartels.[17] A number of factors were involved in this process and two stand out as being of particular importance. The first was control of routes into New York and major terminal facilities on the eastern seaboard. The second was the timing of construction of new rail routes from the anthracite fields to Buffalo, in relation to business cycle peaks and troughs, which proved to be of major long-term commercial significance. At Buffalo, connections could be made both with western railroads and with the Great Lakes vessel trade. Although the establishment of the ICC and the Sherman anti-trust laws made formal combinations illegal, they were continued informally to the end of the period and beyond by the anthracite railroads, through the activities of a 'shell organization' called the Temple Iron Company.[18]

MANAGEMENT OF TRANSPORTATION FACILITIES, THE COAL ESTATE, AND INDIVIDUAL MINE OPERATORS

The senior managers of the anthracite railroads, Presidents, Vice Presidents, and General Coal Agents, were thus involved in tactical and strategic decision-making in a very dynamic and unpredictable business environment. This extended over all geographical scales from individual mines and coal land tracts, through to the full extent of the anthracite distribution region. It also ranged over all temporal scales, from the daily vicissitudes of securing enough empty coal cars for efficient operation of the mines, to long-term strategic planning of the development of a company's coal properties and avenues to market. The time horizon for this strategic planning could extend over a decade or more.

The decision-making also took place against the background of a wide range of constraints. Some of these were fixed and immutable, such as geology and topography. Others were the result of past decisions on land purchase or the location of mining developments and railroad lines, and yet others were of a temporary nature, resulting from financial stringency, labor shortages or production/transport stoppages due to fire, flood, or accident. Hence, management decision-making by major companies, much of it directly or indirectly associated with the formation and operation of cartels, was a key factor in co-ordinating the dynamics of the mining and distribution regions. At the same time, individual coal operators also had a significant part to play in regional development, though their economic influence declined steadily during the period. They were especially important in developing new mining capacity during business cycle upswings and their coal tonnage was of considerable importance to the railroads that competed for their business. As a result, the contractual and financial links they developed with the coal-transporting railroads became major consumers of management energy and decision-making time on both sides.

Approached in this way, the study is ideally placed to utilize the insights of adjustment theory, which focuses on decision-making under constraints. It allows synthesis of the relevant spatial and temporal dynamics, which range from the local scale to that of an area larger than the manufacturing belt, and from days to decades. It also addresses the processes of adjustment to these dynamics, some of which were endogenously 'given' at the national scale, others of which arose endogenously within the two main regions of interest. These processes of adjustment reflected a range of managerial issues. Prominent among these were firstly, the difficulty of matching often rapidly fluctuating demands for coal with the complexities of managing extended transport and distribution systems. Secondly, the problems of deciding when and where to make substantial, long-term, and immovable investment in new mining capacity, in the context of a highly competitive land-ownership and leasing system. In the event, managers found themselves persistently unable to achieve equilibrium between supply and demand, diagnosed variously by commentators as overproduction, overcapacity or overdevelopment of the mines.[19] However, such a finding is not just an empirical inconvenience (or death knell) for the neat theoretical models of the neo-classical economist, rather it is real economics by Marshall's original definition, as 'the study of men (sic) as they live and move and think in the ordinary business of life'.[20] Equally, it is most definitely real regional geography, in the sense of an accurate representation of the responses of corporate and individual actors in real decision environments, whose actions led to the building of the regions in question.

THE ECONOMIC AND THE SOCIAL

It is apparent from the foregoing that the major processes and interactions that will form the substance of this study are very largely in the economic rather than the social domain. This is not to deny or decry the importance of other approaches to the mining industry from labor or ethnic history perspectives. Rather, it is the economic and managerial aspects of the industry that have previously been neglected. As a result, some studies seem to take place in an environment where wider economic forces appear almost incidental to the operation of the industry. In the light of the evidence to be presented here, such a view cannot be sustained. Nor, in the context of 19th-century regional development in the coalfields, can primacy in explanation be ceded to class struggles, ethnic conflicts or unionization, important though these are in both this and other contexts. For example, advances in unionism associated with periodic crises in production and fundamental shifts in technology, as described by Warf[21] for the Pacific Northwest timber industry over a longer time horizon, were simply not observed in the anthracite industry during the restricted time period of interest. The Workingmens' Benevolent Association of St. Clair and the Knights of Labor came and went,[22] but it was not until the rise of the United Mine Workers at the end of the 1890s and the intervention of President Roosevelt during the 1902 Coal Strike[23] that the anthracite coal companies were finally compelled to recognize the right of an external union organization to deal with them on behalf of their 'own' men. Even after this, as Blatz has observed, many of the demands of the men were never met by their employers, such were the entrenched attitudes of the corporate managers in the anthracite industry.[24] In its most extreme form, this intransigent attitude was embodied in the person of G. F. 'Divine Right' Baer, President of the Philadelphia and Reading Railroad Company, who believed that the rights of the working man were best safeguarded not by unions, but by the owners 'to whom God in his infinite wisdom has given the control of the property interests of the country'.[25] Nor was there any greater change in the localized social structure of the workplace than in the external structure of employer-employee relations. The system of contract mining was in place throughout the period. In this system, individual miners employed their own laborers and the amounts paid for specific types of work could vary between veins and even workplaces within the mine, subject in some cases (though by no means all) to negotiation between miner and inside foreman on an almost daily basis.

A number of previous studies have considered a range of other socio-economic issues in the coal fields, ranging from politics and draft resistance during the Civil War to ethnicity and the role of women in anthracite communities.[26] The changing nationalities of different immigration streams had a progressive impact on the spatial and socio-cultural differentiation of settlements and urban districts within the coalfields. However, the impact of these changes, especially those involving eastern and southern European migrants, became more apparent at the very end of the period and on into the 20th century.[27] They had little or no impact on the course of the major region-building phases of the 1860s–1880s, after which the economic structure of the coalfields was largely set, until the segregation of the railroads from their associated coal mining departments, or subsidiary companies, during the 1920s. In the mining region, therefore, the direction of causality tends to run from the economic to the social, rather than the reverse. Industrial activity was unrelated to the earlier agricultural society of the area. It was firmly based on coal and powered by speculation, investment and the growth of the mining and railroad corporations. The deep financial pockets of the latter were able to weather most, if not all, of the inevitable economic fluctuations, as well as the occasional costly vicissitudes of the mining business. It was the perceptions and calculations by mining engineers and managers, of the present and future value of specific coal lands in relation to existing coal estates that governed how the geography of the coalfields progressively came under the control of the major companies.
A combination of management decision-making and entrepreneurial enthusiasm, if not foresight, initially governed where and when new mine developments took place, although the inexorable accounting of the lease royalty system played a progressively larger role in this over time.

Into this industrial system poured successive waves of immigrants from Germany, the British Isles, eastern and southern Europe, and Russia. Hoping for work, though scarcely a better life, in the difficult and sometimes appalling conditions in the mines, their labors tunnelled and blasted the countless miles of underground gangways, chambers, and airways and constructed huge coal breakers strung out along the Appalachian valleys as far as the eye could see. They came when good times prevailed and moved west when they did not, or stayed to endure poverty and privation until better times returned. Much of the same could be said of many of the individual entrepreneurs, for this was an industrial region that came to belong not to them, but to the major corporations. The logic of its development was largely theirs, both in terms of competition and of combination. Their economic power successfully resisted the rise of unions for decades and it took the President of the United States to bring them even temporarily to heel. Social and ethnic integration, and indeed solidarity, was slow in coming against such a backcloth and in the face of such rapid economic development and change. Until after the turn of the century, the production and reproduction of social life in the coalfields was thus largely disconnected from the economic structures of the capitalist corporate culture that drove the region forward. It will therefore not be a major focus of the present study.

However, the more rarefied processes of production and reproduction of the economic structures of the industry, involving mine owners, coal associations, and the operations of the cartels, do merit close attention. Important aspects of these range from considerations of 'regional elites',[28] to questions of industrial embeddedness, untraded dependencies between firms, power relations, interlocking directorates,[29] and communities of financial interest, comprising groups of shareholders of the major corporations.

In terms of the anthracite distribution region, similar issues arose in connection with the wider structure of railroad organizations, such as the Eastern Trunk Line Association of the 1870s, the later regional associations of the 1880s and the informal communities of interest of the 1890s.[30]

To summarize therefore, this study is concerned with the development of two interacting regions, one of very limited area, the other of very wide and changing geographical extent, akin to the 'doily' concept of Allen et al., used to describe discontinuous regions, which contain holes between the connecting links.[31] The key to the overall approach is a focus on the role of business decision-making across a range of temporal and spatial scales. This decision-making took place in the context of business cycle fluctuations that originated at the national level, but whose impacts could be felt back down through all scales to the local. Corporate managers, entrepreneurs, and investors were constantly at work devising adjustment strategies, in the often vain attempt to match coal production and production capacity to market demand, both in the short and the long term. Often the strategies of different economic agents overlapped, confounded one another, or produced unexpected and unintended consequences for groups of producers. These 'interaction' effects necessitated the adoption of yet more counter-strategies and corrective actions. Many of these strategies were profoundly geographical: the acquisition of lands accessible to existing railroad lines, the siting of new mines, and the construction of new railroad avenues to distant markets, while simultaneously capturing as much monopolistic 'line' trade as possible. They were also subject to fixed constraints of geology or topography. As time went on, the locational consequences of past decisions and adjustment strategies also narrowed the range of available decision options for the future. This continued until the patterns of economic activity and associated settlements became almost fixed, in a landscape legacy that survived in large measure until at least the middle of the 20th century.

Chapter 2

RECURRENT AND NON-RECURRENT ECONOMIC FLUCTUATIONS AT THE NATIONAL LEVEL

INTRODUCTION

It is a matter of everyday experience that the path of economic growth and development is not smooth, but is punctuated by periods of relative stagnation as well as periods of accelerated activity. This applies both at the national and international level. Changes at the latter level may or may not be in step with national-level developments, but it is unusual for the two levels to be entirely decoupled.

The 19th-century experience was in essence little different, despite the lack of sophisticated regulation of banks and securities markets. Financial crises were sometimes more extreme than at the present day and the international transmission of economic impulses was slower, at least before the coming of the transatlantic telegraph after the Civil War. However, significant interdependence of national economies, both within the European sphere and between the United States and Great Britain, was already well established by the 1850s.[1]

Two initial questions can be asked about these fluctuations in business activity. The first is over what timespans they occur and the second is whether or not they occur with sufficient regularity to merit being called cycles or waves. The answers to these questions are important both to the detached student of economic growth and the committed participants in the process, the entrepreneurs, bankers, investors, and speculators who will base their decisions on future expectations of likely changes in business activity. Also, it should be borne in mind that patterns which intrigue the scholar may be irrelevant to business decision-makers, since Keynes has already accurately forecast the outcome for us all in the long run.

It was not until 1862 that a tacit acceptance of the inevitability of business fluctuations, coupled with an optimism that better times were just around the corner, was replaced by the first serious attempt to track detailed processes of economic development. In that year, Juglar published his treatise on the periodic return of commercial crises in France, England and the United States.[2] Some years later, in 1884, the British economist W. S. Jevons found evidence of a long wave in British price data from 1790 to 1849. The first 28 years to 1818 were characterized by a general rise in prices, followed by 31 years in which they tended to fall.[3]

Apart from these early pioneers and treatises on specific crises such as Gibbons' fascinating study of the 1857 banking panic,[4] the 19th-century businessman had little to assist him in interpreting his own experience of changing business conditions. Not until the work of Kitchin,[5] Kuznets,[6] and Kondratieff,[7] did a clearer picture emerge of the different timescales over which business fluctuations take place. From these studies, a standard sequence of nested 'cycles' is now generally recognized, although the application of the term 'cycle' to the longer-term fluctuations has been called into question by more recent research.

The recognized cycles are as follows:[8]

i. The Kitchin or inventory cycle, 3–5 years in length.
ii. The Juglar or investment cycle, 7–11 years long.
iii. The Kuznets or building cycle, of 15–25 years duration.
iv. The Kondratieff cycle or long wave, whose proponents claim it to last for 45–60 years.

For candidates to merit the name of cycle in economic terms, two conditions need to be fulfilled. Firstly, there must be a defensible means of identifying them by statistical analysis of time series data. Secondly, it must be possible to specify the economic processes, either singly or in combination, which give rise to the observed periodic behavior. In the next sections of this chapter the evidence for the existence or otherwise of these cycles is reviewed briefly. This is followed by a more detailed examination of the major events associated with either recurrent or non-recurrent shocks to the economic system in the second half of the 19th century. A comprehensive analysis of this topic is beyond the scope of the present study, so the focus will be on those issues which are most pertinent to the development of the mining, railroad, and related industries. The final section of the chapter examines business-cycle-related flows of human and financial capital into the United States, as these had a significant impact on the growth of the latter industries.

ECONOMIC 'CYCLES' OF DIFFERING DURATION

The Kitchin cycle, although it was not the first to be identified, has perhaps the strongest empirical basis and the most convincing economic rationale. In part, this is a result of its short duration, which allows a number of cycles to be examined even when available economic time-series are not of very great length. Kitchin[9] found a pattern of fluctuation in growth rates extending over 3 years or more. The fluctuations appeared to be driven by problems of managing business inventories. Recovery from periods of recession left businesses short of stocks in raw materials, components, and finished goods. Their replacement orders stimulated demand throughout the economy, encouraging the placement of additional orders until supply overshot demand, with a resulting cut-back in orders and so on. The resulting pattern of overshoots and undershoots constituted the cycle.[10] This cycle is what is commonly understood as the present-day 'business cycle'.

The longer Juglar cycle is a function of investment decisions related not to inventories but to fixed assets, such as machinery and equipment. It is not concerned with investment decisions of longer duration, associated with new buildings and the construction of entire new manufacturing plants. Samuelson's well-known multiplier-accelerator model offers a simplified mathematical demonstration of how such a cycle can be generated, if consumers base their purchases on last year's income and producers base their requirements for capital stock investment on the previous year's output. Clearly, other factors such as the cost of borrowing and levels of retained earnings will also have an impact, but the underlying mechanisms are

TABLE 2.1
Timing of 'Cyclical' Peaks and Troughs.[13]

	Kitchin	Juglar	Kuznets	Kondratieff	Currie's Rainfall Cycle
1856		P			
1857	P				
1858	T				
1859					
1860	P				
1861	T		T		T
1862					
1863					
1864					
1865	P	P			
1866				(P) prices*	
1867	T				
1868					
1869	P				
1870	T				
1871			P		P
1872					

TABLE 2.1 *(Continued)*

	Kitchin	Juglar	Kuznets	Kondratieff	Currie's Rainfall Cycle
1873	P	P		P	
1874					
1875					
1876					
1877					
1878			T		
1879	T				
1880					T
1881					
1882	P	P			
1883					
1884					
1885	T				
1886					
1887	P				
1888	T				
1889		P			
1890	P				P
1891	T				
1892			P		
1893	P				
1894	T				
1895	P			T	
1896					
1897	T				
1898			T		
1899	P	P			T
1900	T				
1901					
1902	P				
1903					
1904	T				
1905					
1906					
1907	P	P			
1908	T				P
1909					
1910	P				
1911					
1912	T		P		
1913	P	P		P	

*Peak of 2nd Kondratieff cycle according to Kondratieff,[14] based on price data.

broadly plausible from an economic perspective.[11] Although Juglar himself published his original study in the early 1860s, it has proven relatively straightforward to identify Juglar peaks, if not troughs, running throughout the latter part of the 19th century. These are given in Table 2.1. Other evidence of the cyclical behavior of prices for producer durable goods in the 1890s can be found in Hoffman.[12]

The Kuznets, or building cycle, as it is sometimes called, was the last of the four types to be identified by economists and has, or had, its clearest expression in the US economy. Easterlin[15] quotes Abramowitz's unpublished definition of these cycles as:

> 'swings in the level of growth of a variable with a duration longer than the normal business cycle, but

shorter than the very long swings with which Kondratieff, Schumpeter and others have been concerned'.

In the early work of Kuznets,[16] swings in prices, output, and profitability were emphasized, but later workers such as Abramowitz[17] laid stress on the role of population-sensitive investments such as housing construction. These were in turn related to immigration movements, after a time lag. This approach has been adopted by later workers such as van Duijn,[18] but is contested by Solomou who believes that the early Kuznets approach is correct, while the demographic emphasis is misplaced. Solomou's careful study sheds important new light on Kuznets cycles. He finds them to be more evident in the European economies of France and Germany than was previously thought, and more pervasive, in terms of the economic variables where they can be identified. Their widespread appearance across a number of national economies leads him to reject the argument that migration-related processes are the dominant force behind them. From a detailed statistical analysis, he is able to provide strong empirical evidence supporting their existence. He is also able to explain the differences in the average reported lengths of the cycles—22 years by Kuznets and 14 years by Abramowitz,—in terms of the different statistical methods used by these authorities to identify them. Most importantly, in the present context, Solomou is unable to find empirical support for any of the earlier theories of the Kuznets cycle which are mono-causal, in terms of migration,[19] the stock of money, or sectoral terms of trade. Instead he argues that the swings:

> 'were generated by the economic system in a process that entails a series of shocks feeding through the economy in a system of organic causation'.[20]

Further to this the pattern of growth identified by the label of the 'Kuznets swing' is only sustained over time by a series of comparable shocks. In other words, the swing does not have a simple economic mechanism behind it. It may be recurrent, but it is not cyclical, in the sense of being predictable in advance. Therefore its precise form at any moment in time is dependent on the nature of the most recent shock and the time elapsed since its occurrence.

The existence of the Kondratieff cycle or long wave has been the most contentious of the four possible cycles, not least because its very long duration means that very few such cycles can have taken place since reliable records of economic activity began. Kondratieff himself sought evidence for long waves in both the USA and the major European economies. He also examined a range of commodity production series such as coal and pig iron, in addition to wages and prices. However, it is price data that is usually taken to exemplify the approach graphically (Figure 2.1).

Kondratieff argued that during the rise of the long waves, years of prosperity are more numerous, while depression years predominate during the downswing. He also identified several characteristics of different stages of the long waves. For example:

1. Agriculture tends to suffer badly during the recession phase of the long wave.
2. Numerous important inventions are made in the recession phase, which find large-scale applications only at the beginning of the upswing.
3. Gold production tends to increase at the beginning of a long upswing.
4. During the rise of the long wave, 'high tension in the expansion of economic forces' leads to major wars and revolutions.[21]

The last of these points is of particular interest, as it implies very clearly that the direction of causation runs from the economic to the social. At the same time, it recognizes that once wars have occurred, they have a potent influence on subsequent economic dynamics.[22]

Later commentators have endeavoured to provide a theoretical basis for Kondratieff cycles. Extended treatment of these arguments can be found in the major texts already quoted, especially van Duijn[23] and Tylecote.[24] The former, for example, examines Rostow's theory that the observed price cycles are caused by under- and over-shooting of optimum levels of capacity and output for food and raw materials in the world economy. The under- and over-shooting results from three types of lags inherent in the system. These are as follows:

1. A recognition lattice—the time between recognition of the emergence of a profit opportunity and the making of investment decisions which aim to exploit the opportunity
2. A gestation lattice—the delay while the necessary infrastructure is constructed to allow production to begin

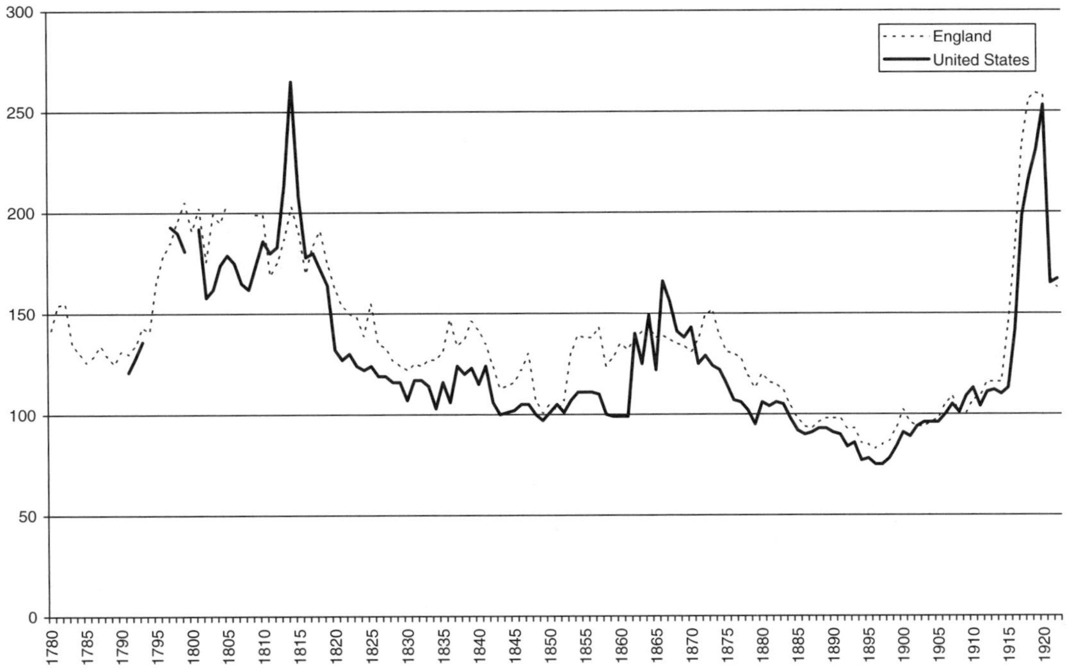

FIGURE 2.1
Kondratieff's Index Numbers of Commodity Prices.

3. An exploitation lag—the time between the onset of initial production and the maximum efficient exploitation of the investment.[25]

It is entirely appropriate to draw attention to these real-world economic processes and they will feature prominently under different empirical guises in later chapters of this study. However, to demonstrate the importance of their operation is one thing, to demonstrate that they necessarily produce a cycle of approximately 50 years in duration is quite another. Tylecote puts forward a somewhat different approach to explanation of the long-wave phenomenon, which stresses the evolutionary nature of economic systems. It therefore rejects the notion of a standard set of economic processes that have been operating unchanged over the last two centuries. Instead, basing his work in part on the French regulation theorists, he develops an explanation in terms of a succession of 'technological styles'. Each style is characterized by the appearance of new 'key factors of production', which are low cost by previous historical standards, have a tendency to become even cheaper, and are potentially all-pervasive.

During the 18th and 19th centuries he identifies three styles. The first was the water style. This was based on improvements in river and canal navigation and technological developments in harnessing water power. From the 1840s this was progressively replaced by the steam transport style—the railroad and the steam engine,—although water power was still important in many industries as late as the 1870s. In each of these two cases, the reduction in transport costs was the key factor, while the railroad also enabled coal to be brought to locations remote from water power at a reasonable cost that declined over time. The third style was that of steel and electricity. Although Henry Bessemer had invented his steel-making process in 1854, it was the beginning of the 1880s before cheap steel made a major impact on markets formerly served by wrought iron. The 1880s also heralded the arrival of electricity as a source of power, light, and communication, following the development of the electric motor, filament lamps, and the telephone in the preceding decade. The fourth style did not emerge until beyond our period, when Henry Ford developed the assembly line in the early 20th century.[26]

Tylecote also dates the onset of the different styles to about 1785, the end of the 1820s, the late 1870s and 1915 respectively, to give inter-style intervals of 45, 47, 38–39 and 63 years, which would broadly fit with Kondratieff timescales. However, he fails to note that although the timing of Kondratieff price swings is closely coincident in both Britain and the United States, the canal era can scarcely be claimed to have begun in the USA until 1817, when construction of the Erie Canal was initiated.[27] His argument is therefore tantamount to giving causal economic power to ideas some 30 years prior to their actual implementation. It is difficult to accord this any measure of credibility.

A second measure of empirical support for long-wave processes offered by Tylecote refers back to Kondratieff's second point about inventions made during the recession phase, which are widely adopted on the upswing. Using Baker's time series of significant patents 1775 to 1965, he identifies one surge in innovations around 1840, just before the start of a Kondratieff upswing, and a second surge from 1875 to 1905. According to the information in Table 2.1, this latter phase of consistently high levels of patents starts at the Kondratieff peak. It continues unabated through the downswing and halfway through the next upswing. The lack of correspondence, between patents and the start of the water style, he attributes to the delaying effect of the Napoleonic Wars.[28] It would therefore appear that the proportion of years of patent data that are at variance with the theory, are as great as those that support it!

A similar conclusion using different data on innovations is reached by Solomou.[29] He can find no evidence of innovation being responsive to long-wave phases. Solomou also undertakes a detailed statistical test of Kondratieff's hypothesis that years of prosperity cluster on the upswing and years of depression on the downswing. Using the Juglar cycle peak years and looking at US GDP figures, he can find no evidence for the existence of a Kondratieff cycle in overall economic output. Turning his attention to price data, he is able to confirm Beenstock's earlier findings that prices have not followed a long-wave pattern. Overall, he concludes that to claim the existence of Kondratieff waves between 1856 and 1937 misrepresents history. Instead, price behavior should be viewed as a series of shocks, followed by long eras of price stability. In the long run, price variations result from the interaction of several forces: real and monetary, aggregate and sectoral.[30]

A number of important conclusions follow from these studies of economic cycles. Firstly, to summarise the above findings, both Kitchin and Juglar cycles can be identified in the historical record, associated with inventory management behavior and fixed-asset investment respectively. The Kuznets long swing can likewise be shown to be present in the 19th century, but according to Solomou it is not a cyclical phenomenon. Rather it is episodic and its apparent persistence is dependent on the processes of adjustment to a series of economic shocks experienced over time. The only question mark that can be raised over Solomou's conclusion follows from Currie's meteorological study, published at a later date. Currie found that the planetary-generated 'luni-solar tide', with an 18.6 year cycle, has a pronounced effect on US rainfall, crop production and hence the US economy in general.[31] This conclusion is especially important for the 19th century, when agriculture was a much larger proportion of GDP than at the present day. The final column in Table 2.1 shows the peaks and troughs for this cycle in the latter part of the 19th century. Earlier peaks were experienced in 1834 and 1852, with a trough in 1843.

The timing of peaks and troughs of Currie's cycles merit closer investigation in a detailed historical context and this will be undertaken in a subsequent section. Nevertheless, even from a cursory examination of the coincidence of rainfall and economic cycles, it is apparent that the possibility of an underlying cyclical process, which has implications for the periodicity of the Kuznets swings, cannot now be discounted.

Although a variety of plausible mechanisms that might contribute to Kondratieff long waves have been proposed, there is no obvious reason why these should produce regular cycles approximately half a century in length, and statistical support for the existence of the long wave is lacking. The only type of evidence linking economic processes to a 50-year periodicity comes from the work of the MIT System Dynamics school in their work on the US economy.[32] The MIT model produced fierce fluctuations in the interaction between the consumer sector and the capital goods sector at intervals of about 50 years.[33] However, as Solomou remarks, 'simulation is one thing, the empirical verification of a long wave something very different'.[34]

If long waves cannot be identified, the observed price fluctuations, in particular, must be episodic shocks followed by periods of adjustment and relative stability. If the shocks have no rationale in terms of economic theory, then presumably they are attributable to external causes. Unlike the Kuznets swing, no relevant climatic effects can be discerned. This leaves the major upheavals of the Napoleonic Wars, the American Civil War, and the First World War, which happen to have occurred at 50–60 year intervals, as the major sources of external shocks. It is interesting to note that on Kondratieff's price index graph (Figure 2.1), sharp price increases are apparent in Britain, whose timing is coincident with the onset of the Crimean War. The overall effect in the United States is much more muted at the time, and sharp price increases are delayed by almost a decade until the onset of the Civil War. Hence, although Kondratieff and a number of his disciples wished to give causal primacy to the economic system, and make wars and strife a function of economic upheaval, it would seem that the weight of more recent analysis would reverse the direction of causality.

These findings are of fundamental importance for the overall approach to be followed in this study. Of the four types of possible economic fluctuations, the longest does not seem to exist, while the Kitchin cycle and Kuznets swing were not identified until the 20th century. The more intellectual class of banker and entrepreneur in the latter part of the 19th century might just have heard of Juglar's work, although the fact that it was written in French would have reduced its accessibility to many people. Certainly few, if any, would have been willing to base significant investment decisions on it, since as the old investment adage states, the past is not necessarily a guide to the future.

With the benefit of hindsight, the present-day analyst may be able to use these cycle concepts to understand more clearly why certain developments occurred when they did. In contrast, the only expectations we can have of 19th-century economic decision-makers are firstly, that they would have some long-term strategies, based on a general optimism about eventual economic growth. Secondly, they would anticipate that the path into the future would be characterized by ups and downs, but the timing, even in a general sense, was unknown and unknowable. Thirdly, that the actual daily grind of economic life was essentially one of reaction, response, and adaptation to changing economic circumstances as they presented themselves through a succession of dull and active markets, punctuated by occasional and severe crises of business confidence. Those who succeeded, were those who most effectively deployed a range of short-term tactics, to deal with the episodic shocks to which they were exposed, without losing sight of longer-term business goals. The combined actions and interactions of the more and less successful actors in the economic drama, be they managers, entrepreneurs, bankers, investors or ordinary labouring people, generated the overall patterns of adjustment to these unpredictable shocks, that we can now observe in the historical record. Many of the decisions involved nvestments in activities such as mines and railroad construction, which could have long gestation lags and were also fixed geographically, once they were sunk or built respectively. As a result, we can have no expectation of a rapid attainment of some kind of economic equilibrium after each shock. Much more likely, the effects of adjustments to one shock will still be in train when the next shock arrives. This suggests that persistent maladjustment of productive capacity to market demand may be observed in the aggregate, even if decision-makers individually respond in a sensible manner to the prevailing circumstances as they see them. Further to this, the persistence of maladjustment may itself become the source of new actions by decision-makers, either individually or as part of a group. Whether these actions prove beneficial or not may depend on the nature and timing of future shocks!

A final point to note in this context is that although the analysis of economic fluctuations at the national level has attracted a very large literature, there is no comparable body of work that examines the impact of these fluctuations on industrial development at the regional level in the United States. Niemi's[35] study of state and regional patterns of manufacturing growth does not employ a business-cycle perspective and important studies, such as those by Borts[36] and Isard,[37] which touch on these issues, are concerned with 20th-century developments. Only the work of Marshall[38] combines an examination of long waves and economic cycles with uneven patterns of regional development in the 19th as well as the 20th centuries, but his argument is developed in the context of the UK, not the USA. This study also confines itself to broad-brush analysis of variables, such as intercensal employment change. It does not deal in detail with specific regions or with the relationships between entrepreneurial activity and economic fluctuations.[39] In consequence, we have remarkably little systematic knowledge about the relationships between cycles, swings, and financial crises at the national level and their effects at more regional and local levels in the 19th-century United States.

The previous discussion has demonstrated the importance of episodic shocks in the process of economic development and the historically contingent nature of the responses of economic actors. The first stage in examining the interrelationships between the national economy and the present case study of the anthracite coal regions is therefore to outline the major features of the observed cycles and shocks, to provide the context for the more detailed argument of later chapters.

EPISODIC SHOCKS, FINANCIAL CRISES, AND THE ECONOMIC IMPACT OF THE CIVIL WAR

Although the Kitchin, Juglar, and Kuznets cycles are associated with different economic processes or combinations of processes, several authorities have argued that they are closely related, nesting one inside the other. Hence Schumpeter took the view that there were three Kitchins to each Juglar, while van Duijn, by slightly arbitrary selection of average cycle durations, suggested a neat arrangement of two Kitchins per Juglar and two Juglars per Kuznets swing.[40] Any simple scheme of this kind is bound to have observed exceptions, but reference back to Table 2.1 does give a measure of support to van Duijn's approach. For example, the peak to peak Juglar cycles of 1856–1865, 1865–1873 and 1882–1889 each contain approximately two Kitchins. However, in the period 1873–1882 the Kitchin and Juglar seem to correspond. For the trough to trough Kuznets swing 1861–1878 the van Duijn scheme seems to hold, whereas, taken peak to peak 1871–1892, slightly more than two Juglars are covered.

Several other short periods of time stand out from the table in terms of the coincidence of peaks or troughs of different cycles. For example, 1861 stands out as a trough year for Kitchin, Kuznets, and Currie cycles, and the Juglar trough (not specified) was doubtless at around the same time also. The more extended period 1871–1873 shows a broad coincidence of cyclical peaks, followed by a similar grouping of troughs in 1878–1880. Thereafter the pattern becomes more scattered, until marked coincidence of peaks is again observed on the eve of the First World War.

It is interesting to compare the timing of these cycles and periods of cyclical coincidence with other measures of variation in business activity, which perhaps approximate a little more closely to the lived experience of the business entrepreneur than to the stylized facts of the economist theorist. The first example, shown in Figures 2.2a and 2.2b, is Ayres' index of American business activity, which is redrawn from a little known study dating from the dark days of the Great Depression.[41] This index is based on a number of price and output series, weighted and rebased to give an overall picture of fluctuations about the average level of business activity over a long span of years. It also predates the 'official' NBER business cycle chronology and some of its peaks and troughs do not correspond exactly with the latter. However, it does indicate noticeable fluctuations on timescales even shorter than that of the Kitchin cycle, which may nonetheless be significant in a short-run business context. Owing to data limitations in the earlier years and the amount of manipulation required to bring quite disparate data series down to a common base, it would be unwise to set too much store by the implied mildness or severity of specific fluctuations on the graph. Rather it is an aid to grasping the general sequence of economic shocks in the second half of the 19th century.

The second example, shown in Figure 2.3, highlights one aspect of 19th-century price behavior, namely monthly anthracite pig-iron prices. These are the original raw data and have specifically not been subject to any smoothing or the application of price deflators, to provide a contrast with the processed data in Figures 2.1 and 2.2a/b. A first striking characteristic of the graph is clearly the inflationary 'spike' in prices in the latter part of the Civil War in 1864–1865. Equally important are the lesser peaks, at approximately decadal intervals from the mid 1840s to the beginning of the 1880s, which correspond closely to Juglar peaks. Thereafter, a long period of secular slow decline in prices set in, ignoring the Juglar peak of 1889. No significant upward tendency becomes apparent until after the Juglar peak of 1899. However, the double peak of the graph at the start of the new century was more a function of the anthracite coal strikes of 1900 and 1902[42] than of any wider business developments.

Taking the graphs together, a number of different periods of growth, depression, and crisis can be identified. In general, the early 1850s were prosperous, following the mild depression of 1848–1849, with minor recessions in 1851–1852 and again briefly at the end of 1854. Pig-iron prices fell from the late 1840s until conditions began to improve in 1852. Better prices were sustained until

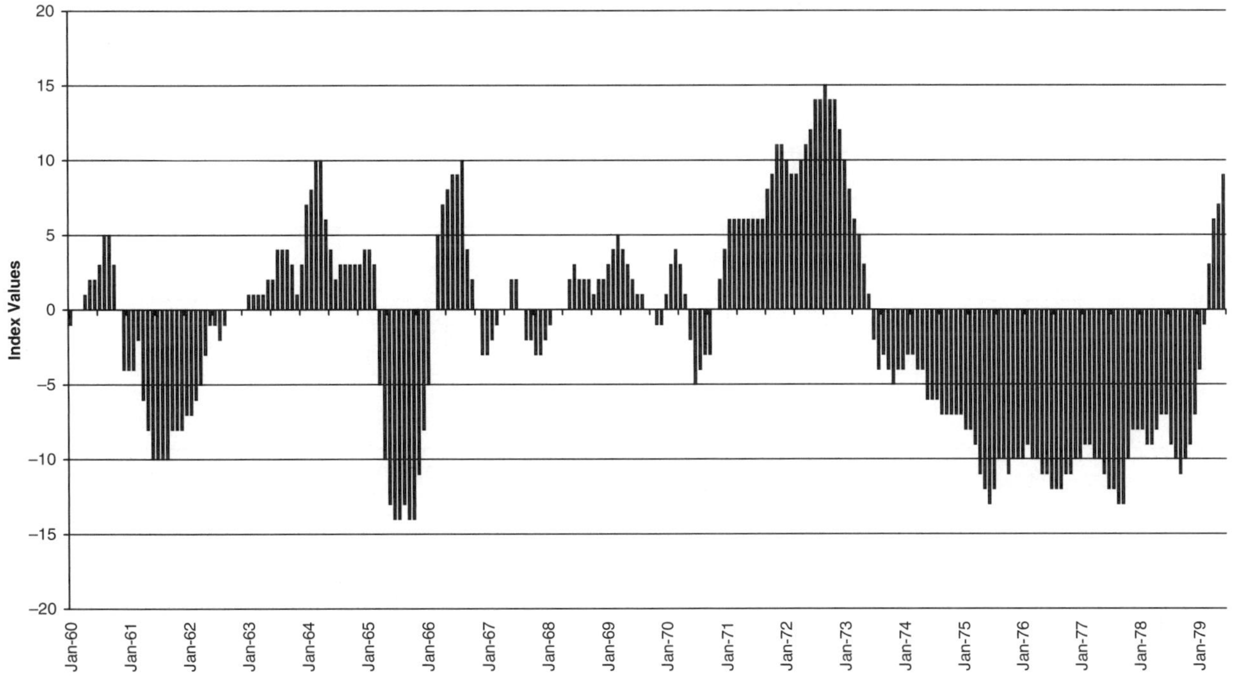

FIGURE 2.2a
Ayres Index Business Activity Fluctuations 1860–1879.

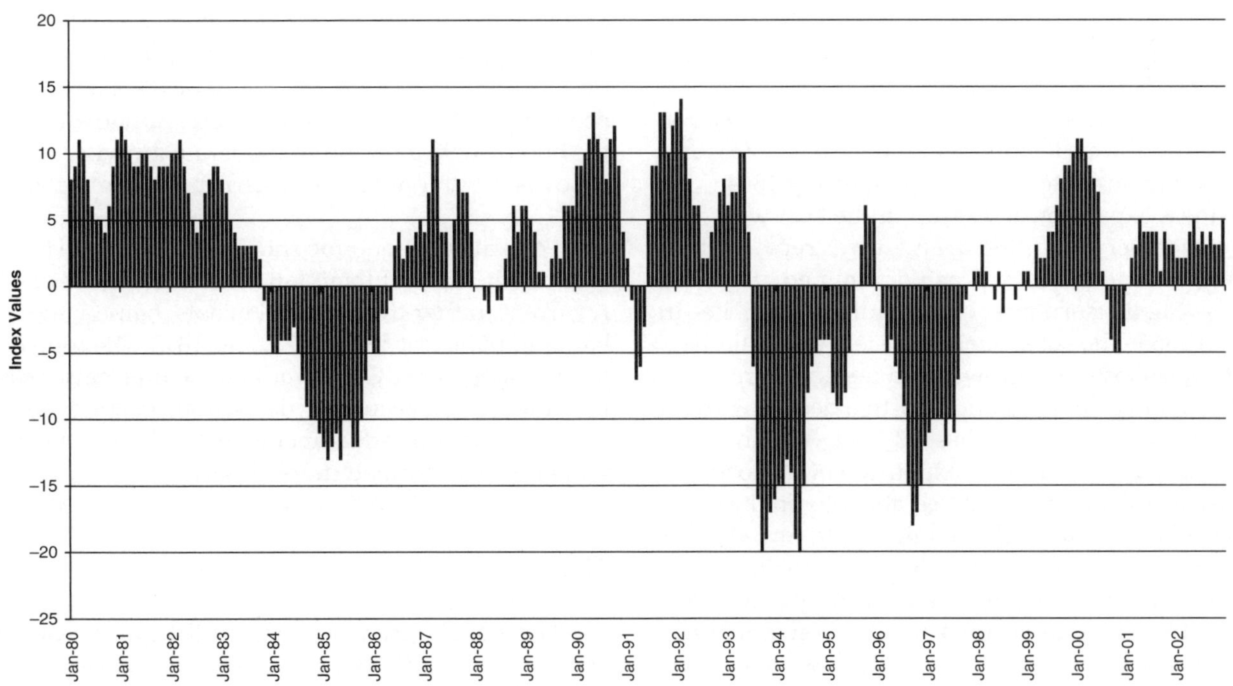

FIGURE 2.2b
Ayres Index Business Activity Fluctuations 1880–1902.

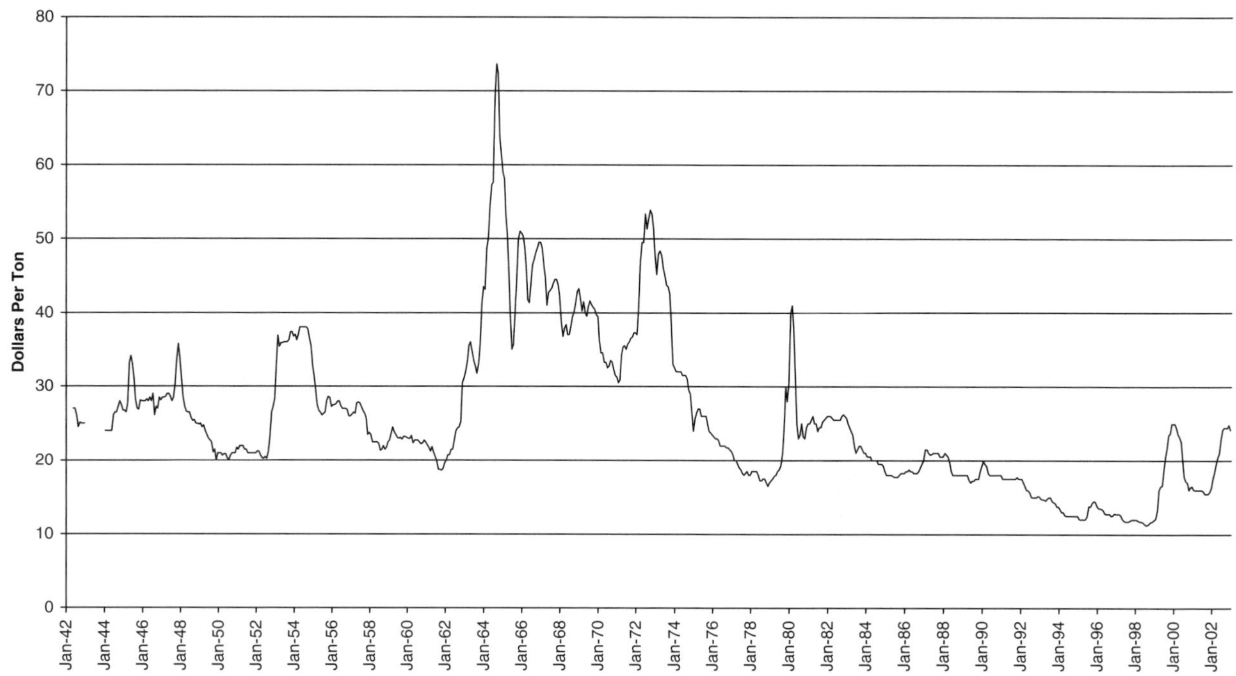

FIGURE 2.3
Monthly Prices of No. 1 Anthracite Foundry Pig Iron in Philadelphia 1842–1902.

1855. Thereafter, prices began to fall again in a step-wise fashion, declining even further following the 1857 banking panic. After the sharp depression following the panic, growth was minimal before output declined again during the depression that accompanied the onset of the Civil War. These adverse conditions were reflected in low but stable prices until war demand and inflationary pressures pushed them up to dizzy heights during 1864. Business expansion as a result of the War was, however, very short-lived, and it was replaced by a sharp decline of business activity immediately following the surrender of General Robert E. Lee in April 1865. The subsequent transient oscillations between short-term growth and decline were reflected in price oscillations with a declining trend, until the expansionary phase of 1871–3 set in. This was closely related to the expansion of railroad construction but it terminated abruptly in the Jay Cooke Panic of 1873. This panic was followed by the longest unbroken period of depressed business activity in the 19th century. It only drew to a close when the return to the gold standard on January 1, 1879 brought a rapid return of business confidence and a further railroad construction boom. This was heralded by a short-term hike in pig-iron prices at the beginning of the 1880s and it ended in the middle of the decade. No further extended periods of depression placed a major restraint on economic development until the onset of the depression of the 1890s, which was marked by the 1893 banking panic. Unlike the uninterrupted gloom of the late 1870s, the 1890s depression had two distinct phases separated by a brief revival in the latter part of 1895. After the economy finally emerged from recession towards the end of 1897 a period of moderate growth followed, with only minor interruption, until 1902.

Four major economic shocks can therefore be identified. Three of them fall into the category of recurrent, rather than cyclical events, namely the banking panics of 1857, 1873, and 1893. The fourth, in the shape of the Civil War and its aftermath, was fortunately non-recurrent. These four major events and the economic adjustments that followed them will now be examined more closely.

THE 1857 PANIC

The inappropriateness of simplistic mono-causal explanations of episodic shocks has already been highlighted in earlier sections, so any discussion must take account of a range of antecedent conditions, as well as the conjunction of a number of factors which led to the specific timing of the panic.

A first monetary stimulus to the economy from the beginning of the 1850s was the discovery of gold in California and Australia. It has been found that an annual increment of 3 percent or more in gold stocks in the period resulted in price rises, and between 1848 and 1857 the increment was in excess of 5 percent.[43] If banks had increased gold reserves they could extend their loan portfolios, especially to the agricultural sector. These favorable monetary conditions coincided with the Currie rainfall cycle peak of 1852, and circumstances therefore encouraged agricultural expansion. This in turn fostered new railroad schemes. The regional impact of these can be seen in Figures 2.4 and 2.5, which show steady investment during the middle years of the 1850s in the eastern Midwest and sharp peaks in investment in 1853–1854 in the Middle Atlantic states and to the west of Chicago. Revisiting the Ayres index of business activity (Figures 2.2a/b), it is tempting to associate monetary factors with the first growth surge of the 1850s and railroad development with the second. Based on the railroad figures and the collapse of investment after 1854, Fishlow has argued that the 1857 panic would have occurred earlier had it not been for events on the other side of the world.[44] In October 1853 the Crimean War began when the Turks declared war on the Russians. Britain and France joined the conflict in March 1854 and it continued until the signing of the Treaty of Paris in March 1856.[45] The period of the war coincided with poor grain harvests in France throughout the period 1852–1856. Similar problems were experienced in Britain in 1852 and 1853. In the winter of the latter year, more foreign grain was imported than during the famine years of the 1840s. Thereafter, war-related demand and inflationary price increases produced very favorable commercial circumstances for US grain producers, especially in western areas, who found a ready export market.[46] Although Fite[47] has argued that the new railroads of the period were built 'in advance of the needs of the country', there can be no doubt that they facilitated this unanticipated growth in the export trade.

Unfortunately for western farmers, this export market quickly dried up after the end of the war, when European and Russian producers rehabilitated their agricultural economies and harvests temporarily improved. In the year to June 30, 1857, Britain imported 8.56 million bushels of grain, but by 1859 this had fallen to under 1 million bushels. Huston supports the business journalists of the period, who argued that these agricultural market

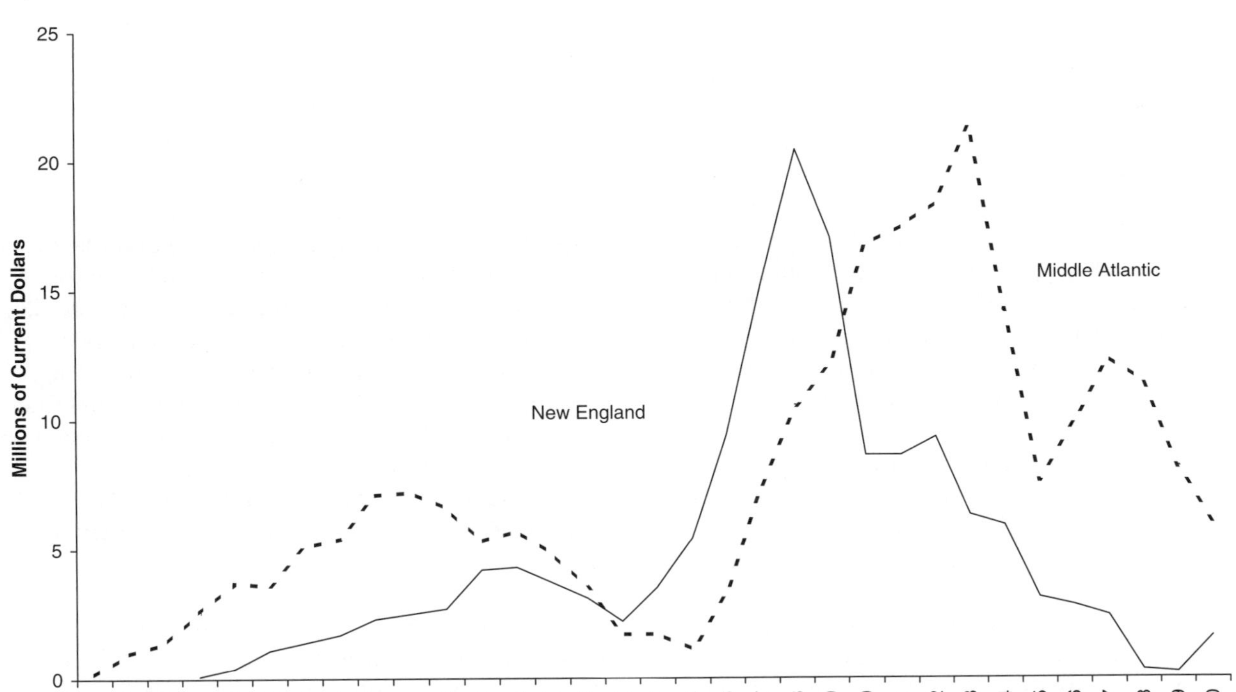

FIGURE 2.4
Gross Railroad Investment in the North-Eastern States 1828–1860.

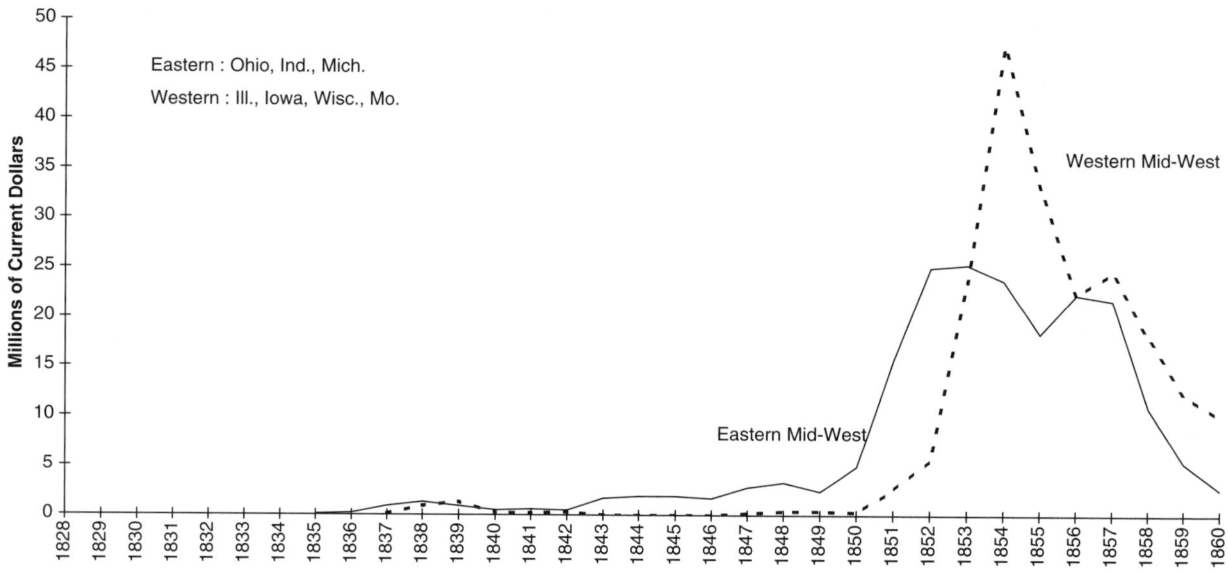

FIGURE 2.5
Gross Railroad Investment in the Mid-West 1828–1860.

changes were at the root of the 1857 crisis and that the banking crisis stemmed from this.[48] However, it is important to stress the role of speculation as a mediator in this process. Speculation was encouraged by rising commodity prices because of war-related inflation. It was also fuelled by a peak in western sales of public lands in 1854–1855. Calomiris and Schweikart also stress the importance of changing expectations of land speculators, especially in Kansas, as a precipitating factor in the panic.[49] Rising grain exports made the new railroads seem attractive investments and rising share prices attracted further speculation. Unfortunately, chronic shortages of gold at the local level, despite the overall increase in supply at the national level, had resulted in many states allowing agricultural banks to hold part of their reserves in securities instead of specie. This placed them at risk of failure in the event of rapid share price declines.[50] A further factor that weakened the banking system was a failure to grasp the full significance of the growth of demand deposits. The required 25 percent specie reserve was linked to notes in circulation, not deposits. Hence if a 'run' on the bank occurred, all depositors could simultaneously demand repayment in gold.[51] Also, banks began making increasing volumes of loans on the basis of these deposits. Gibbons, in his classic treatise on the New York Banking System, records that in 1855 the specie reserves of the New York banks did, in fact, amount to 25 percent of deposit capital, but by 1856–1857 this had dropped to 18 percent and loans were still trending upwards.[52]

Such a state of affairs with rampant speculation and over-extension of credit could not be sustained indefinitely and required only a small reversal to trigger a collapse of confidence. This duly occurred on August 24, 1857 with the collapse of the Ohio Life Insurance and Trust Company, amid allegations of embezzlement by one of its employees.[53] The New York banks immediately began calling in loans to improve their reserve positions. Initially this was at a faster rate than depositors were withdrawing funds, but the tide began to turn on September 26, when the Philadelphia banks suspended specie payments. Three days later the panic hit Chicago and by October 13, despite earlier statements that they would not suspend, the New York banks were forced to follow suit. Resumption of specie payments did not take place until December 14, and trade was paralyzed throughout the country during that interval.[54]

Despite Temin's claim that the collapse was a minor episode in the economic history of the period, it was two years before there was a return to full employment.[55] Likewise Van Vleck[56] reproduces a table from the Banker's Magazine, which shows almost 5,000 banks and mercantile firms failed in 1857, with aggregate liabilities of nearly $292 million.

A further 4,225 firms failed the following year, but the aggregate liabilities were much smaller at $95 million. It is interesting to note that while the worst of the damage in New York State, Pennsylvania, and Ohio was confined to 1857, many of the western states, such as Michigan, Illinois, and Wisconsin, that had boomed in the mid-1850s, suffered even more severely in 1858 than they had in the immediate aftermath of the crash itself. Specific examples in the anthracite fields, both of the impact of the speculative mania of 1855 and of the collapse of 1857, will be discussed in chapter 4.

Recovery in many branches of industry became apparent by 1859 when coal and textile production passed 1856 levels.[57] Improvement in the iron industry overall was delayed until 1860, although this was largely because early recovery in anthracite iron was offset by continued decline in charcoal iron production.[58] It is also interesting to note that, owing to the continued buoyancy of the cotton trade, new railroad construction in the south continued unabated in 1858–1859 and exceeded that of the north during this period.[59] A prosperous South, a fragile recovery in the North and the election of Lincoln were to provide the background for the next unexpected shock, as the storm clouds of secession appeared in the latter part of 1860.

THE ECONOMIC IMPACT OF THE CIVIL WAR

This topic has attracted scholars with diverse viewpoints over a period of many years and it is difficult to do it justice in a brief section here. Good summaries of the major issues can be found in works such as Andreano,[60] Gilchrist and Lewis,[61] and O'Brien.[62] For present purposes, the key aspects of the argument revolve around whether or not the War retarded or accelerated growth. O'Brien[63] has summarized the relevant issues in the form of three hypotheses:

1. The war interrupted the boom under way before hostilities commenced, which would have continued.
2. Since booms always come to an end under normal circumstances anyway, the long-term effect on the overall economy of the premature end of a single boom was marginal.
3. The war removed potential bottlenecks to growth, so that, over time, the economy sustained even higher rates of growth than would have been possible in the absence of armed conflict between the North and the South.

Some of the differing views on these hypotheses can be characterized by reference to the classic papers of Cochran[64] and Salisbury.[65] Cochran bases his arguments on several data series analysed decade by decade, or at 5-year intervals. In the iron industry, for example, he finds 17 percent growth 1855–1860, 1 percent growth 1860–1865, and 100 percent growth 1865–1870. The not-unrelated series for new miles of railroad added for equivalent periods shows figures of 8,721, 4,076 and 16,174 miles added respectively. He therefore argues for a retardation of growth by the war. Salisbury interprets similar figures slightly differently, by comparing the decade of the 1850s with that of 1865–1875 immediately following the war. On this basis he finds, for example, a 50 percent growth in pig-iron production in the first period, followed by a more than doubling of output in the second. For railroad miles, he finds 20,000 miles added in the first period and 40,000 in the second. Hence he stresses the post-war period experienced a substantial boom with growth rates much above those of the pre-Civil-War era.

Unfortunately, in the light of the earlier discussion on business fluctuations, use of such arbitrary time slices clearly has major limitations for interpretative purposes. Reference back to the earlier Figure 2.2a of the Ayres Index for the period 1860–1879 shows a much more complex pattern of growth and recession, where positive and negative influences may cancel out to create an appearance of overall stagnation. The index plots support the notion of a fledgling recovery in 1860 (the first substantial improvement since 1856), which was immediately brought to a close by Lincoln's election in the fall of 1860. Quite a sharp contraction followed, leading to a business panic and the suspension of specie payments in cities such as Philadelphia and Baltimore, but not in New York.[66] Perhaps as a result of absence of suspension in the latter city, this panic could almost be classed as 'forgotten' in comparison to the major upheavals of 1857, 1873 and 1893. However, the number of business failures in the north in 1861 was 5,935, almost 50 percent greater than in 1857, although the liabilities were smaller at $178 million. In 1862 the comparative figure was 1,652 failures, but the rate of failure diminished steadily after the introduction of Greenbacks in April 1862. Some 8 percent of Boston business houses failed, and 5 percent of those in New York met the same fate; a process that was accelerated by the repudiation of $300 million of debt owed to Northern firms by those in the South.[67] The succession of panics over such a short period left a veritable army of unfortunate bankrupts, conservatively estimated at 35–40,000 people in total, to be pursued around the fringes of commerce by their creditors.[68] Such conditions were scarcely auspicious for a return of business confidence, especially when they coincided with military setbacks for the Union in the early part of the war. Nevertheless, a faltering recovery took hold in 1863 after the Battle of Gettysburg. It gathered momentum in 1864 on the back of easy paper money, price inflation, and lucrative government contracts for military clothing, coal and armaments, before coming to an abrupt halt on cessation of hostilities in April 1865. As wartime demand evaporated, prices fell. Ironworks stopped almost immediately to force wage reductions, which matched the falling prices. When the anthracite companies of the Northern coalfield similarly reduced wages, strikes ensued and output fell to a 'dribble'.[69] Hence it is misleading to use 1865 as a start or endpoint of calculations on growth rates associated with the war period. If used as a start point, it will overstate future growth, if used as an endpoint it will under-estimate the extent of expansion in the later stages of the war. Thus, in the period 1860–1864, pig-iron production rose by 23.5 percent, but production fell off by 200,000 tons in 1865[70] to yield the minimal increase 1860–1865 identified by Cochran. The sharp and disruptive contraction in 1865 also explains how apparent absence of growth cited by some authorities can co-exist with the contemporary reports of business prosperity during the later stages of the War in the North.

Further to this, there is well-documented evidence of prices rising faster than wages during the inflationary period of 1863–1864. Fite estimates that a 50 percent price rise in basic commodities was associated with only a 25 percent increase in wages. However, some skilled groups such as iron moulders fared better, with wage increases up to 56 percent.[71] Also, wage increases lagged behind price increases.[72] Further to this, in some industries, such as anthracite mining, there was an element of monopoly over transportation by the railroad corporations. Fite argues that this enabled prices to rise by more than that of benchmark commodities such as flour, where no monopoly existed. So, while flour barely doubled in price, Schuylkill white ash lump coal in Philadelphia, which had sold at $3.28 per ton by the cargo in January 1860, peaked at $10.75 in August 1864,[73] a more than three-fold increase (Figure 2.6). The proportionate increase for anthracite pig iron (Figure 2.3) was almost exactly the same as for coal, even though the latter was only one of the raw material inputs to the smelting process.

Crucially, in the present context, the boom of 1863–1864 was therefore accompanied by excess or windfall profits, despite the discounting effect of inflation. These accrued to capital in greater measure than to labor, to provide funds either for re-investment or speculation. It should also be noted that mining and carrying corporations benefited doubly from higher coal prices and higher freight rates, since the latter advanced almost in proportion to the latter. Between the beginning of 1862 and the end of 1864, for example, freight rates on Schuylkill coal to Philadelphia increased from $1.73 to $5.12 a ton. O'Brien comments briefly on these windfall profits, but only to indicate how difficult it is to investigate how they were used.[74] Neither he nor other authorities examine the role of speculators during this period. Both of these issues will be examined in more detail, in the context of the anthracite industry, in a later chapter. Windfall profits, in particular, played a key role in financing some of the new mining developments of the post-war period. Hence it can be shown that the seemingly contradictory hypotheses about the role of the Civil War can all be seen to play a part in the processes of economic development, provided these are viewed in the context of short-term business fluctuations rather than arbitrary long-term growth averages.

THE 1873 PANIC AND ITS ANTECEDENTS

The sudden and sharp decline in business activity at the end of the Civil War ushered in a period of several years, which was characterized by alternating growth and retrenchment, coupled with oscillating, but downward-trending prices (Figs. 2.2a, 2.3, 2.6). In the case of anthracite coal, prices had dipped down to their pre-war levels by the spring of 1868. Prices of anthracite pig iron, by way of contrast, remained well above 1860 levels until the middle of the 1870s. Also, there is only limited coincidence between the oscillations in the two series. Sharp price increases in coal were associated with market shortages, which followed on from a number of strikes and suspensions, firstly in 1865, as already mentioned, and subsequently in the years 1867–1871. The strikes of 1869 and 1871 were especially notable,[75] the latter cutting off Schuylkill shipments to such an extent that price quotations ceased for several months (Figure 2.6). Coal price rises following the 1869 strike seem to be reflected in the pig-iron price series, but the 1871 strike resulted in no comparable ensuing price rises. This was because of weak demand in the wider economy as a result of a short-lived decline in business activity. Equally, the sharp rise in iron prices in 1872, which coincided with high levels of demand in the railroad sector, did not have any significant knock-on effects to the anthracite coal industry. Production had soared from less than 14 million tons in 1869 to nearly 20 million in 1872[76] and supply was able to keep pace with demand. The developments in the coal industry during this period will be the subject of more detailed examination in later chapters.

As soon as the war was over, investment in railroad construction began to increase again. New railroad miles added increased from 1,177 in 1865 to 7,379 in 1871, the peak year of the period, although construction figures remained high until 1873.[77] However, by 1872 doubts were beginning to surface in the investment community about the profitability and security of railroad bonds. These doubts were based on fears that the railroads in the western half of the country had been overbuilt too far in advance of demonstrated demand for their transportation services. The firm of Jay Cooke and Company had taken on the risky task of financing the Northern Pacific Railroad. This road had spent $15 million by May 1873 on construction, but only 500 miles of track were in operation through sparsely populated areas, and over 1,000 miles of additional track needed to be laid to complete the project. Several attempts by the Company to place large bond issues on behalf of the railroad were unsuccessful in the USA and Jay Cooke was forced to advance to the railroad money derived from short-term bank deposits. The aim was to keep the construction going in the hope of placing the bonds in Europe. However, such a position was unsustainable when the bonds could not be sold, and on September 18, 1873 the Company failed after depositors took fright and started withdrawing their funds. Other companies followed in quick succession and share prices plummeted. The New York Banks deployed the mechanism of clearing-house certificates to settle inter-bank balances and prevent a run on specie. In this they were initially successful, but mounting

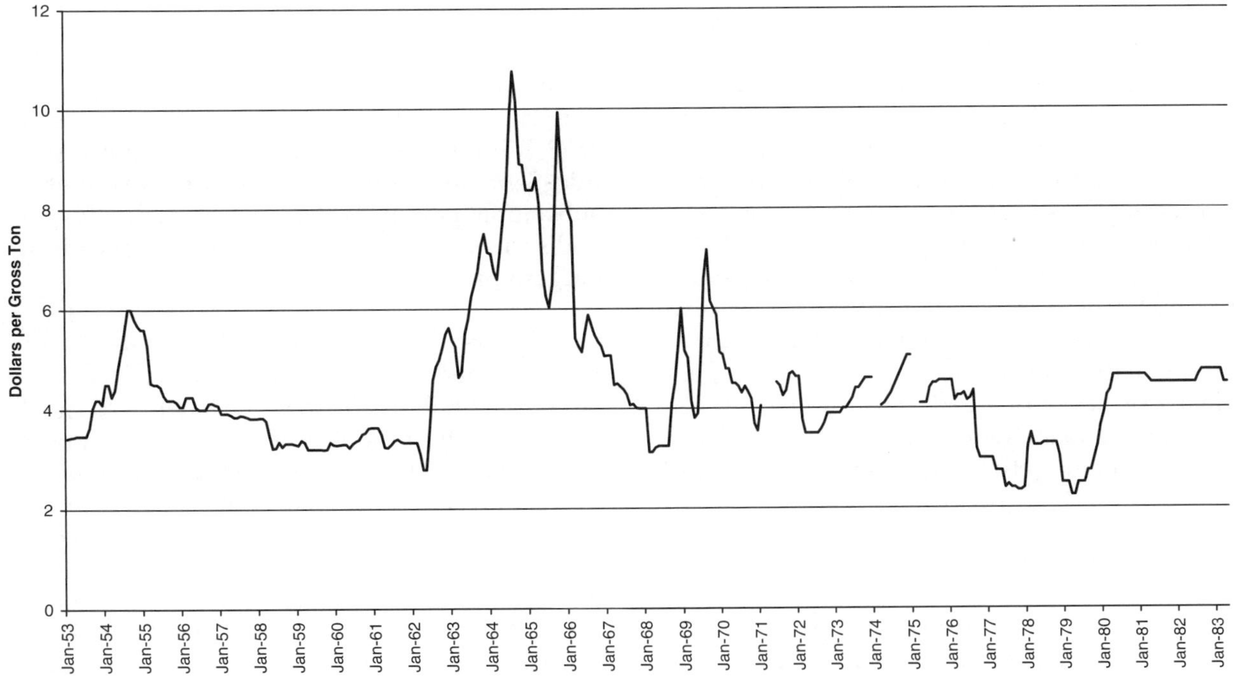

FIGURE 2.6
Monthly Prices of Schuylkill White Ash Lump Anthracite by the Cargo at Philadelphia 1853–1883.

calls for money from interior country banks eventually forced a partial suspension of cash payments, which then spread to other parts of the country, with the exception of California. Fortunately, the suspension only lasted three weeks, but the overall effect of the panic on business confidence was much more long-lasting.[78]

It is apparent from examination of the Ayres Index graph (Figure 2.2a) that business activity had been declining since the beginning of 1873 after two years of solid expansion. So the panic was simply an outward manifestation of a process that was already well underway. Again, as can be seen in a very striking manner from the graph, the contraction that followed was the longest in the history of American business cycles, lasting for more than 5 years until the early months of 1879, when the return to the gold standard and good agricultural prospects prompted a rapid improvement in business confidence. Despite its length, downward price flexibility during the contraction mitigated the fall in output to some extent.[79] This is apparent in the pig-iron price series (Figure 2.3), where prices fell to below 1850s levels in 1878. Anthracite at its lowest, fell to only half 1873 prices, despite repeated attempts to form cartels,[80] and 1878 output, in particular, was well below 1873 levels also. Eventually, the combined reduction in price and output brought many iron, coal, and railroad companies to financial ruin. Bankruptcies exceeded 5,000 in 1873 and the liabilities involved reached $228 million. Five years later, the number of insolvent companies had doubled to 10,478, although the liabilities were only slightly greater than those of the earlier year. Other financial measures paint a similar picture. New York bank clearings fell by 40 percent in comparison to the period before the panic and the average level of railroad stock prices declined by nearly 60 percent. The severity of the contraction affected labor as well as capital. Rising unemployment and corporate opposition to trade unionism culminated in a period of major strikes in 1877 and a reduction in the number of national trade unions from more than thirty to just nine between 1873 and the same year.[81]

THE PANIC OF 1893

In contrast to the 1870s when depression had predominated, the period between the return to the gold standard in 1879 and the end of 1892 was one of several cycles of expansion, punctuated only by the major trough of 1884–1886 and the minor trough that followed the 'Baring Crisis' of 1890.[82] Elsewhere in the world, the Baring Crisis was the start of a recession that lasted until the mid 1890s, but in the United States, serious financial collapse was delayed until 1893. The surge in business activity in 1891–1892 in the United States is generally attributed to bumper crops, perhaps influenced by the 1890 peak of the Currie rainfall cycle, while in Europe crops failed.[83]

When the United States finally joined the rest of the world in recession it did so with unseemly speed in the early months of 1893. The onset of the panic was heralded in February 1893 by the bankruptcy of the Philadelphia and Reading Railroad, the weakest of the anthracite carriers in financial terms and no stranger to the attentions of the receiver.[84] According to the Ayres index, business activity was still in positive territory at this point in time, so the Reading failure can be attributed to internal corporate problems, rather than external factors, but it was the harbinger of problems to come and it unsettled sentiment on Wall Street.[85] Finally, after gloomy news of other failures, both at home and abroad, investors' nerves cracked and May 3, 1893 saw the sharpest fall in the Stock Market since 1884. Wall Street brokerages began to tumble, followed by the banks. Hardest hit were the South and West of the Country, where 153 of 158 banks failed. Pressure built up on the New York banks and they had to resort to the standard mechanism of issuing clearing-house certificates to conserve specie reserves. By early August, a partial suspension was required, as in 1873, but the reserve position rapidly improved by mid-September. This was largely the result of substantial gold imports, to pay for rising exports and securities being purchased by Europeans at bargain prices.[86]

The first depression wave reached its nadir in June 1894 and made a halting recovery which peaked in limited growth in the latter part of 1895. Unfortunately stock market confidence evaporated once again on December 21, 1895 after President Cleveland's message on the Venezuelan boundary dispute. Despite several bank failures and the deployment yet again of clearing-house certificates, there was no overall panic, but business activity still fell off sharply. Recovery was delayed until the spring of 1897, when improved output and exports by the agricultural sector, in the wake of poor harvests in Europe and Russia, brought a welcome return of business confidence.[87]

Throughout the double-pronged depression the tale of woe was shared by capital and labor alike. By the middle of 1894, 192 railroads or 25 percent of the total US trackage were in receivership. Business failures increased progressively in 1894–1896, peaking at over 15,000 in the latter year, but interestingly the liabilities of failures in 1894 and 1895 at $150 million and $159 million respectively, were less than during the worst year of the 1857 panic. About 40 percent of the workforce in the iron industry was unemployed by the end of 1893, although there was some improvement in later years. In April 1894, 130,000 bituminous coal miners struck, but were eventually forced back to work, empty-handed. Unemployment was a serious problem in both the bituminous and anthracite coalfields and was still as high as 15 percent in the latter region in 1897. In addition, Hoffman has calculated that between 1892 and 1897, average earnings of coal miners declined by more than 31 percent, far more than in any other industry. Manufacturing earnings declined by only about 9 percent in comparison.[88] Overall the depression of the 1890s ranks with that of the 1870s in its overall adverse effect on the economy. What the former lacked in duration it made up in terms of severity. In both cases, explanations that have been offered include declining investment in the railroad sector and monetary contraction.[89] In relation to monetary factors, the operation of the gold standard contributed in different ways. In the first case, government-induced reigning in of the money supply made the depression more protracted. In the second, lack of investor confidence in the ability of the government to maintain the gold standard under pressure from the silver interests led from money-market pressure to panic, bank failures, and stock market collapse.[90] Likewise in both cases, improvements in agricultural output were important factors in the restoration of business confidence.

In every one of the economic shocks reviewed, the panic is a lagging indicator of a widening crisis that has already enveloped the business community. Over-extension of credit and rampant speculation were more prominent on some occasions than others, but the weaknesses of both individual firms and the banking system as a whole were ruthlessly exposed on every occasion, despite some of the lessons learned from the 1857 panic. Nor was the US economy isolated from broader international trends and the confidence of overseas investors could ebb and flow with remarkable and potentially destructive speed. Yet in all this, our knowledge of the behavior of investors and speculators and the pattern of survival and failure of individual concerns in specific regions is remarkably limited, even for the 1890s when improved national level data became available. Our understanding of the coping strategies of small and large firms is also minimal. The same applies to different sections of a labor force characterized by growing ethnic diversity. Many of these issues will be examined in subsequent chapters. However, this overview of the economic context of regional development in the anthracite industry would be incomplete without a brief survey of international movements of finance and human capital, which can now be placed within the framework of the major economic fluctuations of the period.

INTERNATIONAL CAPITAL MOVEMENTS

This is a large topic, which has attracted a substantial literature, so it is only necessary to provide a brief review of key points here, as they relate to the importance and timing of major business fluctuations. The following discussion is largely based on information provided by the seminal study of Davis and Cull.[91]

During the 19th century, the United States was the world's largest recipient of foreign investment, yet the population also had a high propensity to save. On average, during the period 1790–1900 only 5 percent of the nearly $60 billion increase in capital stock can be attributed to overseas investment. However, this conceals the fact that at certain periods, foreign investment was a much larger percentage of the total. For example, it is estimated to have been 20 percent in the years 1815–1840, 15 percent during the Civil War decade, and 8 percent during the 1880s.

As we have already seen, decadal averages can conceal a variety of dynamic behavior, so it is useful to examine the year-by-year changes in net capital movements to the USA, shown in Figure 2.7. Positive values imply a net inflow of funds, negative values a net outflow. It is apparent that boom periods prior to panics attracted inward investment, although in the case of the 1857 shock, declining net investment served as a leading indicator of trouble to come. The depressed years of the early Civil War period caused limited capital flight, but an upward, though fluctuating pattern of investment developed when it became apparent that the Union had gained

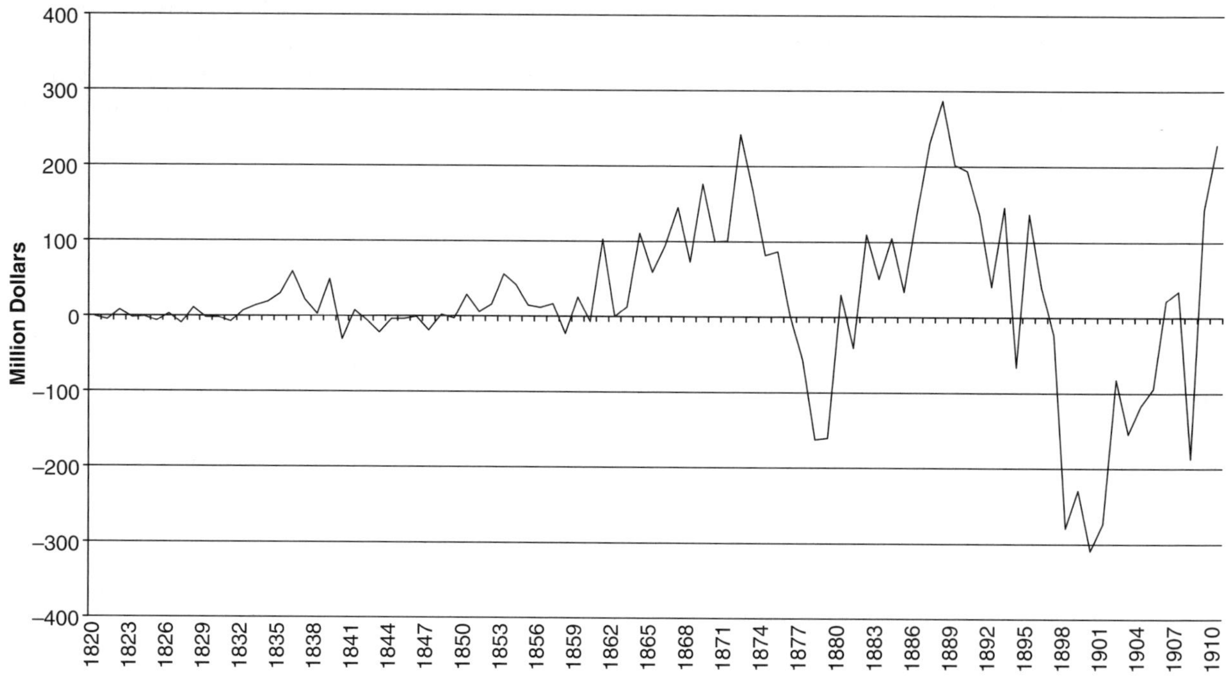

FIGURE 2.7
Net International Capital Movements to the United States 1820–1910.

the upper hand. Following the Jay Cooke Panic, investor confidence evaporated and was not restored until the beginning of the 1880s. A similar effect was apparent after the 1893 panic, but by then the United States was also becoming a major exporter of capital to Canada and South America.

British investors were very keen to acquire American railroad securities.[92] Initially their attention was focused on eastern railroads, such as the Camden and Amboy and the Philadelphia and Reading. Later, interest developed in the Midwestern roads during the period 1865–1873, before moving to the railroads of the South and West. Major railroads such as the Pennsylvania were able to buck this geographical trend, however. Foreign holdings of this railroad's stock rose from 7 percent in the 1870s to more than 50 percent by 1895. Overseas investor interest was not confined to railroads, but extended to the cattle-ranching, brewing, and mining industries. For example, Robert Fleming, of the investment house that still partly bears his name, had a financial interest in the Lehigh and Wilkes-Barre Coal Company at the end of the 1870s.[93] Despite the distance, investors also took a significantly 'hands-on' approach to safeguarding their investments. There were often close links between railroads and individual banking houses. Also, when bankruptcies became prevalent, investors regularly formed investigating committees to check on the management of their assets.

In sum, the overseas investor of the 19th century viewed the United States as an emerging market that presented the usual combination of high risk and high reward. Securities that offered 6–8 percent return during profitable periods, when British Consols provided only 3 percent, were an obvious attraction. As with any risky investment, however, funds would be attracted rapidly when business was booming, but would be withdrawn with equal alacrity during depressions and periods of market volatility. In this way, stock market instabilities could be exacerbated. So it is not possible to consider only the national economy when examining episodic shocks, but the behavior of international investors must also be taken into account.

IMMIGRATION – THE MOVEMENT OF HUMAN CAPITAL

Following the completion of the transatlantic cable in 1866,[94] it might be assumed that while finance capital could now move rapidly between America and Europe, the pace of immigration would lag far behind and show only a muted response to economic fluctuations on the national scene. However, the clear weight of evidence points to a remarkably rapid response of migration streams to economic malaise as well as opportunity, and this rapid response antedates the availability of transatlantic telegraphic communication.

An overview of some of the major characteristics of 19th-century immigration to the USA can be obtained from Figures 2.8 to 2.12. The first figure shows total immigration, and the correspondence of periods of decline with familiar features such as the Kuznets cycle troughs of 1843, 1861, 1878 and 1897–1898 is immediately apparent. In the 1850s, the falloff in immigration after 1854 acts as a leading indicator of the coming panic and it is interesting to note that a Currie Rainfall Cycle peak also occurred in 1852.[95] In 1873 and 1893, by contrast, the decline in immigration begins either in the year following the panic (1873) or in the panic year itself. The apparent sharp drop in 1868, which is visible in several of the series, should be ignored. Owing to a change in reporting arrangements, the data for 1868 only cover part of the year.[96]

While the aggregate picture shows a very close correspondence to economic circumstances, the breakdown by nationality groupings reveals additional dynamics at work. In the case of immigration from Ireland (Figure 2.9) the decisive 'push' factor was the Potato Famine of 1846.[97] Hence the peak immigration preceded the peak of the economic cycle in the mid-1850s. The subsequent response to changes in business conditions is also very limited, with the exception of the 1873–1879 depression, when immigration dropped to very low levels. The modest rate of Irish immigration after the 1880s contrasts sharply with that of Germany (Figure 2.10), where three major phases in the 1850s, 1865–1873 and 1881–1893 are apparent. Both increases and decreases track variations in economic circumstances until the turn of the century, when Germans ceased to contribute significantly to the rising tide of immigration.

The patterns of immigration from the Austro-Hungarian and Russian Empires, and also from Italy are strikingly similar to one another (Figures 2.11, 2.12), but completely different from that of Germany. Very few immigrants came from these source areas until the 1880s and further growth after 1893 was stemmed by the depression of the 1890s,

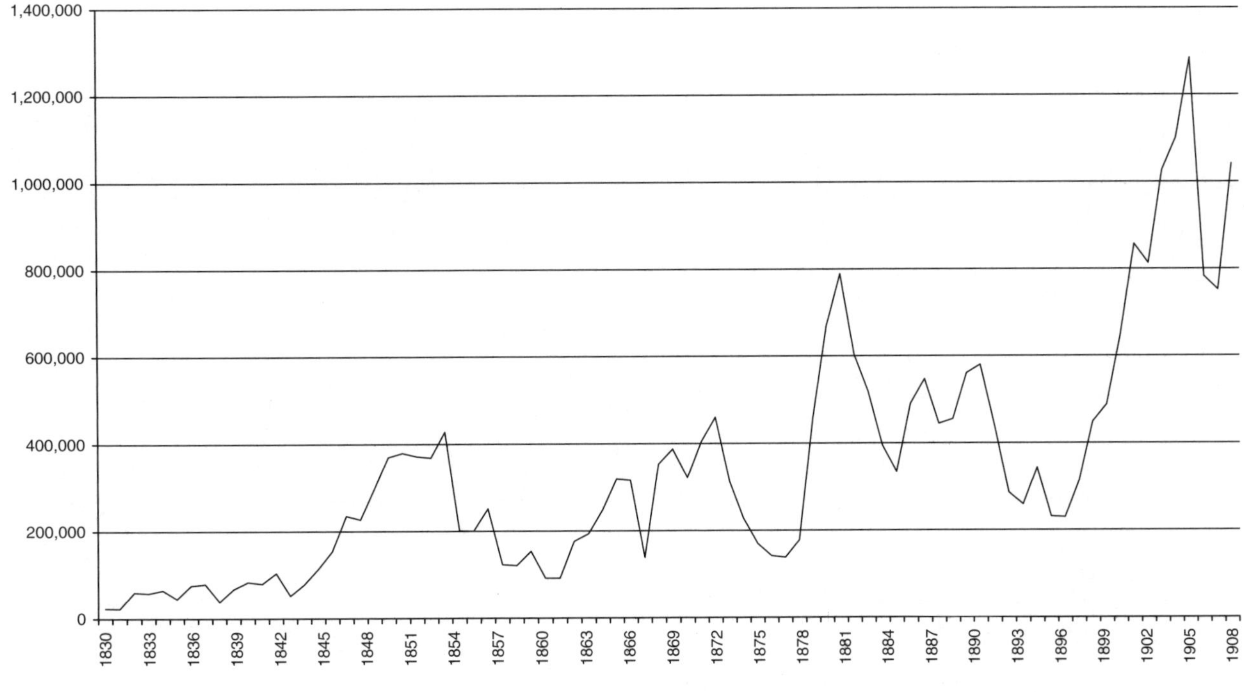

FIGURE 2.8
Total Immigration to the United States 1830–1910.

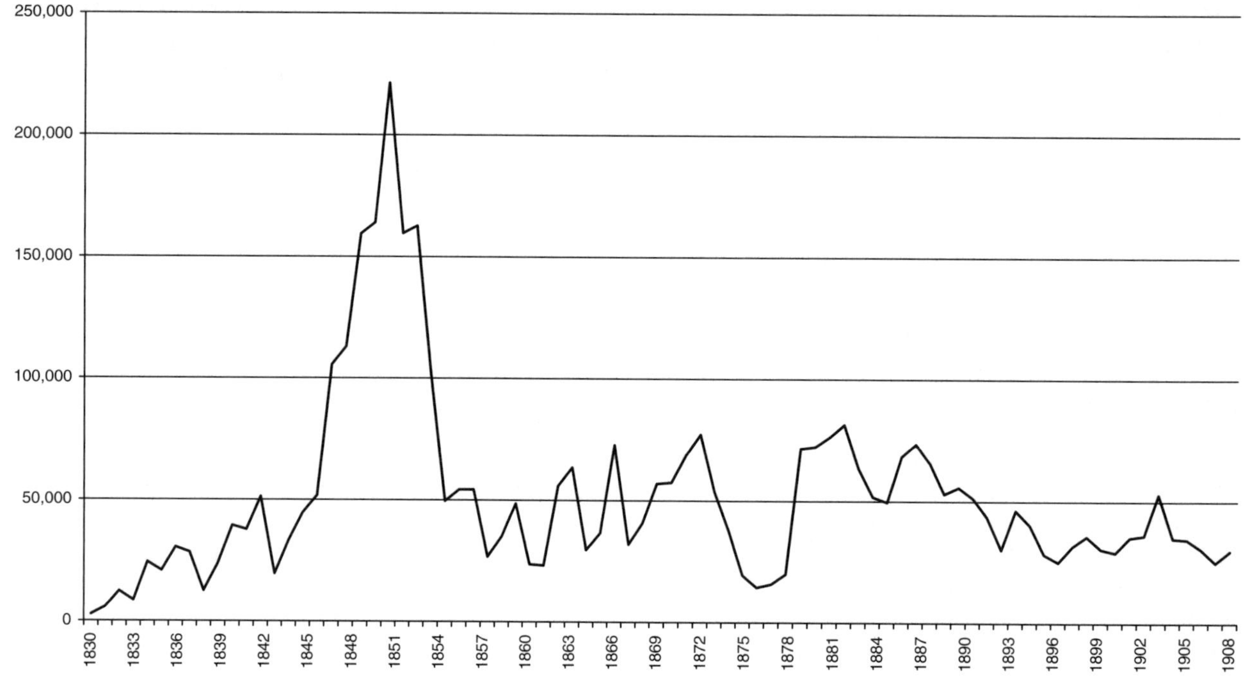

FIGURE 2.9
Immigration to the United States from Ireland 1830–1910.

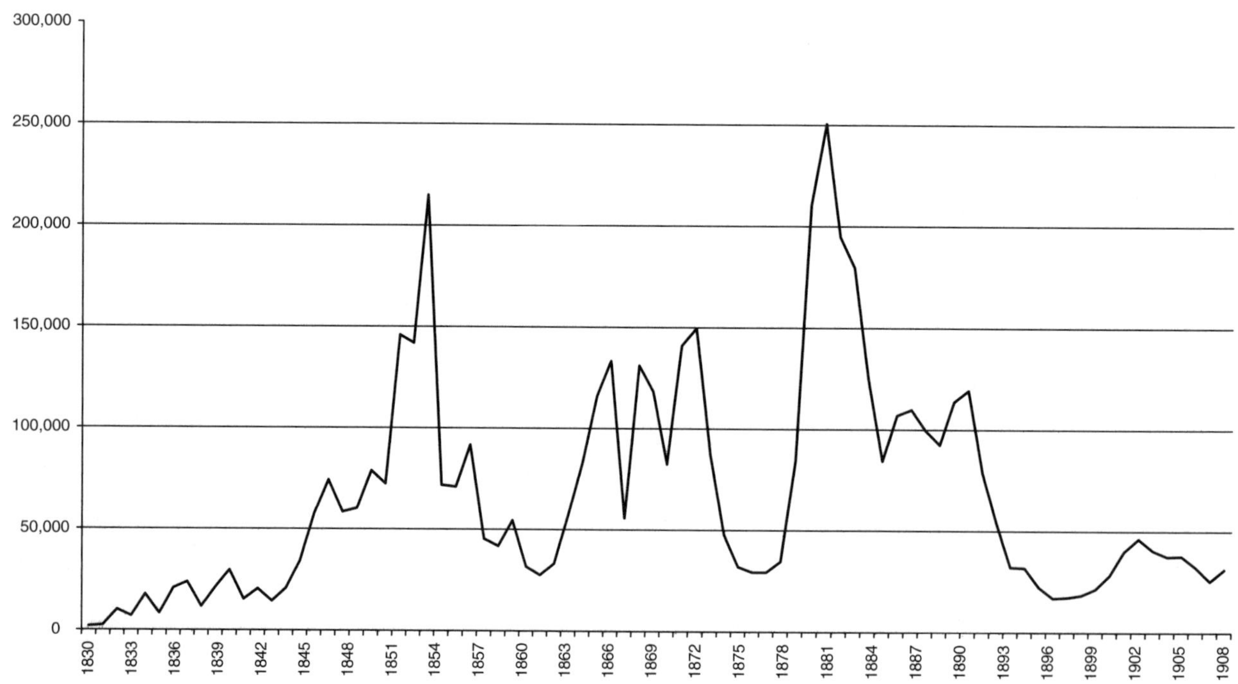

FIGURE 2.10
Immigration to the United States from the German Empire 1830–1910.

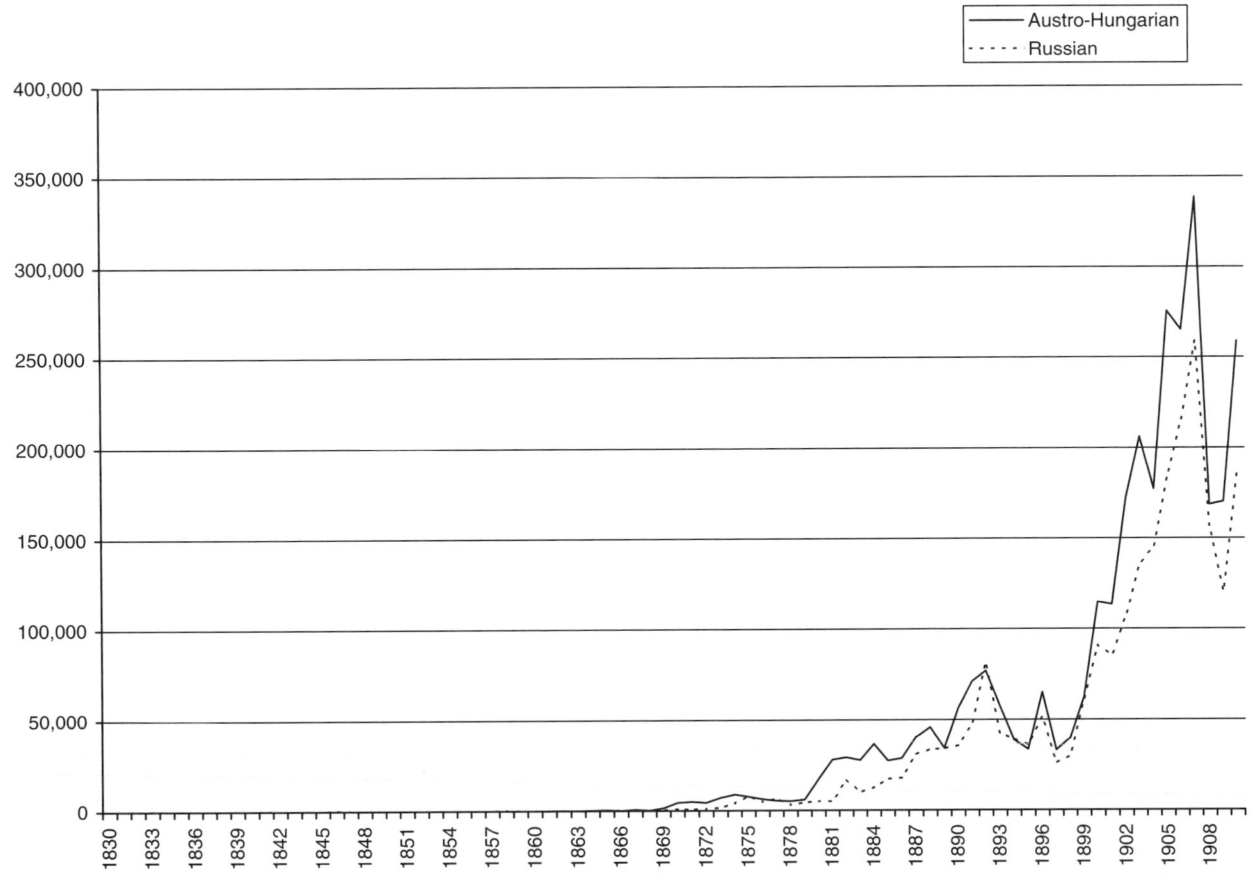

FIGURE 2.11
Immigration to the United States from the Austro-Hungarian and Russian Empires 1830–1910.

until improved business conditions returned in 1899. Leading authorities on the immigration process have listed a variety of historical 'push' factors such as the 1848 revolutionary disturbances in Germany,[98] high birth-rates in Italy in the 1870s and 1880s coupled with improving agricultural incomes to pay the cost of passage[99] and persecution of Jews in Russia in the 1880s and 1890s.[100] Nevertheless, after a detailed examination of the data, Jerome is forced to conclude that migration to the United States was dominated by economic 'pull' factors rather than 'push' factors in the countries of origin. This is a very important conclusion, since it supports the view of Solomou that the Kuznets swings are not initially generated by demographic factors. Migration movements followed improving economic opportunity at a lag which varied from one month to one year and population-related investment in housing construction followed at a further lag.[101] The most likely source of the initial upswings therefore remains the agricultural sector, dependent in turn on the celestial mechanics of the Currie rainfall cycle. Since farming accounted for 50 percent of the labor force of the United States as late as the 1880s,[102] a causal sequence of this kind seems entirely credible.

The impact of these fluctuating migration streams on the anthracite coalfields varied between the different national groupings. The geographical location of the coalfields, which were within striking distance of New York, made them very important as a first destination for incoming migrants and as a staging post prior to movement further west. First to come were the British, Irish, and Germans,[103] and the Potato Famine brought the expected surge in the Irish population. Although the Welsh have attracted scholarly attention[104] and their mining skills were greatly sought after, in any given year the total number of Welsh immigrants was very small compared to the Irish or Germans. Fewer than 60,000 entered the United States between 1820 and 1910[105] and a number had Ohio as their destination rather than Pennsylvania.[106] These groups were followed by the Eastern Europeans, Russians, and Italians in

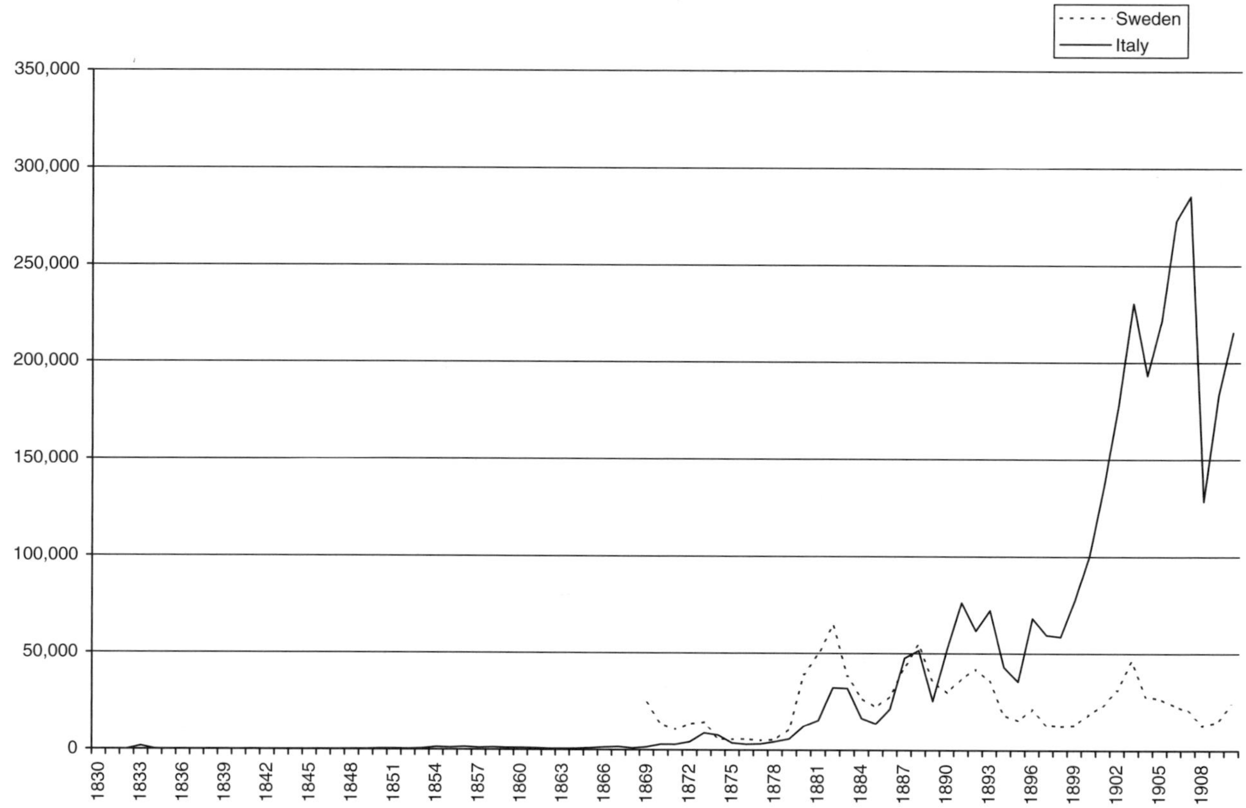

FIGURE 2.12
Immigration into the United States from Italy and Sweden 1830–1910.

the period up to the end of the 19th century, although significant numbers did not begin to arrive until after the end of the 1873–1879 depression, as the earlier figures have shown. Scandinavians never featured strongly in the anthracite mining regions.

Apart from these well-known broad generalizations about the successive waves of immigrants, remarkably little systematic study of exactly when different nationalities arrived in the various mining regions and how long they stayed, has been undertaken. An exception is the study of Welsh mine officials at the start of the 1880s which is currently in progress.[107] Glimpses of the changing composition of the workforce can, however, be gleaned from several different sources. A nationality census of the employees of the Diamond Mine in Scranton was taken in 1857 and some of the findings are shown in Table 2.2.

The numerical superiority of the Irish is noteworthy at this early date, as is the relatively even split between those working inside the mine and those outside. The Welsh in contrast were predominantly skilled miners working underground. Contrary perhaps to the usual stereotype of the Irish laborer, it is interesting to note that of the 119 underground workers, 29 were miners, 69 laborers, 20 drivers and one a carpenter.

TABLE 2.2
Nationality of Employees at the Diamond Mine, Scranton 1857.[108]

Nationality	Inside	Outside	Total
Irish	119	100	219
Welsh	66	20	86
German	19	11	30
English	8	10	18
American	2	9	11
Italian	0	5	5
Scottish	0	1	1
Total	214	156	370

So a number had already moved up the occupational ladder.

A much later summary of changing nationalities at individual though anonymous mines is given in one of the Reports of the Immigration Commission. From contextual material it is possible to surmise that the mines in question may have been in the Mahanoy City area of the Western Middle coalfield. A variety of circumstances are presented. At one mine it is stated that the newer immigrants would work for less than the older employees and gradually displaced them as a result. In a second case, it is argued that there had not been any significant displacement of any nationality by any other. A third example is of particular interest, because it gives more information about when specific groups arrived. The colliery was opened in 1860 and its first employees were the traditional German, Irish, and British nationalities. In 1880 Lithuanians and Poles began to settle in the area and work at the colliery. A decade later in 1890 they were joined by Ruthenians and Slovaks and in 1900 by the Italians. Other sample collieries date the arrival of Poles, Lithuanians, and Slovaks to the early 1870s and that of the Italians to 1890.[109]

These examples are sufficient to give a sense of inward movement of labor to the coalfields, although little is known in detail about onward migration, beyond occasional comments such as the following made by William R. Storrs, General Coal Agent of the Lackawanna Railroad at the start of 1881:

> 'Labour, skilled and unskilled has been, and is now, at this dull season very scarce. A few miners from the old country come and tarry awhile, but not enough to fill the places of those who go away to the coal districts of the west'.[110]

The timing of this quotation from January 1881 is interesting, since the Ayres Index (Figure 2.2b) suggests a brief slackening of growth in the later months of 1880, although it picked up again quickly thereafter. Similarly, total monthly anthracite output was lower in the summer of 1880 and in December 1880/January 1881 than it had been in the corresponding months of 1879/80, but it accelerated in the spring of 1881.[111] Immigration was in the process of recovering from the dark days of the late 1870s, and was just about to begin the steep climb to its 1882 peak (Figure 2.8). So it is difficult to be certain about the nature of the situation William R. Storrs was describing. The first possibility is that it was a short-term reaction to the slightly less prosperous conditions of 1880 after the very large number of days worked by many operating mines in 1879.[112] The second, is that miners were leaving, not just because of limited opportunities in their present employment, but because of the perception that even better opportunities were opening up in the bituminous fields, as the railroad boom of the early 1880s began to take hold. Regardless of the ebb and flow of in- and out-migration from the coalfields, the slow but steady process of occupational mobility continued. Each new ethnic group found itself at the bottom of the ladder on first arrival, but its members worked their way up the social hierarchy and their places on the lowest rungs were taken by others as the years passed. It was thus straightforward for the Immigration Commission to summarize the mining employment situation by 1910, in the following succinct manner:

> 'Managers and superintendents, Welsh; foremen and bosses, Irish; contract miners, Poles and Lithuanians; inside labourers, Slovaks and more recent Poles and Lithuanians, and outside labourers, Slovaks, Ruthenians and Italians.'[113]

CONCLUSION

For the potential investor, entrepreneur, or speculator in the anthracite coal business, the fluctuating nature of the late 19th-century economic environment presented both pitfalls and opportunities. Without the benefit of hindsight and our latter-day statistical analysis, the timing of short cycles and long swings was scarcely knowable and it was therefore difficult to make advance provision against the vagaries of an uncertain future. A significant part of any analysis of business behavior in the anthracite industry must therefore be concerned with the range of reactive and adaptive strategies, which were developed by different groups of economic decision-makers. These strategies were aimed at growth in good times and survival in the bad, in the face of externally generated changes, over which individual firms had no control. However, these strategies themselves could have their own unanticipated 'knock-on' effects, which were internal to the workings of the regional economic system. These effects might, in turn, necessitate further adjustments by decision-makers. In the later chapters of this study, the interplay between these strategies and adjustments and the fluctuating external environment will be examined in detail, as it was this that constituted the overall process of economic development in the anthracite coalfields.

Chapter 3

CONSTRAINTS ON BUSINESS DECISION-MAKING—THE IMPACT OF GEOLOGY, TOPOGRAPHY, AND MINING TECHNOLOGY

INTRODUCTION

In coal mining, as in all other extractive industries, the business decision-making that leads to commercial exploitation of the mineral resource has to take account of a number of environmental constraints. Some of these, such as the location of coal outcrops and the geological configuration of underground coal basins, are immutable. Others, such as topography or drainage patterns may be partially overcome by engineering measures, such as excavations, construction of inclined planes up mountain slopes, or re-routing of small streams. At any moment in time, there will also be a mix of feasible commercial strategies dependent on the current state of knowledge of geological conditions and available technology. The latter may also be subject to what can be termed application and diffusion delays. Application delays may reflect the time required between the 'proof-of-concept' stage of a new technology, as reported in scientific and technical journals, and its actual deployment under a wide range of practical conditions. Diffusion delays are, in a sense, an aspect of application delays, in that effective spread of a new technology may require the physical movement of workers skilled in its use between industrial regions, or at the very least, a series of visits by competent engineers from one district to locations in another where new methods are being used effectively. A classic example of this was the emigration of David Thomas from South Wales to the Lehigh Valley of Pennsylvania in the 1830s, bringing with him the knowledge of how to smelt iron using anthracite, a technology that was to have a major impact on the growth of industrial markets for coal after 1840.[1]

Although it is appropriate to highlight geological and technological factors at this early stage, for convenience of exposition, it should be emphasized that they cannot be treated in an abstract manner, separate from specific geographical contexts. Observed patterns of mining development will also be a function of land ownership or control of mineral rights and accessibility to transport networks by road, rail, or canal, as well as a range of other factors, such as availability of capital. To take an example that anticipates the later discussion, technology and geology may clearly indicate that the best way to mine coal in a certain area is to locate a deep shaft over the center of a coal basin. However, if the land surrounding the preferred mine site is owned by a competing company, it will not be possible for the first company to develop its portion of the coal basin in that way, necessitating an opening in the coal from the outcrop. This may still be profitable in the short term, but will prove costly to both companies in the long run.

With this important caveat in mind, a number of aspects of geology and mining technology will now be examined. These include the geological structure of the anthracite coalfields, the types of anthracite found in the different coal basins and the suitability of these types of coal for differing industrial and domestic markets. Following this, the range of mining methods adopted will be examined, together with the effect of geological and topographical conditions on the location of mine openings and coal breakers. The latter were large surface structures for sizing and sorting coal prior to shipment. The impact of technological change on coal-breaker design over the period will also be considered and contrasted with the relative lack of technological innovation in underground mining work. Finally, engineering solutions to the problem of transporting coal by canal, inclined plane, and railroad across the topographic barriers imposed by the ridge and valley province of the Appalachian Mountains will be examined.

THE GEOLOGY OF THE ANTHRACITE COALFIELDS

Anthracite deposits cover an area of about 484 square miles in eastern Pennsylvania, if the area of the lowest workable coal bed is taken. The coal measures are not continuous, but are divided into four main fields, which themselves contain a number of subsidiary coal basins (Figure 3.1). The Northern or Wyoming Field covers 176 square miles, stretching from Forest City in Lackawanna County to near Shickshinny on the Susquehanna. The Eastern Middle Field is comprised of a number of small basins totalling 38 square miles, with Hazleton as the major town, set on a plateau in the center of the field. The Western Middle Field is conventionally divided into two areas, the Shamokin Division to the northwest of Locust Gap and the Mahanoy Division to the southeast. These two divisions have 94 square miles of coal lands. Finally, the Southern Field, stretching for a length of 73 miles through the counties of Schuylkill, Carbon, Dauphin, and Lebanon, covers 181 square miles between the Lehigh and Susquehanna rivers.

In geological terms, the coal measures are mainly composed of alternating layers of sandstone, shale, fireclay, slate and coal, which overlie the resistant rock of the Pottsville Conglomerate. The maximum remaining thickness of the entire geological column of the coal measures is about 2500 feet. The coal seams themselves are quite well distributed throughout this column at intervals ranging from a few feet to 200 feet or more. Both the intervals and the thickness of the seams vary from place to place. The lower 3–500 feet of the column, from the surface of the Pottsville Conglomerate to the top of the Mammoth or Big Vein, contains the thickest coals with a greater yield than measures higher up. The degree of folding of the strata into basins (synclinals) and anticlinals, the topography and the amount of erosion of the upper measures will therefore determine the actual amount of coal available in any vertical column. This has obvious implications for the economic value of coal lands. However, since the most productive seams are towards the bottom of the column, even basins from which the majority of the coal measures have been removed by erosion may still be very important in mining terms. The greatest thickness of the coal measures is found in the deep basins of the Southern Field where the column is more than 2500 feet deep, with workable veins extending to near the surface. In the Western Middle Field about 1500 feet of measures remain, while in the Eastern Middle Field the upper strata have been removed from the column, leaving only the bottom 700 feet in the deepest basin at Hazleton. Deeper measures are again found in the Northern Field to about 1800 feet in the area west of Wilkes-Barre, but the upper measures have few workable beds in comparison to the Southern Field.[2]

Folding of the measures is most pronounced along the southern boundary of the Southern Field where the strata are contorted into a vertical or even overturned position. In the Northern Field, by contrast, the coal deposits are relatively flat bedded with dips usually less than 30 degrees[3] and, near Wilkes-Barre, as low as 5–6 degrees. On the edge of basins in the other regions the seams may dip at angles as high as 60–70 degrees, with the angles tending to decline as the center of the basin is approached.

CHEMICAL COMPOSITION AND TYPES OF ANTHRACITE

Towards the end of the carboniferous era the coal deposits already laid down in what is now eastern Pennsylvania were subject to mountain folding and metamorphic action to produce the hard coal we know as anthracite. The direction of folding was from SE to NW producing the hardest coals in the east grading westwards into slightly softer coals.[4] True anthracite has a low percentage of volatile hydrocarbons (usually less than 7 percent). Bituminous coal, by comparison, has more than 18 percent volatile matter. In the Scranton area and at the western end of the Southern and Western Middle Fields, the coal has been less subject to metamorphism. As a result, semi-anthracite with 10–15 per cent volatiles is found. This type of coal ignites more easily than ordinary anthracite and was therefore preferred by domestic consumers.[5] By contrast, the harder anthracites of the Lackawanna District around Carbondale and from the Eastern Middle Field were preferred for use by ironworks and other manufacturing interests.

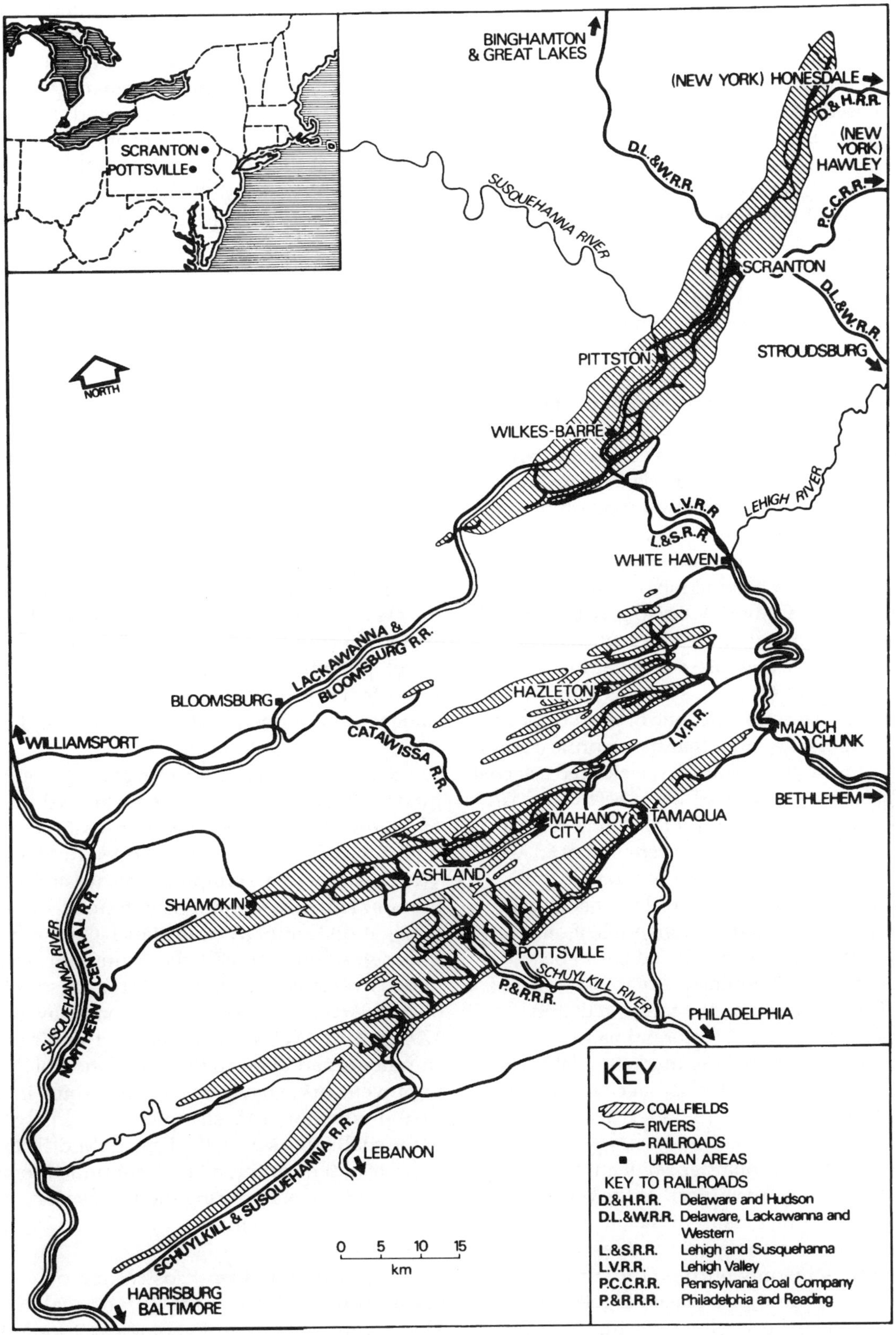

FIGURE 3.1
Map of the Anthracite Coalfields and Major Avenues to Market circa 1867.

Within these broad categories of anthracite there were further distinctions made between the coals found in different districts. Among the most important of these were between red ash and white ash coals, based on the colour of the ash remaining after combustion. Coal seams producing red ash are found both in the lower and higher parts of the coal measures. The higher red ash seams in the Pottsville area, such as the Little Tracy (L) or Gate (M), were very important in the early output of the Southern Field, because they were easily ignited, burned freely in open grates and produced very little ash. Other important red ash coals from the Southern Field included the Lorberry and Lykens Valley coals.[6] In the Northern Field, especially near Plymouth, the red ash coals were largely drawn from the Buck Mountain (B) or Red Ash seam in the lower part of the coal measures. This seam was very important for use in iron furnaces and light blacksmithing work, although it yielded a greater amount of ash than the seams in the upper measures.[7] The most important of the white ash seams, and indeed the most important in terms of production overall, was the Mammoth (E) or Baltimore Vein, to give but two of the names attached to it in the different coalfields. White ash anthracite was also a popular furnace fuel and was well suited to closed stoves. In 1855 an article in the Pittston Gazette, one of the Northern Field newspapers, declared that the calorific output of 2000 lbs of Schuylkill (Southern Field) red ash coal was equivalent to that of 2387 lbs of best white ash for stoves and grates, so pound for pound red ash was cheaper, despite the cost differential of $5.50 a ton as against $4.60.[8] It is important to note that these selling price differentials did not necessarily have an equivalent differential in terms of mining costs, so the type of coal produced could be a major determinant of mining profitability.
Not surprisingly therefore, the terms 'red' and 'white ash' became part of the brand names of different coals, although these marketing labels sometimes diverged from the actual characteristics of the product.[9]

There is one additional aspect of anthracite as a fuel that needs to be introduced at the outset, namely that as a direct result of the hardness of the coal and the low percentage of volatiles, individual pieces of anthracite only burn on the surface.
If different sized pieces are placed together into a grate or furnace, the smaller ones fill the voids between the larger ones and the mass of coal will not burn. Anthracite could not, therefore, be shipped 'run-of-mine' like bituminous coal, but had to be carefully broken and graded into different sizes before shipment, so each boat load or railroad car load would contain pieces of a similar size. This process of preparing different sizes of coal for market necessitated the construction of large wooden, and later iron structures at the mine opening, called coal breakers or 'crackers'.[10] This additional preparation of the product before shipment to market added significantly to the costs of production in comparison to the bituminous coalfields to the west. The evolving technology of the coal breaker is examined in more detail below.

TYPES OF MINING

A number of different methods were employed in the anthracite coalfields for the opening and exploitation of coal seams. These included open quarries or 'strippings', drifts, tunnels, slopes, and shafts. Geological constraints naturally had a major role in determining which methods were feasible, but, as already noted, capital requirements, the pattern of land holding, and technological developments, e.g. in shaft sinking, also had an important role to play in determining which type of mine would be opened at a given time and place.

Some of the earliest mining was by open quarry or 'stripping', whereby a limited thickness of overburden was removed to reveal the coal, which was then worked as an open quarry face. The most notable example of this was the old Summit Hill mines of the Lehigh Coal and Navigation Company near Mauch Chunk (now Jim Thorpe). At Summit Hill, the Mammoth seam is folded over to yield an aggregate thickness of 60–100 ft of coal. It was first worked in a small way as early as 1792 by Cist, Weiss, and others, but significant shipments had to await the completion of improvements to navigation on the Lehigh in 1820.[11] Coal production from strippings, especially in some of the shallow basins in the Eastern Middle Field and in parts of the Southern Field, continued to be important not only during the 19th century,[12] but also into the 20th century.

The earliest and least costly kind of underground mine was the drift or 'water-level'. There were two types of geological/topographical conditions where drifts were the method of choice. In the first, a coal seam ran effectively horizontally through a mountain, which was cut by streams, and an opening could be driven directly into it from any point on the outcrop. In the second, the seam was part of a series of inclined measures, which again had been cut by river action, and the opening was driven into the side of the inclined seam along the strike of the measures.[13] Under these circumstances the location of the drift was effectively determined by the geological/ topographical circumstances. The drift was opened for both types either above the water table ('above water level') or at the level of it, e.g., close to a stream. The workings were so arranged, as they developed, to ensure that all water drained naturally from the mine. This avoided having to install pumping equipment. As the drift extended into the mountainside it became a normal mine gangway, with rails laid so the mine cars could be pulled in and out by mules or horses.

Since the drift was driven in coal and required no pumping or hoisting machinery it was, to borrow Yearley's phrase, 'an appealingly inexpensive' way of opening a mine.[14] Even more importantly, from the perspective of the small independent coal operator, it allowed saleable products to be shipped as soon as the first few timbers of the mine opening had been set, thereby minimizing cash flow problems.

While the conjunction of geology and topography was suitable for drift mining in many parts of the coalfields, in others it was not. Alternatively, where the conditions were suitable, the distance to available transportation routes, such as rivers and canals, might have been too great to make the operation economically viable. Under these circumstances, an early alternative to the drift was the rock tunnel, again driven at water level to avoid drainage costs. The tunnel usually began at the base of a hillside and cut horizontally through a series of inclined measures to open several different veins of coal in sequence. Depending on the length of tunnel required before the first vein was reached, quite a significant amount of capital could be sunk in the enterprise before any return was obtained.

A classic example of this was the Grand Tunnel of Freeman Thomas in West Nanticoke in the Northern Field. This was started in 1828 and required 3–4 years to complete. In the process it consumed all Thomas' own funds and all he could borrow. Indeed, as related by Phillips, the local historian, it was the last shot fired when funds were completely exhausted, that finally broke through to reach the Mammoth Vein.[15] A similar early example was the Old Hacklebarney Tunnel of the Lehigh Coal and Navigation Company, in the Southern Field. This was worked from 1826 until November 1868, when the combination of uneconomic working and an underground fire caused the mine to be abandoned.[16]

If the coal vein descended underground below water level, a slope was driven down the vein from the outcrop and machinery had to be installed both for hoisting coal and for pumping out water. Slope mining was preferred to a shaft opening if the vein outcropped at the surface and dipped at more than 15–20°. After sinking the slope a distance of 100–110 yards, gangways would be opened out to right and left, to make what was termed the first 'lift'. Coal would usually be worked at this level for several years, before sinking the slope a further lift of 75–100 yards, but on occasion slope sinking continued for two or even three lifts before the first lift was fully opened. The more lifts sunk at the outset, the greater the initial capital outlay, as only a small amount of coal was produced in the course of slope sinking itself. However, with two or more lifts working simultaneously, the productive capacity of the mine would be correspondingly greater than when only a single lift was in operation, provided, of course, that the capacity of the hoisting, pumping and breaker equipment was also equal to the task.[17] An interesting example of a mine opening that combined both a slope and a drift is shown in Plate 1 for the No. 3 Lee mine. This mine was at the western end of the Northern Field.

When there were no coal outcrops on the property and the coal veins were too deep to be mined by stripping, shaft sinking was required. From the coal operator's perspective, this represented a major investment which did not produce a return in the very short term. The cost of shaft sinking per yard was much greater than that of slope sinking or tunnel driving. Hoisting and pumping engines, albeit of limited capacity, were required at the outset, and no significant coal could be obtained before the first workable vein had been reached and gangways opened out on either side of the shaft. The sinking process was also slow and specialized work. This was not too serious a problem if the planned shaft depth was only 100–150 feet. However in the deeper parts of the coal basins, from the 1870s onwards, deep shafts in excess of 1000 feet deep were needed. Operations of this magnitude required timescales measured in years rather than months to complete.

Since shafts were both major and immovable investments, it was very important that the best possible location was chosen at the outset. Ideally, this was directly over the deepest part of the coal basin found on the tract to be developed. In the earlier days of the industry before the Civil War, when the detailed geology of many parts of the coalfields was poorly understood, the process of identifying this optimal location could range from educated guesswork to little more than vague conjecture.[18] After the Civil War, with improvements in exploratory borehole drilling apparatus, a more scientific approach could be adopted, although it was still prone to error in districts subject to marked variation in the strata over short distances.[19] A properly located shaft sunk into a basin with several workable veins could expect a long and productive life, although the risk of financial loss from an unexpected quarter could never be ignored. Such was the case at the Buttonwood shaft, near Nanticoke. This was originally opened in 1855, but later closed because of the amount of gas generated and the fires and explosions that resulted.[20]

UNDERGROUND MINING

Once the main mine opening was completed to the required depth or length, gangways in the seams would be driven, usually about 8 feet high and 9–10 feet wide. In gassy mines such as those in the Wilkes-Barre district, the gangways would be up to 14 feet wide to facilitate ventilation. From the gangways, 'chambers' or 'breasts' would be driven at right angles, separated by pillars of solid coal to support the overlying strata. The configuration of these chambers and pillars depended very much on the prevailing geological conditions. By convention, three categories of workings were identified. These were flat workings, those dipping 1–25°, and steeply pitched veins running at angles from 25° to vertical.[21] Depending on the conditions of the roof, the chambers would be opened 8–12 yards wide and the pillars of a similar size.

In very flat workings, as in some of the mines near Wilkes-Barre, the chambers could be driven long distances, often as much as 600 feet. Elsewhere, the inclination of the measures usually restricted the length to 3–500 feet. In the flat workings, mine cars able to hold 2–3 tons of coal could be taken directly to the coal face. If the floor of the chamber rose 5–12°, a smaller wagon called a 'buggy', holding 1/2–1 ton of coal would be used to bring the coal down to the gangway. On inclinations of 12–25° a sheet iron chute would replace the buggy. Above 25°, the coal would be mined on the 'run', falling of its own accord after blasting, into a storage chute constructed at the entrance to the breast. From there it could be loaded into cars waiting in the gangway. However, enough broken coal had to be left in the breast for the miner to stand on, so he could drill the next set of holes for blasting into the coalface above his head. A diagram showing this method in operation in one of the Eastern Middle Field mines can be seen in Figure 3.2. It takes little imagination to appreciate the difficulty and danger of working in these conditions, even before account is taken of the dust, the darkness and the invisible threat of 'firedamp' or methane gas.[22]

It is apparent from the preceding discussion that the chamber and pillar method of mining, as described, would result in up to 50 percent of the coal being left underground. Extracting more coal required what was termed 'robbing' of the pillars, to allow more coal to be taken out until the roof collapsed. This was done when the chambers in a portion of the mine had been worked to their fullest extent, but there was always the risk of inducing a 'squeeze' or subsidence over a wider area than intended, damaging gangways and airways needed for ventilation.

As early as 1866, Daddow and Bannan in their classic treatise entitled 'Coal, Iron and Oil', observed that British mining engineers had developed improved mining methods since the 1840s, such as long-wall working, which greatly reduced the amount of coal left in the mine.[23] This led Yearley, a much more recent commentator, to question why these methods were never adopted, quoting examples of their effective use in England.[24] However, Chance in his well-known treatise on mining methods from the 1880s, answers the case convincingly in relation to long-wall working, because it was not applicable to steep dips, nor could it be used in thick seams. Further to this, he notes that in the thick ten-yard coal of South Staffordshire in England, the percentage of coal left in the ground was about the same as in the Mammoth bed in the anthracite regions.
His conclusion therefore was clear — 'in thick bed mining we can learn little from English engineers'.[25]

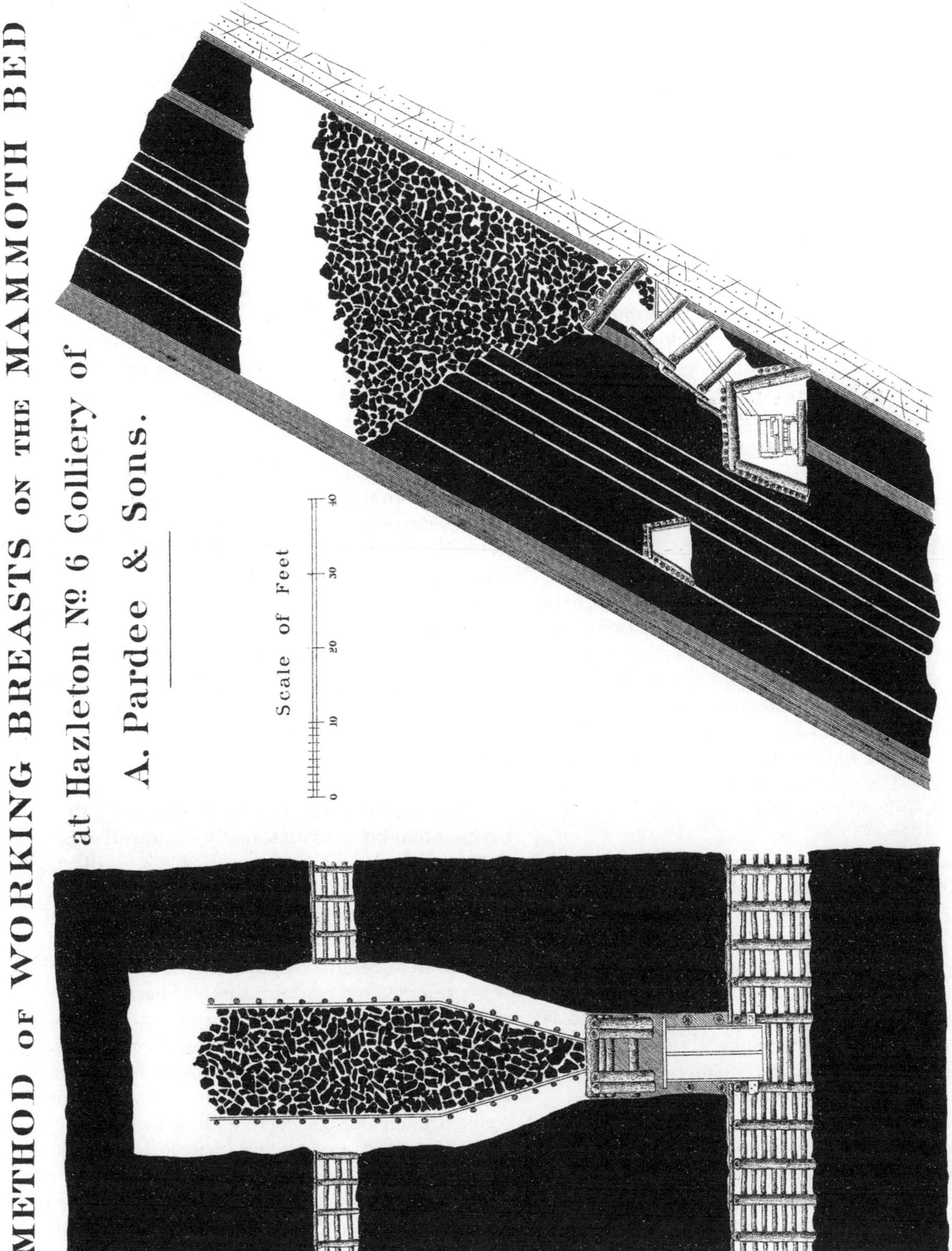

FIGURE 3.2
Method of Working Breasts on the Mammoth Bed at Hazleton No. 6 Colliery.

In other areas of Pennsylvania, such as the Blossburg coalfield to the northwest, long-wall working was adopted,[26] and trials were made in various parts of the anthracite fields of modified mining plans intermediate between chamber/pillar and long-wall working, such as Veith's Boundary Plan.[27] The aim in each case was to maximize the coal taken from the mine, while minimizing the danger to the miners and the likelihood of uncontrolled collapse of the workings.

Overall, during the period of interest in this study, relatively limited progress was apparent in the general approach to anthracite mining underground. For specific tasks, however, such as drilling and haulage, the companies were quite rapid in their adoption of later technological innovations like compressed air drills and electric mine locomotives. Earlier, in the 1870s, the adoption of steam driven ventilating fans had progressively replaced the older and hazardous method of using underground furnaces.[28]

THE COAL BREAKER

In contrast to the relatively unchanging nature of underground work, there were major developments in the surface plant for preparing coal for market, between the 1840s and the turn of the century. In the early days of the industry, the sorting of coal into sizes was rudimentary. Larger lumps were broken with hammers and the smaller coal passed over hard screens. A good deal of fine coal was simply dumped on the refuse pile or 'culm bank'.

The first recorded steam powered coal breaker was installed by Gideon Bast at Wolf Creek in the Southern Field in 1843–4 and the innovation spread rapidly to other parts of this area,[29] although rather more slowly into the Northern Field. In addition to its primary purpose of breaking and sorting coal into the various merchantable sizes, the breaker also served as the means of separating the good coal from the slate, rock and 'bony coal'. The differentiation of the latter from good coal was based on appearance rather than any scientific analysis of heating capacity. In consequence, yet more potentially usable coal was consigned to the refuse pile as unsaleable. The amount of rock and slate brought to the surface with the coal depended in part on the geological conditions. When the coal came from a thick bed on a steep dip, a good deal of slate was intermixed with the coal. With thin beds and flat dips, most of the slate would be left in the mine. Variations in the amount of refuse brought up with the coal resulted in a system of 'dockage', whereby a proportion of the weight of any loaded mine car was deducted from the miner's tonnage output, to allow for the rock and slate. Unsurprisingly, the dockage system was a persistent source of friction between the coal operators and the men.[30]

The basic principles of operation of the coal breaker involved firstly, passing the mine output over iron bars to separate out the largest pieces of coal and rock. The material that passed through the bars was then fed through a revolving circular screen that separated the fine coal from the intermediate sizes. These intermediate sized pieces were broken by cylindrical rolls ('pony' rolls) with iron teeth and the resulting mix of smaller pieces was screened again, to sort it into the various sizes.[31] The sorted coal subsequently moved down chutes to storage pockets for the different sizes, before being loaded into railroad cars. Cross-sectional diagrams of both front and side elevations of a representative breaker from the 1880s, the Hollenback near Wilkes-Barre, can be seem in Figures 3.3 and 3.4. While in the chutes, small pieces of rock and slate were removed from the coal by 'breaker boys' or 'slate pickers', who sat astride the moving mass of coal, controlling the flow with their feet.[32] The breaker boys were usually children of miners, working in cold and very dirty conditions. Their plight was immortalized first in the photographs of George Bretz during the 1880s[33] and later, at the turn of the century, by the camera of Lewis Hine.[34]

The different sizes of coal produced by the breaker had become reasonably standardized commodities by the 1880s (see Table 3.1), although the precise sizes of mesh over or through which pieces of coal would pass varied a little between different companies. The different sizes were utilized in different markets. The largest size, lump was preferred by iron furnaces and early on by railroads. Steamboat coal was used by coastal and ocean-going steamers as the name suggests,[35] although from the 1870s onwards little of this size was made, the trade having been lost to bituminous coals. Broken and egg were widely used for industrial steam-raising and institutional heating, e.g. in hospitals and asylums. Stove and chestnut were the preferred sizes for domestic heating. Pea and buckwheat coal were originally not marketed at all, but in later years their lower cost made them attractive to industrial consumers for general heating purposes. These preliminary observations

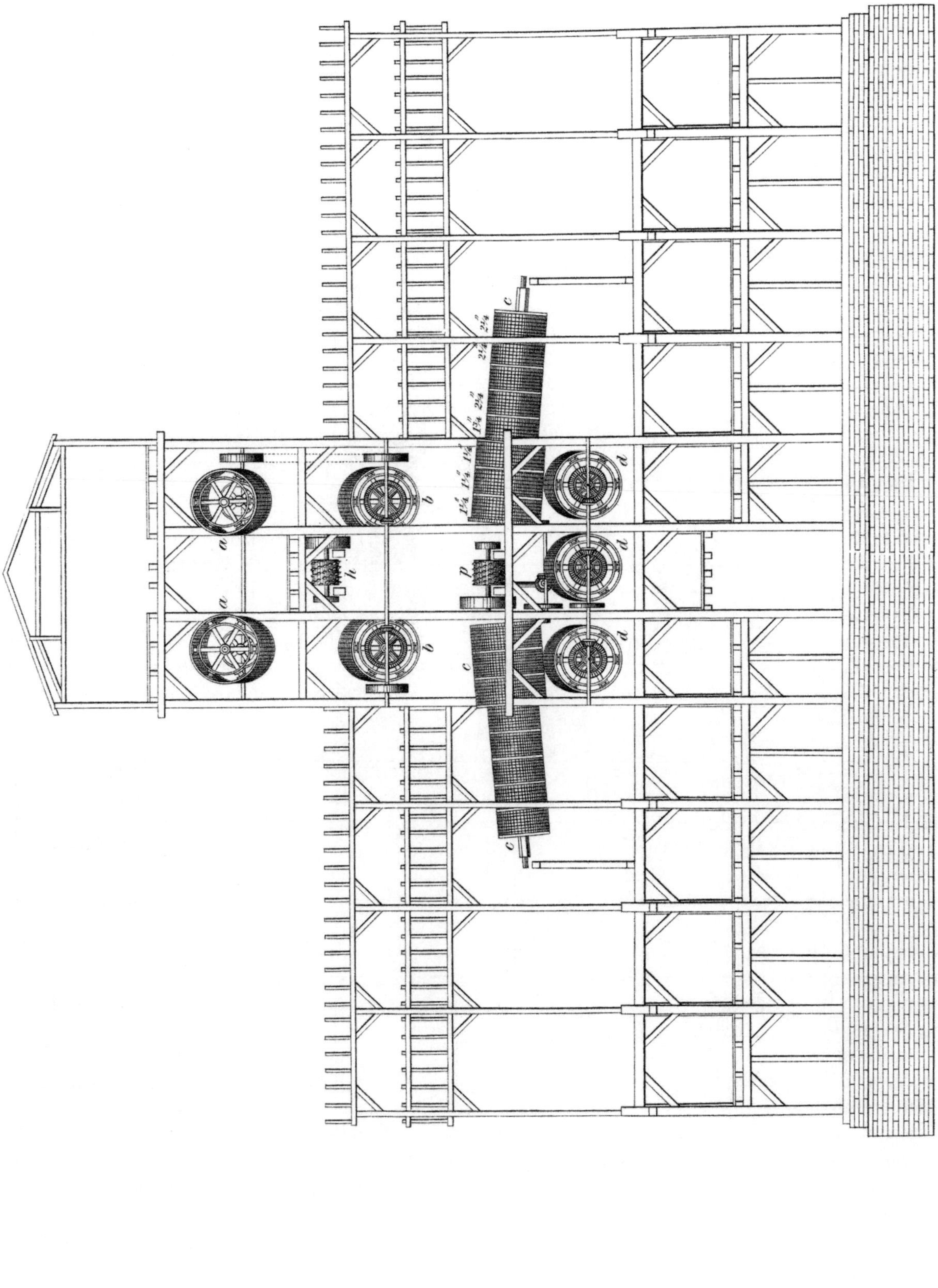

FIGURE 3.3
Front Elevation, Cross-Section of the Hollenback Breaker, Wilkes-Barre.

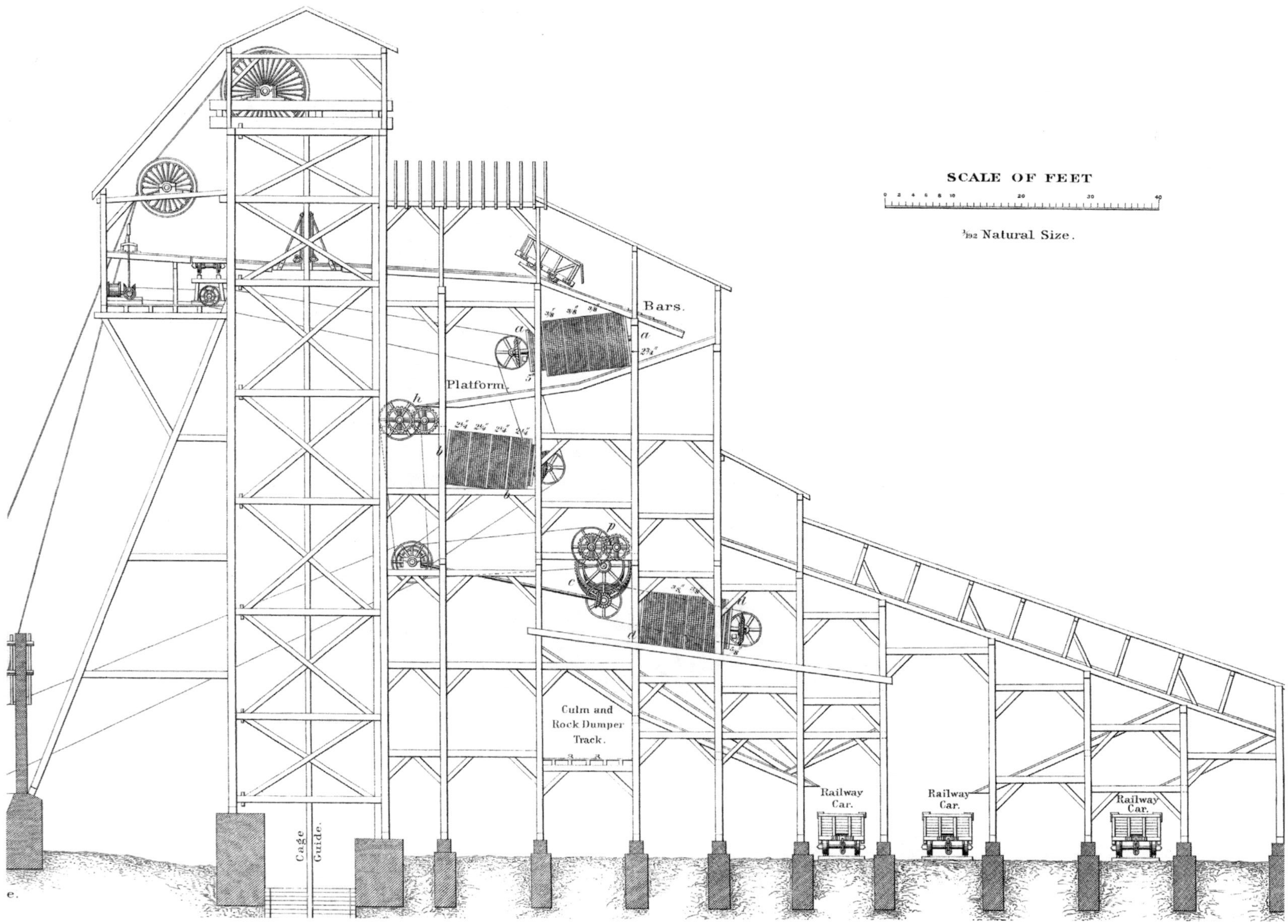

FIGURE 3.4
Longitudinal (Side) Elevation of the Hollenback Breaker, Wilkes-Barre.

TABLE 3.1
Different Sizes of Anthracite in Inches Showing the Range of Variation Between Companies.[37]

Size	Sorting Method	Over Min	Over Max	Through Min	Through Max
Lump	Bars	7.0	9.0	–	–
Steamboat	Bars	3.5	5.0	7.0	7.0
Broken	Mesh/Bars	2.38	2.88	3.25	4.5
Egg	Mesh	1.75	2.25	2.38	2.88
Large Stove	Mesh	1.25	1.88	1.75	2.25
Small Stove	Mesh	1.0	1.25	1.25	1.5
Chestnut	Mesh	0.63	0.75	1.0	1.25
Pea	Mesh	0.38	0.63	0.63	0.88
Buckwheat	Mesh	0.17	0.38	0.38	0.63
Dirt	Mesh	–	–	0.19	0.38

on anthracite markets will be elaborated in later chapters, but it is appropriate to note here that the storage pockets in the breaker were limited in their capacity. If pockets became full, through lack of orders for particular sizes or shortage of railroad cars for loading, the breaker had to stop. Maintenance of a balanced stream of orders for all the sizes and obtaining sufficient empty cars from the railroad were therefore major preoccupations of mine operators and their sales and commission agents throughout the period. Larger companies, which owned multiple breakers, could smooth out fluctuations by allocating shipments among their different works, since each would produce a differing mix of size percentages. Companies involved both in mining and transporting coal could also dump coal in storage yards or 'stocking grounds' along their railroads or canals. A good example of this was the Pennsylvania Coal Company at Dunmore near Scranton.[36] The only major company that seems to have specifically used a stocking ground at the mines for inventory management, to disengage the working of the breakers from the immediate demands of the market, was the Susquehanna Coal Company at Nanticoke in the Northern Field. As a matter of policy they dumped all the coal already prepared in their No. 1 breaker onto the stocking ground, regardless of size. Then they re-processed the coal through the adjacent No. 4 breaker, at a later date, before loading it into canal boats or railroad cars.

As a contemporary observer noted this was 'of great value to the mining operations, acting for them as an elastic spring to keep them in constant working' even though it added 4 cents per ton to the cost of coal production.[38]

Even without reprocessing the coal twice, the necessity of breaking coal added significantly to the cost of anthracite coal production as compared to competing bituminous coals, since not only were breakers expensive to build, they were also expensive to run. For instance, an 1886 report on mining by the U.S. Census Office, recorded that breaker construction could cost between $20,000 and $100,000 for the smallest and largest breakers of the day, which had daily capacities ranging from 500 to 2500 tons.[39] Likewise, Eckley B. Coxe, a well-known mining engineer and coal operator in the Eastern Middle Field, testifying before a House of Representatives Select Committee in March 1888, stated that the cost of preparing coal in the breaker contributed about 20 percent, or 30 cents, to the average coal production cost of $1.50 per ton.[40]

Many developments in breaker technology and design were incremental, the result of experience and experimentation. In 1866, Daddow and Bannan were certainly of the opinion that there was 'plenty of room for improvement'.[41] These improvements took the form of changes to screens, re-design of the teeth on the breaker rolls to minimize the production of fine waste, and labor-saving devices, such as jigs or separators. These utilized both the differing shape of slate and coal particles and variations in specific gravity to try and separate the two types of material. However, in the early 1880s, Chance was forced to admit that of the types available at that time, 'only a few [had] given satisfactory results'.[42] Nevertheless, these were to be the forerunners of much more sophisticated, mechanical separating devices such as the Chance and Menzies Cones, which contributed to the eventual disappearance of manual slate picking in the 20th century. As the equipment became more sophisticated and the demands for increased production more pressing, the modestly sized, often ramshackle, breaker buildings that hugged the valley sides in the 1860s were replaced by huge wooden and later iron structures, 100 or more feet high. These spread across the anthracite fields in their hundreds to produce one of the world's most unique industrial landscapes and a 19th-century tourist attraction in its own right.[43] Good examples of early coal breakers dating from the late 1850s and early 1860s are shown in Plates 2–4. The first of these shows Eaton and Company's breaker at Archbald. Gibson's Breaker (Plate 3) was later renamed Jermyn No. 1. The Alexander Gray Breaker (Plate 4) near Wilkes-Barre is of particular interest because of the method of hauling coal cars directly into the top of the breaker using mule power.

MINE SAFETY AND THE LOCATION OF BREAKERS AND MINE OPENINGS

The terrible disaster at the Avondale mine near Wilkes-Barre on September 6, 1869, which caused the deaths of 110 men and boys, brought the question of mine safety to the forefront of public consciousness, not only in the anthracite coalfields, but across the state and beyond. Prior to that point, limited mine safety legislation had only been enacted to protect miners in Schuylkill County to the south.[44] The disaster resulted from a fire in the shaft, which was communicated to the breaker built directly over the shaft. The rebuilt breaker can be seen in Plate 5.[45] Since the shaft was the only means of exit from the mine at the time, the entire underground workforce were asphyxiated by noxious fumes or lack of oxygen before help could reach them in the aftermath of the fire.

The absence of a second opening became a focus of the subsequent investigation into the disaster, even though there was evidence that the operators, the Nanticoke Coal and Iron Company, a subsidiary of the Lackawanna Railroad, were 'working to secure a second opening as soon as practicable'.[46] As a direct result, new mine safety legislation enacted in 1870 required, amongst other provisions, that every mine had to have two openings or outlets not less than 150 feet apart on the surface, to ensure that the disaster could never be repeated.[47] Importantly, this and the other requirements applied retrospectively to all mines then in operation, as well as to new mines. In the light of the understandable public outcry at the time, and the re-telling of the events by subsequent commentators, it is easy to make an incorrect assessment of the actual nature of mining operations at the time and assume that wilful neglect of miners' safety was the order of the day. While there is no question that many of the older and smaller mining operations throughout the coalfields were badly laid out, poorly ventilated, and had only one means of egress, the same cannot be said of all of the larger concerns, especially in the Northern Field. For example, two weeks after the Avondale disaster, William R. Storrs of the Lackawanna Railroad Coal Department provided a detailed report on second openings from the company's mines to the Railroad's President Samuel Sloan. Of the 14 operating mines listed, only Avondale was without a second opening at the time of the fire. The Taylor mine had two shafts, although they were only 45 feet apart, and the fire risk was recognized, so the company had already begun driving a 2000 foot tunnel in the coal to the surface, to provide a third exit. The Cayuga mine was still under construction and its lack of a second opening was to be remedied by connecting with the adjacent workings of the Von Storch mine.[48]

While the more responsible and solvent companies clearly wasted no time in addressing safety issues highlighted by the disaster, the sudden additional burden of development costs to secure second openings, as a result of the legislation, was a problem to many smaller concerns, although most eventually co-operated in large measure with the demands placed on them by newly appointed Mine Inspectors.[49]

While increased legislative concern for mine safety was essential, from a narrow investment perspective it also raised the barriers to entry into the industry and the time and costs required to open new mines. This was especially the case when shaft workings were sunk in areas remote from coal outcrops and existing workings. Under these circumstances, two shafts had to be sunk and the gangways between them connected, to comply with the law, before production could begin. Some of the implications of this will be examined in later chapters.

A second legislative act of June 30, 1885 also had important implications for mine development and the layout of surface improvements. This act required that no inflammable structure, including a coal breaker, could be built over a mine opening, other than the necessary head-frame for hoisting.

The coal breaker had to be at least 200 feet away, connected to the opening by a trestle, if required. Unlike the 1870 legislation, the act did not apply retrospectively, either to existing breakers or to breakers in course of construction. Perhaps curiously, in the light of the Avondale experience, operators were still allowed to rebuild breakers lost by fire in the same location as before.[50]

A large number of breakers were still positioned above or adjacent to the mine opening at this time, so the important beneficial effects of the legislation took much longer to be felt than in the case of the 1870 act. Once again, however, it added to the costs of building new mine improvements on the surface, because of the additional trestling and machinery that was required to get the coal from the mine mouth into the breaker. It also produced some unexpected legal wrangling. Shortly after the act was introduced, Mine Inspector Blewitt used it to restrain William T. Smith, operator of the Mount Pleasant mine in Scranton, from sinking a shaft as a replacement for, and adjacent to his existing slope, because the new opening was within 200 feet of his breaker. The difficulty arose because the breaker predated the act, but the new shaft did not. The matter went to court and after a complex argument Judge Hand finally ruled in favor of the operator, and allowed the shaft sinking to proceed.[51]

TOPOGRAPHIC CONSTRAINTS ON COAL TRANSPORTATION

While topographic constraints affected the design and layout both of the breaker and associated mining plant, a much more substantial problem was overcoming the steep gradients of the ridge and valley province of the Northern Appalachians to bring a heavy and bulky commodity to market. Initially, the major rivers of the Susquehanna, Lehigh and Schuylkill (Figure 3.1) were exploited through improved navigation and canalization. Subsequently these routes through the mountains were utilized by an ever-expanding network of rail lines, which steadily displaced the canals as the dominant mode of transport.

At certain points in the coalfields, however, areas of higher ground intervened between the river valleys to make communication difficult. To overcome these obstacles, inclined planes were constructed at a number of locations, to haul coal cars up and down steep gradients. Prominent among these were the Gordon and Mahanoy Planes. These were built across Broad Mountain, between the Western Middle and Southern Fields. Of similar importance were the Ashley Planes of the Lehigh Coal and Navigation Company, which lifted coal out of the Susquehanna Valley onto the high ground between Wilkes-Barre and the Lehigh River.

The most complex series of planes were on the gravity railroads of the Delaware and Hudson Canal Company and the Pennsylvania Coal Company, designed to bring coal out of the Lackawanna Valley. The original Delaware and Hudson gravity railroad had eight planes, but by 1866 the system had been extended to include a total of 28 planes. An excellent illustration of the expanded system is shown in Plate 6, which dates from about that time. It shows the Valley Junction terminus of the gravity railroad below Olyphant and the foot of Plane No. 23. The modest coal breaker to the right of the picture is not identified in the caption to the original photograph, but cartographic evidence suggests it may be the first Eddy Creek breaker.[52] By way of comparison with the Delaware and Hudson, the Pennsylvania Coal Company's gravity railroad, built between 1847 and 1850, had a total of 22 planes.[53] Such extensive series of engineering works required constant maintenance and contributed greatly to the costs of transporting coal to market. In consequence of this, both sets of planes were abandoned in favor of locomotive roads before the end of the 19th century, those of the Pennsylvania Coal Company in 1885 and those of the Delaware and Hudson in 1899.[54]

Into this landscape of obvious topographical constraint, but uncertain geological conditions, came a mix of speculators and capitalists from the eastern seaboard cities. Sometimes they bought out local land-owning interests, at other times they formed financial alliances with them. Prominent initially in the promotion of enterprises for improved transportation of coal to market, the mining and carrying corporations they themselves helped create, in turn largely supplanted them, because of their relative financial might and their control over the major railroad routes. In the process, both the fertile agricultural valley of the Susquehanna and the rugged forested mountains to the south and east were transformed from rural idyll on the one hand and wilderness on the other, into one of the most important and dynamic industrial regions in the United States as the 19th century progressed. It is to the beginnings of this industrial transformation that we now turn.

Chapter 4

PRIOR INVESTMENT IN MINING AND TRANSPORTATION INFRASTRUCTURE

INTRODUCTION

The early development of the anthracite coalfields does not lend itself to a straightforward exposition, in terms of a series of well-defined stages of growth. This is in part because of the differing geological and topographical conditions across the region and in part because access to organized transportation, in the form of canals and railroads, came early to some locations and late to others. Hence it is not appropriate to differentiate the 'canal era' from the age of railroads in the traditional manner of the general texts on economic history.[1] A distinction can be made, however, between an early pioneering period prior to the 1820s and a later period up to the start of the Civil War. The pioneering period was characterized by isolated mining activity, absence of any reliable or rapid means of transportation, and limited acceptance of a new product i.e. 'stone coal', in the market place. Although the economic impact of anthracite as a commodity was very limited during this period, enough became known about the possible location of deposits and the possible range of uses for the new product to fuel the imagination of visionaries, speculators, investors, and entrepreneurs. The resulting schemes for the development of the coalfields and of the necessary avenues to market by canal, inclined plane, and railroad inaugurated the later period. These initial transportation projects linked the areas where early mining was already established, to the growing markets of the Eastern Seaboard cities. However, the presence of improved transport facilities within the coalfields stimulated the development of new mining activity on adjacent lands. Improved transportation en route to major markets likewise encouraged the establishment of new industries and associated settlements. The manufacture of anthracite iron was a particularly important example of this. Additional sources of coal and manufactured goods in turn fostered new and competing transportation routes, which further stimulated regional economic development. In analysing the overall process, it is essential to keep in mind both the regional scale geography of the coalfields in relation to their industrial and urban markets, and also the detailed local geography of mines, landholding, railroads, and canals within the coalfields. Initial investment decisions were not arbitrary, but were made based on detailed recognition of the importance of these different geographies as they prevailed at any moment in time. Decisions once made subsequently modified these geographies. New markets or new possibilities for future investment could be opened up. Alternatively, acquisition of lands by one company could inhibit the territorial ambitions of another, or deny competitors access to transportation facilities essential for the expansion of their mining operations.

The next section of this chapter briefly describes the early pioneering period, before examining in more detail the patterns and processes of initial mining and transportation development across the coalfields. This is followed by consideration of the different markets for anthracite as a fuel, in competition with wood, bituminous coal, and charcoal; the widening of markets through extension of the transport network; and the growth of the anthracite distribution industry. This last took the form of growing networks of sales and commission agents, and the establishment of large docks/terminal facilities on the Great Lakes and in the eastern seaboard cities.

THE PIONEERING PERIOD

The first recorded mention of 'stone coal' or anthracite is claimed by Eavenson to have been on a map of the Wilkes-Barre area made in 1762 for the

Susquehanna Company of Connecticut. However, by the 1940s when Eavenson was writing, this map could no longer be found.[2] In 1766 there is a reported complaint by Nanticoke and Mohican Indians of white men robbing their mines in the Wyoming Valley near Wilkes-Barre, and by 1769 anthracite was in use in the blacksmith shop of Obadiah Gore and his brother.[3] In 1776 coal was sent down the Susquehanna from the Rezin Geer mine on the property later purchased by Colonel G. M. Hollenback above Wilkes-Barre. It was taken to Harrisburg and then carted by wagon to the United States Armory at Carlisle, a trade that continued throughout the Revolutionary War.[4] Thereafter, reports of the presence of coal beds and coal mines become increasingly frequent. Records of the Pennsylvania Legislature in 1784 make reference to coal mines, at the location where Pottsville now stands in the Southern Field[5] and Philip Ginter is credited with the discovery of coal near Mauch Chunk in 1791.[6]

Although anthracite met with early acceptance by blacksmiths, accustomed to blowing through the coals with a strong draught, understanding of how to use it as a domestic fuel or for manufacturing purposes in furnaces and for steam raising, took much longer to acquire. The first important development in this context is usually taken to be the discovery by Judge Jesse Fell of Wilkes-Barre in February 1808, of how to burn anthracite in an open domestic grate.[7] Other authorities also cite an earlier successful experiment by Oliver Evans, prior to 1802, that was less widely publicized.[8] The recognition accorded to Jesse Fell may derive from the chronological coincidence of his discovery with the establishment of the first significant commercial mining operation in neighbouring Plymouth, by Abijah and John Smith. Abijah Smith had moved to Luzerne County in 1806 from Connecticut and he bought land on either side of Coal Creek.[9] In 1807 he commenced mining and loaded 50 tons of coal on to an ark, which was floated down the Susquehanna to Columbia during high water in October. However, the difficulty of burning it in the domestic grates then in use for wood, made it difficult to sell. Having been joined by his brother John, they returned to these potential markets in the spring of 1808 with a mason who set up grates in public places to give demonstrations of the use of anthracite for domestic purposes.[10] They used Fell's successful design for these grates.[11] Thereafter a steady trade developed along the Susquehanna and the Smiths shipped a total of 6,000 tons in arks between 1807 and 1820.[12] Later, in 1813, they were joined by two other shippers, who sent 4 ark loads down the river from mines further up the valley at Mill Creek and Port Griffith.[13] Arks were substantial purpose built boats, up to 90 feet long, 16 feet wide and 4 feet deep, coming to a sharp point at both bow and stern. The cost of construction was about $70 and they could carry 60 tons of coal. In the absence of canalization, the floating of such arks down river, either on the Susquehanna or on the Lehigh, was a hazardous business that had to be undertaken during flood conditions. It is not surprising, therefore, that as many as 1 in 3 arks despatched for market failed to reach its destination.[14] Despite the risk of loss, the arks were able to support more than a local riverside trade. In 1812 the Smiths delivered about 220 tons of coal to Havre de Grace, Maryland, a distance of almost 200 miles, where it was loaded onto the schooner Washington, consigned to Price and Waterbury in New York.[15]

In the same year, Colonel George Shoemaker loaded nine wagons with coal from his mine on the turnpike road between Pottsville and Ashland in the Southern Field and transported it to Philadelphia. Unfortunately, some of the inhabitants of that city, to whom he disposed of it on a trial basis, were unable to get it to burn and denounced him as an imposter and a swindler! Fortunately, he also sent one load to Messrs. White and Hazard of the Fairmount Nail and Wire Works, and another to the Delaware County Rolling Mill. White and Hazard's men tried every device to get it to burn in one of their furnaces, but seemingly without success. However, after leaving it in disgust and closing the furnace doors, they returned later to find the entire furnace glowing with a white hot mass inside. Thereafter, when properly handled, anthracite's reputation as a first class fuel became rapidly established. In 1814, Charles Miner and Jacob Cist started 5 arks of coal from their mines near Mauch Chunk down the treacherous waters of the Lehigh bound for Philadelphia. Although only 2 arks survived the journey, all the coal was readily taken by White and Hazard at $21 a ton, as they needed no further convincing of the merits of the new fuel for industrial purposes.[16]

The concentration of a number of pioneering attempts to bring anthracite down to the coastal conurbations in the period 1812–1814 was by no means accidental. The War of 1812 and the subsequent blockade of the Chesapeake and Delaware bays by the British seriously impeded normal coastal commerce. As a result, the price of Virginia coal which had previously been sold in Philadelphia for 30 cents a bushel, had risen to 75 cents – $1 by March 1813 and remained at $1 for the next two years. Hence there was a considerable economic incentive to establish a trade in anthracite, both with Philadelphia via the Schuylkill and Lehigh rivers, and with Baltimore via the Susquehanna.[17]

Until the return of peace in 1815, Jacob Cist, now aided by Isaac Chapman, endeavoured to develop the fledgling Lehigh coal trade, by removing more overburden in the Summit Hill quarry and constructing a yard for building arks at the mouth of Nesquehoning Creek. Cist also kept a close eye on matters of corporate finance, buying up stock in the Lehigh Coal Mine Company which dated back to 1792, to ensure control of its future operations.[18] Despite Cist's tireless efforts as an advocate of the merits of Lehigh anthracite, the ending of the war brought hard times for the coal trade and finance for new developments was hard to obtain. However, the state legislature had become aware of the problems of shipping anthracite caused by navigation difficulties on the Lehigh River. Acts of 1814 and 1817 provided for the establishment of a company to improve navigation and committed $50,000 of state support for the enterprise. Attracted by this prospect of state funds, White, Hazard, and a further partner George Hauto, visited the Summit Hill mine and associated enterprises in December 1817 and were impressed by their findings.[19] In January 1818 they contracted with the Lehigh Coal Mine Company for a 20-year period to mine and ship coal to Philadelphia, and the following March they obtained an act to improve navigation on the Lehigh. The resulting corporation, named the Lehigh Navigation and Coal Mine Company, was to be capitalized at the large figure of $200,000. Unfortunately this large undertaking did not meet with the support of Philadelphia investors, but a revised and more modest proposal for two companies, one to improve navigation, the other to engage in mining, was deemed more acceptable and obtained the necessary financial backing. After two further reorganizations, the two companies were again merged in 1822 under the new charter of the Lehigh Coal and Navigation Company, which was to become one of the largest private corporations in the United States during the 1820s. Josiah White's greatest achievement was the construction of the 46-mile slack water navigation system from Mauch Chunk to Easton that finally made reliable coal shipments from the Lehigh mines feasible.[20]

It is interesting to note that White and Hazard turned their attention to the Lehigh coal trade because of their failure to secure control of the competing enterprise of the Schuylkill Navigation Company, as a result of opposition from merchants in the counties of Philadelphia's hinterland. The original aim of canalization of the Schuylkill from Philadelphia to Pottsville was to provide an outlet for the agricultural and lumber resources of the river valley. At the time the company was chartered in 1815, the coal trade was not large enough to be thought commercially significant, but this early assessment had already been superseded by the time the canal was officially opened to commercial traffic on October 1, 1825. By 1827, over 57 percent of its total receipts were derived from coal shipments.[21]

Thus, by the beginning of the 1820s, the eastern end of the Southern Field in the vicinity of Summit Hill was provided with an avenue to market, and work was in progress to provide a similar outlet via the Schuylkill for the developing mines around Pottsville. Although no equivalent improvement schemes had yet been mooted in the Northern Field, the less capricious nature of the Susquehanna had enabled several coal operators, led by the Smith Brothers, to establish a modest commercial trade down the river. The first shipments by wagon and river transport from the Western Middle Field were not made until 1826, the same year that John Charles, a hunter, discovered coal in the Eastern Middle Field near Hazleton.[22]

EARLY DEVELOPMENTS IN TRANSPORTATION

The pioneering improvements on the Schuylkill and the Lehigh rivers gave the mining areas they served a significant initial advantage in the coal markets of Philadelphia and New York. Aiming to press home this advantage, the managers of the Lehigh Coal and Navigation Company planned to enlarge the Lehigh Canal to a width of 60 feet and a depth of 5 feet, with locks large enough to carry 120 ton boats. These enlargements were completed in June 1829.[23] In terms of the coal trade this was perhaps premature, as only 41,750 tons were carried in 1830, less than all of the tonnage on the Schuylkill Navigation, whose dimensions were much smaller.[24]

The coal merchants and entrepreneurs in the Northern Field quickly realized that they suffered from two major disadvantages in comparison to their competitors in the Schuylkill and Lehigh Regions. Not only did they lack good outlets to the Philadelphia and New York markets, but they were also subject to the vagaries of their circuitous route to the eastern seaboard along the Susquehanna. Their response took the form of two quite separate canal schemes. The first was the Delaware and Hudson Canal to provide an outlet eastwards from the northern end of the Northern Field to the Hudson River and thence to New York. The second was the North Branch Canal, which paralleled the Susquehanna from Pittston in the heart of the coalfield down as far as Northumberland to the southwest.

Coal deposits were known to exist in the Carbondale area of the Northern Field prior to 1812 when William Wurts, a prosperous young Philadelphia merchant, and his brother Maurice decided to invest in coal lands and open up a trade with the Quaker City. In common with the other pioneers of the period, they used arks to bring coal to market via the Wallenpaupack, Lackawanna, and Delaware rivers in 1815. However, the hazards of unimproved navigation on small rivers and the costs of carting combined to make the initial enterprise unprofitable. A second attempt at selling coal in the Philadelphia market in 1822 also failed because of competition from Schuylkill coals, so the Wurts Brothers resolved to turn their attention to the New York market instead. In March 1823, the Pennsylvania legislature authorized Maurice Wurts to improve the navigation of the Lackawanna River. In the following month, the New York legislature granted a perpetual charter to the Delaware and Hudson Canal Company to construct a canal from the Hudson River to the Delaware and thence up the Delaware to the mouth of the Lackawaxen.[25] In 1825, the new company purchased the coal lands of Maurice Wurts and, on July 13 of that year, canal construction began. The whole line of the canal from Honesdale to Rondout on the Hudson was completed by October 1828, and the first coal shipment arrived on February 5, 1829. The canal was large enough for 30-ton boats and required 110 locks.[26] As it was infeasible to build a canal all the way to the mines, in April 1827, J. B. Jervis, Chief Engineer of the Canal Company, was instructed to survey a route for a railroad from Carbondale to the planned canal terminus at Honesdale. Jervis' report indicated that the route could be covered by a combination of inclined planes and almost level sections of railroad. On the latter the new technology of the steam locomotive could be deployed.[27] In January 1828, Horatio Allen, one of the company's engineers, was sent to England to buy 4 locomotives and the necessary railroad iron. The first of these locomotives, the Stourbridge Lion, arrived in New York in May 1829. With Allen in charge it made its first trial run in August 1829, but after a second trial run the following month, it became clear that the engine was too heavy and the road-bed too unstable to support this method of locomotion. Horsepower on the level sections of the railroad was therefore substituted instead.[28] In the ensuing years, coal shipments increased steadily and 111,777 tons were sent to market in 1833.

Although it was mooted at about the same time as the Delaware and Hudson project, the North Branch Canal took significantly longer to come to fruition, partly because it became embroiled in wider political questions about the policy for internal improvements across the State of Pennsylvania. A large public meeting in Wilkes-Barre in November 1822 initiated a lobbying campaign to persuade the legislature to fund improvements to navigation on the Susquehanna River. Not surprisingly, leading protagonists included those with business interests in the coal trade, such as John Smith and Jacob Cist. Cist, in particular, before his untimely death in 1825, supported not only the larger scheme to link the Allegheny and Susquehanna rivers, but also was a prominent advocate of a canal extension along the north branch of the Susquehanna into the Northern Field. In 1827, surveys for a proposed North Branch Canal from Athens near the New York state line down to Northumberland, a distance of about 160 miles, were completed. The cost estimates varied from $1.8 to $2.2 million, depending on the precise route chosen. On March 24, 1828 the legislature finally approved the first phase of construction, after intensive lobbying by the representatives from Luzerne County, one of whom, Garrick Mallery, was later to become a prominent Pittston coal operator. The ground-breaking ceremony for the canal took place near Berwick in July 1828 and construction continued for the next two years on the first stretch, which extended up to Nanticoke Dam. On October 25, 1830, the first three boats carrying anthracite were sent down to Berwick. The next stage of the work involved construction of the Wyoming Division of the Canal from Wilkes-Barre to Pittston. And, in March 1831, a first appropriation of $100,000 was made by the state legislature for this purpose.

On the lower section the canal had been built on the west side of the river, but after much argument it was decided to take the route along the east side which would take it through the city of Wilkes-Barre. The Wyoming Division was completed in June 1834 at a cost of $342,625.[29] It is difficult to argue that the canal provided a major early stimulus to the coal trade, since as late as 1841 only 41,210 tons were carried[30] as compared to 192,270 tons on the Delaware and Hudson at the same date. Indeed, based on his estimates of the early river trade in coal which had reached 25–30,000 tons by 1827, Powell argues that it took nearly a decade for the canal to become the dominant transport route from the western end of the Northern Field.[31] This should not be taken to imply, however, that the canal was unimportant to later development of the mining industry and its evolving geography. A key aspect of this was that for a significant proportion of its length, the canal, unlike all the other anthracite canals, was within the boundaries of the coalfield itself. The implications of this will be examined in more detail below.

Since the other canals were outside the coal measures, the construction of railroad tracks to bring the coal down from the mines was essential not just for the Delaware and Hudson Canal Company, but in other areas also. The most extensive network of 'coal lateral' railroads, as they were known, developed rapidly as feeders to the Schuylkill Canal. These included the Mill Creek and Mine Hill Railroad completed in 1829, the Mine Hill and Schuylkill Haven Railroad in 1831, and the Little Schuylkill Railroad, which had locomotives running on it by 1833, having previously used horses and mules to pull the coal cars.[32] The first gravity railroad to the Lehigh Canal was constructed in 1827 by the Lehigh Coal and Navigation Company to cover the 9-mile distance from the Summit Hill mines to Mauch Chunk. Subsequently, the Beaver Meadow Railroad opened up the coal basin of the same name in the Eastern Middle Field, the 26-mile line being completed in 1836. This was followed by the Hazleton Railroad, chartered in the same year, which connected to it at Weatherly.[33]

In these early enterprises, the railroad was clearly the servant of the canal, but this dominance was very quickly to be eroded. The original charter of the Little Schuylkill Railroad allowed it to build not just coal lateral lines, but to continue down the Schuylkill Valley to the city of Reading, only 54 miles from Philadelphia. A further railroad company, the Philadelphia and Reading was chartered in 1833 to link to it and provide a through route to Philadelphia from the coalfields. Construction on the Philadelphia to Reading line commenced in 1835, but owing to the 1837 panic the Little Schuylkill Railroad never carried out its half of the project. Despite precarious financing arrangements, the line to Reading was finally opened in December 1839[34] and, with the help of the Second Bank of the United States, construction continued northwards, the Little Schuylkill having abandoned its plans for an extension to Reading. The entire line from Pottsville to Philadelphia was finally opened on January 10, 1842[35] and its adverse effect on the level of shipments of coal by canal began to be felt immediately.

The proprietors of the Lehigh Canal were in a more fortunate position in these early years from the viewpoint of competition. The construction of the Delaware Division canal between 1827 and 1832 from the terminus of the Lehigh Canal at Easton down to Bristol gave a reliable outlet to Philadelphia. It also provided a somewhat circuitous route to the New York market via the Delaware and Raritan Canal. Another route to New York was provided by the Morris Canal, which ran in an easterly direction across New Jersey. Construction began in 1825 and was completed from Easton to Newark in November 1831. The canal was subsequently extended to Jersey City.[36]

The first major railroad development, apart from the coal lateral feeders, was undertaken by the Lehigh Coal and Navigation Company itself and therefore aimed to augment the traffic on the canal rather than compete with it. In 1844 the Lehigh and Susquehanna Railroad, as it was named, linked Wilkes-Barre in the Northern Field with Whitehaven on the Lehigh, which by then was the head of canal navigation. Coal was raised out of the Wyoming Valley by means of inclined planes at Ashley below Wilkes-Barre. Not only did the new railroad add to the Lehigh Canal tonnage, it also served to undermine the growing trade of the North Branch Canal by siphoning off potential traffic. Nevertheless, its initial impact was modest, since only 19,590 tons were carried as late as 1849.[37]

A major threat to the Lehigh Canal from railroad competition did not arise until 1853, when construction of the Lehigh Valley Railroad from Mauch Chunk to Easton commenced, under the direction of Asa Packer.[38] Although a comparatively

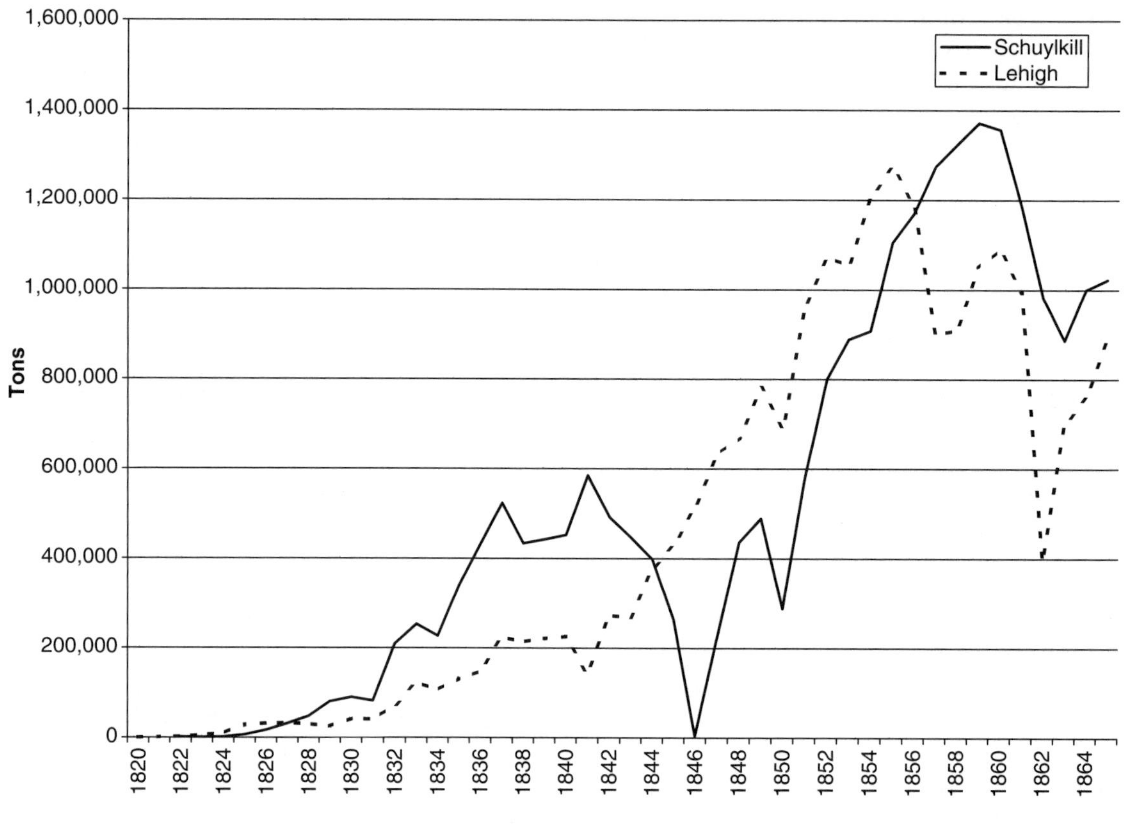

FIGURE 4.1
Anthracite Shipments on the Schuylkill Navigation and the Lehigh Canal 1820–1865.

short line when first completed in 1855, it immediately had outlets to New York via the Morris Canal and the Central Railroad of New Jersey, which had its terminus in nearby Phillipsburg. The latter road had been partly constructed as early as 1838 under the name of the Elizabethtown and Somerville Railroad. However, it was only completed from Elizabethport to Phillipsburg, under its new name of the New Jersey Central, in July 1852.[39]

The two graphs of canal shipments up to 1865 show the differential impact of direct railroad competition during the period. In Figure 4.1, the time series for the Schuylkill Canal shows the benefit of its 'early mover' advantage before the Reading Railroad reached Pottsville in 1842, followed by a dramatic collapse in shipments. The railroad not only competed with the canal but actively cut off its feeder traffic, by making arrangements for the trade of the different mining districts. Only when the two companies agreed to work together to raise tolls and increase income did shipments start to recover.[40] The Lehigh Canal, in contrast, had a much slower period of early growth, largely because of the limited number of operating mines tributary to the canal. When the Eastern Middle Field began to be opened up more extensively in the 1840s, growth accelerated, peaking in 1855, the year of completion of the Lehigh Valley Railroad. The coal tonnage of the latter grew dramatically from a mere 9,063 tons in 1855 to 577,652 tons by 1859. Thereafter the canal went into decline, although the precipitate fall in shipments in 1862 was a result of disastrous flood damage rather than adverse business conditions.[41]

The Delaware and Hudson Canal Company did not experience any equivalent direct competition for coal tonnage from areas tributary to its own gravity railroad in the Carbondale area. Hence its shipments show a continued increase through the period, punctuated only by the aftermath of the 1857 panic (Figure 4.2). The North Branch Canal, which had risen to rival the Delaware and Hudson in terms of shipments by the mid 1850s, likewise suffered from the adverse business conditions of the subsequent

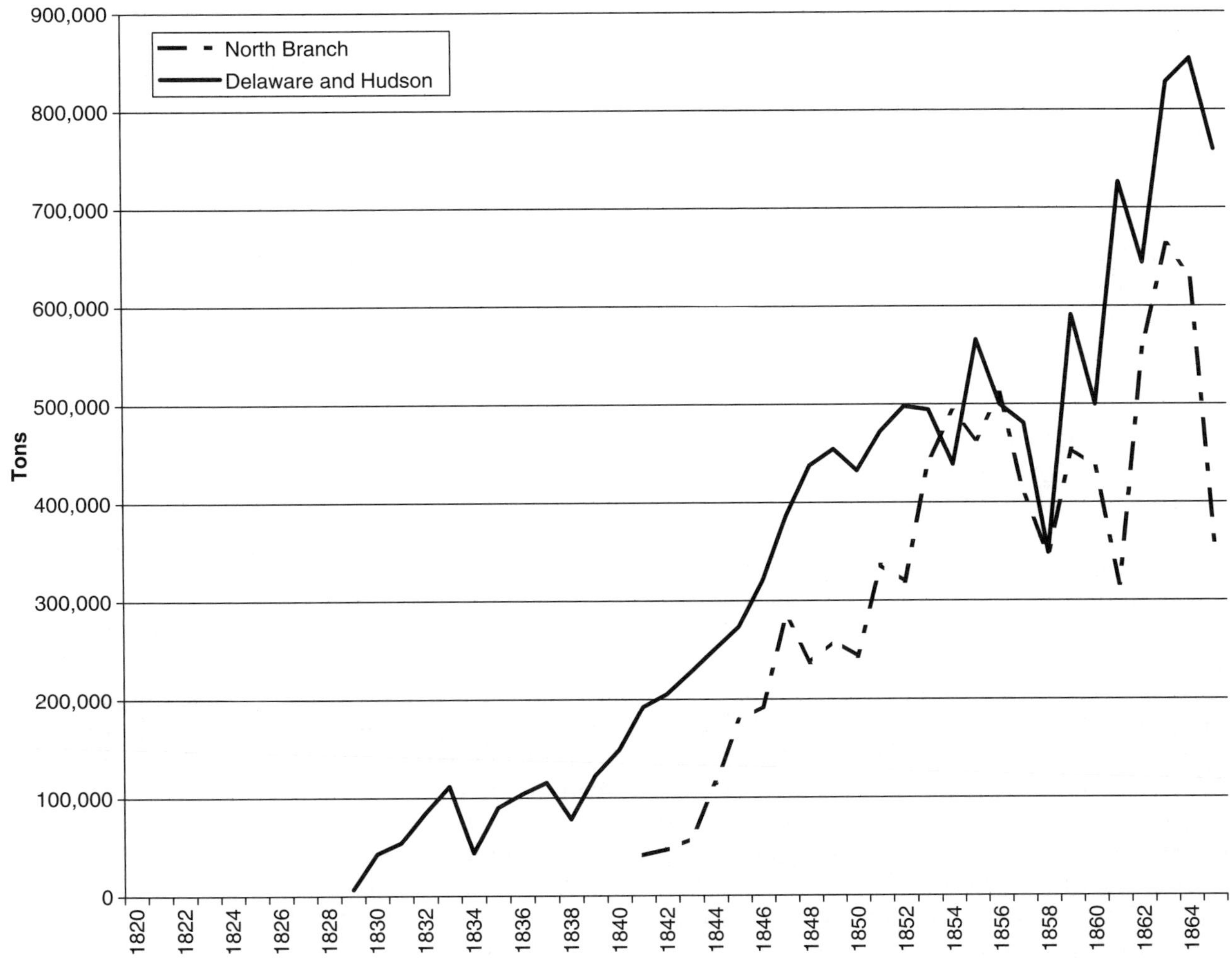

FIGURE 4.2
Anthracite Shipments by Canal from the Northern Coal Field 1820–1865.

years. Recovery was delayed until the Civil War, since the impact of railroad competition, in the form of the Lackawanna and Bloomsburg Railroad, had begun to be felt. This latter railroad was organized on April 16, 1853 and the first construction contract was let in September 1854.[42] By June 1856, track was laid from Nanticoke as far as West Pittston[43] and the first shipments were made in 1858. The extension to Northumberland was completed in June 1860[44] and the Northern Central Railway could then be reached over the tracks of the Philadelphia and Erie Railroad to Sunbury. This provided a through line all the way from Scranton west and south to Baltimore along the Susquehanna Valley. It effectively signalled the start of the canal's demise, since the railroad surpassed it in coal tonnage in its third year of operation. It was also to have a major impact on the location and timing of development of mining on the west bank of the Susquehanna, because of the large areas of relatively untouched coal lands it crossed en route from Nanticoke to Scranton. The Lackawanna and Bloomsburg had been supported both by Plymouth coal interests and the Delaware, Lackawanna and Western Railroad since its inception, but in 1864 the chance arose at short notice for the Pennsylvania Railroad to obtain a controlling financial interest in the company. The minutes of the Pennsylvania record that this came to the attention of the Board of Directors. Instead of taking a prompt decision to secure the Lackawanna and Bloomsburg, they decided to form a special committee to consider what action they should take, but 'with the favorable recommendation of the Board'. Unfortunately, before this committee had time to meet, the

Delaware, Lackawanna and Western stepped in to take control instead![45] This incident of bureaucratic bungling provides scholars of the anthracite industry with an interesting opportunity for counterfactual speculation. Had the Pennsylvania Railroad been quicker or more decisive, the later course of the anthracite industry would undoubtedly have been changed. The Northern Field, in particular, would have had a much stronger orientation towards western markets, at an earlier stage in its development, than actually proved to be the case.

CO-OPERATION AND CONFLICT: THE PENNSYLVANIA COAL COMPANY AND THE DELAWARE AND HUDSON

A further interesting aspect of railroad competition, as it affected the North Branch Canal, related to the development of the Pennsylvania Coal Company. This company opened a number of mines in the Pittston area, several of them in close proximity to the terminus of the canal, but never shipped any significant tonnage over it. The explanation of this lay in the rather complex legal and organizational background to the emergence of the Pennsylvania Coal Company as a major player in the coal trade of the Northern Field. The original Charter of the Company was granted in 1838 and it authorized the building of a railroad to transport coal either to the North Branch Canal or to another railroad. Prior to 1849, it could also hold 2,000 acres of coal land. In that year, it doubled its coal estate by taking over the Washington Coal Company. This had a similar charter, allowing it to own 2,000 acres of coal lands and build a railroad, either to the North Branch or to the Delaware and Hudson Canal.[46] Thus from the outset the Pennsylvania Coal Company had the option of sending its coal shipments by routes other than the North Branch Canal. Indeed, by the time the Washington Coal Company was taken over, the latter had already begun the construction of a gravity railroad quite separate from that of the Delaware and Hudson, to join the latter company's canal at Hawley on the Lackawaxen below Honesdale.[47] The orientation towards the Delaware and Hudson Company and the New York markets was strengthened because several individuals such as James Archbald and Irad Hawley, who were closely associated with the Delaware and Hudson, became heavily involved in the Washington Coal Company. In 1847, Archbald, for example, was appointed engineer with responsibility for construction of the new gravity railroad.[48] President John Wurts of the Delaware and Hudson was also a director of the Washington Coal Company. He was one of the prime movers in the Pennsylvania Coal Company's purchase of 1,700 acres of lands around Dunmore below Scranton, from William and Charles Wurts in September 1847.[49] Recent publications have unfortunately made and perpetuated some factual errors concerning the relationship between the Delaware and Hudson and the Pennsylvania Coal Company at this time, and these need to be corrected. Ruth claims that William R. Griffith, the new president of the Pennsylvania Coal Company, offered to sell to the Washington Coal Company the lands owned by the former below Pittston.[50] The company's minutes tell a different story, however. On October 12, 1848, William R. Griffith was authorized by the Board of Directors of the Pennsylvania Coal Company to purchase from the Wyoming Coal Association a number of parcels of land, amounting to 937 acres 11 perches, for the sum of $134,571.68. Among these parcels was land bought by the Association at an earlier date from William R. Griffith himself, a total of 278 acres 46 perches. Also, Patton,[51] in his recent study of relationships between the Delaware and Hudson and the Pennsylvania Coal Company, claims that Delaware and Hudson interests, including John Wurts, did not approve the merger of the Washington Coal Company into the Pennsylvania Coal Company. Hence they quickly removed themselves from the board of the latter company. This is incorrect on two counts. Firstly, John Wurts was never a director of the Pennsylvania Coal Company, but rather of the Washington Coal Company. Secondly, he and others resigned from the board of this company on January 7, 1848 to enable it to comply with a new State Law on the election of Boards of Directors. This was seventeen months **before** the merger of the Pennsylvania Coal Company and the Washington Coal Company took place. One of the people elected to succeed Wurts was Russell F. Lord, Chief Engineer of the Delaware and Hudson, who remained a director until the merger took effect.[52] By October 1849, rather than falling out amongst themselves, the Delaware and Hudson and the Pennsylvania Coal Company were in fact meeting to discuss an agreement for the sale of the latter's coal by the former.[53]

The final text of this agreement made the payment of tolls by the Pennsylvania Coal Company subject to the earlier agreement of the Wyoming Coal Association with the Delaware and Hudson, dated August 31, 1847.[54] This agreement contained two key clauses. The first was a complex provision, over which the Coal Association and the Delaware and Hudson Company had difficulty agreeing when it was first drafted.[55] It required that if the canal were subsequently enlarged, half of the saving in freight cost per ton by the Coal Association directly attributable to the greater size of the canal, would be payable to the Delaware and Hudson Company as an additional toll. The second clause provided for two arbitrators to be appointed, one by each company, in the event of disagreements arising under the previous clause.[56] The possibility of canal enlargement was not hypothetical, because R. F. Lord had submitted a report on June 16, 1847 to President Wurts, detailing his estimates of the costs of enlarging the canal to take boats of 136 tons. These estimates also included a key figure of 40 cents per ton saving on freight costs that were expected to derive from this enlargement. Interestingly, the report was not, in the quaint phrase of the time, 'spread upon the minutes' until November 17, 1847. So it is possible that some of the other Delaware and Hudson Company directors who had interests in the Wyoming Coal Association and indeed the Pennsylvania Coal Company were not fully aware of these estimates at the time of the original agreement between the Coal Association and the Delaware and Hudson Company in August.

The rationale, both for the moves that led to the formation of the Wyoming Coal Association and for canal enlargement, was clear. While the original canal could be pressed to ship 1 million tons a year with changes in its locks, the gravity railroad could not exceed 500,000 tons capacity, even after improvements were made.[57] Hence the Delaware and Hudson Company could not fill the capacity of its own canal and a second initiative for a further railroad was required. As early as January 1847, the Board of Managers Committee, charged with responding to Wurts' communication of December 1846, was preparing to put a series of questions to James Archbald for his engineer's response. These questions included the comparative costs of gravity railroads, either from Providence (near Scranton) or Pittston, to the Delaware and Hudson Canal. It is not difficult to see, therefore, why the Wyoming Coal Association was brought into being on June 24, 1847. It was a political vehicle for building the new gravity railroad. Five of its seven managers were also on the board of the Delaware and Hudson Company and the others were associated either with the Washington Coal Company or the Pennsylvania Coal Company.[58] If no action was taken, the alternative was the likely future appearance of 'an inconvenient and perhaps dangerous rival'. This would take the form of a railroad connecting the lower part of the Lackawanna coalfield with tidewater, over which the Delaware and Hudson would have no control and from which it could derive no revenue.[59] Knies[60] provides compelling evidence from the surviving James Archbald correspondence for an additional key reason behind the establishment of the Wyoming Coal Association. This was to allow the Delaware and Hudson to circumvent charter restrictions on the amount of coal lands it could hold, thereby extending its control and denying its competitors access to coal resources. Similar manoeuvres by the Delaware and Hudson and other transporting companies were to become a feature of the anthracite trade in later years.

Gravity railroads were 'lumpy' investments. The one to be built by the Washington Coal Company and its successor, the Pennsylvania Coal Company, was to have a larger capacity than the older Delaware and Hudson gravity railroad. Without this new railroad, the canal would be half empty; with it, the old capacity of the canal would be insufficient. Hence substantial enlargement was the only solution to drive down costs per ton, retain market share and, by implication, the level of dividends to which shareholders had become accustomed. Lord's 1847 estimates gave an indicative cost of $1.1 million for these improvements, exclusive of new boats ($400k) and new aqueducts over the Delaware and Lackawaxen Rivers ($300k). It was expected that the enlargement could be sufficiently far advanced by the spring of 1850 to permit larger boats to pass, but it would be May 1851 before the work was completed.[61] The timing proved accurate but the cost less so. By August 1850, over $2 million had been expended on the canal alone and the final cost was estimated to be $2.4 million.[62] Patton gives a similar value for the final costs of the canal enlargement, but argues that half of the amount was for boats.[63] This does not seem to be borne out by the above figures.

Thus the Delaware and Hudson needed the tonnage of the Pennsylvania Coal Company as much as the latter company needed the outlet to tidewater for its coal. Certainly the full tariff tolls payable by the Pennsylvania Coal Company had a dramatically beneficial effect on the profit and loss account of the Delaware and Hudson, whose toll revenues increased tenfold between 1850 and 1853, to a total of $390,000.[64]

The gravity railroad begun by the Washington Coal Company was completed by the Pennsylvania Coal Company in 1850 and shipped 111,014 tons in that year to the canal. The acquisition of the Pittston lands of William Griffith and others by the Pennsylvania Coal Company required an extension of the original gravity railroad from Dunmore to Port Griffith on the Susquehanna, below Pittston. It was also extremely fortunate for the combined Pennsylvania Coal Company/Washington Coal Company because, as Throop observed in his later history of Scranton, the original Dunmore lands of the Washington Coal Company did not prove very profitable when first explored. A 150-foot shaft was sunk but no coal found, the only coal being at the No. 6 opening, where the Company's shops were located.[65] By itself this would not have been sufficient to support a gravity railroad with a capacity to transport 1.5 million tons of coal a year.[66]

In addition to the standard tolls, the Delaware and Hudson expected to receive from the Pennsylvania Coal Company half of its claimed 40 cents reduction per ton in freight costs as a supplementary charge. This was scarcely news in 1850, as it had been clearly stated in the annual stockholders report on March 28, 1848. Provision for it was also in the 1849 shipping agreement with the Pennsylvania Coal Company, which referred back to the 1847 agreement with the Wyoming Coal Association, as mentioned earlier. Immediately after the absorption of the Association by the Pennsylvania Coal Company in May 1851, the transfer of its obligations on tolls etc. was ratified by the Boards of both the Pennsylvania Coal Company and the Delaware and Hudson.[67] With the contractual position seemingly fixed, the Delaware and Hudson not unreasonably expected to begin collecting its supplementary charges. Perhaps the only slight hint that trouble might lie ahead came from a brief entry in the Delaware and Hudson Minutes on May 29, 1851, when the Delaware and Hudson declined to take Pennsylvania Coal Company paper to help the latter company with its finances. However, the Treasurer was empowered to extend temporary loans or cash advances to the Pennsylvania Coal Company to assist its cash flow.

While the Delaware and Hudson Company doubtless considered it premature to request the supplementary payment before the canal enlargement was completed in 1851, the question of these payments, or the lack of them, had come to the attention of the Board by the beginning of 1852. Attention was drawn to it in a letter dated May 8, 1852 from William Musgrave, Vice President of the Delaware and Hudson to John Ewen, President of the Pennsylvania Coal Company (who had resigned as Vice President of the Delaware and Hudson in June 1849). The exchange of letters continued into 1853, the Delaware and Hudson, in particular, attempting to maintain a friendly tone, while pressing its case. This is hardly surprising as the Pennsylvania Coal Company was its best customer, with a coal tonnage that by 1853 exceeded the Delaware and Hudson's own. Equally, the Pennsylvania Coal Company, still a young company building its financial strength, would have recognized its strong bargaining position, and any delaying tactic had the benefit of saving it money. It also regarded the substantial freight cost savings claimed by the Delaware and Hudson as being fictitious as far as it was concerned. Months passed and eventually arbitrators were appointed,[68] but seemingly never met. In the interim, it was business as usual, with the Pennsylvania Coal Company paying its standard tolls but not the surcharge. Eventually, litigation became inevitable. In September 1856, the Pennsylvania Coal Company filed a suit against the Delaware and Hudson for alleged tactics of harassment used in attempting to collect unpaid tolls. The Delaware and Hudson replied with a counter suit in 1857, thereby initiating a long and costly trial, described in detail by Patton.[69] After almost interminable testimony, the hearing finally closed in May 1861, but the verdict was not handed down until 1863. It awarded the Delaware and Hudson Company about $450,000 of the $1.2 million in unpaid tolls it had sought.

The victory for the Delaware and Hudson Company was pyrrhic as well as partial. In November 1863, the connecting railroad from Hawley to the main line of the Erie Railway was completed. By the following month, the Pennsylvania Coal Company anticipated sending 1,000 tons per day via this route to reach tidewater at Newburgh, New York, bypassing the canal completely.[70]

SCRANTON AND THE LACKAWANNA RAILROAD

President Wurts' concerns in 1846 about the appearance of inconvenient and dangerous rivals were perhaps less prophetic than they appeared. Agitation for railroad routes from the coalfield north via Leggett's Gap to Great Bend on the New York state line, and east via Cobb's Gap to the Delaware River, had begun as early as 1825–6. However, investors preferred canal projects in the wake of the success of the Erie Canal and no finance was forthcoming. Later attempts in the 1830s foundered in a similar manner during the depression of 1835–7.[71] The first real impetus to improvements in transportation came from the establishment of the iron-making firm of Scrantons, Grant and Company in Slocum Hollow during 1840.[72] This firm was subsequently renamed the Lackawanna Iron and Coal Company. Slocum Hollow was later to become the city of Scranton and was conveniently situated between Leggett's Gap to the north and Cobb's Gap to the south. Unfortunately, these geographical advantages were yet to be realized. The hard commercial reality of the time was that Slocum Hollow was almost equidistant from the head of the North Branch Canal at Pittston and the start of the Delaware and Hudson Company gravity railroad at Carbondale. Slow travel for some miles over rough wagon roads was required in either direction to bring in raw materials or equipment and export iron products. The high cost of overcoming distance, together with other technological and financial problems, had almost brought the fledgling iron company to its knees by 1846. Fortunately, they were saved by an offer to supply 12,000 tons of T-rail for the New York and Erie Railroad, in time for it to complete its Delaware Division and claim a $3 million incentive award offered by the state. To fulfil the contract, the rails had to be carried by horse and mule teams, first to the gravity railroad for the canal, and thereafter over wilderness roads to various points along the line of the Erie Railroad.[73]

The extreme difficulty of completing this contract spurred the Scrantons into action to improve their transport avenues to market. In 1847 they acquired the old charter of the Leggett's Gap line for the northern route and in 1849 they obtained a new charter for the Delaware and Cobb's Gap Railroad for the route south and east to the Delaware River. This time, with financial backing from John Blair the New Jersey ironmaster and Phelps, Dodge and Company, the project attracted the necessary support and construction of the northern route began in the spring of 1851. It was completed in October of the same year. The topographic obstacles faced on the southern route were much more severe, so construction was slower. The road was finally opened on May 27, 1856, and an arrangement with the Central Railroad of New Jersey enabled shipments to reach tidewater.[74] Prior to this, in 1853, the northern and southern routes had been consolidated into a single line called the Delaware, Lackawanna and Western Railroad.[75] Shipments over the new line grew rapidly, surpassing those of the Pennsylvania Coal Company in 1858 with a total of 683,411 tons.

The different river, canal, gravity and steam railroad routes from the Northern Field produced four distinct concentrations of economic activity in this period. The earliest, linked to the river trade and later the canal, centered around Plymouth and neighbouring Wilkes-Barre, the county seat of Luzerne County, at the lower end of the coalfield. The second area of population growth was Carbondale in the upper part of the field because of the mines and gravity railroad of the Delaware and Hudson. Pittston, at the head of navigation on the North Branch Canal, developed quite quickly in the late 1840s and early 1850s, when growth in the canal trade coincided with the start of mining activities by the Pennsylvania Coal Company and completion of its gravity railroad. Its population increased nearly fourfold to 4,049 from 1840 to 1850.[76] Scranton was the last to develop, though it was later to become the major urban center in the coalfield. Early growth was dependent on the fortunes of the ironworks, until the Lackawanna Railroad began to expand during the 1850s. Elsewhere in the coalfields, Hazleton in the Eastern Middle Field and Pottsville in the Southern Field became the major urban centers, with smaller settlements such as Tamaqua and Shamokin also becoming significant.

CORPORATE CHARTERS AND COAL LAND HOLDINGS

One factor that contributed to the concentration of activity in three of the four areas in the Northern Field was the nature of the corporate organization of the companies involved. The liberal charters of the Delaware and Hudson and the Pennsylvania Coal Company enabled them to own varying amounts of coal lands and engage in the mining of coal as well as its transportation. The same was also true of the Delaware, Lackawanna and Western Railroad.

Initially, the Delaware and Hudson had the largest share, since its amended charter of 1859 allowed it to hold up to 8,000 acres of coal lands.[77] The Lackawanna Railroad Coal Department could initially hold 1,000 acres,[78] while a series of charter amendments and acquisitions of subsidiary coal companies allowed the Pennsylvania Coal Company to build up a substantial coal estate, rivalling that of the Delaware and Hudson. The Washington Coal Company purchase, described above, took its holdings to approximately 4,000 acres, while the property of the Wyoming Coal Association, acquired in 1851, contained 580 acres of coal, and both coal and surface on an additional 3,780 acres. The potential to augment land-holding still further was provided by a later charter supplement of April 14, 1864, which authorised the company to buy another 3,000 acres of coal land.[79]

The only other major company in the industry to have similar rights was the Lehigh Coal and Navigation Company, which from the very early days of the mining industry had held some 10,000 acres of land near Summit Hill. Over the years, it resisted all attempts to have its landholding and mining privileges removed.[80] In the Eastern Middle Field only the Hazleton Coal Company and the Beaver Meadow Railroad and Coal Company had both mining and transporting privileges. These were on a more modest scale, as the two companies owned approximately 3,000 acres of coal land in total.[81] The concentration of large companies with these liberal privileges in the Northern Field contrasted strongly with the situation in the Southern Field outside the territory of the Lehigh Coal and Navigation Company. Here the Reading Railroad and the Schuylkill Navigation Company were both, by charter, 'common carriers'. This meant that in principle they had to provide railroad cars and shipping facilities at the behest of their customers, without discrimination. Since these large companies were also prohibited from investing in mining, the areas tributary to their transportation routes became the province of the small company, or the individual mine operator, working on a small tract leased from a larger land improvement company. These improvement companies became particularly important in the Schuylkill Region after the passage of the 1854 Incorporation Act. Adams has closely examined the relationship between charter legislation at different dates and the expansion of mining in the different coalfields. However, it is apparent from his data that variations in chartering activity closely follow business cycle changes and the expansionary period after 1863, so it is important not to assume that legislation determined the pace of economic development, rather than accommodating to wider economic forces. Also, not all companies chartered for mining purposes, actually engaged in significant mining activity.[82] Later entrants into the anthracite shipment business, such as the Lehigh Valley Railroad and the Central Railroad of New Jersey, were common carriers like the Reading, with no mining rights, as was the North Branch Canal.

Regardless of type of corporate organization, the period up to the mid 1850s in the Northern Field was dominated by gravity railroad and canal transportation. Not until 1863, when the Pennsylvania Coal Company switched from the Delaware and Hudson Canal to the Erie Railway, did the pendulum swing the other way. In terms of the division of shipments between the Lehigh Canal and the Lehigh Valley Railroad, the change in the dominant mode of transport from the former to the latter took place in the same year. In the Southern Field, by contrast, the railroad had become pre-eminent as early as 1845.

COAL PRODUCTION AND REGIONAL GROWTH

It is difficult to make any clear inferences about the impact of the dominant transport mode on the rate of growth of the industry in the different 'regions', as can be seen from Figures 4.3 and 4.4. It should be stressed that, on these graphs, the regions do not entirely correspond with the straightforward geographical breakdown by coalfield. Although the Wyoming Region matches the Northern Field, the Lehigh Region includes not only the Eastern Middle Field, but also that portion of the Southern Field east of Tamaqua. The Schuylkill Region encompasses both the remainder of the Southern Field and also the Western Middle Field. This was the standard classification of the Coal Regions by the anthracite trade in the 19th century and therefore annual tonnages were reported on that basis.

The figures used are those originally collated by P. W. Schaefer for the period up to 1868, thereafter by J. H. Jones, official accountant of the anthracite transporting companies. According to a later coal trade analyst, they were those 'adopted as the most correct' by the trade as a whole.[83] It should be noted, however, that this data series does not exactly match that produced for many years by Benjamin Bannan, editor of the Pottsville Miner's Journal. The Journal contains occasional sideswipes at Schaefer in relation to the accuracy of his statistics, and the disparities probably relate to the

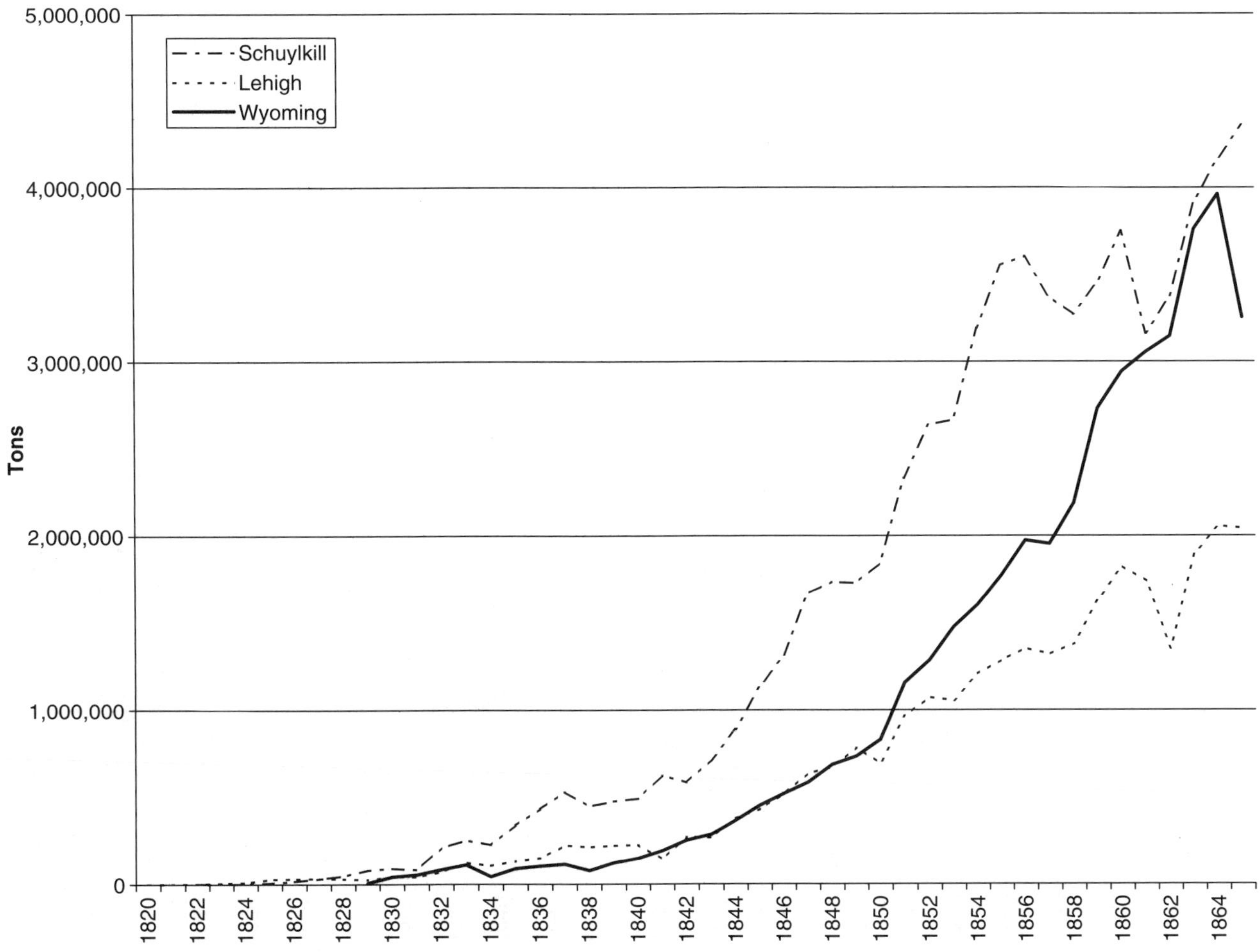

FIGURE 4.3
Total Shipments from the Different Coal Regions 1820–1865.

method of accounting for tonnage carried over more than one railroad on its way to market. It should also be added that the disparities relate very largely to the tonnage of the Schuylkill region. In Bannan's defence, his figures for Schuylkill railroad shipments correspond exactly to those on the railroad's own books.[84] Neither series accounts for the early river trade along the Susquehanna prior to 1840.

Setting aside these minor statistical issues, it is apparent from the graphs that the three regions experienced somewhat different growth patterns in the decades preceding the Civil War. The Schuylkill Region was characterised by growth spurts, followed by periods of relative stasis. The first of these growth spurts is apparent in the 1830s up to the time of the 1837 panic. Proportionately, this was the most rapid period of growth ever experienced by the region, although it was starting from a very low base. In that decade it experienced a more than five-fold increase in shipments. Two subsequent growth phases can also be identified, one in the mid 1840s and the second in the early 1850s. Both were followed by periods of relative business stagnation, that of the late 1850s being particularly marked and protracted. In contrast, the Wyoming Region initially lagged behind Schuylkill, but then experienced growth at an accelerating rate that was scarcely checked until the end of the Civil War period. In the 1840s, shipments increased more than four-fold, again from a very low base. Further rapid expansion of shipments took place in the 1850s and the overall impact of the 1857 panic was much less

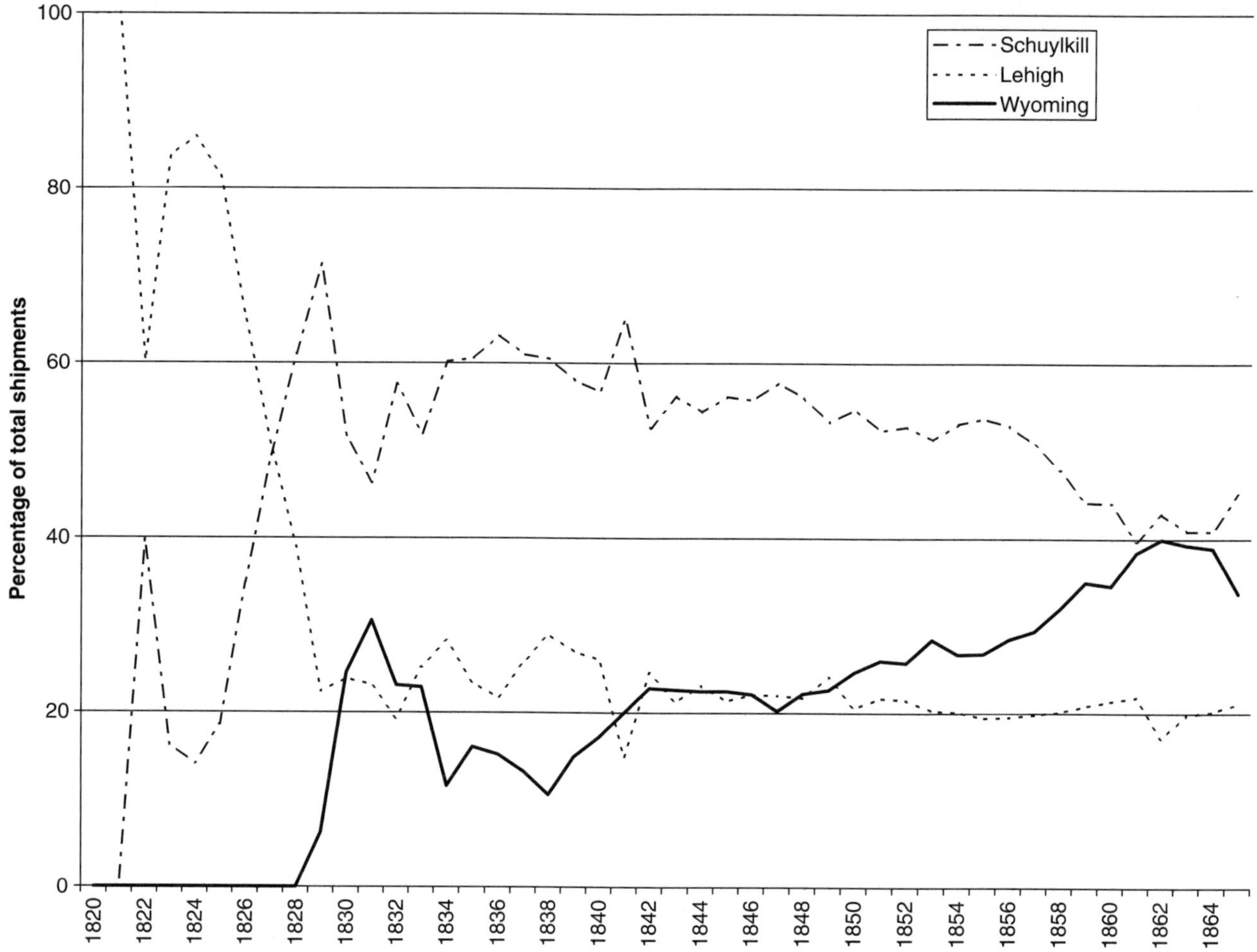

FIGURE 4.4
Percentage Shipments from the Different Coal Regions 1820–1865.

marked than in Schuylkill. The absence of any extended period of stagnation allowed the region almost to catch up with Schuylkill, which, by 1860, was only just maintaining its position as the leading producer. This suggests, for the Wyoming Region, that the stimulus to new mining development generated by the opening of three new avenues to market was the crucial factor in maintaining growth. These three avenues were the Pennsylvania Coal Company gravity railroad to the canal, the Lackawanna and Bloomsburg, and the Delaware, Lackawanna and Western Railroad. In other words, the number of avenues to market and the provision of access to virgin coal lands were more important than the precise mode of transport adopted. In this context, it is worth noting the slightly curious case of the North Branch extension canal. This ran from Pittston north to Athens near the New York state line, where it was to join the Junction Canal to reach Elmira. This should have provided yet another growing outlet to market, but it failed to achieve this goal. A writer in the Pittston Gazette observed that work on it had begun in 1836, was suspended in 1841, resumed in 1849, and was claimed to have been completed in canal reports of December 1853.[85] However, by early 1856 it was still not operational, owing in part to the poor quality of much of the early construction work. The first successful traversal of its entire length by coal carrying boats did not start until November 11, 1856.[86] Unlike the main section (Wyoming Division) of the North Branch Canal, which made a healthy profit for the state, the extension was a financial disaster. Hugely expensive, costing $4.6 million to construct, plagued by maintenance problems and completed too late to establish its traffic prior to the onset of railroad competition, it never shipped more than about 150,000 tons of anthracite in any year. In 1865, severe flood damage led to the operating company being reorganized as the Pennsylvania and New York Canal and Railroad Company, with three directors of the Lehigh Valley Railroad on its board. While retaining its canal privileges, the new company was authorized to construct a railroad along the line of the extension canal towpath. The company operated the canal until the railroad fully absorbed its traffic, leading to closure of the extension in 1872.[87]

The Lehigh Region was different again from the other two regions. It experienced a modest but steady growth in shipments, punctuated only by occasional reversals of business fortune. The most noticeable of these was the reduction in shipments in 1862, attributable to problems on the Lehigh Canal, as has been noted in an earlier section. A comparison of market share of the three regions, shown in Figure 4.4 by their percentage contributions to the overall trade, demonstrates the impact of changing growth rates. Setting aside the fluctuating values of the 1820s, which were based on very low levels of shipments, the overall picture became more stable in the 1830s. The dominance of Schuylkill was apparent from 1830 to 1850 as nearly 60 percent of shipments originated from that region. The Wyoming Region quickly eroded this dominance during the following decade. After performing well during the 1830s the Lehigh Region settled down to a remarkably constant 20 percent share of the trade that persisted until the Civil War and beyond. The changing importance of the different regions to the anthracite trade, as a whole, will be an important factor in the later examination of the development of cartels in the industry.

EARLY DEVELOPMENTS IN MINING

Although in general terms the relationship between developments in mining and associated transportation might be regarded as so obvious as not to necessitate further examination, a more careful consideration leads to a very different view. An analysis of regional development dynamics must consider a range of decision-making and productive processes bound up with this issue. Improved transportation routes might be proposed for a variety of reasons. These included

1. A desire to open up hitherto inaccessible and unexploited coal resources;
2. A requirement to bring coal to new markets;
3. Local boosterism often associated with an anticipated rise in land values;
4. A need to expand the traffic from areas tributary to an existing route before it was diverted onto other routes; and
5. An intention to capture a share of a growing traffic and bring it onto the railroad from a competing mode of transport such as the canal.

As the examples already discussed indicate, however, it could be very difficult to judge the timing and level of investment correctly. It might be premature, available funds may have been insufficient for the task in hand, or promised finances might evaporate in the face of sudden loss of business confidence. Even when finance was not a major constraint, significant time, measurable in years, might elapse between the chartering of a railroad and its opening to traffic. The line of the railroad might be determined in part by the location of existing mining operations, unless it was entering completely virgin territory. Otherwise there might be no traffic to cover its operating costs in the early period of its existence. Equally, it would doubtless have crossed some areas of undeveloped coal lands, immediately raising their potential economic value. Yet new mines, other than small water-level drifts, could not be brought into operation over a very short timescale and they required connecting lateral railroads to be built. Further months or years might therefore elapse before significant new mining capacity came on stream. Few operators would risk initiating the opening of a mine while railroad construction was still at the design stage, in case the project did not come to fruition. Only those with deep pockets, possibly business links with the proposed railroad, or both, could entertain the notion of investing early enough to have the mine ready to ship when the railroad was completed. Even then such investments could not escape the perennial mining risks of fire, flood, gas explosion, and subsidence (mine caves). Such was the lot of the ill-fated Union Iron and Coal Company mine between Scranton and Pittston during the 1850s, which will be examined in more detail below.

A further locational rather than chronological point is that new transport routes through undeveloped coal lands were constrained by topographic rather than geological factors. Hence the existence of a new route could result in mines being opened where convenience of shipping had taken precedence over mining engineering considerations. This could lead to inefficient and wasteful exploitation of limited but accessible coal deposits, followed by early abandonment and poor return on investment. At the same time, any kind of mining development, even if ill-conceived in terms of the geology, added to the available knowledge of coal deposits in specific localities. Likewise, the presence of merchantable coal, once established, reduced the perceived investment risk for those wishing to open other mines in the same area at a later date. This point was more telling in the early years, when understanding of the geology of the coal basins was limited, than in the later decades of the 19th century, when exploratory borehole drilling was undertaken extensively for prospecting undeveloped lands. Thus, it is apparent that the interrelated development of transportation and mining activity introduced several localized forms of geographical inertia into the process of regional development right from the outset. Small-scale mining activity attracted improved transportation facilities, which stimulated further mining investment in areas accessible to the new routes. Expansion of mining then attracted further new routes competing for a share of the growing coal traffic, although existing transportation companies often hindered their progress by refusing to grant rights of way over specific lands or permission to cross existing routes at grade.

Exhaustive analysis of the locational decisions associated with different types of mining activity in all of the coalfields from the 1830s to the beginning of the 1860s is beyond the scope of the present study and would constitute a major work in its own right. However, detailed investigation of company records, manufacturing census data, and newspaper reports from the Northern Field allow a number of general conclusions to be drawn about the pace and nature of developments in this region. This can be compared with summary information provided by Yearley for the Southern Field, although his investigation does not in general provide detailed geographical context. He has also been roundly criticized by quantitative economic historians for stressing the role of the socio-democratic beliefs of the population in the process of mining development. Instead, it has been argued, he should have analysed economic change based on theories of competition and industrial organization and contrasted the growth paths of the different regions.[88] That said, his work does provide a range of insights into the mining economy and its associated cultural milieu. A second valuable study for comparison is that of Wallace.[89] For the most part, this is highly specific to the small town of St. Clair, near Pottsville. Though beguiling in its literary presentation, in essence it uses the technique of generalizing from one small area to the experience of the entire industry, and therefore, in the present writer's view, some of its conclusions have to be interpreted with caution.

Unlike the Southern Field, where the statistical zeal of Benjamin Bannan and the Pottsville Miner's Journal has left a remarkable record of early mining development from 1825 onwards, both manuscript and newspaper records for the Northern Field are very limited in this early period, except for the area around Wilkes-Barre. Although some volumes of Carbondale newspapers for the 1840s are extant, the Pittston Gazette did not begin publication until 1850 and the oldest available issues of Scranton newspapers were published in the mid-1850s.

The earliest map of the Wyoming and Lackawanna Valleys dates from 1830.[90] It was produced by the operator of the Baltimore Mine, from which coal had first been shipped in 1814,[91] and it identifies nearly 100 small mines and outcrop workings, 36 of which are named.[92] Apart from the established mines in Plymouth, mining had also begun in the Nanticoke area, Colonel Washington Lee and George Chanoon having leased the Stiver mine at Newport in 1823.[93] Until the mid-1830s, coal was hauled to the banks of the Susquehanna, but in 1836, Alexander Gray, manager of the Baltimore Coal Company, constructed the first gravity railroad from the mines to the North Branch Canal basin in Wilkes-Barre. Shortly afterwards, the Wyoming Coal Company, comprised of H.B. Hillman, Samuel Holland, and Alexander Lockhart, began mining on the site of what became the Warrior Run colliery in 1837–1838. They also built a gravity railroad 2 miles long down to the Nanticoke Pool, which gave them access both to the river and the canal, and a loading basin which cost $22,500.[94] They shipped their first coal in 1840 and expanded their operations and their network of gravity railroads thereafter. An undated map of their lands, which from the evidence it contains must date from the mid-1840s,[95] shows they, or Holland and Company, had acquired a significant number of tracts of coal lands from below Nanticoke to the town of Pittston, linked to the canal or the Nanticoke Pool by five different short gravity-railroad systems. One of these railroads also connected at the foot of the Ashley planes to the newly constructed Lehigh and Susquehanna Railroad, which carried coal over the mountain to the Lehigh Canal at Whitehaven.

Despite being at the head of canal navigation, mining development in Pittston did not get an initial impetus until 1838, when John Lord Butler and Garrick Mallery opened the Butler mine. Mallery had previously been instrumental in securing funding for canal construction from the state legislature. In a similar manner to the Wyoming Coal Company, they built a gravity railroad 1 mile 800 feet long down to the canal, over which the first shipments were made in 1840.[96] Thereafter, a 'general depression' in all kinds of business resulted in restricted coal shipments until 1843[97] and there were no significant new developments until after that time. In the winter of 1844–1845 Colonel James Johnson sank a shaft near the 'Big Basin' on the canal on a property subsequently taken over by the Pennsylvania Coal Company. Other important new mines in 1845 were those of Samuel Benedict in Upper Pittston[98] and Messrs Bowkley and Price. The former had a short railroad to the canal, while the latter gentlemen built a longer railroad from the mines they had opened on property purchased from Colonel Johnson.[99] Petrillo, quoting the Wilkes-Barre Advocate of February 21, 1844, notes that the Bowkley family had arrived from Staffordshire, England in 1843. They had begun their operations by searching for coal near the canal northeast of Wilkes-Barre. They opened mines, wheeled the coal to the canal at considerable profit, and the area became known as 'Port Bowkley'. As a result, local land prices rose rapidly. The Bowkleys offered one farmer $4,000 for a ten-acre tract along the canal and $60 an acre for more distant lands that required a railroad to reach the canal.[100] Another mine was opened by Benedict and Alton in the summer of 1846. This required a gravity railroad parallel to that of Bowkley and Price to reach the 'Big Basin' on the canal. The property was later operated by the Maryland Coal Company.[101] Further developments ensued until the arrival of the Pennsylvania Coal Company to begin mining on a much larger scale at the end of the 1840s.

Modest coal shipments began from the Delaware and Hudson Canal Company's mines near Carbondale in 1829. They did not exceed 125,000 tons in any year until the end of the 1830s and year-to-year output was rather variable. Steadier growth ensued in the 1840s. Around Slocum Hollow, later to become Scranton, the absence of any efficient transportation routes meant that coal production during the 1840s was restricted to local requirements or those of the ironworks of Scrantons and Grant.

In the absence of any detailed enumeration of the mining industry during this early period in the Northern Field, it is difficult to be precise about either the number or type of mines that were present. Setting aside small outcrop workings largely worked by local farmers, available evidence suggests that drifts and tunnels (e.g. the Grand Tunnel, Holland Tunnel) predominated, with a slowly increasing number of slopes. In terms of shafts, only that sunk by Johnson in Pittston is recorded before 1850. The manuscript manufacturing census of 1850 provides little assistance, since it shows evidence of significant under-reporting and those mines which are recorded, have variable amounts of information attached. Thus, only seven mines have their source of power identified and of these only four make reference to steam power, suggesting the presence of slopes.[102]

Greater detail can be provided for the Schuylkill Region based on Yearley's findings. According to his enumeration, the drift was the most important type of opening in the early years, but slopes increased in significance over time. Thus, in 1839, he identifies 120 drifts and 10 slopes, while, by 1850, there were 122 drifts and 44 slopes.[103]

The decade of the 1850s began slowly for the coal trade of the Northern Field. The Pittston Gazette reported that 'the depression of the iron business during the previous 2–3 years had cut off demand for coal along the Susquehanna'.[104] Business began to improve from 1853 into the middle of the decade, however. By May 1853, the Pennsylvania Coal Company had already shipped 60,000 tons and nine independent operators in the Pittston area had contributed a combined total of 19,261 tons.[105] At the close of the season these same operators had shipped 207,000 tons to the total of 513,000 by the Pennsylvania Coal Company.[106] Although the overall tonnage for the area had increased to almost 800,000 tons by the end of 1856, the increase was all accounted for by the Pennsylvania Coal Company, which had produced 612,500 tons. The number of independent operators had increased from nine to eleven by this time, but their contribution to the overall tonnage had dropped slightly to about

182,000 tons.[107] A trend towards slightly larger and better capitalized companies entering the coal trade was also apparent at this time. Organizations such as the Geneva Coal Company and the Erie and Susquehanna Coal Company took over the leases and property of earlier individual operators, such as D. P. Fuller and Thomas Price (although in the case of the Geneva Coal Company, Fuller retained control as President of the new company[108]). The Geneva Coal Company established its works with a view to both canal and railroad shipments, in order to be ready for the latter when the Lackawanna and Bloomsburg Railroad was completed.[109]

An even greater interest in the coal trade by Philadelphia and New York capitalists and speculators became apparent at the same time in the Wilkes-Barre area. A famous article, written at the end of 1855 by a local correspondent of the North American and US Gazette, describes the energetic development of coal properties by a number of companies. This article is quoted in Harvey's History of Wilkes-Barre and was republished in the Pittston Gazette.[110] These companies included the Empire, William Penn, Kimberton, and Consolidated Coal Companies. Between them, they had acquired 2,300 acres of coal land, stretching from the canal to the Ashley planes, for an investment of about $800,000, and were rapidly pushing forward with new mine developments in time for the 1856 season.

By this time, however, the Scranton area was no longer the laggard it had been during the 1840s. The section of the Lackawanna Railroad running northwards from the town had begun operation at the beginning of the 1850s and the southern portion was in progress, adding to investor confidence. Lewis and Howell, formerly gentlemen of Pittston, began driving three tunnels in Hyde Park in 1851, in readiness to supply customers as soon as the railroad came into operation.[111] Long known as 'Howell's', on the Scranton street that still bears its name, this mine was later to become the Mount Pleasant. Very soon thereafter, the mines of the Lackawanna Railroad's own Coal Department began to come into operation, starting with the Diamond in 1852, the Bellevue in 1854, and the Hampton in 1856.[112] Just as in Pittston, however, the coal business was not to become entirely the domain of the large companies. A Pittston Gazette report, published just a week after the description of new companies in Wilkes-Barre, indicated that similar new developments were also underway just north of Scranton along the railroad. The first of these was the New York and Pennsylvania Coal Company, which was developing a 500-acre property by means of three drifts and a slope. The company had members of the Winton and Rockwell families on its board, and was a well-regarded concern. Nearby were the two new drifts and slope of the Luzerne Coal Company, whose principal owner was Colonel James Johnson, and the burgeoning improvements of Judson Clark and Company. Together, these companies had a capitalization in excess of $800,000 and the rapid expansion of mining activity was having a marked impact on the rate of property development for housing in the Providence area.[113]

Further down the valley, the Union Iron and Coal Company was also preparing, in the contemporary phrase, 'to do a large business'. They had begun shaft sinking and, by May 1856, were building a 1¼-mile lateral railroad to join the Lackawanna and Bloomsburg. Unfortunately, a combination of unexpected events and adverse business circumstances was to lead to the collapse of this concern shortly after it commenced shipments. It is therefore an interesting example of the commercial vicissitudes of the time. In 1855, the company had already spent $103,275 on sinking shafts, opening mines, and building miners houses, of which 41 blocks were eventually constructed. Substantial further expenditure was yet to be made on the breaker, coal and mine cars, and 10 canal boats to use on the Morris Canal from Port Washington. An agreement of March 28, 1856 with the Lackawanna Railroad provided for initial shipments of at least 50,000 tons from September 1, 1856 to the end of the year and 150,000 tons in 1857.[114] Unfortunately just as the company was ready for full-scale operation in early 1857, business conditions took an adverse turn, and the company had to stockpile its coal until more remunerative prices could be obtained. Then a major disaster struck. On June 24, 1857 the breaker and pump house both burned down, cause unknown. Careful managers had taken out £20,000 of insurance cover on the destroyed buildings, spread over several insurance companies in Philadelphia. This spread was important, since one of the companies had just failed in the widening financial crisis of the time. It should be noted that the company's managers were men of impeccable reputation and influential business pedigree, who counted among their number Moses Taylor, the leading New York banker. Despite the standing of the managers in the business community, an entry in the minutes for

December 29, 1857 records that the remaining insurers were refusing to pay their shares of the losses caused by the breaker fire, doubtless because of their own straitened financial circumstances following the 1857 panic, and legal proceedings to recover payment had been instituted. The implications for the mine were severe, as the following minute records:

> 'owing to the disappointment in the receipt of money for said loss and the extraordinary money crisis which had occurred, it was found necessary to arrest all further expenditures for the re-erection of the breaker and orders were accordingly given to the agent to stop all work'.

By this stage, the company had creditors to the extent of about $90,000 and no means of paying. After an unsuccessful attempt to raise money from the stockholders, the personal property of the company was eventually sold at public auction, in partial settlement of outstanding claims. The president was also authorized to compromise the legal actions pending against the insurance companies, to raise any further funds he could.[115] Thereafter, the Union Iron and Coal Company faded into obscurity, but its very valuable lands (almost 2,000 acres) and mine openings, albeit stripped of their fittings and equipment, remained within the orbit of financial interests linked to the Lackawanna Railroad (see Chapter 10). After a gap of almost a decade, in 1866, formal control of the lands passed back to the Steuben Coal Company, one of the Railroad's subsidiaries, at a price 'not exceeding $300,000'. The workings were successfully re-opened and a new breaker built called the Taylor.[116]

Despite its failure to produce significant quantities of coal, the Union Iron and Coal Company is a classic example of the dynamics of new entrants into the coal business. Conceived in the improving business climate of 1854, followed by 2–3 years of investment in mining capacity and connections to major transport arteries, production came on stream to recoup the investment just as the 1857 depression set in and the markets no longer wanted coal at remunerative prices. When ill-timed disaster struck in addition, even the most prudent and experienced management could not spread their financial risk wide enough to avoid the influence of the panic. They and their stockholders were unwilling to dig deep enough into their own pockets to rescue the company in the prevailing adverse business circumstances and it was eventually wound up as an inevitable consequence. Without the fire, the company would probably have survived, albeit with its credit strained to the limit in the early years of operation. A smaller operator, without the business standing of the Union Iron and Coal Company's managers, would possibly have gone into liquidation, even without the fire. All available financial resources would have been committed to establishing the mine, leaving insufficient funds to maintain cash flow during the lean times of the late 1850s.

The expansion of the Delaware and Hudson Company's operations in the northern part of the coalfield had begun in 1843 with the decision to extend the gravity railroad 7 miles south to White Oak Run, where new mines were being opened. Refurbishment of the gravity railroad in 1856 from Carbondale to the coal stocking grounds at Waymart, outside the coalfield to the east, was followed by a further 6-mile extension below Olyphant to Valley Junction in 1858 (Plate 4). Later, a steam-railroad line was constructed on down to the Von Storch lands near Providence. An important aspect of the motivation for these new developments was that many of the old workings of the company near Carbondale were approaching exhaustion and new sources of supply to maintain coal shipments were required.[117] An 1855 report by S. H. Daddow indicates the extent of the Delaware and Hudson's mining activity at that date. They had opened 5 tunnels at Archbald and a total of 9 tunnels, 4 slopes, and 2 shafts in the vicinity of Carbondale, which gave employment to 1,500 men and boys in the mining and shipping of coal.[118]

The extension of their railroad in stages down through the coalfield also provided the opportunity for independent operators to enter the business. In 1856, the company started receiving proposals from new operators such as the Carbondale Coal Company (Jones and Lathrop) and Alva Eaton and Company (Plate 2) who wished to sell them their output.[119] The latter commenced shipping in 1857 and the former apparently not until 1859, according to the draft annual accounts entered into the minutes each March.

The independent and simultaneous expansion of mining investment in four different areas of the Northern Field, linked to an equal number of different avenues to market, helps explain the persistently high rate of growth in output during the

1850s, a rate that the Schuylkill Region could no longer match. Long-term commitment of investment is also apparent in the progressive increase in shaft sinking as a means of opening mines, instead of the cheaper slopes or drifts. At the same time, it should be noted that the geology of the Northern Field was more favorable to the use of shafts than elsewhere in the coalfields. In early 1859, the Miner's Journal lists 10 shafts in the Scranton area alone.[120] Several of these have been mentioned in the brief examples given above. There were also a number of others, such as the Dundee Shaft below Wilkes-Barre,[121] the Ravine Shaft and Shafts Nos. 6 and 7 of the Pennsylvania Coal Company, all in Pittston.[122] By way of comparison, for the Southern Field, H. D. Rogers' Geology of Pennsylvania, published in 1858, records only 4 shafts, but 36 tunnels and drifts, and 66 slopes.[123]

ADOPTION OF THE COAL BREAKER

Since the coal breaker had been invented in the early 1840s, it might be assumed that the expansion of mining activity in the Northern Field from the mid 1840s onwards would have resulted in a rapid spread of this technology across the landscape, as it did in Schuylkill County. The latter had 51 steam-powered coal breakers by 1847.[124] However, this did not prove to be the case. According to Phillips, the first breaker in the Wyoming Valley was built at the Butler mine in 1847, followed shortly after by the Baltimore breaker built by Alexander Gray on Coal Brook in 1848.[125] The different rate of adoption of this new technology between the Northern and the Southern Fields does not seem to have attracted the attention of contemporary commentators, but it still requires explanation. The most likely reason can be traced to the difference in the prevailing transport modes, with canal and gravity railroad combinations dominant in the north and the steam railroad in the south.

Canals and gravity railroads by their nature are locationally constrained by the general topography and cannot readily be modified in position to accommodate access to individual mines. The steam railroad is much more flexible in comparison. Trains of cars in a gravity system drawn by hemp or wire ropes are much less manoeuvrable than when pulled by a locomotive, and the loading arrangements for filling 60- or 120-ton canal boats in a basin are necessarily different than those for loading 6-ton railroad coal cars.

The emphasis in the Northern Field was therefore on what might be termed an intermediate form of coal preparation, where sorting and loading arrangements, rather than facilities for coal breaking were provided. For example, an 1853 report on the new equipment installed on D. P. Fuller and Company's wharf in Pittston makes reference to both coal screens and loading chutes, but not the use of breaker rolls. The screen was 18 feet long and 2½ feet in diameter, worked by the power of a single horse. It could produce a range of coal sizes from steamboat down to pea. The new loading chute was large enough to hold two boatloads of coal at a time. Importantly, these new fixtures cost only $1-2,000, as opposed to perhaps $20,000 for a full-scale coal breaker, a major consideration for a company with a moderate output of 25,000 tons a year.[126]

However, the absence of coal breakers is not predominantly an issue of small companies with limited means, since both the Delaware and Hudson and the Pennsylvania Coal Company used combinations of screens and loading chutes in association with their gravity railroads.[127] Curiously, in both cases it is rather difficult to determine exactly when these two companies introduced standard breakers, as their own surviving records are silent on this matter and the newspaper coverage is sparse for the 1840s and the latter part of the 1850s. There is no doubt that the Delaware and Hudson was aware of developments in breaker technology by 1844, and the James Archbald Papers shed important light on this. These developments included the patented Battin breaker, a different device designed by a Mr. Richardson, who had an interest in the Buck Mountain Coal Company in the Lehigh Region,[128] and Archbald's own idea for breaking coal with toothless rollers.[129] As early as 1844, the Delaware and Hudson had committed themselves to a contract that required coal to be broken, but the surviving Archbald correspondence does not indicate how or where this was undertaken. There is clear evidence that the Delaware and Hudson was building a breaker in 1851 that was expected to be ready by May 1st. James Archbald was to examine it in operation to determine both the cost of breaking and the amount of resulting waste.[130] Unfortunately the location of the breaker was not specified. The first standard breaker of either company whose location and construction date is certain, is that of Racket Brook, built in 1856.[131] A further interesting point about these two companies is that their early breakers were not necessarily located at the mine opening but were often strategically placed to serve several mines. This was the case at Racket Brook.

The mines along the line of the Lackawanna Railroad present a contrast to the developments linked to canals and gravity railroads. The Company's own mines were opened with breakers already installed, the first in the Scranton area being ordered from Cox and Company of Reading by G. W. Scranton at the beginning of February 1852, for the Diamond mine.[132] Others followed at the Luzerne Coal Company and Clark's Mines.[133]

Nevertheless, as late as 1860, the process of breaker adoption was still neither complete, nor clear-cut across the entire coalfield, in terms of railroad versus canal shippers. Table 4.1 reproduces very interesting data from a report in the Pittston Gazette at the beginning of 1861, which lists coal operators in the lower half of the coalfield by location and mode of shipping. In addition, it indicates whether they already had a breaker or whether construction was 'in contemplation'.[134] It is apparent that all the larger shippers (>30,000 tons), regardless of transport mode, had either already built breakers or were considering doing so. On the other hand, established small railroad shippers such as Robert Love, used chutes rather than having a breaker. Love, who worked what became the Jersey mine of the Lackawanna Railroad Coal Department, was the first operator to ship over the newly opened Lackawanna and Bloomsburg Railroad on August 4, 1857.[135] In the context of the table it should also be noted that in 1860 an

TABLE 4.1
Coal Breaking Facilities of Major Shippers in the Wyoming Valley 1861.[137]

Company	Location	Productive Capacity (tons)
1. Companies shipping exclusively by Railroad		
Kingston Coal Co.*	Kingston	15,000
Eno, Shunk and Co.+	Plymouth	20,000
Bennet and Davis	Plymouth	20,000
R. Love	Plymouth	15,000
J. B. Smith	Plymouth	20,000
Grover	Plymouth	15,000
McFarland+	Plymouth	25,000
2. Companies shipping exclusively by Canal		
Harvey	Nanticoke	15,000
Lee*	Nanticoke	40,000
Hollenback*	Wilkes-Barre	40,000
Baltimore Co.*	Wilkes-Barre	100,000
Mordecai*	Wilkes-Barre	20,000
Hillman	Plains	20,000
Hancock	Plains	25,000
Worden	Plains	15,000
3. Companies shipping by both Canal and Railroad		
Lance+	Plymouth	20,000
A. Tompkins+	Pittston	40,000
Maryland Co.	Pittston	20,000
Leyshon	Pittston	18,000
Price+	Pittston	35,000
Ravine Co.*	Pittston	20,000
Freeland	Pittston	15,000
Bowkley*	Pittston	20,000
Junction*	Pittston	80,000

*Mine has breaker already.
+Breaker construction 'in contemplation'.

extension of the Lackawanna and Bloomsburg Railroad was constructed from East Pittston down to the Pittston canal basin. The newspaper observed that 'the completion of this link will greatly enhance the interests of our coal operators, giving them a chance to ship coal in the winter as well as in the season of canal navigation'.[136]

A further factor in the progressive adoption of coal breakers was the impetus of wider domestic markets for coal, where the smaller sizes were preferred. This was in contrast to the industrial markets where, in general, larger sizes were required. The nature of these changing markets in the antebellum period and the impact of new transportation routes to reach them will now be examined.

WIDENING MARKETS FOR COAL

Prior to 1860, there was significant expansion in the range of market sectors where anthracite was used as a fuel. Arguably the degree of widening of markets in geographical terms was more limited, given that the major destinations of coal shipments to the eastern seaboard cities had been established at a very early date.

A number of major market sectors can be identified, including domestic and institutional heating, general manufacturing, iron making, and transportation. These have been examined in detail by Binder[138] so only need to be briefly summarized in this section, to provide context for the later parts of this study.

By the beginning of the 1830s, technical difficulties in the use of anthracite for both domestic heating and cooking had largely been solved and suitable grates and stoves were being manufactured at moderate cost. The inhabitants of New York, who preferred the Schuylkill red ash coals, were consuming over 50,000 net tons of anthracite by 1832–1833 and the quantity used was increasing yearly, as it was in the other seaboard cities. For larger public buildings and institutions, where stoves were inappropriate, Professor Walter R. Johnson of the Franklin Institute had adapted the air furnace for use with anthracite as early as 1825.[139] The initial period of development of the coal trade until the end of the 1830s was thus driven largely by domestic demand, although industrial proponents of anthracite as a fuel were beginning to emerge. This has already been shown in the earlier discussion of White and Hazard's wireworks outside Philadelphia.

Following the first use of anthracite for steam raising in 1825 at the Phoenix Nail Works in Chester County, there were a number of technological developments in boiler design that facilitated the economical adoption of anthracite, such as the use of artificial blasts or steam jets forced into the firebox. Significant sums of money were also spent by companies such as the Delaware and Hudson to promote the use of anthracite for steam raising in the manufacturing industries of New York and New England. Such marketing campaigns were apparently effective, as anthracite shipments into Boston rose from 76,000 tons in 1834 to 248,000 tons by 1847, although the domestic and industrial components of demand cannot be distinguished in these early figures.[140] The ready supply of anthracite in Philadelphia encouraged industrial development in the form of sugar refineries, glassworks, locomotive works, and foundries, among many others.[141] The same could be said of New York, although in each case there are no readily available statistics on consumption by industry type in this period.

However, it was in the making of iron that anthracite registered its clearest and most substantial industrial success during the period. The essential technological pre-requisite for the use of anthracite in blast furnaces, was the invention of the hot-blast, which had been patented in England by Neilson in 1828. The first early experiment to utilize both anthracite and the hot-blast took place at the Valley furnace near Pottsville in 1836, but was not entirely successful. Further experiments followed and, in December 1839, a prize offered for keeping an anthracite furnace in blast for 90 days was claimed by William Lyman of the Pioneer Furnace at Pottsville, thereby ushering in the era of anthracite iron. Although the latter furnace was not subsequently a commercial success, it was rapidly followed by others that were, the most notable being the Lehigh Crane Iron Company. This works near Allentown produced its first iron in July 1840 under the able management of David Thomas, a Welshman and former employee of George Crane, who had first successfully smelted iron with anthracite in Wales during 1837 at the Yniscedwin Works.[142]

In addition to furnaces along the Schuylkill and Lehigh rivers, an important cluster of ironworks developed at Danville in close proximity to some large ore banks.[143] These included the Danville, Montour, and Columbia Iron Works opened in 1840 and 1841. The first two of these were by Biddle, Chambers and Co. and the last by George Patterson, although it was operated by the Groves Brothers.[144] The Danville ironworks were of particular importance to the early development of the Northern Field. Initially they were major customers for the newly opened mines. The Montour Iron Company, for example, contracted for 6–8,000 tons delivered for $1.75 at Danville in the spring of 1844.[145] Later, they began to take a direct financial interest in mining, especially in the Pittston area, where the coal was well suited to furnace use.[146] In a similar fashion, new furnaces in Lancaster County, which benefited from the large ore banks at Chestnut Hill and Cornwall, utilized the hard, durable white ash coal from the Baltimore or Big Vein at Wilkes-Barre. This was supplied by the Baltimore, Diamond, and Black Diamond, coal companies.[147] As noted earlier, Scranton's, Grant and Company commenced construction of their works in late 1840, on Roaring Brook in present day Scranton, to utilize both local coal and nearby ore deposits.[148]

According to Pearse, the average amount of coal used by 8 early furnaces was 1.8 tons per ton of anthracite iron produced, although the first two blasts at the Lehigh Crane Iron Company in 1840 and 1841 used 2.04 and 2.16 tons of coal per ton of iron. Interestingly, later production figures for the same works for 1856–1858 give a very similar average of 2.15 tons.[149] A further independent check, that avoids secondary reporting, can be derived from the surviving blast furnace reports of the Thomas Iron Company for 2 six-month periods at the end of 1858 and 1859. These indicate an average of 2.03 tons.[150] Such figures give good empirical support for the general average of 2 tons of anthracite per ton of iron, employed by both historical and more recent commentators on the iron industry.[151]

Figures for the production of anthracite iron are very scarce during the 1840s, but the total for 1849 is available. Using this figure of 105,950 gross tons, and the multiplier of 2 tons to determine the coal used, implies that 6.5 percent of all coal shipments were used in iron manufacture at that date.[152] By 1859, when anthracite iron manufacture had risen to 421,201 gross tons, the comparative figure was 10.8 percent.[153] Summary figures of this kind are useful, yet they conceal both the temporal and spatial dynamics of market development. In Figure 4.5, all known construction and start dates for anthracite furnaces built prior to 1858 are given, based on data derived from Lesley's Ironworks Directory, augmented by additional information from county histories and other sources.[154] This amounts to a sample of over 90 percent of all anthracite ironworks of the period. From this it is apparent that market development was not uniform over time. After the initial pioneering works of the late 1830s, there were strong surges in the construction of new capacity during the mid-1840s and mid-1850s. Needless to say, in each case, just as the new capacity came on stream, it encountered the depressed business conditions either of the beginning of the 1850s or of the period surrounding the 1857 panic.

The iron industry was always a prized market sector because it took large quantities of lump coal. Since this size did not require breaking, waste was reduced. Use of the next size, namely steamboat coal, for transportation purposes, also developed at about the same time. The primitive technology of marine steam engines and boilers hindered widespread adoption of anthracite in coastal and river steamboats, until Eliphalet Nott designed new patterns of tubular boilers and connected blowers to the fireboxes. Successful trials of converted vessels in 1835 and 1836 proved the effectiveness of these designs and demonstrated the cost savings of anthracite over wood as a fuel. For example, Philip Hone calculated that the second trial, on the steamer Novelty, in 1836, would have cost $240 with wood as a fuel, but the actual cost using anthracite, at $5 a ton, was only $100. Since anthracite was also a much denser fuel than wood, it took up less space on board, so steamers could go longer distances without stopping to refuel. While demonstrating its superiority over wood as a fuel was relatively easy, the same was not true in relation to bituminous coal. Great Lakes steamers were using Pennsylvania and Ohio bituminous coals by the 1840s, and two out of three US Navy steamboats built at that time were also constructed to use bituminous rather than anthracite coal. The case for anthracite was hampered by tests conducted in 1844 by Professor Johnson which purported to show that Cumberland coal was best suited to naval use, although these results were refuted in 1852 by more realistic experiments using marine boilers under actual steaming conditions. These later findings showed anthracite to be cheaper, smokeless (to minimize the

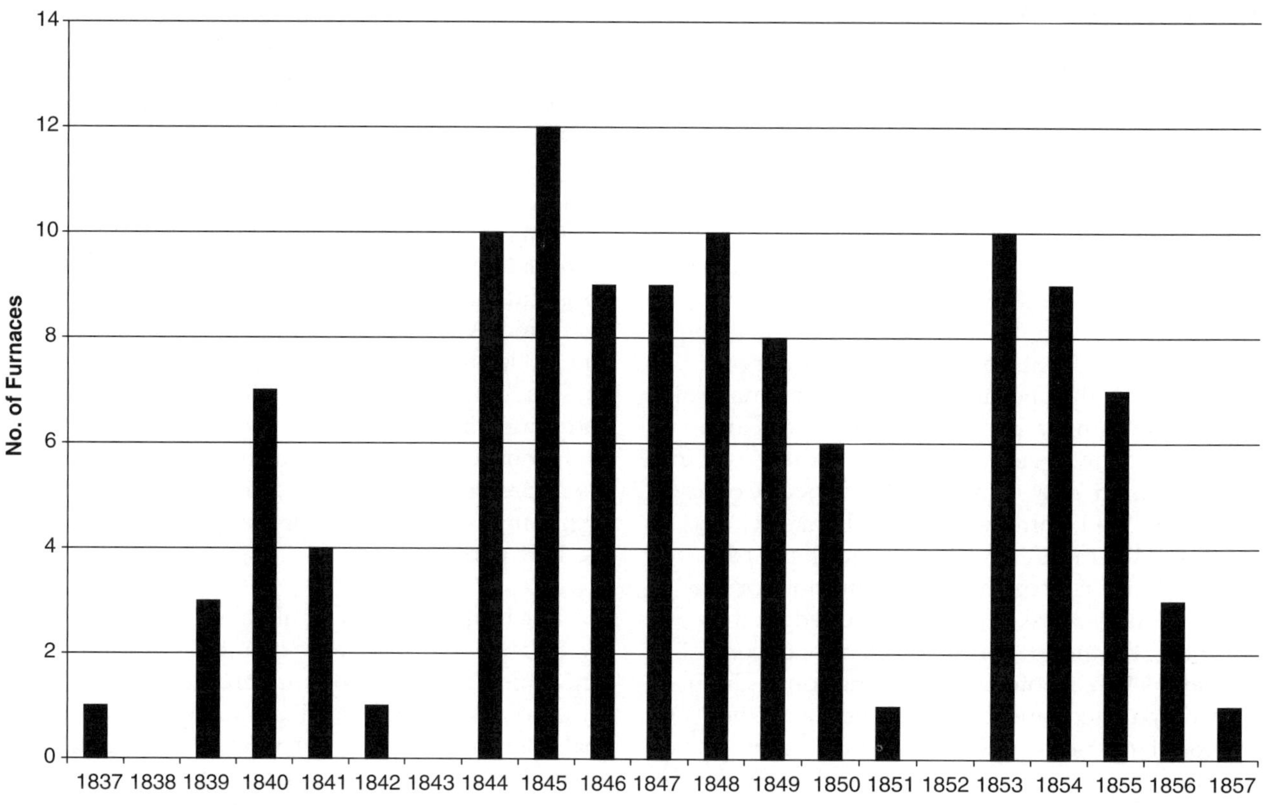

FIGURE 4.5
Construction of Anthracite Iron Furnaces 1837–1857.

chance of detection at sea from smoke trails), and not liable to spontaneous combustion. Its greater energy output per ton also allowed vessels to steam two-thirds longer than Cumberland coal before refuelling. In consequence of this, the Navy shifted to increased use of Schuylkill anthracite until the Civil War, although many of the Old Guard in Naval High Command continued to be advocates of sail against steam power.[155]

A similar shift took place in commercial vessels. Binder quotes a report in De Bow's Review for 1853, which claimed that more than half of the steamers in the US coastal and Caribbean trade were using anthracite, in whole or in part.[156] A slightly earlier report reproduced in the Pittston Gazette in 1851 states that all steamers in the coastal trade used anthracite. Also, of 78 ocean-going steamers listed by Mr. Haswell, US Engineer, 49 were then using anthracite that had not done so in 1850, adding 110–162,000 tons a year to overall anthracite coal use, depending on the assumptions made about daily consumption. A second reported estimate made in December 1851, put the total increase as high as 218,000 tons a year.[157] Given that it was estimated that 150,000 tons of anthracite were being used by eastern steam boats as early as 1840,[158] this suggests that the steamship trade was even more important than the iron trade at the start of the 1850s, in terms of anthracite consumption.

Steamships were not, however, the only sector of the transportation market where anthracite was to become important. On the railroads, as on the ocean wave, technological advances were required before the black diamond could be used effectively. Curiously, these took much longer to achieve than in the case of marine engines. Early experiments with anthracite burning locomotives on the Philadelphia and Reading Railroad proved to be a failure because of poorly adapted boilers, grates, and fireboxes, which resulted in maintenance costs four to five times higher than for wood-burning locomotives. By 1844, therefore, all but one of the company's engines were wood-burning. The necessary technical breakthroughs had to wait until 1851–1852 when James Millholland of the Reading Railroad patented new boiler designs, which were a substantial improvement over the Winans engines developed in the 1840s. These patents were followed by subsequent innovations in firebox design and the use of water-cooled grates. As a result, by 1856 85 percent of the Reading's coal shipments were hauled by anthracite burning locomotives. In 1859, coal use by the railroad had risen to 41,000 tons and wood had reduced to 15,000 cords from a peak of 90,000 in 1847. Competitors such as the Lackawanna Railroad were less advanced, with corresponding figures of 10,200 tons and 34,000 cords.[159] By the following year, however, the latter road was starting to catch up, replacing its old Winans engines with newer technology.[160]

THE GEOGRAPHICAL EXTENT OF MARKETS I: THE EASTERN SEABOARD CITIES

The pattern of major urban centers on the east coast was well established by the 1820s. Hence it is apparent that the construction of canals and the first railroads during the 1830s and early 1840s simply served to strengthen their position as major commercial and industrial centers.[161] Only the progressive development of the anthracite iron and associated industries in the Schuylkill and Lehigh valleys in the 1840s and 1850s provided major new markets that were not primarily within the compass of the tidewater cities.

Statistical data on coal movements to the eastern seaboard have to be interpreted carefully, to distinguish between actual consumption of coal within cities and coal receipts for onward shipment to other destinations. Questions also arise as to what constitutes a city and whether neighbouring towns are included in the calculations. Further to this, the breakdown of published statistics in newspapers, official sources, or company annual reports for the years of interest is not always sufficient to allow shipment destinations to be determined as precisely as is required. Estimates have to be made as a result. Table 4.2 a summarises estimates of coal consumption for major eastern seaboard cities prior to the Civil War. The Boston figures are part of a standard series from the Miner's Journal, but they do not address the question of how much coal was shipped inland by rail from the port of Boston, as opposed to being used in the city and its environs. The Philadelphia data stand in contrast to the figures presented by Binder,[162] which are for coal received rather than consumed. Unfortunately, Binder's figures are not correct, even for coal received. He has added in the entire tonnage of the Delaware Division Canal, but the tables of Jones,[163] clearly indicate that more than half of the total entering this canal was subsequently shipped northwards towards New York on the Delaware and Raritan Canal, rather than south from Bristol to Philadelphia. A further problem with Binder's figures is that he gives the entire tonnages of the Reading Railroad and Schuylkill Navigation as though they were received at Philadelphia. This is also incorrect as, in general, more coal was sold by both companies along the line than reached Philadelphia.[164] The only rider to this is that until 1868, outlying settlements from Philadelphia, such as Manayunk, Nicetown, Belmont, and Germantown, were counted as part of the line trade, not as Philadelphia city shipments. As late as 1866, shipments to these settlements were only about one-third of those to Philadelphia proper.[165] Shortly afterwards, when the Miner's Journal statistics for these locations were merged with those of Philadelphia, their tonnage was approaching that of the main urban area itself.[166] A breakdown of the Philadelphia consumption data is given in Table 4.2 b, using the correct proportion of the Delaware Division shipments, although there is an implicit assumption that coal delivered via this route was consumed in Philadelphia rather than being shipped out again from the port. The railroad and other canal figures follow the Miner's Journal tabulation. Other information makes clear that the line for canal data in this tabulation relates only to the Schuylkill Navigation.

Similarly the railroad data for 1860 is just for the Reading Railroad.[167] It does not include the recently opened North Penn Railroad, which shipped 91,327 tons in that year.[168] Later data[169] suggest that about 22 percent of North Penn shipments were sold on the line, which would give an estimate of 72,000 tons delivered in Philadelphia in 1860. Binder notes that a

TABLE 4.2 a
Estimates of Coal Consumption of Major Eastern Seaboard Cities 1855, 1860.[178]

	Boston	New York	Philadelphia
1855	387,259	2,286,000	1,173,000
1860	584,408	3,655,000	1,134,000

TABLE 4.2 b

Calculation of Estimated Philadelphia Coal Consumption 1855, 1860 (nearest thousand tons).[179]

Route	1855	1860
Philadelphia and Reading Railroad	342,000	306,000
Schuylkill Navigation Company (Canal)	286,000	495,000
Delaware Division Canal	545,000	261,000
North Penn Railroad	–	72,000
	------------	------------
Total	1,173,000	1,134,000

further 100,000 tons a year reached Philadelphia, first via the North Branch, and Susquehanna Canals, and subsequently via the Chesapeake and Delaware Canal. No source for this figure is given and it compares with a total coal tonnage on the latter canal of 193,000 tons.[170] This total includes shipments in both directions and anthracite as well as Cumberland bituminous coal, although the respective breakdowns are not given. However, since Binder also argues that this 100,000 tons was intended for onward shipment to New York, it has not been included in the Philadelphia consumption figures.

Another set of problems with Binder's analysis of Philadelphia's coal receipts concern his statements about the tonnages received at Port Richmond in 1860 and the destination of those shipments. Port Richmond was the coal shipping port of the Reading Railroad in the city. He states that 'by 1860 [tonnage] approached the two million mark, or nearly half the total coal tonnage brought to Philadelphia in that year'.[171] Neither his own quoted sources, nor other tabulations seem to support this view. Although shipments had been in excess of 1.5 million tons in 1855, they declined thereafter and were only about 1.2 million tons in 1860. The two million mark was not reached until the expansionary period of the Civil War in 1863.[172] These issues have been dealt with in detail because the multiple errors described would lead to an inference that about 2 million tons were consumed in Philadelphia in 1860, whereas a more reliable estimate, as indicated in the table, is only just over half that amount.

A final note of caution on Binder's analysis relates to his further observation that 'much of the coal brought to Philadelphia was shipped coastwise to New York'.[173] This is contradicted by his statement later in the same chapter that at least a quarter of the Schuylkill coal deposited at Port Richmond went on to the New York market. This statement is derived from an 1855 source.[174] Fortunately, a different 1855 source, not used by Binder, gives in considerable detail the precise destinations of shipments from Port Richmond in that year. This confirms that the proportion going to New York was exactly 25 percent. It is also interesting to note that other statistics, which bear on this question from the end of the 1860s, indicate that this proportion was relatively constant over quite long periods. The average percentage in the years 1869–1871 was 22.5 percent. Almost exactly half of the 1855 Port Richmond shipments went to the New England states and 72 percent of this New England tonnage went to Massachusetts. These proportions also persisted until after the Civil War.[175]

The ascendancy of New York in the coal trade, as in other areas of shipping and commerce by the mid 19th century, is already apparent from the estimates in table 4.2 (a). Curiously, Albion[176] in his classic study of the rise of New York Port, gives negligible attention to the coal business. Basing his comments on the 1858 report of the New York State Chamber of Commerce, Binder argues that the coal supply of New York averaged two million tons in the years preceding the Civil War. However, two issues have to be considered in relation to this assertion. Firstly, 1858 is an unreliable year from which to extrapolate, as it followed the aftermath of the 1857 panic and the present tabulation suggests a significant increase in shipments above the two million level by 1860. Secondly, it is even harder, in the case of New York than Philadelphia, to identify what is meant in geographical terms, since coal was delivered at a variety of points by different railroads and canals. The Central Railroad of New Jersey shipped to Port Johnston and Elizabethport, the Morris Canal to Newark, the Delaware and Raritan to Perth Amboy, New Jersey and the Pennsylvania Coal Company to Weehawken, New Jersey and later Newburgh, New York via the Erie Railway from Hawley. The Delaware and Hudson sold coal both at

Rondout on the Hudson River and in its own yards in New York City. The vast majority of these shipments would have been classified as reaching tidewater markets in New York Harbor, although a proportion of Delaware and Hudson and Pennsylvania Coal Company coal would doubtless have been shipped north from Rondout along the Hudson towards Albany, rather than south to New York. Unfortunately, precise figures on these proportions have not come down to us, but in 1860 for example, the Delaware and Hudson sold nearly 330,000 tons at Rondout, although the destination was not specified. It also sold a further 210,000 tons, sold on its own account in New York itself, either by the cargo or in retail yards.[177] Table 4.2 c is therefore an attempt to identify the sources of coal coming into the New York Harbor tidewater market. It is partly an exercise in assessing the consistency of figures derived from individual company sources with the overall estimate from the Chamber of Commerce, which is for consumption in the metropolis itself. The notes to the table give the details, but the individual figures for different companies are estimates based on actual tonnages, with deductions for line trade shipments determined from available data.

The major item of uncertainty is the proportion of Delaware and Hudson and Pennsylvania Coal Company coal shipped north rather than south on the Hudson River. An arbitrary, but not unreasonable, figure of 20 percent going north has been used in the table. Regardless of the detailed accuracy or otherwise of any of the figures, the relative contributions of several major transportation routes stand out. The first of these was the Delaware and Raritan Canal, which brought coal both from the Lehigh via the Delaware Division Canal and also very substantially from the Schuylkill Navigation. The latter was shipping a much greater quantity of coal to New York directly in boats than was being transhipped from railroad cars into boats at Port Richmond for the same destination. This practice of direct shipping had started as early as 1829.[181] Secondly, the New Jersey Central played a key role in bringing the tonnage both of the Lackawanna Railroad from the Northern Field and the Lehigh Valley Railroad from the Eastern Middle Field. Thirdly, on the Canal and Hudson River route it is apparent that the Pennsylvania Coal Company was dominant during this period over the Delaware and Hudson, on whom it depended for canal transportation. This situation was not reversed until the early 1860s.

It is noticeable that the resulting aggregate figure for 1860 is much higher than that of the Chamber of Commerce. That said, the proportionate increase from 1855 at 60 percent, compares with the 50 percent increase for Boston, where the figures are not estimates. One factor that tends to inflate the New York figure a little is that some coal was transhipped in New York Harbor onto coal schooners and brigs plying Long Island Sound for the New England seaports south of Boston, such as Providence and New Haven.[182] Some coal would also have been consigned to Boston, but a large proportion of its coal was already being provided directly from Port Richmond, without entering New York Harbor. If the 1855 figures for overall shipments to Boston are compared with the shipments from Port Richmond in the same year, it would appear that only about 21 percent of the Boston total (81,000 tons) was shipped from New York to Boston. The tonnages to other New England ports would have been much smaller than this, so

TABLE 4.2 c

Calculation of Estimated New York Coal Consumption 1855, 1860 (nearest thousand tons).[180]

Route	1855	1860
Morris Canal	262,000	364,000
Delaware and Raritan (from Delaware Division)	141,000	308,000
Delaware and Raritan (from Schuylkill Navigation)	655,000	805,000
Central Railroad of New Jersey	-	790,000
Belvidere Delaware Railroad	-	134,000
Port Richmond	372,000	293,000
Delaware and Hudson	452,000	400,000
Pennsylvania Coal Company	404,000	561,000
	-----------	------------
Total	2,286,000	3,655,000

the overall effect on the New York figures would not have been dramatic.

It is difficult therefore, to escape the conclusion that continued growth in the business activity of Boston and New York between 1855 and 1860, was not mirrored by equivalent expansion in the commerce of Philadelphia at this time, if the coal trade is viewed as an appropriate business 'barometer'. A more detailed investigation of this issue is outside the scope of this study, but one possible explanation might relate to relative rates of business recovery after the 1857 panic. Another key issue is clearly the expansion of new transport routes in the direction of New York, which tended to funnel both coal and other types of commerce away from Philadelphia.

Unlike Philadelphia, statistics for the Baltimore coal trade were not reported for the 1860s in the major sources, such as the Miner's Journal and Saward's Coal Trade Annual. During the second half of the 1850s, shipments averaged a quarter of a million tons[183] and a later comment from Saward indicates that coal received was not transported elsewhere, but was used for local consumption.[184] Coal from the Susquehanna Canal intended for points other than Baltimore on the eastern seaboard simply made its way through the Chesapeake and Delaware Canal without passing through the city.

THE GEOGRAPHICAL EXTENT OF MARKETS II: INTERIOR MARKETS

It is entirely appropriate to emphasize the importance of the eastern seaboard cities as major markets for anthracite. However, it is important not to neglect the steady growth of interior markets in the Lehigh, Delaware, Schuylkill, and Lebanon Valleys. These were largely linked to the anthracite iron industry, but also to the development of regional urban centers such as Norristown, Reading, and Allentown. In 1855, more than 480,000 tons were consumed in the Schuylkill and Lebanon Valleys and a further 280,000 tons in the Lehigh and Delaware Valleys. By 1860, the equivalent figures were nearly 610,000 tons and an estimated 480,000 tons. These totals allow for inclusion of the line tonnage of the Lehigh Valley Railroad, although the breakdown of available figures in 1860 is not as good as for the earlier year.[185] This implies that by the Civil War, Philadelphia's hinterland was as large a domestic and industrial market for coal as the city itself. The consequences were highly significant for coal operators, because the line trade gave higher and more stable prices and profit margins, owing to limited competition. By comparison, the highly competitive tidewater markets produced lower and more volatile prices.

With the exception of upper New York state, other markets for anthracite, such as in foreign ports in the Americas, were an order of magnitude less important than those previously examined. The earliest route for anthracite north into New York state was via the Hudson River and Erie Canal, followed much later by the Lackawanna Railroad that shipped to Binghamton, Syracuse, and Ithaca.[186] A further route north was added by the belated completion of the North Branch Extension Canal in the latter part of the 1850s. Of the million tons of coal shipped by the Lackawanna in 1860, about 75 percent was shipped towards New York City and the remainder north to New York state. A limited amount of this coal reached the Erie Canal at Syracuse, largely for consumption along the line of the canal. A total of nearly 500,000 tons of all types of coal from all sources was carried on the Erie Canal in 1860. Perhaps 300,000 of this was anthracite, judging by the proportions of later years when a breakdown is available. The remainder was largely Blossburg semi-bituminous coal.[187] Unfortunately, regular statistics for anthracite receipts at Buffalo do not begin until 1863,[188] but at that date, of the nearly 500,000 tons of anthracite shipped on the canal, only about 125,000 tons were reaching Buffalo. A further 80,000 tons of bituminous coal were reaching the city by that date, but the majority were by lake shipment rather than from the canal. A portion of the coal received was shipped out to Chicago and other Great Lakes ports. However, in the later 1850s, despite its accelerating urban and industrial growth,[189] the coal business of Chicago was still a modest one. In 1860, only 131,000 tons were received in the city. Nearly 118,000 tons of this came via the Great Lakes,[190] but rapid expansion of this western trade was to begin shortly thereafter (see chapter 8).

Prior to 1860 therefore, the coal trade served steadily growing but geographically static markets, with the exception of the new anthracite iron industry of the 1840s and 1850s. Avenues to market by canal and coastal shipping routes were likewise relatively unchanging until the New Jersey Central Railroad began to funnel large quantities of Lehigh and Lackawanna coal to tidewater at the very end of the 1850s. The earlier completion of the northern division of the Lackawanna Railroad to reach upper New York state did not have an equivalent effect in terms of volume of shipments. Later, however, it was to become one of the key links in new rail connections to Buffalo that enabled the Great Lakes trade in anthracite to develop rapidly, despite the competition from cheaper bituminous coals.

DOCKS AND SHIPPING FACILITIES – PRE-REQUISITES FOR THE ANTHRACITE TIDEWATER TRADE

The combination of rail, canal, and coastal routes for anthracite meant that substantial and efficient facilities for coal storage and transhipment were essential to the conduct of the trade from an early date. The earliest developments of docks and storage yards were undertaken by the Delaware and Hudson at Rondout on the Hudson and by the Philadelphia and Reading Railroad at Port Richmond.[191] Initially, the Pennsylvania Coal Company was unable to follow suit, as its original charter did not allow it to hold lands for depots. In consequence, land it acquired for this purpose in 1850, also at Rondout, had to be held in the name of Irad Hawley until the necessary privileges were granted to it by the state legislature.[192] Development of the land by the building of wharves and offices began in 1851 and early in 1852 the dock property became known as Port Ewen.[193] Hawley was also very active from an early date in pursuing the interests of the company's coal business in the New York area. By October 1849, he had already acquired and developed land for coal docks on the East River at Williamsburgh. Control of these was exercised by the company through a complex financial package of loans and rental agreements, that fully reimbursed him for his costs, but formal ownership did not pass to the company itself, because of the charter restrictions.[194] Very shortly afterwards, the same yards were rented to the Delaware and Hudson, for it to take charge of receiving and selling of all coal of both companies in the New York area.[195] This was part of an overall agreement between the two companies for selling and transporting coal.

As the Lackawanna Railroad did not have an outlet to New York Harbor until 1856, its entrance into tidewater markets was substantially delayed in comparison to its Northern Field counterparts. In that year, its connection to Elizabethport was finally completed, utilizing the tracks of the Central Railroad of New Jersey. It moved quickly during the latter part of 1856 to establish a presence in the New York market, leasing a coal yard on the East River between 15th and 16th Streets from the Manhattan Gas Company.[196] When the Gas Company needed this yard for its own purposes, a lease was arranged for 12 lots on 14th Street from May 1, 1859, at a rate of $1,750 per annum. These lots belonged to Moses Taylor, the New York banker, who was a director of the railroad at the time.[197] Shortly afterwards, this yard had to be doubled in size by the addition of a further 13 adjacent lots. The shortage of space was a direct result of the stockpiling of stove and chestnut coal.[198] The company also opened a yard in Williamsburgh, but a presence in Brooklyn, which the Pennsylvania Coal Company had enjoyed since 1852, was not secured until 1861.[199] Although leasing coal yards aided short-term cashflow, the leases that could be obtained were sometimes for periods as short as 3 years, with the obvious implications for disruption to business at the end of the lease period. Increasingly, therefore, when the financial situation permitted, the companies began buying the dock properties instead. The Pennsylvania Coal Company, for example, purchased part of the land of its expanded Williamsburgh Yard, then held under lease, for $30,000 in 1856. The Delaware and Hudson made a similar purchase of 1,000 feet of waterfront in Weehawken in 1859, for the substantial figure of $53,300.[200]

Apart from the real estate costs of securing waterfront locations, a variety of additional expenditures were incurred in building wharves, piers, and bulkheads, not to mention dredging of channels, trestlework, and hoisting equipment. At Elizabethport, for example, the construction contract stipulated a 10 foot channel, but on examination it proved to be only 5½ feet deep[201] and remedial work was required. A further sum of almost $17,000 was spent on the 'piling ground and fixtures' at the same location in 1858. This included $2,000 for the erection of an engine and rolls for making additional stove coal.[202]

Elizabethport expanded rapidly at the end of the 1850s, with piers and wharves for other companies, such as the Union Iron and Coal Company, in addition to the Lackawanna Railroad itself.[203] However, it did not compare in size to the Port Richmond complex on the Delaware River outside Philadelphia. The Port Richmond site covered 49 acres with 2,272 feet of water front accessible to vessels of 6–700 tons. There were 20 piers and space to stockpile 250,000 tons of coal. A contemporary commentator stated that over 100 vessels could be loaded simultaneously, with coal fed directly from railroad cars down chutes. These enabled a 155-ton brig to be filled to capacity in under three hours. Overall, the size and extensive development of the docks were sufficient to provide an annual shipping capacity of 3 million tons or more.[204]

COAL SALES AGENTS AND COMMISSION MERCHANTS

The larger docks and coal yards represented major investments by the various companies, hence proper management and close supervision of operations was required. To this end, company agents and dock masters were appointed by the Northern Field companies to oversee commercial developments in the major markets and at the key distribution centers. In some of the retail yards, such as in New York, company commission salesmen were also appointed. For example, the Lackawanna Railroad's salesmen at the Williamsburgh and Manhattan yards were paid a basic salary of $500 per annum in 1859, with an additional commission of 10 cents per ton sold at retail. The company had previously decided as a matter of policy that it would not use third-party commission agents in the New York market, but would hire its own staff.[205]

In the more distant markets of upper New York state and New England, a variety of arrangements pertained, depending on the location. The Delaware and Hudson had appointed agents in Albany, Providence, and Boston at an early date[206] and the Lackawanna Railroad made similar arrangements during the 1850s, as it gained access progressively to both western and eastern markets. The importance it attached to the promotion of its coal and the development of its sales and distribution networks is apparent, both from the calibre of people employed and the financial remuneration it was prepared to offer. For the 1859 season, the leading figure of Colonel George W. Scranton was engaged as coal salesman in northern and western markets. A few weeks earlier the company had offered E. A. Quintard the substantial salary of $5,000 per annum to devote all his energies to selling Lackawanna coal in eastern markets. The only stipulation was that he should relinquish that part of his work previously engaged in the marketing of Cumberland coal.[207] In addition to these regional sales managers, as we might now term them, the company employed its own agents in important transhipment ports. Prominent among these was E. R. Holden in Syracuse, an individual whose ability was eventually to lead him to a senior position in the management of the railroad.[208] Elsewhere, as in Binghamton, for example, a local firm was used as the agent, on a commission basis.[209]

However, the sales and distribution arrangements for the coal operators of the Schuylkill and Lehigh Regions were somewhat different from those of the Northern Field companies during this period. Since the Reading and Lehigh Valley Railroads operated as common carriers and did not own coal lands, they did not engage in coal sales until after the Civil War. This function was performed either by the sales agents of the individual smaller coal companies or by commission agents and middlemen. It is more difficult to find surviving evidence of the activities of these agents than it is for the larger companies, but fortunately testimony presented during investigations of the anthracite coal trade in 1871 and 1876 sheds some light on the nature of this part of the business. J. B. Van Dusen provided very useful testimony during the 1876 investigation, because he had been involved in mining and shipping from the Schuylkill Region, as well as acting as agent for the Lehigh Coal and Navigation Company. As a member of several partnerships, he held a wharf at Port Richmond from 1855 until 1872, shipping as far north as Maine and south to Charleston. He stated that prior to the 1870s a great deal of coal was sold on commission, but he also indicated that very few people at Port Richmond were solely middlemen. This implies that most had some involvement in mining ventures also. Commission was usually 25 cents on large sizes and 15 cents on chestnut per ton.[210] Richard Sharpe, who testified before the earlier 1871 investigation, corroborated this information. Sharpe was a prominent Lehigh operator of the firm of Sharpe, Weiss and Company, which had commenced operations at Eckley in 1854. They employed G. B. Newton, later to be a coal operator in the Northern Field, as their Philadelphia sales agent in the 1860s and early 1870s.[211] Newton's job was to provide orders for coal from the line trade of the Lehigh Valley, so the coal did not enter into the lower priced competitive markets of the eastern seaboard as well as having to be handled by the middlemen on commission. As Sharpe himself testified, his company managed to avoid the 25 cents commission at Elizabethport because 'our trade is principally with the consumers'. This indicates that the company had direct ordering arrangements with major customers, such as the Allentown Rolling Mill and the Camden and Amboy Railroad.[212]

Thus, on the eve of the Civil War, a well articulated sales and distribution system for anthracite had been developed along the eastern seaboard for both industrial and domestic retail consumers. The line trade was very important to large and small shippers alike; because the element of monopoly kept prices higher and more stable than those that ruled in the tidewater markets, and the commissions of the middlemen could often be avoided. At the same time, for the smaller operators of the Schuylkill and Lehigh Regions, the commission agents performed an important service. They could consolidate low volume shipments into larger consignments for major consumers like the steamship companies. This enabled the small operators to compete on a more equal footing for major coal contracts with the large companies from the Northern Field, who dominated the New York market in particular. Nevertheless, the advantages accruing to railroads that could engage in both mining and transportation of coal were becoming increasingly apparent, in comparison to those railroads which acted solely as common carriers. Such an imbalance of economic power in an increasingly competitive industry with high fixed costs could not be expected to persist for long. The Civil War itself was to provide one of the triggers for change that was to alter irreversibly the organizational structure of the coal business.

Chapter 5

RAILROAD EXPANSION AND CORPORATE CONTROL I: NETWORK DEVELOPMENT STRATEGIES AND THE ARTICULATION OF THE ANTHRACITE DISTRIBUTION REGION IN INTERIOR MARKETS

INTRODUCTION

It is a daunting task to analyse the business processes and strategies employed by the major anthracite-carrying railroads to expand their spheres of operation and develop new markets for coal. The sheer number of construction projects, traffic agreements, mergers, leases and receiverships involved tends to overwhelm both the investigator and the reader. In consequence, the broader picture can be submerged in a welter of chronological specifics and intricate financial details.

In an attempt to avoid falling into this trap, both this chapter and the next will focus on assessment of the relative success of the different types of business and territorial strategies followed by the various companies. A number of examples of these strategies will be analysed, but without any claim to comprehensiveness of historical or geographical coverage during the period of interest, since this would require a multi-volume study in its own right. In so doing, the opportunity will be taken to examine a number of processes and events that have not been closely studied previously, and also to illuminate the motivations of business decision-makers, based on evidence from corporate records that have not been utilized in earlier studies.

OVERVIEW OF TERRITORIAL AND BUSINESS STRATEGIES

A number of territorial strategies that were followed by the anthracite railroads can be identified. Although there is an element of chronological sequence to them, it is important to recognise that any given company may have pursued several simultaneously. Also, the primary emphasis here is on developments in the transportation network for distributing anthracite beyond the coalfields, rather than in the building of railroad lines within the coalfields. The latter type will be examined in more detail in a later chapter.

The major territorial strategies were as follows:

1. Assertion of control over short-branch or feeder lines connecting the mines to the main transportation arteries.
2. Development of linkages with other short lines to reach the markets of upper New York state.
3. Provision of improved access to western markets via Buffalo and other Great Lakes ports.
4. Opening up of northern markets as far as Montreal.
5. Acquisition and retention of spatial monopoly power to provide a lucrative 'line trade' for coal.
6. Reduction of dependence on other railroads to reach the tidewater markets and New York, in particular.
7. Obtaining control of suitable terminal and shipping facilities at the major transhipment points.
8. Utilization of controlled and linked lines, in conjunction with terminal facilities, to participate in trunk-line pools and compete for through freight from the interior, thereby lessening dependence on coal as a proportion of all shipments.

The first four of these strategies will be examined in this chapter and the remaining topics in the next.

However, there were a number of standard business processes and contractual methods for achieving these territorial goals. These can be divided into two main groupings: methods of controlling other railroads, and approaches to financing and construction of new railroads.

It is appropriate to begin with a brief discussion of these different methods and processes, so their general characteristics can be distinguished from the specific details of actions taken by individual railroads. The examples to be used will stress the nature of contractual arrangements rather than geographical context. They will be referred to again in the later examination of the different territorial strategies, when the emphasis will switch to processes of network expansion rather than contractual details. The next chapter will then assess how effectively contractual arrangements were matched to the geographical requirements of serving different markets, in the context of fluctuating economic conditions, both in the short and the long run.

METHODS OF CONTROLLING OTHER RAILROADS

A variety of different methods for asserting control over other roads can be identified. Among these were:

- Control through stock ownership.
- Control by means of lease agreement.
- Operation under joint-management agreement.
- Consolidation and merger.

The first of these methods was the simplest. It merely required building up a majority stake in a company through purchase of its stock directly, or indirectly through allied financial interests, i.e. wealthy individuals friendly to the controlling railroad. Ownership of a majority of the stock meant that control could be exerted over membership of the Board of Directors. This usually led to senior managers of the controlling road being appointed to the board of what was now effectively a subsidiary company. However, since the corporate organizations were entirely separate, legally the controlling road was not liable for any debts or obligations of the subsidiary. A good example of this method of control was the Northern Central Railway. Originally the Baltimore and Ohio and Philadelphia and Reading Railroads controlled it, but during the monetary panic of 1861 in Baltimore, these companies and other stockholders jettisoned the stock. The Pennsylvania Railroad immediately purchased some 43,614 of these shares and the remainder required to secure majority ownership were purchased in London.[1]

Control by means of lease agreement bound the lessor and lessee railroads into a much tighter contractual framework than majority stock ownership alone. Leases would obviously vary in detail from case to case, but had a number of general characteristics in common. The most important of these was a guaranteed payment by the lessor of dividends at a set level on the stock of the lessee railroad. This level was usually 6 or 7 percent. The lease would be for a term of years, usually 99 years or in perpetuity. In addition, the lessor might guarantee the interest on the bonds of the lessee to enhance still further the security of this form of investment. A lease effectively meant that the lessee railroad became a division of the lessor railroad, and in practice was operated as such. Nevertheless, the two corporate organizations remained legally distinct. In one sense this was a necessity, since if the lease was terminated for any reason, the two companies would have to continue business as separate entities. The legal distinctiveness was also utilized for financing purposes, however. For example, a lessee railroad might have authority under its charter to issue up to $3 million of bonds. At the time of its lease, only $1.5 million might have been issued. At a later date the lessor railroad, which in the interim had funded the construction of new branch lines for the lessee, might call upon the latter to issue a further $500,000 of its bonds and make them over to the lessor. Since the lessor would guarantee these bonds, it could then dispose of them to New York banking houses and other investors, to recoup its outlay on construction expenditure for the lessee.

As a method of corporate control, leasing was very successful and very widely used, so a number of examples will be considered in the ensuing discussion. It was successful because of its attractiveness to stockholders. It gave them a guaranteed long-term return on investments, which would otherwise have been subject to fluctuating or even nil dividends, in the light of changing business circumstances. The financial risk was therefore transferred to the lessor railroad. The lessor had to weigh up the benefits of controlling lines and achieving more effective traffic management against the financial constraints of incurring higher 'fixed charges'. The point about 'fixed charges' was that these were financial obligations, (in this case the dividends on the lessees' stock), which had to be met, regardless of the financial circumstances of the lessor, or the current state of its trade. If the obligations were not met, the lease would be terminated.

A variety of possible arrangements can be placed under the heading of joint-management agreements. The simplest of these was the trackage-rights agreement, which enabled one railroad to use the tracks of another between two points, for a specified period, based on an appropriate charge per passenger or ton-mile of freight carried. More complex forms of agreement might include cases where the companies involved also shared the proceeds of the sale of the freight thus transported. A good example of this was the Jefferson Railroad, opened in 1870, which ran northwards from Carbondale in the Northern Field to connect with the Erie Railway main line at Lanesboro Junction just east of Susquehanna, Pennsylvania.[2] The railroad was operated by the Erie, but the Delaware and Hudson Canal Company had trackage rights to ship its coal over the road at a rate of 1 cent per ton-mile. In addition, however, the two companies shared the profits arising from the sale of the coal thus shipped, after allowing for all expenses, including those attributable to the trackage rights.[3]

The most complex form of joint-management agreement involved two companies each nominating members of their boards to a joint management board. The purpose of this joint board was to allow the two companies to be operated as a single entity, but without the constraints or asymmetrical power relations implied by a lease. An arrangement of this kind was initiated between the Delaware, Lackawanna and Western Railroad and the Central Railroad of New Jersey in 1872, to act as a precursor to a planned merger, which in the event was not carried through.[4]

Merger and consolidation was the ultimate expression of corporate control. In many cases it was the final stage in a process that had begun with stock ownership, traffic agreements or lease arrangements. It resulted in the absorption of one company by another, usually much larger, corporate entity. Stock of the company being merged would be exchanged on a mutually agreed basis for stock of the parent company and all previously existing leases or agreements between the two companies would be terminated. It is worth noting that the replacement of a lease by a merger would be a beneficial step for the larger company, since the exchange of stock would reduce the fixed charges of the latter. The reason for this was that stock, on which dividends had been guaranteed under the lease, would be replaced by stock of the parent company, upon which no such guarantee had been offered.

APPROACHES TO FINANCING AND CONSTRUCTION OF NEW RAILROADS

Acquisition of existing lines, where these reached the required destinations or connecting points, was, in general, a preferred strategy, because it minimized the initial costs of network expansion. If the necessary lines were not extant or had fallen into the hands of competitors, the only alternative was new construction. Railroads could achieve this in a number of ways

- By building the line themselves, financed by retained profits or bond issues;
- By funding other railroads to build the new line; and
- By chartering a new railroad or acquiring an existing unused charter and employing that company as the construction vehicle.

After completion of the new railroad using either the second or third approach, it might be leased by, or merged into, the main railroad that had provided the financing.

Building a new line had the advantages that it assured full control and the choice of the best available route to meet the required business goal. The main disadvantage was the very high cost, which had to be met out of retained earnings or by borrowing. The likely costs depended on the nature of the terrain and prevailing prices for rails and construction labor, but decision-makers were aware of the approximate range they might expect, based on available information from earlier projects. For example, the 63 miles of the main line of the Central Railroad of New Jersey completed from Elizabethport to Phillipsburg had cost nearly $2.4 million by 1852. The easier stretch of the first 35 miles from the coast had cost only $23,000 per mile, but this rose steeply to over $56,000 per mile on the more difficult second stretch of 28 miles, which traversed much higher ground.[5] The connecting Lehigh and Susquehanna Railroad, which stretched from Wilkes-Barre to Phillipsburg, exhibited a similar pattern of costings, when it was constructed by the Lehigh Coal and Navigation Company at a later date. By January 1, 1868, the completed railroad had 145 miles of main tracks and 43 miles of sidings and stood on the books at a total construction cost of more than $10.3 million or $55,000 per mile. The road included the original stretch from Wilkes-Barre to Whitehaven over the Ashley planes, which had been built in the 1840s and had costs of $1.5 million attributed to it by the beginning of 1862. In addition, a further $1.6 million had been invested in rolling stock and equipment for the completed road by 1868. It had been the intention of the Board of Managers to fund the construction by the issue of new stock and 'expected surplus profits', but the costs were much greater than expected and profits much less in the period following the Civil War. As a result, authority to issue a further $5 million loan had to be sought from the legislature. Initial attempts to place the loan in New York or London both failed. In consequence, the financial position of the company was seriously threatened by cash-flow problems, until 'friends of the company outside the board' managed to place the entire loan at 85 percent in the United States.[6]

This last example is instructive because it indicates that the securing of investment capital for new projects, even by old established companies such as the Lehigh Coal and Navigation, could be a source of major difficulty. High costs might exceed the funding capabilities of the company itself, as well as its usual backers. They certainly exceeded the lending capabilities of individual state banks in the region and might require the services of major New York banks to tap into the resources of Wall Street or the City of London. If large and prestigious projects frequently had difficulty securing finance, it was all the more difficult for smaller local railroad projects to attract the attention of potential investors. As a result, the first port of call for a small railroad promoter with a new project in mind was the office of one or more of the major railroads, which might be persuaded to use the fledgling company as a vehicle for construction of one of the links in its expanding network.

A good example of this was the Greene Railroad in New York state, which was a key link between two northward connections of the Lackawanna Railroad. These connecting roads were the Syracuse, Binghamton and New York Railroad and the Utica, Chenango and Susquehanna Valley Railroad. Of these, the former was controlled by the Lackawanna through majority stock ownership from February 1869[7] and the latter by lease dated April 9, 1870.[8] There was a strong tradition of municipal support to attract railroads in New York state[9] and the town of Greene was no exception. It had arranged for a bond issue of $198,700 to aid in building the 8-mile link railroad and was impatient to see it completed. However, it also appreciated that it could not sell its bonds in the financial markets fast enough to enable construction to start in a timely manner. In consequence, agreement was reached with the Lackawanna Railroad for a lease of the Greene Railroad that contained some unusual financing arrangements. In essence the Lackawanna was to act as banker, funding the construction of the railroad in the form of a loan secured on the town's bond issue. Initially the town treasurer made over to the railroad $170,000 of the bonds, retaining $25,000 of the bond issue to be sold as quickly as possible. The proceeds of the first bond sale were to be paid to the railroad by July 1, 1870, at which point the Lackawanna would hand back to the treasurer a further $25,000 tranche of bonds for sale. A second staged payment was to follow on October 1, 1870 and so on until the town treasurer had sold $95,000 of the bonds. The remaining $100,000 of bonds was to be disposed of by the Lackawanna itself, to fund the acquisition of right-of-way for the new road and the first stages of construction. In the event, with the rapid improvement in business conditions during the first half of 1871, this cautious approach to financing was not entirely necessary. By the end of July 1870, $111,700 of the town of Greene bonds had already been sold at par and $26,200 at 95 percent.[10] The road was completed on October 22, 1870.[11]

In other cases, it was deemed appropriate that the major railroad should itself take direct responsibility for chartering a new railroad to complete a chain of lines that formed a new route. The New York and Canada Railroad was a case in point. It was chartered by the Delaware and Hudson to complete a line running up the western side of Lake Champlain, for onward connection to Montreal (Figure 5.1). The key missing section in the route was between Whitehall, the terminus of the Rensselaer and Saratoga RR and Plattsburgh, a few miles south of the Canadian border. The importance of linking the two towns had been recognized by 1866, when the Whitehall and Plattsburgh Railroad was

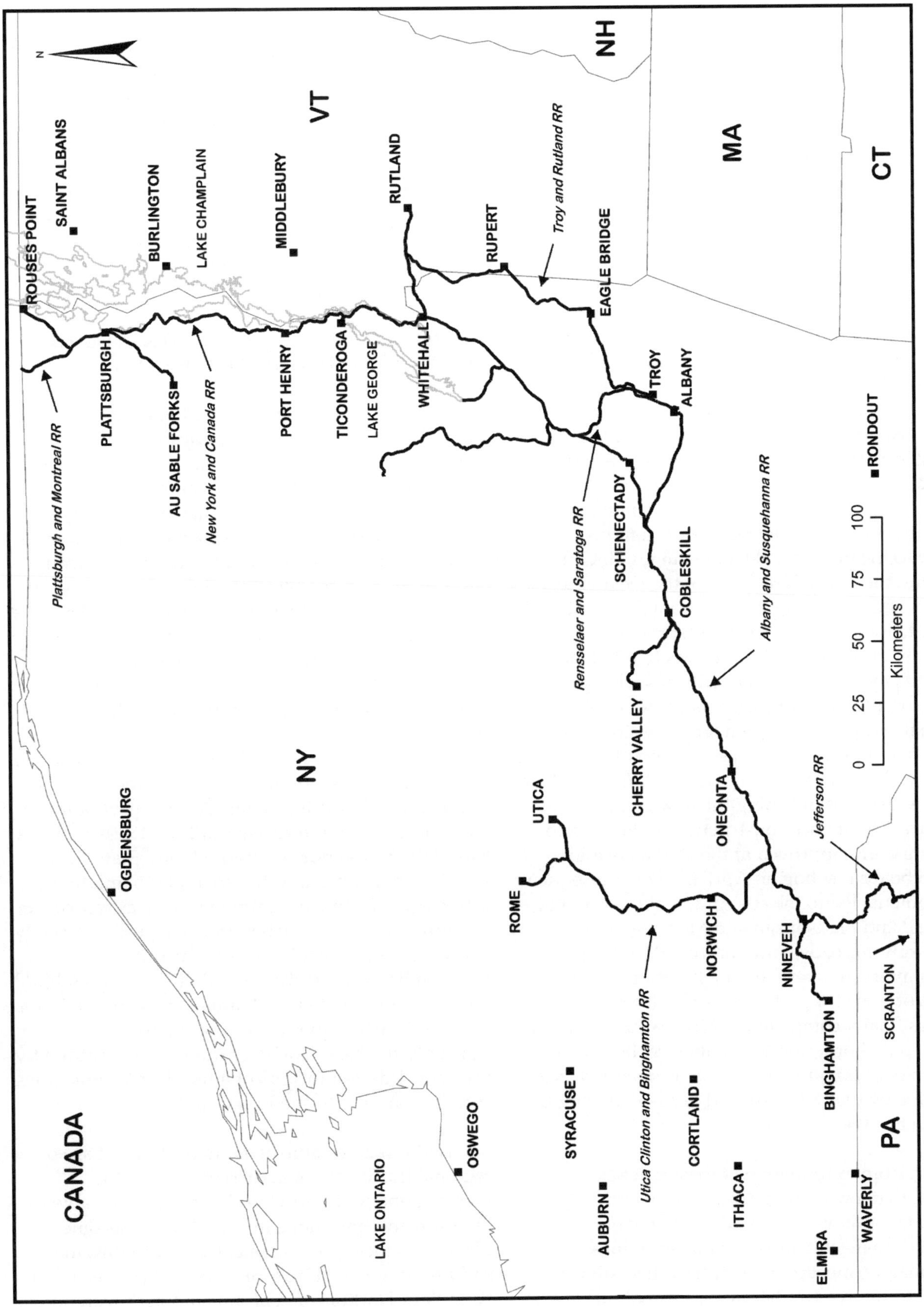

FIGURE 5.1
The New York and Canada and Linked Railroads in Upper New York State.

chartered. A first section of the road, running 20 miles south from Plattsburgh, was completed, together with a separate 17-mile section further south, which ran from near Fort Ticonderoga to Port Henry. Unfortunately, it did not succeed in completing the missing sections and its two stretches of track were finally leased in 1869 and 1890 to the Montreal and Plattsburgh Railroad and the Rutland Railroad respectively. The following year, the latter took over the former to extend its control still further.[12] Clearly dissatisfied with these arrangements, 'parties interested in the Whitehall and Plattsburgh Railroad' approached the Board of the Delaware and Hudson in September 1871 with a proposition for constructing one of the missing sections of the road from Whitehall to Port Henry. This was a logical step, since the Delaware and Hudson had leased the Rensselaer and Saratoga Railroad earlier in the same year.[13] While the 'interested parties' are not named in the Minute Books, Shaughnessy records that early in 1872, Smith M. Weed, prominent citizen of Plattsburgh and member of the State Legislature, lent his support to the proposal. At a meeting with G. T. Olyphant, the Acting President of the Delaware and Hudson, and I. V. Baker, a director of the Whitehall and Plattsburgh and Superintendent of the Rensselaer and Saratoga, Weed was finally able to persuade Olyphant of the benefits of such a project. Much groundwork had clearly been done following the initial proposal in September 1871, since Weed had with him the articles of association of a new company, the New York and Canada Railroad.[14] No time was lost in establishing the new company and the contract to finance construction of the road to Port Henry was approved at the Delaware and Hudson board meeting on April 3, 1872. Subsequent developments led to the absorption of the original Whitehall and Plattsburgh and Montreal and Plattsburgh railroads, now controlled by Vermont railroad interests,[15] and the construction of further sections of track through very difficult terrain. The road was finally completed to Mooers Junction near the Canadian border on November 29, 1875, at a total cost of $8.25 million,[16] and was operated as a leased line by the Delaware and Hudson during the remainder of the 19th century.

Even from examination of these brief examples, it is clear that a wide range of financial and contractual options were available to enable companies to meet their territorial ambitions. Unfortunately, most of these options had significant fixed costs involved, either in the form of interest on bonds or guaranteed dividends on the stock of leased lines. When markets and traffic were expanding rapidly, these increased charges could be readily absorbed. During times of business contraction or intense competition, however, they could become a crushing burden on the profit and loss accounts.

The subsequent sections will examine the interplay of these developments and constraints in reaching the widening geographical markets to which anthracite was shipped.

REGIONAL AND NATIONAL STRATEGIES FOR RAILROAD SYSTEM BUILDING

Chandler's well-known study of the managerial revolution in American business places great emphasis on the importance of the 19th-century railroads as innovators in corporate administration, divisional management, and accountancy practices. It has also established a certain orthodoxy in the treatment of American railroad history that stresses the period of system building from the 1880s until the first decade of the 20th century. This period is clearly characterized by developments at the national level, which culminated by 1906 in seven railroad groupings having control of ⅔ of the country's railroad mileage.[17] Much less emphasis is given to the smaller-scale system-building at the regional level, which was the essential precursor to the later national level developments. A large part of the story of railroad enterprise until the 1870s, especially in the northeastern United States, is a story of short lines and the struggle to link them into more effective operating entities, as our earlier examples have already begun to suggest. Relatively few lines were conceived from the outset as major through lines, apart from the Baltimore and Ohio[18] and arguably the Pennsylvania, although the latter absorbed earlier lines as part of the route.[19] Certainly the New York Central was constituted by the consolidation in 1853 of a number of short lines running parallel to the Erie Canal.[20]

Without an established and profitable base of regional traffic, regional carriers such as the Lackawanna Railroad would not have been able to bid for trunk line status in the 1880s, to become players on the national scale. Further to this, the early moves to acquire large terminal properties on New York Harbor for coal transhipment were to assume an even greater importance in later years, as the volumes of general freight carried to and from the interior grew to become a much larger proportion of overall shipments. Initially, therefore, the motivation for regional-level system-building was to secure access both to supplies of coal and markets for the commodity. The potential this eventually provided for gaining control over the coal industry as a whole will be the subject of Chapters 10 and 11. Once the regional systems were in place, the natural geographical advantages of routes funnelling into the eastern seaboard cities and northwards to Buffalo and the Great Lakes, could be exploited for wider commercial advantage and reduction in the dependency on coal as a single commodity for shipment. These regional-level developments and their subsequent extension to the national level will now be examined in turn.

CONTROL OF FEEDER LINES

The existence of independent feeder railroads, or 'coal laterals' as they were often called, can be attributed in large measure to the fact that the early exploitation of the Schuylkill and Lehigh Regions was based on the canal as the main avenue to market. Canals, by their very nature, could not readily penetrate mountainous terrain to reach the mines directly, so feeder railroads had to be built. Early Schuylkill coal laterals included the Mine Hill and Schuylkill Haven Railroad (MH&SHRR) and the Mount Carbon and Port Carbon Railroad, which brought coal down to the Schuylkill Canal (Figure 5.2). In the Lehigh Region, the early railroads included the Beaver Meadow Railroad and Coal Company and the Hazleton Railroad, whose coal was sent to market via the Lehigh Canal. In a sense, the coming of the competing mainline railroads of the Philadelphia and Reading and the Lehigh Valley in the 1840s and 1850s presented a dilemma for the lateral operations, which had established shipping connections via the canals. They could scarcely forsake their current business partners, yet they could also not ignore the new opportunities for winter shipment presented by the main line railroads. The MH&SHRR solved the problem by dividing its tonnage equally between the canal and the railroad and this proved acceptable over a number of years, against a background of rising coal shipments.

However, after the period of the late 1850s described in the previous chapter, when growth eluded the Schuylkill Region, the attitude of the Reading Railroad towards its independent feeders began to change from one of co-operation towards control. This can hardly be regarded as coincidental. The 'official' rationale for the sudden desire to take over the lateral railroads was given in the 1861 annual report, as follows:

> 'to introduce a more efficient administration, and to reduce the charges for toll and transportation, and thus effect a reduction in the cost of placing coal at tidewater in Philadelphia'.[21]

There was no doubt that reduction of tolls was intended, but it is perhaps significant that the aim was reduction in cost to the Reading, not necessarily reduction in price to the consumer. The reaction of the MH&SHRR to this plan, as stated in its annual reports, put the Reading's actions in a rather different light. They observed that during the period of independence the tolls on the different laterals varied widely, with their own being the lowest. The Reading targeted for leasing, those laterals with the highest tolls. One of its leases contained a clause that dramatically lowered the tolls from the most distant collieries to below the lowest rates that had previously prevailed. The MH&SHRR expressed the view that the sudden change in the competitive position of nearer collieries would probably result in a greater diminution of their output than would be compensated for by increased production at the more distant collieries, to produce an overall unintended reduction in output. By way of contrast, the MH&SHRR claimed it had carefully balanced tolls and motive power charges on its numerous branches, so as to 'maintain the relative ability of the different collieries to compete fairly with each other and to sustain themselves against the rivalry of other regions'.[22]

The reduction of tolls for distant mines, coupled with other developments, suggests a further unstated reason for the Reading's strategy, namely a desire to channel the traffic of the newly developing Western Middle Field southwards down the Schuylkill Valley, rather than eastwards down the Lehigh. In this context it is significant that the Reading leased the Mill Creek and Mine Hill Navigation and Railroad Company in 1861[23] and in the same year began construction of the Mahanoy and Broad Mountain Railroad. This road ran from a connection with the Mill Creek Railroad near St. Clair and extended into the Mahanoy Basin of the

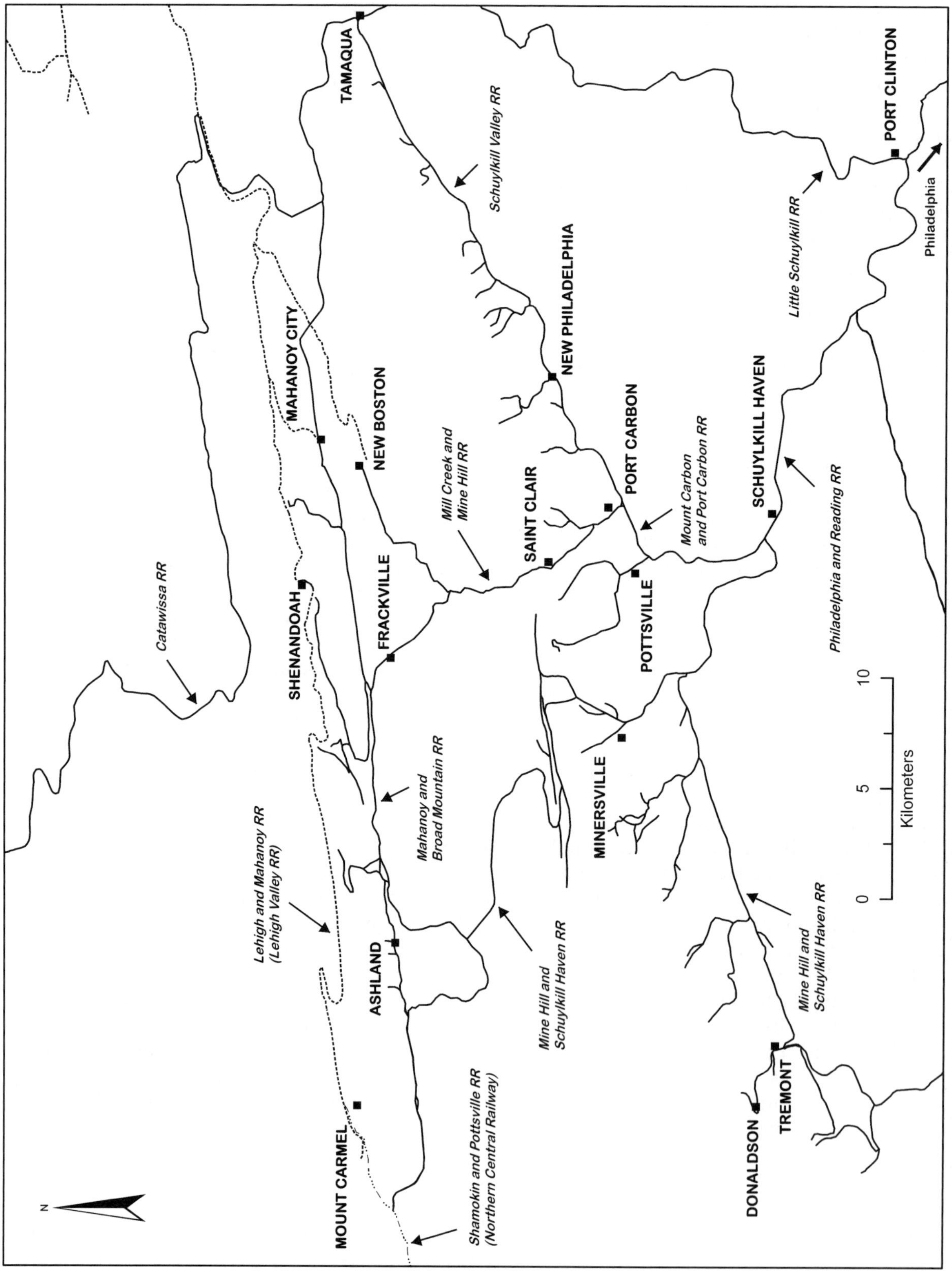

FIGURE 5.2
Major Coal Lateral Railroads in the Schuylkill Region circa 1866.

TABLE 5.1
Details of Leases of Major Laterals and the Schuylkill Canal by the Reading Railroad 1860–70.[30]

Company	Lease Date	Annual Rental	Term of Lease (years)
Mount Carbon and Port Carbon RR	March 5, 1860	$36,250	50
Mill Creek and Mine Hill Navigation and RR Co.	July 25, 1861	$33,000	999
Schuylkill Valley Navigation and RR Co.	July 25, 1861	$29,450	999
Mine Hill and Schuylkill Haven RR	May 12, 1864	$296,000	999
Little Schuylkill Navigation RR and Coal Co.	July 7, 1868	$185,227	93
Schuylkill Navigation Co.	January 1, 1870	$655,000	999

Western Middle Field, via the old line of the Danville and Pottsville Railroad.[24]

Since the Mahanoy and Broad Mountain Railroad encroached on territory that the MH&SHRR regarded as its own, the latter felt it had to respond. It outlined three possible courses of action to its stockholders:

1. Build a line eastwards to the Lehigh Valley Railroad (the Schuylkill Haven and Lehigh Valley Railroad had already been chartered for this purpose in 1856, by interests allied with the Lehigh Valley Railroad,[25] so this may also have been a factor behind the Reading's new strategy).
2. Become subject to lease by the Reading Railroad.
3. Make an agreement with the Schuylkill Navigation Company.

It chose the last of the three options, which involved co-operation on the setting of tolls and a guarantee on dividends for its stockholders by the Navigation Company, but not loss of control of the Railroad. According to its annual report it was not a lease, as later claimed by Bogen,[26] but an agreement between the two companies.[27]

By the following year the MH&SHRR was increasingly unhappy with the manner in which the Reading strategy was unfolding. The reduction of tolls made by the latter company had forced the same on the MH&SHRR, but the tolls on the main line had been increased correspondingly, so all the benefit accrued to the Reading, and none to the mine operators. Further to this, the reduction in income had adversely affected the MH&SHRR, leading it to conclude that the entire Reading strategy seemed 'to have been dictated by a design to damage the prosperity of a company, by whose early enterprise a large part of the mining industry of the Schuylkill Region had been developed'.[28]

Credit must, however, be given to the Mine Hill managers for their initiative in the face of adversity. They promptly leased the Schuylkill Haven and Lehigh River Railroad and agreed to buy its shares at par, to provide the necessary cash flow for construction. This was a serious threat to the Reading, as it would divert the tonnage of its largest feeder eastwards to the Lehigh. It responded, not by taking direct action against the MH&SHRR, but by indirect action against the guarantor of the latter's dividends, the Schuylkill Navigation Company. Since the Reading had already leased three of the feeder lines, it was able to divert tonnage away from the canal and render its operation unprofitable. This brought the canal to the negotiating table with the railroad. The eventual result was a traffic agreement whereby the canal company would receive 45 percent of total coal tonnage during the navigation season, the remainder going to the railroad. In addition, control of the MH&SHRR was to pass to the Reading by a lease dated May 12, 1864, at a rental equivalent to 8 percent on the capital stock. A further condition of the new

TABLE 5.2
Receipts and Expenses of the Mine Hill and Schuylkill Haven Railroad 1867–1869 (excluding lease rental).[31]

Year	Receipts	Expenses	Deficit
1867	$477,563	$635,957	$158,394
1868	$443,393	$525,048	81,655
1869	$477,411	$564,047	86,636

arrangement was that all work on the Schuylkill Haven and Lehigh River Railroad should cease.[29]

In this way, the Reading progressively strengthened its hold over the major transportation routes from the Southern Field and parts of the Western Middle Field. A further lease of the Little Schuylkill Railroad was secured in 1868 and finally the Schuylkill Navigation Company itself was absorbed by lease in 1870, ending competition among routes down the Schuylkill Valley, until the Pennsylvania Railroad intervened in the 1880s. Control was not secured without a significant cost, in terms of fixed charges resulting from the rentals on the leases, as is shown in Table 5.1.

It is very difficult to evaluate the purely financial benefits of absorbing the laterals, because their business is not reported separately in the annual reports of the Reading in the late 1860s and 1870s. However, some of the main account books for the first few years of the Reading lease have survived and from these it is possible to give an indication of how the finances operated under the new management. The relevant figures for the MH&SHRR are given for the years 1867–1869 in Table 5.2.

It would be unwise to attribute too much precision to the above figures, because the operation of the road was progressively integrated with that of the Reading itself, and not all income or costs may have been correctly attributed. Nevertheless, it would appear that the regime of lower tolls made the laterals something of a burden on the parent road, especially when the additional $296,000 of the rental was added to the excess of expenses over receipts. Also, the traffic agreement with the Schuylkill Navigation Company prevented too much of the deficit being recouped on the main line by diversion of tonnage from the canal.

A similar conclusion emerges from an analysis of the impact of the subsequent lease of the Schuylkill Canal in 1870. This analysis is facilitated because, unlike the laterals, the business of the canal is separately detailed in the annual reports of the railroad. In 1871, the first full year of operation of the lease, total canal receipts were $1,077,079, just over a million tons of coal were shipped and the net profit including the lease rental was $18,262. The Board of Managers announced with confidence that 'henceforth it is believed that the lease of the works of the Schuylkill Navigation will be a source of revenue to the Company'.[32] Unfortunately, their initial confidence was not borne out by their subsequent actions. In 1875, the canal lost $130,000. By 1879 the annual loss had grown to $334,000 and shipments of coal had halved from their 1871 levels to only 510,637 tons. This was the result of a policy change after 1871 that led to the canal being used only as a 'dripping pan' to catch the overflow of shipments when the railroad was very busy. It was acknowledged that 900,000 to 1 million tons were needed to make the canal profitable, but it was felt to be in the company's interest to keep as much tonnage on the railroad as possible.[33] However, it can be argued that the 'company's interest' related more to the retention of market share in the anthracite cartel (see Chapter 11) than it did to minimizing the cost of bringing coal to market. In 1879, a ton of coal shipped on the canal yielded an average of 62.7 cents in toll receipts, but the overall ratio of expenses to earnings was only 32.6 percent, even with a modest level of shipments. On the railroad, by contrast, each ton of coal shipped on the main line and laterals yielded 88.2 cents in tolls, but the expenses attributable to coal shipment were 64.2 cents per ton and the overall ratio of expenses to earnings was 68.6 percent.[34] Eventually, the drain on financial resources caused by low utilization of the canal in relation to its substantial lease costs, led to the rental being reduced to $374,582 in 1887. By this time the losses on the canal lease had cost the railroad in excess of $4 million.[35]

Overall, the relatively high cost of gaining control of the laterals was justified by the long-term benefits, in terms of security of tonnage for the main line. At a later period, and in a way that could not easily have been foreseen during the Civil War years, it also meant that the railroad controlled the supply of coal cars directly to the mines. This was to prove invaluable during the operation of the cartel. The case for controlling the canal is much less convincing. The lease may have been a good tactical move initially, to remove the agreement on allocation of traffic between the canal and the railroad, but it should perhaps have been relinquished at the first sign of financial difficulties ahead during the 1873–1879 depression. Since the railroad controlled the laterals already, and the canal era was drawing to a close of its own accord, the Schuylkill Navigation could have been left to wither at the expense of its own stockholders, rather than those of the Reading.

The foregoing example might lead to the impression that the absorption of lateral railroads always took place at the instigation of the main lines and that the laterals were mere pawns in a wider game of corporation expansion and territorial ambition. In at least one important example, however, from the available documentary evidence, this does not appear to have been the full story. The example in question is that of the Beaver Meadow Railroad and Coal Company, which brought coal from the Lehigh Region down along Quakake Creek to Penn Haven and thence by railroad down the Lehigh River valley to East Mauch Chunk, where it transhipped coal onto the Lehigh Canal. Its earlier extension to Parryville, below Mauch Chunk, had been washed out during the floods of 1841 and was not rebuilt.[36] The Beaver Meadow Railroad therefore ran in parallel to the Lehigh Canal down the Lehigh Valley from Penn Haven to Mauch Chunk. The canal continued northwards to Whitehaven, where it connected to the Lehigh Coal and Navigation Company's railroad. This ran via the Ashley Planes to tap the trade of the Wilkes-Barre section of the Northern Field. These arrangements operated satisfactorily until the summer of 1862 when floods severely damaged the Lehigh Canal. The southern section below Mauch Chunk was repaired, but, by May 1863, the costs of this had already reached $566,000. To repair the northern section to White Haven would have cost a further $900,000 and this section had the lightest traffic and the heaviest maintenance costs. A provisional decision was therefore taken to abandon it, but this had several consequences. Firstly, a charter amendment was required, since the charter required the company to make good any damage to any section of the canal. Secondly, the railroad from the Ashley Planes to Whitehaven would immediately become unusable, cutting off the trade from the Wyoming Valley, as would the extensive coal transhipment facilities of the Hazleton Coal Company at Penn Haven. Thirdly, new shipping facilities would be required at Mauch Chunk to replace those abandoned at White Haven and Penn Haven. Fourthly, the only means of replacing the abandoned canal section would be by means of a new railroad from Mauch Chunk to White Haven. The Managers of the Lehigh Coal and Navigation Company moved quite quickly and, on March 4, 1863, a charter amendment was passed which allowed them to replace part of the canal with a railroad[37] However, Asa Packer of the Lehigh Valley Railroad moved even more quickly. In 1857, he had chartered the Penn Haven and White Haven Railroad to build between the two points named, but no work had been started. Seizing the moment of greatest weakness of the Navigation Company immediately following the flood, he placed the road under contract in July 1862, although it was not completed until the summer of 1864. Shortly after completion it was merged with the Lehigh Valley Railroad.[38]

By the time of the stockholders meeting in 1864, the Lehigh Coal and Navigation Company had put under contract the construction of its new railroad from Mauch Chunk to Penn Haven, which was completed in June 1865.[39] It also acquired a further charter amendment to allow it to extend the railroad below Mauch Chunk to Easton, which would enable it to compete directly with the Lehigh Valley Railroad. These developments naturally alarmed the Lehigh Valley Railroad and the Beaver Meadow Railroad. The latter, in particular, tried to stop the navigation company building its railroad and went as far as grading a third track of its own, along the intended route. The navigation company took out an injunction against them to stop this and had the state Supreme Court confirm its obligation to build as far as White Haven, after hearings in February and March 1864. Despite the court rulings, the minutes record 'it was not deemed prudent to rely exclusively on the protection of the courts. We have sought to protect ourselves by permanently occupying the ground'.[40]

Losing out to the Navigation Company in the Supreme Court meant that the Beaver Meadow Railroad's monopoly on rail transportation between Penn Haven and Mauch Chunk would shortly draw to a close. In consequence, the road became increasingly nervous about its long-term position. As a first step, it agreed with the Lehigh Valley Railroad to prorate coal shipments from Penn Haven to Easton, to meet the normal rates between Penn Haven and Mauch Chunk that it expected its new competitor to charge. The minutes of the Beaver Meadow Railroad also record an exchange of letters instigated by the Navigation Company in April 1864, which led to meetings between the two companies. Judge Packer of the Lehigh Valley Railroad was asked to attend when it became apparent that the co-operation of this road was essential to any future agreement. Although the Beaver Meadow delegates urged 'the practicability and importance of a concert of action between the three companies whereby the interests of all parties might be promoted and permanently secured', the conclusion from the round of talks was quite the opposite. It became abundantly clear 'that no form of co-operation or combined action was practicable while three companies were required to assent'.[41]

The Beaver Meadow felt it had reached a point of crisis in its affairs, which endangered its ultimate prosperity. As each day passed, the managers became increasingly convinced that decisive and immediate action was required 'in order to save the company from the fatal effects of a rivalry in which we might prove the weaker party'.[42] By mid-June they had agreed the final destiny of the company in the form of a merger and consolidation with the Lehigh Valley Railroad, dated June 18, 1864. This involved exchanging both common and preferred stock of the Beaver Meadow for Lehigh Valley shares on a one-for-one basis. Together with the capital stock of the Penn Haven and White Haven ($1,415,500) from the subsequent merger of the road, the overall effect was to increase the capital stock of the Lehigh Valley from $3.2 million to $6.6 million almost overnight.[43] It also brought ownership of about 2,000 acres of coal lands held by the Beaver Meadow, although the Lehigh Valley did not enter the mining business at this stage, because the coal lands had already been leased to independent operators.[44]

The circumstances leading to the merger can therefore be traced directly back to the after-effects of the flood of 1862 on the Lehigh Canal, despite the complexity of the intervening events. The Beaver Meadow was not in financial difficulty, it had no debts and its receipts were three times its expenses in 1862 and 1863. Business was booming, with an increase in coal tonnage from 888,221 tons in 1861 to 1,442,706 in 1863.[45] This was in stark contrast to the Navigation Company whose shipments were adversely affected throughout much of the Civil War period by the loss of the upper section of the canal and the shipping facilities at Penn Haven. The Beaver Meadow therefore sought the merger to preserve its stockholders from the ravages of future competition. At the same time, there can be no doubting the skill of Judge Packer of the Lehigh Valley, who utilized events to achieve his strategic plan of a route from White Haven to Easton. By means of the merger with the Beaver Meadow, he strengthened the revenue-earning capacity of the Lehigh Valley and placed it on a more solid footing. This would have been welcomed by coal operators and stockholders alike, since during the war years, it had acquired a bad reputation for its failure to supply sufficient cars and motive power to move all the coal offered to it for transportation.[46]

REACHING THE MARKETS OF UPPER NEW YORK STATE

The early importance of the eastern seaboard markets and the much later development of the Chicago trade via Buffalo can lead to the importance of other markets for coal, such as those of upper New York state, being overlooked. The Erie Canal and its branches had given an initial stimulus to the development of urban centres such as Utica, Rome, Syracuse, Rochester, and Oswego, as well as Buffalo at the western terminus on Lake Erie.[47] It is therefore interesting to note that the original aim of the Lackawanna Railroad was to supply 'the whole of western New York and the country bordering on the lakes, including a part of Canada, with a superior quality of hard or anthracite coal, to be obtained at Scranton and its vicinity'.[48] The orientation towards the eastern markets was a subsequent goal when the initial aim had begun to yield financial success. In the case of the Lehigh Valley Railroad, the situation was reversed. Not content with its eastern outlets towards Philadelphia and New York, it sought to broaden its geographical reach northwards and westwards. As noted in the previous chapter, it achieved this initially by means of the Pennsylvania and New York Canal and Railroad Company, the former North Branch Canal Company, in which it acquired a controlling interest in 1865.[49] The relative success of the commercial and geographical strategies of the Lehigh Valley and the Lackawanna in developing these markets will now be examined.

EARLY LEHIGH VALLEY RAILROAD CONNECTIONS INTO NEW YORK STATE

The approach adopted by the Lehigh Valley was very much in keeping with its traditional policy of either chartering new companies for specific segments of new railroad construction, or progressively obtaining control of independent small lines through purchase of stock and assumption of financial obligations. In either case, a subsequent step would be consolidation and merger into the main railroad itself.

The first aim in the northward expansion of the Lehigh Valley network was for the Pennsylvania and New York to establish a link between Wilkes-Barre and Waverly in New York state, where an interchange with the Erie Railway for westbound traffic could be made. The line was opened to freight traffic in September 1869.[50] Although the idea of a north-south route across upper New York state from the state line to Lake Ontario predated the Lehigh Valley's interest in the region, an early scheme involving the Lake Ontario, Auburn and New York Railroad had foundered at the outset of the Civil War through lack of funding. After the Lehigh Valley's intentions became known in early 1865, the idea was revived, in the shape of the Southern Central Railroad, chartered in November of that year. This new company acquired the graded right of way of the defunct Lake Ontario, Auburn, and New York between Fair Haven on Lake Ontario and Auburn, with the intention of running south through Freeville and Oswego, NY to connect with the Pennsylvania and New York at Sayre on the other side of the Pennsylvania state line (Figure 5.3). The railroad was completed in late 1871, although parts of the route had been in operation since March 1870. A connection was made at Auburn with a branch of the New York Central Railroad. To the north at Weedsport it crossed the Erie Canal and the main line of the New York Central.[51]

A second project sought to extend railroad links northwards from Sayre to Geneva, NY only 97 miles from Buffalo, where another interchange with the New York Central was possible. This route initially involved two short railroads, the Ithaca and Towanda (later renamed the Ithaca and Athens) and the Geneva and Ithaca, which were later consolidated into the Geneva, Ithaca and Athens RR. Connection with the Erie Canal was also possible via Cayuga Lake from the town of Ithaca. Even before the Ithaca and Athens and the Southern Central were both completed in 1871, the chief engineer of the Lehigh Valley, Robert H. Sayre, was emphasizing, the great importance of both to the business of the Pennsylvania and New York Railroad. As a result, an agreement between the Pennsylvania and New York and the Southern Central was negotiated, even before the line was completed. This gave the former effective control of the latter for an initial period of years.[52] Neither the Lehigh Valley, nor its subsidiary, the Pennsylvania and New York, originally held a controlling interest in the Geneva, Ithaca and Athens. Instead, this was in the hands of Ezra Cornell, benefactor and founder of the University that bears his name.[53] The opportunity to obtain control did not arise until March 1875, when the Lehigh Valley, together with other parties, was offered the chance to buy the necessary stock and bonds from Cornell's estate for a total of $550,000, of which the Lehigh Valley would provide $440,000. The offer was duly accepted, and the road became effectively a subsidiary, although it maintained its corporate identity at this stage.

Unfortunately, in strict financial terms, both of the newly controlled companies proved to be a drain on the resources of the Lehigh Valley itself. The Southern Central was asking for money even before the road was completed. Requests for assistance in guaranteeing bonds, providing advances of funds to cover floating debt, or help with meeting fixed charges continued at regular intervals thereafter.[54] By 1880, the patience of the Lehigh Valley was beginning to wear thin and a threat of foreclosure on the property was made, although this was not carried out.[55] By this time, the Lehigh Valley had invested over $1 million in the bonds of the Southern Central and the latter was facing annual interest payments of over $200,000 on its obligations. Since the net earnings of the road available to meet these interest payments ranged only between $91,850 and $163,148 in the years 1879–1881, it could hardly be regarded as a going concern. Without the guarantees on the bond interest payments provided by the Lehigh Valley, the Southern Central would certainly have gone bankrupt. However, the financial pain for the parent was mitigated by the fact that it already owned a portion of the securities whose interest it had guaranteed.[56] Eventually the Southern Central ceased independent operation, when it was taken over by the Pennsylvania and New York in December 1886.[57]

The Geneva, Ithaca and Athens acquisition provides a similar sorry tale of loss and indebtedness. In this case, however, the unsatisfactory state of affairs was not confined to the balance sheet. The Lehigh Valley minutes summed up the condition of the railroad in the mid-1870s as follows:

> 'When the present Company took possession of the road it was entirely out of repair and required the most extraordinary renewals. Between the State Line and Ithaca, the sills and rails were unfit to be used and required renewing for the whole distance'.[58]

This report is perhaps surprising, given the relatively short period the railroad had been in operation, and it indicates the rapidity with which infrastructure could deteriorate, in the absence of a properly executed maintenance program. The Lehigh Valley decided to charge all the renewals to expenses rather than the capital account. This

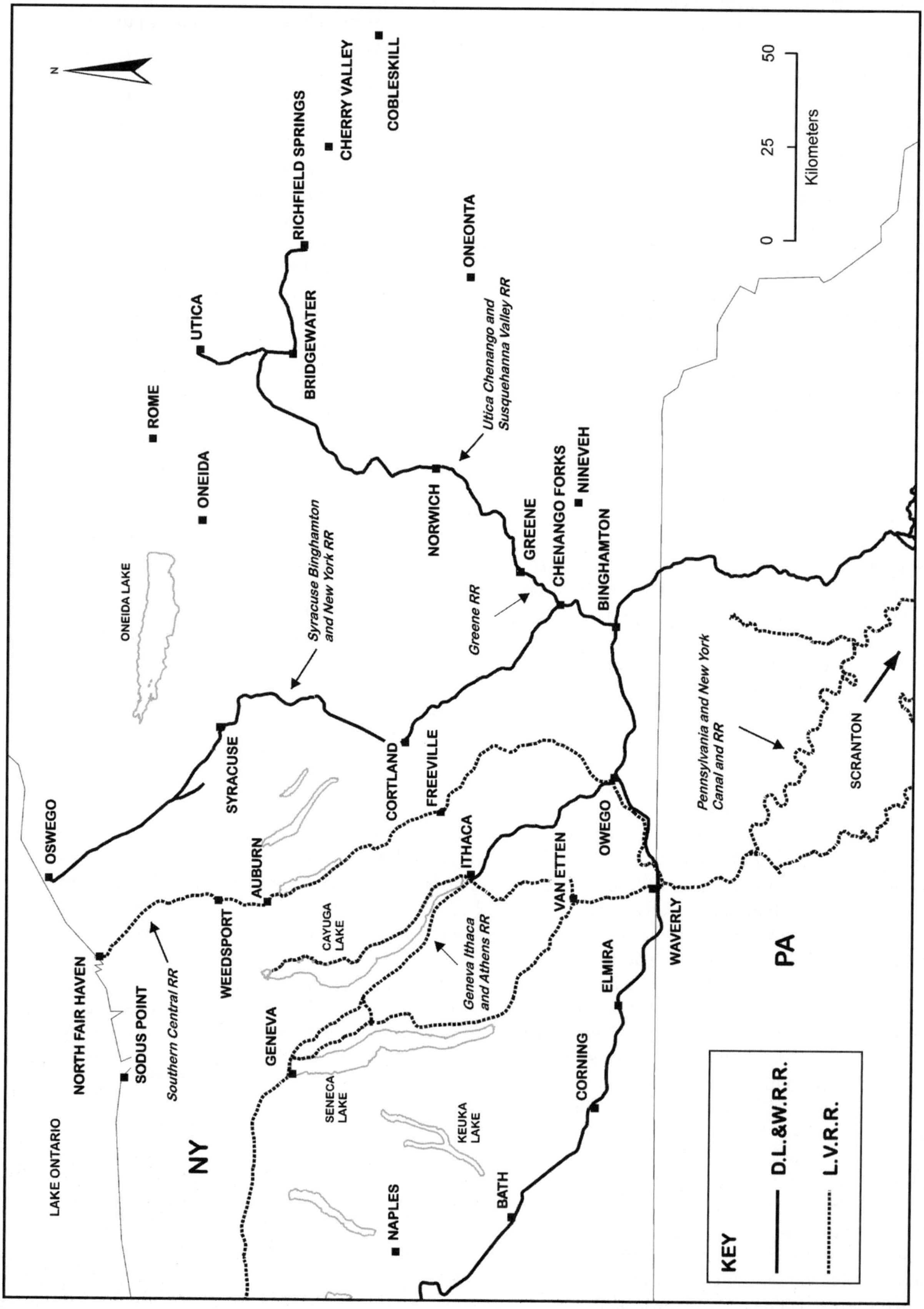

FIGURE 5.3
Northern Connections of the Lehigh Valley and Lackawanna Railroads in New York State.

contributed significantly to the overall losses of more than $470,000 generated by the road in the five years from October 1876 to September 1881, although 50 percent of these losses were attributable to interest payments on bonds and the floating debt. Yet even at $54,000 per annum these interest charges were only a quarter of those of the Southern Central, and given the extraordinary expenditure required following the takeover, the Lehigh Valley believed the overall outcome to be 'quite satisfactory'.[59] Some justification for this statement can be found in the fact that reported overall losses, which had peaked in 1879 at $172,650 had been reduced to $43,570 by 1881. Also, examination of Figure 5.4 shows that a measure of confidence in the performance of the Geneva, Ithaca and Sayre Railroad was borne out in subsequent years, as the increase in coal shipments manifested in 1881 was largely maintained thereafter. Unfortunately, the same could not be said of the Southern Central Railroad.

A further possible reason for the positive outlook on an otherwise poor financial performance was that the primary interest of the Lehigh Valley was in securing profitable traffic on the Pennsylvania and New York Railroad. The latter had financial obligations, in terms of interest charges on bonds and preferred stock, which were larger than those of the New York State lines combined.[60] Hence, if the Southern Central and Geneva, Ithaca and Sayre could approximately cover their costs, while channelling enough traffic north and west to keep the Pennsylvania and New York running to full capacity, they would have served their purpose well. During the boom of the early 1880s, the available evidence supports this interpretation. At the beginning of 1882, a few days after the report on the Geneva, Ithaca and Sayre in the Lehigh Valley Board of Managers minutes, the Pennsylvania and New York reported that 'the road and equipment of the Company have been severely strained during the past year in order to carry the large business'.[61] The same report noted that the second track was being

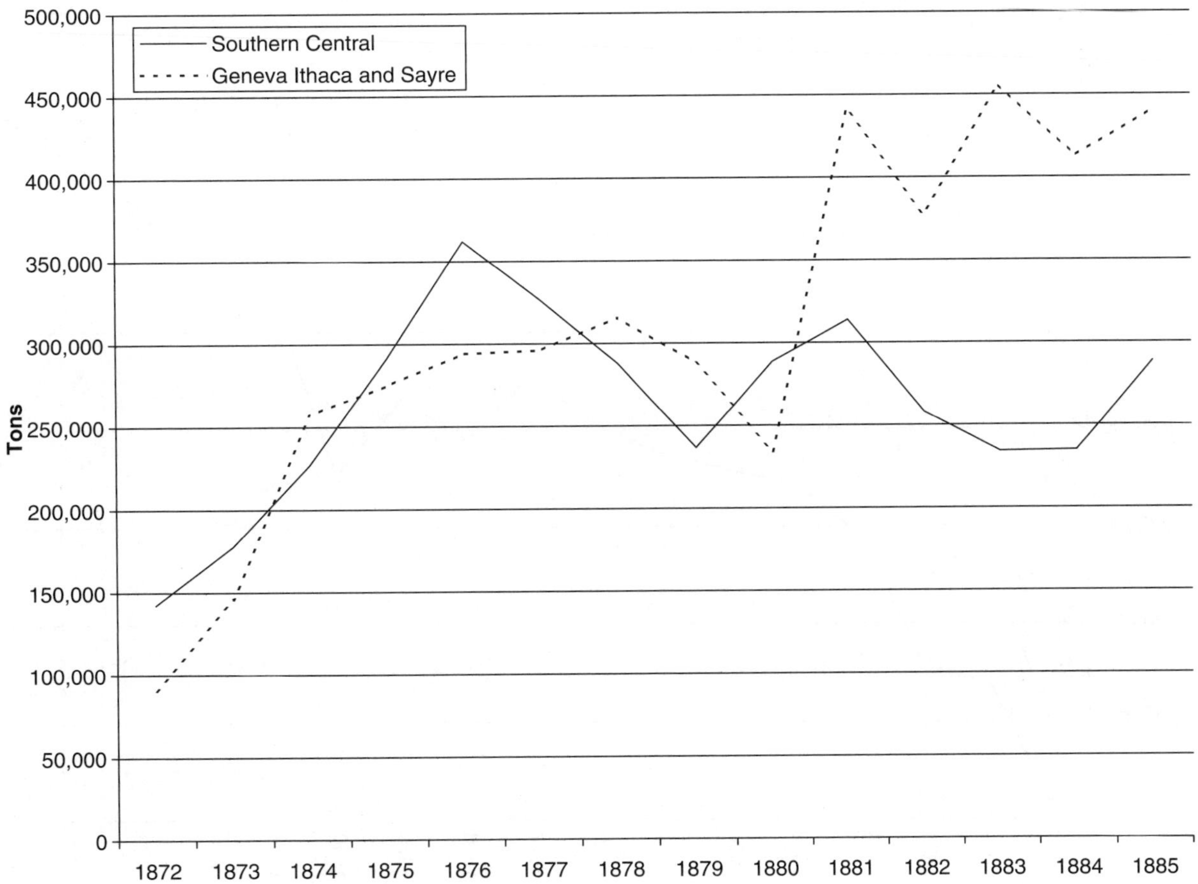

FIGURE 5.4
Coal Shipments of Lehigh Valley Controlled Railroads in New York State 1872–1885.

pushed to completion and a further 30 locomotives were to be purchased. Examination of the anthracite shipment statistics confirms the increase in business. Shipments in 1881 rose to 1,108,057 tons from 705,464 in 1880. Figure 5.5 sets this increase in the overall context of the company's business. Very strong growth of shipments in the early years after the opening of the road was following by a period of slower growth until 1877 and subsequent stagnation prior to the 1881 increase, which was maintained thereafter. The graph of earnings and expenses (Figure 5.6) paints a somewhat different picture of the company's fortunes over the period. Gross earnings in 1880 were only marginally greater than they had been in 1872. Although they increased sharply as the 1882 boom took hold, they fell back almost as rapidly when the 1885 trough approached. In terms of expenses, a slow decline in costs from 1872 to 1879, partly due to wage reductions in the late 1870s, was replaced by a sharp increase as the railroad expanded its transportation capacity in the early 1880s. Such cost increases could be readily absorbed when business was booming. Unfortunately, the increases continued when business began to contract. The inevitable result was a sharp decline in net earnings back down to 1871 levels by 1885. At this level the company had sufficient funds to cover the interest on its bonds, but not enough to give any dividends to the long-suffering preferred stockholders. The latter group had still been recouping the dividend arrears of earlier years as late as 1884.[62]

Since anthracite shipments on the road increased more than four-fold between 1871 and 1885, it may initially be difficult to understand why the financial position had weakened to such a degree by the end of the latter year, even allowing for the increase in

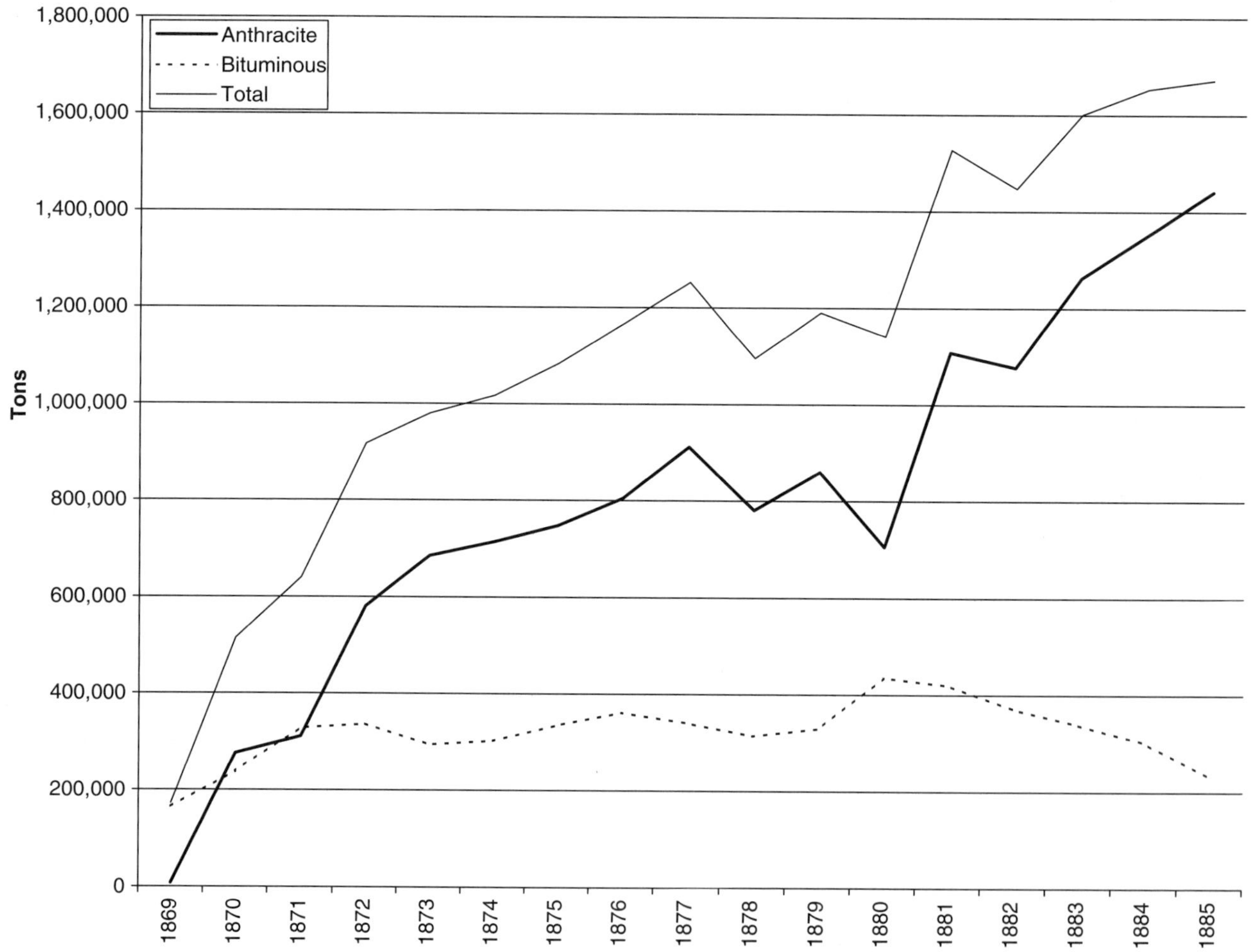

FIGURE 5.5
Pennsylvania and New York Canal and Railroad Company: Coal Shipments by Type 1869–1885.

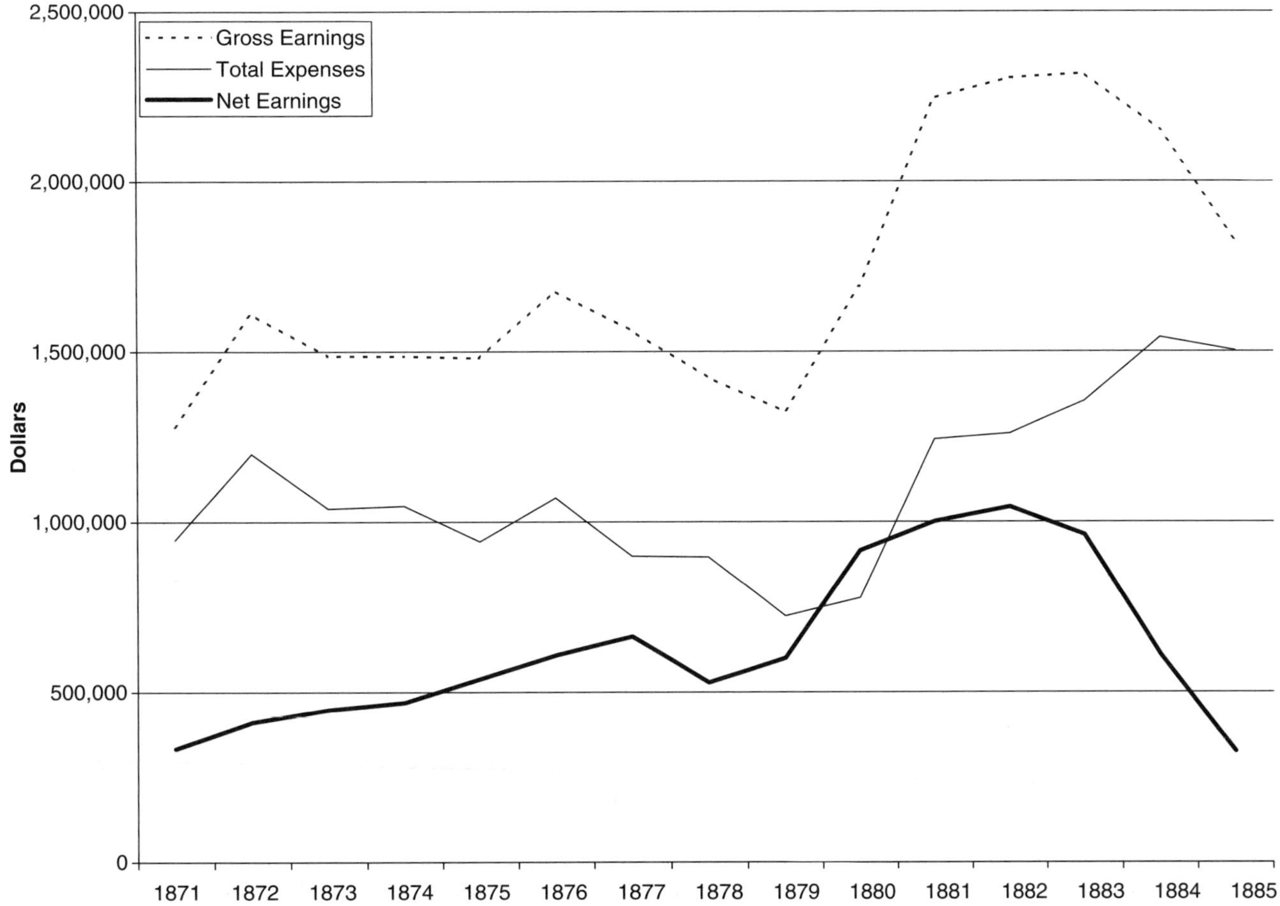

FIGURE 5.6
Pennsylvania and New York Canal and Railroad Company: Earnings and Expenses 1871–1885.

expenses. The answer can be found in Figure 5.7, which compares the freight rates on coal and other kinds of freight with the benchmark of the average freight rates for all types of commodities shipped over the Pennsylvania Railroad during the same period. While the available data for the Pennsylvania and New York do not extend quite as far back as 1871, it is apparent that the route enjoyed competitive advantage both for coal and general freight shipments in the early part of the 1870s, but the advantage was rapidly eroded by the end of the long depression in 1879. By this time, general freight rates had fallen to one-third of 1873 levels. Comparison of this graph with data on different types of shipments from the annual reports indicates that by 1877 continued decline in freight rates was associated with a very sharp increase in freight traffic, since freight earnings rose sharply, despite the reduction in income per ton-mile. This rise in freight income was particularly welcome because it offset a sharp dip in coal earnings. The latter was the result of a combination of declining freight rates and a slight reduction in shipments. Shipment capacity problems allowed a brief recovery of coal transport rates at the beginning of the 1880s, although general freight did not enjoy the same benefits. When rates began to slip again after the 1882 peak, falling to very low levels by 1885 in the case of general freight, the impact on earnings was immediate, and it was exacerbated by the continued relative increase in operating expenses as a proportion of revenues.

Several conclusions emerge from this detailed consideration of the Lehigh Valley Railroad's investment strategy to reach the markets of upper New York state. Firstly, the promoter's dream of a north-south route across the state did not prove to be a resounding business success. Secondly, the attempt by the Lehigh Valley to isolate itself from financial risk by retaining separate operating companies was not particularly successful either. To ensure operating control, it found itself acting as

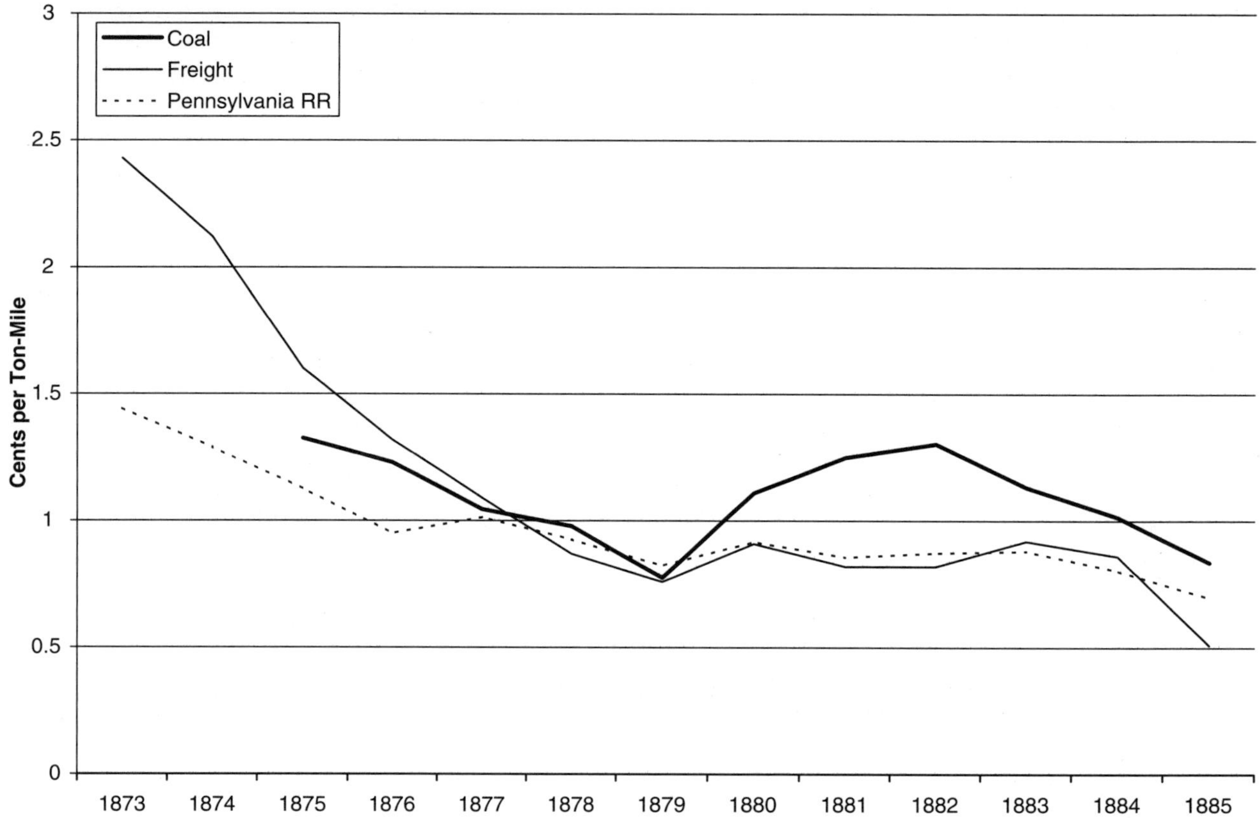

FIGURE 5.7
Pennsylvania and New York Canal and Railroad Company: Freight Rates Compared to the Pennsylvania Railroad 1873–1885.

banker and guarantor of low quality securities. It also found itself responsible for problems arising from high construction costs or inadequate previous track-maintenance regimes on the lines it controlled. Until 1876, the overall strategy made economic sense, because of the benefits that accrued to the Pennsylvania and New York and its preferred stockholders, who received a 10 percent dividend in that year. Thereafter, extended periods of declining freight rates, static or declining shipments, and increases in operating expenses began to eat into its earnings, despite the welcome relief of improved business in the period leading up to the 1882 cycle peak. In later years, with the takeover of the Southern Central by the Pennsylvania and New York and the subsequent leasing of the latter to the Lehigh Valley in December 1888,[63] it becomes more difficult to analyse the separate contributions of different geographical sectors of the railroad's business. Fortunately, however, from 1887 to 1891 monthly reports were made to the Board of Managers relating to the newly constituted 'northern division' of the railroad, comprising the Southern Central and Geneva, Ithaca and Sayre lines. From these it is apparent that static demand for coal over these routes continued to be a feature of the trade at this later period.[64]

It is useful to compare the approach adopted by the Lehigh Valley with that of the Lackawanna, although the absence of any published annual reports by the latter during most of the period of interest, makes it much more difficult to assess the financial implications of the strategies involved.

EARLY LACKAWANNA RAILROAD CONNECTIONS NORTHWARDS

From a purely historical perspective, it is interesting to note that access to markets in upper New York state was occupying the minds of those interested in the development of the Liggett's Gap

Railroad, one of the predecessor lines of the Lackawanna, even before this railroad was completed from Scranton to Great Bend near the Pennsylvania state line, where it connected with the Erie. In June 1849, W.E. Dodge and G.W. Scranton purchased the Cayuga and Susquehanna Railroad, an existing but dilapidated line running from Oswego on the Erie Railway to Ithaca, for $50,000. At that time, there was no coal traffic to route to it, because the Liggett's Gap Railroad was not completed until October 1851. The long-term importance of the Cayuga and Susquehanna lay not so much in the coal business of Ithaca itself, but in the position of the town at the head of Cayuga Lake, which allowed coal to be transhipped onto barges for the Erie Canal and points west. Once coal from Scranton was able to reach Great Bend, it was carried over the Erie to Oswego and then transferred onto the Cayuga and Susquehanna. Although the latter road obtained a charter amendment in 1851, which allowed it to build a new track parallel to the Erie, to connect directly with the Liggett's Gap Railroad, this project was never undertaken.[65] In 1855, the Cayuga and Susquehanna was leased by the Lackawanna Railroad at an annual rental of $68,750,[66] but following the financial problems of the latter after the 1857 panic, this figure was reduced to $54,600.[67]

The next step was to provide an outlet for Lackawanna coal to Syracuse. An important urban and industrial centre in its own right, it was a major transhipment point for the Erie Canal. Also, by means of the Oswego and Syracuse Railroad, a northward rail connection to Oswego on Lake Ontario had been established at the end of the 1840s (Figure 5.3).[68] The Lackawanna could reach Binghamton over the tracks of the Erie from Great Bend, and what was to become the Syracuse, Binghamton and New York Railroad was completed southwards to Binghamton in October 1854, a distance of 79 miles.[69] Despite, or perhaps because of its financial difficulties, the Lackawanna managed to negotiate a trackage agreement with the Syracuse, Binghamton and New York at the beginning of 1858, which allowed it to run coal trains under its own motive power directly to the dumping grounds and pockets in Syracuse, without change of gauge. From the pockets, coal could be transhipped directly from the cars into the 'larger class of boats on the Erie Canal with the greatest possible dispatch'.[70] It is interesting to note that the trackage agreement contained two provisions designed to secure tonnage for the Syracuse, Binghamton and New York, to the detriment of other routes. The first of these routes was the Cayuga and Susquehanna, as the Lackawanna agreed to restrict the tonnage to be sent via Ithaca for transhipment to the lake and canal. Given the fixed rental of the Cayuga and Susquehanna lease, this had potentially significant costs for the Lackawanna and must therefore be taken as an indicator of the importance it attached to securing access to the line into Syracuse. The second route over which restrictions were to apply was the Chenango Canal from Binghamton to Utica. The Lackawanna agreed not to ship more than 10,000 tons a year via the Chenango Canal to points on the Erie Canal west of Syracuse.[71] The advantage of a trackage agreement was that no initial investment costs were required and expenditure was only incurred in proportion to revenues generated, unlike a lease. Also, the railroad owning the tracks was responsible for maintenance costs. This proved useful in 1861 when James Archbald, Chief Engineer of the Lackawanna, made a survey of the Syracuse road and directed them to perform the necessary repair work 'to make transport of coal over their railroad perfectly safe'.[72] The disadvantage of a trackage agreement was that it gave no control over the independent line involved, which left it open to takeover by competitors and blockage of a key outlet to market. At the beginning of 1869, this very nearly happened, just after the missing parts of the link from Syracuse to Oswego, namely a connection in Syracuse and the laying of a third rail on the narrow gauge Oswego and Syracuse Railroad, had been completed. The Lackawanna moved immediately to protect its interests, purchasing a majority stockholding for $751,190, guaranteeing bonds and installing President Sloan of the Lackawanna as president of the new subsidiary, at the modest additional salary of $1,000 a year.[73] A few weeks later, to assure no other difficulties in reaching Oswego, the Lackawanna leased the Oswego and Syracuse Railroad, with effect from March 1, 1869.[74] The only part of the entire route from Scranton to Oswego that was now not under direct Lackawanna control was the 15 miles of Erie track between Great Bend and Binghamton. Other than delays to trains, no particular difficulties with this section of the road had been experienced, although as early as February 1865, the railroad's executive committee had been asked to encourage construction of a railroad from Great Bend to Binghamton, by subscription or other means. Finally, in March 1869, the Valley Railroad was chartered to build the

New York segment of this line. It was leased by the Lackawanna in the following month, and its double track was opened in the spring of 1870. Although short, the line still required significant financial advances from the Lackawanna, amounting to $800,000 for the second track and bridges alone.[75]

For the Oswego, Syracuse and Binghamton route the advantages of a north-south connection had been apparent from an early date. In the case of Utica, farther east and served by good connections on the New York Central Railroad and the Erie Canal, the advantages were less apparent before the Civil War. Nevertheless, once the war had ended, agitation began for a new route to link to the growing system of railroads running west of the city. In December 1865, Articles of Association for the Utica, Chenango and Susquehanna Valley Railroad were drawn up and, by August 1869, construction of the road had reached Sherburne Four Corners. At that location, it connected with the New York and Oswego Midland Railroad, which enabled its passengers and freight to travel a further 12 miles south to Norwich.[76] Its charter allowed it to extend south beyond Norwich as far as the town of Greene, just 8 miles short of a connection with the Syracuse, Binghamton and New York Railroad. The events leading to the closing of this gap by the Greene Railroad in October 1870 have already been discussed in an earlier section. Before the Greene Railroad was completed, the Lackawanna leased the Utica, Chenango and Susquehanna Valley, with effect from May 1, 1870, at a rental equal to 6 percent on the capital stock, which was not to exceed $1.8 million on the date of the lease. However, the line was still not complete, because the section northward from Greene to Norwich had yet to be constructed. The lease made specific provision for the Lackawanna to complete this portion of the road, which was done by December 1870. It would appear from the timing, although the Board of Managers minutes are not explicit on the point, that the sale of $180,750 of stock in the Utica road at 90 percent of par value by the Lackawanna, in July 1870, was used to finance the construction. The sale of stock or bonds, whose interest, principal or both were already guaranteed, was a method frequently used by the Lackawanna for funding construction work on its subsidiary lines.[77] In the same way as with the link from Great Bend to Binghamton, the Lackawanna quickly took steps to remove its dependence on the New York and Oswego Midland Railroad for the section between Norwich and Sherburne Four Corners. Its own track between these points was completed in October 1872. It is interesting to note that the completion of what came to be known as the 'Utica Division', raised important operating problems for the Lackawanna, because the original Utica Railroad was standard gauge, as was the New York and Oswego Midland and the section built by the Lackawanna just mentioned. However the Syracuse, Binghamton and New York to which the Division connected at Chenango Forks, was broad gauge. Taber, the doyen of Lackawanna Railroad historians, speculated that there may have been a third rail at least as far as Norwich, but could find no conclusive evidence to support this.[78] An entry in the Board of Managers Minutes, for March 27, 1874, to which Taber does not make reference, states that part of the Utica Division, including the Richfield Springs branch, which was then narrow gauge, was to be converted to broad gauge. This resolution seems to imply that part was already broad gauge, or had a third rail, but even so, at some point through freight or passengers must have transferred between trains, or specially modified rolling stock, able to operate on both broad and narrow gauges, must have been used. Amusingly for everyone, apart from the Lackawanna stockholders who effectively paid for the work, the decision was reversed in 1876, when the entire road was changed from broad to narrow gauge.[79]

A cursory comparison of the takeovers of the Utica, Chenango and Susquehanna Valley and of the Syracuse, Binghamton and New York, might lead to the supposition that in each case the Lackawanna took the initiative to expand its network of lines. While this is true in the latter case, it does not appear to be so in the former. The Utica road was apparently offered to the Lackawanna under a lease agreement that is not documented, rather than an agreement having been sought out. It is very useful to quote the rationale given by President Sloan for presenting the proposed agreement to the board, for the light it sheds on decision-making at the time. The minutes record, in suitably convoluted prose:

> 'that in view of the fact that the Delaware and Hudson Canal Company had leased the Albany and Susquehanna Railroad and, as he (President Sloan) was informed, had made arrangements for the transportation of coal over the Midland Railroad which in connection with other Roads projected and partially built, was likely to give them access to the Utica Coal Market and to Norwich and other towns between there and Utica—which in his opinion would give them advantages for reaching said

Markets not now possessed by this Company—he had entertained the application.'[80]

Following an examination of the road requested by the directors after the above report, the minutes of the next meeting, on April 5, 1870, indicate that the original proposal for leasing was rejected and replaced by the 6 percent rental arrangement already mentioned.

Sloan's perceptive assessment of the situation in the Utica region and the prompt action of the Board of Managers to forestall the competition from the Delaware and Hudson produced unusual and unexpected results. Two of the projected and partially constructed railroads referred to in the quotation above were doubtless the Utica, Clinton and Binghamton Railroad and the Rome and Clinton Railroad. These were completed in 1871 and 1872 respectively. They were leased to the New York and Oswego Midland Railroad in 1872, at rentals of $70,000 and $25,000 per annum. In November 1872, the Delaware and Hudson undertook to guarantee these leases for the Oswego Midland, but as a result, when the latter went bankrupt in November 1875, the leases were transferred to the Delaware and Hudson.[81] Since the Delaware and Hudson had no lines or subsidiaries connecting to these unanticipated acquisitions, it had no cost-effective means of operating them. In an attempt to minimise expenses, it therefore made an agreement with the Lackawanna for the two roads to be operated in conjunction with the Utica Division. This required construction of a new track near Smith's Valley, between Norwich and Clinton, to link the two systems. The construction costs and operating expenses were divided 70/30 between the Lackawanna and the Delaware and Hudson. The working of the overall system was to be 'at the expense and for the benefit of both parties in proportion to the length of the said roads by each controlled'.[82] Although earlier commentators seem not to have been aware of this, it was also linked to a wider agreement on the division of coal shipments between the two companies destined for the Syracuse and Oswego markets, that could readily be construed by the impartial observer as being in restraint of trade. The agreement continued until April 1883.[83]

In comparison to the Lehigh Valley, the Lackawanna clearly benefited from having a northward route many years before the completion of the Pennsylvania and New York in 1869. It was able to develop its northern markets around Syracuse by means of a traffic agreement that had neither high initial costs nor a fixed rental. It only stepped in to take control when its long-term interests were threatened. Other leases made to guarantee its northern connections, such as the Utica Division and the Valley Railroad, were of roads newly built, without fear of high renewal costs from careless prior management. The old established Oswego and Syracuse Railroad was a profitable, going concern that paid 8 percent dividends in 1868.[84] Only the old Cayuga and Susquehanna lease appeared costly in relation to the returns it brought to the parent company. Overall, by the end of 1870, the New York leased lines had added nearly $290,000 in annual rentals to the operating costs of the Lackawanna, a figure that would have risen to $329,000 if the Utica lease had been running for the full year instead of 8 months. Compared to the net transportation earnings that exceeded $3.5 million, this did not represent a serious financial burden. The Lackawanna policy of requiring leased roads to issue new stock or bonds to cover construction work, which it could sell when required for cashflow purposes, also prevented the accumulation of unwelcome financial obligations. For example, in 1869 and 1870, the Lackawanna expended $407,659 to upgrade the property of the Oswego and Syracuse Railroad, but by the end of the latter year, $400,000 of this amount had already been reimbursed in the form of bonds and other assets received from the railroad.[85] Owing to the absence of published annual reports from the end of the Civil War to 1899, it is not possible to get a clear picture of the profitability of these leased and controlled lines individually, without a detailed analysis of the surviving corporate books of account, which is beyond the scope of the present study. Published information in the 1900 annual report, however, indicates that the Syracuse, Binghamton and New York, at least, was a profitable operation at that time. Earnings of $950,000 and expenses of $557,000 allowed it to meet its bond interest and pay 8 percent dividends.[86]

In the absence of regular accounting reports about the individual lines, the only other possibility is to utilize information on coal shipments. Even here, the reporting of information is sporadic during the latter part of the 19th century. Until the 1860s annual tonnage summaries were published and reported in whole or in part by local newspapers. Thereafter, only the manuscript reports of the General Coal Agent are available. These are

TABLE 5.3
Delaware, Lackawanna and Western Railroad: Distribution of Shipments to Upper New York State.[87]

Year	Syracuse and for Shipment		Ithaca and for Shipment		Utica Division		Oswego		Total Shipments
	Tons	%	Tons	%	Tons	%	Tons	%	Tons
1860	102,465	10.6	28,427	2.9	N/A		N/A		964,489
1868	201,225	13.0	137,655	8.9	N/A		N/A		1,547,267
1876	196,671	9.6	47,756	2.3	124,759	6.1	156,857	7.6	2,054,018
1879	615,268	15.5	56,053	1.4	192,403	4.8	248,042	6.3	3,968,102
1887	496,666	7.9	42,859	0.7	414,361	6.6	313,547	5.0	6,263,298

recorded in the company's letter books and are sometimes reproduced, either in full or in summary form, in the Board of Managers minutes. Table 5.3 shows shipments for sample years 1860–1887 to some of the major points in upper New York state and the proportion that these shipments represented of the entire tonnage of the road. The Syracuse and Ithaca columns contain the additional phase 'and for shipment', which refers to transhipment at these locations for the Erie Canal.

Several points emerge from the figures presented. Firstly, it is apparent that even in the 1860s these markets and distribution points together absorbed a significant proportion of the Lackawanna's overall shipments. This was almost 22 percent in 1868 and it peaked at 28 percent in 1879. Secondly, Syracuse was particularly important during much of the period and there was a very marked increase in shipments to that city, and for the Erie Canal, as soon as the economy emerged from depression in 1879. After the opening of the Lackawanna's new railroad route to Buffalo, however, its relative importance decreased. Thirdly, Ithaca never proved to be a very lucrative destination for coal shipments and the larger tonnage recorded in 1868 was not repeated in any other year for which data are available. Finally, the Utica route and Oswego as a destination both demonstrated a progressive increase in shipments that would seem to justify the investment required to secure control of these routes.

In sum, the early focus by the Lackawanna on markets in upper New York state allowed it to establish itself as a coal supplier in major urban and manufacturing centres such as Syracuse, as well as providing exposure to Buffalo and Great Lakes markets, both over the Erie Canal and via the port of Oswego. A sensible and pragmatic approach to co-operation with, and control over other lines, subsequently gave it a strong presence in the Utica region. Excessive financial obligations were avoided, and the volume of shipments increased to healthy levels in all but the Ithaca marketplace. In contrast, the Lehigh Valley was a later entrant into the market and both its distribution routes, the Geneva, Ithaca and Sayre and the Southern Central, passed through areas where urban and industrial development were much less pronounced. It also acquired badly built and badly maintained railroads, which required substantial investment in comparison to the level of traffic carried and its restricted rate of growth.

ACCESS TO WESTERN MARKETS VIA BUFFALO

The second part of any strategy to ship coal northwards and westwards from the anthracite regions was to reach the port of Buffalo on Lake Erie. An important urban and industrial centre in its own right, it was also the best place for transhipment of coal onto lake steamers, to reach points on the western lakes, and Chicago in particular. Although shipments were modest at the start of the 1860s, as noted in the previous chapter, the Lackawanna Railroad was among the first to recognise the future potential of Buffalo and the Great Lakes markets. In 1861 it secured a yard in Buffalo on a five-year lease, despite the problems of obtaining boats to carry coal there from Syracuse via the Erie Canal at that time.[88] Further land was acquired adjoining the Erie Basin in 1868. This land was later to become the terminus for the New York, Lackawanna and Western Railroad in the 1880s.[89]

Prior to the construction of the Pennsylvania and New York, the Lehigh Valley had no ready means of sending coal to Buffalo, but even before the former road was fully open, it had made a traffic agreement for the Erie to transport its tonnage westwards from Waverly.[90] At this location, large coal pockets were constructed to enable the coal to be transferred from the standard gauge cars of the Pennsylvania and New York to the 6′ gauge of the Erie. Apparently, this agreement also contained the much more ambitious provision for the laying of a third rail from Waverly to Buffalo, 'so that coal cars loaded inthe Wyoming Valley may be dumped on the shore of the lake without break of bulk'.[91] However, this provision was not implemented during the term of the agreement. Under a separate arrangement, dated August 2, 1870, a third rail was laid from Waverly as far as Elmira.[92] Although the Lehigh Valley was still entirely dependent on the Erie to bring its coal to Buffalo at the time, it also recognized the problems at the terminus, where there were very limited facilities for transferring coal to lake vessels for Chicago and other Great Lakes ports. It therefore subscribed 34/40ths of the stock of the Buffalo Creek Railroad. This was a 3½-mile long line designed to connect nearly all the railroads entering Buffalo, including the Erie and the New York Central, with a suitable point on the harbor for the erection of wharves and transhipment facilities. Construction began in 1869 from the connection with the Erie, with a view to reaching the harbor by the spring of 1871. In fact, the initial stretch of the railroad was not completed until June 1870, but was immediately put to good use. In addition to the railroad, the Pennsylvania and New York bought a half-interest in the Lehigh Dock Company property, extending from the Buffalo River to the Blackwell Canal. This provided the possibility of constructing wharves along a total frontage of 2,700 feet on both the river and the canal. By May 1871, at a cost of $42,000, nearly 2,000 feet of trestling had been erected on the Buffalo River, and 58 coal pockets with a storage capacity of 4,500 tons.[93] Two years later, the total investment by the Lehigh Valley in railroad and dock property at Buffalo had reached $336,000, of which $246,000 was in the stock and obligations of the Buffalo Creek Railroad. This was a very high figure given that it was entirely dependent on the good offices of the Erie to bring its coal to Buffalo. Problems were beginning to surface as the tonnage increased, because the Erie was unable to provide a sufficiently regular supply of cars. By the end of 1872, the annual report made clear that a standard-gauge route to Buffalo had become indispensable and the managers of the Erie were said to be proposing laying a third rail on parts of their route and narrowing the gauge on others.[94] Unfortunately, the Panic of 1873 and the ensuing financial problems of the Erie caused it to drag its feet over the matter of the third rail, until an agreement, dated February 18, 1876, was finally reached with the Lehigh Valley and the Pennsylvania and New York. This agreement provided for the laying of a third rail from Elmira to East Buffalo, where a third rail had already been laid. Bogen,[95] quoting a newspaper article, states that the Lehigh Valley advanced the Erie approximately $900,000 for the work. However, the text of the agreement stated that the Pennsylvania and New York was to deliver the necessary rails and joints to the Erie within 3 months of the date of the agreement and the Erie was to lay the rails on both of its tracks within 15 days of receiving them. The rails belonged to the Pennsylvania and New York until paid for in full by the Erie.[96] The agreement must have been adhered to closely, as the laying of the third rail was completed on May 22, 1876.[97] The major provisions for the agreement in relation to traffic were that the Lehigh Valley/Pennsylvania and New York were to transport all their coal, freight, and passengers in their own trains as far as Elmira, and coal between Elmira and Buffalo. The tolls on coal between the last two named locations were split 50/50 between the Erie and the Lehigh Valley.

An evaluation of the benefits of this agreement in 1881 by the Lehigh Valley indicated that, between December 1876 and September 1881, the arrangement had proved to be quite remunerative. The gross earnings had totalled $582,569 and the net earnings $378,972. It had been further calculated that 14.97 percent of the total freight business of the Lehigh Valley system was dependent on the Erie, as compared to 4.82 percent dependent on the New York Central.[98] It is also useful to compare the volume of coal shipments transferred to the Erie by the Pennsylvania and New York with those transferred to the other roads going north across the state. These are shown in Figure 5.8. A strong flow of traffic on the Erie in the early 1870s was followed by a decline after the Panic of 1873, until the benefits of the 1876 agreement initiated an upward trend that continued into the 1880s. By this time, the traffic west across New York State was as important as that going north. It should be noted, also, that the coal shipments on the graph include both bituminous and anthracite coal. In the case of the Erie, nearly all the growth in shipments from 1876 to 1885 was accounted for by the growth of the anthracite trade,

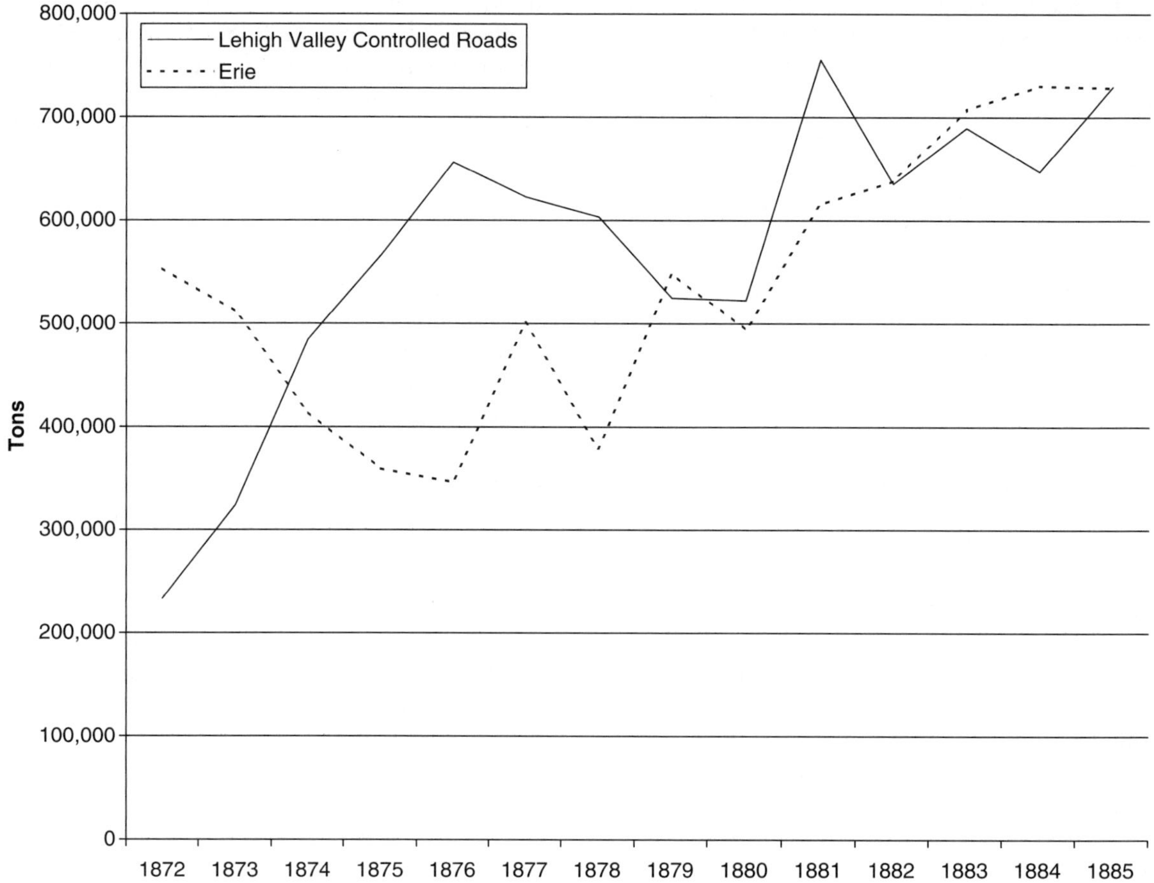

FIGURE 5.8
Coal Shipments across Upper New York State by Roads Connecting to the Pennsylvania and New York Canal and Railroad Company 1872–1885.

since the rate of shipments of Barclay bituminous coal via Towanda was largely static during the period, at an average of about 180,000 tons per annum. For the roads controlled by the Lehigh Valley, the lesser growth trend over the same period concealed a collapse in bituminous shipments from 183,000 tons in 1876 to less than 20,000 tons in 1885. The deficit was made up by a corresponding increase in anthracite shipments.[99]

Following the financial review of the Erie agreement mentioned above, a new agreement for ten years was arranged from May 1, 1882. The Erie granted trackage to points such as Rochester, as well as Buffalo, and received in return 50 percent of actual earnings on coal, freight, and passengers.[100] Despite this close co-operation between the two roads on transportation across New York state, this did not extend to the subsequent development of terminal facilities or arrangements for onward shipment of coal from Buffalo, either by lake or rail. The Lehigh Valley did, however, sell part of its stock in the Buffalo Creek Railroad to the Erie in 1879, to give each a 50 percent share in the company.[101]

Instead, the Lehigh Valley decided to establish its own line of lake steamers to carry coal west to Chicago. It began in a modest way by chartering three vessels for the 1881 season, with a combined capacity of 5,000 tons, at a cost of $55,000. This was quickly followed by a request to buy vessels, and the purchase of two at a cost of $198,000 was approved.[102] By the end of 1882, four more had been purchased to give facilities for carrying 125,000 tons of coal a year.[103] Further substantial expenditure was made in 1881 on the Tifft Farm for wharf development. This 400-acre tract adjoining the Buffalo Creek Railroad had more than a mile of lake frontage on the southside of Buffalo. It cost $255,000 and development of the property began immediately.[104] The overall plan made provision for up to eight linear miles of wharf construction, to be built as the needs of the business required. By 1884, a 100,000-ton coal-stocking trestle had been erected and 6,000 feet of dock front were ready for service, backed by 7.35 miles of sidings to provide standing room for coal and freight cars. Overall investment in the Buffalo docks had risen to $1.23 million.[105]

With the benefit of hindsight, perhaps two aspects of the Lehigh Valley strategy are surprising. Firstly, that it was willing to renew the Erie contract for ten years, rather than five, as had been the case with the previous contract, knowing that it could only expect to derive 50 percent of the benefit of any increases in freight or coal traffic, when the western markets were clearly developing rapidly. Secondly, that it was willing to invest a figure approaching $2 million by the mid-1880s in terminal facilities and lake steamers in Buffalo, while still depending on the Erie both for access to the city and, in part, for motive power to reach it, since part of the company's coal was carried in Erie trains under the new contract.

It was not until 1887 that the plan was first mooted for the Lehigh Valley to build its own through line to Buffalo,[106] and, even then, the plan seems to have been put on hold for a period, perhaps because of the adverse business conditions in early 1888. The first evidence that it had been set in motion was the announcement, in the spring of 1889, that the Geneva, Ithaca and Sayre Railroad had been allowed to default on its bond interest payments, forcing a foreclosure sale. On August 27, the Lehigh Valley bought the railroad at the sale in Elmira, then organized a new company called the Geneva and Sayre Railroad and deeded the old Geneva, Ithaca and Sayre property to it. A few days earlier, on August 23, the Ithaca, Auburn and Western Railroad from Auburn to Freeville had also been sold under foreclosure for $170,000 to the Lehigh Valley. This was in part so some of the tracks could be taken up, because they ran parallel to the Southern Central, and in part so the remaining tracks could be used for a direct line between Ithaca and Auburn. Charters for two other roads had also been secured. The first was for a line between Geneva and Van Ettenville on the Geneva and Sayre Railroad, the second for a line from Buffalo to Geneva. The route of the Geneva and Van Ettenville was located on the high ground between Seneca Lake and Cayuga Lake to avoid the steep grades into and out of Ithaca on the old Geneva and Sayre line, which were unsuitable for heavy through traffic (Figure 5.9). By May 1890, the contracts for grading the Buffalo and Geneva Railroad were let at a cost of $1.25 million. At the same time, the Geneva and Sayre Railroad and the Ithaca, Auburn and Western were merged into the Geneva and Van Ettenville. Shortly thereafter, on June 23, 1890, the Buffalo and Geneva Railroad and the Geneva and Van Ettenville were themselves merged into the Lehigh Valley Railway Company. The latter was another subsidiary of the Lehigh Valley Railroad, which already owned a short line between Buffalo and Lancaster, New York. This series of mergers resulted in a single subsidiary chartered under New York statutes, which controlled all of the Lehigh Valley lines in New York state and also the dock property in Buffalo. Prior to the merger of the two components of the new main line, the Lehigh Valley Railway Company was capitalized at $5.9 million. Following the merger, a mortgage of $15 million was arranged, so the company could issue bonds bearing 4½ percent interest, largely to the parent railroad to cover the cost of advances previously made. Purchasers for $6 million of these bonds had been found, even before they were issued.[107]

In April 1891, notice was given to the Erie that the contract of May 1, 1882 would be cancelled on May 10, 1892. The second vice president, Robert H. Sayre, explained the reason for the 10th of May, rather than the 1st, to the Board of Managers. The plan was to open the new through line to Buffalo the following day, May 11, 1892, as this was the fortieth anniversary of the start of surveys on the original stretch of the Lehigh Valley Railroad from Mauch Chunk to Phillipsburg in 1852. By the late summer of 1891, construction work was progressing well on the Buffalo to Geneva section, and a total of 64.7 miles of track had been laid by early October. Unfortunately heavy embanking work on the line south of Geneva was causing delay and it was becoming apparent that the May 1892 deadline would not be met, although the Buffalo to Geneva section was expected to be completed by the end of 1891, and the branch line to Rochester by January 1892. Inspection of track construction revealed, however, that only 90 percent of the distance had been graded for a double track and a further investment of $250,000 would be required over the winter to complete the work, while the contractors were still on the ground.[108] With consummately poor timing, the entire road to Buffalo was finally opened in September 1892,[109] shortly before the onset of the 1893 panic and the ensuing deep depression that afflicted much of the remainder of the decade.

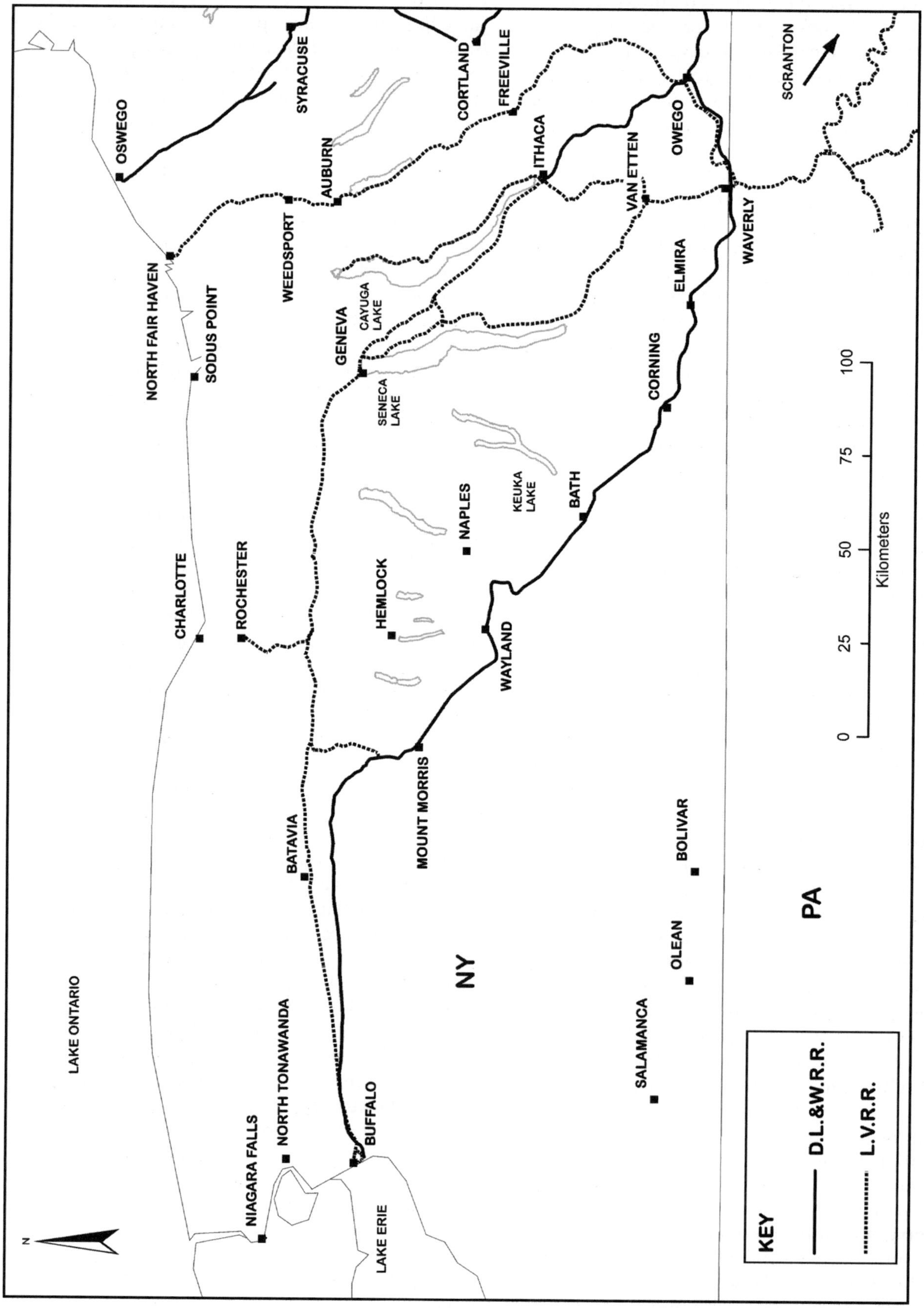

FIGURE 5.9
Western Routes to Buffalo of the Lehigh Valley and Lackawanna Railroads 1893.

The approach to the Buffalo and Great Lakes markets by the Lackawanna Railroad was very different. In the 1860s, when tonnages were still relatively light, the route to Buffalo via Syracuse and the Erie Canal had sufficed. From the beginning of the 1870s, the leasing of the Oswego and Syracuse Railroad provided an outlet to the lakes directly. In consequence, there was no pressing need for an arrangement of the kind adopted by the Erie and the Lehigh Valley. Not until 1879, when shipments both to Syracuse and Oswego jumped 50 percent from the previous year to their highest ever levels, did it become apparent that additional transportation facilities to Buffalo would be required. Unusually, some monthly figures on shipments to Oswego and Syracuse have survived for the period December 1878 to May 1879, and these demonstrate the sudden pressure on transportation facilities very clearly. Even allowing for seasonal variations with the closure of canal and lake navigation, the increase in shipments is still dramatic. In December, Oswego received a mere 2,415 tons and Syracuse 15,605 tons of coal and coal dust. By the following May, these tonnages had soared to 33,647 and 57,624 respectively.[110]

Immediate commercial pressures were not the only factor involved, however. Bogen[111] argues that a group of western railroads having their eastern terminus with the Wabash at Buffalo, were seeking an outlet to the eastern seaboard and viewed the Lackawanna as a suitable candidate. This group of railroads was controlled by Jay Gould. After the collapse of the Lackawanna stock price in late 1876, Gould apparently tried to gain control of the road, but was prevented by Moses Taylor, the principal stockholder. Gould was nevertheless able to exert influence over the future policy of the railroad and served as a director from 1881 to 1888.[112] He was instrumental in securing the eventual co-operation of Moses Taylor and other National City Bank interests, who were well represented on the Lackawanna Board, in chartering the New York, Lackawanna and Western Railroad in August 1880 to build a 207-mile direct line from Binghamton to Buffalo (Figure 5.9). An element of mystery, if not financial subterfuge, still surrounds the manner in which this new line was constructed. The newly chartered railroad did not build the line itself, but made the contract over to a separate company called the 'Central Railway Construction Company'. Following contemporary newspaper reports, Bogen claims that Lackawanna directors had significant stockholdings in the Central Company, which turned the line over to the New York, Lackawanna and Western on completion, at a book value of $22 million. Of this figure, $12 million was represented by first-mortgage bonds of the railroad and $10 million by stock. Bogen further claims, although without giving his source, that the construction cost was only $12 million and that the $10 million stock was distributed to the stockholders of the Central Company, who realized a handsome profit.[113] Unfortunately, this does not entirely match the details of the arrangement as recorded in the Lackawanna Railroad minutes. The division of the capitalization of the New York, Lackawanna and Western Railroad, as between bonds and stock is correct. However, the contract to the Central Company provided for them to receive only $10 million of bonds and $8 million of stock in payment for the construction work. In addition, a further $1.3 million of the bonds were sold to buy equipment for the road. At the point where the railroad was handed over by the Central Company, a total of $19.3 million of stock and bonds had been issued. On October 2, 1882, the New York, Lackawanna and Western was leased to the Lackawanna Railroad for a rental of 5 percent on the capital stock and payment of fixed charges as they became due. Following the lease, the remaining $2 million of capital stock and $700,000 of first-mortgage bonds were issued to the Lackawanna by the leased road, to cover the cost of advances for construction already made. In May 1883, a further $5 million issue of second-mortgage bonds was authorized. Of these, $3.25 million had been issued to the Lackawanna by 1884. In August of that year, accounting for remaining unpaid advances made by the Lackawanna, the total book cost of construction stood at $26,129,744.[114] The only reason this detailed information has come down to us is because of the actions of a dissenting minority stockholder at the annual meeting on February 26, 1884. He had moved a motion for an investigating committee to report on the cost of the new railroad and this report was subsequently placed on the minutes. However, the minutes pointedly note that the stockholder in question had only acquired his shares on February 4 and therefore was not entitled to vote on them! It is also very interesting to observe that a second motion followed the first, which represented a loud rap on the knuckles for the Board of Managers. The motion read that it was:

> 'the sense of the meeting that no further extension of this company should be made beyond Buffalo westerly, either by construction, lease or otherwise,

or if officers of this company consider such extension advisable they are to put it to the stockholders at a regular or special meeting'.

In an atmosphere charged with embarrassment for the Board of Managers, the author of the minutes rather weakly recorded the following observation that 'no objections being made, the above resolutions were seemingly concurred in by those present.'[115] Bogen's newspaper evidence supports the contention that minority stockholders were unhappy about the cost of the lease, although it is unclear why they waited until 1884 to complain, when they had previously approved the lease in 1883. Also, it would appear that the huge profits Bogen says were received by shareholders of the Central Railway Construction Company could not have been as large as he claims, because some of the stock he believed had been issued to them was in fact issued to the Lackawanna Railroad.[116] That said, even though we currently have no conclusive evidence of the actual construction costs, aside from newspaper hearsay, it would appear that substantial profits were made by a select band of capitalists, while the investing public were largely shut out.

Despite the expense of the construction contract, the new road was substantially built and well-suited for high-speed running of through trains.[117] The parent Lackawanna Railroad continued to invest in the road throughout the 1880s, especially in land and terminal facilities in Buffalo and East Buffalo. Some $350,000 was advanced to the New York, Lackawanna and Western in 1885 to purchase additional wharf property in Buffalo from the Union Steamboat Company, and extensive work on new sidings and coal trestles was also undertaken. In 1890, the leased road was asked to assume the financial obligations incurred, amounting to $2,632,750, by the issue of Terminal Improvement Mortgage Bonds.[118]

It might be imagined that these high levels of investment in the new route to Buffalo would have resulted in a rapid re-orientation of the Lackawanna's anthracite business towards the interior rather than the eastern seaboard. However, the data presented in Figure 5.10 indicate both that this re-orientation had begun before the New York, Lackawanna and Western project had ever been initiated and that it took a number of years to come to fruition. The graph is based on data for shipments on the Northern and Southern Divisions of the railroad. The boundary between the divisions was in Scranton. Shipments to that city and all points north into upper New York state were the responsibility of the Northern Division, while all shipments in the other direction, predominantly to New York City, went over the Southern Division. The first point to note is that the temporal pattern of change in shipments in either direction is broadly similar from the 1860s onwards. During the Civil War, the Northern Division carried about a quarter of the company's coal tonnage. Thereafter, it increased to about 30 percent of the total on average, although 1868 was unusually high at 36 percent. Secondly, in 1878, northern shipments began to climb steadily year-by-year, approaching 40 percent by 1880. In 1888, for the first time since the 1850s, northern shipments exceeded those on the Southern Division. Although no annual figures have been located for the 1890s, more detailed weekly reports of the planned distribution of coal north and south have survived for several periods, especially in the later part of the decade. Several examples are shown in Table 5.4. From these it is apparent that the shipments northwards assumed even greater importance at this later time than they had during the 1880's. In 1898, at the peak of the autumn shipping season over the Great Lakes, shipments to Buffalo alone accounted for more than half of all coal shipped by the Lackawanna. Taking account of other northerly destinations, such as Syracuse and Oswego, the overall proportion carried through New York state rose to two-thirds of all shipments. Only with the close of lake navigation did shipments to Buffalo fall off to a low level, to be replaced by increasing shipments to the urban areas of New Jersey and New York City, via Washington and Hoboken, both in New Jersey. This dovetailing of demand at different times in the markets served by the two divisions of the railroad was clearly a valuable means of evening out potential production peaks and resultant capacity problems in the mining regions themselves.

It is apparent therefore that the Lackawanna Railroad was able to develop its markets for coal in upper New York state and to the west, via Buffalo, in an effective and progressive manner. Initially, the connections to Syracuse and Utica provided access to both urban and industrial centres. Syracuse also provided an outlet to Buffalo via the Erie Canal, and Oswego provided an important transfer point for coal onto Great Lakes vessels. The railroad's Scranton coal was therefore established as a commodity in a range of markets, even before the surge in western demand at the end of the 1870s

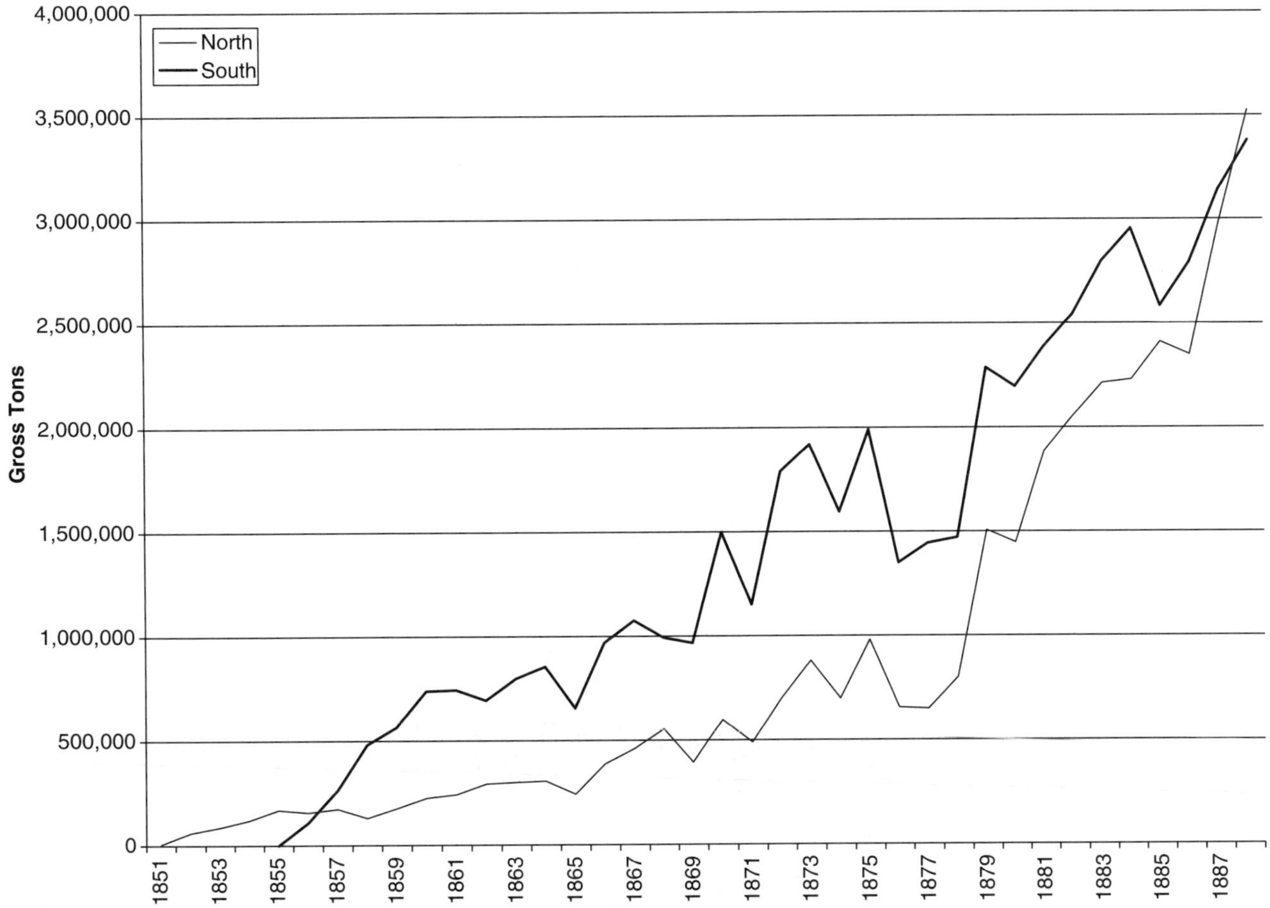

FIGURE 5.10
Delaware, Lackawanna and Western Railroad: Distribution of Anthracite Shipments 1851–1888.

provided the impetus for further developments. Rapid and well-timed construction of the direct line to Buffalo placed the railroad in a very strong position. It was able to derive substantial financial returns from the generally buoyant conditions of the 1880s, and its resulting healthy balance sheet allowed it to weather the relative stagnation of the 1890s.

OPENING UP OF NORTHERN MARKETS AS FAR AS MONTREAL

Traditionally, coal sales by the Delaware and Hudson Canal Company had been almost exclusively to tidewater locations,[120] but the need for geographical diversification of markets had become apparent by the time of the Civil War. Although initial actions had begun to be taken several years earlier, this new strategy was not clearly advocated either in the annual reports of the company or the Board of Managers minutes until 1867. At this time, President Olyphant set before the board a key letter, at once reflective and forward looking in terms of company policy. His argument ran as follows:

> 'We should aim to meet our competitors on at least equal terms, in every market where they may appear, asserting and maintaining our ability to furnish our full proportion of the quantity needed at every point of consumption; and to enable us to do this, we must construct, or aid others to construct, all necessary connecting roads'.[121]

The earliest recorded action in accordance with this evolving strategy dates back to the end of 1863. At this time, discussions were held with the Erie about possible arrangements for shipping coal westwards, in the event that a connecting road were built northwards from the coalfield to the

TABLE 5.4

Planned Distribution of Total Daily Production (tons) of Delaware, Lackawanna and Western Railroad Coal Department Company's Coal for Sample Weeks in 1898.[119]

Week of	Buffalo	Syracuse	Oswego	Utica	Northern Total	Port Morris and Washington, NJ	Hoboken	Southern Total	Line Trade	Transport Dept.	Total	Buffalo %	N %	S %
June 20, 1898	5,555	1,575	1,500	1,200	9,830	1,095	2,755	3,850	600	1,820	16,000	34.7	61.4	24.1
July 18, 1898	9,400	1,075	1,300	1,150	12,925	1,545	1,750	3,295	600	5,200	22,000	42.7	58.8	15.0
Aug. 6, 1898	7,650	1,275	900	1,080	10,905	2,025	1,675	3,700	900	2,295	17,800	43.0	61.3	20.8
Sept. 3, 1898	13,325	1,325	1,315	950	16,915	1,700	2,700	4,400	650	1,800	25,000	53.3	67.7	17.6
Sept. 24, 1898	9,000	1,650	900	1,800	13,350	1,915	4,120	6,035	730	1,700	25,000	36.0	53.4	24.1
Oct. 29, 1898	5,950	2,350	1,350	2,460	12,110	5,105	4,095	9,200	760	1,630	25,000	23.8	48.4	36.8
Nov. 26, 1898	7,000	1,970	1,900	2,320	13,190	4,280	1,450	5,730	760	1,750	25,000	28.0	52.8	22.9
Dec. 24, 1898	3,500	2,000	0	2,500	8,000	5,790	4,375	10,165	885	950	20,000	17.5	40.0	50.8

N.B. Columns may not sum to total as coal was assigned to 'surplus' in some weeks.

Pennsylvania state line. No definite agreement was made, but early in 1864 a sub-committee was appointed to make arrangements for the Jefferson Railroad to construct this connecting line to a point on the Erie at, or near, Lanesboro. The road was not actually built at this time, but a later minute indicates that the route was laid out and rights-of-way were progressively secured by the Canal Company's subsidiary, the Northern Coal and Iron Company.[122]

Part of the reason for the delay was doubtless attributable to the restricted nature of the company's charter, granted in 1823 by the New York State Legislature, which had not conferred general railroad rights. This was remedied by a charter amendment of May 9, 1867 which allowed the company to construct, own, and maintain railroads in New York state and make contracts with other railroads.[123] A further reason for moving relatively slowly following the first tentative contacts with the Erie in 1863, was that there were at the time no other serious candidates for connecting roads that might take traffic north-eastwards towards Albany, as opposed to westwards towards Buffalo. The fledgling Albany and Susquehanna Railroad, intended to connect the former city with Binghamton, had made little progress in the first twelve years of its existence, as a consequence of financial problems, and had only a few miles of track in operation. However, in that year it managed to secure financial support from the state legislature, provided it could complete 83 miles of track to Oneonta. This evidence of backing from public funds encouraged private investment and construction was able to proceed thereafter at a much more rapid pace. By the beginning of 1865, the railroad had reached Cobleskill and it was extended to Oneonta by August.[124] Further construction in the direction of Binghamton brought it within striking distance of the Pennsylvania state line, making a future connection with the Delaware and Hudson a realistic proposition.

Discussions between the two companies must have begun at about this time, as the minutes record the finalization of an agreement between them in February 1866. Under this agreement, the Delaware and Hudson was to subscribe to $500,000 of Albany and Susquehanna Railroad second-mortgage bonds. In return for this investment, the latter railroad would complete its road to Nineveh near Lanesboro within a year, and undertake to carry anthracite coal in its broad gauge cars. Evidently, the Delaware and Hudson intended to revive its project to extend a road northwards from Carbondale, and a committee was formed in May 1866 to look into the matter. However, the committee was discharged the following month and the project dropped again, possibly because of cost, although the reasons were not recorded. Adverse business conditions in 1867, and the resultant collapse in profits, dampened enthusiasm for major capital projects until early 1868, when yet another committee looked into the question of railroad extension, but it also reported in the negative. At a cost of $2.5 million, it was not felt that the company should proceed, 'in view of the many more important and pressing claims on their attention and means'. However, if anyone else should wish to foot the bill they would be happy to entertain a lease on fair terms.[125] It is not known if the committee's report was entirely accepted by the board, as they would have known that the Albany and Susquehanna was already completed beyond the planned junction point at Nineveh, leaving the tantalizing gap south from the Pennsylvania state line. Also, by April 1868, the board had approved expenditure of $81,000 for the coming season, to lay a broad-gauge third rail on the company's road within the coalfield and on the mine laterals. This suggests plans were already being made for the future arrangement with the Erie, which was finalized on September 2, 1868. Under this contract, the Erie was to build the line to Lanesboro Junction under the charter of the Jefferson Railroad, but the Delaware and Hudson was to provide much of the funding required, through its purchase of $1.5 million par value of Boston, Hartford and Erie Railroad bonds at 90 percent. The purchase money was to be paid out at the same rate as construction costs were incurred. The subsequent operating arrangement was not a lease, as might have been expected. Rather, the Delaware and Hudson acquired trackage rights over the railroad, paying the Erie 1 cent per ton-mile plus half of the net profits on the sale of the coal after making due allowance for all expenses.[126] While this railroad was still under construction, an extraordinary fight for control of the Albany and Susquehanna Railroad developed, shortly after its completion to Binghamton in January 1869. Erie interests represented by the notorious speculators Jay Gould and Jim Fisk endeavoured to gain control of the road, firstly by share purchase, then by legal writ, and finally by pitched battles between rival gangs of railroad employees along the line. Eventually the state militia had to be called out to restore order and the matter was finally settled by the courts in the interest of the original management, the judgement being handed down in January 1870. Perhaps sensing the need to shelter behind a larger but more civilized body corporate, the Albany and Susquehanna entered into a perpetual lease by the Delaware and Hudson in February of the same year, having previously declined the offer of a lease by Gould and Fisk.[127]

At this stage, the Delaware and Hudson was still pursuing its earlier strategy of meeting competition in a range of markets, not just those in the direction of Albany. In March 1869, like the Lehigh Valley and the Lackawanna, it purchased a property on Buffalo Creek in Buffalo for use as a coal depot, at a cost of $149,500. A further $50,000 was spent the following year to acquire 330 feet of riverfront adjoining the original purchase. Also in 1870, an agreement was made with the Northern Central Railway for the transport of coal south in the direction of Baltimore. However, the orientation to northern markets began to assume an increasing degree of importance when the Rensselaer and Saratoga Railroad was offered to the company for lease in 1871, the eventual lease itself being dated May 1, 1871.[128]

The final link in the chain of railroads to the north, the New York and Canada, has already been discussed as an example in the early part of this chapter. The timing of its construction, which began in March 1872 and was completed in November 1875, was a classic case of a project conceived in the enthusiasm of buoyant business conditions, but whose financial obligations still had to be met when prospects deteriorated after the 1873 panic. Even the staid commentator responsible for the centenary history of the Delaware and Hudson was led to note how remarkable it was that the company had been able to proceed with this project, when the aftermath of the panic had so greatly impeded railroad construction elsewhere in the United States.[129]

Overall, the leased lines became an increasing financial burden on the Canal Company as the 1870s progressed, because growing rental bills were not matched by corresponding increases in earnings. The Albany and Susquehanna lease required 7 percent dividends and interest to be paid on the stock and bonds of the company, which stood at about $2.5 million and $4.5 million respectively, at the time of the lease. However, just like the Lehigh Valley Railroad, the Delaware and Hudson had acquired a railroad whose track and equipment were in bad condition. This necessitated substantial investment in improvements and upgrading, for which $2.5 million of additional stock and bonds had been issued by the end of 1876, although this further increased the annual cost of the lease, from $490,000 in 1871 to $683,121 in 1876. The Rensselaer and Saratoga lease was even more onerous, since from the second year of its operation, the rental amounted to 8 percent on stock, in addition to payment of interest on outstanding bonds. Although it appears the road was in better condition than the Albany and Susquehanna when it was leased, a further 7,629 shares were issued between 1871 and 1876 to cover certain types of improvements. Other investment, such as the $308,000 spent on laying steel rails, was not covered by issuance of stock or bonds, since this was not authorized under the terms of the lease. Again the net effect of issuing more stock was to increase the rental charge from $671,848 in 1872 to $756,339 by 1876. The financial arrangement was somewhat different for the New York and Canada Railroad. Its construction cost of $8.25 million was largely represented on the books by a $4 million or £800,000 loan whose interest was payable in sterling and in gold at Baring Brothers in London, and by $4 million of stock. The Delaware and Hudson guaranteed principal and interest of the gold loan, while also holding almost all of the $4 million in stock. Initially, at least, the Delaware and Hudson received 70 percent of the gross earnings for operating the road, leaving 30 percent as an effective rental paid to the New York and Canada. In 1876, this latter sum amounted to $119,831.[130]

The size of the fixed charges during a depression era made losses almost inevitable and the full sorry picture is shown in Figure 5.11. Prior to 1876, the annual reports do not reveal the worsening position overall and not until 1883 are the details of the performance of the individual lines made manifest. Even then, the rentals attributable to each are not given and the leased lines are lumped in with the railroads in Pennsylvania to give a more favorable impression of overall operations. Only the 1877 report of the stockholders investigating committee reveals the detailed breakdown of earnings, expenses, and rentals in the early years. From this report it is clear that the major losses, after taking account of fixed charges, were being generated by the Albany and Susquehanna. The Rensselaer and Saratoga was initially profitable, but subsequently began to sink into deficit, as the

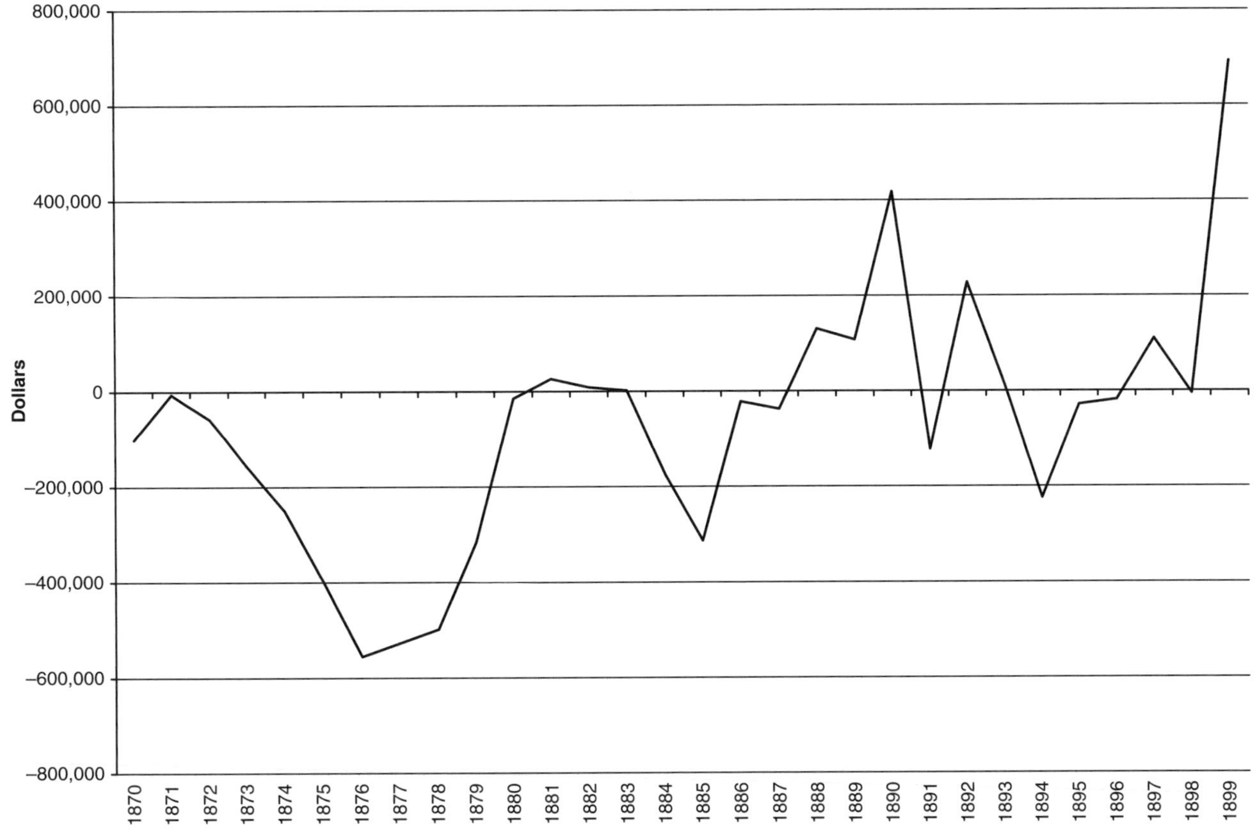

FIGURE 5.11
Delaware and Hudson Canal Company: Annual Profits and Losses on Leased Lines 1870–1899.

impact of rising fixed charges was felt against a background of static earnings.

It might have been expected that the 1880s would bring welcome relief and a return to prosperity, but in the case of the leased lines this meant only a break-even condition until the beginning of the 1890s. Fixed charges continued to creep upwards from $1.6 million in 1876 to $2.4 million by 1891, before levelling off. Part of this increase was accounted for by the proportion of increased earnings on the New York and Canada. The early promise of the Rensselaer and Saratoga was not fulfilled, largely because of minimal earnings growth in the last two decades of the 19th century. By contrast, the Albany and Susquehanna was able to demonstrate significant improvement in performance, especially in the 1890s when earnings increased but expenses were kept in check. The New York and Canada continued to disappoint as an operating unit, its 1899 revenues being lower than those of 1890. Only the earnings of the Jefferson Railroad, where no fixed charges were payable, seem to have been a consistent source of revenue to the company as a whole (see Figure 5.12). In total, over the period 1870–1899, the leased lines made a net loss in excess of $2 million.[131]

Newspaper columnists and investors were not slow to criticise the policies of the Canal Company in relation to the leased lines, and indeed its financial arrangements more generally. The New York Tribune made adverse comment on February 21, 1877,[132] and three days later a minority stockholder brought suit against the company, claiming it was hopelessly insolvent and bankrupt. President Dickson vigorously and successfully rebutted the charges, but the damage to shareholder confidence was done and the stock price fell from 125 to 50 in the space of a few weeks.[133] The Board of Managers were so concerned by these developments that they offered a $5,000 reward, in conjunction with the Law Committee of the New York Stock Exchange, for information 'resulting in the conviction of any

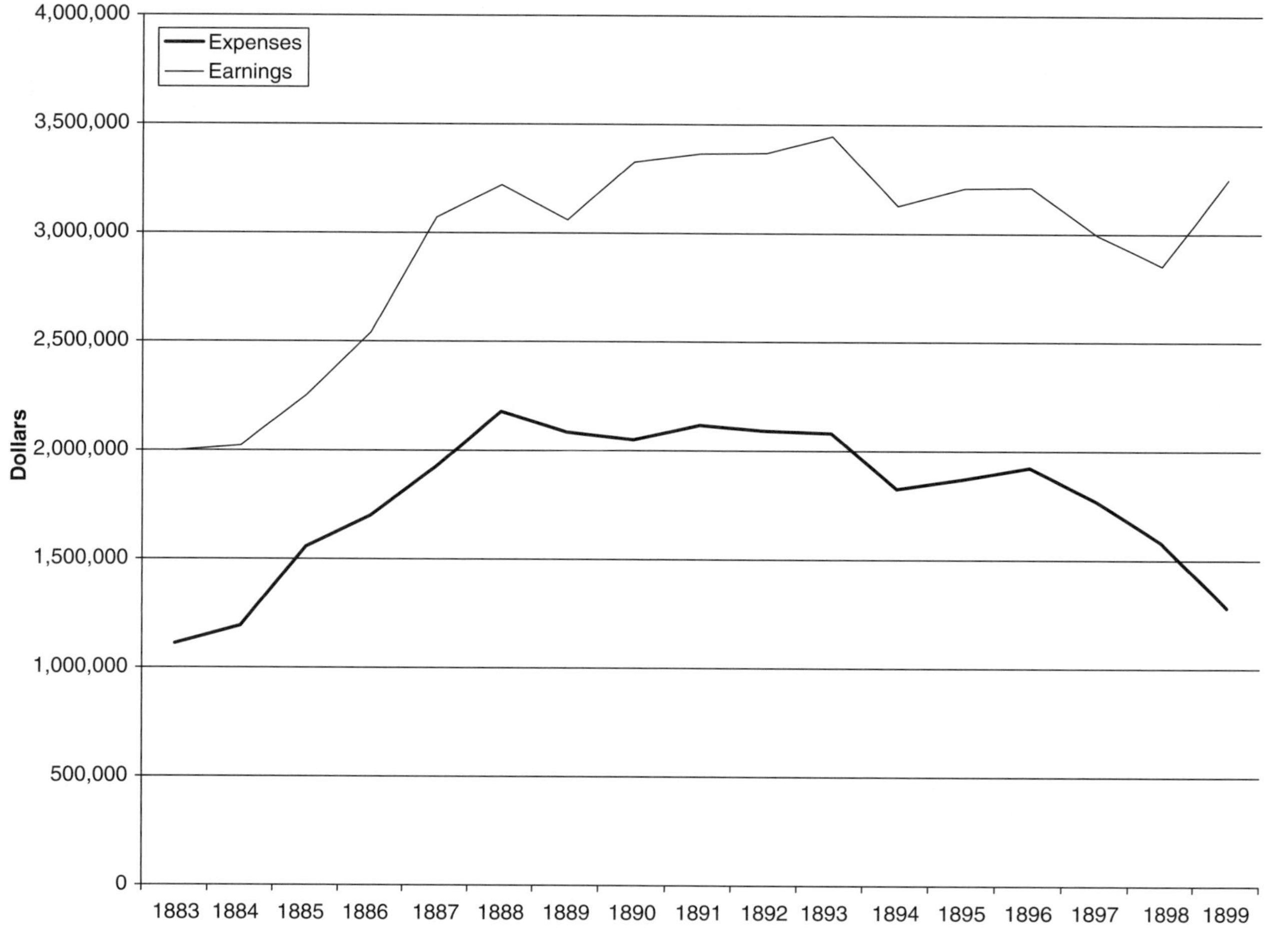

FIGURE 5.12
Delaware and Hudson Canal Company Railroad Operations in Pennsylvania: Earnings and Expenses 1883–1899.

person engaged in circulating the damaging reports in regard to the credit of the Company'.[134]

Despite this robust response to criticism, the company could not escape the scrutiny of its own stockholder investigating. This committee was appointed on May 8, 1877.[135] The observations of the committee seem, with hindsight, to have been quite fair and balanced, in respect of weighing up the exigencies of the current financial depression against the long-term strategic benefits of actions that had been taken by the Board of Managers. Several of these observations are of particular interest in the context of the present study. Firstly, it was acknowledged that the Albany and Susquehanna lease was expensive. However, if the company had not secured it at that time, there was a danger a competitor would have leased it. Secondly, it was argued that the real benefit from the Albany and Susquehanna and Rensselaer and Saratoga leases was that they provided an extended market for the company's coal, a market that was not subject to competition and which was already taking one-third of the company's product. Hence, although a substantial loss was made on the leased lines in 1875, it was greatly outweighed by the profit of $1,080,061 on the coal sold on them. Thirdly, it was admitted that the construction of the New York and Canada was premature, but at the time the iron industry was prosperous and 'the demand for Lake Champlain ores seemed limitless'.[136] Unfortunately, losses on the leased lines of $1.5 million by 1876 were to climb still further to $2.87 million by 1879, large enough to threaten the financial viability of the entire enterprise. It is not surprising, therefore, that continued attacks in the press were made against the Delaware and Hudson in the latter part of 1877. These attacks centred on the excessive fixed charges being paid on the leased lines, as would be expected, but also on claims of unwarranted over-valuation of depreciating assets, and payment of dividends out of borrowed money rather than genuine earnings.[137]

Fortunately, the general improvement in business after 1879 came to the rescue of the company and profitability was restored during the 1880s. Nevertheless, the leased lines continued to be a drain on the company's resources, since, despite a near doubling of earnings between 1880 and 1890, expenses more than kept pace. The northern markets may have been free of competition from other anthracite carriers, but the growth of demand was moderate at best, and during much of the 1890s demand was static. By focusing its attention northwards rather than westwards to Buffalo, the Delaware and Hudson effectively consigned itself to a more limited role in the anthracite industry than its previous proud history might have led it to expect.

It is apparent from examination of these different case studies that both the timing and the location of new railroad developments had important interactions with business-cycle changes and the differential growth of different areas of the eastern and central United States. In this context, the expansion of the western trade via Buffalo to Chicago proved to be particularly important. The other key market sector was the east coast tidewater cities. The different strategies used to exploit these markets will be examined in the next chapter, before making an assessment of the relative financial success that accrued to the different companies, based upon their overall mix of development strategies across the anthracite distribution region.

Chapter 6

RAILROAD EXPANSION AND CORPORATE CONTROL II: TIDEWATER MARKETS, TRUNK LINE CONNECTIONS AND COMPARATIVE ECONOMIC PERFORMANCE

THE RAILROAD PUSH FOR NEW YORK HARBOR

As Chapter 4 has already made clear, New York had become the magnet for eastbound coal shipments from a very early date. By 1860, receipts of coal in the harbor by rail, canal, and coastal freighter were more than three times those of its nearest rival, Philadelphia, although 8 percent of the coal arriving in New York had been re-shipped from Port Richmond. Two-thirds of the estimated coal receipts arrived by canal or river barge and a quarter by railroad, the vast majority of it over the Central Railroad of New Jersey to Elizabethport (Table 4.2c). Initially, the existence of the latter railroad was a great boon to the fledgling and cash-strapped Delaware, Lackawanna and Western and Lehigh Valley railroads, as it provided a key market outlet for the anthracite mines tributary to their respective lines. The 1860s also ushered in a very profitable period for the New Jersey Central as the dominance of the canal was progressively eroded by superior railroad technology. This can be seen in Figure 6.1, which shows the increase in its coal shipments during the period, distinguished by whether they originated in the Lehigh Valley or on the line of the Delaware, Lackawanna and Western. Growth was particularly rapid in the immediate post-bellum period, with shipments almost doubling between 1862 and 1868. As the railroads originating in the coalfields grew in ambition and financial stature, however, they began to appreciate the importance of good terminal facilities on New York Harbor and to chafe at the lack of control over their key avenue to market across New Jersey.

As usual, there were several strategies that could be adopted to address these issues. The first was to take control by lease of existing railroad or canal routes, which fed into the New York area. The second was to construct an entirely new route and thirdly, there was the possibility of entering into alliances or traffic agreements with other companies that fell short of outright control or ownership. Owing to its importance in the coal trade generally, it was to be expected that each of these strategies would impinge on the fortunes of the Central Railroad of New Jersey, and this indeed proved to be the case.

LACKAWANNA RAILROAD STRATEGIES

The first railroad to make a move was the Delaware, Lackawanna and Western. Despite its existing good relationship with the New Jersey Central, it wished to be assured of independent control of a route into New York. Thus, on December 31, 1868, it leased the Morris and Essex Railroad at a rental of 7 percent on the capital stock. Although quite costly, in terms of fixed charges, the lease was a sensible one from the Lackawanna's viewpoint. Extending from Hoboken and Newark to Phillipsburg on the Delaware River, the Morris and Essex intersected the Southern Division of the Lackawanna at Washington, NJ, a short distance from Hampton Junction, where the latter railroad interchanged traffic with the New Jersey Central. The Morris and Essex was already double-tracked over much of its length and the planned Boonton branch would provide a second route into Hoboken via the important manufacturing centres of Paterson and Passaic, NJ. This branch was eventually

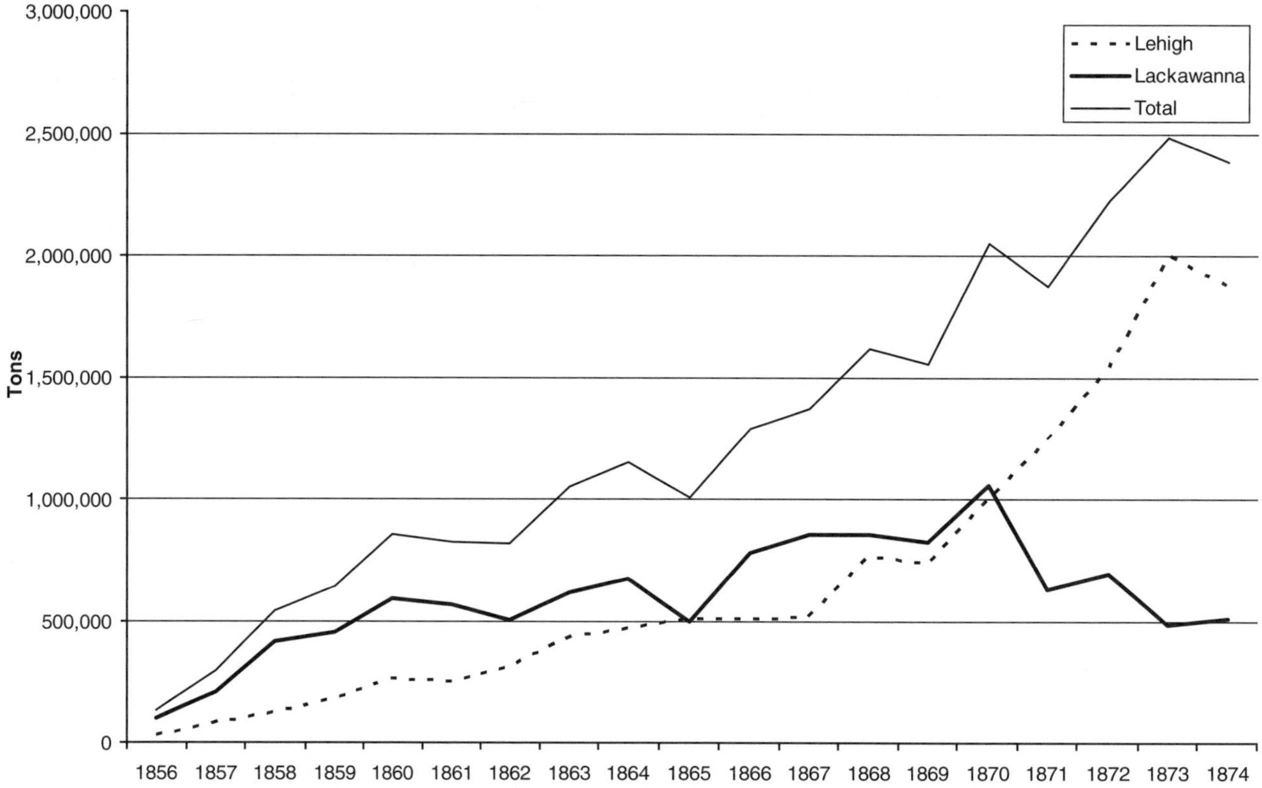

FIGURE 6.1
Central Railroad of New Jersey (Central Division): Annual Coal Shipments 1856–1874.

completed in September 1870. It is also interesting to note that the Lackawanna was not the first company to have contemplated a lease of the road. A lease had been negotiated by the ill-fated Atlantic and Great Western Railroad in 1865, but financial difficulties of the latter in 1866 prevented the lease provisions being implemented. Also, in 1868, Jay Gould, president of the Erie, had been in discussion about a lease, but the matter was dropped and the more favorable terms offered by the Lackawanna were taken up.[1]

Under the lease, the Lackawanna was required to build the Boonton branch and to complete the double tracking of the main line to Phillipsburg. No time was lost in attending to these matters and, by April 1869, more than $45,000 had already been spent on securing rights of way for the Boonton branch. However, a report to the Board of Managers also indicated that securing the right of way for the last ten-mile stretch of the road east from Paterson to Hoboken was going to be an extremely expensive proposition. The sum of $200–$250,000 would have to be set aside for this purpose. In expectation of the large increase in the company's business at Hoboken once the new line was completed, a further 260 feet of waterfront, south of the property they owned, was secured for $550,000, payable in convertible mortgage bonds of the Morris and Essex Railroad.[2]

An important operational difficulty in merging the traffic arrangements of the two companies was that the Morris and Essex was standard gauge, while the Lackawanna was broad gauge. Laying of a third rail had been the means of circumventing the difficulty for shipments over the New Jersey Central to Elizabethport. However, in the summer of 1868, prior to the lease being negotiated, such a remedy for traffic consigned to points on the Morris and Essex had been rejected in favor of a converse plan for laying a third rail on the Southern Division of the Lackawanna, to enable narrow gauge cars to be hauled to Scranton.[3] Following the execution of the lease, this plan was reversed yet again. A third rail was to be laid during the construction of the Boonton branch and additional stretches of third rail to link with it were laid from Washington to Port Morris and Dover, NJ in 1869–1870.[4] The implication of this was that the Boonton branch would immediately become the main route for coal shipments to Hoboken, bypassing the older route via Morristown and Newark (Figure 6.2).

A further implication of the growth of coal shipments was that increased congestion through the Bergen Tunnel outside Hoboken was to be expected. According to the Lackawanna minutes, the Erie held direct rights to use the tunnel, originally built by the Long Dock Company, and the Lackawanna claimed these same rights under an 1859 agreement between the Long Dock Company and the Hoboken Land and Improvement Company. However, there were issues over the scheduling of trains and whether or not the Lackawanna was required to pay tolls for use of the tunnel. After some months of disagreement between the Lackawanna and the Erie in the latter part of 1870, Jim Fisk of the Erie brought matters to a head in December, with his customary subtlety, by blocking the tunnel with locomotives at either end. The State Governor intervened to have these obstacles removed, as they were blocking a public highway, and the matter was referred to the courts. Following a ruling in January 1871, which upheld the right of the Lackawanna to pass trains through the tunnel, an agreement between the companies was reached. Each was to pay half of the cost of upkeep of the tunnel and the Lackawanna was to pay the same proportion of $100,000 a year, to the Erie, as its traffic bore to the total traffic through the tunnel. The Erie could require the Lackawanna to abandon use of the tunnel by January 1, 1881 on three years' notice, but to do so would involve the Erie in a financial penalty of $116,550.[5]

Given this agreement and the likelihood of increasing delays as coal shipments continued to grow, the Lackawanna had already decided by September 1871 to build a new tunnel for its own traffic, close to the mouth of the existing tunnel.[6] The contract was let in August 1873 at an approximate cost of $800,000, with an optimistic timescale of 21 months to completion. In the event, owing to labor troubles and problems of water seepage, it required a further two years to complete the job.[7] The tunnel was finally completed in May 1877. A first special train ran through it on May 12 and regular service was inaugurated on May 24. At the same time, the Lackawanna notified the Erie that it was abandoning the use of the old tunnel under the 1871 agreement.[8] A minor but interesting aside to this construction saga is that the main contractor for the tunnel was John McAndrew of Scranton.[9] The construction accounting vouchers of the Lackawanna Railroad Coal Department show that this same contractor was also involved in a number of projects for masonry work at new coal breakers in the Scranton area between 1867 and 1873.[10]

Almost from the outset, the Central Railroad of New Jersey had been unhappy about the lease of the Morris and Essex Railroad by the Lackawanna and this had a major impact on the relationship between the two companies in the early 1870s. This was hardly surprising, since it could only have an adverse effect on Lackawanna shipments over its road to Elizabethport, despite the existence of the perpetual traffic contract between the two companies, dating back to 1854.[11] By the beginning of 1871, the New Jersey Central was complaining that the Lackawanna was sending most of its traffic over the Morris and Essex in claimed violation of the old contract, and threatening to sue for damages.[12] However, this statement points up a significant problem with discussions of the relationship between the two companies during this period by previous authors. Since the Lackawanna issued no printed annual reports during the period, reliance has, in the past, been placed on the version of the saga put forward in the annual reports of the New Jersey Central. Fortunately, the manuscript records of the Lackawanna reveal another side to the story, contrary to Taber's assertion that the Lackawanna's views were 'not recorded for posterity'.[13] In the first place, surviving Lackawanna account books give a breakdown of the relative value of coal sales at Elizabethport and Hoboken throughout much of the period, although there are gaps in 1871–72. These data are plotted in Figure 6.3 and indicate clearly that in 1870 barely 20 percent of Lackawanna coal shipments were going to Hoboken via the Morris and Essex, so, at best, the New Jersey Central was overstating its case at the time.

The long strike in the first part of 1871 must have delayed further consideration of the matter. The next reference to it came in the form of an announcement that instead of pursuing the matter in the courts, as the New Jersey Central had previously threatened, the two companies had agreed on March 16, 1872 to operate their two concerns in close collaboration, as a 'joint estate'. All receipts and expenses were to be joint and equal dividends were to be paid on each company's shares. Initially, the

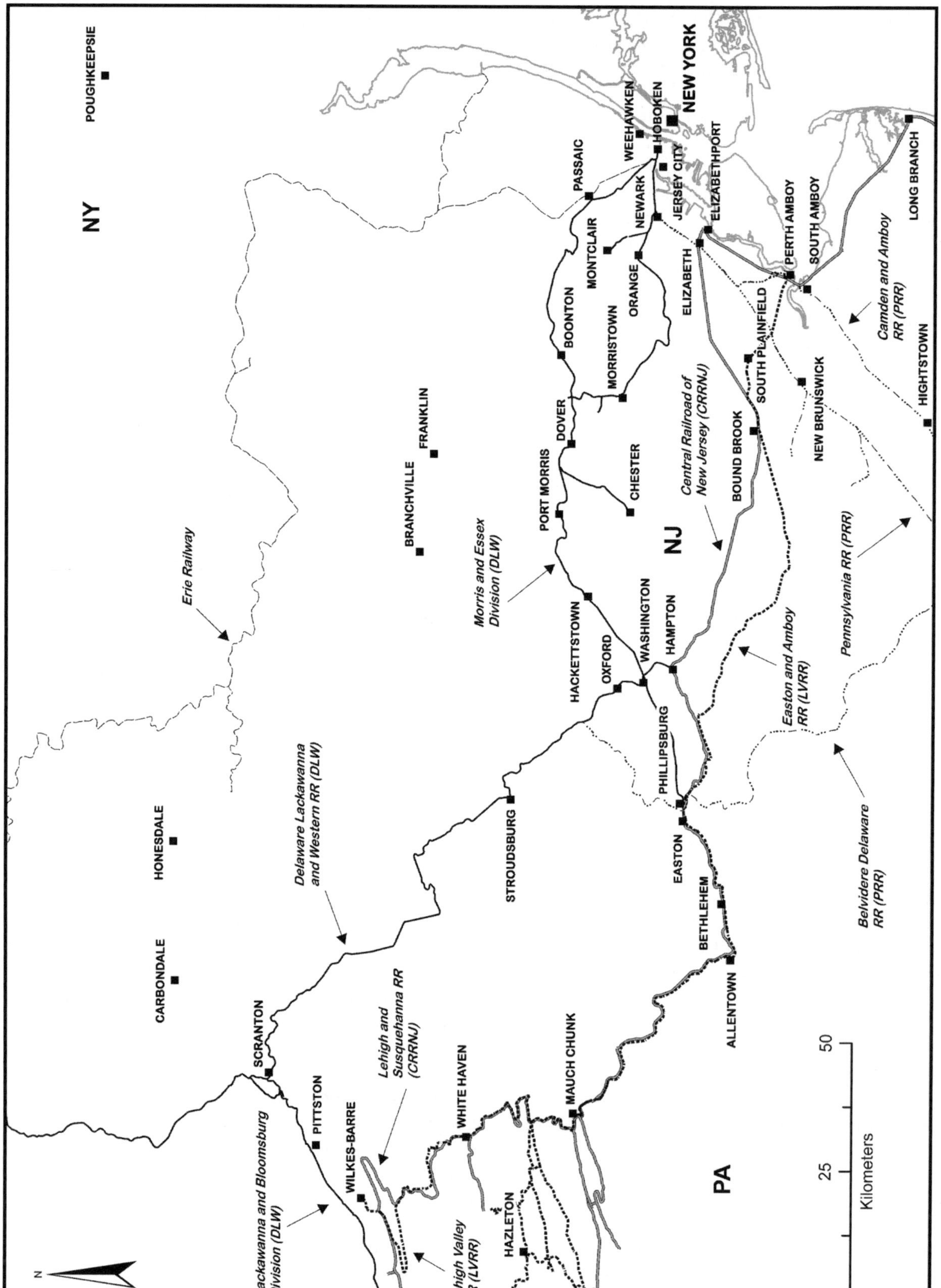

FIGURE 6.2
Competing Railroad Routes into New York circa 1875.

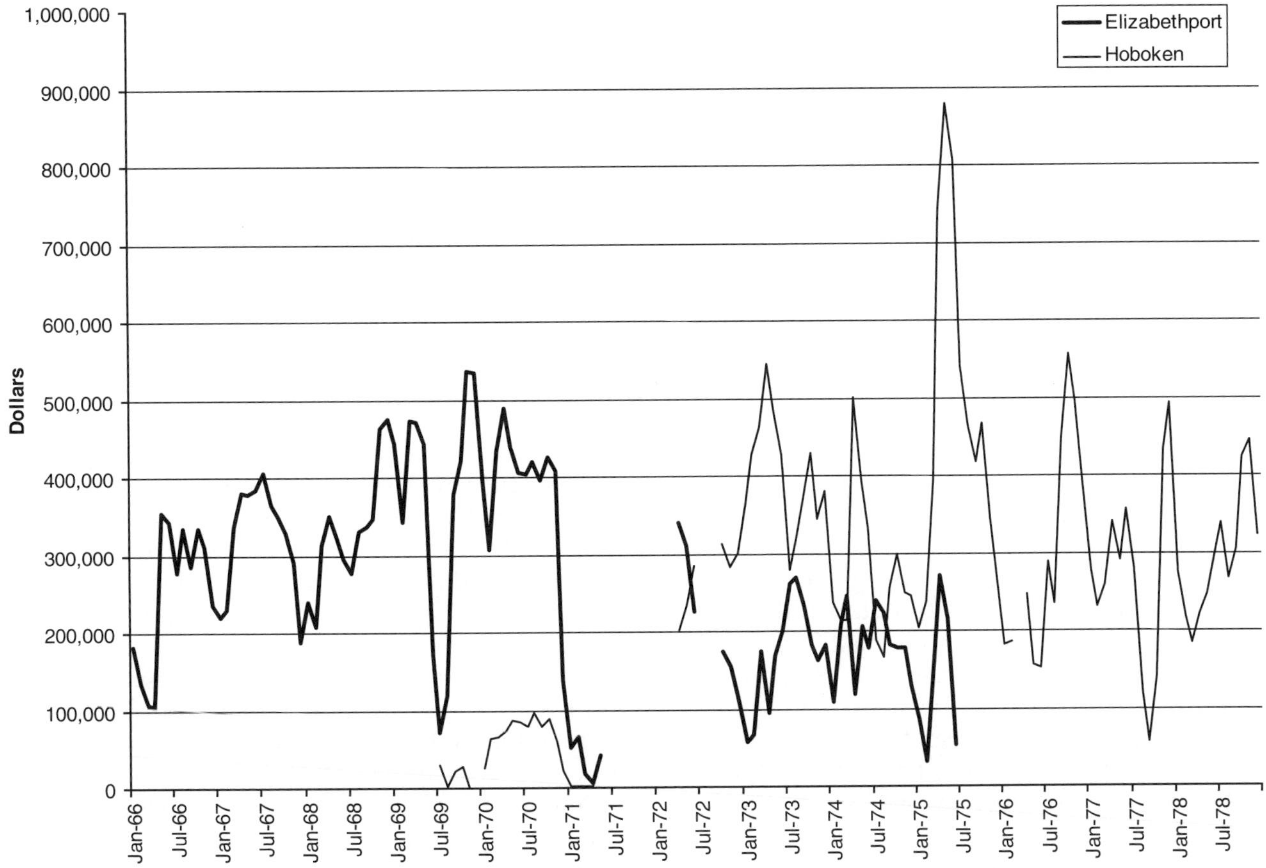

FIGURE 6.3
Delaware, Lackawanna and Western Railroad: Coal Sales in Eastern Seaboard Ports 1866–1878.

respective boards and officers were to manage the ordinary business of each company, subject to the general direction of a 'Joint Committee' made up of the two presidents and five directors from each side. The longer-term aim was to effect a 'perpetual union of interests' and consolidate fully the operations of the two companies to simplify management and improve efficiency.[14]

Such a turn-about, when relations between the two companies appeared to have soured, seems very surprising when first encountered, and previous authorities have offered no convincing explanation of this. Detailed examination of the almost illegible correspondence of President Sloan of the Lackawanna to General Coal Agent Storrs in Scranton has, however, unearthed an important clue. A letter from Sloan at the end of 1873 indicates that the idea of a merger, involving not only the Lackawanna and the Jersey Central, but also the Lehigh Valley, was not new. It had originally been mooted by Sloan and Johnston of the Jersey Central prior to the lease of the Morris and Essex in 1868. Unfortunately, such a proposal had not found favor with Judge Packer and Ario Pardee of the Lehigh Valley. As a result, substantial outlays of many millions of dollars were made by the Lackawanna on the Morris and Essex lease, by the Lehigh Valley on a route to New York Harbor (see below), and by the Jersey Central on leases of its own. Sloan's remarks indicate that he regarded this as a case of personal ambition taking precedence over the interests of stockholders, and he closes his letter with an exasperated cry of 'What mistakes!'.[15] Given the private nature of this correspondence to a trusted senior manager and friend, there is little reason to suppose these are anything other than the firmly held views of President Sloan and that his intentions were honourable in trying to achieve the best outcome for his stockholders. Hence, the 1872 merger plan simply revisited an earlier joint proposal by the two companies and appears to have been, initially at least, a genuine attempt to avoid a damaging split between two companies, which had collaborated profitably over many years. That said, there were other pragmatic reasons on the Lackawanna side for maintaining cordial business relations with the Jersey Central, not least because it still needed access to the docks at Elizabethport. As a Board of Managers resolution recorded, the 'present facilities of the Company at Hoboken are inadequate for the shipping of all of the coal of this company brought to tidewater and of the coal of other parties which it is the duty of this company to transport'.[16]

Although no formal records of the proceedings of the Joint Committee appear to have come down to us, the New Jersey Central later claimed that it had been wracked by dissension and clash of corporate cultures from the outset. The root cause, however, probably lay in the differing financial strength of the two companies at the time the 1872 agreement was signed. According to the New Jersey Central, the Joint Committee found that the Lackawanna was in such a straitened financial condition that it demanded the issue of $3 million of convertible bonds by the latter company. The distinct impression was given that the New Jersey Central felt it had been misled over the health of the Lackawanna prior to the March 1872 agreement. As the Annual Report avers, 'without charging deception it is safe to say that no such state of urgent indebtedness had been exhibited to our committee'.[17] Company records of the Lackawanna report that on May 9, 1872 the president had identified the need to either issue bonds or sell some of the securities owned by the company. Interestingly, there is no mention of this having been demanded by the Joint Committee, although the latter approved the bond issue at the end of May.[18]

While the New Jersey Central was quick to scold its partner company for apparent misrepresentation of the financial facts, the situation was actually more complex than might initially appear. The first point to note is that the Lackawanna had a large Coal Department at the time, whereas the New Jersey Central was only just moving into the mining, as opposed to transportation business. Secondly, average coal prices fell by more than one-third between October 1871 and June 1872, from $5.51 to $3.51 in the case of Delaware and Hudson Coal sold in New York.[19] With the benefit of hindsight, this was hardly surprising. The strike in the first part of 1871 had restricted output to such a degree that the wants of the market in the second part of the year could scarcely be met and high prices were maintained. In 1872, with no strikes or restriction of output by producers, shipments continued at the high levels of the second half of 1871. Falling prices ensued as markets were flooded with a plentiful supply of coal, and it is fortunate that the economy was in an expansionary phase at the time, according to the Ayres Index (see Chapter 2) or prices would undoubtedly have fallen even further. Thirdly, sharp falls in prices usually affected the miners of coal more severely than the shippers of coal, although freight rates would tend to decline also. The net effect of this was that, although the New Jersey Central bemoaned the state of the trade in its 1872 Annual Report, it still managed to generate net earnings of $2.88 million, which left it a small surplus of nearly $35,000 after payment of all fixed charges, the rental on the Lehigh and Susquehanna Railroad, and a 10 percent dividend to stockholders.[20] The Lackawanna was not so fortunate. The company's main account journals reveal that a healthy $4.84 million in net transportation earnings was partly offset by a substantial $1.33 million loss in the Coal Department. This was also after the funds from the $3 million bond issue had been deployed. Since its fixed charges were higher than those of the New Jersey Central, it found itself with insufficient funds to pay a full 10 percent dividend and therefore had to pass the dividend for six months at the end of the year.[21] In the light of these circumstances, the Lackawanna was probably almost as surprised as the New Jersey Central at the rapid deterioration in its financial position, which would have been much less apparent at the beginning of 1872, when negotiations over the March agreement were underway, than would have been the case later in the year.

The emergency bond issue by the Lackawanna can scarcely have enhanced its reputation in the investment community and it was an uninspiring start to the joint arrangement. If the number of resignations by Lackawanna directors is any indication, the atmosphere in the joint committee must have been tense and difficult from the outset. The New Jersey Central objected to the Lackawanna's long-standing practice of selling coal by monthly auction in New York, although these auctions were continued with consultation. The Central's policy of publishing annual reports was at odds with the Lackawanna's culture of secrecy and the former claimed that the Lackawanna was to have taken immediate steps to change to standard gauge, when instead it began to press for broad gauge to be extended from Elizabeth to Jersey City!.[22] It should be noted, however, that none of these issues were mentioned in the text of the March agreement.

Nevertheless, the joint committee persevered with its work and, by the summer of 1872, produced proposals for the committee and management structure of the planned joint company. Behind the scenes, however, financial pressures and tensions must have continued to mount on both sides. Both cause and effect of this can be observed in Figure 6.3. From the spring of 1872, a rapid decline in Lackawanna coal sales in Elizabethport is observable, while there was a matching rise in sales in Hoboken in the latter part of the year. The specific economic rationale behind this only becomes clear when the circumstances in the coal trade outlined above are linked to the provisions of the old 1854 agreement with the New Jersey Central. Under this agreement, coal was shipped to Elizabethport at a **fixed rate** of 1¼ cents per ton-mile.[23] Since the rate was fixed, under conditions of rapidly falling prices in tidewater markets, this had the effect of transferring all the loss of revenue to the Lackawanna, because the Central's revenues were dependent only on tonnages moved. In consequence, it was essential to minimize shipments to Elizabethport, thereby upsetting the Jersey Central, to avoid the additional reduction in cash flow. The diverted tonnage could then be shipped over the Morris and Essex at proportionately lower freight rates, thereby preserving a greater percentage of the tidewater price for the Lackawanna (both transportation and coal departments), yet still enhancing the receipts of the Morris and Essex to help in meeting the fixed charges on its lease. In justification for its actions, the Lackawanna later argued that the contract of 1854 had been merged into the agreement of March 1872, although again the wording of the latter agreement did not explicitly state that this was the case.[24]

The tensions reached breaking point in September 1872 when it became apparent to the Lackawanna that the Jersey Central was intending to declare a full 10 percent dividend without the agreement of the Joint Committee. The latter road would have been aware that the financial circumstances of the Lackawanna did not permit it to follow suit. This would mean that the companies were not abiding by the equal dividends clause of the March agreement. In response, the Lackawanna issued a Bill in Chancery to restrain the Jersey Central from paying a dividend 'until the earnings of the two companies should warrant paying a dividend upon the stock of both' and the payment had been approved by the Joint Committee.[25]

The problem was compounded in late October when the Lackawanna issued a further injunction to prevent the Jersey Central issuing $3 million of its own convertible bonds to pay for coal land purchases, even though the Joint Committee had approved this purchase. Apparently the concern here was that the Lackawanna might find itself with some liability for these bonds as a result of the March agreement, so an appropriate legal escape clause needed to be added to the paperwork surrounding the bond issue.[26]

Shortly after this, relations between the two companies must have broken down entirely. In early December, the Jersey Central formally accused the Lackawanna of ceasing entirely to comply with the 1854 agreement. In consequence, they were commencing a suit against the Lackawanna to obtain compensation for present and future losses. Further to this, in a letter dated December 13, 1872, President Johnston informed President Sloan that from December 31 no coal would be transported over the Jersey Central in broad gauge cars. This provoked a vehement outburst from the Lackawanna Executive Committee, claiming the resolutions and proposed actions of the Jersey Central were:

> 'illegal, unjust, unnecessary, unreasonably wasteful, injurious to the joint interests of said two companies and as contrary to, and violative of the letter and spirit of the Agreement for unity of business, profits and losses, made between the said two companies and dated March 16, 1872'.[27]

An intense bout of litigation followed and, despite the protestations of the Lackawanna, the Jersey Central managed initially to get the March agreement voided on a technicality.[28]

However, the Lackawanna succeeded in persuading the State Legislature of New Jersey to pass an act, on March 7, 1873, confirming and validating the agreement despite the earlier court ruling. Unfortunately, the opposition of the Jersey Central's stockholders to any further moves towards consolidation had become apparent by this stage, and it appears that a final truce was called, the attempt at union having ended in complete division.[29] Yet the Jersey Central still needed coal traffic, the Lackawanna still did not have sufficient shipping capacity at Hoboken for all its coal, and by this time, the price weakness of mid-1872 had abated. Business pragmatism therefore took over and the intimate relationship, which had so badly soured, was replaced by an arms-length traffic agreement, dated May 9, 1873. This superseded all previous agreements and provided well-defined escape routes for both parties. Under the new arrangement, the Jersey Central was to retain the third rail to Elizabethport for five years, if required, and transport whatever coal the Lackawanna provided, up to a maximum of 600,000 tons in any one year. The freight rates were unchanged from the earlier agreements at 1¼ cents per ton-mile.[30] As Figure 6.3 shows, there was a modest revival of traffic to Elizabethport under the agreement, although the heavier shipments at peak periods were being routed to Hoboken. In the weakening business climate following the 1873 panic, it was less and less in the Lackawanna's financial interest to maintain docks and terminal facilities at both locations. Hence, at the end of October 1874, a resolution was passed to terminate the contract with the New Jersey Central on April 30, 1875, giving six months notice, as required by the 1873 agreement.[31] In terms of loss of revenue, this timing was rather unfortunate for the Jersey Central, as it was only able to derive very limited benefit from the long strike of 1875 in the Schuylkill Region. This resulted in greatly increased shipments by the Lackawanna and other Northern Field companies in the spring and early summer of 1875. The commercial parting of the ways was fixed and final, as the Jersey Central tore up the third rail to Elizabethport, leaving the Lackawanna docks idle and inaccessible to broad gauge cars.[32] President Sloan was unconcerned. The arrangement had lasted long enough for the Lackawanna to establish its shipping facilities at Hoboken and the 1873 agreement was simply the endgame, as far as he was concerned. His views on the former business partner were summed up in correspondence to W.R. Storrs — 'I am glad we are done with them'.[33]

Although shipment data for the Lackawanna by destination are rather fragmentary, the surviving information indicates that Hoboken continued to be an important destination for coal throughout the 1870s, with tonnage reaching nearly 1.24 million in 1879. At this time, these tidewater shipments comprised 54 percent of the total tonnage sent south from Scranton. By 1887, however, the tonnage to Hoboken had declined by nearly a third to 841,000 tons, and, as a share of southbound shipments, it had fallen by half to only 27 percent of the total.[34] This suggests a growing importance of shipments, to points in industrial New Jersey along the line of the Morris and Essex. It is not surprising, therefore, that the last major single investment by the Lackawanna in the 19th century, to provide for additional terminal facilities, was the purchase of land adjacent to its Hoboken terminal in Jersey City in June of 1881. This waterfront property was acquired for $1.8 million from the Jersey Shore Improvement Company, the entire sum payable in bonds of the Morris and Essex Railroad.[35]

President Sloan was entirely justified in lamenting the amount of resources devoted to the development of the Morris and Essex Railroad, as shown by his statement quoted above. At the same time, control of a key avenue to tidewater, which owned prime real estate in New York harbor, proved to be essential to later success as a trunk line to Buffalo. Unfortunately, the absence of any readily available figures, even after brief annual financial statements began to be issued in the late 1880s, makes it very difficult to determine how quickly the Morris and Essex became cost-effective. Limited information can be gleaned from the Lackawanna accounts and other sources, however. There can be no doubt that establishment of the road as an efficient coal carrier was a very expensive proposition. By December 1869, the Lackawanna had advanced $5.37 million on it, followed by a further $4.95 million in 1870, of which $1.64 million was for grading, $537,000 for equipment and $83,000 for piers and docks. Of this $10 million total, $5.89 million had been reimbursed in cash and $3.88 million had been covered by the issue of Morris and Essex stock.[36] As more stock was issued, the rental under the lease began to climb from its 1870 figure of $1.22 million, which represented at the time some 64 percent of the entire rental bill of the Lackawanna for leased lines.[37] Shortly afterwards, the Jersey Central advised that the railroad was 'very nearly' meeting the cost of the rental out of its earnings, although the high cost of the rental was regarded as one of the drawbacks of the 1872 joint agreement.[38] By 1875, the total amount of stock reached $15 million, in addition to $15.4 million of bonds, and it continued to be increased by further issues thereafter. By 1900, the lease rental on the Morris and Essex had more than doubled from its 1870 figure to $2.68 million, as compared to $1.67 million for the New York, Lackawanna and Western Railroad. Rental of these two lines comprised almost 85 percent of the entire charges of $5.15 million for leased lines at the time, a figure that was little changed from that of a decade earlier.[39]

LEHIGH VALLEY RAILROAD STRATEGIES

In general, the other anthracite carriers, who initially lacked routes under their own control, managed to achieve their aims with a lesser degree of friction and ill feeling than that which developed between the Lackawanna and the Jersey Central. Perhaps the most ambitious of the schemes in the early post-bellum period was that of the Lehigh Valley Railroad. By 1871 the Company was aware that it had been suffering for some time from 'the want of an independent outlet to tidewater'.[40] To reach Philadelphia, it utilized the Lehigh/ Delaware Division canal system and the North Penn and Belvidere Delaware Railroads. The equivalent routes for New York were the Morris Canal, the Morris and Essex Railroad and the New Jersey Central. The latter was the largest forwarder of coal originating on the lines of the Lehigh Valley Railroad in the years immediately following the Civil War. So it was a matter of some sensitivity when the Jersey Central leased the competing Lehigh and Susquehanna Railroad in 1871.[41] The graph shown earlier, in Figure 6.1, reflects the growing importance of this new feeder line to the Jersey Central, as the Lehigh Valley Railroad's contribution to the Central's overall 'Lehigh' tonnage peaked in 1868 and 1870 at about ¾ million tons.[42]

There were two elements to the scheme adopted by the Lehigh Valley to obtain control of its avenues to New York. The first was to lease the Morris Canal in 1871; the second was to embark on the construction of an entirely new railroad line from Easton to tidewater at Perth Amboy, New Jersey (see Figure 6.2). It can hardly be imagined that the Lehigh Valley saw a prosperous future ahead for canal technology, but the Morris Canal lease offered two important advantages for the development of its eastbound coal trade. The first and most strategic of these was that it owned a 60-acre terminal basin in Jersey City with 1500 feet of frontage on the North River, directly opposite New York. A second, more pragmatic reason was that control of the canal would facilitate the speedy unloading of coal trains into waiting boats at Phillipsburg, thereby increasing the tonnage that could be carried over the main line of the railroad. The lease was not especially cheap, as bond interest and dividends at 6 and 7 percent on different classes of issues had to be paid annually, amounting to $256,000 per annum. In addition, a number of outstanding debts to major banks totalling $166,687 had to be settled by the railroad at the time of execution of the lease.[43]

There can be no doubt that the issue of securing good real estate for the construction of wharves and piers for the New York coal trade was uppermost in the minds of the railroad's Board of Directors at this time. In the summer of 1871, 378.76 acres of land at Perth Amboy were purchased at $400 an acre, for a total outlay of $151,504, payable in several instalments. This was followed later in the same year by the acquisition of additional tracts adjoining the original purchase.[44]

The availability of two alternative sites for the future development of terminal facilities, one in close proximity to the centre of New York, the other opposite the southern tip of Staten Island, may have left the Board of the Railroad uncertain about the best way to proceed in the short term. As evidence of this, the initiative to build the Easton-Amboy link was preceded in 1871 by the purchase of a controlling interest in the New Jersey West Line Railroad, formerly the Passaic Valley and Peapack Railroad. The shares were acquired by Asa Packer in a personal capacity before being turned over to the railroad at cost, although available sources do not indicate the length of time taken by him to obtain the shares.[45] Originally chartered in 1865 through a series of charter supplements, the road had acquired by 1870 the right to build, from a connection with the Newark and New York Railroad in Newark, to Milford on the Delaware River south of Phillipsburg. The route crossed the Morris and Essex Railroad at Summit, but the only section of track actually built during this period was a short stretch westward from Summit to Bernardsville.[46] The same Board of Managers' minute that records Packer's purchase of the shares also indicates that the plan was for the Lehigh Valley to aid in the completion of the road, to provide 'a through connection for our business'. However, this plan was not carried out and the minutes contain no further reference to the matter. Bogen claims that the Lehigh Valley lost control of the road to the Morris and Essex[47] but does not give a source for this information, and Archer[48] repeats the claim. However, this interpretation seems to be incorrect. The Lehigh Valley held a controlling interest, so could not simply 'lose' it to the Morris and Essex. Taber notes that, in fact, the railroad ran into financial difficulties during the mid 1870s and was eventually sold in 1878 to R. G. Rolston, who was acting for the Lackawanna Railroad.[49] This implies the Lehigh Valley had changed its mind about building a through line along this route, and hence withdrew financial support for the New Jersey West Line.

The rapidity with which a more southerly route was adopted suggests this change of strategy must have happened quite quickly. In March 1872, the Lehigh Valley purchased the old 1858 charter of the projected Perth Amboy and Bound Brook Railroad, and it also secured a new charter for a railroad from Bound Brook, near Piscataway, to Easton. As soon as the subscription books were opened to buy stock in the latter railroad, the Lehigh Valley secured 11,000 out of a total of 20,000 shares to ensure control of the railroad. Later in the year, the two roads were consolidated to form the Easton and Amboy Railroad, almost all of whose 25,000 shares were taken by the Lehigh Valley.[50] The construction timetable for the railroad was ambitious and no time was wasted in moving the project forward. The aim was to complete the road by April 1, 1874. To that end, six of the heaviest sections of construction work were placed under contract in April 1872, and a further 25 heavy sections in August. The contracts were staggered with a view to all being completed simultaneously. A large force of men and horses was deployed, as well as four steam-powered excavators. Unfortunately, the planners had not allowed either for the impact of the 1873 panic or the problems that arose in building the Musconetcong Tunnel. After the panic, light construction work was suspended to save money, as it could be completed rapidly when required, but the heavy work was pushed forward to avoid further slippage in the project schedule. The tunnelling problems were to delay the overall completion of the line by an additional 9–12 months. To the end of 1873, approximately $2.8 million had been spent on the road, 30 percent of which was for right of way and real estate.[51] By the end of the following year, some 50 miles of track were laid, the grading was nearly complete, and the tunnel was the last remaining obstacle to completion of the route. Over the majority of the distance, the road had already been graded for double track. Finally, on May 28, 1875, the new road was in sufficiently good order for a coal train of 125 cars to pass over its entire length. One month later, the road was formally opened. To that time, the construction cost, including terminal facilities, was almost $8.5 million, although the final cost had risen to over $10 million by the end of 1877.[52]

Extensive work on the development of wharf and pier facilities at Perth Amboy had begun in 1873, confirming this location as the preferred destination for the majority of the Lehigh Valley's tidewater coal and freight shipments. The plan called for extensive trestlework and tracklaying, initially to two coal piers, with a third being added later. Some 800,000 feet of southern pine lumber for the trestlework were contracted for, in 1873 alone, for this purpose. Two freight piers, each 800 feet long, were also to be built. These were completed by the end of 1876. The channels adjacent to the piers had been dredged to a depth of 16 feet, and there was provision for the stocking of up to 200,000 tons of coal.[53]

Although the Lehigh Valley proudly announced, in 1876, that the Easton and Amboy Railroad was 'entirely free of debt',[54] the $10 million expended on it began to weigh heavily on the parent company's treasury when coal prices and tolls fell to all-time lows in 1877. In consequence, a $10 million mortgage was arranged for the Easton and Amboy on all its property. A total of $5 million in 6 percent bonds resulting from the mortgage was given to the Lehigh Valley in part payment for construction outlays. The remaining $5 million of these costs was represented by stock of the Easton and Amboy, also held by the Railroad.[55] Unlike the subsidiary railroads in New York, the Easton and Amboy began to pay its way from the start. Shipments in 1877 were 60 percent up on the previous year, gross earnings exceeded $1 million, yet costs were only $522,000, to provide a sound return of 5½ percent on the entire cost of building the road.[56]

The rapid switching of its main-line tonnage on to the Easton and Amboy had major effects on the distribution of tonnage to other connecting roads, although these effects were not felt equally by the different companies, as can be seen in Figures 6.4 and 6.5. The initial brunt of the change was borne in dramatic fashion by the Jersey Central, whose shipments of Lehigh Valley Railroad tonnage collapsed in 1875. In the short term, the Morris and Essex was apparently a beneficiary of this change, but it shared in the decline shortly afterwards. While the present discussion is primarily concerned with shipments to New York, it is also instructive to compare the fortunes of the two connecting roads going in to Philadelphia, namely the North Penn and the Belvidere Delaware, shown on the two graphs. Although shipments over the North Penn declined after the 1873 panic, the sharpest fall came in 1880,

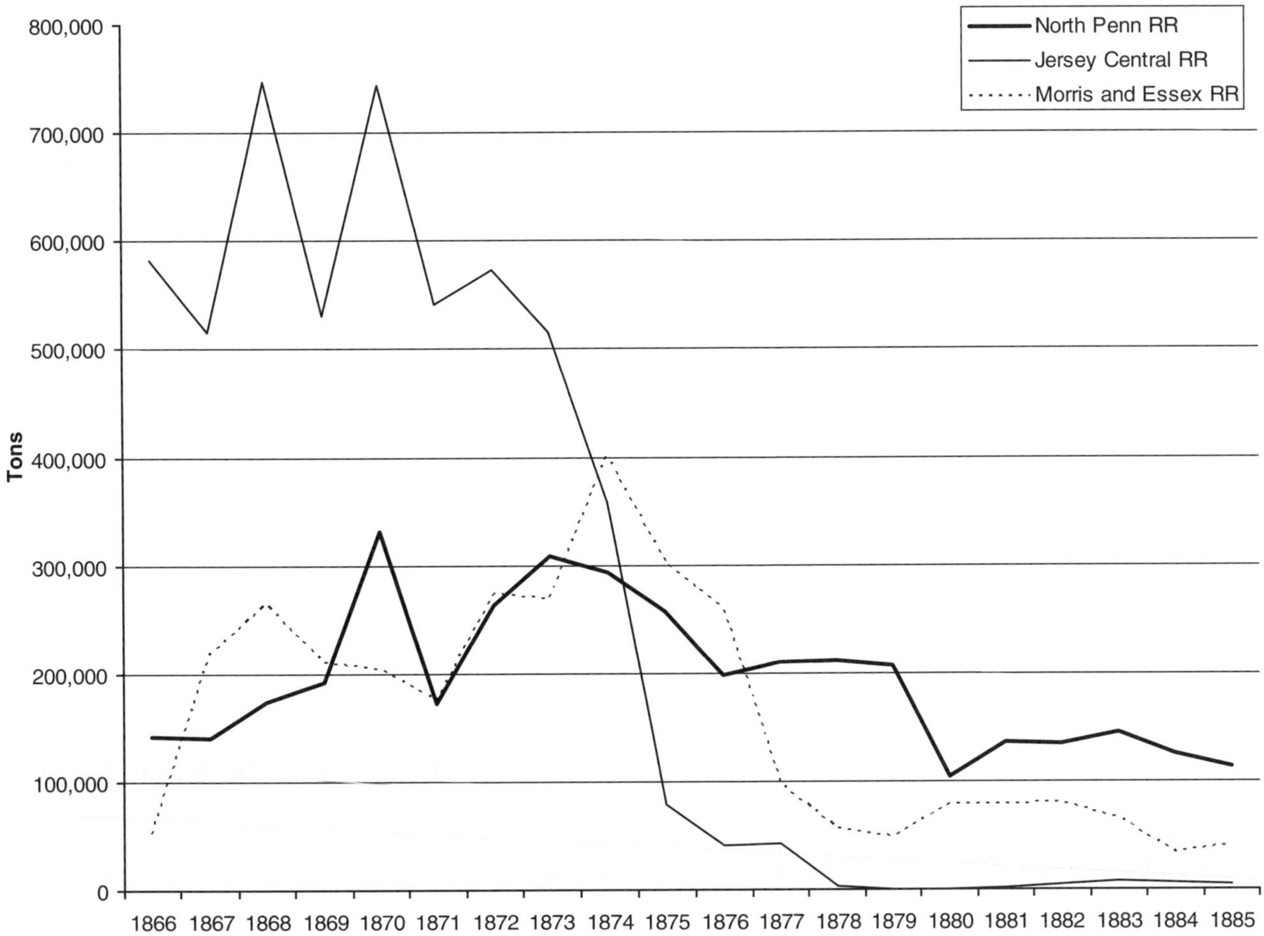

FIGURE 6.4
Lehigh Valley Railroad: Connecting Avenues to Market Experiencing a Decrease in Shipments 1866–1885.

when shipments halved after the Reading Railroad leased the road.[57] From Figure 6.5 it is clear that tonnage was diverted on to the Belvidere Delaware Railroad, which had been under the control of the Pennsylvania Railroad since 1871.[58] This switch can hardly be regarded as fortuitous. Relations between the Pennsylvania and the Lehigh Valley were sufficiently close that they attracted the scrutiny of others in the trade. As President Sloan observed in 1873, 'I am rather suspicious of the Lehigh V[alley] and the Penna R.R. They are very friendly. They both go to Amboy'.[59]

At the same time, there was no love lost between the Reading and the Pennsylvania. So the Lehigh Valley would doubtless have been happy to take any steps that would find favor with its powerful and long-standing ally, yet which did not have any adverse impact on its own business. One final point to note from Figure 6.5 is that the traffic on the Morris Canal actually increased between the 1860s and the 1880s, although it remained at relatively modest levels. This is in marked contrast to the steady decline of tonnage on the Schuylkill Navigation during the same period.[60] The reason for this was that the lease was perpetual and there was no easy way for the Lehigh Valley to wriggle out of it, despite several attempts. It was therefore necessary to keep some traffic moving over the canal to offset expenses. The lease was to continue for over half a century, until finally cancelled in 1922.[61]

The dock and terminal properties acquired and developed during the 1870s at Perth Amboy and Jersey City served the Lehigh Valley well for a decade or more. In the late 1880s, it became apparent

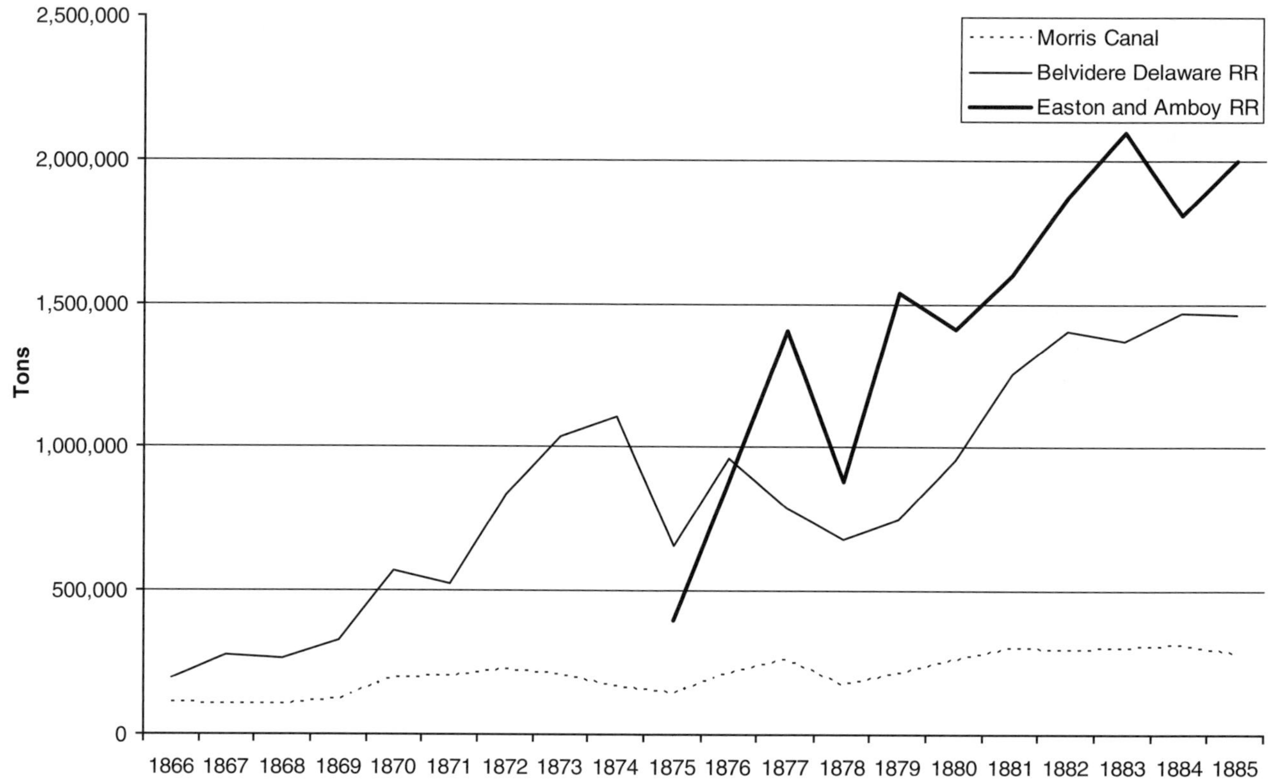

FIGURE 6.5
Lehigh Valley Railroad: Connecting Avenues to Market Experiencing an Increase in Shipments 1866–1885.

that further investment was required and a proposal was made in 1887 to spend $1.5 million on additional wharves in Jersey City. The high level of shipments to Philadelphia over the Belvidere Delaware could also not be ignored. During 1890, several tracts of Delaware River front in the 6, and 18th Wards of the city were purchased, by means of complex financial arrangements, for a total cost exceeding $650,000.[62]

The Jersey City property assumed increasing importance in the late 1880s and 1890s, but the Lehigh Valley lacked a direct railroad entry into the Morris Canal basin. Initially, this was achieved by a combination of connecting line construction and a trackage agreement with the Jersey Central. This enabled tidewater freight shipments to be diverted from Perth Amboy to Jersey City after December 1888. Through construction of additional connecting lines and control of the strategically important National Docks Railway, a short line running along the waterfront south of Jersey City, the Lehigh Valley finally obtained its own direct route from South Plainfield on its New Jersey Division (formerly the Easton and Amboy Railroad) to Jersey City in 1899. A further short connection from the National Docks Railway gave it access to the Constable Hook oil refineries. It was also able to build its own dock facilities at that point to handle freight shipments to New England, thereby reducing congestion at the Jersey City terminal.[63]

TRAVAILS OF THE JERSEY CENTRAL

The prosperity of the Central Railroad of New Jersey up to the early 1870s had been based on the coal tonnage it transported to tidewater for the Lackawanna and Lehigh Valley Railroads. The leasing of the Lehigh and Susquehanna Railroad in 1871 should have ensured it of continuing financial success, even though the Lackawanna Railroad was increasingly siphoning off tonnage to Hoboken. However, the additional impacts of the long depression that followed the 1873 panic, and the sudden loss of half a million tons of

Lehigh Valley shipments transferred to the Easton and Amboy Railroad, contributed to the railroad's growing financial problems. As a result of this loss of tonnage, earnings fell from $4.67 million in 1873 to $3.28 million in 1875. This was almost entirely the result of a decline in receipts, since operating expenses varied little between the two years. Foolishly, the railroad continued with its policy of paying 10 percent dividends. Yet even a cursory glance at its financial accounts for the year showed that this was an unwise course to follow. The dividend cost over $2 million, the annual rental of the Lehigh and Susquehanna exceeded $1 million and the debit balance on the interest account stood at $658,243. These three items alone significantly exceeded the net earnings and large expenditures were continuing on new rolling stock and new branch lines in New Jersey, although the latter were covered by the issue of almost $5 million of bonds[64]. In a perfectly mistimed move in late 1873, the railroad had also acquired two coal properties, those of the Wilkes-Barre Coal and Iron Company and those of the Lehigh Coal and Navigation Company in the Northern Field, and leased the latter's canals and mining operations in the Lehigh Region.[65] At the other end of its line, similar expansionary zeal exhibited itself. The railroad had originally begun acquiring tidewater dock properties as early as 1860 by purchasing the capital stock of the American Dock and Improvement Co. for $1.5 million. This company had rights to a 2½-mile stretch of waterfront south of the Morris Canal basin in Jersey City, with a total land area of 1,000 acres. By 1876, an estimated $6 million had been spent on initial development of part of this property, represented by $3 million in bonds and the same amount in stock.[66] These escalating obligations, coupled with lack of restraint on new investment at a time when coal demand was slipping and prices weakening, led to eventual receivership in February 1877, after attempts to find a rescue plan acceptable to holders of the company's securities had failed. The problem was compounded by the separate receivership of the Lehigh and Wilkes-Barre Coal Company, the subsidiary into which all its newly acquired coal interests had previously been merged. Substantial difficulties were experienced by the Receiver Judge Lathrop in securing the necessary support for a financial re-organization plan and averting the break-up of the railroad through surrender of the Lehigh and Susquehanna lease. However, agreement was eventually reached in 1878 by all parties, which gave the company some financial respite until 1880.[67]

Fortunately, the upturn of 1879 brought relief from the commercial stringencies of previous years, and this relief was afforded to the Jersey Central as well as its competitors. Shipments jumped from 2.26 million tons in 1878 to 3.83 million tons in 1879, nearly all of it originating on the line of the Lehigh and Susquehanna Division. In the latter year, however, a new source of shipments became available. The Reading Railroad had always shipped coal to New York by water from Port Richmond, but it lacked its own rail route into New York and use of other railroads cost it 85 cents a ton to reach the metropolis. In 1879, President Gowen saw an opportunity to remedy this problem and cut the cost of shipments. He leased both the North Penn Railroad, running northwards out of Philadelphia, and the recently completed Delaware and Bound Brook Railroad, for a period of 999 years. The latter road, linked to the Delaware Branch of the North Penn Railroad, gave him a connection to the Jersey Central, which would allow coal to be shipped to Elizabethport.[68] The Reading leased the old abandoned wharves of the Lackawanna Railroad at Elizabethport for $10,000 a year over three years, in the first instance.[69] Shipments were small in the first year, but rose to half a million tons in 1880 and over ¾ million tons by 1882. This reduced costs for the Reading and provided much needed revenues for the Jersey Central. Such a beneficial alliance between the two carriers led President Gowen to try and gain greater control over the Jersey Central by means of a lease. He secured the support of the Vanderbilts, but his plans were bitterly opposed by Jay Gould and the Pennsylvania Railroad. He finally succeeded in making the lease on May 29, 1883, although the lease proved more of a financial liability than an asset when business conditions worsened in 1884–1885.[70]

It was hardly surprising that the Pennsylvania Railroad should oppose a new direct route from Philadelphia to New York Harbor, since this provided a significant new competitor to its own route between the two cities. In fact, the Pennsylvania had both a water route, via the Delaware and Raritan Canal, and a rail route, over the former Camden and Amboy Railroad which extended to tidewater at South Amboy, New Jersey. These two routes formed a major part of the United Canal and Railroad Companies of New Jersey, an organization that originally had interchange traffic agreements with the

Pennsylvania for carrying of its New York-bound shipments. The United Canal and Railroad Companies had been leased by the Pennsylvania in 1871, to secure its route into New York, but on very expensive terms. The Executive Committee of the United Companies had indicated that they were 'unanimously of [the] opinion that not one member of our Board will entertain any proposition to lease at less than 10 percent clear of all deductions'. In fact the terms were 10 percent on the capital stock of $19.49 million and interest on the bonded debt of $16.94 million. The route to New York was so important that even the mighty Pennsylvania had to agree on this basis if it wanted to gain control.[71]

Despite the high costs, both of lease rentals and new line construction work, there can be little doubt that investment by the different railroads in improving their access to the New York market was worthwhile in both the short and the long term. Traffic levels quickly rose to the point where receipts were at least sufficient to cover costs. Over a longer time span, control of a route into New York was essential to becoming a player in the trunk line arrangements that began to develop from the late 1870s onwards. The same argument applies over the long term to the expenditure on real estate for terminal facilities and the construction of wharves and docks. When expenditures by the Reading in Philadelphia, totalling approximately $3 million[72] between 1867 and 1875, are also taken into account, the overall investment was well in excess of $10 million at tidewater between 1860 and 1880. However, in the short term, many of the waterfront properties were much larger than actually required for prevailing levels of business, and undoubtedly represented a drain on corporate finance, despite their unquestionable long-term strategic importance. It should also be remembered that the railroads became quite heavily involved in coastwise shipping in conjunction with their tidewater outlets, just as they did on the Great Lakes. This required the construction, lease, or purchase of tugs, barges, and coastal freighters. The Reading Railroad is a good example of this, because the costs and earnings from this aspect of their coal business are clearly set out in the annual reports over a period of years. In the early 1870s, they invested $2.5 million in vessels. Despite the high cost, this proved to be one of the few ventures from the Gowen years to yield solid and quick returns. Between 1872 and 1889, the Reading steam colliers generated $2.95 million in net earnings, despite the loss of several vessels in treacherous weather conditions off the New England coast.[73]

TIMING, TRUNK LINES, AND DIVERSIFICATION

It is apparent from the previous chapter and the foregoing discussion that the timing of investment to capture a share of western markets via Buffalo was more varied than that diverted towards securing control of tidewater routes to the eastern seaboard. The latter was particularly concentrated in the period up to the 1873 panic and just beyond, while new railroad building in the direction of Buffalo continued into the 1890s. A minor rider to this is the entry of the New York, Susquehanna and Western and the New York, Ontario and Western Railroads into the picture from the 1880s onwards, but these two carriers never achieved the same stature as the major players in the anthracite trade.

There are two key questions that have never been considered in the literature, and which are omitted from Bogen's interesting conclusions about the growth of anthracite railroads.[74] The first is the effect of the interaction between business cycles and the timing of investment in eastern and western connections on earnings growth and the ability to participate in trunk-line arrangements. The second is the extent to which the financial performance of the railroads varied over the period, in terms of operating ratios and the ability to accumulate surpluses. This financial performance was partly a function of the quality of infrastructure on leased roads, as has been discussed in earlier sections. It was also dependent on the nature of the terrain traversed, the rate of growth of shipments over time, the level of fixed charges and other financial encumbrances, and the extent to which each railroad could diversify its traffic beyond coal into a general freight business, to maximise the use of its expensive facilities and motive power. This last was clearly closely bound up with the success in participation in trunk-line arrangements. One further factor, the effect of the anthracite coal combinations or cartels, will be examined as a separate topic in Chapters 10 and 11. The issues to be considered here provide important information relating to the context of cartel decision-making. They are also crucial to an understanding of the interactions between the railroads and their mining subsidiaries. Some of these issues are touched on by Bogen, but the absence of published annual reports from the Lackawanna, and the variable nature of the information presented in the annual reports of other companies, prevented him from making any systematic comparisons of the kind required.

Access to the archives of the major players, such as the Lehigh Valley, the Reading, and the Lackawanna, has now made this possible to a much greater degree, although a number of gaps in available information still remain. Since a good deal of detailed empirical information about specific initiatives by the different railroads during the period has already been presented in earlier sections, the focus here will be on comparative examination of the temporal dynamics of aggregate measures of company performance.

At the simplest level, a comparison of transportation receipts of several of the major railroads (Figure 6.6) gives an indication of the relative rate of growth of overall traffic during the period. Several major points are immediately apparent on examination of the trajectories of the different companies. The first of these is the dominance of the Philadelphia and Reading (P&RRR in the figure) during much of the decade of the 1860s, followed by progressive erosion of its relative importance thereafter. Not until the turn of the century did it begin to outpace its rivals once again. Secondly, the Delaware, Lackawanna and Western (DL&W) accelerated strongly in the boom period of the early 1870s, but suffered more heavily than its competitors in the depressed conditions of the latter part of that decade. Thirdly, the Lackawanna experienced an extended period of sustained growth during the 1880s, barely faltering during the depression of 1884–5, while its competitors were more adversely affected. Fourthly, all the companies peaked in terms of earnings at the beginning of the 1890s, before the 1893 panic ushered in an extended period of flat revenues during the long depression that followed.

Finally, the Lehigh Valley (LVRR) trailed the other companies from the 1870s onwards. As a point

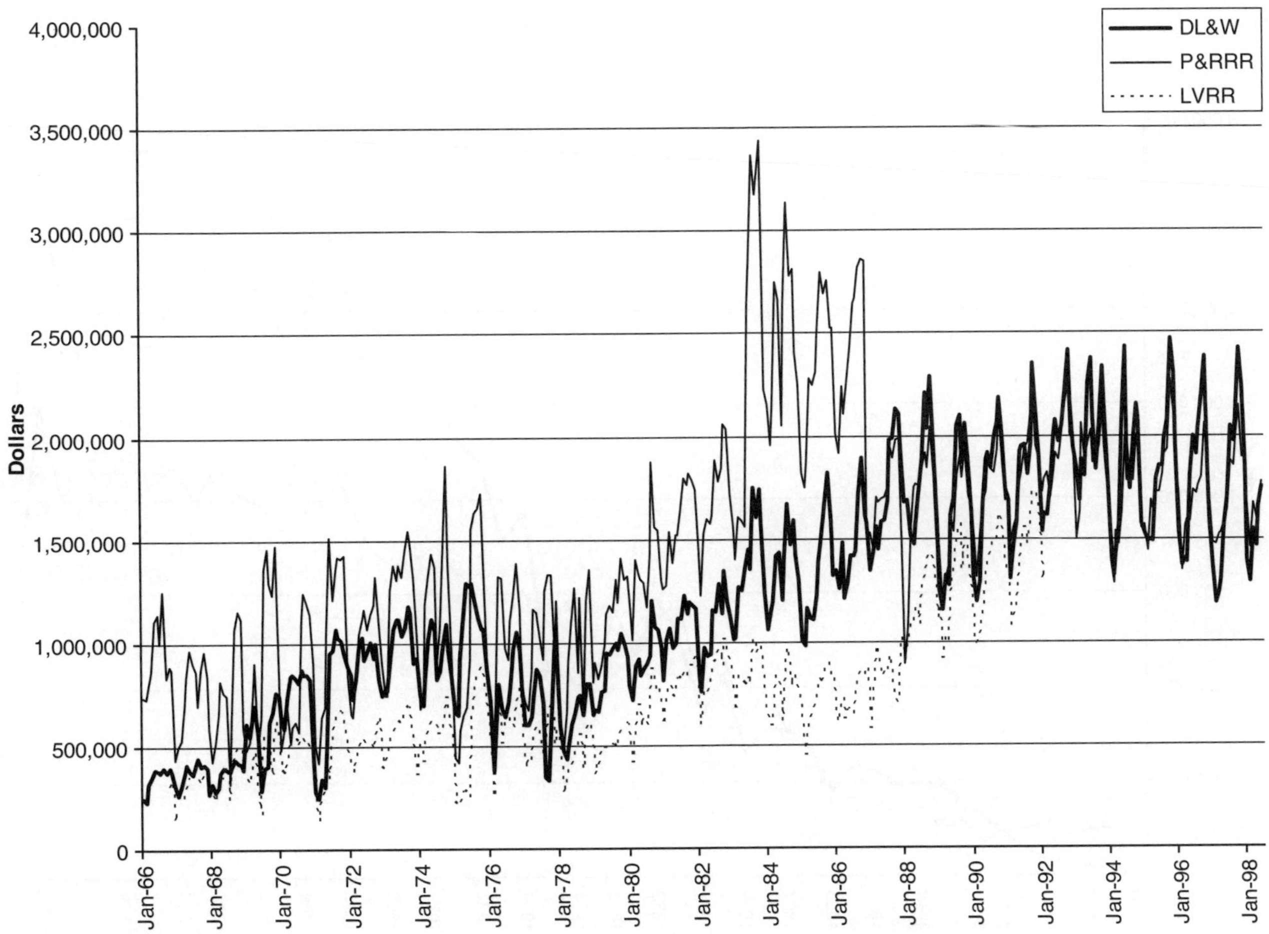

FIGURE 6.6
Monthly Transportation Receipts of Major Anthracite Railroads 1866–1898.

of clarification, it should also be noted that the apparent steep rise in its receipts in 1888 is due to the incorporation of the Pennsylvania and New York Railroad into its reported figures rather than a sudden improvement in business performance!

Further light can be shed on this first aggregate measure of gross receipts by examination of comparative anthracite shipments for several of the companies (Figures 6.7 and 6.8). The first of these shows the Lackawanna against the Delaware and Hudson (D&H). The similarity in many of the minor peaks in the series is explained either by the adverse impact of strikes that affected both companies, or the beneficial effects of strikes in other coal regions that benefited them equally. Thus, in 1875 and 1887–1888 these Northern Field companies increased their output to compensate for labor troubles in the Lehigh and Schuylkill Regions, while the converse was true in 1877. All were uniformly adversely affected by the major event of the 1902 coal strike.[75] Prior to the 1880s, the Lackawanna had held an output advantage in the period 1858–1865, while the Delaware and Hudson marginally gained the upper hand from 1868–1872. However, a major performance gap opened up in the Lackawanna's favor from 1879. This became most pronounced in the 1880s and was largely sustained throughout the remainder of the period of interest.

A somewhat different picture emerges from the comparison between the Philadelphia and Reading and the Lehigh Valley. Here the gap between the two companies narrowed in the 1880s as compared to earlier and later periods, and neither company showed the sharp increase in shipments, experienced by the Lackawanna in particular, during the 1880s.

The outperformance by the Lackawanna Railroad during the decade following the 1870s depression is therefore the most striking feature of this initial stage in the analysis. The excellent timing of the construction of the New York, Lackawanna and Western Railroad and its rapid completion, allowed the Lackawanna to derive considerable

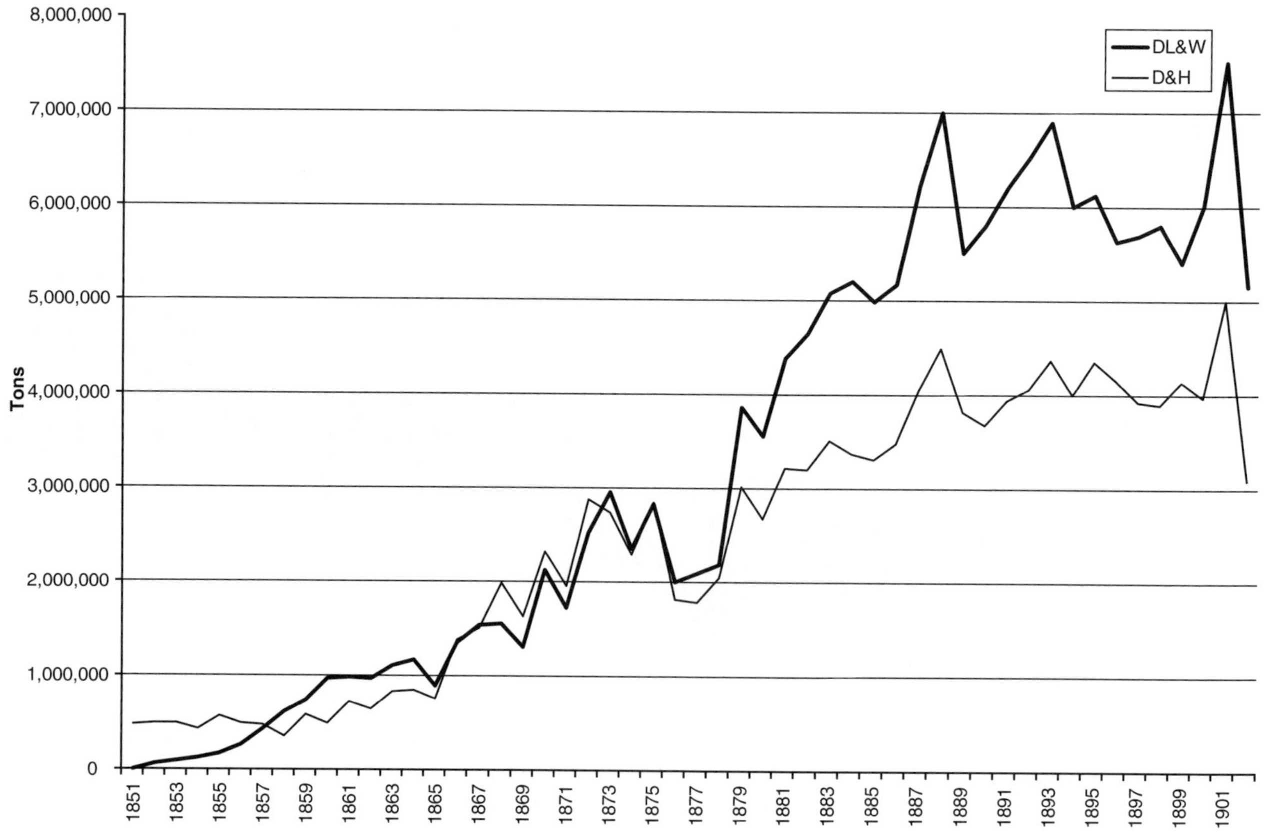

FIGURE 6.7
Major Anthracite Railroads: Comparative Annual Shipments 1851–1902.

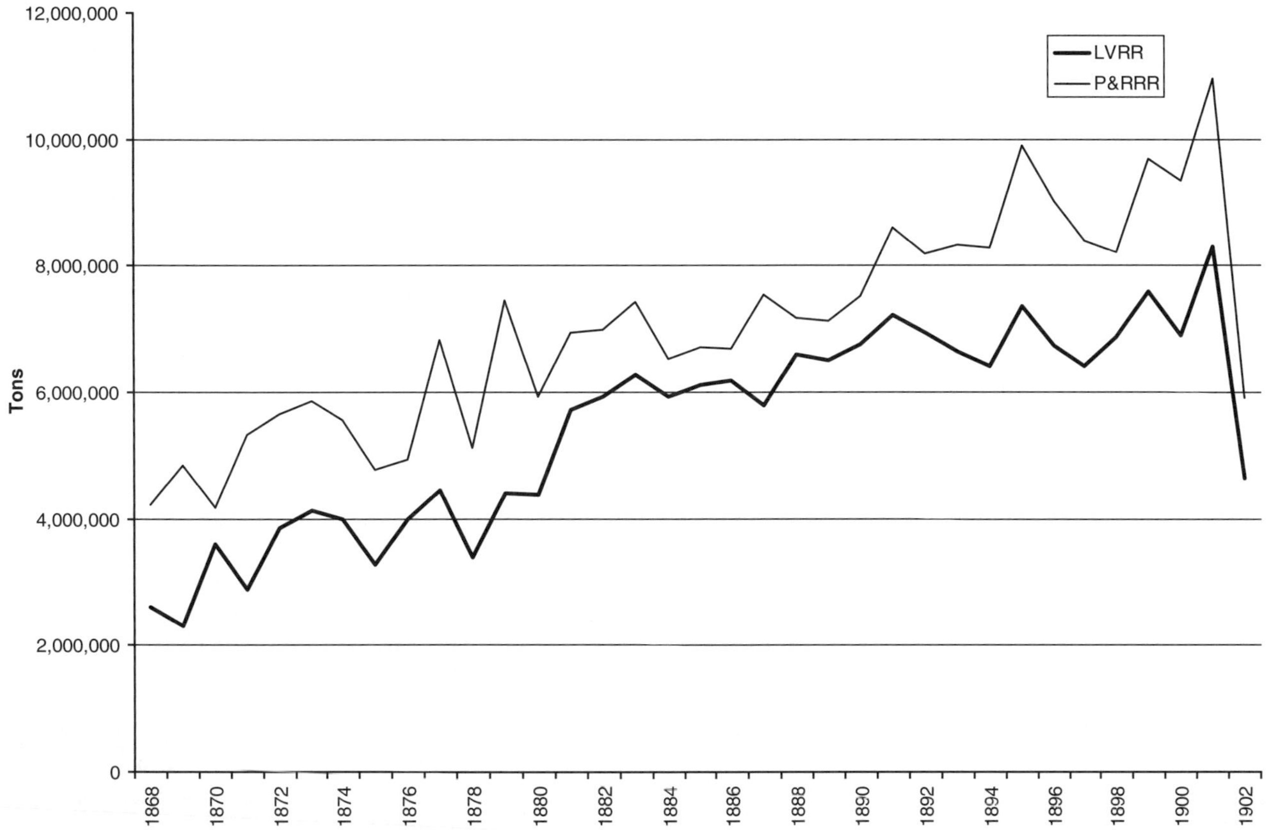

FIGURE 6.8
Major Anthracite Railroads: Comparative Annual Shipments 1868–1902.

commercial benefit from the improved business conditions of the 1880s and to reduce its dependence on the competitive eastern seaboard markets. That said, the earlier transfer of terminal facilities to Hoboken from Elizabethport and the development of northern connections via Binghamton and Syracuse must also have played their part, since the first sharp increase in coal shipments predated the completion of the through line to Buffalo.

By contrast, the timing of two of the major new investment projects by the Lehigh Valley Railroad appears to have been very poor. The Easton and Amboy project was begun in 1872 shortly before markets peaked in 1873 and was completed in 1875, as the long depression began to take hold. In a similar fashion, its own line to Buffalo was completed in 1892, shortly before the 1893 peak, and the subsequent cessation of growth in traffic, which lasted until the late 1890s. Only the construction of the Pennsylvania and New York Railroad can be regarded as timely. Its completion in 1869 provided an outlet for sharply increased shipments of Lehigh Valley Railroad coal to the north and west in the expansionary period up to 1873. Shipment levels were broadly maintained during the 1870s depression, before increasing again in the early 1880s, albeit at a much slower rate than the Lackawanna (see Figure 5.5).

Less well-known and less ambitious than the schemes of the Lackawanna and the Lehigh Valley were the Reading's plans to extend its western connections, beginning with the lease of the Catawissa Railroad in November 1872.[76] In the first full year of operation of the lease, some 320,000 tons of coal were shipped over it, rising to 500,000 tons by 1880.[77] This initiative was complemented, in the early 1880s, by the construction of the Shamokin, Sunbury, and Lewisburg Railroad, which linked to the Jersey Shore, Pine Creek and Lewisburg Railroad for onward connection to Buffalo, via the New York Central. Bonds to finance construction of the Jersey Shore were jointly guaranteed by the latter, in partnership with the Reading and the Corning, Cowanesque and Antrim Railroad. This new route was specifically intended to allow the Reading's Coal and Iron subsidiary to reach 'the large Northern and Western markets for anthracite coal which it has so long desired to obtain entrance to'.[78] Unfortunately, post 1882, the Reading's annual reports no longer provide a breakdown of shipments, but weekly tonnage reports from the Secretary's office indicate that, by the summer of 1884, tonnage over the Shamokin, Sunbury and Lewisburg branch was running at 15,000 tons a week, as compared to about 5,000 over the Catawissa and Williamsport Branch.[79] The Catawissa branch was also extended in 1883 to connect it with the Jersey Shore Line.[80] Useful though these developments were, the Reading was clearly behind even the Lehigh Valley in developing its western connections, and the most important of these came on stream just as the boom years of the early 1880s were drawing to a close. It is thus hardly surprising that the Lehigh Valley was able to narrow the gap in shipments between the two roads during this period.

However, the growth of anthracite shipments is only part of the overall picture of the transportation business and account must also be taken of other types of freight and passenger traffic. Figures 6.9–6.11 show the proportion of receipts derived from coal, merchandise, and passengers for each of the railroads over the period of interest. Good data are available for the Reading and Lehigh Valley for an extended time span, but unfortunately the Lackawanna data are reported more sporadically, even when recourse is made to the original account books. Nevertheless, data have been located for a sufficient number of years for the overall pattern to emerge. From the graphs, it is apparent that the Reading led the way in diversifying its traffic away from the very high level of dependence on coal, which had characterized its early operating years. Although all three of the roads earned about 70 percent of their revenues from coal

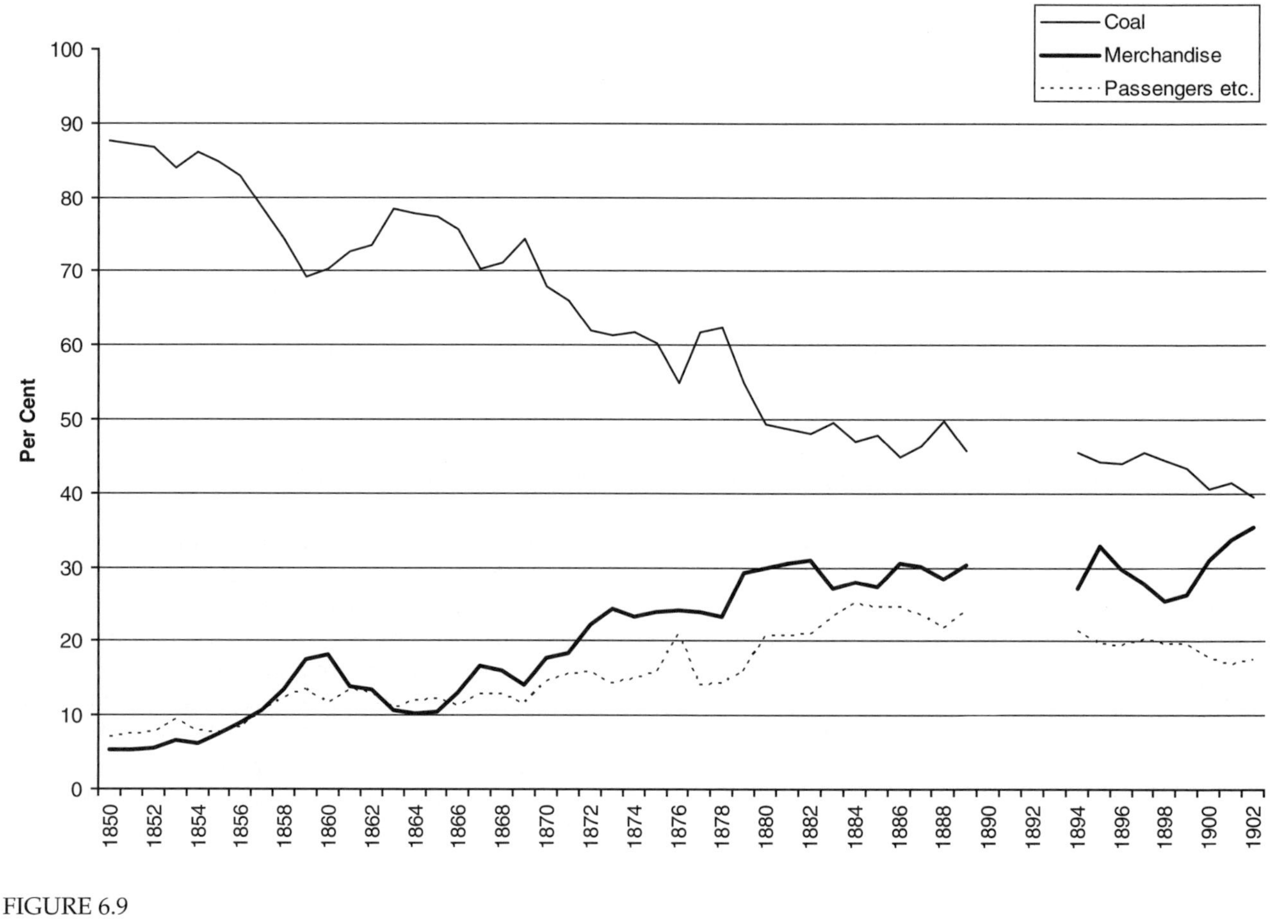

FIGURE 6.9
Philadelphia and Reading Railroad: Percentage of Receipts from Different Traffic Types 1850–1902.

FIGURE 6.10
Lehigh Valley Railroad: Percentage of Receipts from Different Traffic Types 1856–1902.

in 1870, the Reading was able to reduce this proportion to 50 percent in the ensuing decade, through a combination of increases in both passenger and merchandise revenues. The Lackawanna achieved a lesser reduction, to about 60 percent, by the start of the 1880s. This was almost exclusively due to an increase in passenger revenues in the populous parts of suburban New Jersey traversed by the Morris and Essex Division. Coal receipts remained extremely important during the 1880s and were only overtaken by combined merchandise and passenger revenues at the end of the decade. Most of the increase was due to the rise in merchandise receipts from 20 to 30 percent of the total. However, given the large increase in coal shipments during the decade, for merchandise shipments to rise by this proportion implies an extremely large growth in the latter type of traffic. Once again, the Lehigh Valley provides a contrast. Its dependence on coal lasted much longer than the other two railroads, hovering around 70 percent or more from the days of the Civil War until the mid-1880s. Only when the Pennsylvania and New York Railroad figures were included, did the latter road's large merchandise traffic begin to dilute the preponderance of coal shipments on other parts of the Lehigh Valley system; so the level of receipts from coal did not fall below 50 percent of the total until the early 1890s.

Several more detailed points can be made, based on these findings. Figure 6.12 shows the monthly shipments of merchandise by the Reading Railroad over much of the latter half of the 19th century. This striking time series contains a number of features worthy of note, although space only permits a brief mention of these here. The first is the absence of any obvious stimulus to general economic activity in the Philadelphia region during the Civil War years, followed by a surge of growth up to the 1873 panic.

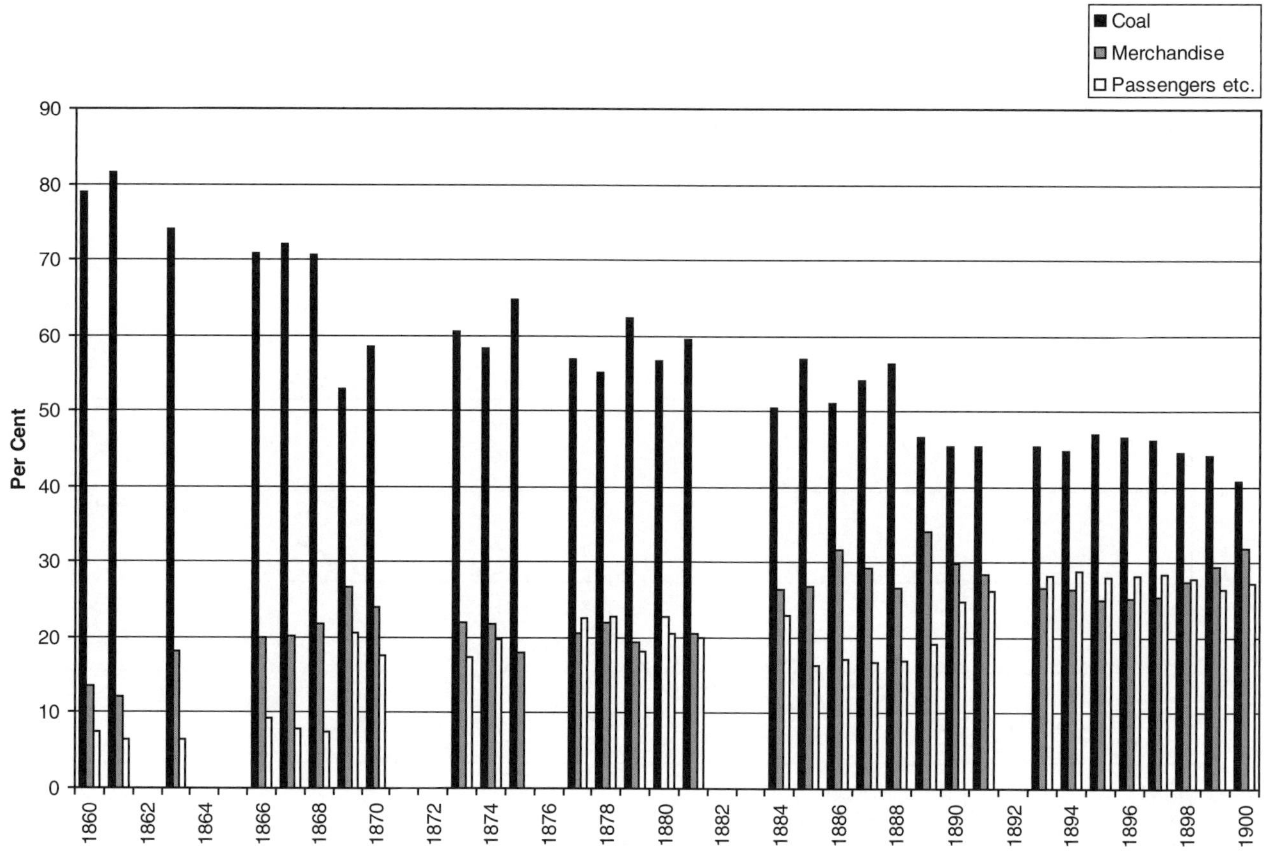

FIGURE 6.11
Delaware, Lackawanna and Western Railroad: Percentage of Receipts from Different Traffic Types 1860–1900.

Secondly, the depression years of the 1870s are characterized by absence of growth more than a dramatic reduction in business activity. This is in marked contrast to the depression of the 1890s when some months experienced activity 50 percent lower than the corresponding periods in earlier years. Thirdly, a marked increase in seasonality of traffic is apparent in later years, no doubt a function of increasing dependence on trade via Buffalo and the Great Lakes. Fourthly, in the summer of 1879, there is a sudden 50 percent rise in merchandise traffic over the space of a few weeks. In some measure this will have been a function of the increase in business confidence, following the return to the Gold Standard at the beginning of the year, but no such sudden increase is apparent in the data for the Lehigh Valley or the Lackawanna (not shown). It is also useful to note that no equivalent surge is apparent in coal shipments for the railroad, so it is not a general upturn in business that is being observed. It would seem therefore that this can only be fully explained as a result of the lease of the Bound Brook and North Penn Railroads in May 1879, which must have produced an immediate and sharp jump in through shipments to New York Harbor. The subsequent expansion of western connections would have fuelled further development of traffic during the rest of the 1880s.

Finally, there is one aspect of the overall temporal pattern of transportation receipts (Figure 6.13) that remains unexplained, namely the sharp increase in receipts for the Reading at the very end of the 19th century, which was not experienced by the other two roads. This increase had two components. The first was a very sharp increase in merchandise traffic, which almost doubled from 8.3 million to 16.4 million net tons between 1897 and 1902. The second was a less dramatic though still marked increase in coal shipments, starting in 1898. The rapid increase in merchandise traffic has scarcely been mentioned in previous studies, as

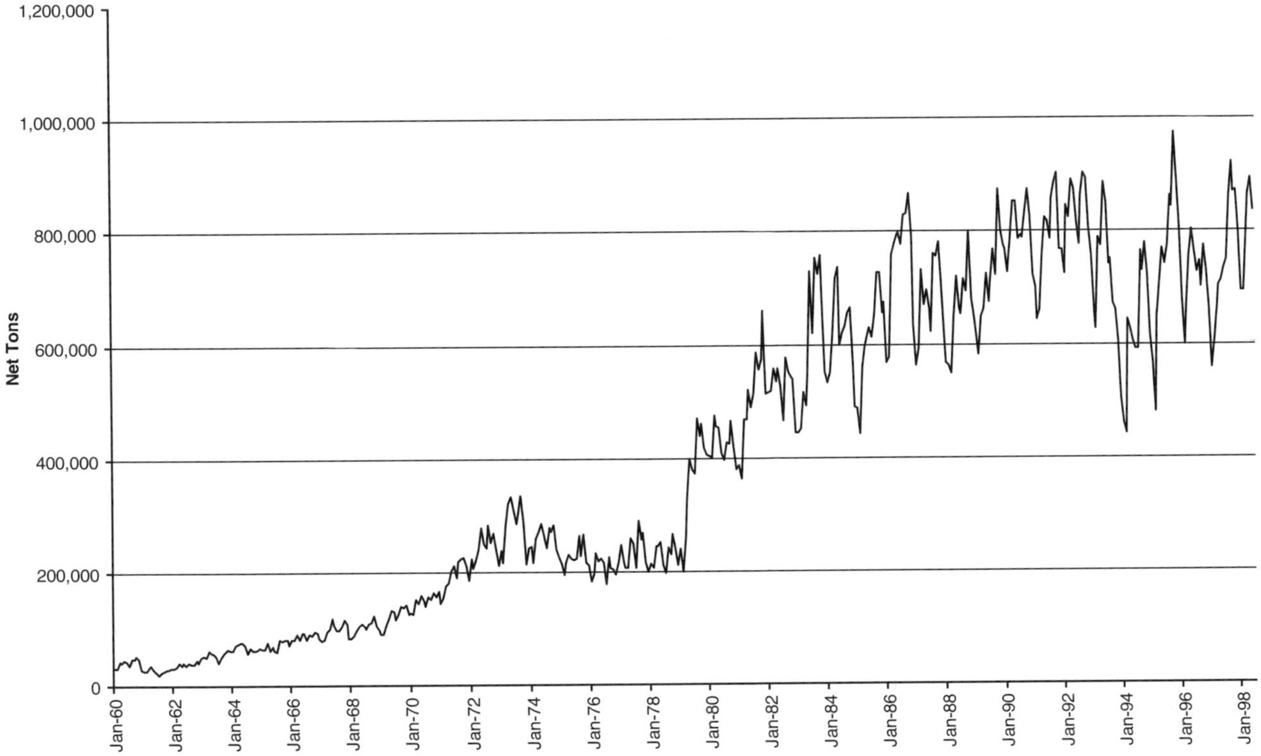

FIGURE 6.12
Philadelphia and Reading Railroad: Monthly Merchandise Shipments Paying Freight 1860–1898.

Bogen's table of shipments 1897–1923 uses decadal intervals for the earlier years, thereby obscuring the pace of the upturn.[81] Neither he nor the Railroad's own published reports offer any clear explanation for the observed change, although the annual reports at the turn of the century make several comments about the importance of export traffic from the Philadelphia wharves and the need to improve shipping facilities there.[82] The increase in coal shipments is significant, because it is almost entirely due to growth in bituminous coal tonnage rather than anthracite. The 1901 anthracite tonnage of 10.5 million was only marginally greater than it had been in 1892, whereas the corresponding bituminous tonnages were 6.09 and 1.79 million tons, a more than three-fold increase.[83] In 1902 the Reading made agreements to carry soft coal eastwards for the Buffalo, Rochester and Pittsburgh and the New York Central Railroad and this traffic grew steadily thereafter, eventually surpassing that of anthracite by the time of World War I.[84] Expansion of this traffic was a final vindication of the Reading's approach to westward extension of its connections, since neither the Lackawanna nor the Lehigh Valley were transporting significant bituminous tonnage over their routes from Buffalo by 1900.

The experience of the Lackawanna, as shown by the respective proportions of coal and merchandise traffic, is instructive, because of the light it sheds on the role of the Trunk Line Pool in the development of the railroad's business. Bogen quotes Ringwatt in stating that the Lackawanna was first recognized as a fully-fledged competitor for trunk-line traffic east and west in 1877. However, it did not make a major bid for trunk-line traffic until after completion of the line to Buffalo in the early 1880s. Even then, according to the financial press of the time, it preferred to compete in the open market for a period, with the aim of securing a better bargaining position to enter the pool at a later date. When Commissioner Fink of the Trunk Line Pool finally persuaded President Sloan to join in 1884, the road was allocated 12.6 percent of the westbound traffic.[85]

It is interesting therefore to relate these events to the actual levels of merchandise earnings recorded

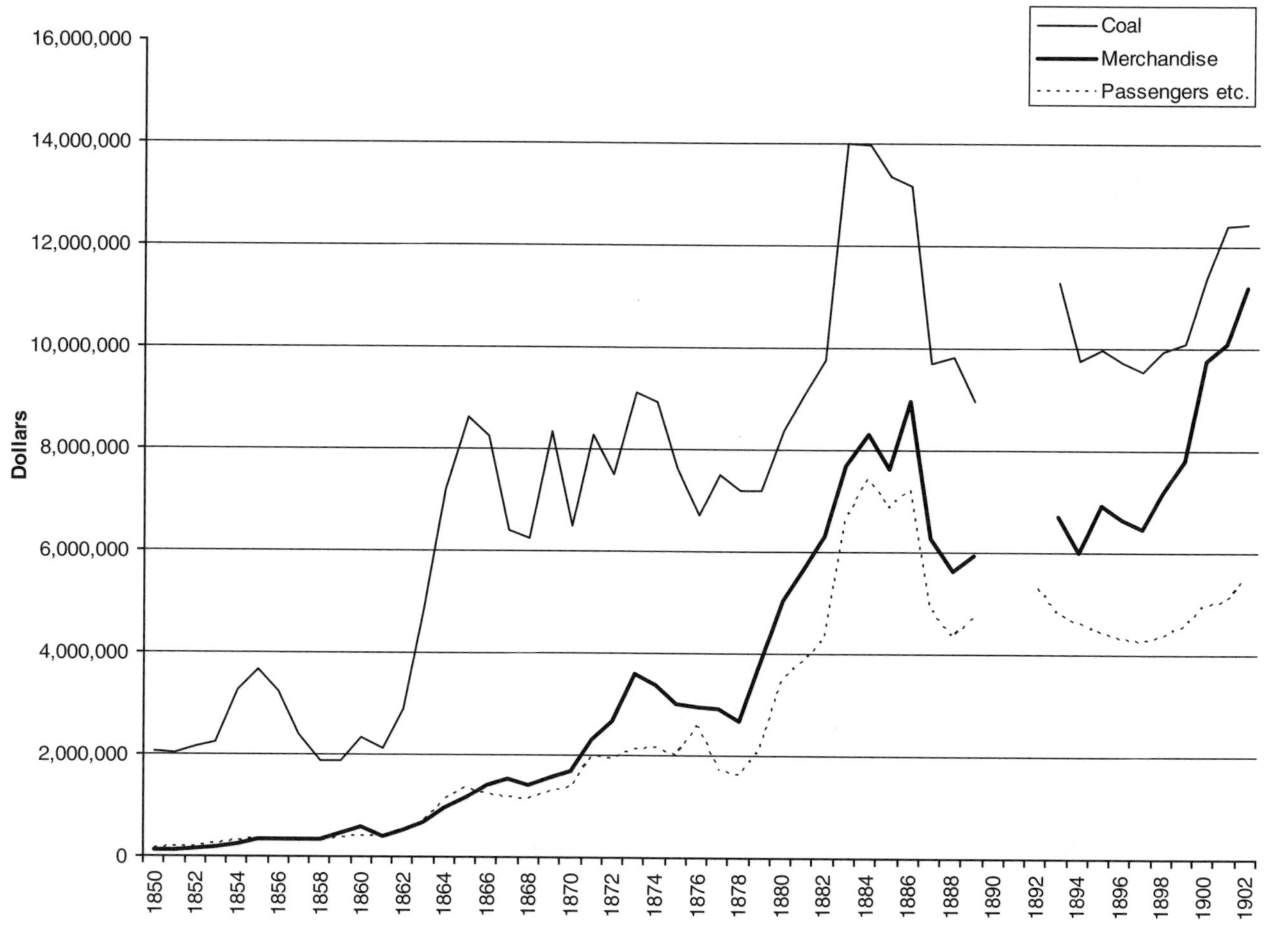

FIGURE 6.13
Philadelphia and Reading Railroad: Receipts from Different Traffic Types 1850–1902.

in the company's books, since coal and merchandise traffic were deliberately merged together by the railroad when reporting its business to the Commonwealth for the annual reports of the State Auditor General. In 1877, the merchandise earnings were modest at $1.7 million but had grown to $2.5 million by 1880. At this time, they were approximately half of the equivalent figure for the Reading. In 1884, the year of entry to the Trunk Line Pool, they had climbed to $4.3 million, reaching $6 million by 1887, to almost equal the Reading at $6.3 million. Thereafter, however, little or no growth was experienced before the turn of the century, and earnings fell back during the depression of the 1890s. Indeed, it should be noted that passenger earnings in the mid-nineties exceeded those from merchandise for an extended period, unlike the Reading or the Lehigh Valley.[86] Thus it is apparent that much of the work of developing a general freight business was undertaken without the help of the Trunk Line Pool. The completion of the through line from Buffalo to Hoboken earlier than its major rivals allowed it to build up coal and merchandise traffic faster than they did, catching up to its most senior competitor, the Reading, in dramatic fashion, as has been seen in Figure 6.6. It should be noted that the apparently large increase in Reading earnings 1883–1886 shown in that figure is, of course, the effect of the short-lived lease of the Central Railroad of New Jersey.

COMPARATIVE ECONOMIC PERFORMANCE

Although earnings growth is a useful measure of increasing volumes of business, the efficiency with which the traffic is carried is equally important to the profitable running of the railroad. The simplest measure of this efficiency is the operating ratio, the ratio of gross expenses to gross earnings. As this ratio approaches unity, the costs of running the road absorb a larger and larger proportion of the earnings derived from the traffic moved. This implies either a lack of efficiency in operation or very high maintenance costs on the railroad infrastructure, or both. In addition, extended periods of operation at a level well below capacity will tend to produce inefficiency, since high fixed costs have to be spread over a smaller amount of earnings.

Annual operating ratios for a number of the major anthracite railroads are plotted in Figures 6.14 and 6.15. In Figure 6.14, earnings and expenses data for lines leased by the Delaware and Hudson are not reported until 1876. Comparison of the two figures suggests that the Lackawanna and the Reading both operated in a cost-effective manner in the early 1860s but there was a sharp reversal with spiralling costs in the mid-1860s. This was brought under control quite quickly by the Reading, but took much longer in the case of the Lackawanna. A detailed examination of the reasons for these changes are beyond the scope of this study, but factors may include disparities in the rate of reduction of salaries and freight tolls in the deflationary period after the Civil War, or rapid expansion in infrastructure and rolling stock, which created 'teething troubles' and management problems. In the mid-1870s, it appears that the Lehigh Valley was most successful in controlling costs, followed by the Lackawanna. During the early 1880s, the latter road experienced increasing costs as operations began on the new line to Buffalo, but a period of very efficient operations followed as traffic volumes grew rapidly during the 1880s.[87] The Reading presents a somewhat similar picture during the same period, aided no doubt by the watchful eyes of the receivers during a large part of the decade. Overall, the Delaware and Hudson lines were slightly less efficiently run than those of the Lackawanna, although there was convergence in the 1890s. While both the Lehigh Valley and the Reading experienced a decline in efficiency in the late 1880s, once again the latter was able to recover its position by the turn of the century.

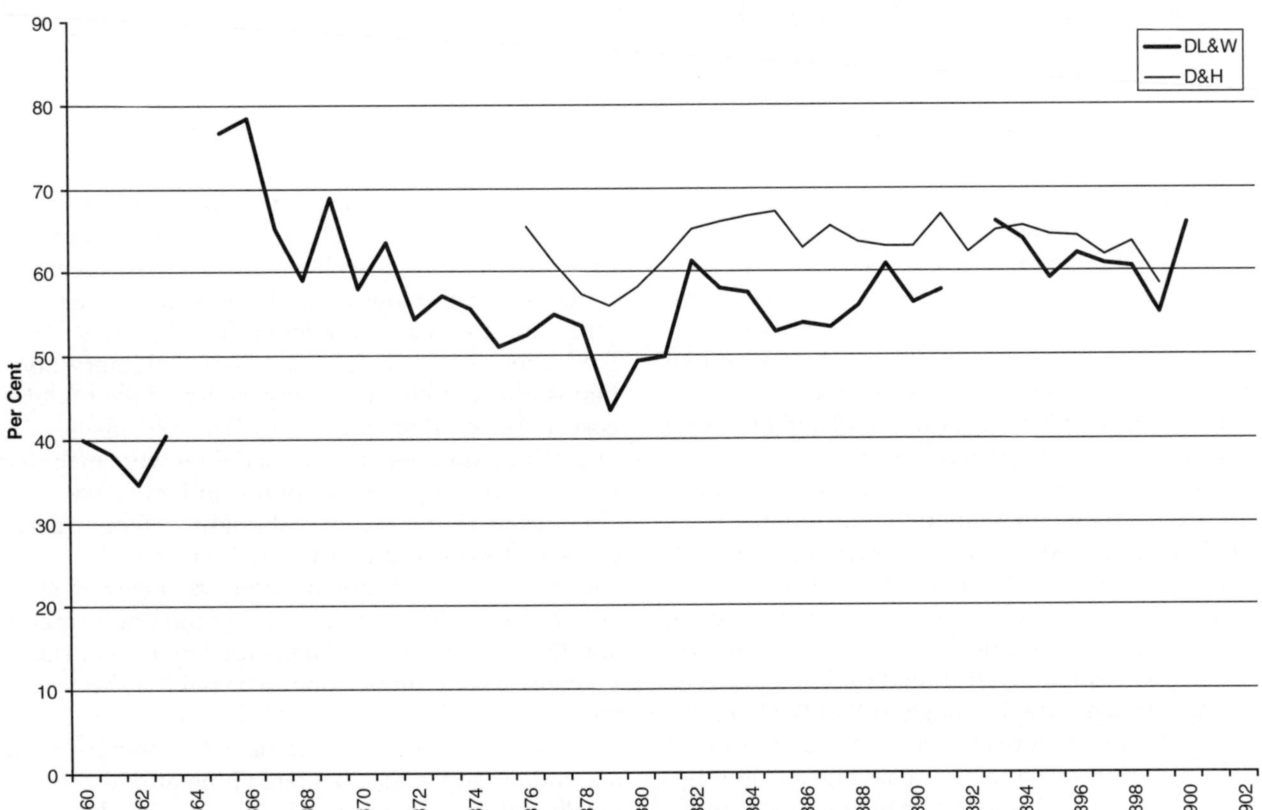

FIGURE 6.14
Major Anthracite Railroads: Operating Ratios 1860–1902.

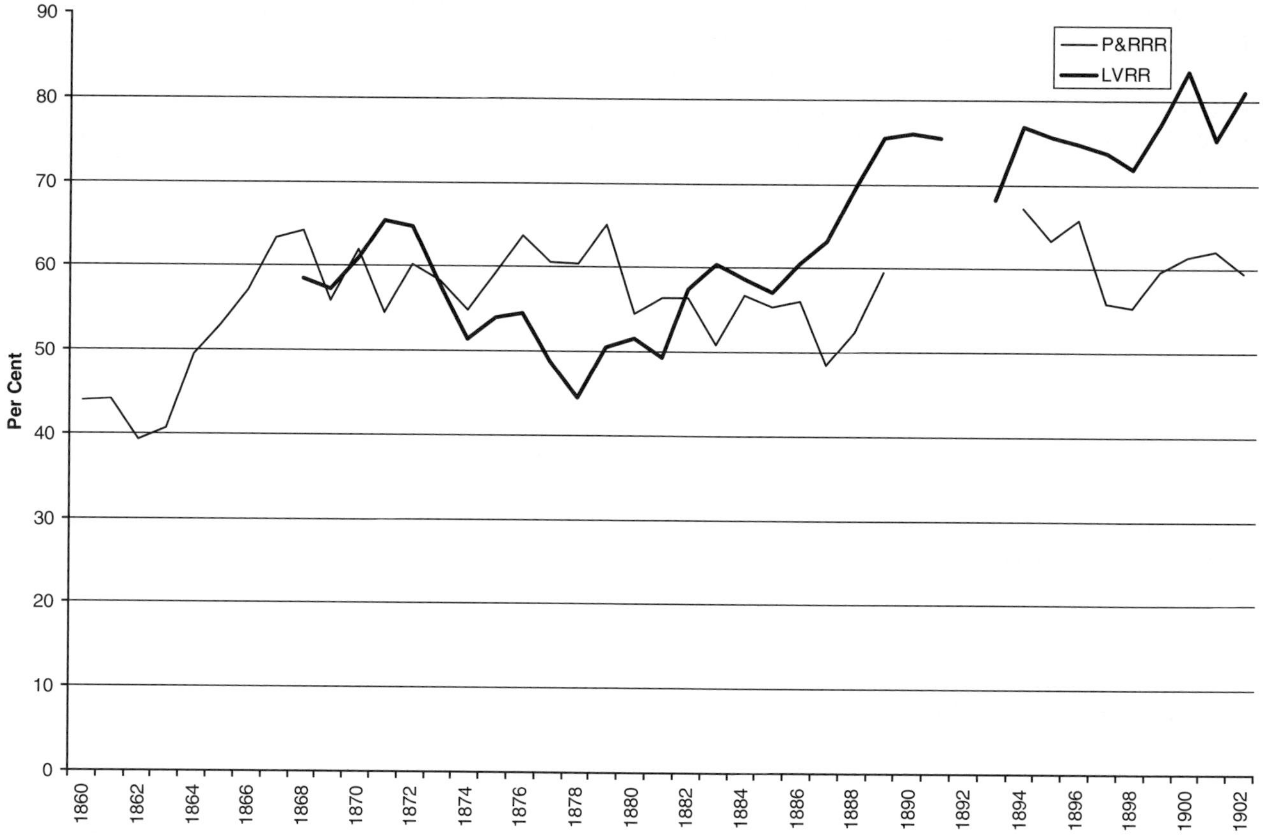

FIGURE 6.15
Major Anthracite Railroads: Operating Ratios 1860–1902.

(Figure 6.15) Only the Lehigh Valley stands out in later years as persistently failing to keep costs in check in relation to revenues. Indeed, the trend in its operating ratio indicates progressive deterioration of economic performance right up to the end of the period of interest. The timing of the onset of serious efficiency decline also suggests that the Pennsylvania and New York Railroad and its lines in New York state were particularly at fault in this regard. One implication of this is that maintenance costs were too high in relation to the volume of traffic being carried. A further interesting aspect of the Lehigh Valley's management can be seen in Figure 6.16. that disaggregates the operating ratio by traffic type. While questions may remain about how consistently costs were attributed to the respective earnings streams, the data suggest that the Lehigh Valley was rather more efficient in handling its coal traffic than its merchandise or passengers. On this basis, it is not surprising that its overall operating ratio worsened sharply in the late 1880s, when the northern lines with their heavy freight traffic were merged in to the main railroad's accounts.

Once transportation earnings and expenses were known, the final elements in determining company profitability were miscellaneous earnings, such as rents from real estate; miscellaneous costs, such as taxes; and the crucial matter of 'fixed charges'. Under the latter heading were included interest on bonds issued by the company itself and the rental costs of leased lines. Since the earnings and operating expenses of the leased lines were included in the main company's accounts under normal circumstances, the rental cost might well have been earned. However, if that was not the case, the shortfall had to be made up from other earnings. Otherwise the lease obligations would not be met and the lease itself could be voided by the courts. Likewise, bond interest had to be paid or the receivers could be called in. Clearly, dividend payments to stockholders of the main company were dependent on the surplus remaining after these legally enforceable financial obligations had been covered. Failure to pay dividends would reduce the financial strain on the company, but could have a dramatically adverse effect on its stock price

FIGURE 6.16
Lehigh Valley Railroad: Annual Operating Ratios by Traffic Type 1868–1885.

(see Chapter 11). This could make the company vulnerable to take-over if its stock was bought at low prices on the open market.

Given the differing ways in which information was reported by different companies and year-to-year changes by individual companies, it is extremely difficult to give an entirely accurate, consistent, and comparable series of data on the financial performance of the different anthracite railroads in the absence of strict and strictly enforced standards of account reporting. In addition to the obvious problems, there are further difficulties arising from several sources. The first of these is inconsistent treatment of depreciation. Companies frequently failed to make full provision for depreciation in lean years, and then attempted to 'catch up' with large provisions in better years. Secondly, especially in the 1870s, the extent of loans and underwriting of losses at subsidiary coal or coal companies was often far from transparent. Thirdly, mysterious and sudden appearances of 'floating debts' could occur. These were borrowings from banks, or occasionally other railroads, which were not secured by a mortgage on the company's property, but on which interest had to be paid. Finally, extraordinary charges for items such as changing track gauge or repairing flood damage could also be made. In the light of these factors, the specific value of any individual year's results should be regarded with circumspection because of the extent to which it was open to manipulation. However, over a period of several years, experience during the 1870s showed that it was impossible to conceal mounting losses, despite numerous subterfuges, accounting infelicities, and myriad attempts at converting floating debt into funded debt (bonds).[88] Hence, over time, the broad patterns of changes in net earnings, surpluses, and fixed charges did begin to show the overall health of the company in question. It is in this light that the earnings figures should be interpreted.

As in earlier parts of the analysis, three of the railroads have data available for an extended continuous period, long enough for the relevant trends to emerge. In the figures (6.17 to 6.19) for each of the railroads, it has proved possible to chart broadly comparable financial indicators, although there are some differences in detail that need to be explained briefly. For the Reading, the net earnings figure is transportation earnings plus real estate and investment income, net of all taxes and deductions other than interest payments (on bonds or floating debt) and rentals of leased lines and canals. Interest and rentals comprise the 'fixed charges'. Unfortunately, inconsistent reporting over the 35-year period means that, prior to 1880, the rentals of coal lateral lines and other leased lines are incorporated into net earnings, whereas canal rentals are not. Thereafter, both are included in the fixed charges. Hence, both net earnings and charges are lower than they should be for consistency up to 1879, but the gap between the two data series is correct across the graph! For the Lackawanna (Figure 6.18), the absence of annual reports means that a simple measure of transportation earnings net of expenses has had to be used, as a full audit of the accounts to incorporate all taxes and extraordinary items is not possible. The fixed charges are interest payments and rentals throughout, so both series are fully comparable, although the earnings figure slightly overstates the funds actually available to apply to fixed charges. Finally, for the Lehigh Valley (Figure 6.19), the net earnings figure is again for transportation earnings, but excluding taxes and depreciation. These, together with losses on coal and canals, are lumped into an overall figure for charges, to which bond interest has been added. It should be noted that depreciation was not charged in the straitened economic circumstances of the late 1870s but substantial charges were made under this heading in the years 1880–3, totalling almost $3 million.[89]

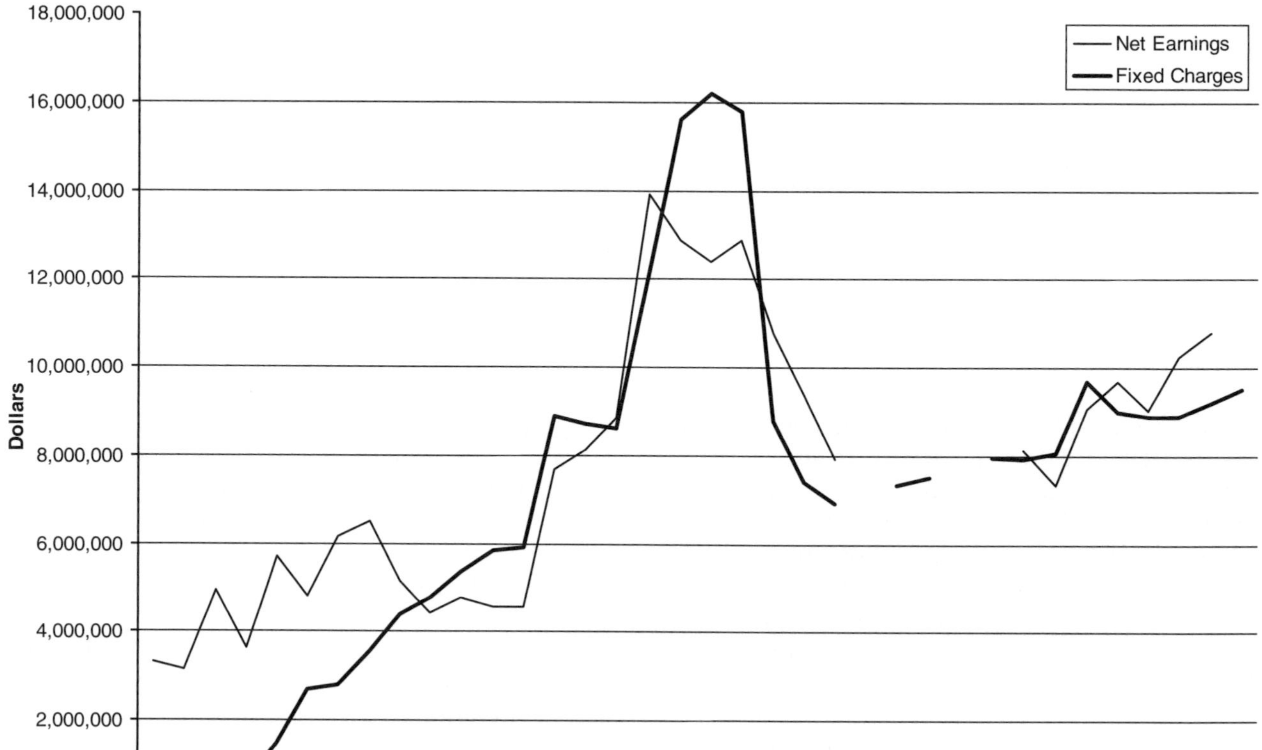

FIGURE 6.17
Philadelphia and Reading Railroad: Indicators of Financial Performance 1867–1902.

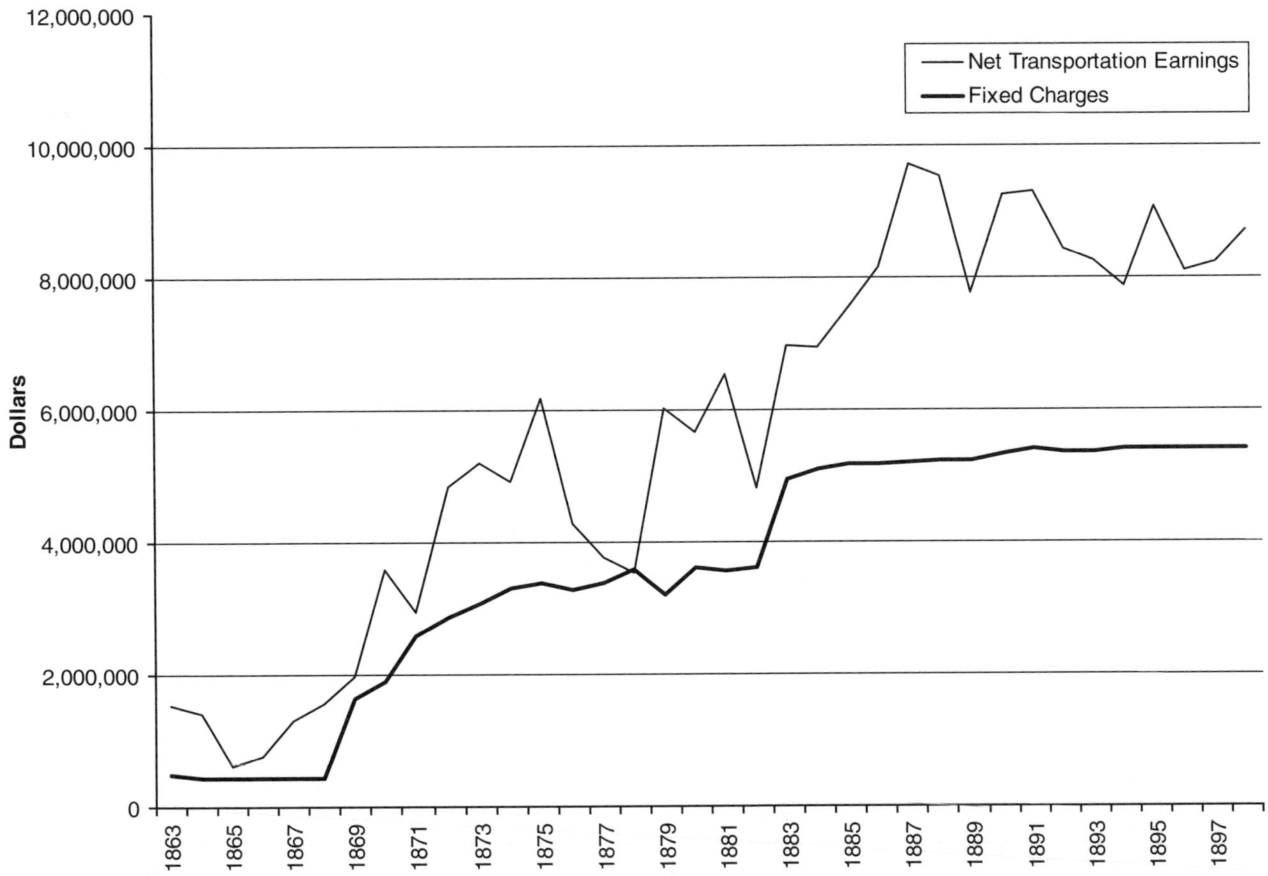

FIGURE 6.18
Delaware, Lackawanna and Western Railroad: Indicators of Financial Performance 1863–1898.

The aim of the graphs is to give an indication of the maximum amount of funds likely to be available either for distribution to shareholders or for retention as a reserve fund. This amount is represented by the gap between the two lines. Provided the earnings line is uppermost, the company has available funds. If not, a deficit has been incurred, with potentially serious consequences. One final point on comparability is that the effect of any coal subsidiary company has been removed from the Lackawanna series. It has partly been removed from the Reading series after 1879 when the accounts were properly separated, but effects continue in relation to bond interest payments. Unfortunately, the Lehigh Valley accounts are rather opaque and it is difficult, except at the end of the period, to distinguish cash transfers to and from the Lehigh Valley Coal Company from the general railroad accounts given in the annual reports. Figure 6.19 should therefore be read in conjunction with the later Figure 13.4, although the data on which the latter figure is based are not drawn from the Railroad's annual reports.

Despite these caveats, the graphs are able to highlight key aspects of the differing financial fortunes of the transportation operations of the companies over a lengthy timespan. The Reading graph shows the railroad emerging from the 1860s in a strong financial position. However, this position reversed with alarming speed when depressed earnings from the mid-1870s coincided with rapidly rising bond interest payments, which were largely the result of large-scale purchases of coal lands (see Chapter 10). A substantial deficit built up in the latter part of the 1870s, financed initially by creation of a large floating debt. The company was finally forced into receivership in 1880 under the weight of its financial obligations. This was despite the tireless efforts of its silver-tongued president, Franklin B. Gowen, to talk up its credit to the dwindling band of

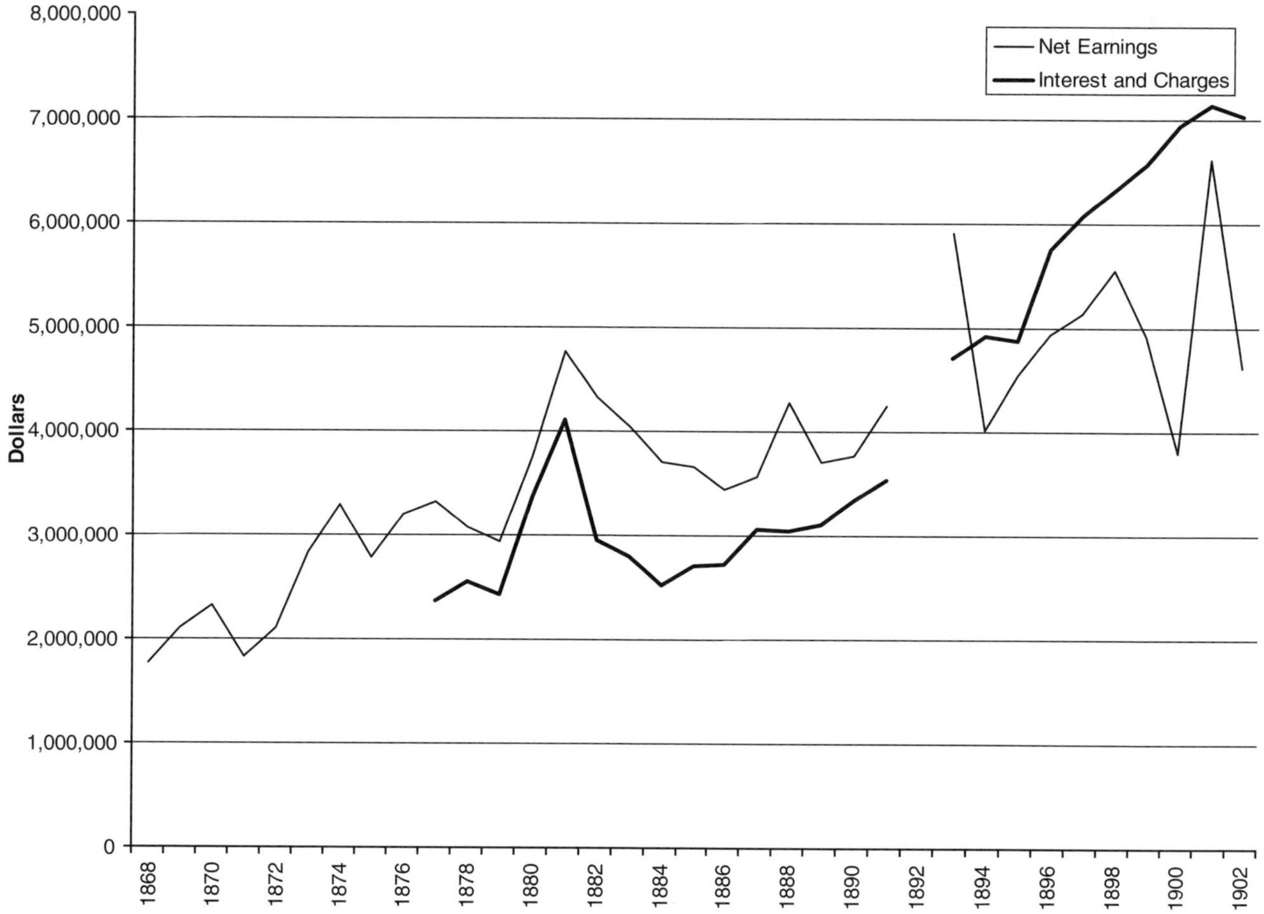

FIGURE 6.19
Lehigh Valley Railroad: Indicators of Financial Performance 1868–1902.

financiers on both sides of the Atlantic who would listen.[90] This first receivership was terminated in 1882, but the adverse business conditions of 1884 brought the company back into receivership for a second time. As the graph shows, the lease of the Central Railroad of New Jersey, consummated in 1883, merely added to the company's financial woes, rather than alleviating them, before it was terminated by the Central's own receivers on January 1, 1887. Extensive financial reorganization and debt re-financing was undertaken during the second receivership, with a view to eliminating the floating debt and reducing fixed charges by $1.5 million a year.[91] Despite the success of the reorganization plan, the company had lost the opportunity of the growth period of the 1880s to build up reserves and, when the Panic of 1893 struck, receivership followed again, accompanied by yet another financial reorganization plan in 1895.[92] Only at the beginning of the 20th century did the railroad really begin to generate a satisfactory cashflow. By 1901, the reconfigured Reading Company had managed to build up a reserve of $3.8 million. However, a chastened and conservative management was quick to observe that this could not be made available for distribution to shareholders without weakening the company.[93]

Until the start of the 1880s the Lehigh Valley Railroad was seemingly in a much more satisfactory financial position than the Reading (Figure 6.19), despite the annual reports being remarkably uninformative about the condition of the company's accounts, in contrast to the state of the bridges along its line! Its coal land acquisitions had been on a much more modest scale and it had not engaged in an orgy of bond issuance on falling markets. Thus, in 1879, its annual bond interest payments were $1.56 million, only one-third of the Reading's total.

Similarly, its small floating debt of less than $350,000 was entirely eradicated by the following year. In contrast, the Reading had accumulated a very substantial floating debt of $7.55 million by 1879, which had further increased to $10.33 million by 1881.[94] Although the Lehigh Valley benefited from the boom of the early 1880s, thereafter a period of relative decline in net earnings set in. Unfortunately, subsequent improvements in earnings were halting at best and very variable on a year-to-year basis. This may have been a result of complacency or lack of strategic foresight on the part of the management, in addition to the burden of expensive maintenance on its Northern Division. The Erie arrangement for shipments to and from Buffalo was profitable enough that the Company initially seems to have failed to appreciate that, without its own through line from Buffalo to New York, it was not a major player in the trunk-line game. By the time the realization dawned in 1887, it was too late to benefit from the expansionary phase of the 1880s. As noted earlier, the through line was completed just as traffic growth stalled, when the depression of the 1890s set in. However, the provision of expensive new capacity without increased traffic made for high costs, as the worsening operating ratio has shown. Further to that, the new developments had resulted in a 37 percent increase in bond issues and a 47 percent increase in capital stock between 1881 and 1893, while net earnings had only risen by 16 percent, and they were to fall again in the aftermath of the 1893 panic.[95] The corollary of this is obvious from the graph. Fixed charges began to outpace earnings growth, leaving no surplus funds to distribute to shareholders in the form of dividends for an extended period. Given these compounding failures in the timing of capital investment and in strategic positioning within the industry, it comes as no surprise that, by 1895, the credit of the company had reached a relatively low point. Yet further substantial investment in new equipment was still required for efficient operation of its expanded network.[96] The Lehigh Valley's failures of judgement and poor timing were matched only by those of President McLeod of the Reading, whose abortive lease of the former road in 1892 cost the latter over $1.4 million in the space of only six months. The credit as well as the share price of the Lehigh Valley were only rehabilitated when J.P. Morgan and Company stepped in to institute a new era of conservative accounting practice and a prolonged program of investment in the railroad's equipment and infrastructure.[97]

A quite different scenario from either of the other two is presented by the Lackawanna Railroad in Figure 6.18. In terms of fixed charges, two periods of rapid increase are apparent. The first of these, at the end of the 1860s and the beginning of the 1870s, was associated with the lease of the Morris and Essex Railroad and the expansion of control in New York state. The second was in the early 1880s, associated with the lease of the New York, Lackawanna and Western Railroad. It is interesting to note that in each case, an increase in financial obligations was followed rapidly by substantial increases in earnings, so the latter could be absorbed without damaging the company's financial position. It is apparent therefore, that the timing of new initiatives under the careful hand of President Sloan was excellent, and maximum benefit was derived from the expansionary phases of the early 1870s and early 1880s. The benefit also appears to have been cumulative. The dark days of 1877, when business confidence bottomed out and strikes plagued the Northern Field,[98] caused funds potentially available to shareholders (the gap between the two lines on the graph) to evaporate completely, and dividends had to be suspended. However, despite the adverse impact on the share price (see Chapter 11), reserves accumulated in the earlier years prevented the company spiralling into debt, unlike the Reading, and dividends were resumed in 1880, increasing thereafter.[99] By the time of the next severe contraction, following the 1893 panic, a solid base of net earnings well in excess of those required to meet essentially static fixed charges, had been created over a period of years. In addition to this, efficient infrastructure and absence of strikes minimized the adverse impact of the business contraction, as compared with the late 1870s. It is worth observing that the composition of the fixed charges was also very different between the Lackawanna and the Reading. The former had the highest level of rentals of leased lines, which rose to about $5 million a year, to the consternation of its minority stakeholders, as we have already seen. Bond interest, however, was at a very modest level, falling below $250,000 per annum in the early 1890s.[100] In contrast, the Reading had bond and other interest payments that rose as high as $6.6 million in 1885, before financial reorganization plans began to affect a measure of reduction in this item. Rentals of leased lines and canals were about two-thirds of those of the Lackawanna (excluding the short-term effects of the Jersey Central lease), before dwindling traffic volumes on the Susquehanna and Schuylkill canals forced a progressive reduction in lease rentals for these companies to very low levels.[101] The burden of canal rentals was made much worse, because in general, their net earnings were far from sufficient to cover their costs, unlike leased railroad lines.

CONCLUDING COMMENTS

This analysis of overall company financial performance is deliberately incomplete, because the aim has been to distinguish as far as possible the operations of the transportation departments from the coal departments of the railroads. Without the latter, it is premature to discuss annual surpluses and long-term strategies for re-investment of retained earnings. This topic will be taken up in detail in Chapter 13. Nevertheless, it is already very striking how different were the management responses to fluctuating economic circumstances by companies transporting coal and similar types of commodities across similar terrain, often from neighboring mines, to the same markets east and west. Injudicious and hasty action, such as that of the Reading in overextending its coal estate too quickly, was speedily punished by the 1870s depression, and the adverse financial repercussions were visited on successive Boards of Managers for a generation. Laggardly and complacent management, poor investment timing, and the squandering of a previously sound financial position by the Lehigh Valley Railroad likewise received their just reward in a progressive deterioration of the company's infrastructure, credit, and standing in the market. Only the Lackawanna, by careful strategic acquisitions, consummate timing, and conservative, if secretive accounting practices, was able to present a picture of financial health that grew more robust with every passing year. Humbled by financial problems in the aftermath of the 1857 panic, it met each subsequent period of dislocation and major depression from a position of growing strength. No better testament to the skill and judgement of President Sloan and his managers can be given.

Chapter 7

WAXING AND WANING MARKETS I: SECTORAL SHIFTS IN THE USE OF ANTHRACITE

WAXING AND WANING MARKETS: INTRODUCTION

At the beginning of the period, as discussed in Chapter 4, when the bituminous coal fields of the interior were not well developed and transport of their product to eastern markets was expensive, anthracite held sway both as a domestic and an institutional heating fuel and for a wide variety of industrial uses. Prominent among these were the smelting of anthracite iron and steam raising on the railroads and in steamboats. Anthracite was also widely used in stationary engines to drive powered industrial machinery, and for specialized types of manufacturing where a high level of heat output was required, such as in the production of glass or the evaporation of brine to make salt.

The next three chapters examine a number of market changes experienced by the industry in the latter part of the 19th century. Several different types of changes can be identified. These include sectoral shifts, as anthracite gained or lost ground in particular industries; geographical shifts, associated with the rise of new urban markets and widening of the distribution region; and price substitution effects, as different groups of consumers sought cheaper fuels to meet their heating or steam raising requirements. A convenient classification such as this is necessarily somewhat artificial, since relative price movements affected industrial decision-making or the ease with which different types of fuel could be introduced into new geographical markets. Similarly, a sectoral shift might be accompanied by changes in the location of manufacturing plants, rendering them more or less accessible for the receipt of anthracite coal. Nevertheless, the classification is useful for the purpose of exposition, provided these caveats are borne in mind.

In terms of the adjustment processes that this changing business environment engendered within the anthracite industry, there are a number of issues to be explored. At the broadest level, there is the question of whether these market shifts increased or decreased the cyclical sensitivity of the industry. For example, it might be expected *a priori* that a progressive reduction in sales to the iron industry, coupled with increasing sales to domestic consumers, would dampen business-cycle effects on coal output, since demand from the latter group of consumers was affected more by the march of the seasons than wider economic conditions. A second question is whether all parts of the Coal Regions experienced these market changes to the same degree. Anthracite was not a uniform commodity, and individual mines had differential access to certain coastal or interior markets, depending on the railroads that served them. Hence it would be reasonable to expect differential responses at the sub-regional level, and even on a firm-by-firm basis. It is undeniable, however, that one of the key overall impacts of market shifts was to reduce the sale of larger sizes of coal and increase that of progressively smaller sizes. This necessitated an increased amount of coal breaking, which resulted in growing volumes of coal waste and a reduction in profitability.[1] Again, however, it is important to investigate whether these changes were gradual over long periods of time, or whether they were concentrated around periods of economic dislocation, such as the 1873 panic. Equally, were all major producers affected to the same degree at any given time or were there variations, between parts of the Coal Regions, that were large enough to have an impact on the financial viability of firms in the medium term? Campbell's preliminary survey of changing uses of anthracite in the late 19th century provides some useful clues in examining these questions,[2] but a systematic analysis of both printed and archival data sources is needed for a fuller picture to emerge. This analysis forms the subject matter of the ensuing chapters.

SECTORAL SHIFTS IN THE USE OF ANTHRACITE

The major market sectors for anthracite were already well established by the onset of the Civil War, but, over time, the relative importance of these sectors began to change. It is appropriate to examine the pattern of demand changes across these sectors individually, in order to assess the impact on market volatility and the preferred mix of coal sizes required by industrial or domestic consumers. The different sectors will be examined under the following headings:

- the anthracite iron industry;
- the transportation sector;
- other industrial uses;
- public utilities and institutional heating; and
- domestic consumption.

Owing to the complexity and the widening geographical extent of the anthracite coal trade, the straightforward collection of statistics about these different sectors became impossible, especially in the later years. Information from numerous sources must therefore be employed to build up a coherent picture of the different market sectors. This will necessarily introduce issues of comparability in reporting between sources, and even within sources over extended periods of time, so these need to be borne in mind in relation to the ensuing discussion.

THE ANTHRACITE IRON INDUSTRY

To understand the evolution of the anthracite iron industry in the latter part of the 19th century, it is necessary to examine it in both a national and regional context. Further to this, the changing fortunes of the industry can be divided into two phases. The first of these is a period of growth until the early 1870s. The second is a period of stasis, then decline in absolute terms, after 1880, following the recovery from the depression of the mid 1870s. At the national level, these trends can be seen clearly in Figures 7.1 and 7.2. These figures show the comparative output of pig iron classified by fuel type. It should be noted that the scale on the second graph is a factor of 10 larger than on the first graph, so the changes of smaller absolute magnitude in the early period can still be identified. It is apparent from Figure 7.1 that the localized anthracite iron industry more than held its own against the raw bituminous/coke iron industry of the rest of the nation until 1874. Thereafter, anthracite iron slumped more severely during the depression and was overtaken by the bituminous sector, which was increasingly fired by Connellsville coke.[3] During the more favorable business conditions of the 1880s, anthracite iron output oscillated around the 2 million ton mark, although its overall market share declined steadily after the 1885 depression. The graph also suggests that the surge in coke iron output in the late 1880s was quickly followed by sharp retrenchment in the anthracite sector. This began in 1891 and continued through the entirety of the depression of the Nineties.[4] The graphs are based on the standard classification of fuel types maintained in the statistical reports of the American Iron and Steel Association,[5] but, in the later years, they cannot be taken purely at face value, as anthracite was increasingly mixed with coke in eastern furnaces. Unfortunately, no detailed statistics of the varying proportions of anthracite and coke used have come down to us, but several pointers to the nature of these changes can be found. According to the Coal Trade Journal, anthracite furnaces first began using coke during the 1875 anthracite strike and continued using it in varying proportions thereafter.[6] By 1890, although the amount of 'anthracite' iron made was nearly 2.45 million tons, the Journal reported that only 279,184 tons were made with anthracite alone, the rest with anthracite mixed with coke.[7] Although this period of decline has not been very closely studied by commentators on the iron industry, several factors seem to be involved. As early as 1877, comparisons between the economy of coke and anthracite in blast furnaces were being made in the coal trade press. One particularly interesting discussion was a critique of comparisons between the two fuels, made originally by John Fulton, General Mining Engineer of the Cambria Iron Company. He found that the Isabella Furnace in Pittsburgh used 1.07 tons of coke to make a ton of pig iron, while the Glendon Furnace in the Lehigh Valley required 1.29 tons of anthracite to do the same. However, it was noted that the Isabella was a somewhat larger furnace and it used Lake Superior ores, which were richer than the magnetic ores from New Jersey. Also, chemical analysis of anthracite indicated that on average it contained 90 percent fixed carbon, whereas the

FIGURE 7.1
United States Pig Iron Production by Fuel Type 1860–1879.

figure for Connellsville coke was only 87.5 percent. These points suggested that coke would not, in fact, produce any economies of operation over the use of anthracite, in the Glendon furnace. Significantly, however, the Isabella was producing over 700 tons per week, compared to 289 tons at the Glendon Furnace. This large output disparity was out of proportion to either furnace size or quality differentials. To explain this a further factor was invoked, namely the 'greater rapidity of driving'.[8]

'Hard driving' as it came to be known,[9] required a large blast furnace and progressive increases in the heat and strength of the blast, by the use of improved stoves and more and better blowing engines. Unfortunately, anthracite as a fuel did not facilitate hard driving, because of a limit on the achievable rate of combustion. This derived from the fact that the pieces of anthracite in the furnace only burned on the surface. Coke, on the other hand, because of it porous, honeycomb structure, has a much higher surface area, which permitted a greater degree of potential acceleration of combustion, faster heating of the furnace charge and greater output. This characteristic of coke led it to find favor with eastern ironmasters, despite its high cost to transport the works, according to 1890 figures provided by Temin. However, furnaces originally built for anthracite were not as well adapted for hard driving as the western furnaces built specifically for coke. Hence, only partial economic benefits could be derived from changing to coke, or a coke/anthracite mixture, unless the furnace was completely rebuilt on the western pattern.[10]

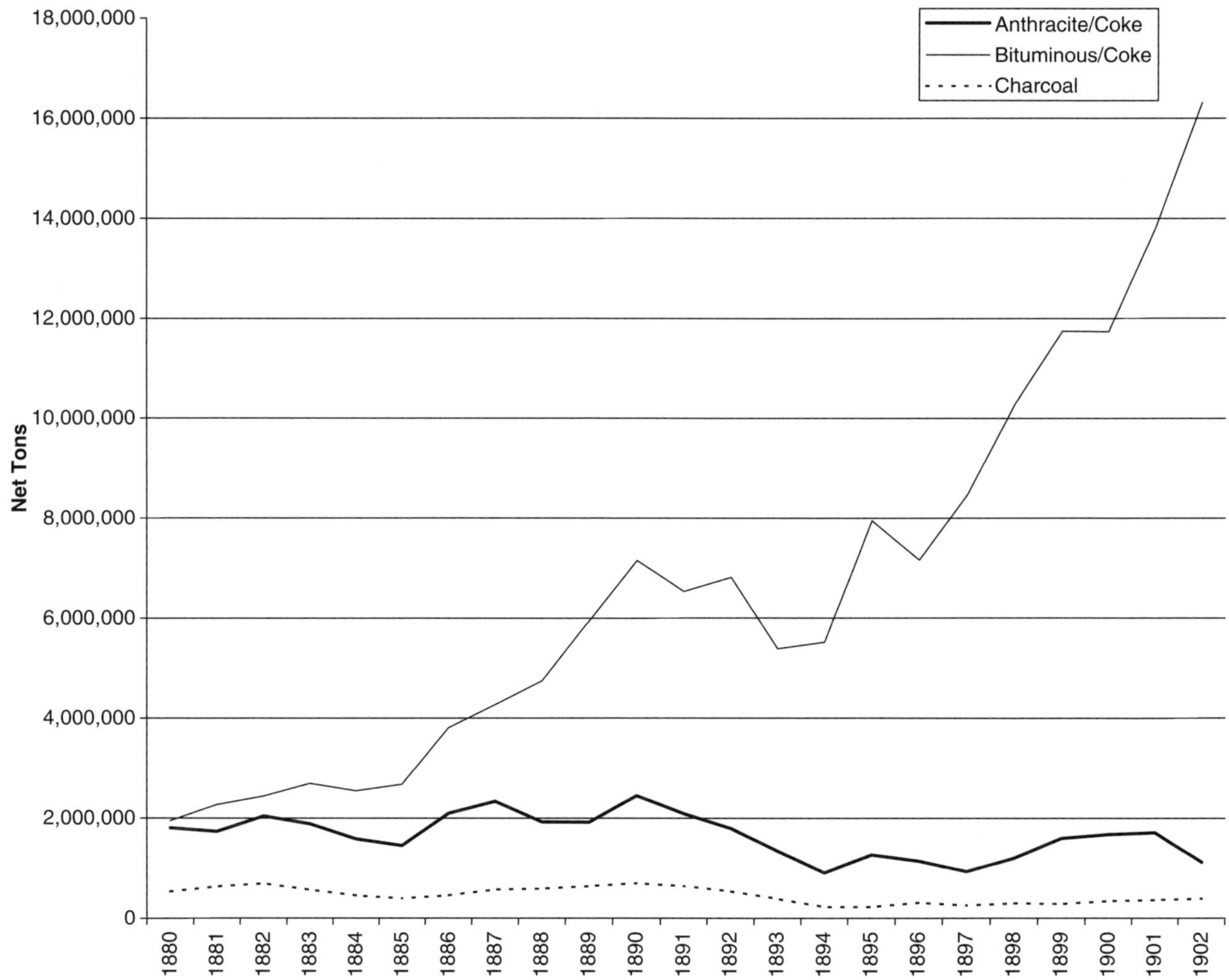

FIGURE 7.2
United States Pig Iron Production by Fuel Type 1880–1902.

Temin's argument, which is summarized in the preceding paragraph, is based on rather limited evidence and it collapses what may well have been a complex process of technological adaptation and economic adjustment, extending over a number of years from 1875 to 1894, into an apparently straightforward process. Unfortunately, a detailed examination of this issue is beyond the scope of the present study. Nevertheless, Figure 7.3 gives an indication that the anthracite iron producers did not fade away rapidly in the face of competition from Andrew Carnegie and the other coke iron producers of the Pittsburgh region. The anthracite and anthracite/coke iron industry produced a substantial majority of Pennsylvania's mineral fuel irons from the 1850s up to 1879. As late as 1887, anthracite was still ahead in the production tables, with an output of almost 1.93 million tons, as compared to 1.74 million from bituminous/coke furnaces. The key to the eventual demise of the eastern producers, however, lay in the fact that the anthracite/coke output required 147 furnaces, while the bituminous/coke output was made by only 75 furnaces.[11] Since it enabled a larger yield from generally newer, bigger furnaces, the benefit of coke lay in reducing cost by spreading expenses over a larger tonnage. The Pittsburgh Commercial Gazette emphasized the point even more strongly:

> 'With coke as a fuel, in furnaces of the same size fifty percent more iron can be produced than by the Anthracite process'

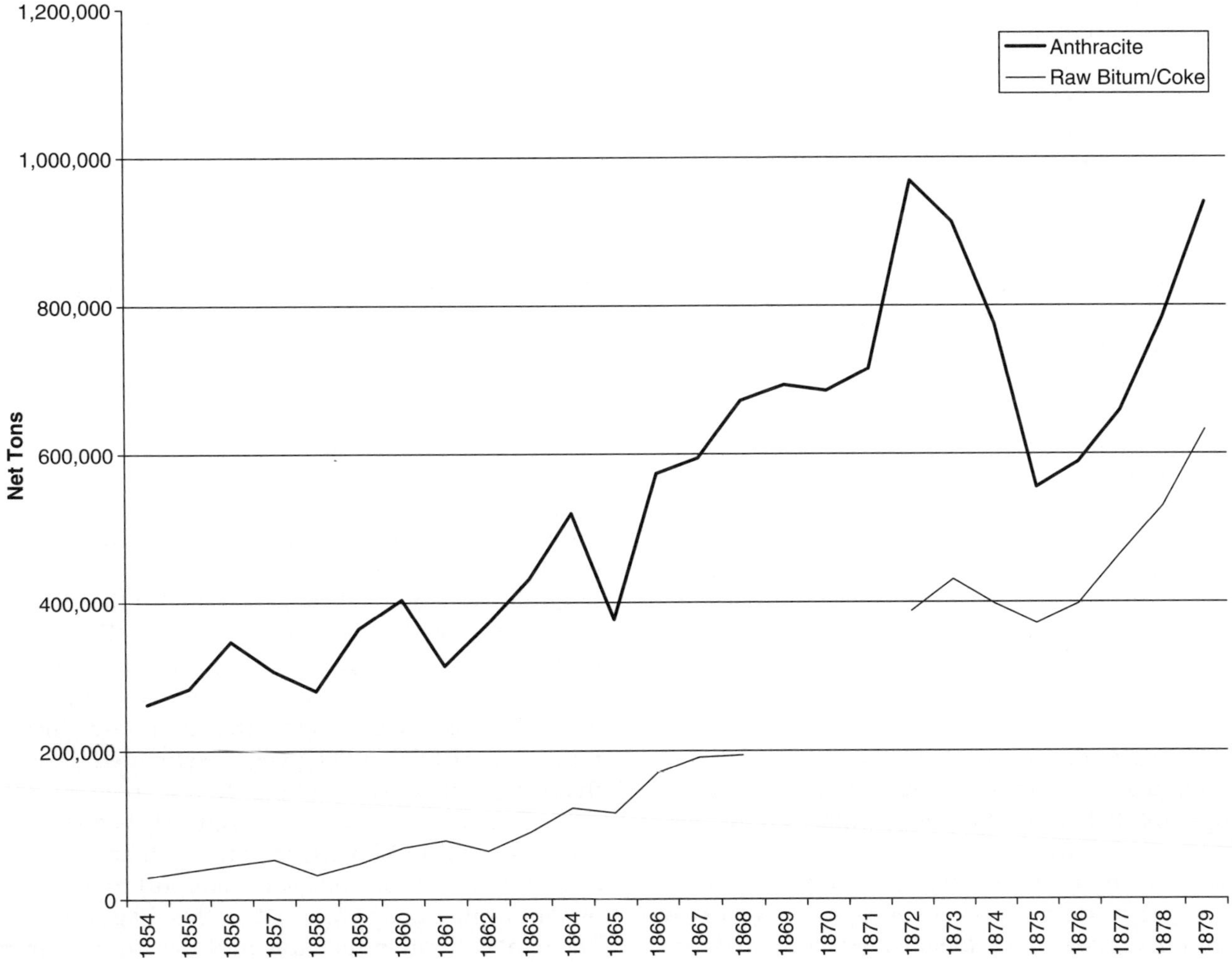

FIGURE 7.3
Pennsylvania Pig Iron Production by Fuel Type 1854–1879.

The quote continued with boosterism of a most prophetic kind, based on the economic benefits of coke as a fuel:

> 'When the capacity of our furnaces and mills has been steadily increasing in the face of the unprecedented depression of the past five years, and when our manufacturers have stood up so bravely and maintained their trade against such adverse circumstances, all talk about Pittsburgh losing its prestige as the great centre of the iron trade is idle'.[12]

In terms of the temporal dynamics of market shares in the iron industry, Iron Age placed the onset of competition for anthracite from coke as early as 1873, when falling iron prices necessitated strict economy in production. Since coke was a low cost fuel at source and the large outputs of the Isabella and Lucy furnaces in 1874 became widely known, coke began to attract attention in the East. After the 1875 strike, some use continued in anthracite furnaces and it increased again in 1879, but the proportion used was very variable. In a few cases it was as much as ⅔, but in general it only ranged from one-tenth to a quarter. Significant benefits were obviously obtained, however, as there were reports of much higher furnace yields than had previously been obtained using anthracite alone.[13]

In roads made by coke into anthracite furnaces were accompanied by progressive improvements in technology, which had reduced the amount of anthracite required to produce a ton of iron from about 2 tons up to the time of the Civil War[14] to 1.29 tons at the bigger furnaces, such as Glendon, by 1876. Thereafter, further minor improvements continued to be made.[15] Even in prosperous times, these two factors in combination would have slowed the increase in anthracite utilization in the iron industry. When coupled with the sharp decline in the iron business in the late 1870s, however, the adverse effect on coal sales was very marked. One contemporary commentator, writing in 1877, put the reduction in demand since 1875 at 1.5 million tons.[16] This is substantially more than can be accounted for by the blast furnace sector alone, but it should be remembered that anthracite was also used in rolling mills and foundries. It is difficult to find information on foundry use, but the 1880 census reports over 500,000 tons of anthracite being used by working mills in 1879/1880, as compared to 3.9 million tons of bituminous coal and a negligible amount of coke.[17] The same source gives figures for blast furnace consumption by fuel type and this can be compared with Birkinbine's figures for 1884, as shown in Table 7.1.

These figures are of particular value, as they distinguish coke from raw bituminous coal. It is apparent that anthracite was still more important than coke as a furnace fuel in 1880, but was rapidly overtaken in the ensuing years. Proportionately, however, its decline was not as rapid as that of raw bituminous coal. Additional temporal context for these figures can also be provided by the data shown in Figure 7.4 for anthracite furnaces in blast 1874–1900. The data are indicative only, because of their incompleteness, and the fact that extant information is not always available for the same month in each year. Nevertheless, the combination of cycles and an overall downward trend after 1882–1883 can clearly be seen. This culminated in near collapse of much of the industry in the 1890s, when only a few of the larger and more efficient operations were able to remain in blast. Although a good proportion of firms were still in active production in 1888, a report in Iron Age early in 1889 commented on the signs of decay in the industry in eastern Pennsylvania, which it attributed to high freight rates on the railroads, rather than competition from more efficient coke-fired furnaces.[19]

Changing furnace technology, more efficient fuel utilization, and increasing admixture of coke make it somewhat difficult to determine the proportion of total anthracite shipments destined for the iron industry over extended periods of time. Nevertheless, an attempt is made in Table 7.2 to derive indicative estimates, using reasonable assumptions about fuel utilization.

The pig-iron production and anthracite shipment figures are drawn from the standard trade series, so attention must be focused on the 'multiplier' column. For 1860, the standard multiplier mentioned above is used, while in 1884 the multiplier is calculated from the actual anthracite use figure, which is obtained from Table 7.1. The multipliers for 1870 and 1880 are chosen to reflect improving technology by the former year, and increased coke utilization by the latter year. The multiplier of 1.5 in 1880 gives an estimated usage that is very close to the census figure given in Table 7.1, although the census figure is based on the 12 months to May 31, 1880 rather than the calendar year. This lends some support to the use of a multiplier of approximately this magnitude. The final column in the table indicates that the iron trade, as opposed to steel manufacture, broadly maintained its importance as a market for anthracite throughout the first half of the period of interest, and only went into decline during the 1880s. Even then, it is possible that Birkinbine's figure for anthracite use by furnaces in 1884 was rather low, since Saward's Coal Trade Annual asserted that 11.1 percent of anthracite shipped in 1885 was used in iron and steel manufacture, and that 70 percent of the increase in overall shipments during 1886 went to this sector. Alternatively, it may be that the proportion of anthracite used in rolling mills and steel manufacture rose sharply

TABLE 7.1
Blast Furnace Fuel Use 1879–1884.[18]

		Anthracite	Bituminous	Coke	
1879–1880	Blast Furnaces only	2,615,182	1,051,753	2,128,255	(units given as 'tons')
1879–1880	Total in iron and steel manufacture excluding foundries	3,322,498	5,659,055	2,277,555	(units given as 'tons')
1884	Blast furnaces only	1,534,688	326,986	3,697,275	(gross tons)

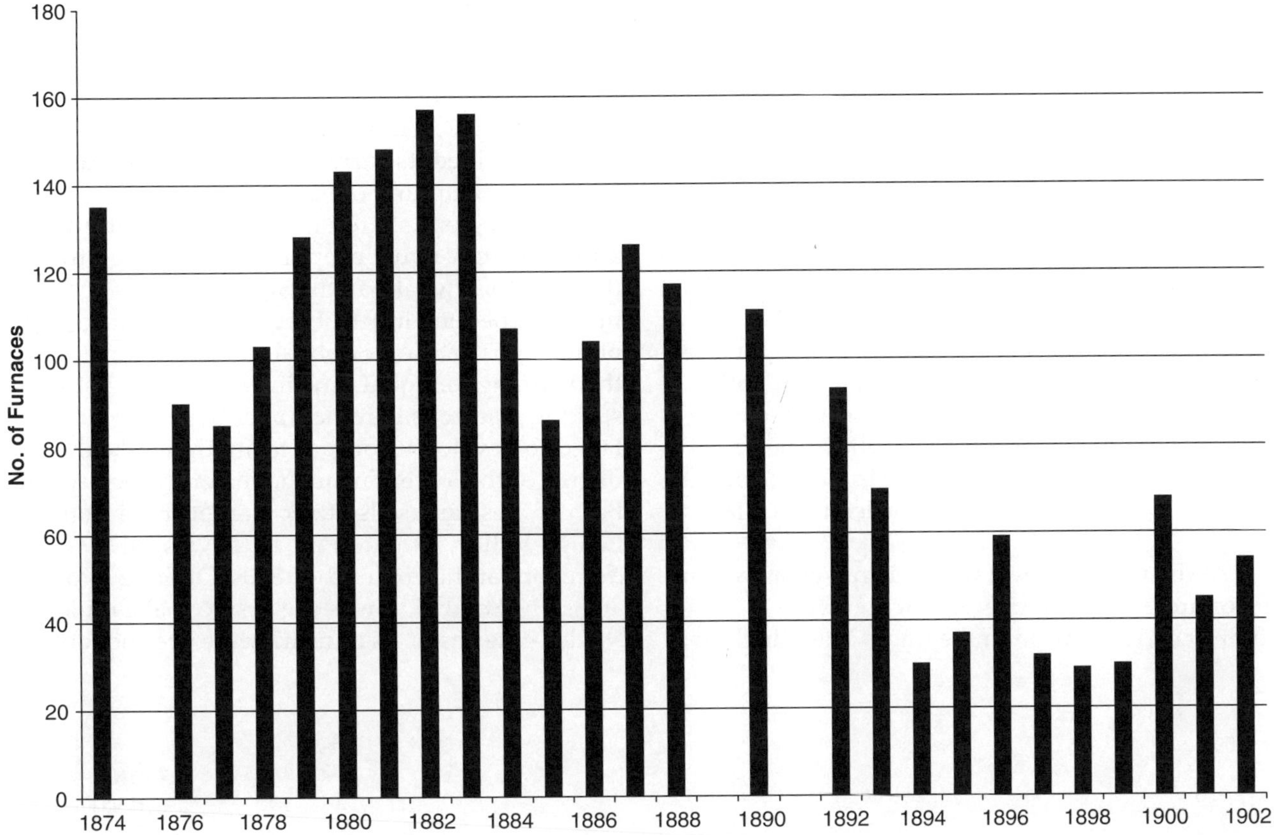

FIGURE 7.4
Anthracite Furnaces in Blast 1874–1902.

during the 1880s. This proportion was only equivalent to 27 percent of the blast furnace utilization in 1879–1880, according to calculations from the figures in Table 7.1. Support for this contention comes from separate figures in the weekly Coal Trade Journal, which reports the use of anthracite in iron manufacture alone, at about 5 percent of total shipment in 1885.[20] This is the same percentage as found for 1884 in Table 7.2, based on actual, rather than estimated figures.

In addition to the temporal dynamics of demand in the anthracite pig-iron industry, it is necessary to consider the geographical dynamics, both at the state level and in more localized regional areas. Beyond the borders of Pennsylvania, anthracite iron production was mainly concentrated in New York state and New Jersey, although there were limited developments in Maryland and New England, in locations where the coal could readily be brought by rail, canal, or coastal vessel. According to Iron Age, prior to the

TABLE 7.2
Estimates of Anthracite Use in Blast Furnaces 1860–1884.

	Production of anthracite pig iron (tons)	Multiplier	Estimated (E) or actual (A) anthracite use in furnaces	Total anthracite shipments	Estimated % of total shipments used in blast furnaces
1860	524,563	2	1,049,126(E)	8,513,123	12.3%
1870	930,000	1.75	1,627,500(E)	16,182,191	10.06%
1880	1,807,651	1.5	2,711,477(E)	23,437,242	11.57%
1884	1,586,453	0.97	1,543,688(A)	30,718,293	5.03%

widespread adoption of coke in the mid 1870s, anthracite was also being shipped to furnaces in Michigan and Wisconsin. Examination of Figure 7.5 shows that, although New York was second in importance to Pennsylvania, its production never exceeded about 30 percent of the output of the latter. Production in New York grew very rapidly in the late 1860s, but did not recover as quickly as Pennsylvania from the depression of the 1870s. New Jersey output was worst affected by this depression, falling by three-quarters from its 1872 peak to a mere 25,000 tons in 1876. Although figures for later years are scarce, data for 1887 indicate that both Pennsylvania and New Jersey had been able to increase their output substantially above the peaks of the early 1870s, but the production in New York state, totalling 217,585 tons, barely exceeded that of 1869.[21]

At the regional level within Pennsylvania, four major areas of anthracite production were commonly recognized in the trade. These included the Lehigh and Schuylkill Valleys and the upper and lower portions of the Susquehanna River basin. The upper portion centred round Danville, while the lower included the area around Harrisburg. From Table 7.3, it is apparent that the Lehigh Valley had maintained its early lead in anthracite iron manufacture up to the time of the Civil War, while the other regions shared the remaining output in approximately equal proportions. Although the data are somewhat patchy, an attempt has been made in the remainder of the table to capture some of the highs and lows of business conditions, rather than aiming for equal time intervals. The first striking feature is that the upper Susquehanna failed to develop its output to anything like the degree of the lower Susquehanna or the Schuylkill. Both of these areas also took market share from the Lehigh Valley, although the latter was still dominant at the end of the 1880s. Of these two areas, the Schuylkill was the first to make rapid strides in terms of its output, while production

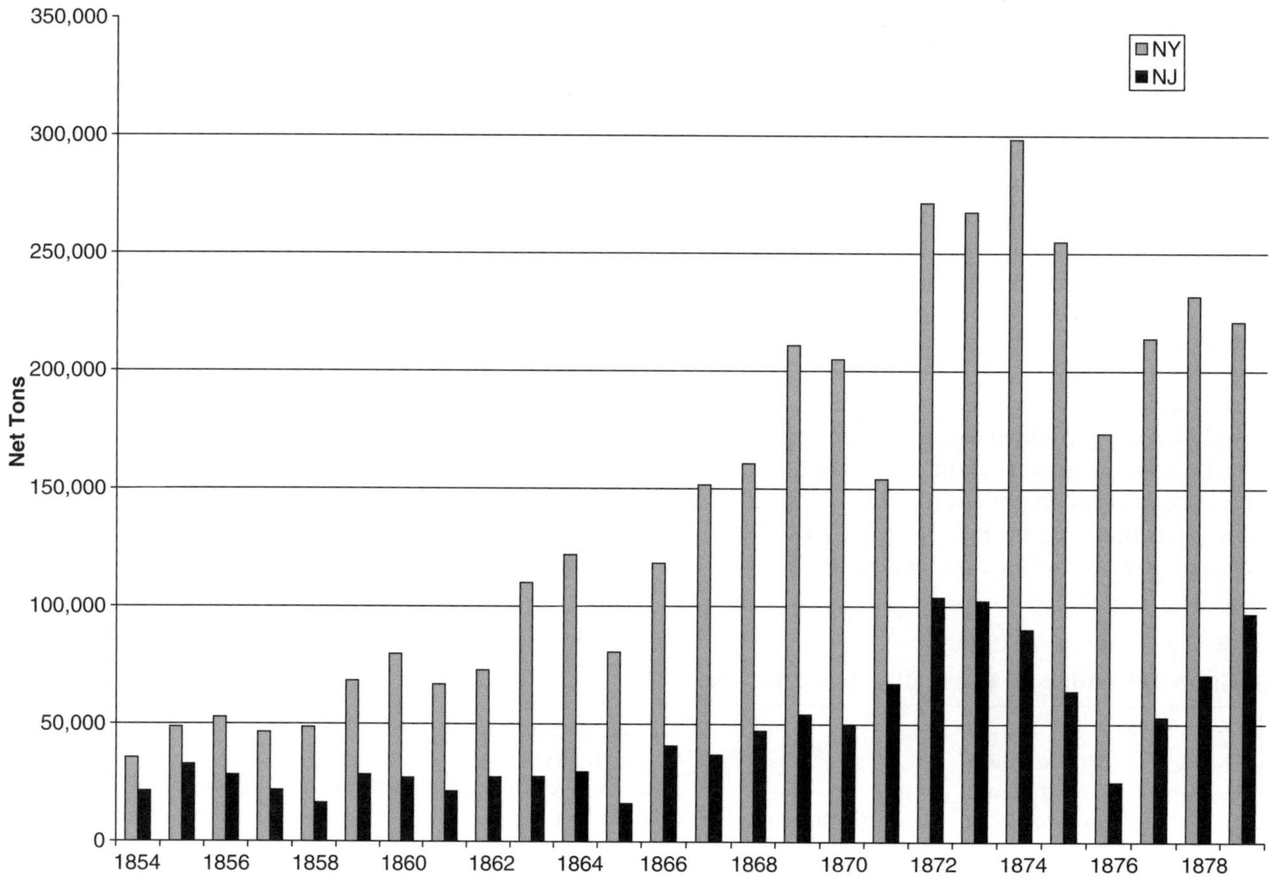

FIGURE 7.5
Anthracite Pig Iron Production: New York State and New Jersey 1854–1879.

TABLE 7.3
Anthracite Iron Production by Pennsylvania Region 1864–1887 ('000 net tons).[22]

	Lehigh	Schuylkill	Upper Susquehanna	Lower Susquehanna
1864	214	113	109	119
1869	302	147	120	118
1873	390	236	129	157
1876	261	145	79	103
1879	456	192	126	166
1885	374	259	90	247
1887	723	520	165	521

surged in the lower Susquehanna during the 1880s. Although the growth of the output in the Schuylkill Valley could not match the Lehigh in absolute terms, it grew proportionately almost as fast between 1864 and 1872. Both regions more than doubled their production in this interval. Over the shorter period, 1869–1872, the growth rate in Schuylkill of 58 percent actually exceeded the 49 percent growth in the Lehigh region. This can be attributed, at least in part, to a change in policy by the Philadelphia and Reading Railroad which served the Schuylkill Valley. In the earlier period it had adopted a laissez-faire attitude to the development of industry, and iron manufacturing in particular, within the orbit of areas traversed by its lines. The Miner's Journal railed against this 'miserable policy' of the railroad,[23] but President Gowen subsequently inaugurated a change in approach that fostered development instead. This included offering to loan bonds of the railroad for one half of the cost of erecting new iron works and manufacturing plants. This not only increased coal shipments, but also the movement of iron and miscellaneous freight.[24] Further examination of the effects of this policy will be undertaken in a later section on the line trade, but a good sense of how far-sighted it proved to be, can be seen in Figure 7.6. This shows the growth of shipments of pig iron and iron ore on the railroad up to 1888, the last year when a breakdown of commodity shipments is provided in the annual reports. It is also interesting to note that even neighboring valleys did not respond to business-cycle changes in exactly the same way. This can be determined from iron production changes in the Lehigh and Schuylkill valleys during the 1870s, a period for which a complete run of data is available. It appears that the Lehigh Valley mirrored changes in the wider economy very closely, quickly absorbing falls in demand and responding rapidly to improvements in business conditions. The Schuylkill Valley in contrast, suffered from greater rigidity in output. It experienced a sharp fall in production between 1874 and 1875 and showed only limited signs of recovery by the end of the decade. This sharp fall is suggestive of large blocks of capacity being taken out of blast. Coal shipments to locations such as Phoenixville, where there was a large ironworks, lend support to this hypothesis. In 1872, over 100,000 tons were shipped to this point by the Reading Railroad, but shipments fell to only 11,400 tons in 1876.[25] It may be that the Phoenix Iron Company had taken the opportunity of adverse business conditions to shut down almost entirely for a period, as they were reported to in 1876, to be building a new rolling mill on their site, which also contained three blast furnaces.[26]

Overall, it is apparent that despite the inroads progressively made by coke as a furnace fuel, the iron and steel industry remained an important market for anthracite producers until the latter part of the 1880s. Even the new technology and entrepreneurial energy of Andrew Carnegie and the other Pittsburgh iron barons required a generation to overhaul, in output terms, the established merchant iron-masters of the eastern part of the state. This fact has not always been fully recognized. When demand for iron was strong, the iron works were favored customers of the mines, because they required substantial and regular shipments of the profitable large sizes of coal. However, as has been seen, the cyclical fortunes of the iron industry could also change rapidly. This necessitated sharp and protracted reductions in coal shipments and produced adverse impacts on mining profits.

THE TRANSPORTATION SECTOR

Fuel for railroads and steamboats was a second market sector that was of longstanding importance

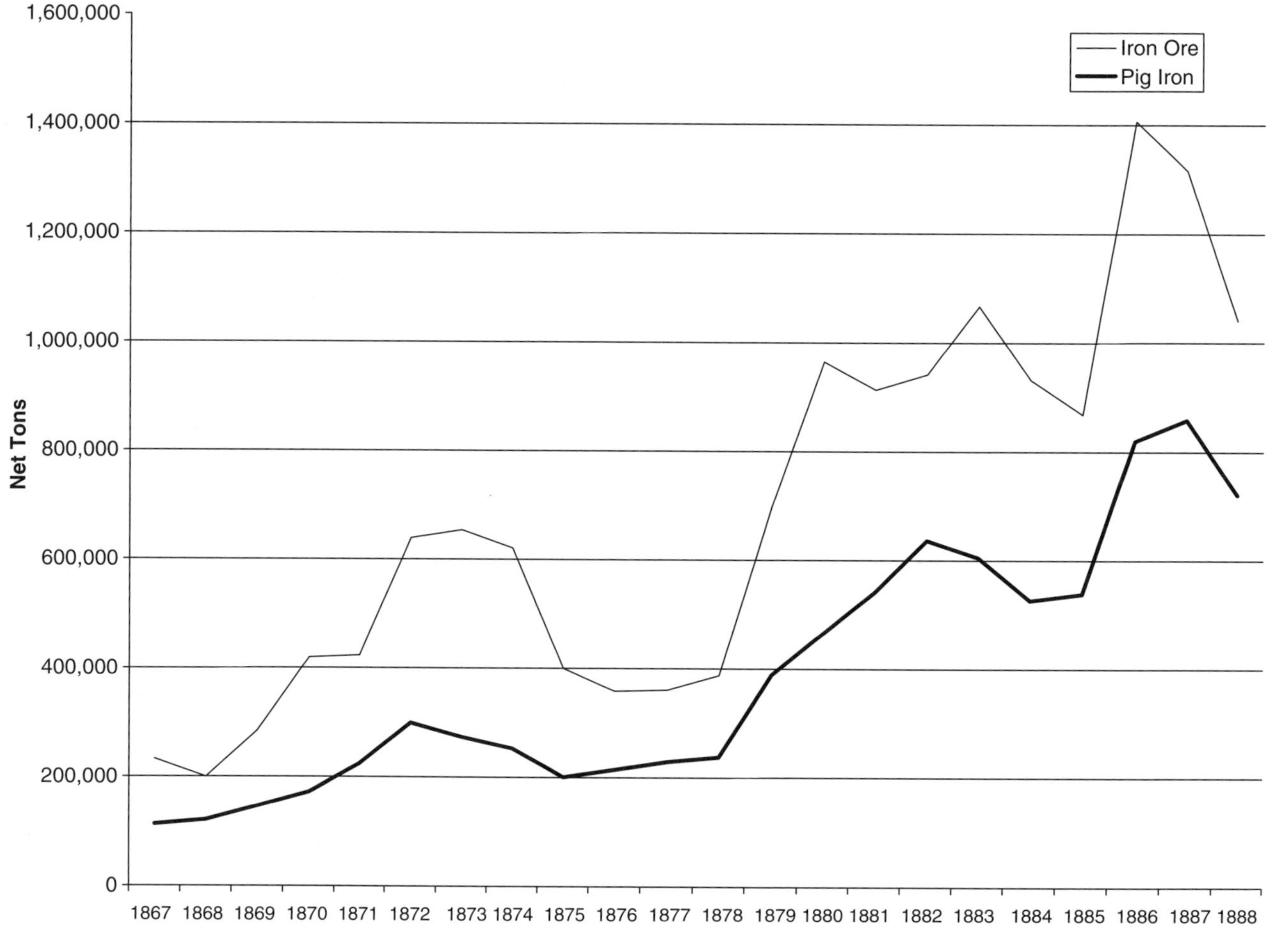

FIGURE 7.6
Philadelphia and Reading Railroad: Shipments of Pig Iron and Iron Ore 1867–1888.

to the anthracite trade. Unfortunately, in the later period, information to track patterns of anthracite use in this sector is even more difficult to find than in the case of the iron industry, as there was no national body equivalent to the American Iron and Steel Association that collected pertinent statistics. Recourse must therefore be had to the records and reports of individual companies, but these in turn vary widely in the extent of information provided about fuel use. In this section, both inter-and intra-urban railroads will be examined first, followed by steamships and ferry companies.

In the 1860s and early 1870s, the railroads were a prized market for the coal companies because of their preference for the profitable larger sizes of coal, especially lump and steamboat coal. These sizes required little or no breaking and hence minimized waste in production. A good example of sales to the railroad sector is provided by Sharpe, Weiss and Company. This firm was a prominent Lehigh Region concern based in the village of Eckley. In 1867, just over 30 percent of the output of their Council Ridge mine was sent to railroads, amounting to 34,973 tons.[27] The major recipients were the Camden and Amboy Railroad, the New Jersey Central, the North Penn and the Lehigh Valley.[28] Railroads, however, were somewhat fickle customers. The Camden and Amboy, for instance, was noted for despatching orders to several mines when their supplies ran low, in case any one company could not deliver expeditiously. Often, however, they received too much coal as a result and cancelled future orders until further notice. Individual railroads were also subject to take-over by larger concerns and this could result in a change in commercial arrangements, to the detriment of their original suppliers of coal. This happened to Sharpe, Weiss and Company in 1873. At this time the Pennsylvania Railroad, which controlled the Camden and Amboy by that stage, decided to switch its coal supplies to one of its other subsidiaries, the Susquehanna Coal Company, whose mines were located at Nanticoke in the Northern Field.[29]

Changes of this kind adversely affected sales of the larger sizes of coal, but Sharpe, Weiss and Company endeavoured to insulate themselves against them by securing a financial interest in various ironworks in the Lehigh Valley.[30] Although their lease of the Council Ridge mine expired at the end of 1874, their successors, Leisenring and Company, were able to benefit from expansion, both in the railroad and anthracite iron sectors in the early 1880s, to generate high levels of output of the large coal sizes at that time. This can be seen in Figure 7.7, which also shows very clearly how broader cyclical changes at the regional and national level had a marked impact on individual companies.

As traffic on the eastern railroads increased and motive power became heavier, to meet the demands for traction placed upon it, the cost of locomotive fuel began to occupy the attention of railroad managers and superintendents. In particular, the possibility was raised of using the smaller sizes of coal and even the waste or culm, which had previously been thrown on the dirt bank. Articles about this appeared in the trade press in the late 1870s[31] and the issue was given added impetus by Platt's 1881 report to the legislature on waste in anthracite mining.[32] Experimental modifications to locomotives led to the development of innovations such as the Wootten firebox, Wootten being sometime general manager of the Reading Railroad. By 1883, it was reported that the new firebox had

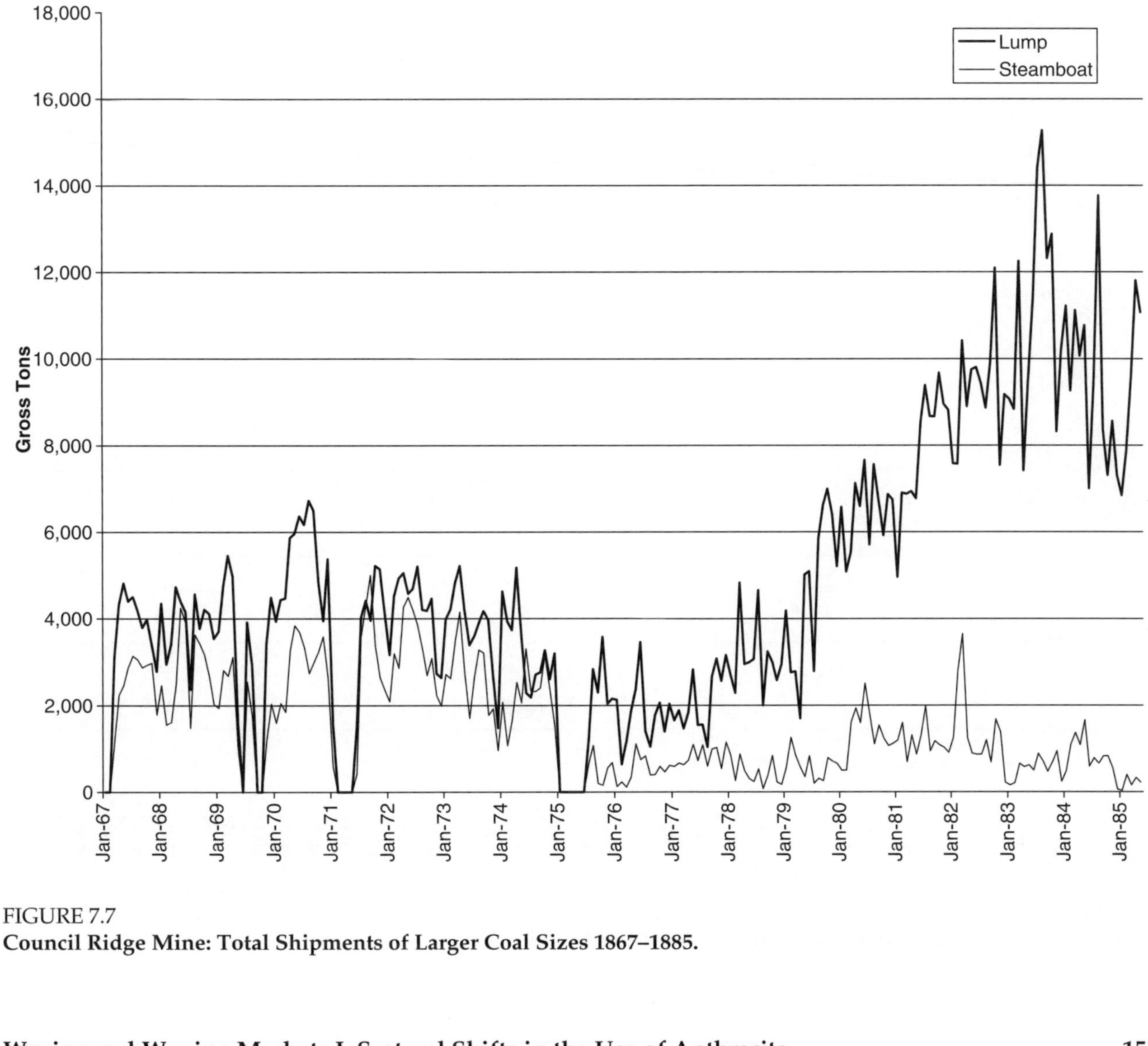

FIGURE 7.7
Council Ridge Mine: Total Shipments of Larger Coal Sizes 1867–1885.

enabled 83 percent of the Reading's main line coal tonnage to be moved by engines burning waste anthracite. A total of 171 engines used over 180,000 tons of this cheap fuel. However, the cost saving of $378,000 was not solely due to lower prices. The engines burned the waste at a lower temperature than lump or steamboat coal and this greatly reduced the expenditures necessary for engine maintenance.[33]

Unfortunately, detailed data series showing the use of anthracite by railroads are very difficult to find, with the exception of the Reading Railroad. In the latter case, coal for company's use is distinguished from general coal tonnage in the company's accounts from 1868.[35] This data series is plotted in Figure 7.8. The upward 'blip' in the series in the mid-1880s is, as usual, attributable to the lease of the Central Railroad of New Jersey, which was terminated on January 1, 1887.[36] The breakdown between anthracite and bituminous is also identified, for coal used by the company, in the annual reports from 1874 to 1896. Until the mid-1890s, the overall figure plotted on the graph can be regarded as indicative of anthracite usage, since the bituminous consumption was very small, usually of the order of 5,000 tons a year. In 1894, bituminous use started to climb and, by 1896, the last year when the breakdown is given, it exceeded 140,000 tons, as compared to almost 830,000 tons of anthracite.[37] Since the general freight tonnage of the anthracite railroads, as well as coal shipments, grew substantially in the later years, as previous chapters have shown, it is to be expected that fuel use by railroads would increase as a proportion of overall anthracite consumption. This assumes, however, that the additional fuel requirements to transport larger quantities of coal and freight were not fully offset by increased efficiency in fuel use. This could have derived from improvements in engine design and better handling of rolling stock in longer trains, which were pulled by more substantial motive power. The annual reports provide clear evidence of year-on-year efficiency gains of this kind.[38]

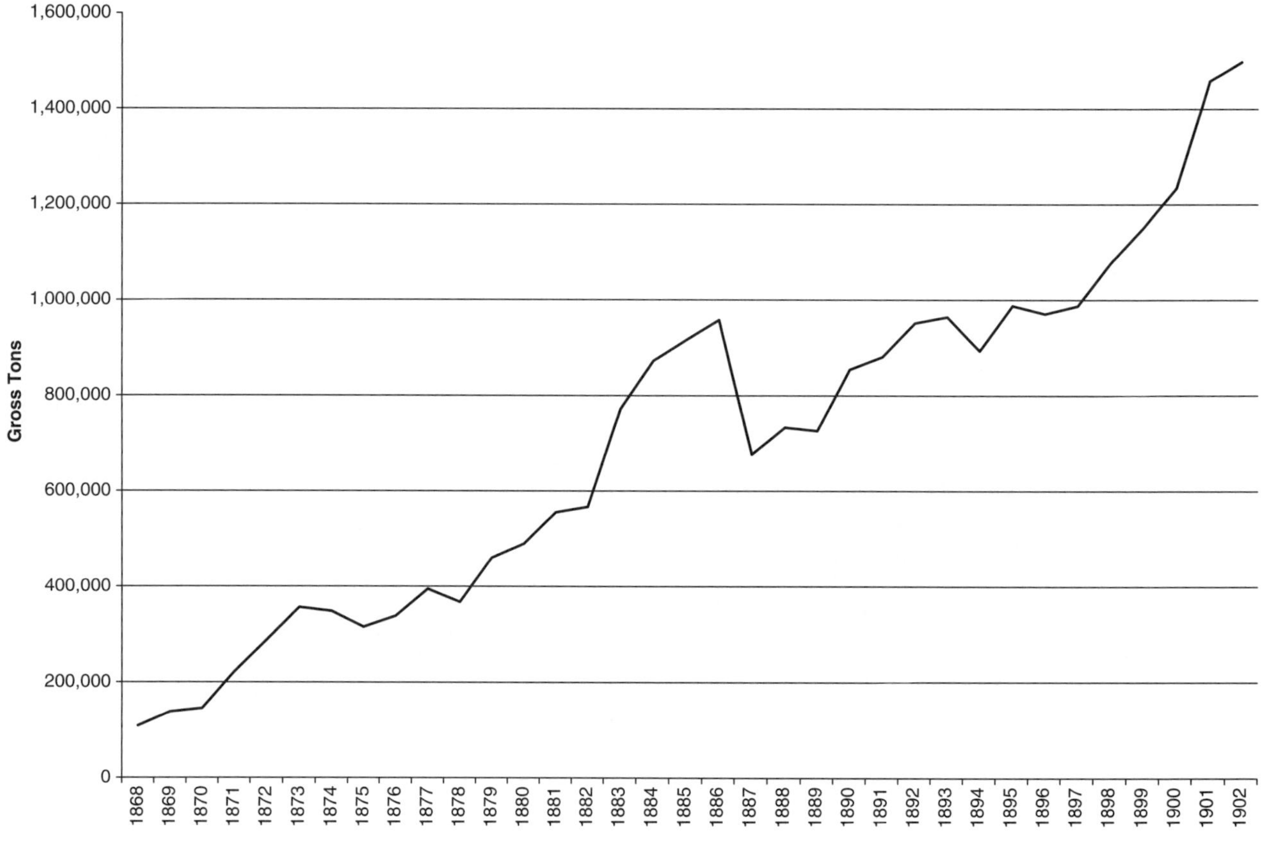

FIGURE 7.8
Philadelphia and Reading Railroad: Coal for Company's Use 1868–1902.

Also, new and doubtless more efficient engines were purchased, sometimes in large numbers. A good example of this was the purchase by the railroad of 47 new freight engines and 143 passenger engines, as soon as it was freed from the iron financial grip of the receivers in 1888. These engines were purchased from the Baldwin Locomotive Works.[39]

Simple calculations of the percentage of total anthracite shipments that were consumed as company's coal on the Reading Railroad bear out the argument. In the late 1860s, ignoring the small contribution from bituminous coal, coal for company's use never exceeded 1 percent of total anthracite shipments. By 1874, when precise anthracite figures are given for the first time, the percentage had risen to 1.67 percent. In 1882 it reached 1.93 percent and peaked in 1890 at 2.37 percent. Thereafter it slipped back slightly to 1.92 percent in 1896, as the bituminous contribution began to rise significantly. Though individually modest, it must be remembered that the Reading was only one of at least six major railroads that made substantial use of anthracite. A clue to the usage by the New Jersey Central is given by the 1886 percentage for the Reading of 2.96 percent, when the figure for the latter included the former, suggesting its usage was about 50 percent of that of the Reading. Early in 1893, it was reported that the Reading and CRRNJ had begun advertising in their timetables that they used anthracite exclusively in their engines, to ensure cleanliness and comfort. Unfortunately, the Lehigh Valley Railroad Annual Reports do not specify company use of anthracite until a very late date. In 1896 the railroad used just over 750,000 tons, compared to the 829,000 tons of the Reading, which equated to about 1.75 percent of total shipments. Shortly after this, the Lehigh Valley began reporting substantial consumption of bituminous coal in its engines, in excess of 700,000 gross tons per annum, and anthracite consumption slipped back to 4–500,000 tons.[40] Data for the Lackawanna Railroad are even sparser. There is no information prior to 1900, but by that time it was consuming about 824,000 tons per annum.[41] Although no breakdown is given, it can be assumed that this was very largely anthracite, as the Lackawanna was not noted for its transportation of bituminous coal, its 1,900 shipments being only 280,716 net tons. Confirmation of this can be found from a turn-of-the century summary of coal use by the anthracite railroads, in the light of high bituminous coal prices at the start of 1900. The Lackawanna Railroad was reported to be extending the use of the small sizes of anthracite wherever possible, particularly in new engines. The Reading, in contrast, was burning bituminous coal by this time, in locomotives whose fireboxes had been designed to burn pea or buckwheat coal. Bituminous coal had never been a major fuel source for the New Jersey Central, but the Lehigh Valley was a heavy user on the western end of its roads. Its eastern lines used bituminous mixed with the smaller sizes of anthracite, obtaining thereby 'a very economical and satisfactory fuel'.[42]

Other railroads, such as the Delaware and Hudson, provided no information on company use of coal during the period, which makes precise identification of overall railroad usage difficult. Nevertheless, the four railroads previously examined consumed perhaps 5–7 percent of overall anthracite shipments up to the late 1890s. This suggests that the total use by major eastern railroads probably did not exceed about 10 percent of the overall output, at its peak, and by the end of the century bituminous coal was increasingly being used by some of the carriers. Assuming that the pattern of change in the Reading's figures was broadly mirrored by other railroads, it is important to recognise, however, that in terms of the period of interest, the railroad sector actually increased in importance as a market between the 1860s and the 1890s. This was in part because the same technological advances that allowed cheap anthracite culm to be burned in locomotive fireboxes also prevented early substitution of low cost bituminous coal.

Although it did not operate on the scale of the major regional railroads, the intra-urban Manhattan Railway Company also became an increasingly important market for hard coal in the latter part of the 19th century. The company rose to importance in 1879, when it leased the two existing elevated railroad companies in New York and united them under a single management, headed by the financier Jay Gould. The intra-urban transit business grew rapidly in New York. Some 60 million passengers were carried in 1880 and 186 million a decade later.[43] Although the trains were light and fuel-efficient, the volume of traffic meant that the Manhattan Railway became a major customer for anthracite. Consumption rose from 140,000 tons in 1884 to 200–225,000 tons of broken coal a year by 1893. At this time, the Philadelphia and Reading Coal and Iron Company had the coal supply contract which had previously been held by the Lehigh Valley Coal Company.

Substantial unloading facilities were built at 2nd Avenue and 129th Street in New York to transfer the coal supplied by vessel from Port Reading.[44] Not to be outdone, by 1891, the Lehigh and Wilkes-Barre Coal Company was supplying boatloads of stove coal to the Brooklyn Elevated Railroad.[45]

The second important component of the transportation sector was the steamship trade, although this, in turn, had two largely distinct branches. The first was the transatlantic lines and the other was the coastal steamers and ferryboat companies. The latter were concentrated in the area of New York Harbor and Long Island Sound. Many routes provided connections from the metropolis to the coastal towns of New England.

The transatlantic lines were particularly fickle customers. As they were large users of coal, their primary interest was in minimizing cost, regardless of source. The routes they followed also gave them the option of coaling on either side of the Atlantic. They did not share the Navy's concern over being seen at a distance, so the availability of smokeless fuel was not a major consideration. In trying to track changing demand for anthracite from the steamship trade, the absence of any statistics on sales is even more of a problem than in the earlier period. Hence the investigation has to rely in the main on isolated information from company records, government investigations of the trade, or inference from coal shipment data.

In the 1860s, there is no doubt that the transatlantic trade was making some use of anthracite. Records of the monthly Scranton auction sales run by the Lackawanna Railroad in New York indicate that the Atlantic Steamship Company was buying anthracite in 1869.[46] However, this segment of the trade does not appear to have been a major market, because of competition from England. Indeed in 1872, the Philadelphia and Reading Railroad expressed the hope that the high prices of coal then prevailing in the old country would induce owners of transatlantic steamers to get more of their coal from the United States than had previously been the case.[47] There is little evidence, however, to support the view that this hope proved well-founded.

The more promising part of the trade was the coastal steamship and ferry companies, for whom the purchase of British coal was not a realistic option. This market had been important since the early 1850s, as already noted in Chapter 4. It continued through the Civil War period, until the advent of cheap bituminous coals began to make inroads into the market domination that anthracite had previously enjoyed. For example, in 1861, the Lackawanna Railroad made contracts with a number of steamboat companies, including the Norwich and New London, the Hartford and the New Haven, and the Fall River Line.[48] Likewise, Sharpe, Weiss and Company supplied nearly 2,000 tons in 1867 to the Camden and Philadelphia Steamboat Ferry Company, and a further 14,452 tons to other coastal steamboat lines.[49] Unfortunately, falling prices of bituminous coal and the repeated interruptions to supply caused by strikes in the early 1870s, made it increasingly difficult to hold the trade. As Richard Sharpe, director of the company, observed when questioned at an 1871 inquiry about the effects of strikes on his customers:

> 'They have gone to other fuel, some are burning soft coal; we find it a difficult matter to induce them to return; the steamships are using bituminous; the rolling mills are using bituminous almost all together, or semi-bituminous'.[50]

Referring back to Figure 7.7, it is apparent that the 1875 strike signalled the end of the trade in the steamboat size, at least as far as the Council Ridge mine was concerned. However, the evidence of a single mine's records cannot be taken as necessarily indicative of trends in the marketplace as a whole. The steamboat size of coal did not vanish altogether, nor did the steamship and ferry operators abandon anthracite wholesale for bituminous coal.

Evidence for this is provided by Table 7.4, which lists the annual coal supply contracts won by the Lehigh and Wilkes-Barre Coal Company for three years in the late 1880s. Both large and small contracts for steamboat coal were still being asked for by customers, although it is apparent that the slightly smaller sized broken coal was then in greater demand, together with the appropriately named pea-sized coal. Using only the minimum contract totals for each year, these steamboat companies purchased at least 6.7 percent of the company's output in 1887 and 8.7 percent in 1889. Available evidence would also suggest that a number of these commercial relationships with the steamboat trade persisted over many years. For example, the Lehigh and Wilkes-Barre was still

TABLE 7.4
Lehigh and Wilkes-Barre Coal Company: Coal Supply Contracts for Steamship Companies.[51]

	1887	1888	1889
Old Colony Steamboat Co.	60,000 (Broken)	75,000*(Broken)	80,000* (Broken)
Brooklyn and New York Ferry Co.	25,000 (Broken) 10,000 (Pea)	40,000 (Stmbt)	35,000 (Broken)
Providence and Stonington Steamship Co.	12,000 (Broken) 35,000 (Pea)	40,000 (Pea)**	45,000 (Pea)
Hartford and New York Transportation Co.	10–12,000 (Pea)		
East River Ferry Co.	12,500 (Stmbt)		
Greenpoint Ferry Co.	3,500 (Stmbt)	6,000 (Stmbt)	6,000 (Stmbt)
Metropolitan Ferry Co.		12,500 (Stmbt)	10,000 (Broken)
Nassau Ferry Co.		3,000 (Stmbt)	3,000 (Stmbt)
Newport and Wickford RR and Steamboat Co.			2,000 (Broken)
Communipaw Coal Co. for Iron Steamboat Co.			10,000 (Egg)
Minimum Contract Totals	168,000	176,500	191,000

*Delivered at North River alongside. Coal required at Fall River and Newport to be added to total.
**Unspecified amount of broken coal also required.

supplying the Providence and Stonington Steamship Company in 1891 with about 50,000 tons a year, and adding large new customers, such as the Savannah Line.[52]

The Lehigh and Wilkes-Barre Coal Company was fortunate to have mines both in the Wilkes-Barre area and at Honey Brook in the Lehigh Region, which produced coals well-suited to steam raising. It could therefore take on large contracts that required a regular supply, as the product of different mines could be substituted, if one of its mines ran into production difficulties. Smaller operators would not be able to compete for these large contracts, as they could not guarantee supply in the event of fire, flood, or mine cave. That said, the range in contract sizes from 3,000 tons to 80,000 tons, in Table 7.4, suggests that steamship companies varied in the degree of risk they were willing to bear from any one supplier. This also implies that a number of companies remained involved in supplying the steamship trade throughout the 19th century.

OTHER INDUSTRIAL USES

Returning to the 1885 assessment by the Coal Trade Annual of different markets for anthracite, just under 20 percent of shipments were said to be directed towards 'miscellaneous uses', including coal for locomotive fuel.[53] Given the percentages already calculated for locomotives and steamboat use, it is apparent that not all companies could have been involved in supplying both these segments of the trade to the same extent. Otherwise, the implication would be that virtually no anthracite was sent to the general industrial sector, which was included in the 'miscellaneous uses' heading above. In fact, the available evidence points to a number of important industrial uses, some of which were port or terminal-based industries on the eastern seaboard, while others were specific to the line of particular railroads. Examples of the former include sugar and oil refining, while the salt industry would come into the latter category. Many of the general industrial users were quick to follow the trend towards smaller sizes of coal, on account of their lower cost in relation to heating value. The sugar refining industry became a particularly heavy user by the late 1880s. A good example is Matthiessen's American Sugar Refining Company, which was willing to take at least 25,000 tons of pea and culm sizes from the Lehigh and Wilkes-Barre Coal Company in 1889. By 1891 the contract had grown to 10,000 tons of buckwheat a month. Unfortunately, this proved too much for the Lehigh and Wilkes-Barre to supply, while maintaining a consistent quality of product. They were forced to abandon the contract in the later part of the year, owing to the 'unmerchantable condition' of their coal.[54] Indeed the standard of coal preparation became so bad at the time that an exasperated sales agent had to castigate the superintendent of the Wyoming (Wilkes-Barre) Division of the company in no uncertain terms:

> 'The coal you are sending from Stanton at present is simply infamous in quality. You would be much better off if you pitched it into the dirt heaps'.[55]

The oil refining industry on the eastern seaboard also provided a ready market for anthracite from the 1880s onwards. While some companies such as Columbia Oil of New York used the egg size,[56] the largest customers, such as Standard Oil at its refinery in Bayonne, New Jersey, took progressively smaller sizes as time went on. In 1891 they were taking consignments of pea, but by 1893 they regarded buckwheat as their standard size, although they were also burning mixtures of 90 percent pea and 10 percent bituminous coal. Further to this, they were obtaining good results with even smaller types of coal, such as buckwheat No. 3, or rice coal. According to J. Dunbar Wright, the company's coal purchasing agent, Standard Oil regarded themselves in some respects as 'pioneers in the utilization of sizes of coal at one time considered of little value'. They took extreme care in fuel utilization and watched every detail on a monthly basis, in terms of the cost of different sizes and mixtures.[57]

The salt industry provides an interesting contrast to the refining industries, which were clustered around tidewater points, and New York harbor in particular. Its location was obviously determined by the requisite geological formations and significant quantities of fuel were required for evaporating purposes. The industry had initially developed prior to the 1860s in upper New York state and the Lackawanna Railroad had been keen to bring its coal into use by the industry before the Civil War.[58] The war itself gave added impetus to the industry, 9 million bushels being produced in the Syracuse area in 1862, a 100 percent increase over the 1860 output. However, production declined somewhat thereafter, probably as a result of competition from the Saginaw area of Michigan.[59] Anthracite also had to compete for the salt trade with the Blossburg semi-bituminous coal, which was well placed geographically to serve customers in upper New York state. Indeed, the salt producers were not slow to recognise the importance of securing adequate and regular coal supplies. To this end, the Onondaga Salt Company went as far as working its own mines in the Blossburg coal region.[60] In later years, the Lackawanna Railroad found an important group of customers among the salt companies of Wyoming and Livingston Counties in upper New York state. These included Royal Salt and Leicester Salt of Mount Morris, and the Warsaw Salt Company, in the town of the same name. Companies such as these sought to purchase culm and coal dust to minimize costs, although the Le Roy Salt Company of Le Roy, NY, was also interested in the purchase of stove coal.[61]

The burgeoning industrial centres of New Jersey also provided important markets in types of manufacturing that were unrelated to the refinery industries already discussed. However, the financial pressures were very much the same, namely to generate the maximum amount of heat or steam power at minimum cost. As a result, they sought the smaller sizes of coal in substantial quantities. Two types of manufacturing that featured prominently in the Lehigh and Wilkes-Barre order book included the zinc smelting industry and the textile industry. For example, the New Jersey Zinc and Iron Company in Newark took 3,000 tons of buckwheat coal and 16,000 tons of culm in 1888, while the Passaic Zinc Company took up to 10,000 tons of buckwheat in the same year and an equivalent quantity of culm. The following year, the latter company had increased its usage to 12,000 tons of each of the two preferred sizes. Textile mills such as the Raritan and Bound Brook Woollen Mills also sought buckwheat, but in somewhat smaller quantities than the zinc industry.[62]

The limitations of surviving records have necessarily restricted the range of examples to the business of a small number of companies. However, enough information has been gleaned from the faded letter-books and contract records to indicate that industrial markets across a range of sectors remained important to major anthracite shippers until the end of the period, even if they did not dominate the industry by that stage. The greater cyclical sensitivity of industrial markets, especially in the iron and related trades, transmitted economic impulses to the mining sector. This was more pronounced in the period prior to 1890 than subsequently, because of the waning influence of anthracite iron. Hence the growing significance of the institutional and domestic heating sectors tended to buttress the anthracite industry against the worst excesses of the depression of the 1890s in particular, although the impact of this period of business contraction was still substantial. While industrial demand was more cyclic over a 3–5 year time-span, it is important to note that, on a month-to-month basis during the year, it was more regular than domestic demand, which was obviously concentrated in the colder seasons. This had important implications for stocking coal to meet peak demand at certain times of the year.[63] This issue will be taken up further below.

The more domestically oriented sections of the market will now be examined, beginning with the utilities, which in some ways occupied an intermediate position between the industrial and the domestic components of overall demand.

UTILITIES AND INSTITUTIONAL HEATING

Sources of information for these markets are even more scattered and fragmentary than for general manufacturing. The major utilities in question were water and gas companies. In addition, there were private companies in the larger cities, which provided steam heating to blocks of apartments or office buildings. This latter type of company has not previously received any attention in the literature.

Large municipal water and sewage companies were valuable customers, because of the need for steam to drive their pumps. Both larger and smaller sizes of coal were employed. For example, in 1884, the Brooklyn Water Works asked for bidders for coal to drive its pumping stations. The coal merchants F.A. Potts and Company won the contract. This company often acted as middleman for the Lehigh and Wilkes-Barre Coal Company. They offered 19,900 tons of Plymouth Red Ash broken and egg coal at $4.25 a ton. Their Red Ash coal had been the most expensive of the bids per ton. Yet its steam raising power was so much greater than the cheaper coals offered by the other bidders, that it proved to be the most cost-effective fuel.[64] Similarly, in 1886, the Philadelphia Water Department bought 34,500 tons of pea coal, while, in 1888, the Brooklyn Works negotiated a further contract with the Lehigh and Wilkes-Barre Coal Company for 24,000 tons of broken coal. Other contracts awarded to the latter company in the following year included the Consolidated Gas Company of New York and the U.S. Illuminating Company, which took 10,000 tons of broken and 12,000 tons of pea respectively. A decade later, Consolidated Gas was still a customer, so some of these commercial relationships were clearly of an enduring kind.[65]

It should be noted that 'producer gas' for heating and illumination was largely generated from bituminous coal, and the gas coals of western Pennsylvania, in particular. Anthracite did not compete strongly with these fuels, owing to the low percentage of volatiles that it contained. However, it could be made into what the Europeans called 'power gas', by means of the Dowson process, to provide heating power for industry. In 1892, for example, some 415,000 tons of coal were used for making gas in New York. Of that total, only about 25 percent was anthracite. By the late 1890s, experiments were also being reported from Scranton in the use of waste anthracite from the culm banks to produce 'power gas', but it had not yet been developed into a commercial process at that time.[66]

Steam heating of commercial buildings was a particular market niche that favored anthracite, because of its high calorific value. Some of the companies involved became substantial enterprises. Prominent among them was the Manhattan Steam Heating Company, which provided heat and steam power to 70 buildings in the dry goods district of New York. The fuel used in 1884 was a mixture of 80 percent anthracite pea and dust and 20 percent soft coal, at a price of $2.60 per ton. The company employed some 65 engines and 20 furnaces, which consumed in excess of 7,000 tons of coal per annum.[67] Firms such as this would be able to deliver a service cheaply through bulk purchases of coal on contract, but they depended on dense concentrations of customers in confined areas of cities. Large institutions, which generally had to operate their own boiler plants, included hospitals and insane asylums across the northeast. As would be expected, the coal companies were solicitous in trying to minimize disruption of supply to such customers, especially during strikes and periods of especially cold weather.

Government buildings and installations were a further type of potential customer. In this context, the Coal Trade Journal recorded an amusing tale about the supply contract for government buildings in Cincinnati in 1892. Although the contract was for the broken (or grate) size of anthracite at $5.11, the supplying agent had difficulty obtaining sufficient hard coal to fill it. When he tried substituting coke as an emergency measure, all manner of stories about his collusion with, and partiality towards soft coal suppliers began to circulate. The imputation made was that he was trying to introduce replacement fuels by a back door route.[68]

Unfortunately, the surviving evidence about the size of the commercial and institutional heating market as a component of overall non-industrial demand is too meagre for any precise statements to be made about its importance. Nevertheless, it shared with home heating a strongly seasonal component. This added to the level of demand fluctuation throughout the year that had to be accommodated by the anthracite shipping companies.

DOMESTIC CONSUMPTION OF ANTHRACITE

In his reflective article in 1892 on trends in the anthracite industry, J.S. Harris, president of the Lehigh Coal and Navigation Company, observed that despite the inevitable fluctuations in any commercial business, the growth of the anthracite trade overall had been remarkably steady. He attributed this to three major factors:

1. the increase in population;
2. the decrease in the wood supply; and
3. the decrease in freight rates.

Although these changes were also of benefit to bituminous coal producers, he was in no doubt that anthracite was the better fuel for domestic use, and all other purposes where the smoke and dirt associated with the burning of soft coal needed to be avoided.[69] In general, it appears that the market shared his view, since just over 65 percent of anthracite was being used for domestic consumption by 1885, according to the statistics from the Coal Trade already quoted in relation to other sectors.[70] However, it is unclear whether this figure includes any of the institutional customers discussed in the previous section.

The traditional sizes used by the domestic market were stove and chestnut. In the early 1870s, the Reading Railroad estimated that each member of the community using anthracite as a fuel required one ton per annum, as a rough guide. However, there was a caveat that consumption in centres of manufacturing, or lake ports opened by new railroads, tended to exceed this figure. More specifically, in 1874, the marked increase in domestic use was credited with offsetting the fall in industrial demand in the aftermath of the 1873 panic.[71] A similar observation was made by the Miner's Journal, which claimed that in the decade following the Civil War, domestic consumption had doubled.[72] By the early 1890s, rather different rules of thumb pertaining to domestic consumption were being employed. The trade press argued that small suburban towns required 2 tons per inhabitant per year, and only a very small proportion of this was soft coal. Further to this, the larger the city, the larger the amount consumed per capita. While part of this was attributable to factories, as opposed to domestic consumption, it was also a result of the greater availability of steam heating 'and other luxurious habits'.[73]

However, increasingly profligate consumption by the moneyed classes in major urban centres was not the only significant shift in domestic demand. The early 1890s also began to register a shift in the size of coal used, to include the slightly smaller pea size. Individual households could not employ the large-scale furnace technology that allowed manufacturing enterprises to switch from larger to very small sizes. Hence any shift could be expected to be of limited extent. Nevertheless, in a market as large as the domestic sector, even a small shift in consumer preferences could have a marked impact on aggregate demand. The trade press duly provided appropriate advice for domestic customers considering a change to a smaller size:

> 'Pea Coal can be used with satisfactory results, both for cooking and heating, if care is taken to keep a thin fire and supply sufficient draughts'.[74]

Several years later, advice of this kind seemed to have been heeded, since an 1897 report on the coal trade of Philadelphia referred specifically to the growth in the use of pea as a domestic fuel.[75]

Although anthracite was unquestionably a better fuel for domestic purposes than bituminous coal, there was a brief period in the late 1870s when the possibility of using crushed coke as a serious alternative was highlighted in several articles in the trade press. One interesting report from 1877 noted that in Cincinnati the cleanliness of anthracite was increasing its use on the one hand, while the introduction of crushed coke was reducing it on the other.[76] A very long haul was required to bring anthracite to such an interior point, which resulted in a high price of nearly $10 a ton at the time. The fact that anthracite was making any headway in such a marginal market is, therefore, perhaps as telling as the finding that it was suffering competition from other fuel sources. A further discussion of the potential of coke was published in Iron Age in 1879 and the merits of the fuel were examined. It was certainly clean, with a high calorific output. Although it was less suitable for open fires than anthracite, it could be burned in stoves previously used for bituminous coal, provided care was taken. The major difficulty was that coke from the oven tended to break into awkward shaped pieces. So it had to be crushed or broken before it was suitable for sale to domestic customers. The equipment required to do this was not as sophisticated as an anthracite breaker and could be erected at the retail distribution point. In general, however, the dual requirement both to manufacture the coke and prepare it for domestic consumption seems to have militated against it becoming a widespread competitive fuel for anthracite, except in localized areas. A good example of the latter was Altoona, where certain coal dealers fell out with the anthracite companies and established a regular coke trade as an alternative supply, at a rate of 12–15 twelve-ton cars a week.[77]

Further discussion of the expansion of domestic demand in major urban centres and the competition with coke and bituminous coals will be found in the next chapter on the geographical expansion of markets. Sufficient has already been said, however, to indicate that a variety of progressive shifts in the demand for different coal sizes took place in the latter part of the 19th century. Some sizes were specific to individual market sectors; others were in demand from several directions. Available evidence is not sufficient for any kind of year-by-year table of use by different sectors to be drawn up. Yet the overall impact of changes in aggregate demand did register on individual mines, companies, and areas of the coalfields, by way of the proportion of their shipments over time that fell into different-size categories. Information of this kind has survived for many companies. It is also available over longer time-spans than detailed data on shipments to specific market sectors and is a useful partial proxy for it. Such information allows an examination of the varied experience of different parts of the anthracite industry. It also permits investigation of the range of adjustments in output mix of coal sizes that were required to maintain market share over the longer term. Detailed analysis of these latter issues will be undertaken in Chapter 9.

Chapter 8

WAXING AND WANING MARKETS II: THE CHANGING GEOGRAPHY OF MARKET POWER

INTRODUCTION

Previous chapters have examined the ways in which the anthracite railroads sought to extend their reach. Initially, this was by developing connections to the eastern seaboard markets. Subsequently, it involved participation in the western trade, much of which was routed over the Great Lakes from Buffalo. However, it is now necessary to investigate in more detail the relative effectiveness of the geographical strategies adopted by the different companies. These strategies were aimed at maintaining, if not enhancing, their market share in the anthracite coal trade. The emphasis is placed here on market share, rather than overall financial performance (which has already been evaluated in chapter 6), because it was the former that was of greatest significance to the large shippers, in relation to the operation of anthracite cartels. The latter are the subject of later chapters.

These geographical strategies can be analysed under several main headings:

- line versus tidewater shipments;
- the continuing importance of the eastern seaboard cities;
- the Great Lakes ports and the expansion of western markets; and
- meeting the competition from soft coal in the interior cities.

These will now be examined in turn.

LOCAL MONOPOLY AND COMPETITIVE PRESSURES: THE LINE TRADE VERSUS TIDEWATER SHIPMENTS

As the name suggests, the line trade for anthracite was made up of customers who were located along the major canal and railroad lines en route to the eastern seaboard markets. Since the individual lines followed separate routes for at least part of the distance, this introduced a varying element of spatial monopoly into the marketplace. Down the Lehigh Valley, for example, by the beginning of the 1870s, there were tracks of both the Lehigh and Susquehanna Railroad and the Lehigh Valley Railroad. These were in addition to the canal, although the latter was declining in importance by this time. Hence, a significant measure of actual or potential competition remained in relation to the coal business of the ironworks in the valley. By contrast, along the Schuylkill River, the Reading Railroad had reinforced its monopolistic position by leasing the Schuylkill Navigation Company's Canal in 1870,[1] to become the sole transportation provider in this area. This position was not challenged until the Pennsylvania Railroad opened its Schuylkill Valley branch in 1886.[2] A similar level of monopoly power was held by the Lackawanna Railroad along much of its Morris and Essex Division and by the Delaware and Hudson on the New York and Canada Railroad.

Spatial monopoly power carried with it both benefits and obligations. The benefits included higher and more stable prices, as well as protection from competing suppliers, either of anthracite or of lower priced bituminous coals. The major obligation was to maintain coal supplies to long-standing customers, even during periods of coal shortage, since these customers had no means of obtaining coal via alternative transportation routes. A clear example of the financial benefits to a mining company of a healthy line trade can be seen in Figure 8.1. This shows the prices received by J. Leisenring and Company at Mauch Chunk for stove coal destined either for the line trade or for tidewater markets at Elizabethport. Since the contribution of railroad tolls has been removed from

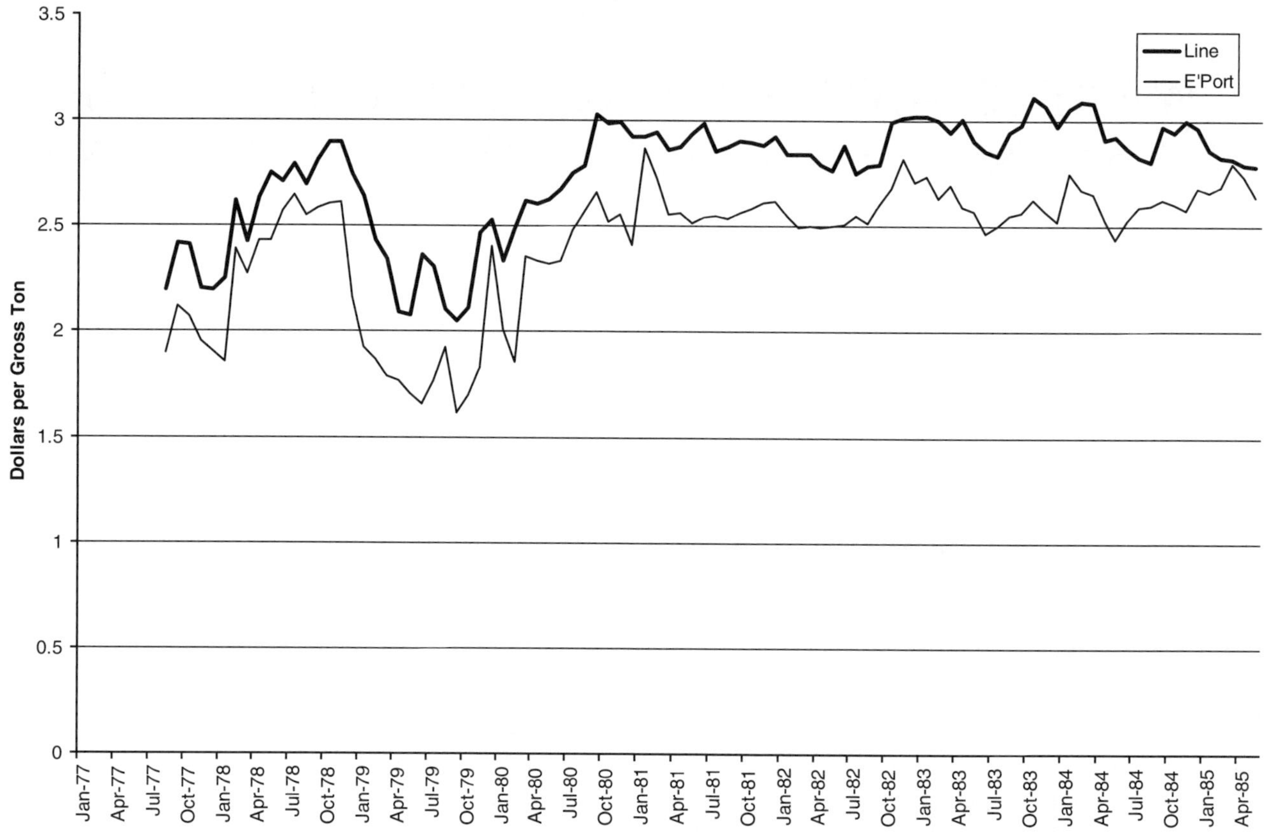

FIGURE 8.1
Council Ridge Mine (J. Leisenring and Co.): Comparative Prices Received at Mauch Chunk for Stove Coal Consigned to the Line Trade and Elizabethport (Tidewater) 1877–1885.

the prices shown, it is apparent that coal for the line trade generally netted the company an average of 25–50 cents a ton more than that destined for Elizabethport. A price differential of this size was undoubtedly sufficient to make the difference between overall profit and loss in lean years.

This is confirmed by the testimony of Richard Sharpe to the State Committee on the Judiciary in 1871, to which reference has already been made in Chapter 4. Sharpe, it will be recalled, was one of the proprietors of the predecessor firm to Leisenring and Company at the Council Ridge mine. The substantial line trade built up for Council Ridge coal saved both Sharpe and Leisenring from having to pay the 25 cents a ton commission on sales at Elizabethport. Sharpe also added that when coal prices were $5 a ton at Elizabethport they could only obtain a very small profit at that point.[3]

Figure 8.2 for Leisenring and Company shows the shipment figures for line and tidewater coal for the same period as the price data in the previous figure. The regularity of the line trade contrasts strongly with the rapid fluctuations experienced in tidewater markets, which indicates the desirability of maximizing the former rather than the latter. Rapid variations in output were difficult to manage in the mining business, because of the implications for availability of labor and railroad cars for shipment.

The Delaware and Hudson, which had the advantage of being both a miner and shipper of coal, was forceful in extolling the merits of the line trade. A Board of Directors minute from 1873 states that sending 2 million tons of coal to tidewater had not paid its costs and expenses, whereas only 800,000 tons consigned to the line trade had brought a handsome profit of $450,000.[4] Interestingly, the company's corporate history also claimed that the

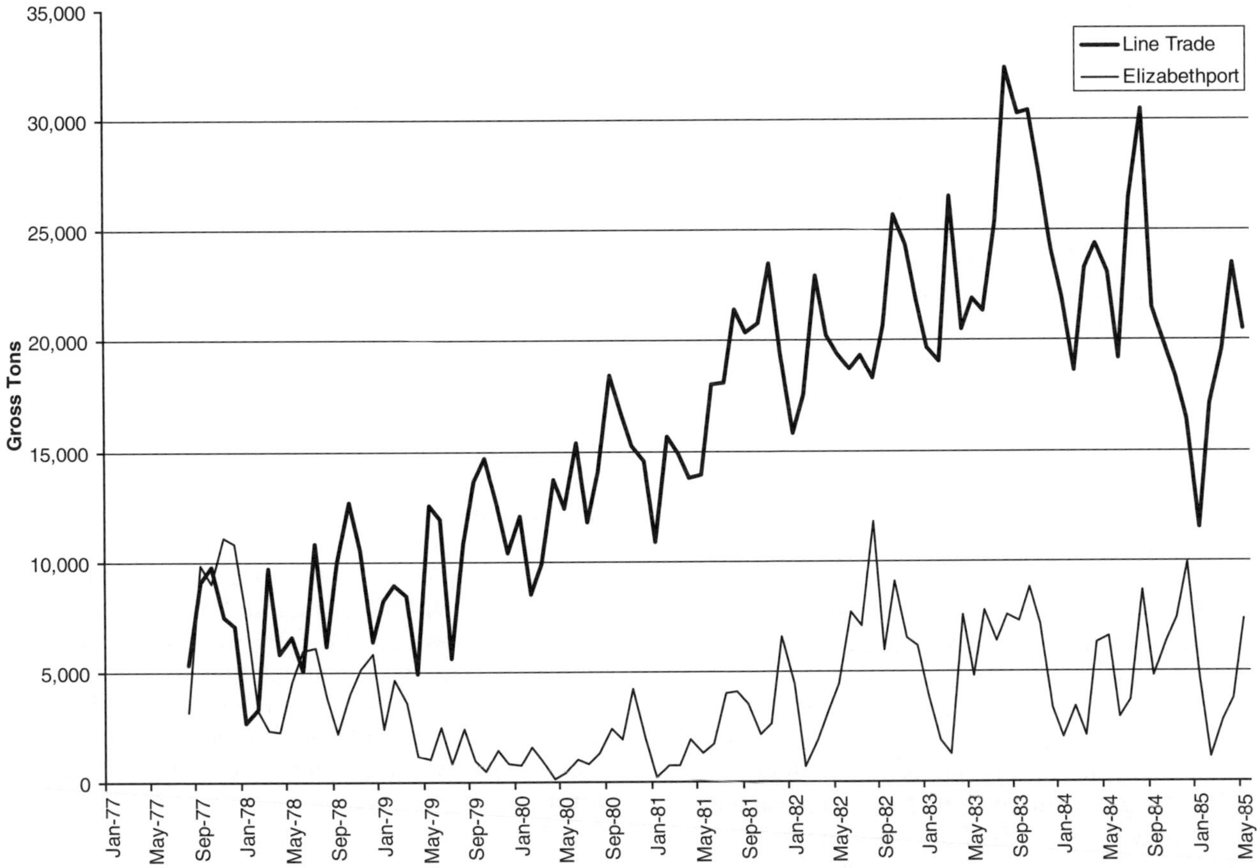

FIGURE 8.2
Council Ridge Mine (J. Leisenring and Co.): Distribution of Shipments 1877–1885.

losses incurred on the leased lines in New York State were made up by the profits of selling coal to the line trade along these roads.[5] As a result of these profits, a growing proportion of the company's coal was diverted away from the competitive tidewater points, as can be seen from Table 8.1.

TABLE 8.1
Percentage of Delaware and Hudson Shipments to the Line Trade and Tidewater Points 1869–1876.[6]

	Inland (Line Trade)	Tidewater
1869	18.5	81.5
1870	30	70
1871	38	62
1872	39	61
1873	40	60
1874	38	62
1875	45	55
1876	42	58

Although the increases in the trade in percentage terms are impressive, the shift is even more striking in terms of actual tonnage. In 1869 the line trade of the Delaware and Hudson in New York state was a mere 128,000 tons, but by 1873 this had soared to almost 899,000 tons. The Lackawanna Railroad also expanded the line trade along its northern routes through New York state by 84 percent, to about 833,000 tons during the period. The equivalent figure to the south for its Morris and Essex Division was 67 percent, in the context of an overall increase of 53 percent in anthracite shipments from 1869 to 1873.[7]

In the earlier section on the anthracite iron industry, it was noted that prior to the appointment of F. B. Gowen as president of the Reading, the latter railroad had adopted a laissez-faire attitude to the development of the line trade of the Schuylkill Valley. However, this was replaced by a much more pro-active approach to fostering industrial development in areas tributary to the railroad at the beginning of the 1870s. The beneficial effects of this became apparent quite quickly, as can be seen from Table 8.2.

The cumulative impact of these different initiatives by the anthracite railroads was considerable. Over the same period 1869–1873, the line trade for anthracite in the states of New York and New Jersey almost trebled from 1.323 million to 3.628 million tons. Across the entire industry, the line trade nearly doubled from 4.2 to 8.3 million tons, while tidewater shipments grew more slowly from about 9 million to an estimated 13 million tons.[9]

Unfortunately, detailed statistics for the breakdown of the line and tidewater shipments for the whole industry are not available for the latter part of the 1870s and tidewater shipments are not reported again on a regular basis until 1882. It is interesting to note, however, that the reported figures for tidewater shipments during the 1880s and much of the 1890s scarcely rise above 1873 levels, despite an increase in overall anthracite shipments from 21 million tons to 45 million by 1900.[10] A significant portion of this increase was taken up by the growth of the western trade and increased use of all rail routes to New England,[11] rather than vessel shipments from New York Harbor.

The significance of the line trade continued to be recognized throughout the period by the coal-trade press. For example, in 1889, the Coal Trade commented that 'the line trade of some of the anthracite coal roads is of great importance'. It also noted that in one particular week, of 145,000 tons sent from Pottsville by the Reading, only 62,000 tons was destined for tidewater, while the Jersey Central regularly sent 60,000 tons to the line trade for every 65,000 tons shipped to tide.[12] A further example can be seen in Figure 8.3. This shows the way (line) and through (tidewater) shipments of anthracite on the Belvidere Delaware Railroad, a route which was absorbed by the Pennsylvania Railroad in 1871.[13]

TABLE 8.2
Line Trade of the Schuylkill and Lehigh Valleys 1869–1873 (Thousands of Gross Tons).[8]

	Schuylkill Valley	Lehigh Valley
1869	1,037	996
1871	1,223	1,180
1873	1,787	1,438

MARKETS IN THE EASTERN SEABOARD CITIES

PHILADELPHIA

Analysis of the pattern of shipments prior to the Civil War, which has already been undertaken in Chapter 4, has indicated the central importance of the eastern seaboard cities to the early development of the anthracite trade. In the case of Philadelphia, this meant that the major transportation arteries bringing coal to the city, namely the Reading Railroad, the Schuylkill Navigation and the Lehigh/ Delaware Division Canals, were already well established by 1860. Although the number of routes to New York was larger than to Philadelphia at that date (see Tables 4.2b and 4.2c), the trade over the New Jersey Central was in its infancy, while new lines such as the Morris and Essex, and the Easton and Amboy were yet to be constructed. These new lines would bring the potential both for increased shipments and for greater competition in the New York Harbor market.

Perhaps surprisingly, it becomes increasingly difficult to get precise figures on coal shipments to Philadelphia and New York as the 19th century progresses. In part, this is due to the complexity of the respective market places, and in part it reflects the difficulty of delineating the metropolitan areas, especially in the case of New York and the neighboring cities in New Jersey. Or, as the Coal Trade bluntly put it in 1887, speaking of the New York coal business:

> 'From the peculiar methods of receiving and distributing coal in this vicinity it is almost impossible to give the definitive figures'.[14]

In the later years, therefore, reliance must increasingly be placed on estimates from the trade press and the figures must be regarded as indicative, rather than exact.

The limited number of routes into Philadelphia and the increasing dominance of the Reading Railroad enables quite a good picture of the development of this market to be obtained for the 1860s and 1870s, before the railroad ceased to report the detailed destinations of coal shipments in the

FIGURE 8.3
Belvidere Delaware Railroad/Division: Way and Through Shipments 1864–1887.

early 1880s. Figure 8.4 shows coal shipments by the railroad from 1861 to 1882. It distinguishes between coal destined for local consumption in the Philadelphia area, and coal consigned to Port Richmond for onward shipment by vessel. Several particular features of the trade merit further comment. Firstly, there is an absence of any sustained growth in shipments from Port Richmond after 1863. Shipments to Philadelphia itself seem approximately to double in the period 1863–1872, with little change thereafter during the long depression of the 1870s. However, some of this increase in the 1860s is attributable to a change in the method of statistical reporting, as has already been noted in Chapter 4. Table 8.3 provides further context for these figures. This gives the Miner's Journal estimates for Philadelphia local consumption and Philadelphia tidewater shipments, broken down by transportation route.

By 1874, almost 62 percent of the shipments from Port Richmond were destined for the New England states and, of these, 75 percent went to Massachusetts.[16] Separate figures from the Coal Trade indicate that more than 578,000 tons of anthracite were shipped direct to Boston from Philadelphia in that year. This total was equivalent to 28 percent of the total shipments from Port Richmond. Thereafter, shipments to Boston increased progressively until the early 1880s, when over 1 million tons were shipped. This amounted to 50 percent of the Port Richmond shipments, although some of the coal would probably have come from the Greenwich Point Terminal of the Pennsylvania Railroad, in addition to Port Richmond.[17]

Data on the Philadelphia coal trade are hard to locate during the 1880s. An 1889 trade estimate claimed that 2 million tons of anthracite and 0.5 million tons of bituminous coal were being used, supplemented by a small local trade in coke.[18] Later information from the 11th census would prove this to be an underestimate, although the census figures did not become available until the mid-1890s. When they were eventually released, the census data

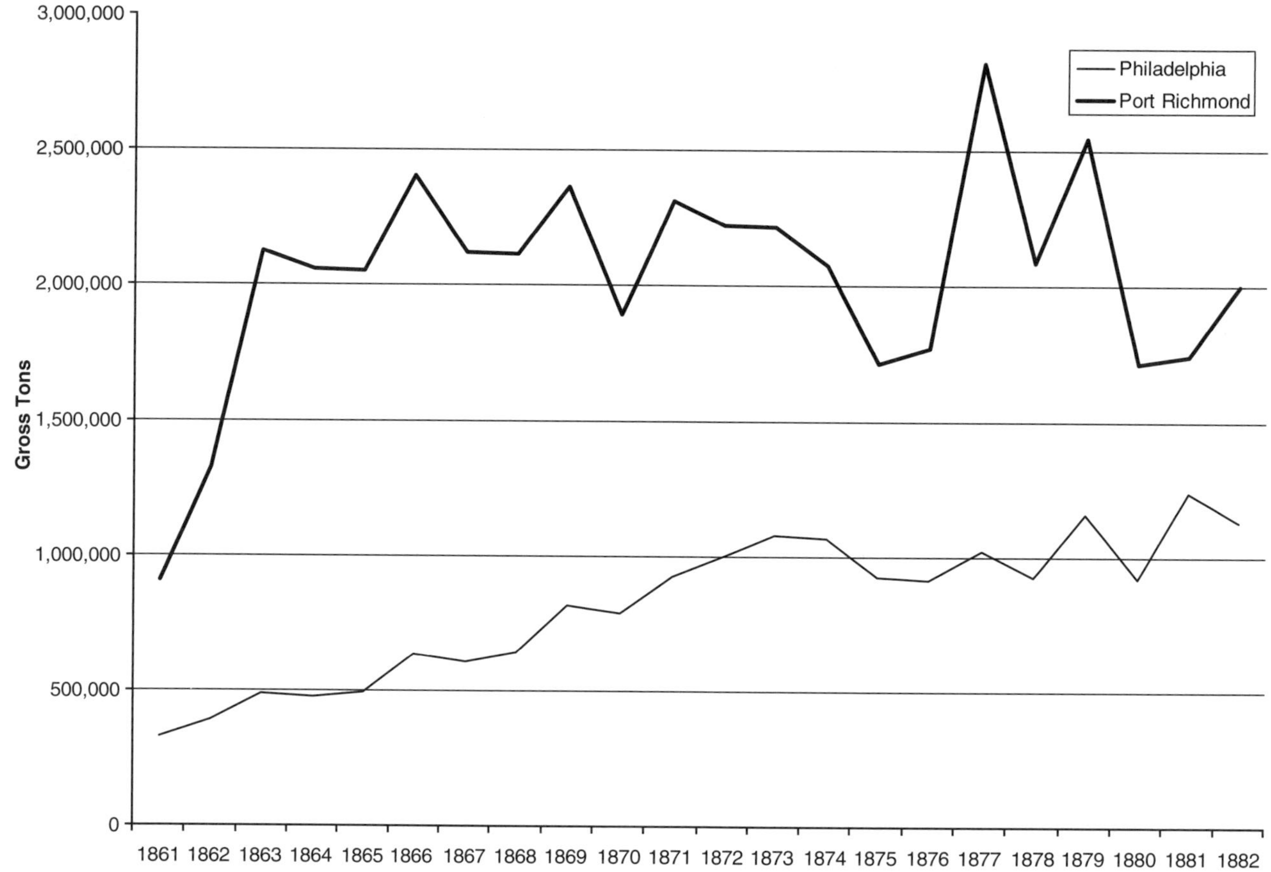

FIGURE 8.4
Philadelphia and Reading Railroad: Anthracite Shipments Consigned to Philadelphia and Vicinity and to Port Richmond 1861–1882.

indicated that Philadelphia's 1 million inhabitants were consuming 3,188,094 tons of anthracite and 919,187 tons of soft coal at the start of that decade.[19] These figures are of particular interest, because they indicate that soft coal had not, in fact, increased its share of the local market since 1869. Also, the overall growth of demand for coal had been relatively modest, as it had taken two decades to double from the 1869 figures.

During the 1890s, several changes became apparent. The Pennsylvania Railroad's Greenwich Point Terminal became increasingly important for the shipment of soft coal and coke, and more Beech Creek soft coal, in particular, began to be shipped from Port Richmond. Total shipments from the two locations reached 5 million tons and continued to grow. By 1896, the Pennsylvania was loading 3 million tons of bituminous and gas coal and 0.5 million tons of anthracite at its piers.[20] Its shipments became so large that plans for a new pier, capable of loading four large steamers at a time, were announced in 1900. Such developments prompted the trade press to assert that Philadelphia was set to become the chief coal exporting port of the eastern seaboard and it was likely to 'regain its old time prestige as a coal shipping center'.[21] In the same year, local consumption reached 5 million tons, two-thirds of which was still anthracite. It is apparent from the trade reports that the rising flood of soft coal being routed to the eastern seaboard was being held in check by the progressive shift to smaller and cheaper sizes of anthracite for steam raising, and the widespread adoption of pea coal for domestic use.[22]

NEW YORK

Between the Civil War years and the 1873 panic, the coal trade of New York experienced a progressive shift in favor of railroad shipments. Tonnage brought via the Delaware and Raritan

TABLE 8.3
Miner's Journal Estimates of the Philadelphia Coal Trade 1869–1873.[15]

Route	Local Consumption 1869	Tidewater Shipments 1869	Local Consumption 1873	Tidewater Shipments 1873
Anthracite				
Reading Railroad	898,663	2,279,376	1,075,256	2,217,315
Schuylkill Navigation	205,185	386,629	259,964	366,816
North Penn Railroad	171,296	0	176,654	74,182
Lehigh/Delaware Division Canals	160,000	25,011	240,000	72,984
Sub-totals	1,435,144	2,691,016	1,751,874	2,731,297
Combined Local and Tidewater Total (Anth.)		4,126,160		4,483,171
Bituminous				
Cumberland and Broad Top	300,000		150,000	
Gas Coals	140,000		225,000	
Grand Total (Local Consumption)	1,875,144		2,126,871	

Canal from the Schuylkill and Lehigh Regions began to decline. This was especially the case from the former region, while railroad tonnages over the New Jersey Central and the Morris and Essex Railroad showed very marked growth. Only the Delaware and Hudson Canal, among water-borne carriers, managed to increase its share of shipments in the first part of the period. Even here, however, its relative position began to slip at the beginning of the 1870s. This was attributable to increased line shipments to its northern markets, the details of which have been examined in the previous section. The changing pattern of shipments to New York via the Schuylkill Navigation can be seen in Table 8.4.

It should be noted that the peak shipments of 1866 were still below the equivalent figure for 1860 (table 4.2c). Further information about sources of coal for the metropolis in 1869 can be found in Table 8.5. This presents data originally reported in the Miner's Journal, with the addition of the shipments over the Schuylkill Navigation/Delaware and Raritan Canal and the onward shipments from Port Richmond. It is possible that there is a small element of double counting associated with these figures, but if so, it cannot be precisely determined from the available data.

A further comparison of the 1869 figures with the equivalent, but unfortunately incomplete data for 1873, shows the marked trends in the trade even

TABLE 8.4
Anthracite Shipments (Gross Tons) to New York by the Schuylkill Navigation Company 1863–1870.[23]

Year	Shipments	Year	Shipments
1863	525,145	1867	465,261
1864	578,706	1868	535,279
1865	591,672	1869	316,389
1866	701,120	1870	246,783

TABLE 8.5
Transportation Routes for Anthracite Delivered to New York and Vicinity 1869.[24]

Route	Shipments
Delaware and Hudson Canal	1,254,480
New Jersey Central–Lehigh coal	610,003
New Jersey Central–Lackawanna coal	810,896
Pennsylvania Coal Co.	775,611
Morris and Essex Railroad	360,066
Morris Canal	129,895
Lehigh Coal via Delaware and Raritan Canal	247,345
Belvidere and Trenton Railroad	455,684
Sub-total	4,643,980
Schuylkill Navigation	316,389
Re-shipped from Port Richmond	611,457
Grand Total	5,571,826

more clearly. Although the Delaware and Hudson had increased its shipments by about 25 percent at the later date to more than 1.57 million tons, the shipments over the New Jersey Central had doubled in the space of four years, while those over the Morris and Essex had soared to almost 1.64 million tons.[25] These figures include shipments to tidewater points in New Jersey such as Newark and Jersey City, as well as those destined for New York City itself.

Following an unfortunate dearth of information for the latter part of the 1870s, the Coal Trade Annual begins to provide useful estimates of the business of the New York market from the beginning of the 1880s, and increased amounts of information are available for some of the later years of the century. By 1881, it was estimated that the New York area consumed about 5.5 million tons of coal. Of this total, about 3.5 million was anthracite, the majority for domestic use. About one million tons were used for steam raising in factories, ferry and river boats, and coastal steamers. By this time the elevated railroads were using about 150,000 tons per annum. The bituminous total included 300,000 tons of gas coal for lighting purposes, but only 100,000 tons of soft coal for domestic use, this sector of the market being entirely dominated by anthracite.[26]

A decade later, in 1891, a more detailed geographical breakdown of coal use in the metropolitan region is quite revealing. New York City proper was consuming about 3.5 million tons of anthracite, little changed from five years earlier, whereas bituminous consumption for all purposes had forged ahead over the same period, from 1.375 million to 2 million tons. Other important adjacent urban markets included Brooklyn, Jersey City, and Staten Island. These locations consumed 1.8 million, 350,000, and 200,000 tons of anthracite respectively. In comparison, Brooklyn took only 200,000 tons of bituminous coal. The grand total of at least 5.85 million tons of anthracite was somewhat inflated by a protracted strike among coke suppliers in the early part of the year. According to trade sources, this added 100,000 tons a week to the demand for hard coal during that period.[27]

Fortunately, the Coal Trade Annual, which reported the business of 1901, provides a very full summary of the state of the New York coal trade towards the end of the period of interest. It is not made clear if Jersey City is included in the figures, but Brooklyn was, of course, formally part of the New York metropolis by this date. An estimated 15 million tons of all types of coal were used or supplied in the city and its environs at this time. Of this total, 7 million tons were anthracite, 5.5 million tons were bituminous, and the remainder presumably specialized types of gas coals and cokes.[28] The detailed breakdown of these totals into major sectors, shown in Table 8.6, is of particular interest. It shows the concentration of bituminous coal use into steam raising requirements for factories and steamships, while penetration of the domestic sector remained minimal. However, the figures for anthracite reveal a continuing diversity of uses and market sectors.

The well-established domestic sector retained its dominance, taking over 55 percent of shipments (categories 1 and 2 in the table), but commercial, institutional, and industrial heating/steam raising still comprised nearly 24 percent of the total (categories 3 and 7 in the table). In terms of transportation uses, anthracite had progressively

TABLE 8.6
Coal Use by Major Category in New York and its Environs 1901.[29]

Category No.	Category Type	Consumption (Million tons)
Anthracite		
1	Domestic, private houses, small stores	2
2	Flats and apartment houses	1.9
3	Hotels, factories, institutions	1.35
4	Gas manufacture	0.45
5	Elevated and surface railroads	0.4
6	Steamboats and shipping	0.4
7	Department stores, office buildings	0.3
Bituminous		
1	Steam trade, factories, etc.	2.6
2	Steamship bunkering	2.4
3	Heating, stores, offices	0.2
4	Blacksmith trade	0.1

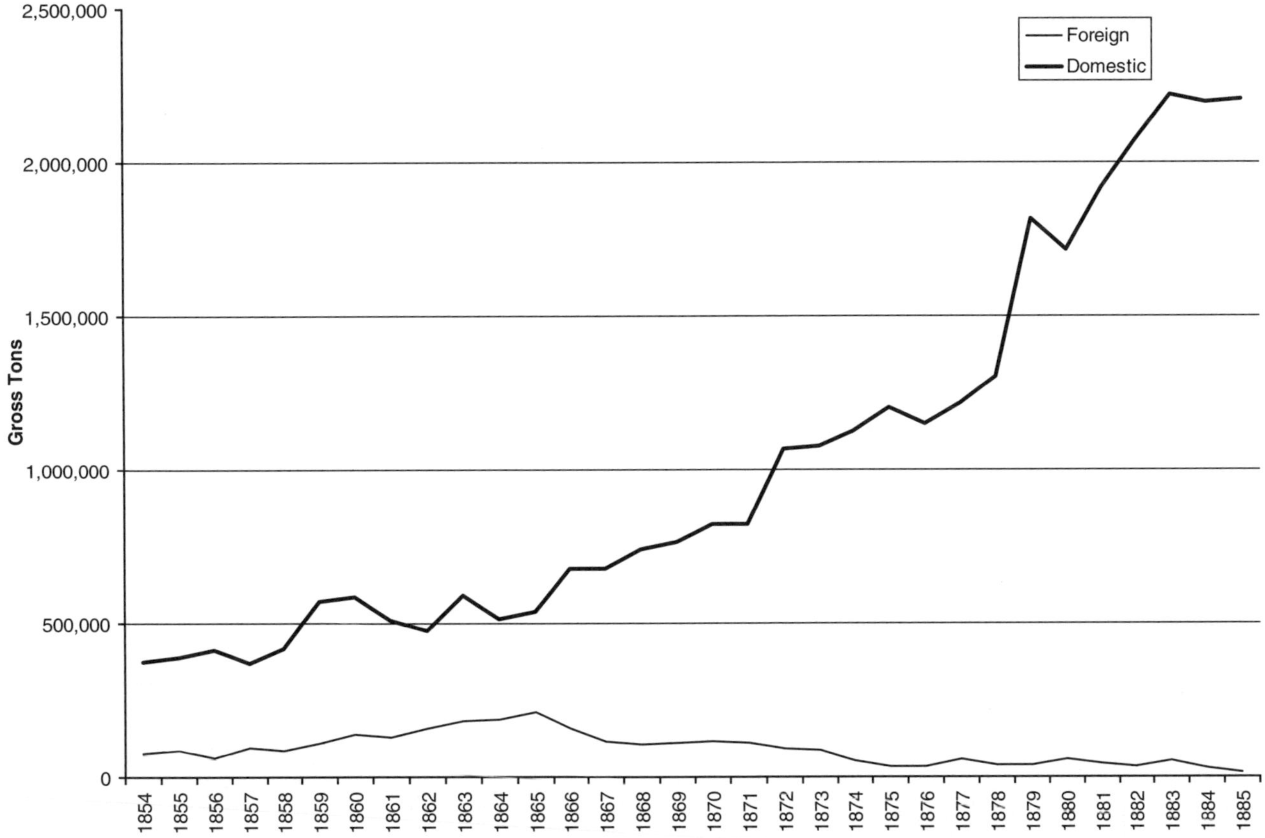

FIGURE 8.5
Boston Coal Receipts 1854–1885.

lost ground to bituminous in the steamboat trade over the last two decades of the 19th century, but some of this loss had been recouped by the trebling of use on elevated and surface railroads since 1881. These statements do not fully support the Coal Trade's own assertion in 1889 that 'hard coal (was) almost entirely a domestic fuel'.[30] They also indicate why the impact of the 1902 anthracite coal strike on the city was so great, since nearly every household and commercial building would have been affected. Problems were also exacerbated by the difficulty of storing any significant quantity of coal in the 'populous portions' of the city, because high real estate values made lots too expensive to devote to this class of land use.[31] Interruptions to fuel supply were never welcome news for consumers and the epilogue to the 1902 strike in the trade press was to inject a note of foreboding into the 20th century experience of the anthracite business:

> 'Probably more soft coal will be burned this year [1903] as a result of its introduction in new quarters during the strike'.[32]

BOSTON

In contrast to New York and Philadelphia, the Boston coal trade had a more cosmopolitan flavor to it from an early date. In part, this was because its rail links to the coalfields in Pennsylvania, Maryland, and the Virginias were much less well-developed than those to the other major urban centres on the eastern seaboard, and distances were much greater. Conversely, however, it was much more accessible from the Nova Scotia coalfields. Also, established trading links with Great Britain allowed coal to be brought from the British coalfields as ballast in transatlantic cargo vessels. Statistics for Boston coal receipts traditionally included coal destined for points on the railroads emanating from Boston, in addition to that delivered within the city and surrounding settlements, such as Brookline, Cambridge, and Somerville. Figure 8.5 shows the growth of this trade from the 1850s onwards. 'Foreign' coal included coal from Great Britain, but was largely made up of Nova Scotia coal. 'Domestic' coal was

FIGURE 8.6
Boston Coal Receipts by Port of Origin 1874–1885.

almost exclusively anthracite until after the Civil War, when Cumberland bituminous coal began to displace the Nova Scotia coals.

Shipments from the Cumberland region were followed in the 1870s by Clearfield bituminous and then Pocahontas steam coals in the ensuing decade.[33] Between 1874 and 1885, data on Boston receipts by port of origin are available (Figure 8.6). Throughout this period, Philadelphia remains the major source for anthracite and Clearfield coal, while shipments of Cumberland coal from Baltimore remain at a modest level until 1884. The most noticeable feature of the graph is the very sharp and sustained increase in shipments from New York and other points in 1879. Although disaggregated data are not available, it is likely that a significant factor in this was the opening up of the rail connections to New York by the Reading Railroad over leased lines in that year. These connections have been examined in detail in an earlier chapter.

The picture becomes clearer in 1888 when the Boston trade statistics begin to differentiate between receipts of anthracite and bituminous coal, as shown in Figure 8.7. It is apparent that anthracite dominated the market until the mid-1890s. Indeed, it gained market-share at the expense of bituminous until 1895, and it was not until the following year that a sustained increase in soft coal shipments began. It should be noted that anthracite receipts for 1900 and especially 1902 were very adversely affected by the strikes of those years.[34] One aspect of the trade that is not apparent from the statistics

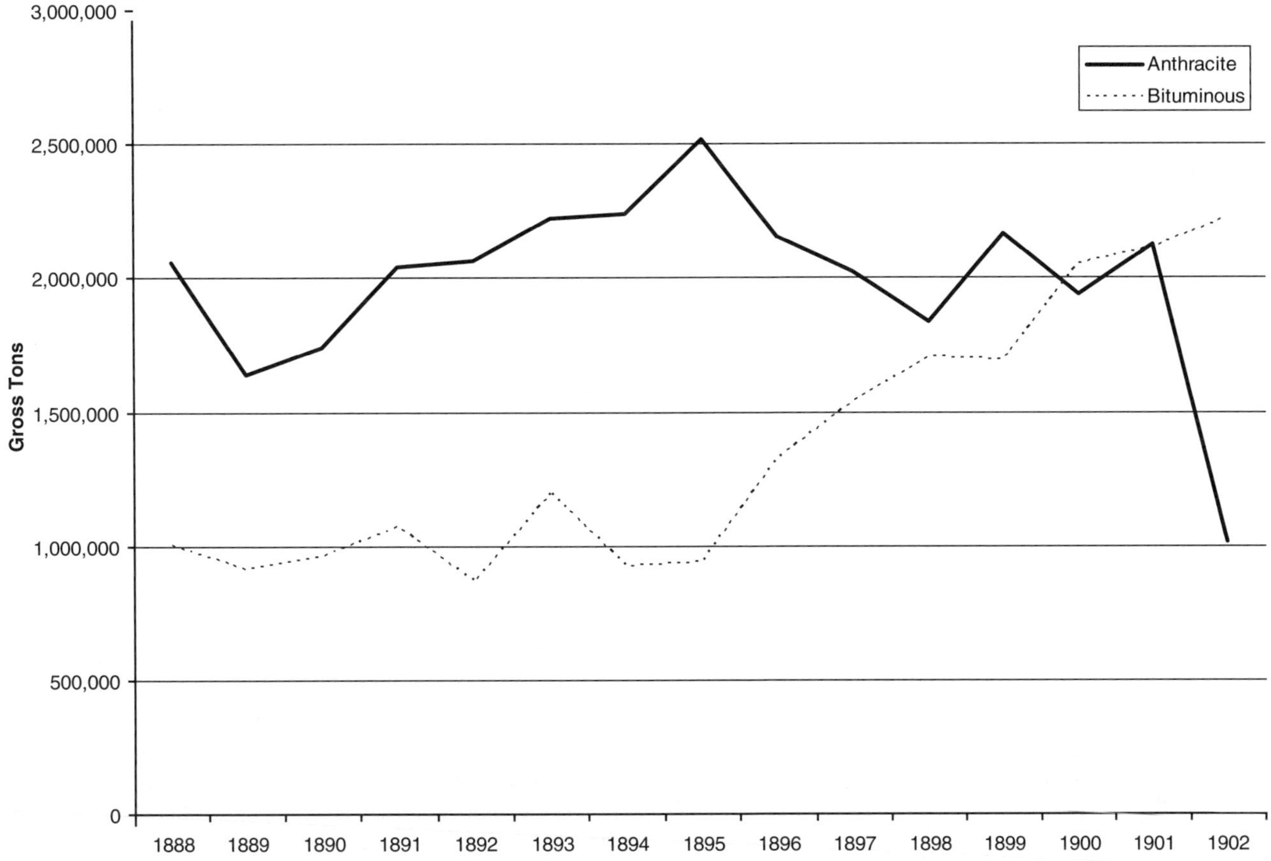

FIGURE 8.7
Boston Coal Receipts by Type 1888–1902.

alone was that the continued growth in demand for anthracite may have been as much a consequence of the short supply of bituminous coal as it was of any preference for the premium quality fuel. The reports in the trade press contain frequent references to the short supply of soft coal despite brisk demand for the fuel. These supply problems seem to have been a direct result of transportation difficulties from the bituminous coalfields. For example, in 1889, it was reported that 'the delay in the shipping ports, caused by the lack of transportation facilities from the mines, kept the supply at this end (i.e. Boston) very short'.[35] Or again, in the latter part of 1892, 'the movement on the railroads to tidewater was wretched and coal came along very slowly'.[36] Even as late as 1899, problems of this kind persisted. An initial soft coal shortage developed at the end of 1898 after a fleet of vessels carrying soft coal sank in a storm. The resulting coal famine was perpetuated by car shortages at Baltimore, Norfolk, and Newport News, which led to detention of vessels. Later in the year, the same problems arose again when the railroads ran out of both cars and available motive power. Vessels were detained for 3–4 weeks awaiting cargoes and Boston experienced its worst ever shortage of soft coal.[37] The problems continued into the following year and a veteran of the trade likened the circumstances to the Civil War era: 'one was able to recall the war times, when one could sit in the office and have a customer come in and almost beg for a car of coal'.[38] The tables were turned in 1902 when the strike led to severe shortages of anthracite. One noticeable feature of this period can be observed from Figure 8.7. This indicates that the shortfall in anthracite was not made up by corresponding increases in bituminous coal, suggesting continuing supply problems on the domestic front. However, the graph does not show receipts of foreign coal, which were reported separately. As a result of the supply difficulties from the United States coalfields, a major opportunity arose for foreign suppliers in Nova Scotia, and even

Great Britain, to enter the market. Foreign coal receipts were a trivial 14,778 tons in 1898, but by 1900 the total had risen to 551,817 tons. During the 1902 strike this figure almost doubled again to just over 1 million tons, of which 301,520 tons were from Great Britain. Hence the anthracite shortfall was almost entirely made up from foreign coal imports. A limited attempt to bring in British anthracite was also made, but customers disliked the coal, because it was not properly prepared into sizes and had to be broken up before use.[39] A second point was that customers continued to ask for Pennsylvania anthracite until prices climbed to a heady $15 a ton. At this point, cost considerations overrode personal preferences and they switched to soft coal instead. However, when the strike ended, customers went back to anthracite so quickly that the surge in demand for vessels to bring it to Boston prevented bituminous shippers from obtaining the necessary transportation for their own coal. According to Chamber of Commerce Statistics, in 1902, 96 percent of anthracite and almost 95 percent of bituminous coal of US origin was still being brought to the city by water. Thus, only a very small proportion of the total, in each case, was transported overland by railroad.[40]

In summary, there is no doubt that Boston remained a very important market for anthracite throughout the period, although the dominance of the latter was challenged by soft coal at the end of the 1890s. The city and its environs consumed about 20 percent of the anthracite used in New England in 1900, almost 2 million tons out of an estimated total of 9.9 million. Of this total, the great majority was thought to be used for domestic heating, but an estimated 1.25 million tons was used for steam raising. The experience of Boston was mirrored across the myriad manufacturing towns and cities of New England. The region as a whole consumed an estimated 10.6 millions tons of soft coal, only a little more than the anthracite total.[41]

BALTIMORE

Unlike their New England colleagues, the merchants of Baltimore apparently eschewed any careful attempt at documenting the progress of the coal business of that city. This caused the Coal Trade to lament that it was one of the most difficult places about which to obtain information—'people who should know apparently do not, and they profess the utmost indifference in regard to the value of the collection of statistics'.[42] Despite these problems, information on the coal shipments reaching the port is available for the two main railroad carriers, as shown in Figure 8.8. The Baltimore and Ohio brought bituminous coal to the wharves at Locust Point from areas such as the Cumberland region. At a later date, shipments also came from the Elk Garden and Fairmont fields in West Virginia and Somerset County in Pennsylvania. The Northern Central Railway initially brought anthracite from the western end of the hard coal regions to the piers at Canton. Subsequently, it also carried bituminous coal from the Clearfield region.[43] In the first part of the period, the Susquehanna Canal was a second important source of anthracite shipments but, in common with the other canal routes, its significance declined rapidly from the 1870s onwards.[44]

Two aspects of the trade are difficult to quantify for Baltimore. The first is the total amount of coal consumed in the city and its environs; the second is the proportion of that total that was anthracite. Consumption is hard to define because, not only were there large shipments of Cumberland and West Virginia coals from the port to other coastal locations along the eastern seaboard and to New England in particular, but, in later years, a large overseas export trade in coal also developed. In the late 1880s the export trade was limited to about 50,000 tons a year or less, but the total began to increase markedly at the turn of the century, reaching an estimated 600,000 tons in 1902.[45] An indication of the significance of coastal shipments can be obtained by comparing coal receipts at Baltimore with onward shipments to Boston. In 1882, for example, almost 1.6 million tons of bituminous coal were carried to Baltimore over the Baltimore and Ohio Railroad. Of these, 1,046,709 tons were from the Cumberland region. Onward shipments to Boston in the same year were 262,639 tons and, as noted above, Boston accounted for about 20 percent of the New England coal trade.[46] However, the Northern Central Railway also brought significant quantities of soft coal to Baltimore by this date. Prior to 1875, it only shipped anthracite to Baltimore at a rate of 200–250,000 tons per annum, but, in that year, the first shipments of Clearfield bituminous coal were made south to Baltimore over the road.[47] Unfortunately, thereafter a clear distinction is not maintained between the proportions of hard and soft coal in overall coal shipments over the Northern Central. Despite this, the relative absence of fluctuation in

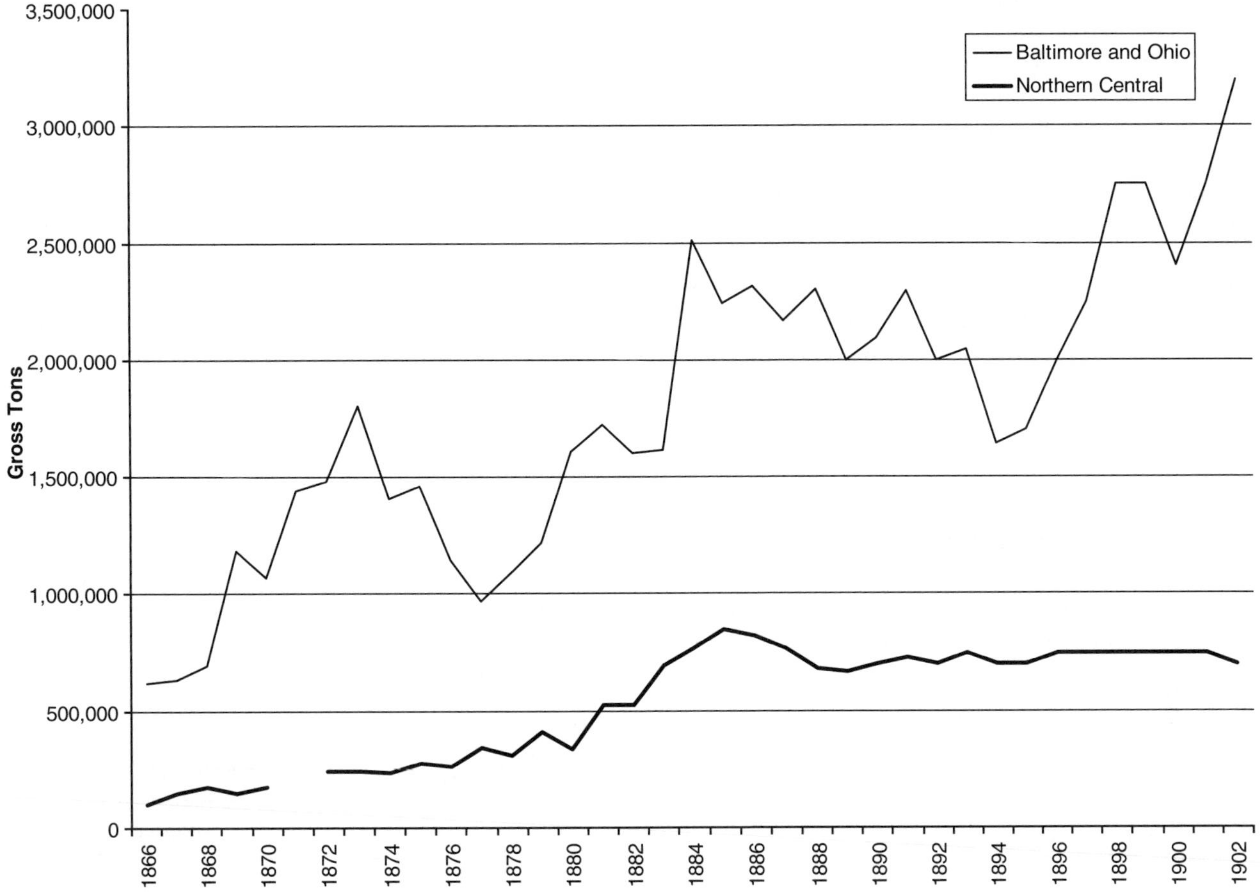

FIGURE 8.8
Baltimore Coal Receipts by Major Railroad Carrier 1866–1902.

shipments over this road from the early 1880s allows a reasonable indication of the proportionate shipments to be made. It appears that there was very little growth in the anthracite trade after the 1870s, since the trade reports for both 1892 and 1902 estimate that about 300,000 tons were brought to Baltimore out of a total of about 700,000 tons, the remainder being soft coal.[48] Unlike soft coal, however, the anthracite was all destined for local consumption, and there was no significant onward shipment.[49]

The final component of the Baltimore anthracite trade was the coal shipped over the Susquehanna Canal to Havre de Grace at the mouth of the Susquehanna. From there, some shipments were towed by tug to Baltimore, while others went east to Philadelphia by way of the Chesapeake and Delaware Canal.[50] Initially, the trade to Philadelphia dominated, but, by 1857, shipments to the two cities were on an equal footing, judging by data on the distribution of canal boats to the two destinations.[51] The total shipments on the canal reached about 230,000 tons by the late 1850s, suggesting a little over 100,000 tons per annum were reaching Baltimore by this means.[52] Although later figures are hard to obtain, there does not appear to have been any very marked expansion of the canal trade to Baltimore thereafter, although it remained important for longer than might have been expected. As late as 1884, almost 245,000 tons were still being carried on the canal, a reduction of about 43,000 tons from the previous year.[53] Shipments continued at roughly this level until 1888. Thereafter, they declined sharply as operations on the canal were progressively wound down, until they finally ceased in May 1894.[54]

In conclusion, to take an overview of the coal markets of the eastern seaboard, it is apparent that anthracite remained a fuel source of major importance to the cities of New York and Philadelphia throughout the period. Although domestic uses predominated, it continued to find markets for industrial steam raising and transportation purposes. In the case of Boston, anthracite continued to increase its market share until the mid 1890s, when bituminous coals from a variety of locations began to be adopted to a rapidly increasing extent. Only Baltimore, at some distance from the anthracite fields and a focal point for bituminous coal shipments from the interior, saw stagnation in demand for hard coal in the later decades of the 19th century.

THE GREAT LAKES PORTS AND THE EXPANSION OF WESTERN MARKETS

The rise of the Great Lakes coal trade was closely bound up with the westward spread of settlement across the country and the concomitant growth of Chicago, which rapidly assumed a pre-eminent position among midwestern cities.[55] The port of Buffalo on Lake Erie similarly rose to a position of dominance. It was a key destination and lake transhipment point for the railroads emanating from both the anthracite and bituminous coalfields of Pennsylvania. It rapidly outstripped potential rivals such as Cleveland and Erie on Lake Erie and Oswego on Lake Ontario. Although Chicago and Buffalo were the main centres of the coal trade, rapid growth in coal receipts was also experienced by Milwaukee on Lake Michigan and at a later date by Duluth on Lake Superior.

Figure 8.9 shows the pattern of growth in the coal trade of Chicago and Buffalo during the latter part of the 19th century. The equivalent data for Boston are provided on the same graph, by way of comparison. Several salient points are immediately apparent. The first is the marked surge in the

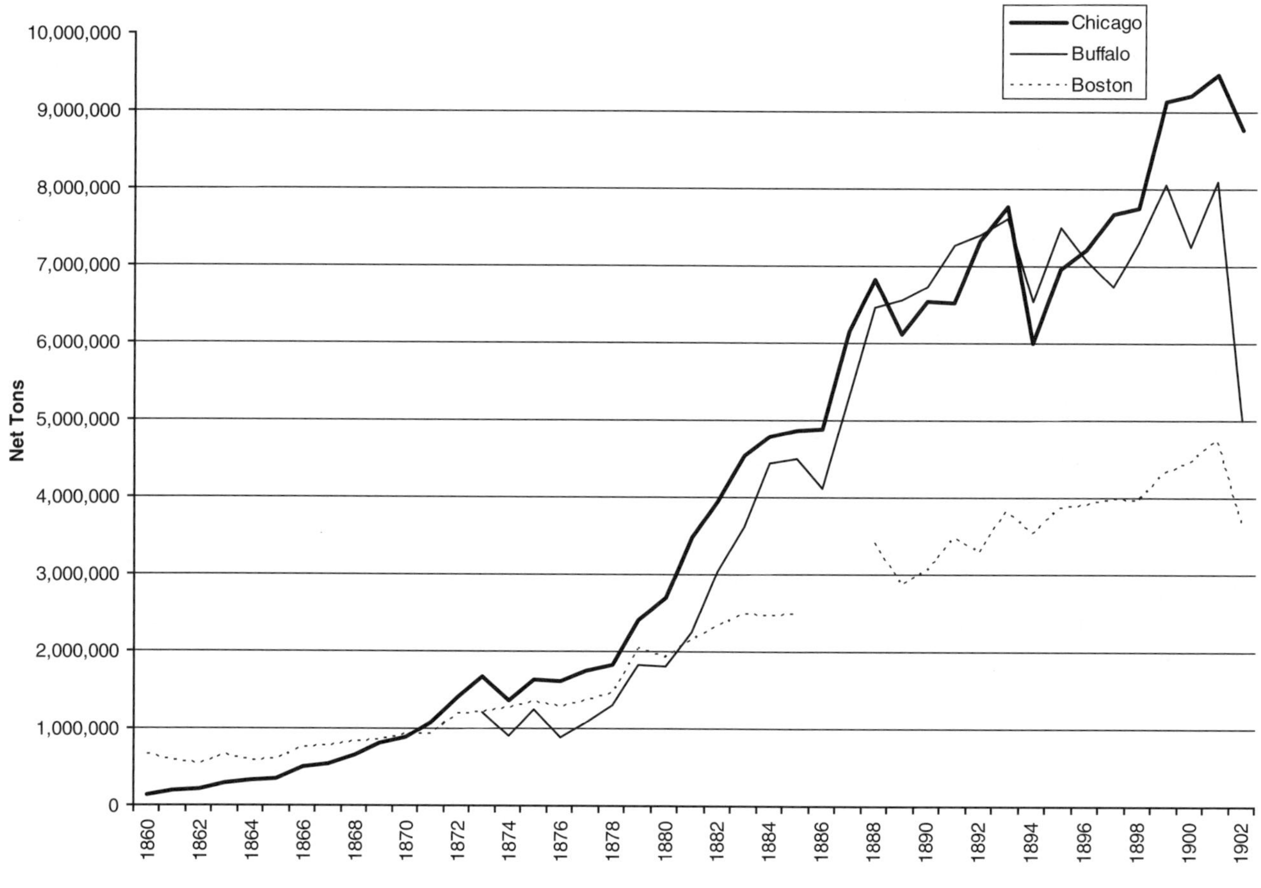

FIGURE 8.9
Coal Receipts of Major Cities 1860–1902.

Chicago trade during the 1880s, followed, after a short time lag, by Buffalo. In contrast, the Boston trade grew much more slowly, especially in the early years of that decade. Although the greatest increase in absolute terms for Buffalo and Chicago took place at that time, it should also be noted that the decade of the 1860s showed a much greater proportionate increase of nearly 7-fold, albeit from a much smaller absolute base. The second point to note is the impact of both major and minor periods of recession in the wider economy on overall rates of growth in coal receipts. For example, the relative standstill during the 1873–1878 depression was followed by a phase of rapid growth until the end of the 'railroad boom' and the onset of the 1884–1886 depression. In this context, it is interesting to observe that recovery in the 1870s apparently began in 1877, even before the return of broader business confidence in 1879. During the investigation into labor troubles in the Anthracite Regions in 1887–1888, F. E. Saward, the experienced trade commentator, argued that the low prices of hard coal during 1877, in the absence of any combination between the shippers at that time, was a key factor in the opening of new western markets for hard coal.[56]

The surging growth of the 1880s was not replicated in Buffalo during the 1890s. Shipments at the end of the latter decade only slightly exceeded those of 1893 and were very adversely affected by the 1902 anthracite strike. Judging by the graph, the shock of the 1893 banking panic to the commerce and industry of Chicago was more severe than in the case of Buffalo. However, 1895 saw the beginning of a sustained recovery in the former city, with very strong growth in 1899 and only limited impact of the 1902 strike. The figures for Buffalo confirm the significance, for the different anthracite railroads, of the time when they reached the city over their own tracks. This point was originally highlighted in an earlier chapter. The Lackawanna arrived in 1882, to derive full benefit from the growth of western markets. The Lehigh Valley arrived ten years later, only to find that growth in shipments would be elusive or absent for a decade or more to come.

From examination of the graphs, it is also tempting to imagine that the growth of the trade in each of the two cities was very closely linked, and nearly all Buffalo shipments were destined for Chicago. However, this was not in fact the case. The data for each of the cities need to be disaggregated further for the actual trade relationships to become more apparent.

DEVELOPMENT OF THE BUFFALO COAL TRADE

Authorities differ somewhat on the early development of the Buffalo coal trade. One of the standard local histories claims that, in the 1850s, the coal trade of the city was confined to soft coal, brought to fuel the local foundries and machine shops. A figure of 60,000 tons for the year 1852 is generally quoted.[57] However, an earlier and possibly more authoritative source states that 25,000 tons of anthracite was brought to Buffalo in that year. This was part of a total of 67,487 tons of hard coal sold 'at the west' by J. J. Albright, who was the Coal Sales Agent for the Lackawanna Railroad in the North and West. This latter figure was believed to encompass almost all the anthracite sold to western markets at that time.[58] By the beginning of the 1860s the anthracite coal trade began to expand steadily. The Lackawanna Railroad acquired a coal yard in the city in 1861,[59] and, in the same year, it formed the Western Coal Association with J. Langdon and Co. and the Pittston and Elmira Coal Co.[60] In 1868, total receipts of coal reached almost 300,000 tons.[61] From the following year onwards, a breakdown of coal receipts in the city is available on an annual basis, as shown in Figure 8.10. It should be noted that these figures from the trade press do not always coincide with those from other sources. For example, in 1883 Perry Smith's data on hard coal receipts suggest a total of over 4.5 million tons, whereas the data used in Figure 8.10 contain a value for 1883 which is less than half that amount. However, detailed examination of railroad annual reports and accounts indicates that while the trade figures may be conservative, the Perry Smith estimates are much too high.[62]

Despite the possibility of some underreporting, the overall picture is quite clear. From the beginning of the 1870s until the end of the century, anthracite dominated the coal trade of Buffalo. The expansion of the hard coal business was especially marked during the 1880s, but it stagnated throughout the 1890s. The soft coal trade made very few inroads, in terms of market share, until the turn of the century, although it derived some benefit from the precipitous decline of anthracite shipments during the 1902 strike.

From the beginning of the period until the start of the 1880s, anthracite shipments to Buffalo by canal formed an important, though declining, proportion of the trade. The strongly cyclical nature

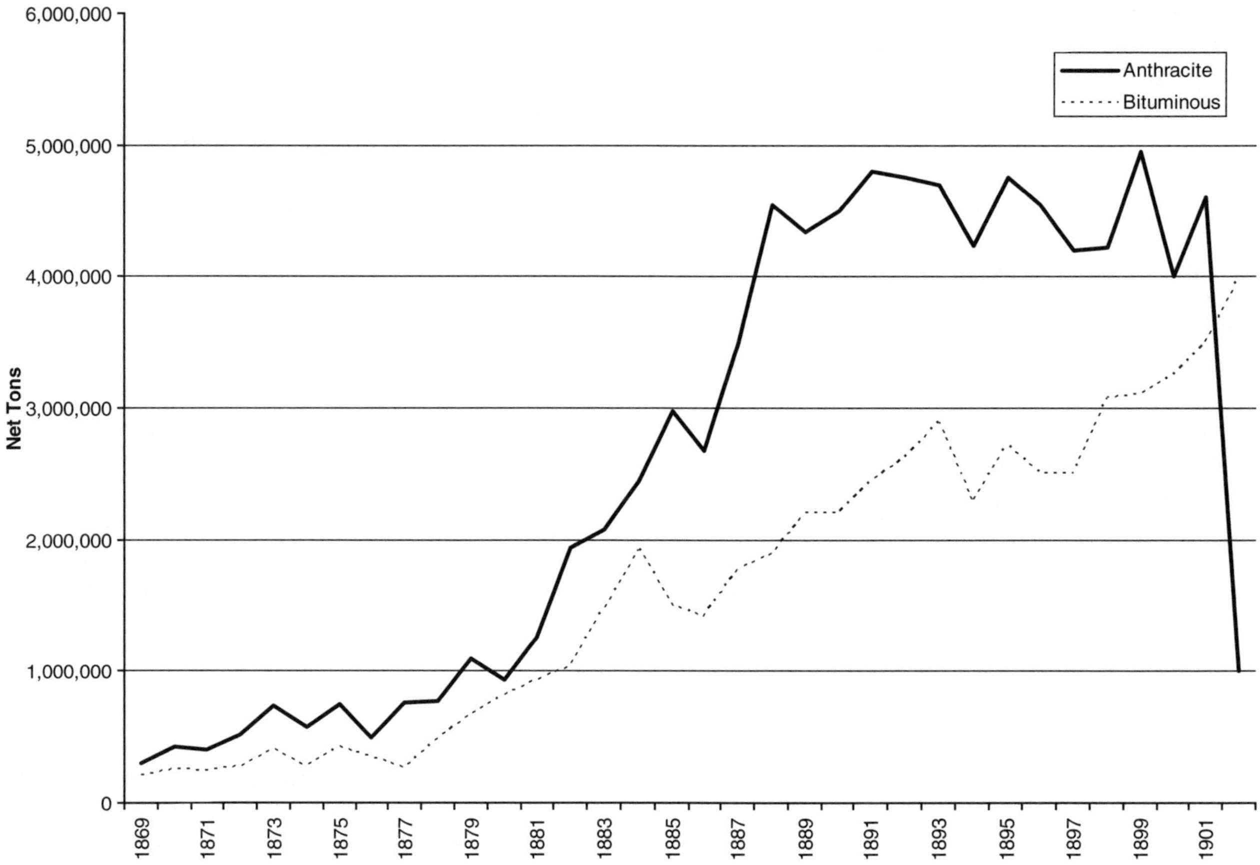

FIGURE 8.10
Buffalo Coal Receipts by Type 1869–1902.

of these shipments, shown in Figure 8.11, corresponds broadly with the periodicities of the Juglar cycle described in Chapter 2. It also indicates that the canal was used as an 'overflow' transportation system, to handle surges in demand that could not readily be met by existing railroad capacity. After the arrival of the New York, Lackawanna and Western Railroad in 1882, however, the Lackawanna Railroad, which had previously made westward canal shipments from Syracuse, had an alternative and faster railroad outlet, and the canal trade rapidly went into terminal decline.

Buffalo benefited from further additional rail connections at the beginning of the 1880s. These included the New York, West Shore and Buffalo Railroad, which chose Buffalo as its western terminus; the New York, Chicago, and St. Louis trunk line; and the Buffalo, New York, and Philadelphia. In the words of the Coal Trade Journal, the latter 'made Buffalo the lake outlet and entrepot of the Baltimore and Ohio'.[63] However, these rail connections did not generate the same adverse effects on lake shipments as they had on the canal trade. The percentage of total anthracite receipts by railroad and canal that were forwarded on by lake oscillated between 39 percent and 69 percent during the 1870s. After peaking at 70 percent in 1883, lake shipments then settled down to an average of about 54 percent of the total over the next nineteen years to 1902.[64] In contrast, the amount of soft coal forwarded from Buffalo by lake was very small. In 1885, for example, out of about 1.5 million tons brought to the city, only 10,000 tons was sent on by lake.[65] No doubt this was a function of competition from other shipping points on the Great Lakes such as Cleveland and Toledo, which had shorter rail hauls from the other coalfields and were closer to Chicago by water. The proximity of the interior coalfields of Illinois and Indiana to the midwestern metropolis also made it difficult for distant soft coal suppliers in Pennsylvania to compete with them.

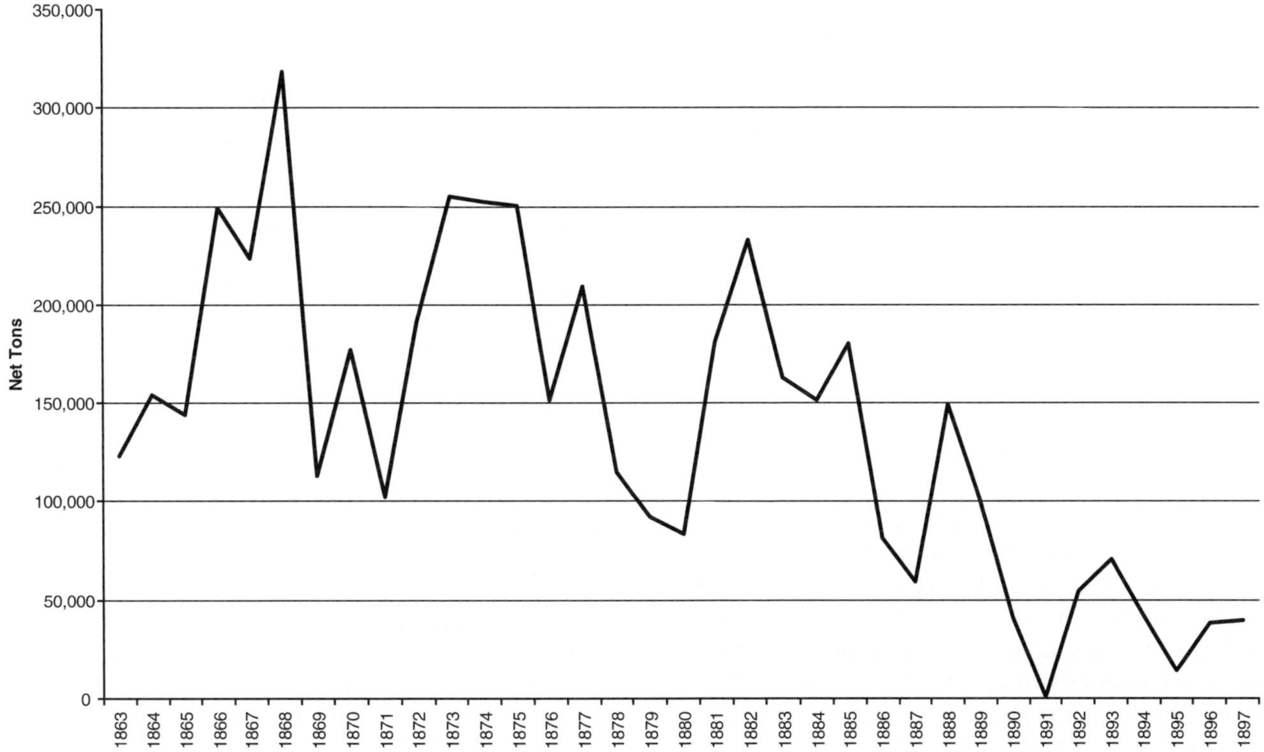

FIGURE 8.11
Buffalo Anthracite Receipts by Canal 1863–1897.

The economics of lake transportation of anthracite raise some interesting questions about the conduct of the western trade. In the mid 1880s, lake freights between Buffalo and Chicago were as low as 50 cents per ton in the spring and summer months, rising to $1.00 or more in the final stages of the navigation season in November.[66] Published freight rates by rail between the two cities at the same time were much higher. For example, the 'special anthracite coal tariff' published by the Lake Shore and Michigan Southern Railway in 1885, indicated that the charge was $2.00 per ton, although major shippers doubtless received some rebates and drawbacks on these figures.[67] It did not make economic sense for the anthracite railroads to try and extend their own rail lines to Chicago, given the presence of a number of pre-existing lines. Even the mighty Pennsylvania only achieved this by means of take-over and leasing of other roads.[68] Yet the construction of lake transhipment facilities and their own fleets of lake coal carriers was within their means, and this policy was pursued with vigour, as indicated in an earlier chapter. In addition to giving the companies control of the transportation of anthracite all the way from the mines to the final destination city, it had another effect, which has not previously been documented. It is clear from trade reports that the major shippers operated a tight system of control over the western trade before the passage of the Interstate Commerce Commission Act in the latter part of the 1880s. This control was exercised through the 'Western Anthracite Joint Committee', a successor to the old Western Coal Association, with expanded membership. This committee allocated westbound tonnage between the different interests on a monthly basis. Further to this, it set prices in the different markets. A particularly informative report on the October 1885 meeting of the committee makes clear that prices were set regardless of whether the coal was received at the relevant destinations by rail or by lake. For example, the prices in Chicago, Milwaukee, Racine and Duluth were set at $5.80 per ton for egg coal and $6.20 for stove coal in November of that year.[69]

There are several implications from this finding. Firstly, put one way, the lake trade was subsidizing the railroad trade, since the margin for the transporting company mining its own coal was much larger for water-borne shipments. Put another way, the lake trade allowed anthracite to be made available in the west at lower prices than would otherwise have been the case, thus expanding the potential market area for the coal. Secondly, given the freight rate differential, it is reasonable to ask why the lake trade did not continue to increase its share of shipments at the expense of rail traffic. The fact that it retained its share during a period of very rapid expansion of westward shipments indicates that lake shipping facilities were also expanding rapidly.

However, given the time required both to construct new vessels and build major dock improvements, it is likely that the companies would have been hard pressed to expand their facilities any faster than they actually did. They may also have been reluctant to invest too heavily in facilities that could only be used during the months available for navigation, since the bulk of coal had to be carried by rail. Thirdly, there was advantage in having multiple transportation routes available. Although the all-weather advantage of railroads partially offset their higher costs, they were not immune to serious problems and traffic delays. Car shortages frequently arose. An interesting case in 1885 was attributed to a combination of car detentions on sidetracks in many areas, low levels of eastward grain shipments and high levels of miscellaneous freight forwarding westward. Low grain shipments were a problem because they meant there was no east-bound back-haul traffic to complement west-bound coal shipments, and it did not pay the railroads to haul empty cars one way at the then prevailing rates of freight. Westward freight forwarding also reduced the available pool of cars for shipping coal.[70]

This example illuminates the close interrelationship between the eastward grain trade and the westward coal trade, even though, in 1885, this relationship did not work as intended. The proper operation of the joint trade was admirably described by Eckley B. Coxe of the Coxe Brothers Mining Company in his remarkable testimony before the House of Representatives Committee investigating labor trouble in the Anthracite Regions in 1887–1888:

> 'The western farmer does not buy coal until he has got his harvest in; he then hauls the grain to the elevator, and buys a load of coal. The car that takes his grain to Chicago brings back a load of coal to the elevator. The vessel that takes the grain to Buffalo brings a load of coal from Buffalo. The boxcar that takes coal to Buffalo from the mines may be loaded with grain to New York and Philadelphia, or to our mines'.[71]

Bradsby, in his 1893 History of Luzerne County, went as far as referring to the adoption of boxcars for the coal/grain trade as a veritable revolution:

> 'with full loads each way transportation is so cheapened that anthracite is being used all through the west in competition with the bituminous coals which underlie many of the farms of those who use it'.[72]

The railroads hastened to participate in the new manner of conducting business. In 1884 for example, the Reading Railroad built a new depot just west of Shamokin, on its Shamokin, Sunbury and Lewisburg branch, for transferring coal from coal cars into boxcars for the western trade. The depot had a bin storage capacity of 2,000 tons and could load 100–125 10–12 ton cars per day. This was much quicker and more cost-effective than the previous manual methods of coal transhipment.[73]

To return to the activities of the Western Joint Committee, it is useful to examine details of their tonnage allotments by lake and rail. These have fortunately come down to us for certain months in 1885. For example, the November 1885 figures are listed in Table 8.7. These are based on expected total shipments of 800,000 tons for the month and they

TABLE 8.7

Tonnage Allotments by the Western Anthracite Joint Committee for November 1885.[74]

Shippers via	% Allotment	Equivalent tons
Delaware, Lackawanna and Western Railroad	30.45	243,600
Lehigh Valley Railroad	19.40	155,200
New York, Lake Erie and Western Railway	23.13	185,040
Pennsylvania Railroad	17.63	141,040
Philadelphia and Reading Railroad	9.39	75,120
Totals	100	800,000

TABLE 8.8
Buffalo Coal Trestles 1884.[77]

Trestle Owner	No. of Storage Pockets	Capacity of Pockets (gross tons)	Annual Handling Capacity of Trestle and Dock in tons
Lehigh Valley Railroad	64	6,000	300,000
Delaware, Lackawanna and Western Railroad	39	4,000	350,000
Delaware and Hudson Canal Co.	57	4,800	220,000
New York, Lake Erie and Western Railway	30	2,000	200,000
Pennsylvania Coal Co.	26	2,800	200,000
Buffalo, New York and Philadelphia Railway	38	3,500	200,000
J. Langdon and Co.	22	1,500	50,000

show the dominant position of the Lackawanna Railroad as a supplier to the western markets at that time. Not only did it have the largest allotment, but data for preceding months indicate that it had also exceeded that allotment between May and September of 1885 by a much larger percentage than any of the other railroads. This suggests it had the greatest capacity to respond to demand fluctuations in the western markets. As a result, it had actually shipped more than 32 percent of all the anthracite going west during the five months in question.[75]

It is useful also to compare the 1885 allotments with data on the coal-handling capacity of the different loading trestles in Buffalo given in Table 8.8. Again, at the time, the Lackawanna was pre-eminent in annual handling capacity, but the Lehigh Valley's Tifft Farm development was still under construction. The apparent absence of the Reading Railroad is noticeable, because it was waiting for its own docks to be built, as part of a development by the New York Central along the Niagara River. In the meantime, coal originating from its Mahanoy mines in the Western Middle Field, was being handled by the Buffalo, New York and Philadelphia Railway. The coal was actually brought in by W. L. Scott and Company under their name. The latter company also handled the anthracite of Coxe Brothers in the Eastern Middle Field and soft coal from the Clearfield and Pittsburgh districts.[76]

By the end of the century, the major players in the Buffalo market, namely the Lackawanna and the Lehigh Valley Railroads, had retained their positions (Table 8.9). However, the Philadelphia and Reading and Coxe Brothers had relegated the Delaware and Hudson and the Pennsylvania Coal Company to a more minor role at this stage. It should also be noted that the figures do not contain a breakdown of coal type, so it is possible that a limited amount of soft coal is included in the totals given. Buffalo's pre-eminence as a coal-exporting center somewhat overshadowed its significance as a local consumption point. In the mid 1880s, about 300,000 tons of anthracite were being used in the city[78] and a significant quantity of bituminous coal in its foundries and rolling mills, although precise figures are hard to locate. The remaining anthracite was forwarded by rail, the total exceeding a million tons in 1885. For a short period at around this time, further details of the destinations are also available. Detroit took about 100,000 tons a year, Chicago 3–400,000 tons and St. Louis about 50,000. The trade

TABLE 8.9
Lake Shipments from Buffalo by Major Company 1899.[79]

Company	Tons
Delaware, Lackawanna and Western Railroad	644,681
Lehigh Valley Coal Co.	472,915
Philadelphia and Reading Coal and Iron Co.	368,243
Williams and Peters	273,356
Coxe Bros and Co.	255,086
E. L. Hedstrom	197,477
Pennsylvania Coal Co.	154,048
Delaware and Hudson Co.	131,187

to way-points in Michigan and Indiana rose steadily, exceeding 300,000 tons in 1887, and destinations in Illinois outside Chicago received more than 200,000 tons in the same year.[80] Northern rail shipments to Canada also assumed a growing level of importance. By 1900, some 383,000 tons of anthracite and nearly 2.2 million tons of soft coal were being forwarded in this direction.[81]

A particularly interesting feature of the Buffalo coal trade was the close links between some of the smaller shipping firms and individual mines within the anthracite fields. Some of these connections were long-established, others developed at a later date. For example, J. Langdon and Company, who had been in the business since 1858, handled the product of several mines in the Shamokin region. They also operated mines in that area on a leased basis during the 1870s, namely Big Mountain, Hickory Ridge, and Henry Clay No. 1.[82] In the early 1860s, they had operated the Gaylord and Sweatland mines near Plymouth. Likewise, the Pittston and Elmira Coal Co. operated the Seneca, Ravine, and Twin Shaft breakers near Pittston.[83] In the early 1880s, a further spate of take-over activity gave other Buffalo firms direct control of the output of specific mines. Moser and Hoole, under the trading name of the Northern Coal Co., took over the Eagle Shaft, and G. R. Wilson took over the Tompkins Shaft. Both of these were in the Pittston area. Similarly, Andrew Langdon, of the Chicago coal-shipping firm of Langdon, Richardson and Co. established himself in Buffalo and acquired the Grassy Island Coal Company, operating in Winton Borough, together with the Enterprise in the Wilkes-Barre area. Finally, W. H. Davis and Co., a company organized in 1882, acquired the Fairmount Coal Company in Pittston to expand their existing lake and rail trade in anthracite.[84]

Given the close links between Buffalo and Chicago in terms of supply and demand for coal, it is not surprising that Andrew Langdon was only one of numerous entrepreneurs and firms with important commercial connections in the two cities. In 1878, W. L. Scott and Co., who were based in Erie but were an important Buffalo shipper, took over the Chicago coal firm of Dewey and Co. This latter was the oldest coal concern in the city, having been established in 1854.[85] A second major firm, that of Meeker, Hedstrom and Co., played a very important role in the development of both the coal and iron business of Chicago. The senior member of the firm, A. B. Meeker, came to Chicago in 1857, and had close business connections with Judge Packer of the Lehigh Valley Railroad. In addition to his involvement in the anthracite trade, he also constructed the first blast furnace in Chicago, was President of the Menomonee Iron Company and also of the Joliet Iron and Steel Works, which opened in 1869.[86] The other leading member of the firm, E. L. Hedstrom, resided in Buffalo and was responsible both for their local trade in that city (amounting to 75,000 tons per annum by the mid 1880s) and onward western shipment of 300,000 tons. Prior to the establishment of the agency in which he was a partner, Hedstrom had also handled the Buffalo trade of the Lehigh Valley Railroad. Subsequently, he concentrated on the sale of anthracite from the mines around Scranton. He had originally become involved in the coal business during the 1860s. Like the Langdons, he had dabbled in leasing of anthracite mines directly, one example being the Gaylord in Plymouth, for a short period in 1871–1872.[87]

THE CHICAGO ANTHRACITE TRADE

The Lackawanna Railroad was the earliest of the major anthracite railroads to be represented in Chicago in its own right, although precise details are hard to find. There is evidence that it owned property on Clark Street in the city by the late 1860s, which it apparently sold for nearly $70,000 in 1869. Another property was purchased in 1870 for $36,000, to be shared with Robert Law, Jervis Langdon, and the Pittston and Elmira Coal Company. Much later, in 1883, a substantial property was acquired on the Calumet River for $85,000.[88] Shortly after this, the Delaware and Hudson also established an agency, by taking over the long-standing firm of King and Bogle, retaining Walter Bogle to run the business. Fortuitously, the timing of the purchase coincided exactly with the bottom of the depression in May 1885 and the company was able to develop its business thereafter in a healthy fashion.[89]

Further examination of the growth of the Chicago coal trade during the period reveals quite a complex interplay between supply and demand for different types of coal and the use of alternative modes of transport. Until the beginning of the 1870s, lake receipts predominated (Figure 8.12). Prior to 1870, there is no breakdown available by type, but, in that year, approximately two-thirds of the tonnage was anthracite and the rest bituminous. Railroad shipments were negligible until after the Civil War when the Chicago and Alton Railroad and, to a lesser extent, the Pittsburgh, Columbus and

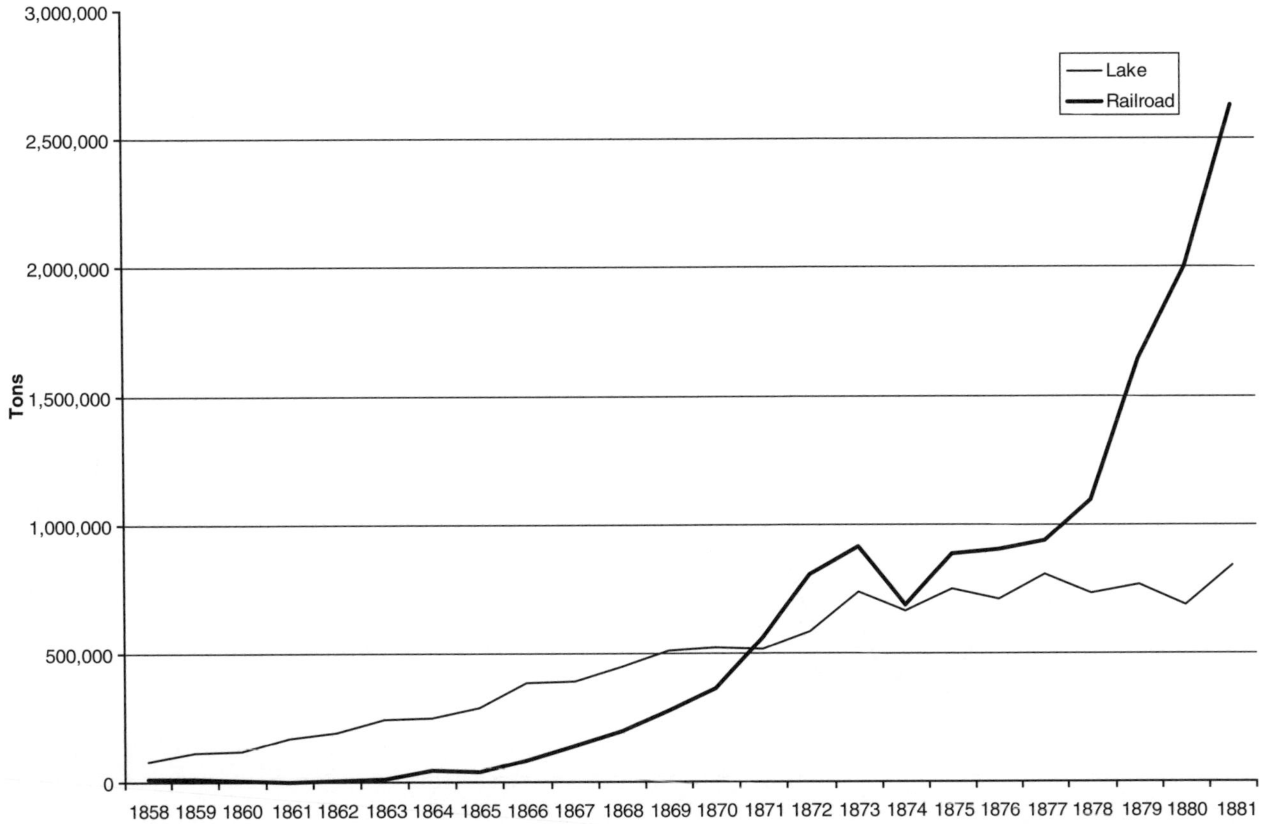

FIGURE 8.12
Chicago Total Coal Receipts by Mode of Shipment 1858–1881.

St. Louis and the Pittsburgh, Fort Wayne and Chicago railroads began to emerge as the major coal shippers.[90] Anthracite faltered after the 1873 panic and receipts declined from nearly 549,000 tons in 1873 to 325,500 tons by 1878. The deficit was made up by increased lake shipments of bituminous, while railroad shipments of the latter type of coal remained essentially static throughout the depression of the 1870s. The recovery of business confidence in 1879 brought an extraordinary surge in bituminous receipts in the city, which continued unabated until the depression of 1884–1886. This surge brought rapidly increasing traffic volumes to the railroads only, as lake shipments of bituminous coal declined progressively after 1878, falling to less than 54,000 tons by 1889.[91]

A rather different scenario unfolded for anthracite, as indicated in Figure 8.13. Lake shipments almost doubled between 1880 and 1884. Initially, this appears to have diverted traffic from the railroads, although receipts by this mode of transport began to recover after a time-lag of about two years. Improving business conditions after 1886 brought a short-term increase in railroad coal receipts, but this was almost immediately reversed by a sharp climb in lake traffic, which suppressed the railroad trade until the end of the century. Lake statistics are not reported after 1891, so shipments from Buffalo to Chicago are included on the graph as a surrogate measure, to indicate the approximate trajectory of the lake trade during the 1890s. The difference between Chicago receipts and the Buffalo shipments was very largely made up by shipments from the port of Erie, delivered to the docks by the Philadelphia and Erie division of the Pennsylvania Railroad. Lake anthracite shipments from Erie trebled between 1887 and 1895 from 157,000 to 480,000 tons per annum.[92]

Evidence of the sharp reversal in the fortunes of the anthracite trade by rail to Chicago after 1887 is difficult to locate. However, the rationale probably lies with the question of lower freight rates by lake, discussed above in connection with the Buffalo trade. The large increase in soft coal

FIGURE 8.13
Chicago Anthracite Receipts by Mode of Shipment 1872–1902.

receipts in Chicago during the 1880s must have put significant downward pressure on prices. Anthracite, as a premium commodity, would not have been immune to this pressure. Maximizing shipments by the cheapest means of transport would have been an obvious response to keep the price differential with soft coal as small as possible. There is certainly clear evidence of this downward pressure on prices, especially as the downturn of the 1884 depression began. A Chicago report in the trade press of the time made particular reference to the oversupply of soft coal in the city. The effect of this was to force down prices, send smaller firms to the wall and consolidate the business in the hands of a few corporations 'backed by railroad influences'.[93] However, at the same time, in a remarkable precursor to present-day concern with globalization, a quite difference reason for prices ruling low was also advanced:

> 'A prominent Chicago coal operator states that the western market for coal and its price is likely to be affected by the low price at which the farmer is selling his stock of wheat and the low price at which he must expect it to be sold, if he can sell it at all in the markets of the world, in competition with wheat from India and Australia'.[94]

Somewhat similar circumstances arose in 1892, as the Chicago market began to falter after several years of growth, prior to the panic of early 1893. At the beginning of 1892, there was a large glut of both anthracite and bituminous coal on the market. It was reported that over 5,200 loaded coal cars were standing on the tracks of the Chicago and Eastern

Illinois Railroad, so impeding the conduct of its business that it threatened to withdraw car service until they were cleared from the tracks.[95] The problem was admirably summarized by one of the dealers:

> 'Chicago is a regular dumping ground for soft coal. All that a mine operator cannot sell within a radius of 20 or 30 miles, he ships to Chicago where it will fetch something. His profit is in his local trade, and he can afford to lose something on what he sells here'.[96]

The Coal Trade Annual made a similar point slightly more crudely:

> 'The "rush" of coal to this market was enormous, as many mine operators believed the absorbing powers of Chicago were illimitable, the inevitable result of which was a glut which lasted for months and margins were lost sight of entirely in the desire "to get rid of the stuff" '.[97]

Early in 1892, the establishment of the 'Great Combine' in anthracite, as it was termed by the trade press, became known. This was formed as a result of the leasing of the Jersey Central and Lehigh Valley Railroads to the Philadelphia and Reading. Initially, President MacLeod of the Reading claimed that the object of the new organization was not to increase the price of coal, but to equalize it.[98] There was some substance to this claim, as there was a long-standing complaint that Chicago was paying relatively less for its hard coal than eastern cities.[99] However, it progressively became clear that MacLeod had rather more than fairness to his far-flung customers in mind. First came strict enforcement of the circular price of $5.00, so coal was no longer offered from the dock at $4.75, as it had been at the beginning of the year. Then, tactics reminiscent of the late lamented F. B. Gowen began to be employed. As an initial step, the Combine purchased the Silver Creek and Morris Coal Company for $550,000 at the beginning of March. This company had acted as western distributor for the Reading Coal and Iron Company in the preceding two years. The property included four coal yards in the Twin Cities, a transfer dock at West Superior and three docks in Chicago itself . Then followed a steady series of price rises that brought the wholesale price up to $5.50 ($6.50 retail) by the beginning of May, and $6.10 by September.[100] Not surprisingly these rapid price increases in anthracite produced a corresponding increase in demand for soft coal. The message for the corporate managers was spelled out bluntly by an industry observer:

> 'There is now no question that the combine have about ruined the anthracite trade west of the Missouri. The continued advances have antagonized both dealer and consumer. Business is very quiet throughout that region and few orders are now being placed at September prices. In many places the percentage of Eastern Anthracite consumed will be reduced fully 50 per cent. Many consumers have discarded their hard coal burners and will use the better grades of soft coal'.[101]

The annual summary of the trade amplified these points further to indicate that the high prices of Pennsylvania anthracite also allowed these western consumers beyond the Missouri River to substitute Colorado semi-anthracite on competitive terms.[102] Examination of relative prices in the Chicago market indicates the significance of the anthracite price rises, because they were not matched by an equivalent rise for soft coal. For example, at the end of August, when anthracite was $6.00 (per net ton), Pittsburgh coal sold in the range of $3.00–3.15, Indiana Block Coal $2.35–2.40, Wilmington (Illinois) coal at a very low figure of $1.50 and the other premium fuel, Connellsville coke, at $5.05. It is interesting to compare these prices with figures for earlier and later years. In 1878, for example, anthracite ranged from $5.75–6.75, when the high quality Briar Hill bituminous sold for $5.00 and Wilmington coal for $3.00 for much of the year. By 1901, the price of anthracite was unchanged from 1892 levels, Pittsburgh bituminous lump was $3.40, Indiana $2.00–2.25, and Wilmington $2.40. Long gone were the days of the Civil War, when a handful of dealers monopolized the trade and anthracite was sold as high as $20.00–22.00 a ton.[103] In the face of such long-term price differentials between anthracite and plentiful, cheap soft coals, it is not surprising that anthracite consumption in Chicago grew more slowly than that of bituminous coal during the last two decades of the century (Figure 8.14). Despite this, it is also apparent from the graphs that cyclical variations in coal demand in the city were very largely absorbed by the soft coal sector. Hence the Chicago anthracite trade, with its regularity of demand and relatively stable prices, formed a marked contrast to the more volatile markets of the eastern seaboard cities. This was doubtless an important factor in explaining the willingness of the major companies to invest so heavily in shipment facilities for the western trade, and especially by lake from Buffalo. The plot of onward shipments of anthracite in Figure 8.14 also suggests that there was a large measure of truth in the above quotation about the hard-coal trade west of the Missouri River. It does indeed appear that shipments westwards from Chicago fell off sharply after the price rises of the combine took effect in 1892–93.

THE SMALLER GREAT LAKES PORTS

From the mid-1880s onwards, detailed figures are available for the destinations of lake anthracite sent from Buffalo to ports other than Chicago. Although these give only a partial picture of consumption, they reveal some interesting differences between ports closer to Buffalo and those more distant. For example, shipments to Toledo peaked at just over 100,000 tons per year in 1892, but fell away to negligible proportions by 1902. The Detroit trade, which never significantly exceeded 40,000 tons, peaked a little earlier and declined to less than 4,000 tons by 1897.[104] Other more distant points such as Gladstone, Michigan and Green Bay, Wisconsin both had their largest shipments of approximately 55,000 and 38,000 tons respectively in 1893, before a slow decline set in.[105] However, the most important lake destinations, apart from Chicago, were those at the greatest distance by railroad, suggesting that water transport remained competitive for longest where the hauls were the most extended. The ports in question were Milwaukee and the rival shipping points of Duluth and Superior. A comparison of shipments to these destinations with those to Chicago can be seen in

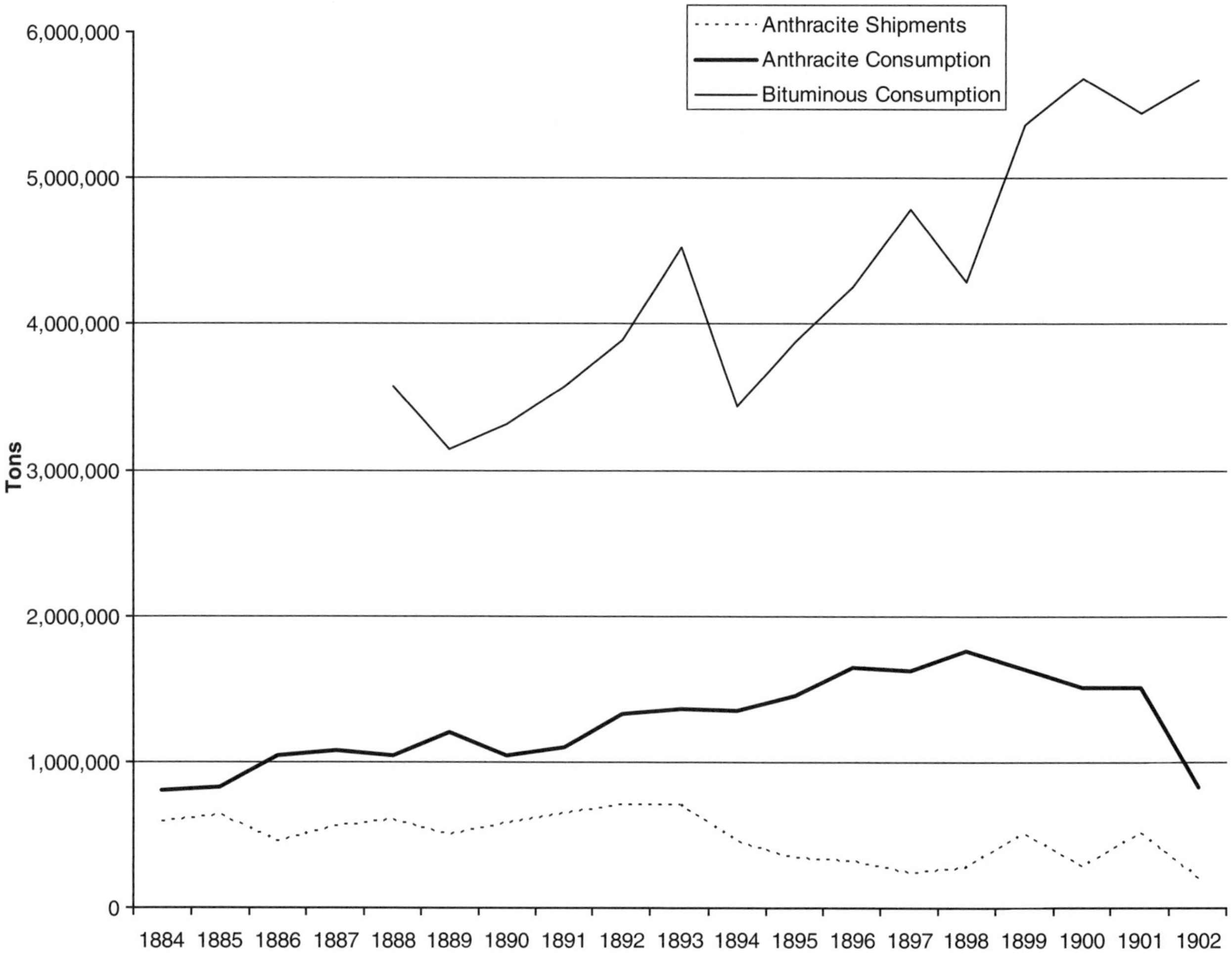

FIGURE 8.14
Chicago Coal Trade 1884–1902.

Figure 8.15. It is apparent that, taken together, their total receipts by the early 1890s matched those of Chicago, and the proportionate growth of shipments to Duluth and Superior, in particular, was much greater in the years 1885 to 1899, than that of the midwestern metropolis.

Turning to these smaller ports, the coal trade of Milwaukee was initiated in 1851, when R. P. Elmore entered the business, bringing a 600-ton cargo of inferior quality soft coal to the city. It was apparently sold off in small quantities and purchasers of ton lots 'were looked upon as capitalists worth cultivating'.[106] Prior to this, occasional small consignments of anthracite had been brought from Buffalo, usually packed in hogsheads, for the purpose of melting pig iron.[107] The trade remained at very modest levels over the next twenty years, but rose sharply in the years preceding the 1873 panic, as can be seen in Figure 8.16. Despite the unfortunate gap in the figures in the middle of the series, it is apparent that the boom years of the early 1880s ushered in a period of sustained growth in the coal trade as a whole. This lasted until 1899, apparently unaffected by the depression of the 1890s. As in the case of Chicago, the breakdown of receipts by type indicates that the expansion of the 1880s was anthracite-driven, while continued growth after 1895 was the result of increases in demand for soft coal. A trade report reflecting on the worsening business outlook of 1883 indicated the relative importance of hard and soft coal at that time. It stated that the rolling mills of the city usually consumed 75–80,000 tons of soft coal, but these and other manufacturing establishments had been compelled to run on limited time during the year, so only 30,000 tons were actually used in the iron business. Overall, approximately 300,000 tons of anthracite were received by dealers, and 45,000 tons of soft coal. A further 100,000 tons of bituminous were used in flour mills, brewing, grain elevators, and

FIGURE 8.15
Buffalo Coal Trade: Major Lake Destinations for Anthracite 1885–1902.

rolling mills, not counting coal for locomotives. The vast majority of coal received during the mid-eighties was shipped by lake, while rail receipts varied—60–83,000 tons per annum.[108] By 1893, the major players in the hard-coal trade were the Philadelphia and Reading Coal and Iron Co., Coxe Brothers, and the three major Northern Field companies, namely the Pennsylvania Coal Company, the Lackawanna, and the Delaware and Hudson. The last of these was represented by R. P. Elmore and Co., who were still active in the trade after more than forty years. The Lackawanna was represented by the North Western Fuel Company.[109]

The coal trade on Lake Superior was concentrated initially on Duluth. It began much later than on Lake Michigan and did not follow on rapidly from the opening of the Sault Ste. Marie Canal in the mid 1850s. The recorded reminiscences of one of the pioneers of the Duluth coal trade, B. S. Russell, indicate that shipments to Duluth commenced in 1871, following a visit in May of that year by J. J. Albright, now of the Delaware and Hudson, and C. M. Underhill of Buffalo, who represented other anthracite interests. The intended market was the Twin Cities of Minneapolis and St. Paul and points to the south of them. The aim was to encourage the adoption of anthracite in place of wood, which was the only available fuel in the region prior to that time. The first coal was brought in 3–800 ton capacity schooners and the onward rail freight to St. Paul was $1.50 a ton. However, adoption of the new fuel was slow at first, and much of the 8,000 tons brought in the first year remained unsold until the following season. In 1872, construction of a ship canal enabled vessels to reach the inner harbor and docks were constructed to facilitate unloading of the coal. Russell took responsibility for handling coal at Duluth. He also shipped it on to St. Paul to the other members of his syndicate, comprising the firms of Hill, Griggs and Co. and Garrett and Johnson.[110] In subsequent

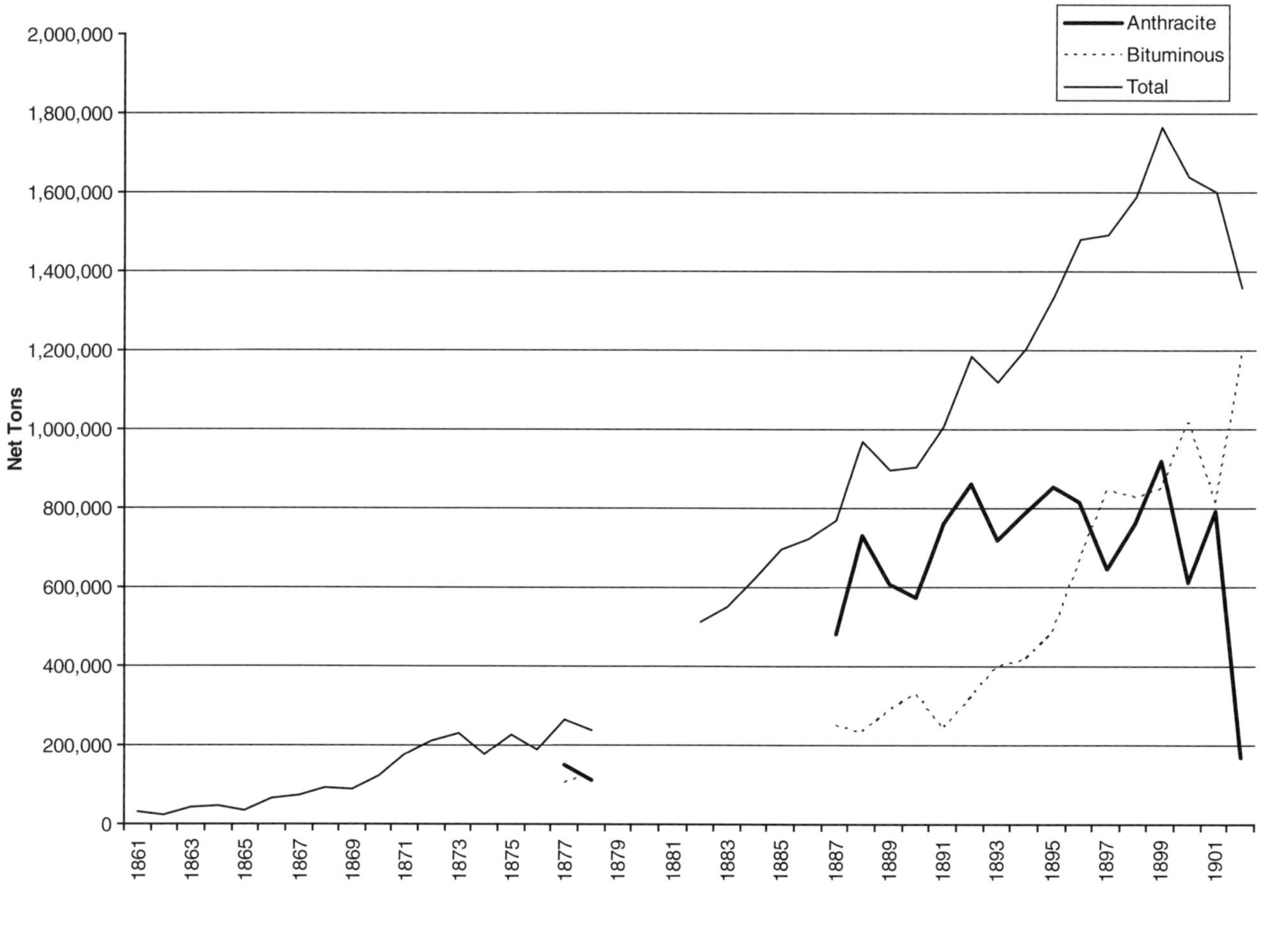

FIGURE 8.16
Milwaukee Coal Receipts by Type 1861–1902.

years, the commercial links between the Twin Cities and the Great Lakes shipping ports continued to develop. In 1876, James J. Hill, E. N. Saunders and others established the North Western Fuel Company in St. Paul. This company secured the north-western agency of the Lackawanna Railroad and progressively built up a very large trade in Scranton anthracite. About 1880, Hill sold his interest to W. L. Scott and Co., to turn his attention to railroading. He was eventually to become President of the Great Northern Railway and the 'biggest man in the Northwest'.[111] By 1893, the North Western Fuel Company owned docks in Duluth and West Superior, in addition to those in Milwaukee mentioned above. Its annual trade of 2 million tons included large shipments of both hard and soft coal from Pennsylvania.[112]

Another linked concern, called the Pioneer Fuel Company, also had a large 50-acre dock property in Duluth, where it could obtain cars for interior shipment without any switching charges. This company was incorporated in 1885 to handle the Minneapolis business of the North Western Fuel Company and had its headquarters in that city. It had a second large dock at Gladstone, Michigan. The company handled anthracite originating from the shippers over the Erie Railway, via Williams and Peters in Buffalo, and the Lehigh coal of Coxe Brothers.[113] Eckley B. Coxe underscored the importance of his connections with the Great Lakes ports, and the importance of the storage facilities at the docks, in his 1888 testimony, as follows:

> 'The farmer out west practically buys his coal almost exclusively in the months of August, September, October and November. But we cannot furnish all the coal needed then. The consequence is we are obliged in the months of May, June, July and August to pile up at Chicago, at Duluth, at Milwaukee and other shipping points, a large amount of coal, amounting to hundreds of thousand [sic] dollars in value, because we know when the farmer begins to move his grain he will want the coal'.[114]

And again:

> 'We shipped a great deal of coal to Minneapolis last year. That went to Buffalo by rail, from there by water to Duluth, and then to Minneapolis by rail, where it was delivered to consumers'.[115]

Interestingly, the anthracite operators were by no means the only people to foster a close association with the grain trade of the Great Plains. The coal docks in Duluth were carefully sited close to the grain elevators 'to make sure of a good car supply'.[116] In Minneapolis, one of the largest customers for anthracite was the Peavey Elevator Company. This was not because of its own demand for steam raising, but because it engaged in its own lucrative coal supply business. As the Coal Trade Journal observed:

> 'It sells coal to the farmers at the two hundred points where its local branch elevators are located; the wagon that brings in grain takes some coal back and thus the two branches run together in a very profitable way'.[117]

Other major coal companies with docks in Duluth or nearby Superior/West Superior included the Ohio Coal Company, which represented the Pennsylvania Coal Company and the Delaware and Hudson, the Lehigh Coal and Iron Company, representing Lehigh Valley Railroad interests, and the St. Paul and Pacific Coal and Iron Company, who were originally agents for the Philadelphia and Reading. The Lehigh Coal and Iron Company boasted the largest dock in Superior, but fell on hard times following the 1893 panic and ended up in the hands of the receiver. This may have accelerated the decision by the Lehigh Valley Railroad to build and operate its own docks in Superior, which had a capacity of 200,000 tons.[118] In similar fashion, the Reading Railroad also decided to operate independently of its former agents, as its business grew.[119]

The proximity of Duluth and Superior and the expanding coal and iron ore trade at each location, led to intense local commercial rivalries that impeded the accurate collection of statistics. The position was further complicated by vessels entering their papers in Duluth, but discharging their cargoes in Superior.[120] Nevertheless, the available data indicate that Duluth was the more important coal port during the 1880s, but Superior gained ground thereafter. Duluth held off the challenge from its neighbor, in terms of anthracite shipped from Buffalo, but was eventually surpassed in overall coal shipments. By 1900, two-thirds of the coal brought to the head of the lake went through Superior and only one-third through Duluth, although the latter shipped more than twice as much iron ore.[121] Such were the size of the coal-handling facilities in Superior/West Superior that in 1892 the Coal Trade Journal proclaimed the port to have surpassed Chicago, to become the 'greatest coal market in the

West'.[122] This may have been slightly premature, since it was based on returns from the shipping companies that appear optimistic in the light of other data. Nevertheless, the upward trend continued, despite setbacks in the mid 1890s (see Figure 8.17). By 1900, individual companies had docks 1,200–2,000 feet long able to unload multiple lake steamers simultaneously and store up to 500,000 tons of coal.[123]

The other Lake Superior ports of significance for the coal trade were Port Arthur and Fort William. These were on the northern shore in Canada. Much of the hard coal that reached cities like Winnipeg was brought from these ports by means of the Canadian Pacific Railroad. It was sold directly from the cars without being unloaded into storage.[124]

COMPETITION FROM SOFT COAL IN THE INTERIOR CITIES

Apart from the Twin Cities, where anthracite was largely brought by lake, followed by a short overland haul by rail, most other interior cities received their coal largely by rail. The cities in question included points relatively closer to Pennsylvania, such as Cleveland, Cincinnati and Louisville, as well as those more distant, such as St. Louis.

All of these cities also had more local supplies of lower-cost soft coal. Nevertheless, in several instances, modest levels of anthracite consumption were eventually achieved, which together formed a useful extension to overall markets for hard coal. Cleveland is a case in point. By 1879, its anthracite receipts of about 76,000 tons were only 5 percent of bituminous receipts. Relatively steady growth followed until 1893, when more than 250,000 tons were received [Figure 8.18]. This represented a very small increase in market share to 7.2 percent of bituminous receipts, but this share subsequently declined again during the remainder of the 1890s.[125] Despite its location on Lake Erie, Cleveland did not participate in onward shipments of anthracite by lake, although 15–25 percent of anthracite receipts were being re-shipped by rail during the early 1890s. Hence the remainder must have been used for local consumption.

FIGURE 8.17
Duluth/Superior Coal Receipts 1882–1902.

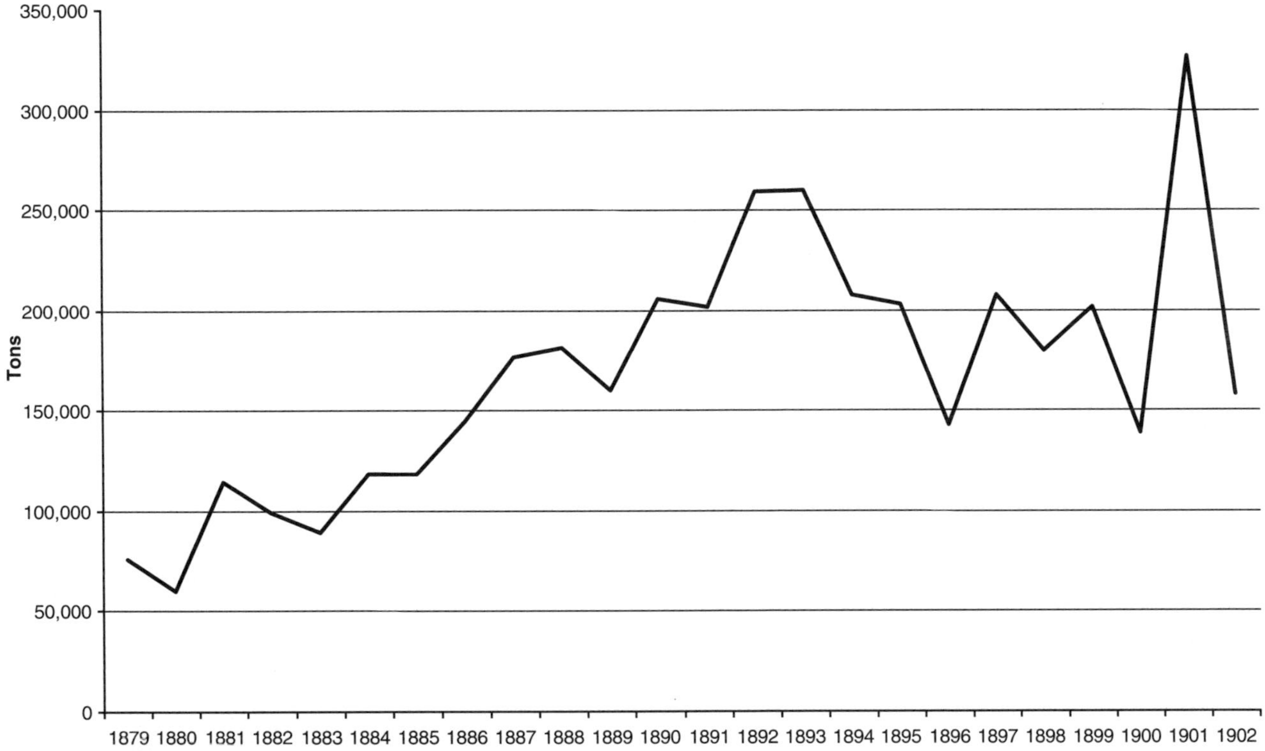

FIGURE 8.18
Cleveland Anthracite Receipts 1879–1902.

St. Louis provides a somewhat similar picture. In 1879, a mere 40,000 tons of anthracite were received, compared to almost 1.7 million tons of soft coal. However, the market for hard coal expanded until 1892, when its market share peaked at 7.8 percent of the bituminous total. Thereafter, although receipts increased slightly, in terms of market share, anthracite entered a period of relative decline. The main routes for bringing hard coal to the city were via the Wabash and Pennsylvania Railroad systems.[126]

Cincinnati and Louisville never became significant markets for anthracite. The former city was taking about 40,000 tons a year by the early 1890s, but the latter never exceeded about 20,000 tons during the period. By 1895, a trade report on Louisville confirmed that 'the output of Kentucky's mines [was] holding the local markets.' So not only was there little scope for expansion of the anthracite trade, but soft coal shipments from other areas, such as Pittsburgh, had also gone into decline.[127]

To summarize the overall picture of the development of the western and interior anthracite trade, it is apparent that two major factors enabled anthracite markets to extend more widely than perhaps the hard-coal trade had a right to expect, given the number of intervening opportunities for western customers to buy lower-priced soft coal. The first of these was the efficient and comparatively short rail hauls to Buffalo from the coalfields, over lines largely under the control of the hard-coal carriers themselves. The second was the persistence of low freight rates by water across the Great Lakes, initially to Chicago and Milwaukee, and subsequently to Duluth and Superior. Maintenance of low freight rates was facilitated by the burgeoning shipments of Lake iron ores, which enabled cargo vessels to carry this key raw material in one direction and fuel in the other. A similar synergy existed between westward coal shipments and eastward grain transportation. The continued growth of anthracite shipments to western destinations half a continent away from the mines, was a testimony to the degree of customer preference for the premium fuel and to the energy and persistence of the sales agents for the different companies. It also indicated the willingness of the latter to invest in the necessary docks and handling facilities to supply new markets, wherever they could be found.

Chapter 9

WAXING AND WANING MARKETS III: REGIONAL SHIFTS, PRICE BEHAVIOR, AND THE CHANGING SIZE COMPOSITION OF ANTHRACITE PRODUCTION

The accumulation of innumerable long-term investment decisions in specific geographical markets, short-run production decisions, and tactical responses to meet the competition from bituminous coal resulted in discernible shifts in the last two decades of the 19th century in the pattern of shipments to different regions. Changes in price behavior and in the proportion of different sizes of anthracite sent to market also took place. In the light of these changes, it is appropriate to investigate, as far as available data allow, the extent to which all companies participated equally in these long-term trends. Some, by dint of market inertia, managerial skill, or quality of product may have been able to buck the overall trends, at least temporarily, securing thereby a measure of economic advantage. These topics will now be examined in more detail.

REGIONAL SHIFTS IN ANTHRACITE CONSUMPTION

It is always difficult to obtain statistics that have been calculated on a consistent basis over long periods of time. This is especially true when a geographical breakdown is required. Fortunately, for two periods in the 1880s and 1890s, such statistics were collected and reported for the regional distribution of anthracite consumption, although no original documents have survived that might allow their correctness to be checked. The available regional breakdown is quite a broad one. The three most important categories are the middle Atlantic states, which included Pennsylvania, New York, and New Jersey, followed by New England and the 'western' states. Other more minor categories include the southern states, Canada, and the Pacific Coast. Figure 9.1 shows the relative shifts in importance of two of the three major regions over the period, for the years when the data were collected. It is apparent that the regional share of the 'western' states almost doubled between 1882 and 1894, from 7.6 to 14.9 percent. This growth was very largely at the relative expense of the middle Atlantic states. However, at the end of the century, the latter region began to reassert itself and the proportionate share of the 'western' states registered a slight decline. These changes need to be seen against the backdrop of an overall 50 percent increase in anthracite shipments during the period covered by the graph. The continued dominance of the cities of New York and Philadelphia, together with their industrial hinterlands, is a striking feature of the period. Equally interesting is the fact that New England's regional share, which is not shown on the graph, remained remarkably constant in the range 15–17 percent. The southern states consistently took about 4 percent of shipments. This was slightly more than Canada, whose proportion increased from 2.1 percent in 1882 to 3.6 percent in 1901. Shipments in the latter year totalled 1.93 million tons.[1] In this context, the cities of Toronto and Montreal present an interesting contrast. Until the late 1890s, Toronto consumed more anthracite than bituminous (Figure 9.2). In the latter city, coal of US origin, which was effectively all anthracite, found a ready but static market. Increasing demand from the city was almost entirely met by growing receipts of Nova Scotia bituminous from the 1880s onwards (Figure 9.3).

While general statistics on the regional distribution of shipments are hard to obtain, it is even more problematic to find equivalent information for individual companies. In the case

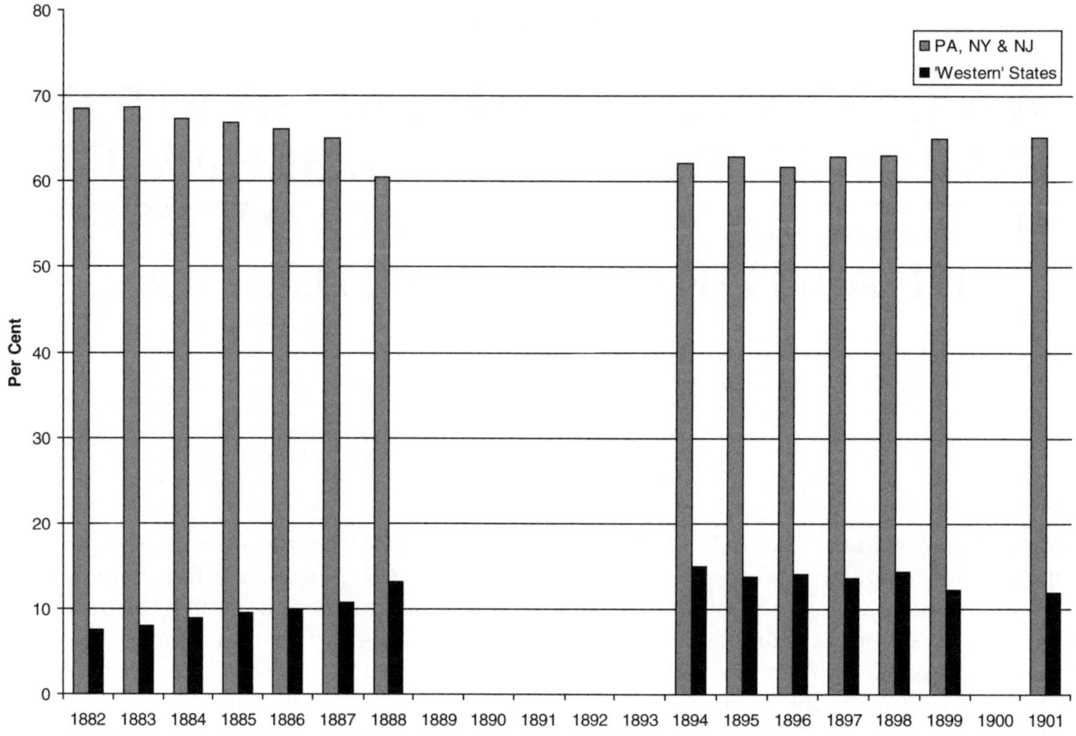

FIGURE 9.1
Anthracite: Percentage of Total Shipments by Destination Region 1882–1901.

FIGURE 9.2
Toronto Coal Receipts by Type 1887–1902.

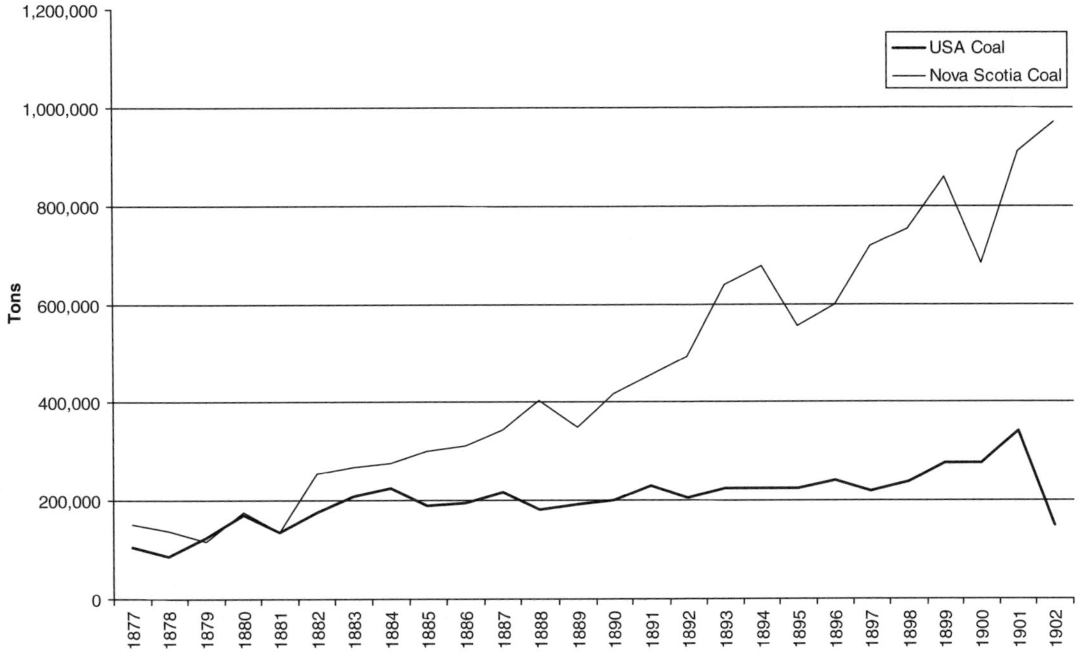

FIGURE 9.3
Montreal Coal Receipts 1877–1902.

of the Lackawanna Railroad Coal Department, this has proved possible, but only by delving deep into the company's accounts. Even then, the data are reported in some years and not others. Table 9.1 shows coal sales by the company in a number of major cities over the period 1882–1893. Collectively, these sales account for more than 95 percent of total sales in each of the specified years. The total sales figures in question are given in the bottom row of the table.

TABLE 9.1
Coal Sales in Major Cities by the Lackawanna Railroad Coal Department.[2]

	1882		1886		1890		1893	
	Tons	%	Tons	%	Tons	%	Tons	%
Hoboken, NJ	980,753	28.6	603,805	16.3	850,916	18.2	1,093,220	20.5
Washington, NJ	303,393	8.8	306,725	8.3	367,442	7.8	401,651	7.5
Scranton	474,124	13.8	562,282	15.2	673,399	14.4	813,502	15.2
Albany	-	-	61,487	1.7	108,518	2.3	105,737	2.0
Utica	219,173	6.4	221,753	6.0	344,712	7.4	412,028	7.7
Syracuse	359,128	10.5	362,348	9.8	469,721	10.0	505,507	9.5
Oswego	175,760	5.1	141,689	3.8	152,370	3.3	154,651	2.9
Buffalo	353,581	10.3	344,593	9.3	551,529	11.8	602,996	11.3
Toledo	116,585	3.4	186,102	5.0	218,151	4.7	248,419	4.6
Cleveland	-	-	24,745	0.7	58,053	1.2	85,939	1.6
Chicago	208,459	6.1	314,874	8.5	414,139	8.8	507,588	9.5
Milwaukee	-	-	114,977	3.1	221,378	4.7	177,583	3.3
Duluth/Fort William	-	-	50,517	1.4	3,892	0.1	567	0.0
Lake Superior	-	-	-	-	101,567	2.2	117,038	2.2
Montreal/Quebec	51,030	1.5	85,075	2.3	74,254	1.6	81,493	1.5
Boston	72,091	2.1	145,545	3.9	-	-	-	-
Worcester	-	-	95,517	2.6	35,261	0.8	-	-
Total Sales to all points (Tons)	3,432,169		3,695,955		4,682,929		5,343,809	

Several points emerge from examination of these figures. The first is that, in terms of the broad regional breakdown of shipments, the category for Pennsylvania, New York, and New Jersey tends to be over-represented (83.5 percent in 1882 and 76.6 percent in 1893), compared to the industry average, which is shown in Figure 9.1. The same applies in the western region. Secondly, the corollary is that the company had essentially withdrawn from the New England market by the early 1890s, leaving the business to the Reading Coal and Iron Company and the Lehigh and Wilkes-Barre Coal Company, among others. Thirdly, local industry and domestic demand in the Scranton area remains of considerable importance throughout the period. Finally, sales in Chicago increased by about 2.5 times, outstripping Buffalo sales, which grew more in line with the overall increase in tonnage. It is also possible that some Buffalo coal was re-shipped to Canadian points after being sold by the company, but this cannot be verified. Overall, these figures confirm earlier evidence about the relative dominance of the Lackawanna Railroad in the western trade. This was a result of its 'first mover advantage', in securing its own rail outlet to Buffalo prior to the Reading and the Lehigh Valley.

PRICE BEHAVIOR

OVERVIEW

At the outset, it is useful to set price behavior during the period of interest within a much longer historical context. This can be seen in Figure 9.4, which shows the standard series of annual prices in the century following the opening of the Pottsville trade in 1825. After a period of rapid price fluctuations in the early years, prices settled into a band roughly between \$3 and \$5, from the mid 1840s until World War I. The two most marked deviations from this were the inflationary period of the Civil War and the bottom of the depression of the 1870s. Such a graph tends to support the hypothesis referred to in chapter 2, that the 19th century was characterized by relative price stability, punctuated by occasional inflationary and deflationary episodes. However, the annual figures conceal sharper month-to-month fluctuations in the eastern seaboard market cities. For example, Figure 9.5 gives the monthly prices of Schuylkill white ash lump coal from 1853 to 1883. This shows that the monthly peaks during inflationary periods were significantly higher than appears to be the case from the annual figures and the troughs slightly lower. Sharp price changes in 1869 and gaps in the early 1870s reflect

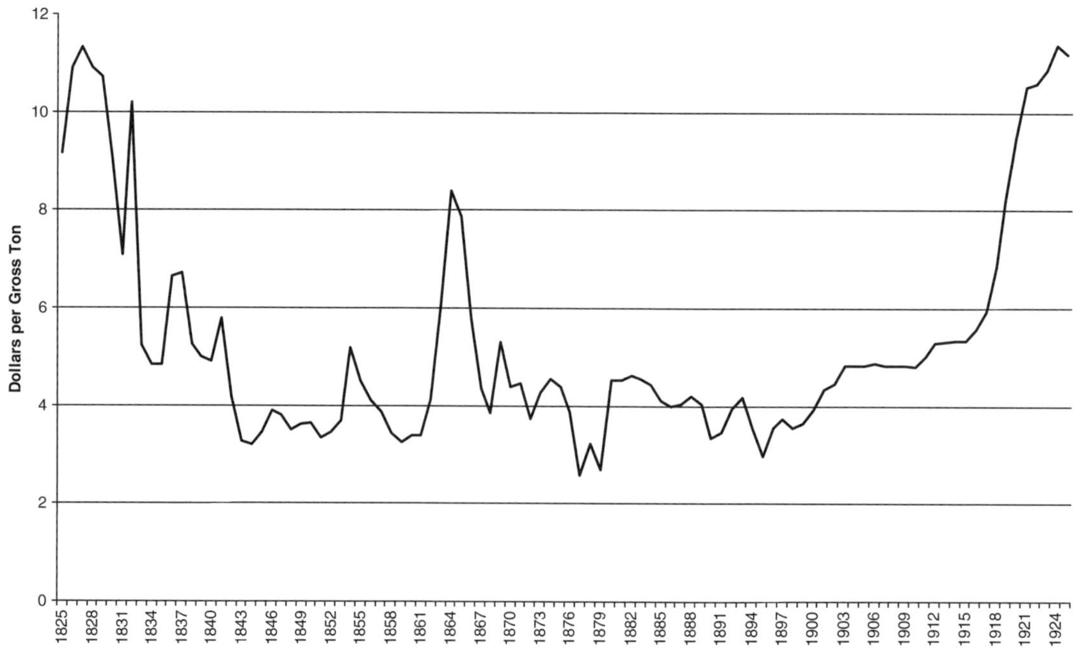

FIGURE 9.4
Average Annual Price of Anthracite Coal 1825–1925.

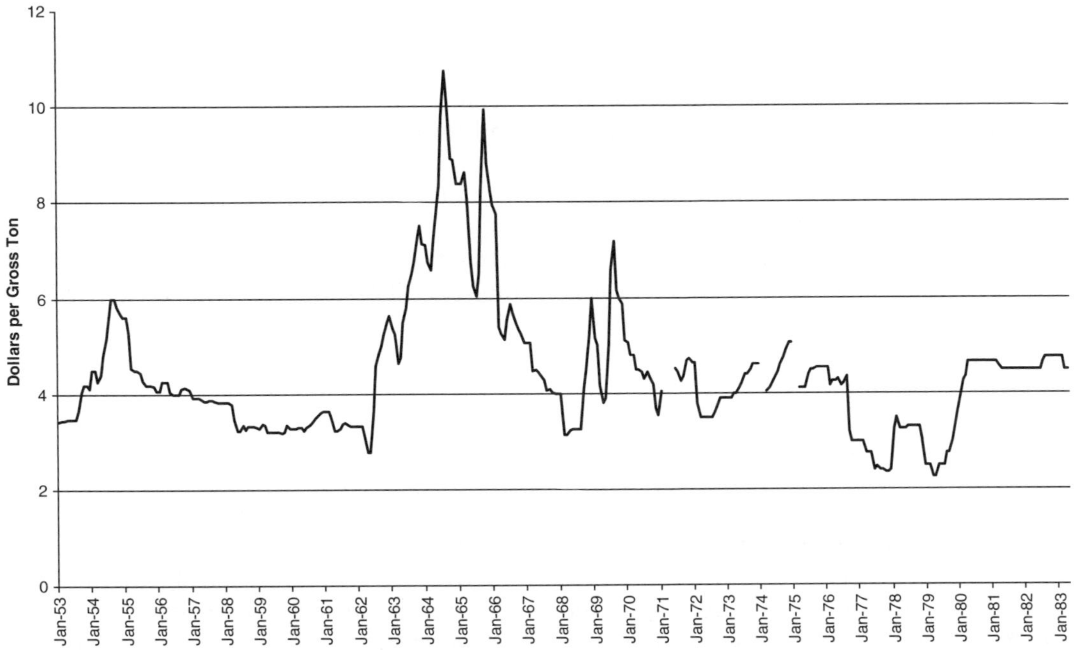

FIGURE 9.5
Monthly Prices of Schuylkill White Ash Lump Anthracite by the Cargo at Philadelphia 1853–1883.

strike activity. Confirmation of these sharp monthly peaks can be found in Figure 9.6, which shows the average prices realized by the sale of stove and chestnut coal at the 'Scranton coal auctions' of the Lackawanna Railroad, as they were known in the trade. These auctions, held in New York, were initiated during the Civil War and finally discontinued in 1879, although they had also been

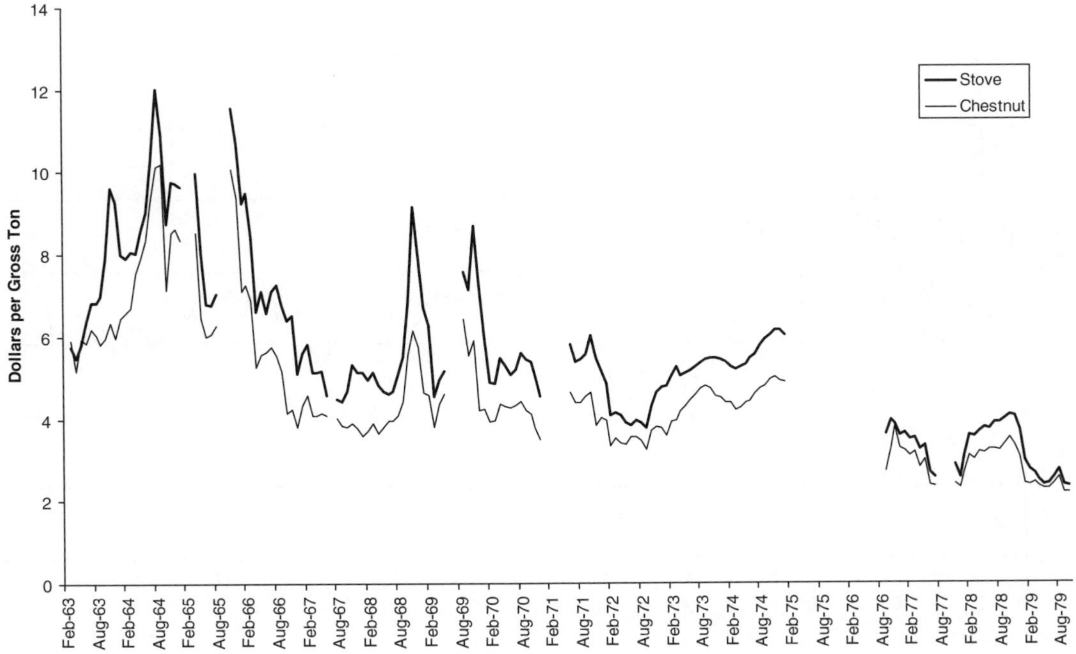

FIGURE 9.6
Scranton Coal Auctions 1863–1879: Comparative Prices of Stove and Chestnut Coal.

suspended for a period during the mid 1870s. They were described by one contemporary commentator as 'the best barometer to indicate the prices of anthracite coal'.[3] In the absence of any complete tabulated record of the auctions, these price series were pieced together, largely from reports in a variety of newspapers of the period. Some of the small gaps in the series are the result of strike activity, others from incomplete reporting. Nevertheless, it is clear that prices of stove coal, the premier domestic size, briefly reached $12 in 1864, nearly 50 percent higher than the annual average would indicate.

Unfortunately, it is very difficult to find consistent monthly price series that span all, rather than part of the period of interest. The best data available for the middle of the period come from the Delaware and Hudson Canal Company (Figure 9.7). This series overlaps with that from the records of the Lehigh and Wilkes-Barre Coal Company, although the latter series has a gap, in the mid 1890s, during the period when it was temporarily taken over by the Reading (Figure 9.8). In the Lehigh and Wilkes-Barre series, it is particularly tempting to observe the troughs of two Juglar cycles in 1886 and 1895.

The apparent 'double-dip' trough of the late 1870s also reflects other factors. These include the deflationary period prior to the return to the Gold Standard in 1879 and the presence or absence of the anthracite combination.

Taking all these series together, it is possible to identify several different periods of price behavior, while not forgetting the underlying influence of business fluctuations of different periodicities. These periods are as follows:

1. Civil War inflation and post-war price adjustment — 1862–1872
2. Partially effective combination to combat depressed markets — 1873–1879
3. Shipment quotas and attempts at price stability — 1880–1886
4. Covert co-operation and fluctuating fortunes under the Inter-State Commerce Commission and Sherman Anti-Trust Acts — 1887–1902

The first of these has, as its major feature, the expected inflationary rise in prices, peaking in August 1864, which corresponded to overall price

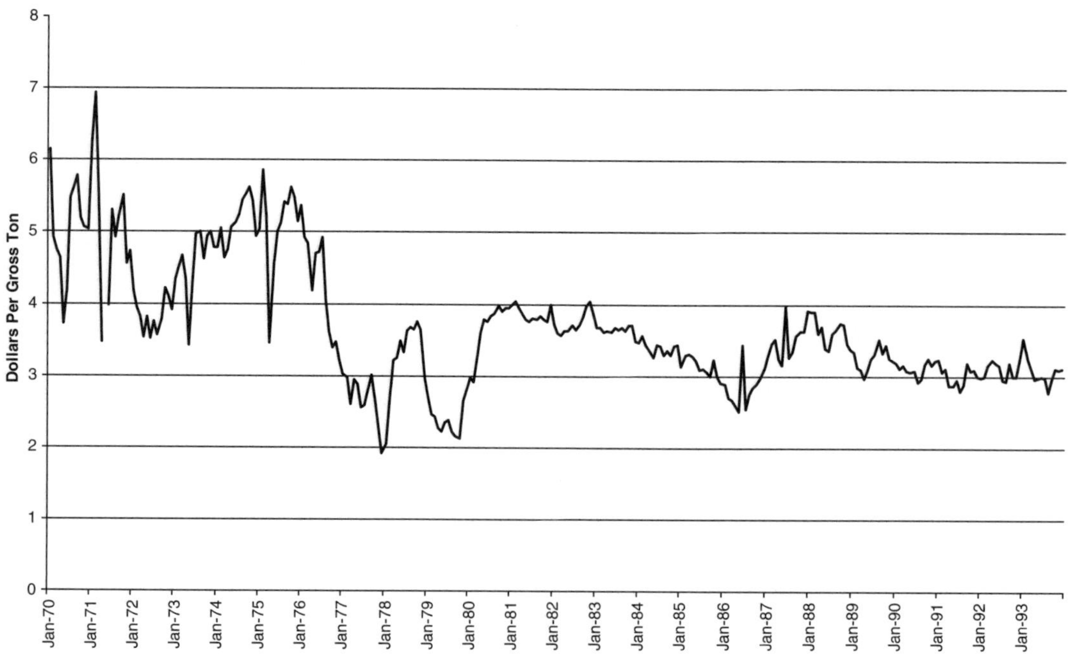

FIGURE 9.7
Delaware and Hudson Canal Company: Average Wholesale Prices for All Sizes of Anthracite F.O.B. New York 1870–1893.

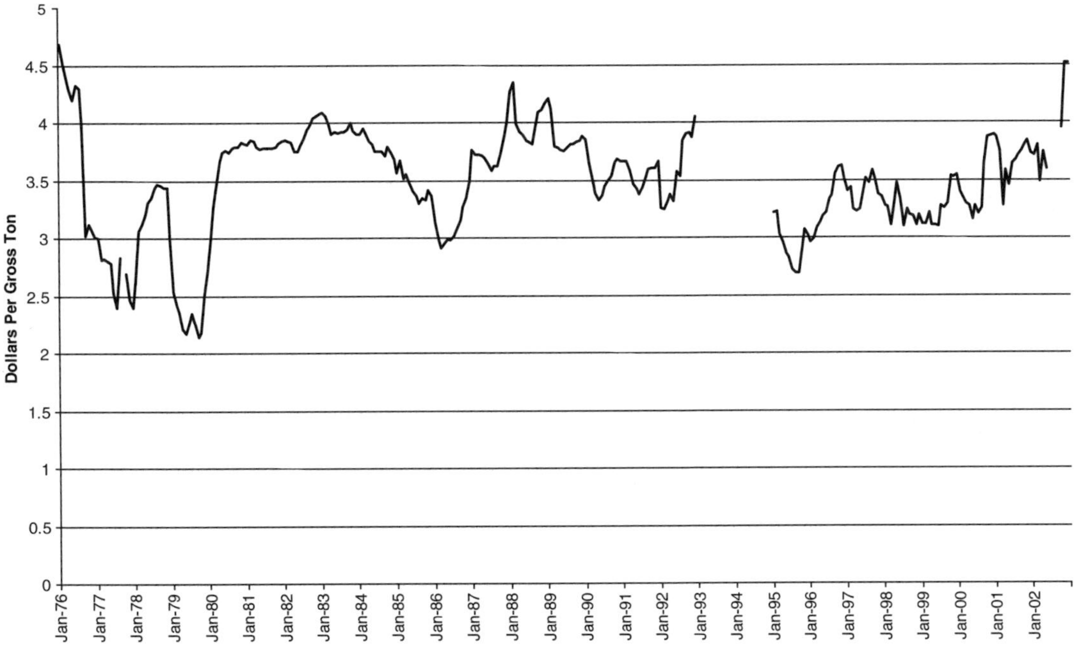

FIGURE 9.8
Lehigh and Wilkes-Barre Coal Company: Average Prices of All Sizes 1876–1902.

movements in the national economy during the Civil War. However, secondary peaks during 1865, at the end of 1868, and in the late summer of 1869 reflect the impact of strikes in different parts of the coalfields. The effect of these deviations from normal market conditions can be seen in Figure 9.9, which compares monthly prices of anthracite and Cumberland bituminous. Although the slightly patchy data for the latter track anthracite prices quite closely from 1858–1864, the later secondary peaks in the anthracite data are not present for the bituminous fuel. It is also noticeable that the good quality and limited supply of Cumberland coal allowed its quoted prices to remain stable and slightly higher, on average, than anthracite from 1866 to 1873. It should also be remembered that small additional freight charges from Baltimore would be required to make it more comparable to the Philadelphia anthracite prices. This would seem to imply that competition from bituminous coal was not a major factor affecting anthracite prices on the eastern seaboard, during this period.

During the second and third periods, the activities of the 1870s anthracite combination, in its various forms, dominated attempts to control or manipulate prices, either directly or indirectly, via coal supply restrictions. These topics will be dealt with in detail in Chapter 11. However, it is interesting to note that upsurges in combination-related activity in 1886 and 1895 correspond very closely with periods of severe price weakness (Figure 9.8). The increasing degree of price fluctuation visible in the latter graph between 1887 and the late 1890s is understandable, given the legislation to counter restraints on trade and the formation of trusts. In the light of this, the progressive price declines of 1884–5 merit closer attention, since the combination was in force, government regulation by legislation was not yet in place, but the stable prices of the early 1880s could not be maintained. An explanation for this price behavior will be sought in the next section, in terms of growing competition from a variety of sources of bituminous coal supply.

During the final period, under the influence of legislation, attempts at market control had to become more subtle. These involved a range of devices, from collusion between sales agents to the 'community of interest' at the end of the century.[4] Other factors also entered into the equation, such as the depression of the 1890s, which, despite its overall severity, did not produce as serious a decline in prices as had been experienced during the 1870s. Likewise the impact of labor troubles in 1887–1888

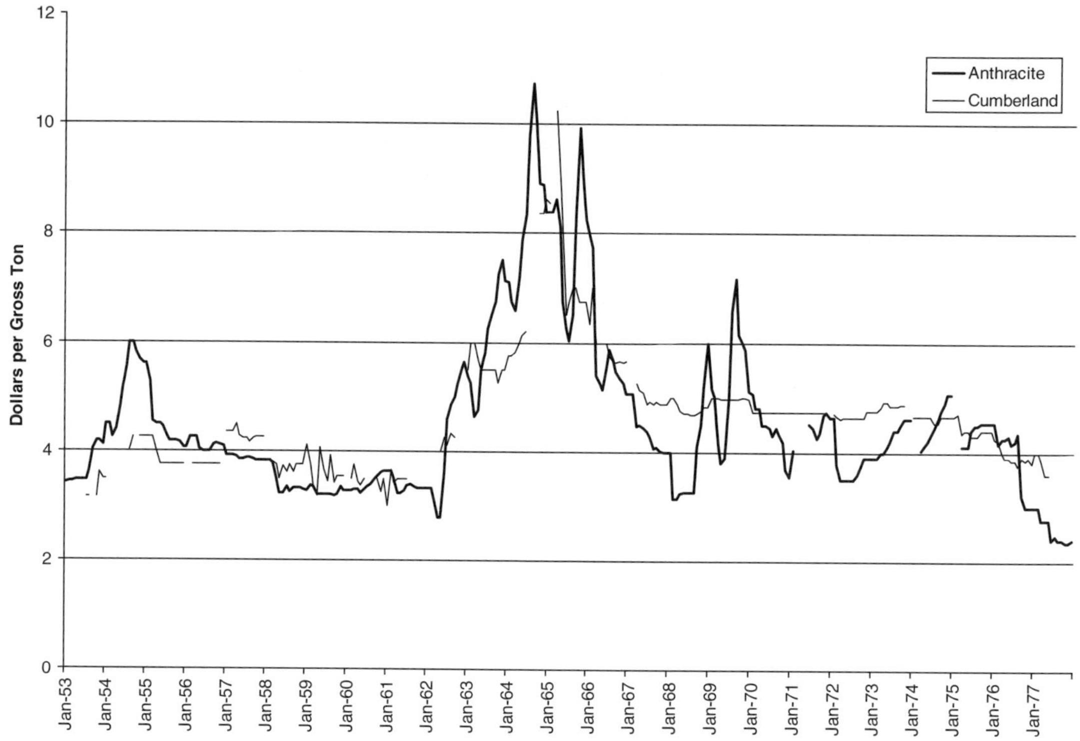

FIGURE 9.9
Monthly Prices of Schuylkill White Ash Lump Anthracite at Philadelphia and Cumberland Bituminous at Baltimore 1853–1877.

and 1900 failed to generate the amplitude of price fluctuations occasioned by similar events in the 1860s. It took the all-out struggle in 1902 between the operators and the men to bring about marked price increases, right at the end of the period.[5]

A second aspect of price behavior is the extent to which different regions, different companies, and different coal sizes experienced similar patterns of price changes over extended periods of time. Comprehensive comparisons are difficult because of the data limitation on the availability of long-time series, but a number of general points can still be made. In Figure 9.10, the prices for stove coal at the Scranton coal auction sales are plotted against those for stove coal from the Lehigh Region at the shipping point on New York Harbor. The closeness of the correspondence for coal from the two different regions is very noticeable. However, there is an interesting tendency for the auction prices to lead the Lehigh figures in any fall by a very short time and to drop slightly lower into the bottom of price troughs. Such results would tend to confirm both the earlier quotation about the role of the auction sales and also the competitive nature of the market for anthracite on the eastern seaboard.

An alternative method of comparing prices is to examine the series for different companies. In Figure 9.11, the data from earlier graphs are re-plotted to show this for the Delaware and Hudson and the Lehigh and Wilkes-Barre Coal Company. These prices are the average for all sizes and, in the case of the latter company, they are prices f.o.b. Port Johnston, the shipping point of the Jersey Central Railroad on New York Harbor. It is possible that the curious spikes in the 1886 and 1887 figures for the Delaware and Hudson are errors in the original printed tabulations, which had not been noticed, because they have not previously been plotted. Until 1882 there is a very close correspondence between the two series, although the Delaware and Hudson appears to experience slightly higher peaks and deeper troughs when fluctuations occur. Thereafter, a distinct gap, which is often as much as 50 cents a ton, opens up between the two series and this is maintained for at least a decade. Price differentials of this magnitude, if not swallowed up by higher royalty payment on leases [see Chapter 12], would be of considerable significance for the profitability of mining operations, when multiplied up across the entire output of a company. In this particular case, the prices shown for the Lehigh and Wilkes-Barre cover the output from its mines in the Wilkes-Barre and Plymouth areas, but not those in the Lehigh Region.[6] Its Plymouth coals were particularly sought after for domestic use and they carried a significant price premium over its own Wilkes-Barre output, as can be seen from Figure 9.12. This helps to explain the overall price differential with the Delaware and Hudson. Although the latter also had mines in Plymouth and Wilkes-Barre, as well as producing 'Lackawanna' coal from the Carbondale area north of Scranton, in general its anthracite was not quite of the choicest kind.

The remaining type of price differential was a function of the different sizes of coal. Comparison of Figure 9.6 and Figure 9.13 shows that price differentials for the smaller coal sizes of stove and chestnut persisted from the beginning of the period until 1900. There is, however, an indication, from the 1880s onwards, that at times of peak overall demand, chestnut prices would climb to match those of stove coal, the preferred domestic type. In contrast, when demand weakened, as in 1884, chestnut prices would fall away more quickly than those of stove. Consignments of pea coal were never offered for sale at the Scranton auctions, but it is apparent from Figure 9.13 that pea coal sold at a discount of approximately $1 a ton to chestnut coal. This discount persisted for two decades and only widened slightly at the very end of the period. Prices for the still smaller buckwheat size were not reported until 1891, when a discount of about $1 compared to pea prices was offered. Interestingly, by 1900, the discount, rather than widening, had narrowed significantly to about 30–40 cents per ton, on average. In some months, the differences were even smaller.[7] At the other end of the size spectrum, broken and egg coal followed very similar price trajectories. These, in turn, mimicked the pattern of chestnut coal prices in Figure 9.13 very closely.

Supporting evidence for these company-specific trends and differentials can be found from the wider market also. Data from the Anthracite Coal Operators Association in 1897, shown in Table 9.2,

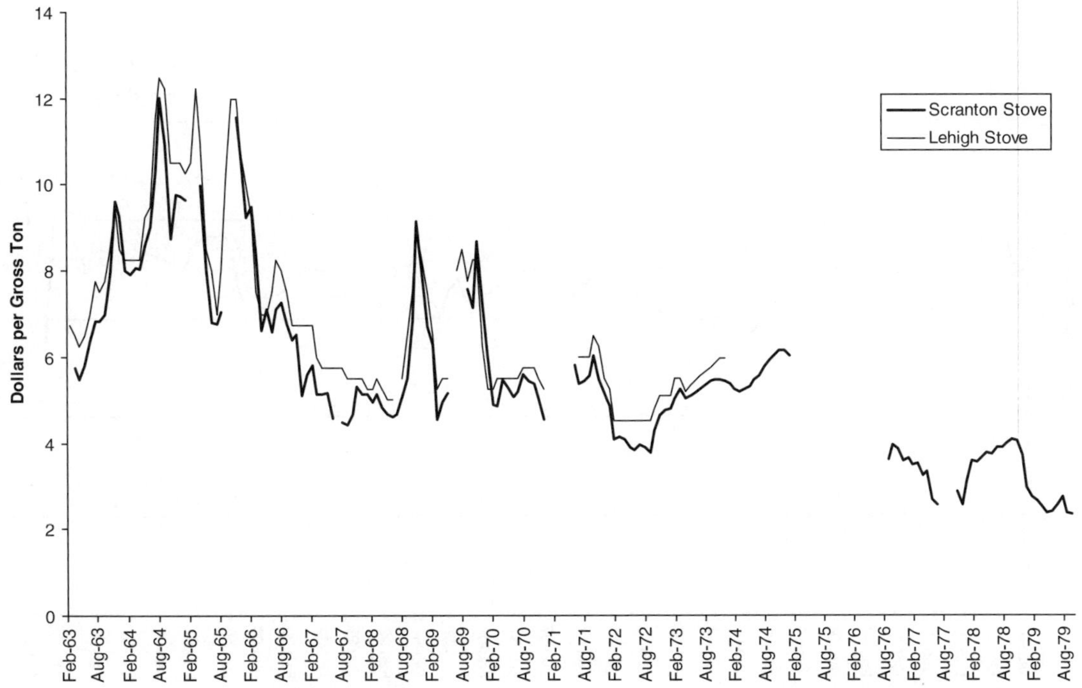

FIGURE 9.10
Comparative Prices of Stove Coal 1863–1879.

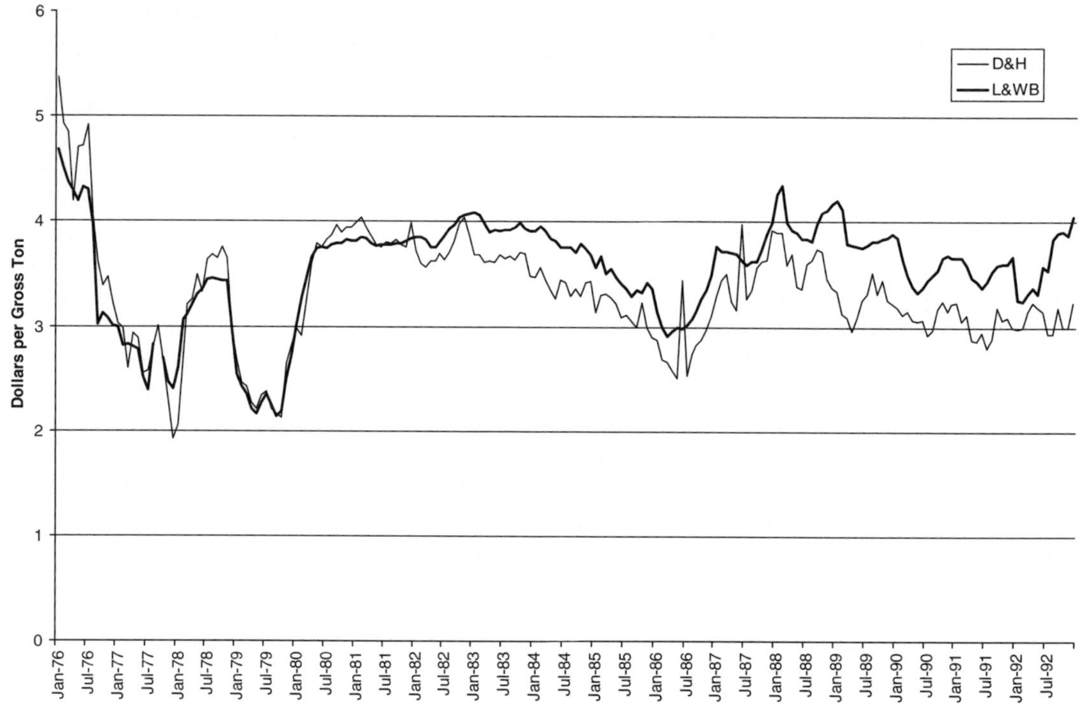

FIGURE 9.11
Comparative Average Anthracite Prices by Company in New York 1876–1892.

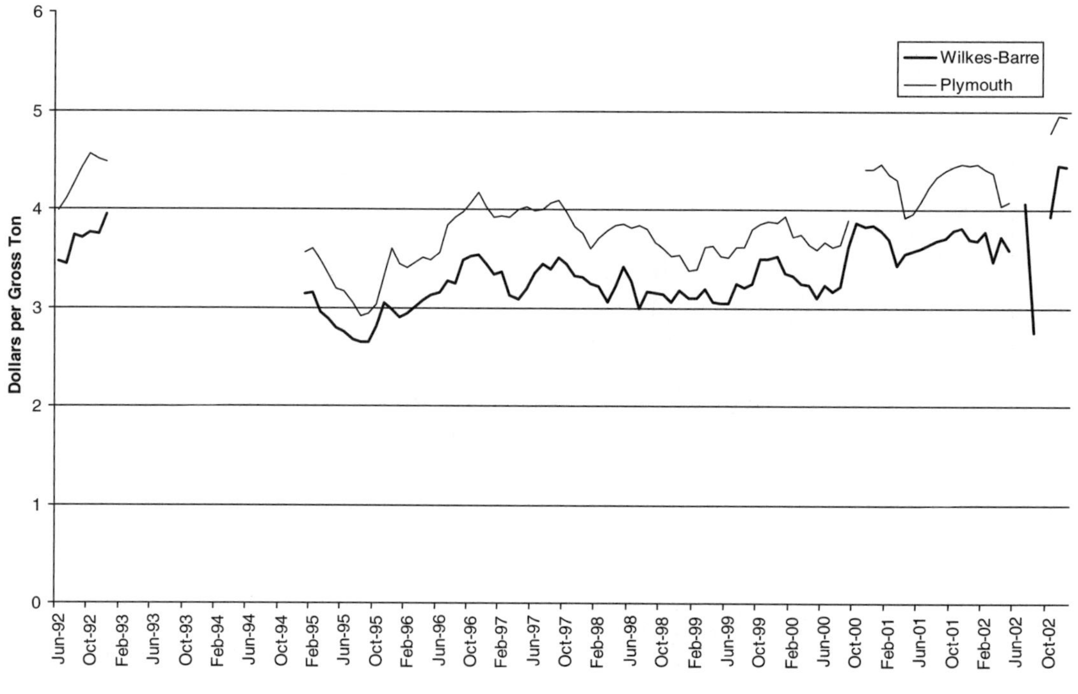

FIGURE 9.12
Lehigh and Wilkes-Barre Coal Company: Average Prices by Origin of Coal Shipments 1892–1902.

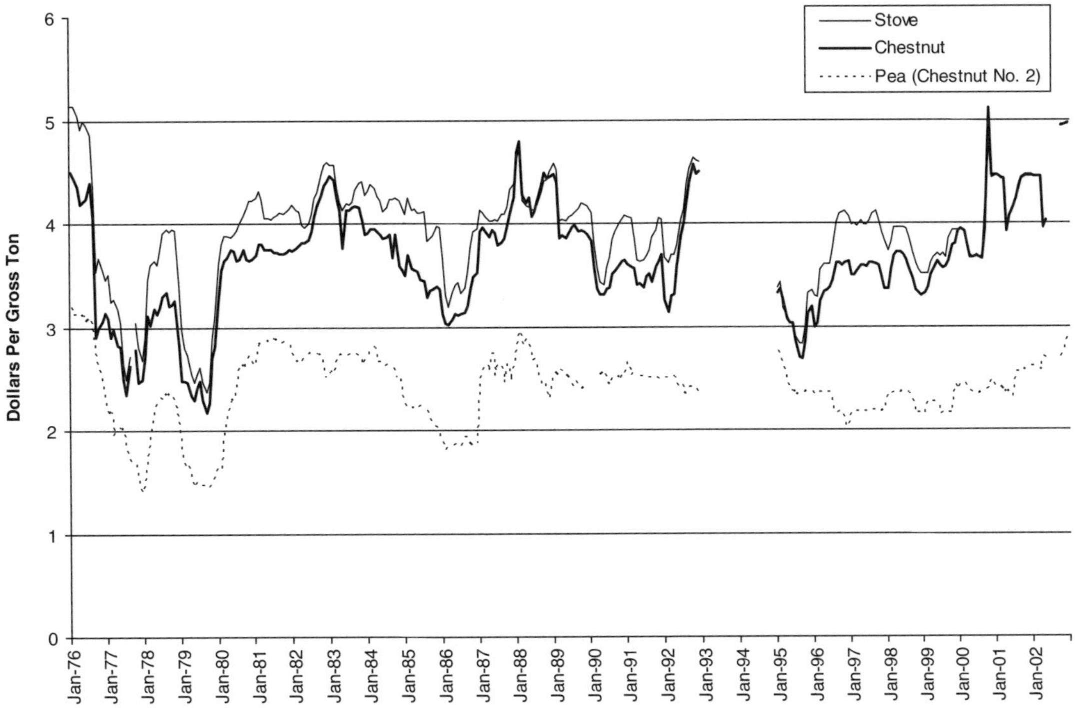

FIGURE 9.13
Lehigh and Wilkes-Barre Coal Company: Prices of Smaller Sizes 1876–1902.

would tend to confirm that the Lehigh and Wilkes-Barre figures for these sizes are quite representative.

PRICE COMPETITION FROM BITUMINOUS COAL

In the early part of the period of interest, the limited development of the interior bituminous coalfields and the lack of good, low-cost transportation to the eastern seaboard meant that anthracite did not experience serious competitive pressures from soft coal at this time. The only possible exceptions were the availability of Cumberland coal from Baltimore and Nova Scotia coal from Canada. The latter was only really a significant factor in the Boston market and the supply of the former was limited, as noted earlier. Not until the late 1870s and early 1880s did eastward shipment of bituminous coal start to become large enough to threaten the position of anthracite in the coal trade of the tidewater markets. It is surprisingly difficult to find convenient price series for bituminous coal in the eastern ports to provide a direct comparison with anthracite, but it has proved possible to build up several series of limited duration for both Cumberland and Clearfield coals. The first of these is given in Figure 9.14. It presents an annual version of the monthly data given in Figure 9.9, but it extends the series through the key period of the 1880s. The purpose is less to provide an exact price comparison (since the coals were quoted at different shipping ports) than to examine the timing of price trends. In the latter part of the 1870s, there is a distinct impression that depressed conditions and low prices in anthracite were putting downward pressure on Cumberland bituminous prices. However, at the start of the 1880s, when anthracite prices rose sharply on re-establishment of the combination, Cumberland prices registered a

TABLE 9.2
Average Annual Prices for Different Sizes of Anthracite at Tidewater in 1897.[8]

Size	Price ($)
Broken	3.373
Egg	3.713
Stove	3.900
Chestnut	3.601
Pea	2.215
Buckwheat	1.820

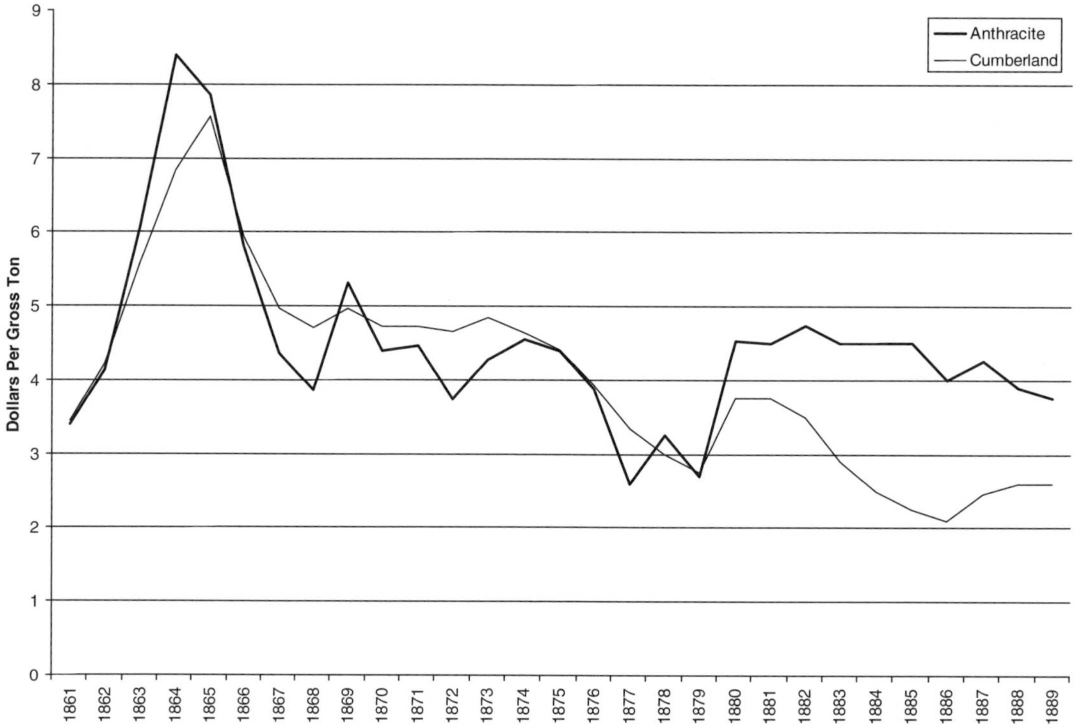

FIGURE 9.14
Annual Prices of Schuylkill White Ash Lump Anthracite at Philadelphia, and Cumberland Bituminous at Baltimore 1861–1889.

smaller rebound. As soon as the boom faltered in 1882–83, it appears that roles were reversed and anthracite started to follow Cumberland in a downward trajectory. It should be remembered also that the monthly anthracite prices, especially for the smaller sizes discussed in the previous section, show a much more marked decline than is apparent from the annual data. It is dangerous, however, to make causal inferences from a single graph. The reason for this is apparent from Figure 9.15. The data plotted here, unlike the previous figure, are clearly comparable in absolute terms, because of the proximity of Elizabethport and South Amboy as shipping points on New York Harbor. The Clearfield bituminous coal was brought from central Pennsylvania over the Pennsylvania Railroad. It showed the same pattern of substantial price cuts after 1881 as the Cumberland coal. This was prior to any significant change in the level of anthracite prices, although price cuts in the latter in the late 1870s had been matched immediately by reductions in Clearfield prices. Hence the downward trend in bituminous prices in the early 1880s can be identified in different sources of supply brought to seaboard by different railroad routes. It is therefore necessary to extend the investigation to determine if one or other of these, or indeed different coalfields again, had a dominant influence on changes in supply that could have initiated a downward price spiral.

Figure 9.16 shows the annual shipments from the Clearfield and Westmoreland regions in central and western Pennsylvania respectively, together with those from the Cumberland region. It is apparent that the latter had a rather chequered rate of progress during the period of interest, since its early shipment peak in 1873 was not surpassed again until the late 1880s. It is therefore difficult to attribute a continuing effect on prices to coal from this source. In contrast, the years 1874–1884 were a period of dramatic growth for the Clearfield region. Its shipments rose almost five-fold to more than 3 million tons during that interval. The growth was also concentrated after 1878. Together with the more modest growth in Westmoreland County, also served by the Pennsylvania Railroad, this degree of expansion in the bituminous trade over a short period was sufficient to explain the price reductions experienced in the seaboard markets. By contrast,

FIGURE 9.15
Annual Prices of Lehigh Lump Anthracite at Elizabethport, NJ, and Clearfield Bituminous at South Amboy, NJ 1872–1889.

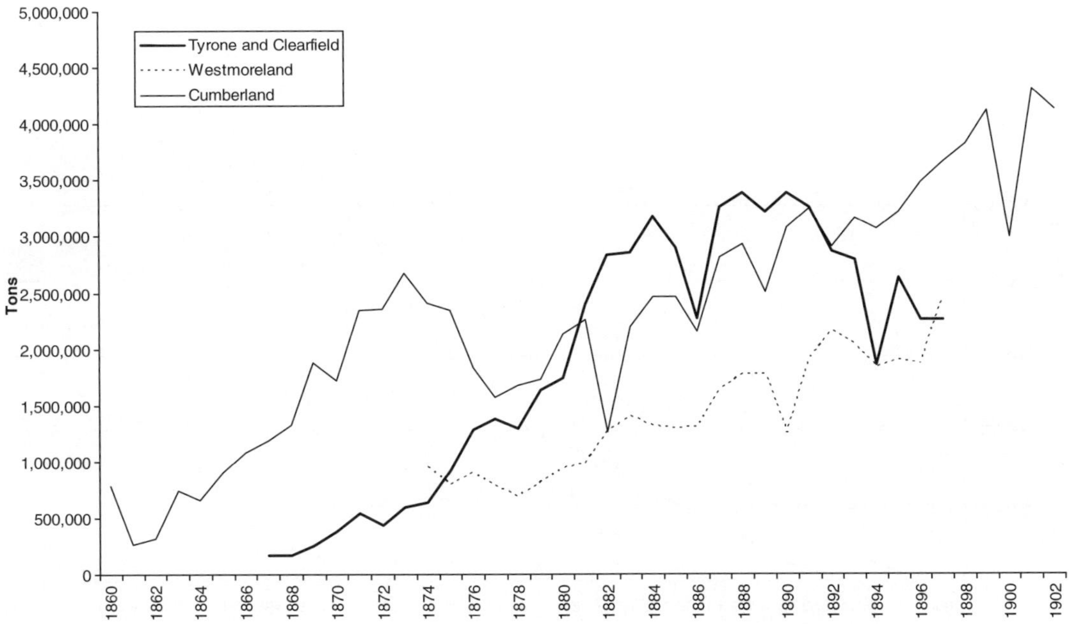

FIGURE 9.16
Growth of Shipments from Bituminous Coal Regions in Pennsylvania and Maryland 1860–1902.

expansion in coal production elsewhere in Virginia and West Virginia (Figure 9.17) did not take place early enough to have an impact on prices in the period currently being examined.

It should not, however, be imagined that bituminous shippers were operating in concert against the anthracite trade. Rather, it appears that each field regarded itself as being in competition with all other sources, anthracite or bituminous, that had a foothold in any given market. For example, in 1877, the Coal Trade quoted the Alleghenian to claim:

> 'When the price of Cumberland coal is within a reasonable margin in comparison with other coals, there need be no fear for our prosperity, *for no other coal can equal ours*' (phrase italicized in original).[9]

In a similar vein, early 1884 brought press reports of attempts by the Pennsylvania Railroad to remove railroad competition for coal traffic between Baltimore and Philadelphia. The aim was to prevent Cumberland coal underselling Clearfield coal in the latter market. (Clearfield coal was brought to Philadelphia as well as to South Amboy by the Pennsylvania Railroad.) John Norris, Editor of the Philadelphia Record, commented on the events of the period in his testimony to the investigation into labor troubles in 1887–1888. He claimed that, in 1884, Cumberland and Clearfield coals displaced about 600,000 tons of anthracite, forcing the anthracite coal companies to lower their prices. He further asserted that competition from soft coal had 'always been one factor in protecting the manufacturers and consumers of the East from the exactions of the anthracite coal combination'.[10] In the light of the evidence just presented, it is a little difficult to agree with this last assertion. Indeed, prior to the 1880s, it could equally be argued that the higher prices charged by the combination facilitated the development of the eastern bituminous trade, which otherwise would have been rendered uncompetitive, because of high freight rates.

By the mid 1880s, however, the competition from bituminous coal had become a force to be reckoned with. The Reading Railroad found it necessary to explain to its stockholders how improved rail links with western Pennsylvania were allowing bituminous steam coals to reach New York and compete with broken and pea coal.[11] Likewise,

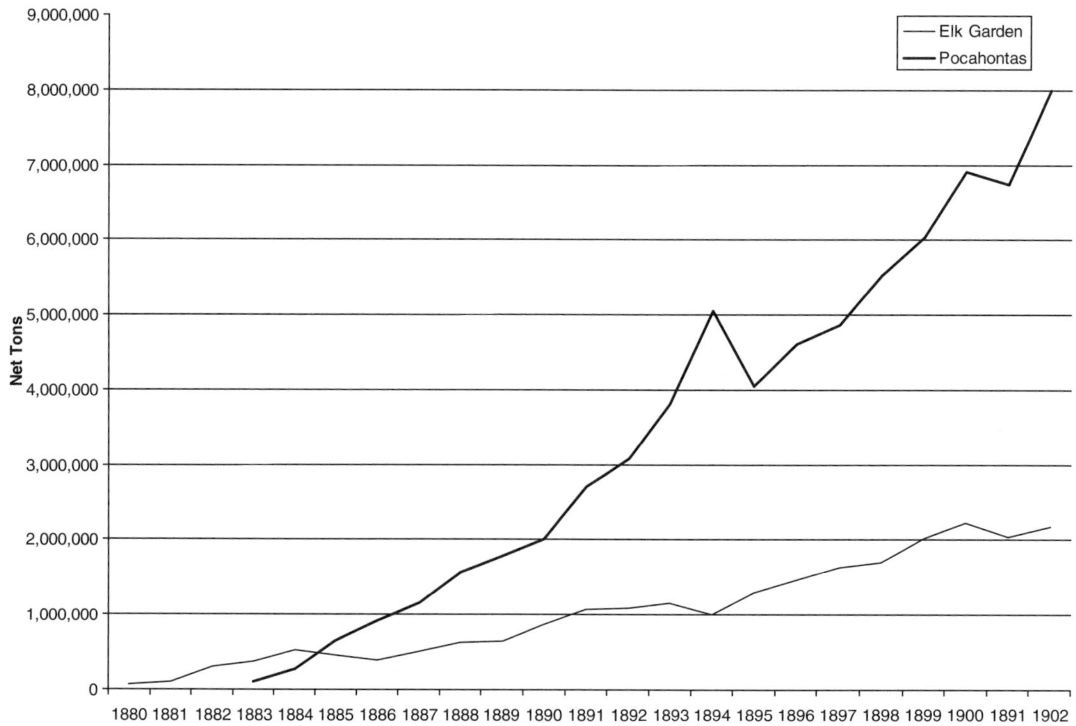

FIGURE 9.17
Growth of Shipments from Bituminous Coal Regions in Virginia and West Virginia 1880–1902.

Eckley B. Coxe averred in 1888 that soft coal competition was preventing any increases in anthracite prices, and Joseph S. Harris, then president of the Lehigh Coal and Navigation Company, stated that price variations in anthracite resulted in customers oscillating between hard and soft coal for their supply. Prior to 1885, when anthracite prices were relatively high, consumers switched to bituminous, but then 'a considerable amount of the trade' returned to anthracite in 1885–6, when prices of the latter fell sharply.[12] Some seesawing in price competitiveness seems to have remained a feature of the trade until the end of the period, judging by Figure 9.18. It should be noted that the anthracite prices used in this graph are quoted circular prices rather than actual prices received. They therefore exhibit fewer fluctuations. The Georges Creek coal came from the Cumberland region, but these are New York rather than Baltimore prices, and the coal was at a relative disadvantage, despite its high quality, because of greater transport costs. Nevertheless, despite the use of rather crude annual data, there is an indication that the depression of the nineties had a more adverse effect on bituminous coal markets in New York than was the case with anthracite.

Apart from direct price reductions, the only other option available to anthracite producers facing bituminous competition was to increase shipments of lower-priced smaller sizes.[13] Unfortunately, this meant that not only would more coal with a low profit margin be sold, but also more breaking of coal would lead to an increase in coal waste, until improved breaker technology could be developed. The nature of the changing proportions of anthracite sizes being shipped will now be examined in this context.

CHANGING SIZES OF COAL

The evidence in the previous section shows striking similarities in price movements between different companies and coal regions, even though quality variations were reflected in persistent price differentials. On this basis, it might be expected that experiences of changing mixes of coal sizes marketed

FIGURE 9.18
Annual Coal Prices by the Cargo Alongside in New York Harbor: Lowest Circular Prices for Anthracite (Chestnut and Larger Sizes) and Best Georges Creek Bituminous Prices 1860–1901.

throughout the period would also be broadly similar. However, a detailed examination of the patchy available data leads to a rather different conclusion.

As early as 1867, companies in the western portion of the Coal Regions, such as the Summit Branch Railroad and Coal Company, which sold coal predominantly to the domestic market from its Williamstown colliery, were shipping none of the lump and steamboat sizes and 94 percent of production was concentrated in the egg, stove and chestnut sizes.[14] Shortly afterwards, the Miner's Journal lamented that the Schuylkill Region as a whole could not sell all the lump coal it could produce. This resulted in large amounts of waste when the coal was broken down into smaller sizes.[15] By contrast, in 1868, the Lackawanna Railroad was shipping almost 30 percent of its output in the lump and steamboat sizes and the Pennsylvania Coal Company nearly 45 percent, two-thirds of the latter amount being lump.[16]

Therefore, not only did companies start out with differing mixes of sizes in their overall output, but the temporal dynamics of how the size proportions changed also varied much more than has previously been known to be the case. A good example of a company in the Lehigh Region, which was a major supplier to the iron trade, is once again the Council Ridge mine at Eckley. This was worked by Sharpe, Weiss and Company until 1874 and thereafter by J. Leisenring and Company, as noted previously. As Figure 9.19 shows, in the early part of the period for which data have survived, lump coal varied between 30 and 40 percent of the total and steamboat made up about 20 percent. As the 1870s depression in the iron trade began to bite, shipments of lump coal declined, only to recover to earlier levels during the railroad boom of the early 1880s. As subsequent examination of other data will reveal, such high levels of lump shipments were not found in other parts of the coalfields by the mid-1880s. They are testimony to the continued high levels of anthracite use in the Lehigh iron furnaces up to that time. Unfortunately, no such recovery was experienced by the steamboat size. The impact of steamship lines starting to switch to bituminous coal, because of lower prices, was becoming apparent by 1872, and the trend accelerated thereafter.[17] Following the disruption to supplies during the 1875 coal strike, former customers for this size deserted hard coal and were never to return.

In the early years, from 1867 to 1874, the large proportion of sales in the lump and steamboat sizes minimized the amount of coal processed through the breaker, reducing costs, minimizing waste, and maximizing profits. When the markets for these sizes declined in the mid-1870s, more coal had to go through the breaker and shipments of the broken,

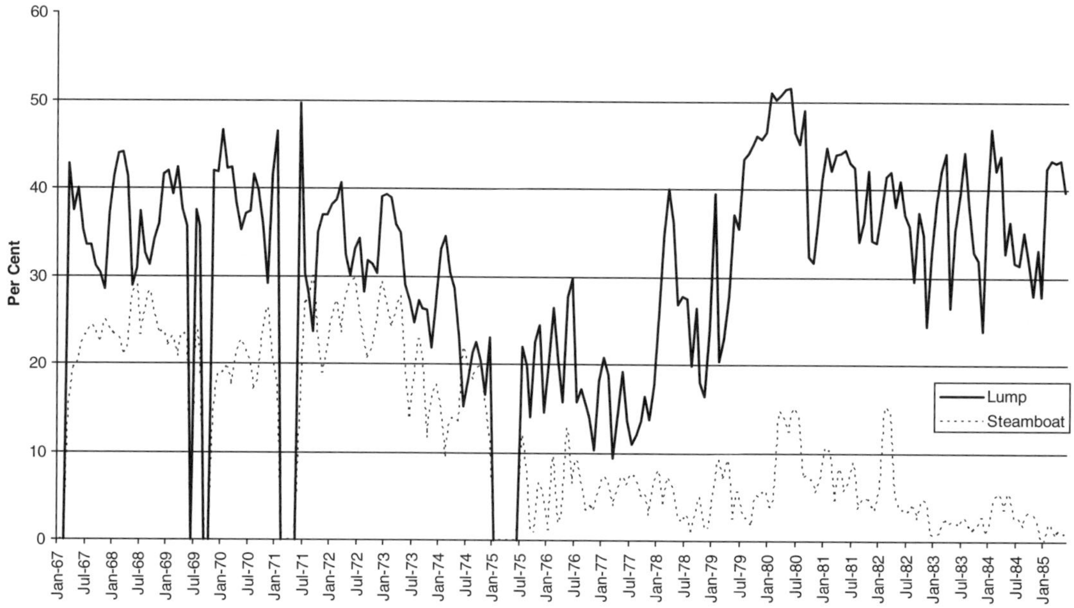

FIGURE 9.19
Council Ridge Mine: Percentages of Shipments in Larger Sizes 1867–1885.

egg, stove, and chestnut sizes increased accordingly. More pea coal was also a by-product of increased breaking, as Figure 9.20 indicates, but the larger volume of resulting waste culm had to be consigned to the dirt bank. Interestingly, the smaller buckwheat size, first marketed in the 1870s, only found customers in periods of weak demand, despite its very low price. The annual summary data for the Lehigh Coal and Navigation Company for 1872–1885, shown in Figure 9.21, provide an interesting comparison with the Council Ridge mine, since both were at the eastern end of the coalfields. The former was in the Southern Field, the latter in the Eastern Middle Field. The Coal and Navigation Company found its markets for lump coal declining as rapidly as Sharpe, Weiss and Company in the early 1870s, but, unlike the latter, there was no significant recovery in the early 1880s, and lump shipments resumed a steady decline. At the other end of the size spectrum, pea coal assumed a greater importance than at the Council Ridge mine in the early 1880s, but shipments declined proportionately thereafter. There is, however, evidence that pea coal increased in importance again in the latter part of the 1880s.[18]

Yet another pattern of behavior is displayed by the very comprehensive monthly and annual data available for the shipments of Plymouth and Wilkes-Barre coals by the Lehigh and Wilkes-Barre Coal Company between 1876 and 1902. The shipments in question were all made to tidewater at Port Johnston. Figure 9.22 indicates that the largest sizes were of limited importance by the time available records begin in 1876, and their significance continued to decline thereafter. The 1885 depression brought an abrupt end to lump production that was briefly compensated by modest increases in sales of the steamboat size. The largest fraction of the company's coal at this time was shipped in the broken size (Figure 9.23). These changes tally with the table of shipment contracts made by the company, which was discussed in an earlier chapter (Table 7.4). However, once again the impact of major economic dislocations is apparent, this time from the 1893 panic. Despite the unfortunately timed gap in the data, it is still apparent that there was a sharp drop in the percentage of broken shipments at this time, from which only a partial recovery was made during the remainder of the decade. Stove and chestnut shipments oscillated around 25 percent and 15–20 percent respectively in most years. The striking changes in the proportionate output of pea coal can be seen in Figure 9.24, which plots the monthly fluctuations in this size. A similar slow decline in output of that size to that experienced by the Lehigh Coal and Navigation Company is observed throughout the 1880s, presumably because sales were buoyant of the larger, more expensive,

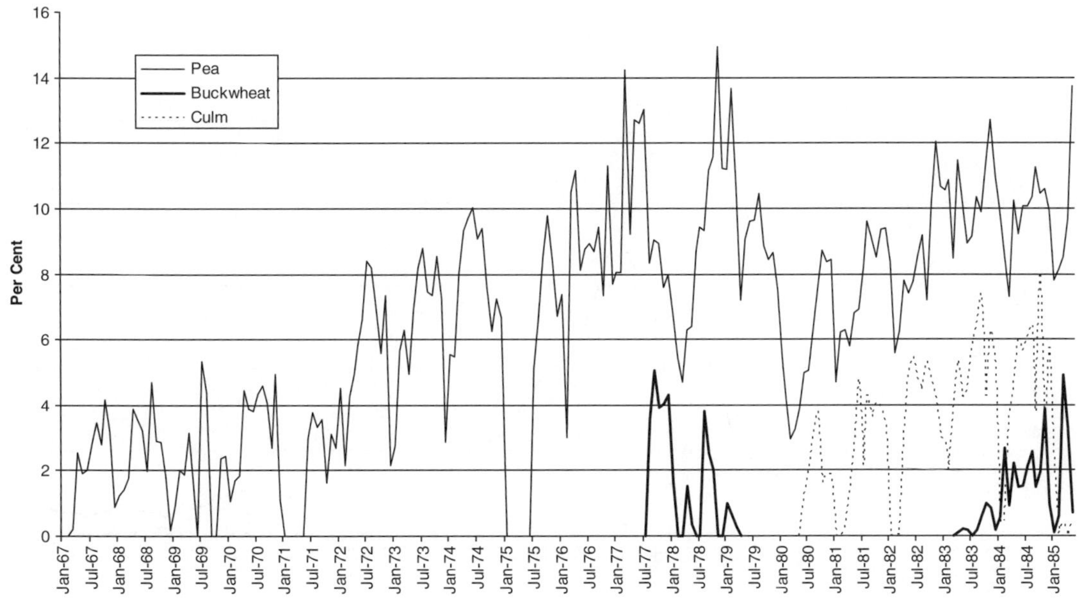

FIGURE 9.20
Council Ridge Mine: Percentages of Shipments in Smaller Sizes 1867–1885.

FIGURE 9.21
Lehigh Coal and Navigation Company: Percentages of Lump and Pea Sizes Shipped 1872–1885.

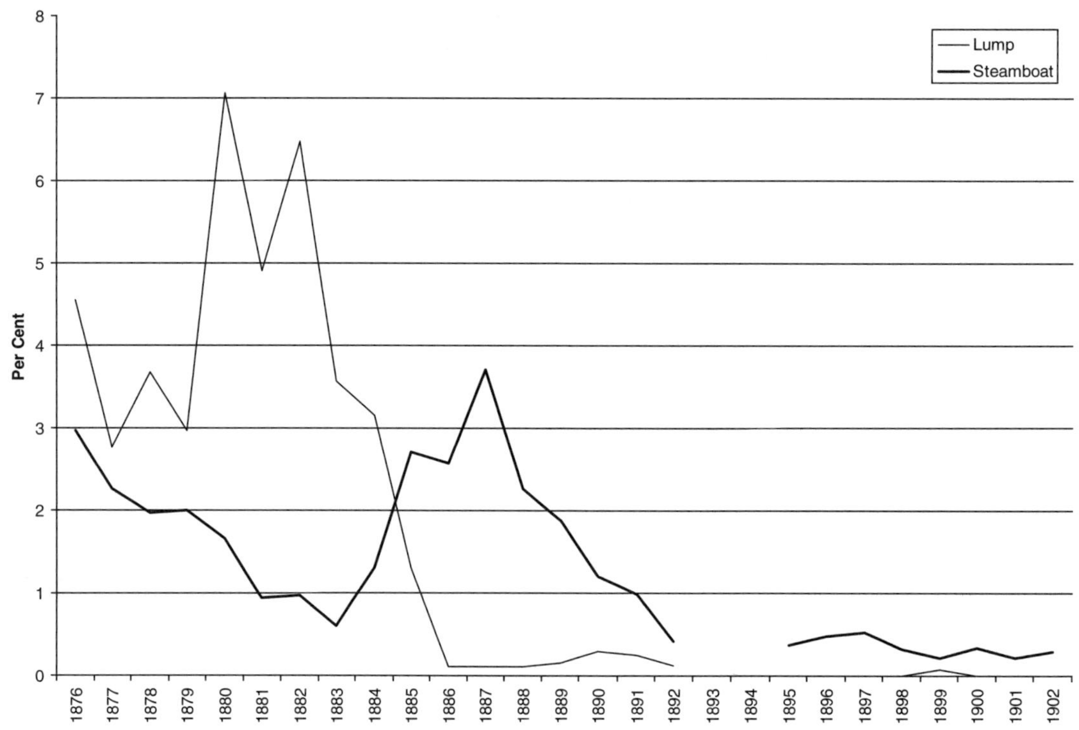

FIGURE 9.22
Lehigh and Wilkes-Barre Coal Company: Percentages of Larger Sizes Shipped 1876–1902.

FIGURE 9.23
Lehigh and Wilkes-Barre Coal Company: Percentages of Sizes Shipped 1876–1902.

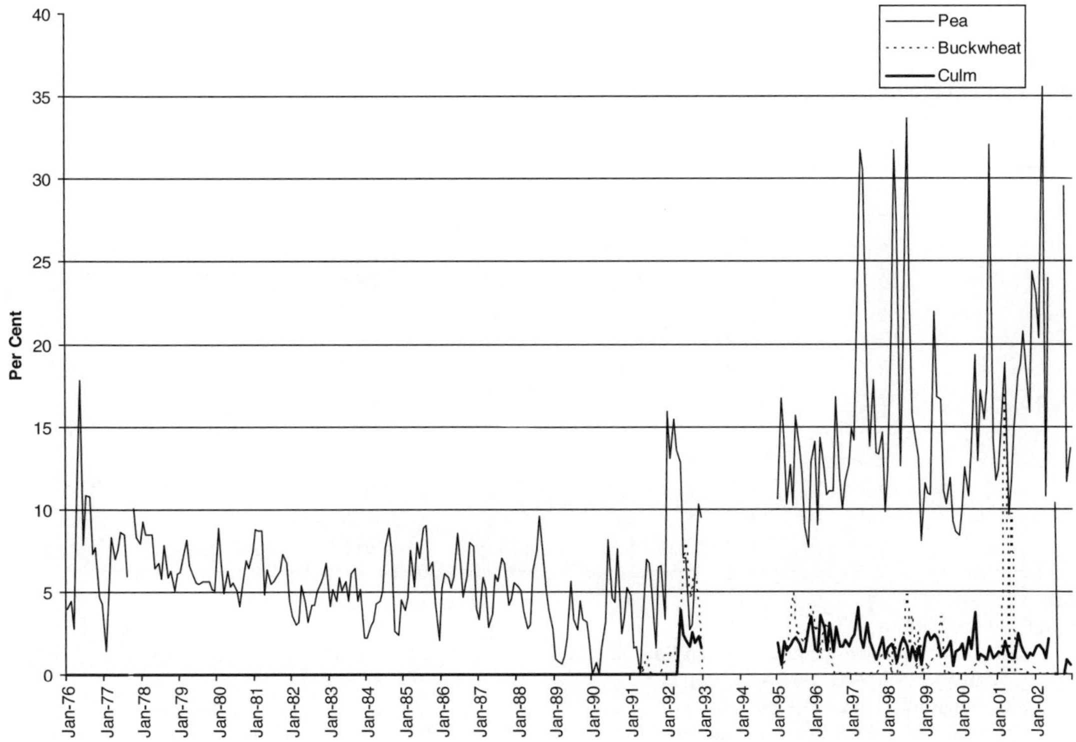

FIGURE 9.24
Lehigh and Wilkes-Barre Coal Company: Percentages of Smaller Sizes Shipped 1876–1902.

and also more profitable sizes. However, as business conditions began to deteriorate, prior to the 1893 panic, demand for pea coal suddenly began to increase. This demand continued unabated for the rest of the decade. Corroboration that this surge in the market for pea coal was industry-wide, rather than company-specific, comes from Mr. Storrs of the Lackawanna Railroad. In January 1893, he reported that there was unprecedented demand for pea coal, far in excess of the company's ability to supply it. More graphically, he declared:

> 'Manufacturers who have always used it and whose margins will not allow a higher priced size of coal, widows and orphans who beg for a living, the high and low, the rich and poor, all want it but it can't be had'.[19]

Judging by the data in Table 9.3, the rise of buckwheat coal, from the 1880s onwards, is at least as significant as the increase in sales of pea coal. Overall, the progressive disappearance of the larger sizes and their replacement, in terms of tonnages shipped, by the very small sizes, is very apparent from the table. Equally clear is the fact that not all companies shifted their production percentages at the same time. The Reading Coal and Iron Company, despite the shortage of data outside the decade of the 1880s, appears to have been the leader in this trend, (although it still had large steamboat shipments). In contrast, the Lackawanna was something of a laggard. It was still shipping significant amounts of lump coal, especially for railroad supply, until the beginning of the 1890s.[21] Output trends for the Lackawanna are even more distinctive, if production from its own mines is distinguished from that of the individual operators tributary to its road. These disaggregated figures (not shown in the table) indicate that the company produced more lump from its own mines in 1887 (16.3 percent) than it had in 1868 (15.3 percent). As late as 1898, it could still, on occasion, produce more than 10 percent lump from this source.[22] Other corroboration of the growing importance of the smaller sizes can be found in a report by Heber S. Thompson, who was Engineer of the Girard Estate Coal Lands in the Western Middle Field.[23] Thompson found that, while the percentage of chestnut coal produced by the estate's mining lessees between 1871 and 1891 had fluctuated between 11–16 percent, pea coal output had risen from 4.1 to 11.9 percent over the period. Similarly, buckwheat, which had first been shipped about 1878 from the area, increased from nearly 2 percent in 1881 to 9.6 percent a decade later.[24] The trend to the smaller sizes continued thereafter. By 1899, the

TABLE 9.3
Comparative Percentage Shipments by Size for Major Companies 1879, 1887, and 1898.[20]

Year/Company	Lump	St.	Br.	Egg	Stove	Chest.	Pea	Buck.
1879								
DL&W	15.5	3.4	12.5	9.8	29.5	17.4	11.9*	0.0
L&WB	3.0	2.0	33.5	15.1	22.2	18.2	6.0	0.0
SCCo.	19.7	0.0	11.2	16.4	28.8	16.3	7.6	0.0
1887								
DL&W	12.9	0.2	10.2	12.8	26.8	25.0	7.6	3.8
L&WB	0.1	3.7	30.7	19.6	24.2	16.6	5.2	0.0
P&RC&ICo.	2.2	10.3	14.9	13.5	21.9	12.6	17.7	6.9
SCCo.	6.3	0.0	8.8	14.8	31.6	16.8	12.7	9.0
1898								
DL&W	5.2	0.1	6.1	14.5	24.3	25.5	12.8	11.1
L&WB	0.0	0.3	18.5	21.6	21.9	17.8	16.5	1.2
SCCo.	4.3	0.0	9.9	13.07	25.2	20.1	13.4	13.9

*Pea and Dust
Key to Company Abbreviations in Table
DL&W Delaware, Lackawanna and Western Railroad
L&WB Lehigh and Wilkes-Barre Coal Company
SCCo. Susquehanna Coal Company
P&RC&ICo. Philadelphia and Reading Coal and Iron Company
NB. Percentages may not sum to 100 percent because of rounding and omission of other very small size categories used by some companies and not others.

Girard Estate percentages for the above sizes had increased to 22.6 percent, 13.1 percent, and 17.5 percent respectively. In addition, buckwheat No. 2, or rice coal, first marketed in 1895, now accounted for 3.05 percent of the total.[25] This new size was even more important in the case of some other parts of the coalfields. By 1900, the Lehigh Coal and Navigation Company shipped 20.3 percent of its output in this size, a larger percentage than for pea or buckwheat.[26] For 1901, the overall size averages for the industry were reported by the Anthracite Coal Strike Commission and a comparison given with 1892. These figures are reproduced in Table 9.4 and further compared with the 1892 data for the 'largest producer of anthracite in the world', which other data indicate was the Reading Coal and Iron Company.[27]

These figures confirm the progressive replacement of the very large sizes by the very small, while the intermediate sizes were much less affected. Comparison with Table 9.3 indicates some of the range of experience of individual companies compared to the industry average. The seemingly anomalous listing of 'steamboat' coal by the Reading Coal and Iron Company in 1892, when it is entirely absent from the industry average for the same year, suggests that the company's steamboat size was merged with the lump category by William Ruley, who compiled the overall industry figures reported by the Coal Strike Commission.

A further point to note about the comparative experience of different companies is that there were considerable benefits, in terms of efficiency of operation, in keeping the size percentages produced relatively constant. Rapid fluctuations would cause some of the storage bins in the breaker to be filled, while others were empty. In the early days this necessitated stoppage of the breaker, in later years, excess quantities of sizes not required by the market would have to be stockpiled.[29] In this context, the monthly shipment figures for the Lackawanna Railroad and the Lehigh and Wilkes-Barre Coal Company for 1896–1899 are revealing (Figures 9.25 and 9.26). The graph for broken coal shows that the latter company experienced rapid variations in shipment percentages both on a monthly and a year-to-year basis. A slower overall downward trend with more muted fluctuations is apparent for the Lackawanna. The picture for pea-coal (Figure 9.26) is ever more striking. The Lackawanna percentage was remarkably constant over an extended period, while the Lehigh and Wilkes-Barre experienced sudden sharp peaks in output. Such differences probably reflect, in part, the smoothing effect of more stable western markets. These took a substantial proportion of Lackawanna shipments, while those of the Lehigh and Wilkes-Barre were all destined for Port Johnston. Here they were at the mercy of the vicissitudes of competitive eastern seaboard markets. The differences may also indicate the overall superiority of the Lackawanna in managing its coal business.

CONCLUDING COMMENTS

It is apparent from the detailed analysis in this and the two preceding chapters that there was a complex interplay of both sectoral and geographical

TABLE 9.4
Comparative Size Percentages 1892 and 1901.[28]

	Reading Coal and Iron Company 1892	Industry Total 1892	Industry Total 1901
Lump	1.6	9.3	4.1
Steamboat	8.3	-	-
Grate (Broken)	13.1	11.2	8.3
Egg	12.1	14.5	13.1
Stove	20.9	23.1	19.7
Chestnut	15.4	17.9	19.1
Pea	14.5	12.6	14.1
Buckwheat	12.8	9.5	14.7
Rice	1.3	1.9*	6.9*

*Listed as 'smaller than buckwheat No. 1' in the original tabulation.

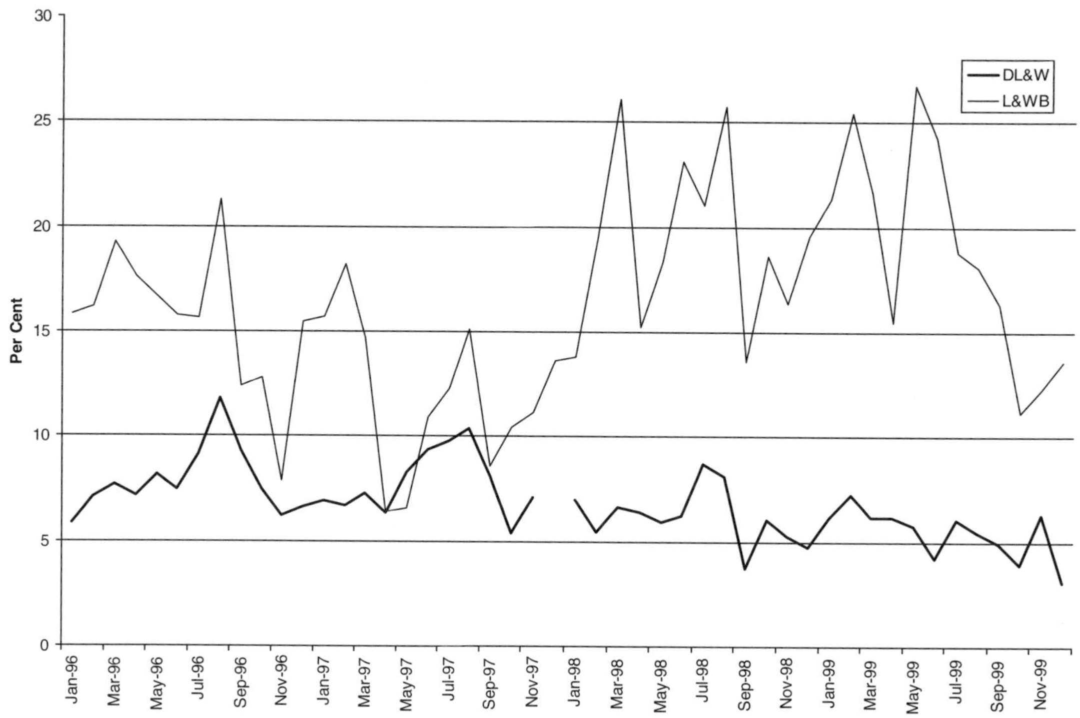

FIGURE 9.25
Comparative Percentages of Broken Coal Shipped: 1896–1899.

changes in markets for anthracite. Coupled with competition from a variety of types of bituminous coal, these changes found summary expression in the shifting mix of coal sizes shipped to market over the period. The differing experience of individual companies in this context reflected variations in their engagement with a range of market segments, both on the eastern seaboard and in the continental interior. However, changes in the size composition of total output were not entirely neutral, in economic terms, for the firms involved. In the first place, leaving aside the vexed issue of dockage for rock and slate in the mine cars, the miner was paid on the basis of the coal hoisted out of the mine, not on the basis of the tonnage finally shipped from the breaker.[30] Hence, other things being equal, the greater the amount of waste after breaking, the less economical the mining operation. Maximizing the tonnage shipped in the large industrial sizes, which did not pass through the breaker rolls, would minimize the waste due to the breaking process itself. That said, in the earlier years, when demand for these larger sizes was greater, it used to be the case that buckwheat, pea and even chestnut would often 'be thrown very freely into the dirt bank when there was a slack demand for them'.[31] This was the worst of all possible outcomes, because the cost of breaking had already been expended on these sizes, which were then thrown away. If they could be loaded onto cars and sold without depressing the prices of the larger sizes, this would obviously provide a net gain to the mine operator, as well as additional tolls to the railroad.[32] Prior to the 1890s, it is important not to overstate the importance of the smaller sizes, however. Buckwheat and smaller sizes were not shipped by the major companies prior to the late 1870s, and then only in limited or negligible quantities in the following decade. It should also be noted that market forces or decisions by mine operators were not the only factors affecting the size distribution of coal output. In some locations, where the coal was softer or more friable because of geological conditions, there was only a limited amount of large coal to begin with. Similarly, in some of the deeper veins, crushing and movement of the strata would have the same effect.[33] In his well-known article on the anthracite trade in 1892, Joseph S. Harris stated that about 20 percent of the coal hoisted from the mine was formerly consigned to the culm bank, but, more recently, utilization of the smaller sizes had reduced this proportion considerably.[34] The problem of waste was still

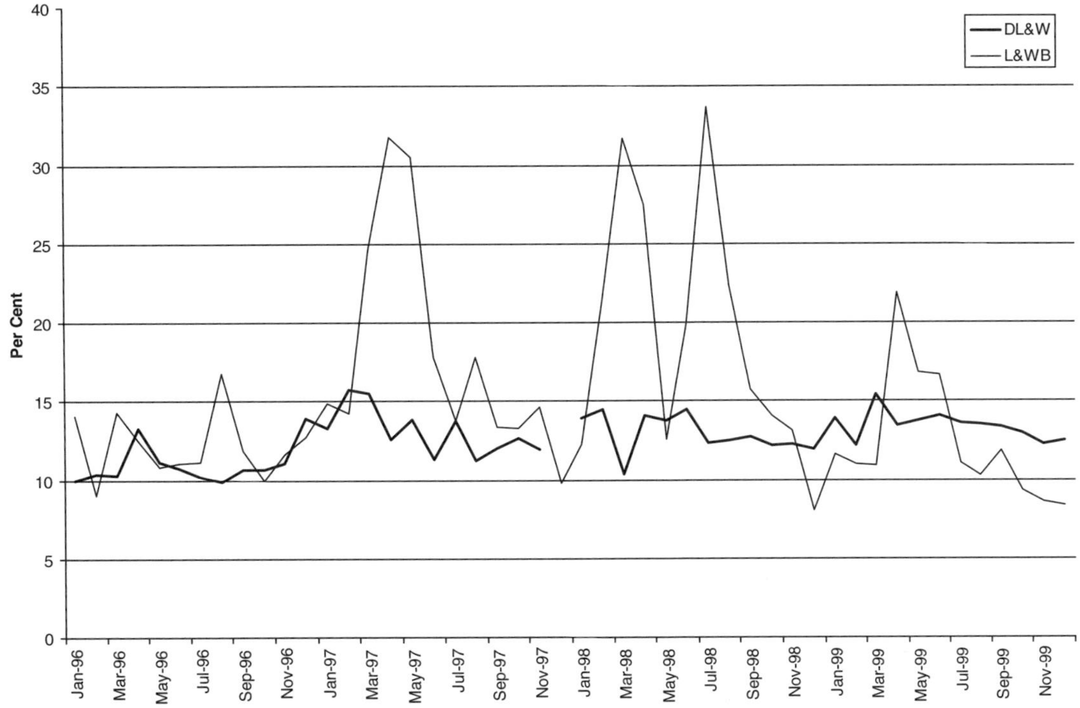

FIGURE 9.26
Comparative Percentages of Pea Coal Shipped: 1896–1899.

deemed to be sufficiently serious, however, for a commission to be appointed to investigate the matter. The 1893 report of this commission was the successor to an earlier investigation of the same issue by Franklin Platt for the state legislature, at the beginning of the 1880s.[35]

By the turn of the century, it could no longer be argued that failure to utilize the smaller sizes was regrettable, wasteful and, almost incidentally, an unwelcome lost opportunity for additional profit. Instead, increased costs from deeper mining and a greater requirement for pumping,[36] had made realization of additional revenue from the smaller sizes essential to the commercial viability of anthracite production as a whole. The economic argument was also presented rather differently than in the 1870s. Average costs were calculated across a greater number of sizes, with the addition of buckwheat No. 1, No. 2, and No. 3. Unfortunately, the small sizes were being sold at less than the average cost of production, owing to the low prices required to meet bituminous competition for steam raising. In 1902, it was estimated that No. 3 realized only 30 cents a ton for the operator, No. 2 about 40 cents, No. 1 about 85 cents and pea coal about $1.30 (these are net receipts, not profits). Were it not for the higher prices received on the larger domestic sizes, which were perceived as subsidizing the small sizes, it would have been uneconomic to mine at all.[37] Since the percentage of total shipments above pea size had fallen from 66 percent in 1892 to only 60 percent by 1901, the scope for this cross-subsidy was steadily disappearing. The only way forward was further cost-cutting by means of continued improvements in breaker machinery and underground haulage. Wage reduction was no longer a feasible option in the face of the growing power of the United Mine Workers Union.[38] Under these circumstances, the mine operators found themselves progressively squeezed by rising costs and static or falling prices. This was especially true of the smaller concerns. In contrast, the major anthracite railroads, with their vastly greater financial resources, were able to exert much more control, both over sources of coal supply and delivery of coal to consumers. In the former endeavor, they were eventually to achieve a very large measure of success. In the latter they were to be opposed by a growing array of political and legislative devices. These were initially set in motion at the state level and subsequently at the national level. The railroads countered by replacing early and more blatant attempts at combination with the subtle, almost covert, harmony of operation represented by J. P. Morgan's community of interest, that developed towards the close of the period. It is to these changing financial and organizational arrangements between the railroads that we now turn, prior to a detailed examination of the mining economy and its relationships with the transportation sector.

Chapter 10

CORPORATIONS, COMPETITION, AND THE RISE OF THE CARTELS I: PRECURSORS AND PRE-DISPOSING FACTORS TO INDUSTRY-WIDE COMBINATION

INTRODUCTION

From the very early days of the anthracite coal trade it had been recognized that adoption of different modes of industrial organization would have an impact on the future development of the mining industry. The 1834 report of Senator Samuel J. Packer's investigating commission promoted the free pursuit of individual enterprise, while recognizing the benefits that had been brought by incorporated companies. However, it railed against the evil of giving corporations both mining and transporting privileges, which would give them the 'power to lock up at pleasure the resources of a whole valley or community'.[1] Adams has traced the subsequent development of corporate organization in the mining industries of Pennsylvania and Virginia, and he has highlighted the differing roles of general purpose, i.e. template, charters, and special charters as legislative instruments in this process.[2]

By the beginning of the period of interest, unincorporated individual enterprises were already disappearing from the scene in the anthracite coalfields, a trend that was accelerated by the Civil War. Increased costs and financial risks attendant on deeper mining, a greater requirement for pumping, and more sophisticated coal breaker technology had raised the barrier to entry, in terms of requisite starting capital, to the point where the necessary means were largely beyond the reach of individuals. In 1865, even in the Schuylkill Region, almost 50 percent of the tonnage was mined by 52 corporations,[3] although both the Reading Railroad and the Schuylkill Navigation Company acted as common carriers. As discussed in Chapter 4, the Wyoming Region was already the province of the three main mining and carrying corporations, the Lackawanna, the Delaware and Hudson, and the Pennsylvania Coal Company. In the Lehigh Region, there was a mixture of companies that had both mining and transporting privileges, such as the Beaver Meadow Railroad and the Lehigh Coal and Navigation Company, and those that did not, such as G.B. Markle and Company. The Lehigh Valley Railroad, like the Reading, acted as a common carrier.

From this point, and with this particular disposition of economic power within the coal regions, it is necessary to examine the factors that led to the next stage of evolution in industrial organization, or what might be termed metaorganization. This took the form of a series of cartels or combinations in the trade that aimed, in a variety of ways, to control coal output, price, or both. This stage was to dominate the industry during the last three decades of the 19th century and beyond, until government anti-trust legislation finally began to chip away at the united front represented by the anthracite-carrying railroads.[4]

Quite detailed descriptions of the basic structure and operation of the major industry-wide combinations from 1873 onwards have been provided by Jones' classic study, which draws on an earlier investigation by Virtue, and in Bogen's examination of the anthracite railroads.[5] More recently, Schlegel has focused on the role of Franklin B. Gowen in the establishment of combinations, and Campbell has provided some additional insights into cartel operation during the period of the early 1880s.[6] However, much less attention has been devoted to analysis of the economic circumstances prevailing in the period prior to 1873 that may have pre-disposed the industry to move in the direction of new forms of organization. This analysis will now be undertaken and, in the process, it will become clear for the first time that the antecedents of industry-wide combination can be traced much further back in time than has previously been realized.

More specifically, the investigation will be conducted under three main headings:

- early combination activity prior to 1873,
- overview of pre-conditions for the first industry-wide combination, and
- accelerated acquisition of coal lands.

In the course of elaborating these issues, it will become apparent that combinations were complex reactive processes of industrial meta-organization. They were closely linked, not only to changes in general economic conditions, but also to the perceived need by companies to protect past infrastructural investment and to guard future market share and position in the trade against actual and potential competitors. These processes had important geographical implications, both within the coalfields and the wider anthracite distribution region. They also had a number of unanticipated consequences for the mining industry as a whole. The latter will be the subject of a later chapter.

EARLY COMBINATION ACTIVITY PRIOR TO 1873

The conventional historical view, first put forward by Virtue and echoed by Jones, is that, prior to the combination of 1873, any attempts at combination were local and temporary, although the former does draw attention to the pooling arrangements between the Reading Railroad and the Schuylkill Navigation Company to control the coal traffic from the Southern Field down the Schuylkill Valley.[7]

However, a closer examination of early sources reveals that interest, from miners and shippers of coal to western markets, in controlling and regulating the trade, without regard to the consumers, can be dated as far back as 1856. In December of that year, the Lackawanna declined to participate in an arrangement with shippers over the North Branch Canal to set prices 'mutually' at a profitable level.[8] This call for price fixing followed several months of declining business activity after the peak in mid-1856, a fall that presaged the financial panic of the following year. Although the business prostration after the panic appears not to have produced any more documented attempts at combination, the persistence of stagnant economic conditions that marked the beginning of the new decade brought further proposals forward. The first of these came from Thomas Kimber of the Williamsport and Elmira Railroad in January 1860, and it was closely followed by a further proposal from Messrs. Lawrence and Langdon of Elmira. In what seems to be have been a third and final attempt to get a western market combination off the ground, a joint proposal from Kimber and Langdon was eventually rejected by the Lackawanna Railroad, not because it would act in restraint of trade, but because there was no mechanism for preventing the Pennsylvania Coal Company and the Delaware and Hudson spoiling the party, by shipping to the same markets themselves![9]

Nevertheless, the idea must have taken root, since on January 3, 1862, in the worst phase of the depression that accompanied the onset of the Civil War, a contract to establish a 'combination' was signed by a number of parties to the western coal trade.[10] Although no copy of this contract has been found to date, other evidence indicates that it included semi-bituminous companies from the Blossburg Region, such as the Tioga Improvement Company and the Fall Brook Coal Company, as well as the Lackawanna, Langdon and Company, and presumably what became the Pittston and Elmira Coal Company, who were also parties to the Anthracite Coal Association, formed in Buffalo in 1861.[11] The latter company, which eventually operated the Seneca, Ravine and Twin mines in Pittston, was referred to as 'Mercur and Frisbie' during the Civil War period, but shortly thereafter the same E.N. Frisbie emerges as the president of the Pittston and Elmira Coal Company.[12]

Although there is no evidence that the Delaware and Hudson joined the combination at this time, such arrangements must have been part of a wider movement to find ways of regulating the trade in adverse business conditions. The Delaware and Hudson certainly appointed a committee of directors in 1862 to meet with like committees from other companies, in the hope of identifying 'reasonable and proper measures to preserve the coal trade in a healthy condition and secure such prices as will be renumerative to all concerned'.[13]

Specific action by the company was to be delayed until the beginning of the 1870s, however. Reflecting on the first two years of the combination, the superintendent of the Fall Brook Coal Company, felt moved to heap fulsome praise upon it for the benefits it had brought. Such testimony to its initial success is unsurprising, however, given the steady improvement in overall business conditions during those same years.[14] The soft coal interests remained largely focused on Buffalo, where they had important markets in the manufacturing sector. However, the hard coal arm of the combination quickly extended its reach to Chicago, under the lengthy name of the 'Scranton, Pittston, Wilkes-Barre, and Shamokin Coal Company of Pennsylvania', run by its petulant agent Robert Law.[15] Fortunately, some angry exchanges between the participants in the Chicago combination in April 1867, with Law caught in the middle, reveal the basis on which it operated. Half the revenues from coal sold were due to the Lackawanna and a quarter each to Langdon and the Pittston and Elmira Coal Company, provided they all had 'coal in the market'.[16] Unfortunately, at that time, the newly appointed General Coal Agent in Scranton, W.R. Storrs, had insisted initially on remittances being sent in proportion to the coal shipped to Chicago, thereby incurring the wrath of the other parties to the agreement. Fortunately, Frisbie was confident he could sort Storrs out and show him the error of his ways, asserting forcefully in his letter of reply, 'I'm sure we can knock it into you when we meet'.[17]

Despite these internal frictions, the combination at the time seems to have achieved its aim of strengthening its hold on the Chicago market during a period of weakening demand and falling prices. As Robert Law observed:

> 'There is a terrible weeping wailing and knashing (sic) of teeth by the coal dealers and speculators towards the combination on a/c of $9 per ton'.

And again:

> 'The revulsion and panic is terrible among coal dealers. Many are selling out and no more competition hereafter will be encountered by us.'[18]

At the same time, the Buffalo arm of the combination was busy tying dealers along the line of the Erie Canal into annual contracts for coal, while reducing their margins.[19] Unfortunately for the combination, however, the going was proving much tougher in the more accessible upper New York state than it was in distant Chicago. Other 'Pittston and Lackawanna' companies (i.e. the Pennsylvania Coal Company and Delaware and Hudson) were also appointing special sales agents along the canal from Albany to Buffalo, 'with private circulars at very low rates'.[20] Large consumers of coal were becoming suspicious of the combination's attempts to foist prices on them[21] and, most serious of all, the New York State Legislature was beginning to take an unwelcome interest in the workings of the combination. It did this by means of an investigating committee that met in Albany in March 1867, summoning company presidents, coal agents, and dealers to give an account of themselves, although no action against the combination was ever taken.[22] This was also not the first time that the combination had attracted critical public attention. The Luzerne Union, a Wilkes-Barre newspaper, reminded its readers that the operations of the combination had been brought to the attention of a grand jury in Rochester, but unfortunately no trial or hearing had ever been held. The newspaper also claimed that the Lackawanna, the North Branch Canal (formerly owned by Mr. Mercur and his partners), and the Northern Central Railway (on behalf of Mr. Langdon) were operating in concert to control shipments from the Northern Field north and west, taking traffic in the approximate percentages of 60, 20, and 20 respectively. While such claims seem well-founded, a further claim that no mines other than those owned or leased by the Lackawanna could ship to Syracuse or to New York is not supported by the tonnage records in the railroad's own books.[23]

Not to be outdone by its neighboring state in terms of investigative zeal, Pennsylvania also set up a senate committee in 1867 to examine the problem. This committee held sessions in Harrisburg, Philadelphia, Pottsville, and Wilkes-Barre. Rufus J. Bell, represented the private coal operators and the public and he declared that:

> 'Such is the absolute despotism and such the system of terrorism exercised by these various railroad and canal companies over private coal land owners and the public that a number of witnesses were deterred from appearing before the committee'.[24]

Despite this, Bell felt that enough evidence had been presented about abuses of monopolistic control over avenues to market that the railroads should be forced to forfeit their charters, having, in his words, converted public highways into private thoroughfares.[25] A particular example of this in 1867 was the structure of freight rates on the North Branch Canal, now owned by the Pennsylvania and New York Canal and Railroad Company. Coal cleared to points north of the Athens Pool (i.e. through freight to Elmira) was charged at a prohibitive rate of 10 mills per net ton per mile for small shippers, but a drawback (rebate) of half this rate, or 5 mills, was made for bulk shippers (over 100,000 tons per annum). This effectively prevented small shippers from using the avenue to market. With consummate irony, if not downright hypocrisy, the published toll circular also prohibited combination by small shippers to pool their tonnage and gain access to the rebated rate schedule![26]

Following this spell of legislative investigation and media outrage, the activities of the combination seem to have moved out of the spotlight for a period. No formal strictures ever appear to have been placed upon it. However, it may be that the unwelcome attention it received in relation to its business practices caused some of the worst excesses in its behavior to be moderated, although there is little direct evidence of this.

Several points emerge about the combination in the ensuing years. Firstly, it was based on a long-term contract, since 1868 correspondence refers to the expiration of the 'Western District joint contract' at the end of April 1871.[27] Secondly, not all the partners enjoyed the same degree of financial stability. At the end of 1867 and the beginning of 1868, serious financial problems experienced by the Pittston and Elmira Coal Company threatened the continuance of the combination. This was compounded by a strike among their men, probably related to non-payment of wages. The company's difficulties were attributed to low coal prices and sudden stringency in the money markets, which made discounts difficult to obtain, causing them severe cashflow problems. Among the solutions proposed were for the Lackawanna to buy their interest in the coal yards in Chicago and Rochester and even in some of their lands in Pittston. Although the final details of their financial recovery arrangements have not come down to us, the company did manage to weather the storm, as later correspondence indicates that coal transportation arrangements resumed later in 1868. The Pittston and Elmira's tonnage tally for the year finally reached nearly 110,000 net tons, of which 26,496 tons was shipped to Ithaca and 52,992 tons to Syracuse. Financial rescue for their business partner was clearly regarded as essential by the Lackawanna, as John Brisbin declared, 'to permit the existing arrangement to be broken up is in my judgement the most wanton, wicked suicide'.[28] Thirdly, the relationship between the combination and the Delaware and Hudson (more so than the Pennsylvania Coal Company), seems to have undergone a progressive shift over the period. In 1868, the impact of the Erie Canal agents for the latter companies mentioned above began to be felt in earnest. Complaints were made of the Delaware and Hudson undercutting Lackawanna markets, first in Utica and subsequently by the substantial amount of 65 cents a ton in Syracuse.[29] The avowed 'free trade policy' of the Delaware and Hudson was deemed to be the root cause of the problem, and the only perceived motive from the Lackawanna's perspective was 'to destroy our margins'. Indeed, these actions were believed in themselves to justify the necessity for continuing the joint arrangements of the combination.[30] However, at the same time, a shift in the attitude of the Delaware and Hudson started to become apparent. They put forward a proposal to begin working in harmony with the combination, instead of in competition with it, although no documented rationale for this change in policy has come down to us.[31]

By the beginning of the following year, this newfound spirit of co-operation was formalized when the Delaware and Hudson joined the Western Coal Association, making contractual agreements with J. Langdon and Co. and the Northern Central Railway. These rather complex agreements had a number of key provisions. In the case of the former company, these provisions were as follows. Firstly, the Delaware and Hudson agreed not to sell coal at its Plymouth and Wilkes-Barre mines, destined for any point on the Northern Central and its connections, to any other party than Langdon. Secondly, Langdon was to provide the necessary cars for shipment at the mines. Thirdly, he was to pay for all the costs of mining, preparing, shipping and selling the coal, plus 60 cents per gross ton to cover royalty costs and profit for the Delaware and Hudson. Fourthly, the latter company agreed to provide a minimum of 100,000 tons and a maximum of 250,000 tons under the agreement.

The five-year agreement with the Northern Central was for the joint selling of up to a further 250,000 tons. The companies were to split the net profit after allowing for all production and selling costs, royalty at 40 cents a ton, and incidental expenses. Equal shipments were to be made each year under the two contracts.[32] In this way, the Delaware and Hudson received a guaranteed profit margin on half its coal sent over the Northern Central, while Langdon was assured of a regular coal supply and absence of competition for the sales of these coals over a wide market area. Further formalization of the combination arrangements for 1871 took place in November 1870, when it was agreed that the Delaware and Hudson, J. Langdon, and the Northern Central would be entitled to 50 percent of the western coal market, while the Lackawanna and the Pittston and Elmira would have the other 50 percent.[33]

While the combination was being strengthened in one direction by these agreements, growing competition from other transportation companies in the western markets was undermining the monopolistic advantages of the Coal Association. The weakest member of the Association, the Pittston and Elmira, also became a source of concern, as its high quality coal was an attractive target for the competition wishing to increase shipments from the Northern Field. In both these contexts, the Lehigh Valley Railroad was the major threat. Increased shipments to Buffalo by the road and also the Philadelphia and Erie Railroad, coupled with the low prices they offered, forced the Association into price reductions to avoid losing market share.[34] At the same time, the Lehigh Valley was taking a close interest in the Pittston and Elmira because of the proximity of the Pennsylvania and New York Canal and Railroad Company's tracks to its mines. However, Storrs was of the view at the time that the Pittston and Elmira preferred to stay in the Association because of the arrangement for western shipment. Nonetheless, a certain nervousness about the loyalty of the latter company persisted, although they never in fact deserted the association, as long as operations continued in their name.[35] Unfortunately the company did not survive the aftermath of the 1873 panic. Warning signs were already apparent by September 1873, when their men refused to continue mining until paid for the previous month's work, and, by mid-1874, buyers for the property were being sought. The Lackawanna looked over the property, declined to pay the asking price of $600,000, then vacillated over a figure of $550,000, knowing that the Delaware and Hudson were also interested at this level.[36] Finally, in what appears to have been a complicated break-up of the Pittston and Elmira's assets, 438 acres of its lands were indeed sold to the Delaware and Hudson on December 17, 1874 for $300,000, but its remaining property, which included the Seneca, Ravine, and Twin mines was initially assigned to one Patrick Lynch on behalf of the company's creditors. Subsequently, in 1876, Lynch sold the property in question to the Pittston Coal Company, whose president, F.C. Dininny, was Secretary of the neighboring Butler Coal Company, although the Pittston Coal Company is recorded in the Mine Inspectors' Report as working the Seneca, Ravine, and Twin mines from 1875 onwards.[37] President Sloan of the Lackawanna clearly miscalculated, as he had expected the Delaware and Hudson to purchase the entire property and continue to send its output over his road, but in the event, shipments were transferred to the Lehigh Valley and were lost to the Western Coal Association.[38]

The progressive alignment of the interests of the Lackawanna and the Delaware and Hudson after 1869 was not limited to sales of coal in western markets. Late in 1870, the latter appointed a committee to confer with the Lackawanna more generally on matters pertaining to acquisition of coal lands and leases and the mining and transportation of coal. Unfortunately, legal advice from the Company's counsel found that it would be illegal for the two companies to enter into such a close arrangement and the matter was dropped at that time.[39] Such an initiative needs to be seen in the context of wider attempts at developing a combination that went beyond the Northern Field. These attempts can be traced as far back as March 1869, when the Lackawanna Board rejected as impractical a proposed combination among the 'principal coal interests' whose purpose was to regulate temporarily the supply of coal to be sent to market.[40] It is interesting to note that such proposals predated the appointment of F.B. Gowen to the presidency of the Philadelphia and Reading Railroad in April 1869.[41] They also coincided with a sharp fall in coal prices back to 1867 levels, after they had been artificially raised by the strike of 1868 in the Schuylkill and Lehigh Regions (cf. figure 9.10). However, after 1869, no record of any further attempts at uniting all the major anthracite railroads, as opposed to just the Lackawanna and the Delaware and Hudson, can be found until 1872, when it appears the discussions began that were to lead to the 1873 combination.

Hence, contrary to the received wisdom, prior to 1873 there were longstanding combination arrangements by a small number of companies, to control significant portions of the anthracite coal trade in western and northern markets. It is no accident that such moves originated among corporations with both mining and transportation privileges, since only they had the ability to control both coal supply and access to markets. Neither the Reading nor the Lehigh Valley railroads, which took their 'common carrier' status very seriously at the time, were in a position to exercise an equivalent degree of control, a point not lost on incoming president, F.B. Gowen. However, certain additional developments were required before concert of action between the main coal interests could become a reality, and it is to these that we now turn.

OVERVIEW OF PRE-CONDITIONS FOR THE FIRST INDUSTRY-WIDE COMBINATION

In his classic paper, Virtue identified six pre-conditions for the establishment of a general combination of anthracite producers in the early 1870s. These were as follows:

1. A productive capacity in the industry that was 5–6 million tons larger than the 16 million tons of output that the market would take at paying prices
2. Six major transportation companies controlling the majority of coal lands and effectively all the avenues to market
3. Each major company burdened with debt and anxious to maximise its market share
4. A large number of individual operators each trying to increase their output
5. A substantial coal surplus on hand and rapidly falling coal prices
6. A united labor organization, strong enough to stand up to the operators

These are clearly all important issues in understanding the structure and evolution of the anthracite industry, and matters pertaining to them will occupy much of the remaining argument of this book. Hence only a number of preliminary observations will be made at this stage.

The first point about overcapacity and excess production was a recurring theme from the early days of the industry, although it assumed its greatest significance during the combination era. If indeed the regional economy were governed by the iron laws of supply and demand, it might be reasonable to expect, at least in a simplistic view, that initial overcapacity problems would soon be eradicated, as the price mechanism led to equilibrium between these two forces. On the other hand, if overcapacity problems were to become chronic, equilibrium considerations would have little to offer by way of explanation. Instead, an investigation of technical and strategic decision-making by major companies in relation to capacity planning would be required. Looking ahead, this is the approach that will actually be adopted in later chapters, which examine the development of the mining economy. A further issue that needs to be raised in this context is what exactly is meant by the term 'overcapacity' in an industry so subject to seasonable fluctuations in output and the unpredictable risks of accident, fire and flood.[42] A clear definition is obviously required as a precursor to any attempt to determine quantitatively whether such over-capacity was actually present.

The second point concerns control of coal lands and avenues to market. The importance of control of the latter to successful enforcement of a combination has already been discussed in the previous section. The processes by which companies secured control of coal lands is a large topic; hence the next section is devoted entirely to it. However, since coal lands were long-term economic resources that required substantial capital and time to develop, the question must be raised as to whether coal land ownership, of itself, would enable companies to operate more effectively in combination. If indeed this was the case, how did this work out in terms of the cartel mechanism? This latter point will be addressed in the next chapter.

Under the heading of debt burden, there is a danger of later commentators engaging in ex post rationalization of circumstances that were not actually present at the time the combination was initiated, although they were doubtless partly foreseen by the more far-sighted of managers. Abundant evidence about the financial condition of the anthracite railroads during the period has been presented in Chapter 6. While debt burdens weighed very heavily on certain companies in the latter part of the 1870s, this was not the case early in the decade when the discussions about the combination first took place (see next chapter), remembering that these predated the 1873 panic. Even the Reading, whose plans for coal land acquisition were much more grandiose than its competitors', was still in the process of building up its coal estate at this time.

The effects of its increased financial obligations would take several years and a major depression to wreak havoc with its accounts. In contrast, the desire, first to safeguard and then to maximize market share, is not in doubt.

The role of individual operators is closely bound up with the views of the transportation companies about their 'common carrier' status and the existence or otherwise of contractual arrangements for the shipment and sale of coal. Transportation companies with a strong sense of their former status would endeavour to move all the coal offered to them along their lines by the individual operators, as far as available rolling stock would allow. This would obviously enable the operators to try and maximize their own sales under prevailing market conditions. However, as soon as carriers began to view their ownership of the means of transportation as giving them the power to control, at least in some measure, the economic behavior of operators tributary to their lines, the overall role of the latter would become less important. This power relation could be reduced to a very simple mechanism. Since, with very few exceptions, mines did not have space for stockpiling coal, the breakers could only run as long as railroad cars for shipment were arriving to take the coal away. Otherwise, the breaker storage pockets would fill up in a very short time and the mine would have to stop. Hence, any deliberate rationing in the delivery of empty cars to the mines, when rolling stock was actually available, would give the transportation company control over coal output. The other means of controlling the production of individual operators was by means of provisions in various types of coal purchase and shipment contracts with the transportation companies. These usually required specified amounts of coal to be mined and shipped within defined time limits. Both the car rationing and the contractual mechanisms will be examined in detail in later chapters.

In turn, groups of individual operators found solidarity and unity of action through the formation of Coal Exchanges and Associations, especially in the Schuylkill and Lehigh Regions. Good examples include the Mahanoy Valley and Locust Mountain Coal Association formed in 1867, and the Anthracite Board of Trade formed in Schuylkill County in November 1869.[43] At the time, however, these were more a response to the threat from labor unions than they were a response to discriminatory practices by transportation companies. In general, it would be equally easy to argue that the presence of independently minded coal operators was a hindrance to the formation of an effective combination, because of the difficulty of controlling their output, as it would be to argue that the downward pressure they would put on prices, through competitive forces, would necessitate a combination to restrict coal supply and provide price support.

Virtue's fifth pre-condition concerns coal surpluses and rapidly falling prices. The immediate question here relates to the timescale involved, a point that Virtue does not address. For example, occasional overstocking of markets was common, so is the relevant timescale one of weeks, months, or years? It seems more useful to take a step back and examine general business conditions and price behavior in the months and years preceding the 1873 combination, to gain a clearer picture of the relevant dynamics. Figure 10.1 picks up the business cycle theme first developed in Chapter 2, to show the variations in business activity from 1866 to 1873. As before, positive values of the index reflect broad expansion in the economy, and negative values a process of contraction. From the figure, it is apparent that a sharp oscillation following the end of the Civil War was followed by a period of modest, short fluctuations between 1867 and 1871. Expansion was more concentrated in 1869, and contraction in late 1870 and early 1871. Thereafter, a relatively sustained growth period persisted until the beginning of 1873, when a declining growth rate presaged the onset of the 1873 panic. It might reasonably be expected that market overstocking would occur during downswings, accompanied by declines in prices. However, reference back to Figure 9.10 makes such a relationship hard to substantiate. In the first place, price behavior over the period needs to be seen in the context of the deflationary response to the ending of the Civil War period of high inflation. Secondly, an overall trend of price reduction, back towards ante-bellum levels, is punctuated by several upward price 'spikes'. Each of these, in 1868, 1869, and to a lesser extent 1871, is the product of market distortions arising from strikes. These strike effects can be seen more clearly in Figures 10.2 and 10.3. A year-end seasonal dip in output is to be expected, but the sharp dips in mid-1868 and 1869 can be identified clearly in the first of the two figures. The 1870 and 1871 strikes appear as extensions of the normal seasonal dip, as they affected the opening months of the year. Comparison with Figure 10.3 shows that the 1868 strike only had a substantial effect in the Schuylkill

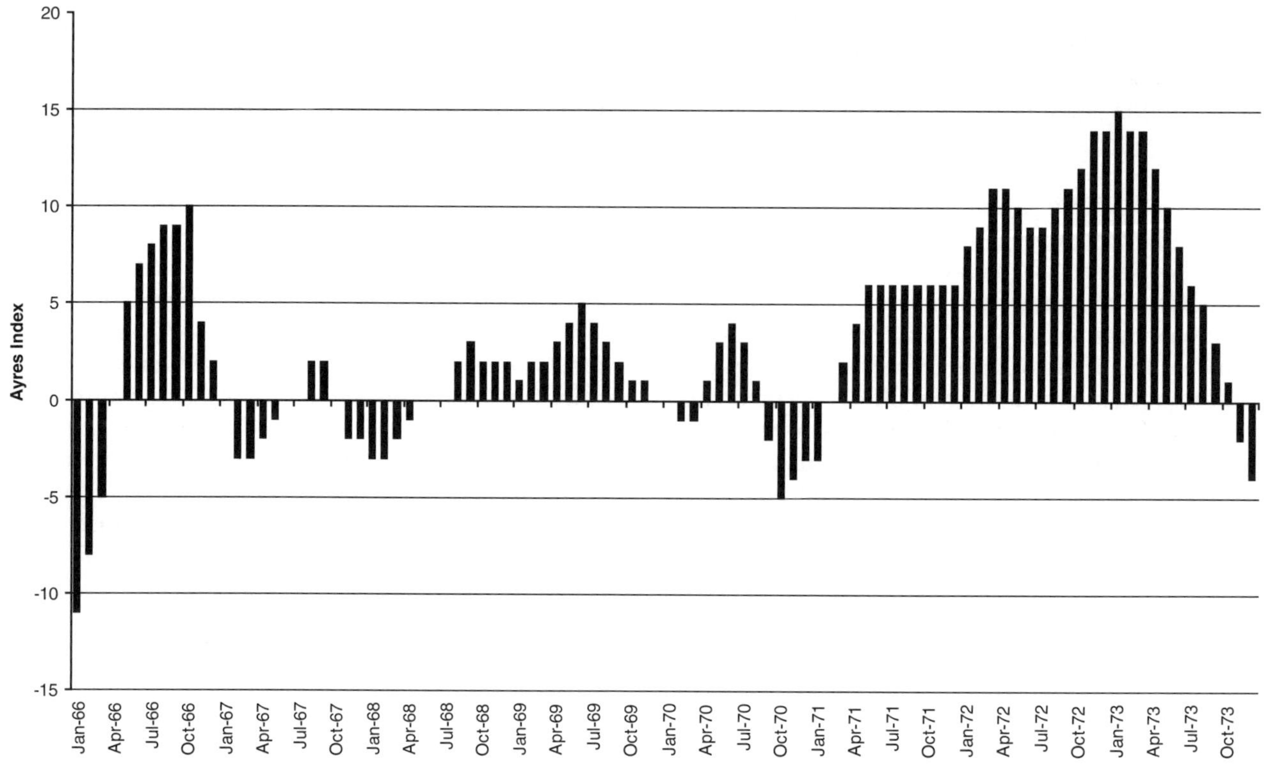

FIGURE 10.1
Monthly Business Conditions 1866–1873.

(Philadelphia and Reading RR – P&RRR) and Lehigh Regions (Lehigh Valley RR-LVRR), while the 1870 strike was only partially effective and then only in Schuylkill. Strikes in the Wyoming Region, by contrast, when they did occur, were more effective in cutting off output than was the case in Schuylkill, as the shipment figures for the Lackawanna Railroad indicate (Figure 10.3). Following strikes, a period of 'catch-up' in shipments followed, and high prices persisted until the shortfall in supply was largely made up, at which point they generally fell precipitately. The catch-up period for the Reading and Lehigh Valley Railroads in 1871 is particularly interesting, because in neither case were the railroads able to sustain shipments at these levels during the ensuing year, unlike the Lackawanna, despite the favorable business conditions.

Hence, major price variations during the period were predominantly a function of artificial market scarcity as a result of strikes. Sometimes these strikes corresponded to periods of business contraction, as in 1868 and late 1870–early 1871. However, the 1869 strikes took place during more buoyant conditions. These strikes were a follow-up to those in 1868 and were dominated by the question of the 8-hour law, rather than specifically business circumstances, a point that will be examined further below.[44] It is also useful to note that, from the operators' perspective, if markets were dull, strikes were not necessarily quite as adverse circumstances as might at first appear. Working the mine well below capacity, or suspending frequently for short periods when demand was slack, resulted in mining costs that were 15 to 25 cents a ton higher than if the mine was running full.[45] Also, in the catch-up period when mines could operate near capacity, selling prices were also much higher than they would have been, which tended to increase profits still further. Unfortunately, these eventual benefits had to be offset against the significant costs of keeping the mine open and free from water during the strike, when no cash flow was forthcoming. So, deep pockets were required to survive a strike of extended duration. A version of the same argument might be thought to apply to the men, since when they did work during the catch-up period, employment would be continuous, at least for a limited period, and wages higher. However, in the view of the Miner's Journal, steady work at lower

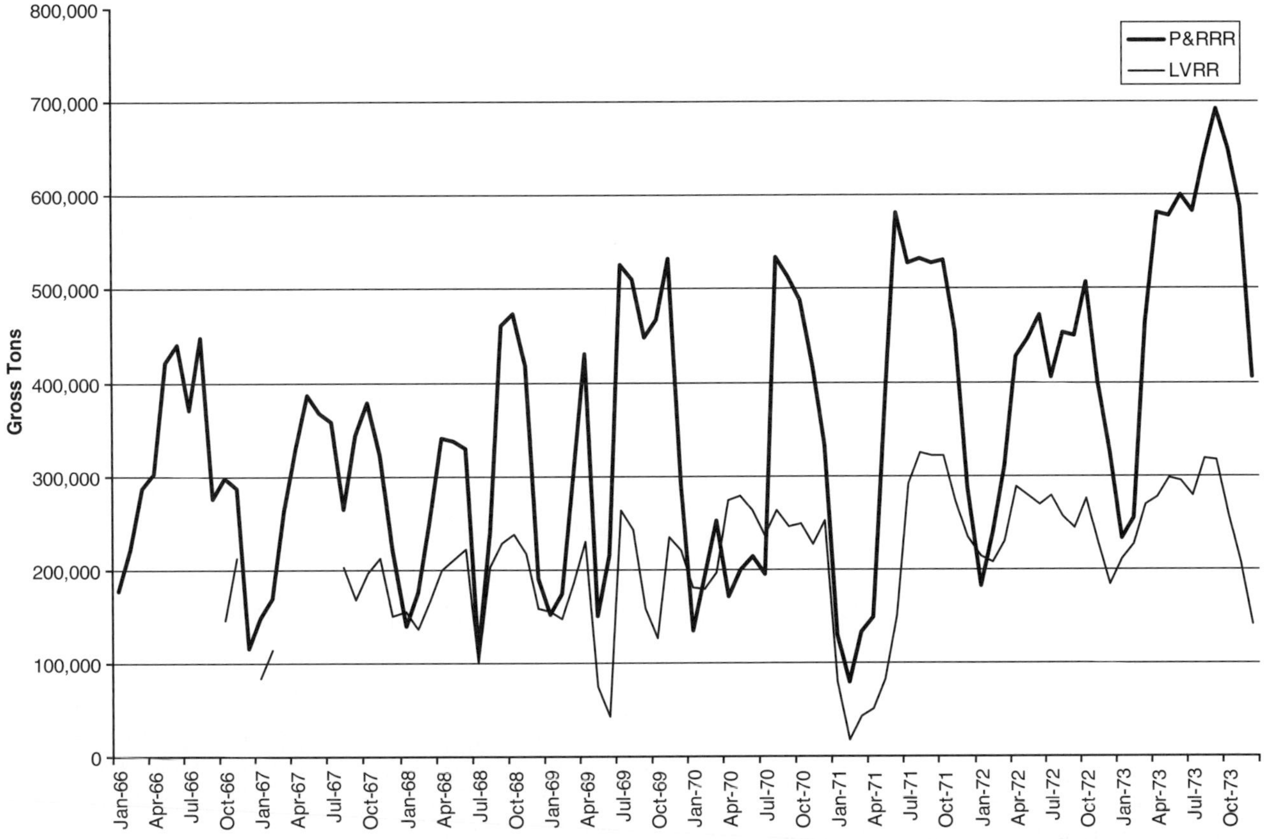

FIGURE 10.2
Monthly Coal Shipments By Railroad 1866–1873.

wages was much more desirable than bursts of higher-paid work.[46]

It is therefore difficult to characterize the period as one of 'rapidly falling prices'. This would only really apply to the period between November 1865 and June 1867, when the auction price of Scranton stove coal fell from $11.75 to $4.58. In contrast, ignoring the strike-related fluctuations, the price in September 1867 was $4.42. The lowest price reached subsequently, before the onset of the combination, was $3.78 in September 1872, although it recovered to $4.31 in the following month.[47]

Similarly, a close examination of coal market reports during 1872 was undertaken to see if the claim about large surpluses on hand could be substantiated. In the early part of the year, there is some evidence of coal being stockpiled at Port Richmond in Philadelphia, but the major cause appears to have been a shortage of vessels to transport it to other eastern seaboard destinations.[48] This seriously curtailed the output of the Schuylkill Region, for a period, causing the Miner's Journal to observe that the Delaware and Hudson, by itself, shipping to New York, was nearly matching the entire Schuylkill Region in terms of production.[49] However, by the end of May, the vessel shortage had disappeared, coal was moving freely, and it was reported that the stock of coal in the market was not large compared with the quantity going forward.[50] Apart from the occasional period of expected market dullness during the summer, there were no large-scale accumulations of coal for the remainder of the year. A November report indicated an absence of overstocking and an expectation that the market could be 'pretty clear' by the time it opened in the spring of 1873.[51] What had become all too apparent instead, was that the Schuylkill wage agreement for 1872 between the operators and the miners' union had been set at too high a level in relation to prices ruling during the year. The same applied to railroad/canal tolls. This meant that to compete effectively in New York and New England markets, Schuylkill operators had either to work with minimal profit margins or even losses. The only

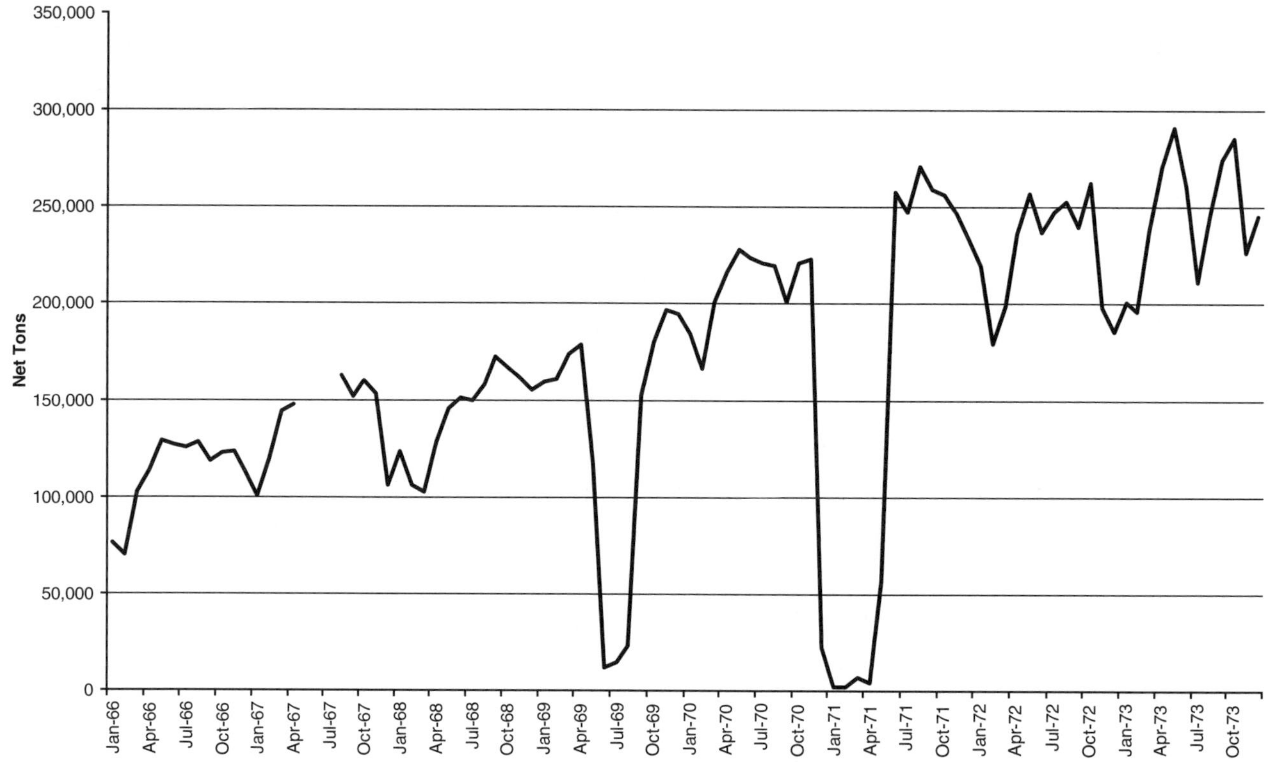

FIGURE 10.3
Monthly Coal Shipments: Delaware, Lackawanna and Western Railroad 1866–1873.

alternative was to suspend operations.[52] It was reported in June that one-third of mines in the Mahanoy area had stopped as a result. These stoppages partly explain why Schuylkill output remained at little over half its actual capacity, even when the vessel shortage disappeared.[53] At the end of July, when the full extent of the problem became apparent, the Reading Railroad made limited reductions in rates on coal to Port Richmond and also in canal tolls for coal consigned to New York.[54] Unfortunately, the damage was already done. Major customers had placed most of their orders elsewhere for the year, leaving the Schuylkill operators with the rump of the trade, at prices that were much less remunerative for them than the Wyoming Region companies, which had not adopted the same wage scales.[55] The implication of these findings is that there was no fundamental problem of protracted market overstocking in the period of interest, which affected all regions to such an extent that concert of action was required to eradicate it.

Virtue's final claimed pre-condition for the combination was the pressure of a strong, united labor organization. There is no need here to rehearse all the details of unionization in the immediate post-bellum period, as this has already been carefully examined by Aurand.[56] Instead, some of the major issues will be highlighted as they bear on the questions at issue here. Attempts at unionization can be traced back at least as far as the Bates Union of 1848, but it was not until 20 years later that labor in the coalfields found common cause around the implementation of the 8-hour law. This stipulated that 8 hours should be a normal working day, except where a contract for a longer day existed. The demand from the miners and laborers was for recognition of the law and a shortening of the working day, without, of course, a corresponding reduction in pay. The operators naturally refused such demands and the Schuylkill and Lehigh Regions came out on strike, although the majority of the Wyoming men declined to join them.[57] The strike ended in August 1868 with a compromise of a 10 percent increase in pay and continuance of the 10-hour day. However, during the strike, plans were laid for the formation of a union spanning the anthracite counties, to be called the Workingmen's Benevolent Association

(WBA), after the earlier eponymous organization local to the town of St Clair, near Pottsville.[58] Interestingly, the primary purpose of the Association was to enable the operators and miners to control the coal market and rid the trade of low prices caused, in their view, by overproduction. They also introduced the idea of the basis system of wages. Under this system, a minimum wage level was to be defined, associated with a specific market price for coal. If coal prices rose above the price specified, wages would advance by a defined incremental amount for each incremental increase in prices. This sliding scale resulted in advances or reductions of wages as prices fluctuated. However, if prices fell below the basis minimum, the agreed minimum wage level was still maintained. Under these conditions, the operators could find their profit margins evaporating rapidly. The only option to avoid working at a loss would be to suspend the mine until prices improved or transportation tolls were reduced, which would amount to the same thing from the operators' perspective.[59] As indicated above, it was the setting of a basis minimum wage, at too high a level for market conditions in 1872, that resulted in suspension of work at many Schuylkill mines during that year.

A very clear example of the labor perspective that linked the 8-hour movement to the subsequent policies of the WBA can be found in a reported speech of Thomas Williams in Wilkes-Barre to a cheering crowd of strikers in July 1868. He argued that one of the objects of the 8-hour strike was to prevent the market being overstocked. If mines ran 10 hours a day through the summer and fall, he felt this would happen, while, if the hours were reduced to eight, demand and prices would increase and the end result would be better both for employer and employee.[60]

A perennial problem for the early union organizers from the Schuylkill and Lehigh Regions was the general suspicion of the Southern Field men by those in the Northern Field. As W. R. Storrs reported to President Sloan at the height of the 1868 disturbances 'our miners have no confidence in the Schuylkill men and will not join them if they can avoid it'.[61] This was not wishful thinking by the General Coal Agent, since the men themselves voted not to make a stand on the 8 hours question shortly thereafter.[62] The other problem faced by the union was the implacable opposition by the three major Northern Field companies to the basis sliding scale of wages adopted elsewhere. Apparently it was also easier for the three companies to agree to work in concert to resist wage demands than it was for them to work in harmony on prices and shipments of coal.[63] Storrs gave his reasons for opposing the basis very succinctly in his account of a meeting with a committee of miners in May 1869. The men had come asking for a 10 cents per car advance, which Storrs declined. He argued:

1. The basis took away control of the business to work or be idle.
2. It was difficult to calculate a basis price for the many markets served.
3. As merchants in the sale of coal, the company wanted no partners, but took all the chances of profit and loss. On the other hand, as producers of coal, the company would pay the full market rate for labor in each department.
4. There were so many contingencies incident to the basis arrangement that it would not prove satisfactory to the men.

The men responded that with a basis 'they would not think of suspension but would stand true on low as well as high prices'.[64]

Some Northern Field operators, such as Charles Parrish and others in the Wilkes-Barre area, did adopt a version of the basis scale. As a result, by agreeing also to advance wages, they were able to get their men back to work in the latter part of June 1869, after only a short strike.[65] The Lackawanna men continued to make requests for a basis agreement, but the company declined. Eventually, after agreeing to get the same advance as the Pennsylvania Coal Company men, the Lackawanna miners started going back to work at the end of August, but without a basis agreement.[66] As circumstances were to prove, avoiding the strictures of the basis system gave the Northern Field companies more flexibility in responding to changing business conditions, as Storrs correctly foresaw in his first point to the men. Most certainly, it prevented a situation where the company would have to work at a loss to fulfil order obligations incurred earlier in any given year. At the same time, basis-related wage considerations continued to feature as part of labor negotiations thereafter, not least because both the Lackawanna and Delaware and Hudson owned mines in the Plymouth and Wilkes-Barre areas, at the lower end of the Northern Field, where many of the neighboring concerns were running under basis agreements.

It is apparent, therefore, that in this pre-combination period labor was at least as active as capital, and arguably more so initially, in trying to restrict coal supply, if not to the wants of the market, then to what the market would bear, while yielding wages deemed satisfactory by the men. This potential alignment of interests between capital and labor, in terms of advancing prices for coal, led to an interesting situation at the outset of the 1871 strike. The operators were so keen to reduce output, because of falling demand, that they offered to pay basis wages in January 1871, if the union could make the strike general![67] However, as strike losses began to mount, the demands of the men continued and the damage to the market share of the Schuylkill Region from inflexible and excessive basis settlements became increasingly apparent, especially after 1872. Collaboration between capital and labor began to appear increasingly illusory, and Storrs' other concerns about control and the desire to avoid having partners in the business came to the fore. In this way, the question of the combination and its operation became increasingly a matter of economic power and control, a point that will be echoed in the next section in coal lands. As Gowen wrote in his Annual Report in relation to the unions, after the end of the 1875 strike:

> 'The determined stand taken during the struggle by the Company and by the individual coal operators, has resulted in permanently rescuing their property from the arbitrary control of an irresponsible trades union, which for so many years has interfered in and frequently dictated its management'.[68]

Hence, there can be little doubt that the rise of unionization encouraged the railroads and the individual operators to find common cause against the perceived threat to the management control of their businesses, as well as that of the trade as a whole. At the same time, there is evidence that, in the early stages of discussion about the proposed combination in 1872, divergent views on how to deal with labor issues were one of the sticking points. As Storrs remarked in February 1872:

> 'An arrangement among operators before some irresponsible ones are bottled up or shut out would not be lasting. A combination with diverse theories on the labor question will be good for nothing'.[69]

The 'diverse theories' were not specified, but they presumably related to whether or not companies should accede to union requests for adoption of a basis system. There were clearly other issues as well, however. The following day, Storrs also made reference to a scheme for division of tonnage between the major interests, apparently proposed by Charles Parrish, which had not found favor either with President Sloan or with Dickson of the Delaware and Hudson.[70]

From the perspective of unionization, therefore, the relationship between capital and labor cannot initially be seen as one of implacable opposition between two antagonistic interests. Yet attempts at an operation to control the trade and restrict coal supply were eventually to founder on the key issue of perception, both by individual operators and the railroads, that others were attempting to dictate the terms of their business. This problem was to echo down the years to the 1902 coal strike and beyond.[71]

Reflecting more widely on this brief background critique of Virtues' six pre-conditions for the combination, it is apparent that questions of market power, market share, and control over decision-making lay at the heart of the negotiations between the major interests that led to the formation of the 1873 combination and the subsequent attempts at concerted action. However, this combination movement must be seen in the light of the decade of prior experience of similar smaller- scale operations, by the Lackawanna and its associated companies, in relation to the western trade. Mention should also be made in this context of the failed merger between the Lackawanna and the Jersey Central in 1872, which has been discussed in a previous chapter. Although detailed documentary evidence is sparse, it seems clear that this move was the tail-end of a more ambitious earlier plan to control the eastern trade from the Wyoming and Lehigh Regions and divert the Schuylkill trade away from the Reading. This earlier plan also included the Lehigh Valley Railroad. One indication of this can be found buried in the minutes of the Lackawanna's Executive Committee in 1869. The committee approved a proposal for the Morris and Essex, the Jersey Central, and the Lehigh Valley to acquire the charter of the Union Central Coal, Iron and Railroad Company, each of the three guaranteeing an equal proportion of the latter company's bonds. The railroad was to be built from Schuylkill County to connect with the Lehigh Valley Railroad at Lehighton.[72] According to the Miner's Journal, preliminary construction work on the new line had begun in the latter part of 1868, but had been suspended temporarily. A few weeks later, the same newspaper was extolling the merits of an alternative plan for a through railroad from Pottsville to Perth Amboy. It claimed that the proposed line of the Union Central was at too high an elevation to receive coal from mines then in operation, unless it was hoisted by inclined planes. The proposed line would also be twice as expensive to build as the alternative route, and the grades were against the direction of heavy traffic.[73] These reasons may explain why the project was apparently not pursued, as no further references to it have been located.

This proposed 'community of interest', to borrow the later term, between the Lackawanna, Lehigh Valley, and Jersey Central cannot be dismissed as a mere whim. A much later letter, written by Samuel Sloan, indicates that it had originally been proposed by Moses Taylor and himself, and that Packer and Pardee of the Lehigh Valley had been involved in the discussions.[74] The events described indicate that this plan must have dated at least back to near the beginning of his presidency of the Lackawanna.

As a final postscript to the pre-1873 era, it is useful to contrast the approaches of the two junior railroad presidents, Messrs. Sloan and Gowen. The former wished to manage the trade through transportation contracts, to control coal supply and mergers to reduce costs and eliminate over-building in the railroad sector. This might be termed a 'banker's approach', with the strong background influence of Moses Taylor. In contrast, Gowen, at least in his early days, appeared to espouse a much more liberal free-market policy. His remedy for overproduction, as set before the 1871 committee of Investigation, was not the periodic suspension mechanism beloved of the WBA, but increased demand at lower prices. Steady work, he insisted, would reduce the price of coal at tidewater by at least 75 cents, thereby increasing the market area by 100 miles and providing the necessary demand to absorb available supply.[75] No doubt his remarks were aimed at presenting his company's stance in a manner likely to appeal to the committee. Yet it seems clear that more of a shift, in the policy of the Reading in the direction of combination-oriented policies, is apparent in the years prior to 1873 than was the case with the Northern Field companies, where such notions were already well-established in the corporate mind. This change was especially apparent in relation to the acquisition and management of coal lands, a topic that will now be examined in detail.

ACCELERATED ACQUISITION OF COAL LANDS

The traditional view of the accelerated process of coal land acquisition by the major companies is that it was 'probably' initiated by the Delaware and Hudson, using excess profits from the boom years of the later Civil War period to bolster its strategic position within the trade.[76] Unfortunately the documentary evidence does not substantiate this view, although the length of time it has stood unchallenged is itself a reflection on the obscurity of the records required to disprove it! In fact, as the following discussion will demonstrate, the rush to acquire coal lands was actually initiated in a complex and almost clandestine manner during the darkest days of the Civil War, by interests aligned with the Lackawanna Railroad. Beginning with this company, the interplay between land acquisition, charter rights, and competitive pressures perceived by the different railroads will now be traced.

THE LACKAWANNA RAILROAD

Recourse to that final arbiter of the occurrence of land transactions, the county courthouse deedbooks, reveals that as early as the summer of 1863, interests acting for the Lackawanna Railroad, via the person of Percy Pyne, began acquiring significant bodies of coal lands in Hanover Township near Nanticoke, below Wilkes-Barre. Pyne subsequently sold these lands, totalling over 1,000 acres, to his father-in-law, Moses Taylor, the New York banker and major Lackawanna shareholder, in November 1863.[77] The latter, in association with J.J. Phelps, also acquired a variety of other land tracts in the same area at about the same time. One particular group of these covered about 1,820 acres and encompassed lands that originally belonged to the old Wyoming Coal Company, the Hanover Coal Company, and the Dundee Coal Company. These two bodies of coal lands were sold to the newly formed Nanticoke Coal and Iron Company in July and December 1864, for the sums of $183,177 and $382,675 respectively.[78] It is interesting to note that Taylor, as a matter of policy, did not make direct personal gain from transactions involving property held in trust for others.[79]

The Nanticoke Coal and Iron Company was the first of several companies to be used by the Lackawanna for holding coal lands over and above its charter entitlement, a corporate device that was later to be adopted by the other transportation companies. Other Lackawanna subsidiaries included the Steuben Coal Company, which acquired lands formerly held by the old Scranton Coal Company, when the latter was merged into the former in 1867, and the Granby Coal Company, organized in the same year.[80] In 1866, the Steuben Coal Company had 821 acres of coal lands in comparison to the 4,388 held by the Nanticoke Coal and Iron Company. There were also other lands and leases held directly by Moses Taylor and J.J. Phelps in the interest of the Lackawanna, including the Central Coal Company lease in Scranton, which, at the time, was held by Phelps, but not recorded.[81] Further to this, Percy Pyne was apparently holding almost 2,000 acres of the defunct Union Iron and Coal Company's property that he had bought at successive sheriffs' sales in 1859 and 1861. Hence, both corporate and personal vehicles were being used to extend the company's influence over coal property. Tardy or incomplete paperwork was also used as a deliberate device to obfuscate the financial linkages back to the Lackawanna. While this may have been sufficient to allow state legislators in Harrisburg to turn a blind eye, the local grapevine was keeping up with the developments. In August 1867, the Luzerne Union carried an article about coal land purchases, noting that the Lackawanna had 'through individuals who ostensibly do not act for the company, purchased largely of coal lands in this valley'.[82]

It is clear from surviving correspondence and accounts that this proliferation of coal companies increasingly proved to be an administrative headache for the Lackawanna, because of keeping track of appraised taxes on capital stock, deeds, lease documents, and the like.[83] This was exacerbated when the company found it needed to initiate developments on the lands of these supposedly independent concerns, which unfortunately lacked the necessary funds in their own right. This necessitated accounting transfers from the Lackawanna itself. A good example of this was the ill-fated Avondale mine, described in Chapter 3. Development of this mine by Lackawanna interests began in 1866, initially under the control of the Steuben Coal Company.[84]

The easiest way to overcome such problems was to amalgamate the smaller companies as far as possible, but legal opinion had found that an 1867 Act authorizing the consolidation of oil and other mining companies did not extend to the coal industry.[85] Undeterred, John Brisbin wrote briskly to the Hon. L.D. Shoemaker in Harrisburg, to provide him with the draft text of an act that he wanted passed in the current session of the legislature 'without fail', to ensure that coal companies were included in merger legislation.[86] Following successful resolution of this legal problem, a stockholders meeting was called in July 1868, to agree to the merger of the Steuben and Granby Coal Companies with the Nanticoke Coal and Iron Company. Thereafter, the latter was run even more closely in conjunction with the main business affairs of the railroad itself, and all major investments in its mines were directly approved by the Lackawanna's Board of Managers.[87] Finally, by agreement dated June 23, 1870, the Nanticoke Coal and Iron Company was itself merged into the Lackawanna at a book cost of $3 million. This merger brought in the new Taylor breaker, on the lands of the old Union Iron and Coal Company, the re-built Avondale breaker, the Scranton Coal Company's mine, and the other new improvements in the Scranton area that were to become the Archbald (Plate 7) and Sloan breakers. The mines were valued at a convenient round figure of $500,000, the remainder being for the immensely important coal estate that had been built up over a period of years, largely through the persistent efforts of Messrs. Taylor, Pyne, and Phelps.[88]

Despite the considerable effort expended on these nominally independent, but actually subsidiary companies, the Lackawanna was not neglectful of its own charter rights as a further means of extending its coal estate. The company's original authorization had been to hold a meagre 1,000 acres. Although this had been increased in stages to 5,000 acres by 1867, it was still felt to be insufficient in comparison to its competitors, given that the Delaware and Hudson, with a smaller capitalization ($10 million as opposed to $15 million), was able to hold 8,000 acres.[89] As a result, in 1867, John Brisbin, then president of the Lackawanna, formulated a detailed case to the legislature for further extension of the railroad's charter privileges. Despite the obvious element of special pleading, it merits quotation at some length because of the insight it gives into the railroad's own assessment of the interdependency between charter privileges, special legislation, and processes of regional economic growth:

> 'To say that none of usthe railroad companies ever ought to have had theselandholding rights is simply

to say that these improvements ought never to have been made and that the Commonwealth and her citizens would have been better off without them, for no sensible man will pretend that they ever could or would have been made without the right to hold lands and deal in coal. I leave it for you to answer the question whether it would have been better had those companies never been born.

If these rights to hold lands and mine coal were taken from the companies I have namedthe Lackawanna, Delaware and Hudson, and Pennsylvania Coal Company, their property would within a twelvemonth pass into the charge of a receiver, and not less than ten millions of dollars which they now annually pay out be withdrawn from circulation in this valley'.

And again:

'To deny to these Corporations sufficient lands for present use, and to afford a reasonable warrant for further investment, does not only injure them but the people of the district, and detracts from the revenues of the Commonwealth and induces the Companies to "whip the Devil around the stump" by controlling the stock of separate coal companies in virtue of special legislation authorizing them to hold the stock'.[90]

From a latter-day perspective, we can only be amused by the effrontery of justifying one method of circumventing charter restrictions on the basis that it is better than a different method that has already been deployed! Yet it remains the case that the range of land acquisition strategies used by the Lackawanna, coupled with powerful pressure on the legislature when required, enabled the company, by the early 1870s, to assemble approximately 25,000 acres of the richest and most accessible coal lands in the Anthracite Regions. Contrary to Bogen's assertion that it was undertaken in self-defence,[91] this approach was carefully planned from an early date with the backing of key New York capitalists, and it was the other companies who found themselves needing to respond in a defensive manner.

THE DELAWARE AND HUDSON

The Delaware and Hudson was the first of the Lackawanna's competitors to respond. Initially, it might appear that this company had no pressing need to acquire additional lands, since the particular circumstances of its early founding and subsequent development meant it was blessed with a larger coal estate than some of its major competitors, as noted earlier. However, on closer inspection, several decades of mining the more accessible coals, coupled with a short-term approach to investment, had brought the company to a point where readily available productive capacity was close to exhaustion. This fact, as President Olyphant later reflected candidly to his managers, had become apparent to him shortly after his appointment in 1858, and was unfortunately 'a secret only to the members of this Board'.[92] To address this problem, he stated that he had embarked on a policy to 'buy lands, open new mines, provide facilities for the proper preparation of our coal, and generally to extend and increase the capacity of our works', although he had not revealed the full scope of his intended restoration plan to the board at the time. He also emphasized that the foundation of permanent prosperity had to be laid 'broad and deep in the possession of an adequate supply of coal-producing lands'.[93] While hindsight may have lent greater clarity to the articulation of his policy than was possible at the time, a reading of these statements might lead one to imagine that the Delaware and Hudson embarked upon significant land acquisition beginning at the end of the 1850s. However, the company's records do not entirely bear this out. According to the company's minutes, in 1859 they did agree to buy 3,500 acres in Luzerne, Wayne, and Susquehanna Counties already owned by G.T. Olyphant, the recently appointed president. This was made possible by a charter supplement that had increased their authority, to hold lands, from 5,000 to 8,000 acres. The timing suggests that Olyphant may have been holding the land in trust for the company, in a similar manner to Moses Taylor for the Lackawanna at a later date.[94] Apart from this, however, between 1857 and the beginning of 1865, only a limited number of coal tracts were bought, such as one in 1859 for $24,721 and two from Edward Jones in 1864 for nearly $40,000 each. The Eddy Creek shaft was subsequently sunk on part of the latter property.[95] Instead, implemented policy prior to the end of the Civil War seems to have focused more on the extension of railroad lines down the Northern Field, to increase the company's access to new coal territory, and to channel coal traffic in the direction of the canal. A second area of activity was the construction of ten new coal breakers.[96] Part of the reason for the initial concentration on railroad activity was that the company's original charter did not authorize it to engage freely in railroad, as opposed to canal transportation, so specific legislative authorization had to be obtained for new developments of this kind.[97] The first step in the process of railroad expansion was the extension of the company's gravity railroad, with a locomotive powered section from Olyphant down to Providence, now a suburb of Scranton, to connect with the Von Storch coal property. Although the managers approved the project at the end of 1859, it was not completed until the summer of 1863.[98] Given the timing of this new construction, the next stage in the Delaware and Hudson strategy is not difficult to comprehend. Having reached Scranton, the next step was to secure, in a circumspect manner, both coal lands and railroad access further down the Susquehanna Valley to funnel coal traffic over the company's works.

This was achieved in a rather complex fashion, suggesting that while the Delaware and Hudson may have had a good strategic sense of its long-term goals, the tactical manner of achieving them was handled in a somewhat reactive way. As Bogen notes, in 1864, three companies, which combined railroad and mining privileges, were incorporated in the Wyoming Region by 'interests friendly to the Delaware and Hudson'.[99] The first of these, the Northern Coal and Iron Company, could hold 3,000 acres of coal land and had authority to construct lateral railroads, up to 20 miles in length, to connect with any existing or future railroad to be built in Luzerne County. A later 1866 charter supplement also allowed it to build tracks in Wayne or Susquehanna Counties to connect with the Delaware or Susquehanna rivers. The company was clearly intended to be a vehicle for the Delaware and Hudson from the outset, since its two largest shareholders were G.T. Olyphant and Thomas Dickson of the latter company.[100] However, unlike the Lackawanna's Nanticoke Coal and Iron Company, chartered a few months after it in 1864, it lay effectively dormant for several years. Instead, early activity focused more on the remaining two companies, the Union Coal Company and the Howard Coal and Iron Company. By 1866, the former owned coal mines in Plymouth, on Mill Creek, east of Wilkes-Barre, and apparently also at Locustdale in the Mahanoy Region, although the latter is not listed as part of the assets of the company in a later sale.[101] The driving force in the company was the president, E.A. Quintard, a prominent coal industry figure, but not someone wholly identified with the Delaware and Hudson. Hence, relations with the latter were not initially quite as close as in the case of the Northern Coal and Iron Company. The activities of the last of the three companies, the Howard Coal and Iron, are clouded in relative obscurity. However, the important point about this company, in relation to the present discussion, is that it had charter privileges to build a railroad connecting the Nanticoke Railroad near Wilkes-Barre with the terminus of the recently completed Delaware and Hudson extension at Providence.[102]

The next step in this complex process was an agreement between the Union Coal Company and the Delaware and Hudson, dated June 28, 1866, for the former to build a railroad from its mines near Wilkes-Barre to the terminus of the latter in Providence. In return, during 1867, the Union Coal Company was to provide 200,000 tons of coal to be shipped to the Delaware and Hudson for carriage on the canal. Failure of the Coal Company to abide by the agreement would result in the Delaware and Hudson taking over its mines, to work them at 30 cents royalty per ton.[103] Further to this, the managers of the latter company agreed to subscribe for $250,000 of the mortgage bonds of the Union, to help finance the construction of the road, which was to begin immediately following ratification of the agreement.[104] Later in 1866, the Union made a second contract with the Lehigh Coal and Navigation Company, lessee of the Nanticoke Railroad, whereby the Lehigh Company was given exclusive rights to run passenger and freight trains and also its own coal trains, over the Union road provided these did not impede the Union in meeting its obligations under the earlier agreement with the Delaware and Hudson. The contract also makes clear that the Union was making use of the charter privileges of the Howard Coal and Iron Company, of which it had become the majority stockholder, to build the railroad, at least in part.[105] This confusing situation was resolved in March 1867 by a merger of the Howard and Union Companies.[106] At this time, the Delaware and Hudson had a token presence among the shareholders of the Howard Coal and Iron Company in the shape of J.J. Albright. More significantly, to protect the Canal Company's investment in the Union's mines, he had taken over as general superintendent of these works prior to this date.[107]

Work on the new railroad began in the late summer of 1866, in the usual manner, with survey teams and acquisition of rights of way. Work on grading the line was in progress by the fall of that year. It seems that work began simultaneously at both ends of the planned line, south from Scranton and north from Wilkes-Barre under the direction of Charles Parrish. By April 1867, only the gap in the middle near Pittston remained to be closed and a formal opening in May was confidently expected. Unfortunately, serious damage to the roadbed from a freshet on the Susquehanna delayed the completion until June 1867.[108]

To this point, the Delaware and Hudson seems to have been largely content with the progress of its strategy to funnel coal to its canal from the Wilkes-Barre district and evinced no pressing desire to control the new railroad as such, only its coal tonnage. In the weakening business climate of the latter part of 1867, however, matters began to unravel in an unsatisfactory manner. With a suitable degree of understatement, Thomas Dickson reported to his board that Quintard had discovered that 'it was easier to make calculations upon paper, than to produce coal'.[109] The latter had also realized that he would need to ship a much larger tonnage, than the 250,000 tons allowed for in his revised contract with the Delaware and Hudson, if he was to recoup his investment in new railroad construction. He therefore applied to the Canal Company for an increase in his permitted tonnage. Dickson's response was perhaps, at first sight, a little surprising. Rather than jumping at the chance of higher shipments immediately, he insisted that a change could only be made if the Union Company secured more lands, and he proposed they try to consolidate with the Wilkes-Barre Coal and Iron Company to achieve this.[110] This would seem to have been a high-risk strategy, since a merger with a competing concern would dilute the Canal Company's control and might have led to diversion of tonnage over alternative routes to market.

The explanation for it may, in fact, lie with the little-known Susquehanna and Delaware Railroad project. The Wilkes-Barre Coal and Iron Company had secured the charter for this proposed route from Wilkes-Barre to the Delaware River, but actual construction had not begun.[111] Had this project come to fruition, there would have been direct competition for the Union Company's railroad, whereas a merger would have relieved the Wilkes-Barre Company from the necessity of building it.

Negotiations between the Wilkes-Barre Company and the Union were apparently initiated, but were then broken off for reasons that are not documented. In all probability, given the timing involved, these were related to the acquisition of a majority of the stock of the former company by the Lehigh Coal and Navigation Company, which already held $300,000 of Union Coal Company bonds.[112] The Susquehanna and Delaware Railroad project was shelved.

A side effect of these negotiations was, however, to prove of fundamental importance to the Delaware and Hudson. President Wilson, of the old established Baltimore Coal Company, caught wind of what was going on and offered to sell his company to Quintard of the Union. Quintard, in turn, applied to the Delaware and Hudson for financial assistance to complete the purchase and an initial agreement was made, dated August 20, 1867, for a consolidation of the Union and Baltimore companies.[113] Before all the provisions of this agreement were completed, however, it was replaced by a separate sale of the Baltimore Company directly to the Delaware and Hudson, for $1.75 million, on October 15, 1867. The Baltimore property was then temporarily leased back to the Union Coal Company for a 6-month period.[114] It was a very valuable estate of about 2,000 acres, 1,750 of which were believed to be underlain by coal. There were three breakers and the coal was of the highest quality, with an established name in the market. By means of the various agreements, the Delaware and Hudson had also increased its contractually required tonnage from the Union and Baltimore properties, now to be worked in tandem, to a figure of 500,000 tons per annum. At first sight, this was a highly satisfactory outcome.[115]

Unfortunately, with the continuing weakness of the coal trade and the high costs of its mining and railroad investments, the output of the Union Coal Company did not reach its contractually required levels and its cash flow became impaired. News leaked out, late in 1867, that certain members of the company had advanced it funds to keep it afloat, but these had not been repaid, and a forced sale of the company's assets was imminent. Thomas Dickson and attorney A.T. McClintock, acting for the Delaware and Hudson, bought the personal property for $147,000 at the Sheriff's sale in early 1868. At about the same time, the Delaware and Hudson invoked the modified original agreement, dating back to 1866. This allowed it to take possession and operate the company's mines on a royalty basis, in the event that they defaulted on provision of coal for transportation by the canal. By this point, the mines in question included the Mill Creek, the Laurel Run, and the old Wilcox Mine.[116] The Chauncey mine was apparently sold separately to J. Stickney, formerly assistant president of the Wilkes-Barre Coal and Iron Company.[117]

On the same day (January 21, 1868) that the Delaware and Hudson took control of the Union coal property, it also purchased the lands and mines of the Boston Coal Company near Plymouth, in a secret deal worth $1,575,000 to the sellers. This estate covered about 1,100 acres and $500,000 was paid for the Coal Company's land. The remainder was paid for tracts owned by others, for the extensive improvements, and for a bridge across the Susquehanna. Since it contained 68 feet of coal seams above the base of the Red Ash seam, about 500 feet below ground level, this estate was regarded as an even more valuable purchase than the Baltimore property. The secrecy was no doubt occasioned by the fact that the mines were being operated at the time by the Lackawanna Coal Department, under a 20-year lease that did not expire until the beginning of the 1880s. Hence the Delaware and Hudson had quite literally bought the ground from under their competitor's feet! Over the following few weeks, yet another property close to the Boston estate was purchased for $606,000 from the Plymouth Coal Company.[118] These, and other lands, such as those bought for $200,000 from J.C. Fuller in December 1870, also in Plymouth, were placed under the ownership of the Northern Coal and Iron Company.

Although the Delaware and Hudson now effectively controlled both the Union and Baltimore properties, it continued to operate them for a period under the name of the re-organized Baltimore Coal and Union Railroad Company, and Quintard remained as president. On December 1, 1868, the relationship was formalized by an 18-year lease that confirmed the responsibility of the Delaware and Hudson for paying the interest on the Union Coal Company bonds. It also set a more realistic annual requirement to ship 300,000 tons of coal per annum from the coal properties. In July 1871, the structure of subsidiary companies was further simplified when the Baltimore and Union was merged with the Northern Coal and Iron Company. The latter, in turn, was leased to the Delaware and Hudson in December 1873.[119] In this convoluted and complex manner, the Delaware Hudson finally achieved its original strategic aim of acquiring substantial coal reserves in the richer lower part of the coalfield around Plymouth and Wilkes-Barre, together with a railroad that connected them to its other transportation facilities in the upper part of the field. Nevertheless, it is clear that Vice-President Thomas Dickson had set the whole process in motion, because of his fears about the extent of the lands that had already been taken over by the Lackawanna.

THE LEHIGH COAL AND NAVIGATION COMPANY AND THE JERSEY CENTRAL

The actions of the Delaware and Hudson in the lower part of the coalfield had immediate repercussions. The initial agreement for the Union Coal Company to acquire the Baltimore in August, 1867 'excited the apprehensions' of the managers of the Lehigh Coal and Navigation Company.[120] The reason for this was clear. The Baltimore had previously shipped all its eastbound coal over the Navigation Company's Nanticoke Railroad, and its tonnages in that direction had been expected to increase significantly after the imminent completion of the Lehigh and Susquehanna Railroad. Diversion of the tonnage onto the Delaware and Hudson Canal was therefore a major blow in commercial terms.[121] An immediate response was essential, to safeguard coal properties tributary to their lines, otherwise their large investments in the Nanticoke and Lehigh and Susquehanna Railroads would be rendered unprofitable. To this end, they negotiated the acquisition of a majority shareholding in the Wilkes-Barre Coal and Iron Company, which operated a number of collieries near Wilkes-Barre and Nanticoke. It owned 3,000 acres of choice coal lands and leased a further 2,500 acres. The deal involved a share exchange that necessitated the issuance of over 17,000 new shares by the Coal and Navigation Company, with a par value of $857,200. The transaction gave the company control over a very large and valuable coal estate for a very low price, as compared to the amounts paid by the Delaware and Hudson for the Baltimore and Boston Coal Company properties. Unfortunately, what should have been an excellent business deal proved to have very adverse short-term consequences. The untimely death of one of the key participants in the share exchange threw a large amount of the newly issued stock on the market. This precipitated a rapid fall in the company's share price late in 1867, which in turn plunged it into a temporary financial crisis! Nevertheless, viewed over a longer timescale, this was a key strategic acquisition for the company, that added to its existing holdings in the Lehigh Region and the 800 acres it had previously purchased near the Nanticoke Railroad in 1864.[122]

Control of the Wilkes-Barre Coal and Iron Company was unaffected by the lease of the Navigation Company's Lehigh and Susquehanna Railroad to the Jersey Central in March 1871.[123] The Coal and Iron Company continued to expand its property holdings, especially in the Plymouth area. In a short period in 1871–2, it acquired a number of important collieries and their operating leases, including the Lance, Gaylord, Dodson, Nottingham, and Washington breakers. At least some, if not all of these, were acquired in conjunction with the Navigation Company, since, in early 1872, it had yet to be decided which of the two companies was to have the Nottingham and Washington recorded as assets on its books. It is interesting to note that these latter two properties, in addition to the Lance and the Gaylord, had previously been offered to the Lackawanna, which had turned them down as being overpriced.[124]

Despite the indirect control over coal supply given to it by the lease of the Navigation Company, the Jersey Central still felt it essential to place 'their most important branch of traffic beyond the reach of contingency of any sort', by starting to acquire coal lands in their own right.[125] They preferred the very best quality lands in large blocks, even if this meant paying higher prices as a result. By the end of 1871, through counting in all the coal lands of the Navigation Company and the Coal and Iron Company, they were able to claim a coal estate of 21,500 acres of the best coal lands in the Wyoming and Lehigh Regions.[126] Their actual control of this estate was greatly strengthened in 1873, when the Navigation Company sold its Wyoming lands to the Honey Brook Coal Company in the Lehigh Region, which was directly controlled by the Jersey Central. A new company, the Lehigh and Wilkes-Barre Coal Company, was formed from the merger of the Honey Brook and the Wilkes-Barre Coal and Iron Company. This new concern also leased the Navigation Company's Lehigh Region estate in the vicinity of Summit Hill on December 31, 1873. The sale of the Wyoming lands netted the Navigation Company $4.7 million, valuing the coal lands at $700 an acre, and the lease provided for a minimum rental of $500,000 per annum.[127] Table 10.1 shows the distribution of coal and timber lands forming part of the coal estate at this time. It is apparent from the table that the coal lands were very equally spread between the Wyoming and Lehigh Regions. The entire estate was valued at over $17.5 million. Of this, $4.2 million represented mining improvements and 'personal property', i.e. coal cars, machinery, etc. These improvements included 37 slopes, 13 shafts, and 26 breakers with a productive capacity of 3.1 million tons.[129] By 1876, a further 3,000 acres of owned and leased lands in the Wyoming Region had been acquired, three additional breakers had been constructed, and the whole property stood on the books valued at $22,437,000.[130]

TABLE 10.1
Coal Estate of the Lehigh and Wilkes-Barre Coal Company 1873.[128]

Coal Properties	Status	Acreage Underlain by Coal	Timber Land	Total Acreage
Wilkes-Barre Lands	Owned	2,320	580	2,900
Newport and Hanover Lands	Owned	5,600	1,173	6,773
Plymouth Lands	Owned	600	200	800
Wilkes-Barre Lands	Leased	1,325	500	1,825
Plymouth Lands	Leased	1,030	0	1,030
Sub-total: Wyoming Lands		10,875	2,453	13,328
Honey-Brook Lands	Owned	3,925	8,075	12,000
Lehigh Navigation and Greenwood Lands	Leased	6,925	2,075	9,000
Sub-total: Lehigh Lands		10,850	10,150	21,000
Additional Timber Lands	Owned		8,000	8,000
Grand Total		21,725	20,603	42,328

THE LEHIGH VALLEY RAILROAD

The purchase of the Baltimore Coal Company by the Delaware and Hudson in 1867 had a similar impact on the directors of the Lehigh Valley Railroad as it had had on those of the Coal and Navigation Company. The 1867 Annual Report drew attention to the moves by the Wyoming Region companies to absorb and control both coal lands and existing mines, to secure present and future coal supply. Initiation of such activity by competitors in the Wyoming Region was of particular concern because the company did not want its new route north along the Susquehanna to find itself starved of coal traffic.[131] The policy adopted to combat this perceived threat was to have important geographical repercussions. At the 1868 annual meeting, stockholders authorized the company to secure 'a proportion of the coal trade from each region, by the purchase of interests in other companies owning lands on or near our several branches'.[132] Since these branch lines extended into the Western Middle Field as well as the Northern and Eastern Middle Fields, territorial acquisition took place across a much wider area of territory than with the other companies examined thus far. It also brought the race to acquire coal lands into contact with areas that the Reading Railroad regarded as tributary to its lines.

The published policy was quite specific about acquiring an interest in companies that already owned lands, because the railroad itself had no authority under its charter to hold coal lands or engage in mining, unlike the other companies discussed previously. However, stating policy is one thing and meeting business requirements quite another. In fact, several important tracts of land were initially acquired in the name of Asa Packer, president of the Railroad, to circumvent charter restrictions, in the manner to which the industry was now becoming accustomed. These tracts included the Everhart and West Pittston lands in the Northern Field and the Trevorton and Russell and Beebe Lands at the western end of the Western Middle Field.[133] A full list of the lands acquired in the different fields between 1868 and 1873 by purchase, lease, or through control of the stock of other companies is given in Table 10.2. A detailed analysis of surviving lease records and later company tabulations has been used to provide comparative information to that for the Lehigh and Wilkes-Barre Coal Company in Table 10.1. The key point is the distinction between the overall acreage of the estate and the actual acreage underlain by coal measures, as it was the latter that controlled the price. The field totals for the Eastern and Western Middle Fields closely match the summary data given in the 1873 Annual Report, and the discrepancy for the Wyoming Field is most probably the result of a combination of later re-survey and omission of the lands of the Mineral Spring Coal Company, that were only acquired in 1873. However, comparison of Table 10.2 with the Annual Report data, which were later reproduced by Bogen, indicates that the latter are misleadingly reported by the railroad as 'coal lands', when the totals given include extensive areas of timber land of very limited value.[134] Taking the coal estate proper, it is apparent that the Lehigh Valley concentrated its purchases in the Northern and Western Middle Fields, where it wished to expand its operations, rather than in the Eastern Middle Field, much of which was naturally tributary to its lines. However, it is important to note that the coal lands in the Western Middle Field, purchased directly, rather than through stock control of other companies, were largely undeveloped and relatively inaccessible from the then current railroad network. The exception to this was the Trevorton estate, where limited mining development had been carried on for some years.[135] In contrast, the lands purchased or leased in the Northern Field (Wyoming Region) were, for the most part, already developed by mining improvements, or were adjacent to lands where mining was in progress. The relative lack of development and inaccessibility of part of the estate, together with the very limited amount of land in the Wilkes-Barre/Plymouth area, which was richly endowed with coal seams, partly explains the low average cost of the lands, at about $410 an acre. This compares with a figure of about $910 an acre for the Lehigh and Wilkes-Barre, based on the data in Table 10.1. Such figures should be regarded as indicative rather than exact, because of the possible sources of error in the calculations, which are explained further in the relevant note.[136] Nevertheless, they would tend to confirm the view of the Lackawanna that the Lehigh and Wilkes-Barre purchased at least part of its property at higher prices than they themselves were willing to entertain.

Comparison of the acreages of the different companies, by the time the 1873 panic brought speculative investment largely to a close, is quite informative. The combined Jersey Central interests led the field with nearly 22,000 acres of genuine coal lands, rising to nearly 25,000 by 1876. The Lehigh Valley was strongly placed with nearly 19,000 acres of geographically diversified coal estate, while the

TABLE 10.2
Lehigh Valley Railroad and Predecessor Companies to the Lehigh Valley Coal Company: Coal Land Acquisition 1868–1873.[137]

	Status	Acreage Under Coal	Timber Land	Total Acreage	Cost*
Wyoming Region					
Coal Properties					
West Pittston	Owned	389	276	665	$160,000
Burrough's	Owned	104	0	104	$250,000
Yatesville	Owned	823	33	856	$400,000
Everhart	Owned	601	942	1543	$550,000
River Warrants	Owned	221	0	221	$58,000
Lacoe and Shiffer	Leased	469	96	565	
Dorrance	Leased	487	0	487	
Hollenback	Leased	429	0	522	
Other small tracts	Leased	522	0	522	
Lands of Companies Controlled by Stock Ownership					
Luzerne Coal and Iron Co. (All stocks)	Owned	730	31	761	$498,615
Franklin Coal Co. (Majority of stock)	Part leased, Part owned	388	918	1306	$180,600
Mineral Spring Coal Co. (Majority of stock)	Leased lands	609	381	990	$112,284
Sub-Total Wyoming Region		5772	2677	8449	
Mahanoy and Shamokin (Western Middle Field)					
Coal Properties					
Trevorton	Owned	1326	609	1935	$350,000
Russell and Beebe	Owned	645	426	1071	$250,000
Brown and Thompson	Owned	865	0	865	$300,000
Brobst	Owned	320	0	320	$50,000
Lands of Companies Controlled by Stock Ownership					
Coal Ridge Improvement and Coal Co.	Owned	1339	974	2313	$543,942
Locust Mountain Coal and Iron Co.	Owned	2025	1975	4000	$549,448
New York and Middle Coalfield RR and Coal Co.	Owned	770	517	1287	$287,734
Delano Land Co.	Owned	2494	2735	5229	$2,000,000
Sub-Total		9784	7236	17020	
Lehigh Region					
Lands of Companies Controlled by Stock Ownership					
Humboldt Coal Co. and Ashburton Lands	Owned	648	5180	5828	$415,000
Lands in Name of Lehigh Valley Railroad					
Beaver Meadow	(acquired 1864)	1094	436	1530	$75,149
Hazleton Coal Co.	(acquired 1868)	1404	685	2089	$930,325
Sub-Total		3146	6301	9447	
Grand Total		18702	16214	34916	7,885,948 (excl. Beaver Meadow)

Delaware and Hudson, at the time of the 1877 stockholders investigation, owned 20,000 acres and leased a further 5,000. However, the proportion of these that were timber lands is not stated.[138] Precise figures for the Lackawanna at this time are hard to locate, but in February 1872 the company controlled about 12,500 acres of land underlain by coal, (approximately 20 percent of which was leased lands), and a further 5,400 acres of timber lands.[139] Although additions to their landholding continued thereafter, these figures from the company's own records indicate that the contemporary newspaper estimate of 25,000 acres of controlled lands, cited by Bogen, was somewhat exaggerated.[140] Bringing up the rear was the Pennsylvania Coal Company, which already had authorization to hold in excess of 11,000 acres by the end of the Civil War, as noted in chapter 4. On this basis, it had rather less of an urgent need to acquire further coal lands. Nevertheless, in pursuance of a steadier purchasing policy, it acquired almost 1,000 acres near Pittston in 1867 and a further 500 acres in the following year.[141] While these figures give a very good impression of the manner in which these companies secured their future coal supplies in the immediate post-bellum period, it should be noted that they do not include the acreages owned or leased by individual operators, who were bound to the transportation companies, for varying lengths of time, by different types of coal-shipment or purchase contracts. This has a very important bearing on the later evolution of the industry and will be examined in detail in a later chapter.

As Table 10.2 indicates, the methods used by the Lehigh Valley to acquire coal lands, in the absence of any charter authority, differed somewhat from those employed by the Lackawanna and the Delaware and Hudson. The purchase of a controlling interest in the stock of mining companies was the preferred instrument in the early period, as the quote from the 1868 Annual Report signalled. The Lehigh Valley did not initially set about chartering companies in its interest through friendly parties, nor did it particularly press the companies it acquired through stock ownership into making further land acquisitions on its behalf. One important exception, however, was the Luzerne Coal and Iron Company, originally chartered in 1866. The earlier West Pittston Coal Company apparently took over this charter in 1869 and, the following year, the Lehigh Valley acquired its entire capital stock of 12,000 shares. Nearly all of these came from E.C. Knight of Philadelphia, who was also involved in sugar refining and was later to become president of the Jersey Central.[142] The Lehigh Valley then used this company to operate other mines, such as the Burroughs and Henry collieries, bought from H.N. Burroughs in June 1871. The former mine suffered a disastrous cave some weeks after its purchase, and the breaker was eventually abandoned in June 1872.[143] However, the lands on which the breakers were built remained in the name of Asa Packer.[144] The other company of particular interest was the Delano Land Company, which, unlike the others, was specifically chartered for the benefit of the Lehigh Valley Railroad. Its purpose was to hold a large body of lands between Delano and Mahanoy City, and the Lehigh Valley wished to guarantee the principal and interest of its bonds. Unfortunately, part of the lands in question lay in Schuylkill County and, under 1869 legislation, railroad companies were specifically prevented from providing such guarantees in that county. (This was originally a mechanism to safeguard the commercial interests of the Reading Railroad). However, the Lehigh Valley managed to get an exception clause inserted into its charter, thereby permitting the guarantee and flouting the original legislation.[145]

Flouting a single act of the Pennsylvania Legislature was one thing, but flouting the new state constitution that was due to come into effect on January 1, 1874, was quite another. The new constitution provided *against* special charters and *in favor* of general purpose incorporation laws. Fearing that its lack of rights to own coal lands and to mine and sell coal would cause insuperable problems under the new constitution, the Lehigh Valley sought out and purchased the favorable charter of the Green Land Company for $8,515 in June 1872. A supplement to the company's charter allowed it to hold unrestricted quantities of land and 'purchase, sell, transport and mine coal and manufacture iron and sell the same'.[146] During 1873, all the lands held in the names of company officers were transferred to the Green Land Company, to prevent any difficulties when the new constitution came into effect. On February 9, 1875, the Luzerne Coal and Iron Company was merged into it also. Finally, on November 13, 1875, the merged companies were renamed the Lehigh Valley Coal Company, with a capital stock of $650,000. All of this, together with a mere $6 million of floating debt, was held by the parent railroad.[147]

Wholehearted commitment to the business of securing long-term coal supply through acquisition of coal lands was less in evidence on the part of the Pennsylvania Railroad than the other participants in the anthracite trade, as will become more apparent below. Reasons for this are not difficult to identify. The Pennsylvania itself was not reliant on anthracite for a major portion of its business, indeed its interests probably lay more with the competing bituminous shippers in the western part of the state and too close an identification with one group of its customers could have led to commercial difficulties. However, since 1861, it had held a controlling interest in the stock of the Northern Central Railway, which was heavily dependent on anthracite shipments from the western end of the coal regions, as was the Pennsylvania Canal, which it also controlled. Acquisition of lands tributary to these transport routes by competing railroads, such as the Lehigh Valley, would have resulted in tonnage being diverted eastwards, hence the Pennsylvania felt compelled to act, to preserve its local business interests.[148]

Action by Pennsylvania Railroad interests can certainly be traced back to 1868, since a report by the Harris Brothers on the Trevorton lands in that year, which has been referred to earlier in connection with the Lehigh Valley, was actually prepared for the Northern Central Railway, not the former road.[149] In the event, the Pennsylvania did not buy these specific lands. It is possible that the Pennsylvania moved even earlier than this in relation to the chartering of the Pittston Railroad and Coal Company on April 11, 1867. This company was the precursor to the Susquehanna Coal Company. However, it is difficult to argue that major stockholders such as Parrish and Quintard were wholly identified with the Pennsylvania interest, based on earlier discussion of their activities, although Parrish was heavily involved in the operations of the North Branch canal over an extended period.[150] Even if the Pennsylvania interest was not dominant at the outset, it very quickly became so. The election of directors in February 1869 brought a cohort of Pennsylvania Railroad officers on to the board including J. Edgar Thomson, Thomas A. Scott, and G.B. Roberts, with Isaac J. Wistar elected president. A month later, the company's name was changed to the Susquehanna Coal Company. It had a wide-ranging charter that allowed it both to hold up to 5,000 acres of coal lands and to construct a railroad to Hawley, by-passing the Delaware and Hudson Canal. It could also issue stock and bonds up to a total of $10 million.[151] A sale of stock gave control of the company to the Pennsylvania Canal and subsequent sale of the bonds of the latter, taken as payment for the stock, gave the Susquehanna Coal Company the financial means to start purchasing coal lands.[152] The first tract acquired was the 1,500-acre Lee Mine tract at Nanticoke Dam, for $1.05 million in 1869.[153] Mining improvements had been in operation on this tract for many years prior to this. Further large purchases took place at the end of 1870, after the company had arranged a $2 million bond issue, with principal and interest guaranteed by the Pennsylvania Railroad. Apparently, the sellers would not accept the bonds in payment for land without this guarantee.[154] A total of 2,122 acres were bought from Col. Washington Lee and his son for $440 an acre. The lands lay between the Lee Mine tract and Shickshinny. Other major purchases in early 1871, on the other side of the Susquehanna, included the 600-acre Harvey Colliery of Jameson Harvey, and nearly 200 acres of the New England Coal Company property, where the Grand Tunnel mine was located. Including mining improvements and personal property, these coal estates cost $160,000 and $110,000 respectively. This phase of acquisition of coal lands in the Northern Field was largely completed by the end of 1871, and summary figures for the end of 1873 indicate that 5,833 acres of lands had been acquired. If the bonds used to acquire much of the estate were valued at par, the lands cost an average of $554 an acre. Valuing them at the then prevailing market price gave a lower average figure of $470.[155]

The Pennsylvania Railroad acquired a second body of coal lands near Hazleton for a cost of $295,000. The undeveloped tracts totalling 2,129 acres cost only $180 an acre, because the extent of the coal veins on the property were not known. Initially, it was thought that nearly all the property was underlain by coal and an early geological report by J. P. Lesley gave the railroad to understand that the important Mammoth Vein was found on most of the lands. Unfortunately, a later detailed survey undertaken by mining engineer Irving Stearns in 1884, based on extensive borehole drilling, showed this early information to be false. In fact, the extrapolations from the early partial geological information proved to be grossly in error, only a very small proportion of the property was underlain by workable seams and the lands were very largely devoid of value for mining purposes.[156]

The third group of coal lands were a rather more successful purchase. They were acquired to secure the transportation, of the sought-after Lykens Valley and Shamokin coals, south over the Northern Central Railway. At exactly the same time as the Pennsylvania moved its directors onto the board of the Susquehanna Coal Company in 1869, it also secured a controlling interest in the Summit Branch Railroad and the Lykens Valley Coal Company. The lands of these companies comprised 9,200 acres of coal lands and a further 3,000 acres of farmland. There were also four mines and breakers, including the Williamstown Colliery. Records of the share ownership of these companies in 1873 by the Pennsylvania suggest that control was acquired at a cost of approximately $1.6 million, or $179 an acre, a very low price considering the premium quality of the coals from this part of the anthracite fields.[157] After several years of separate operation, the two companies were effectively merged into a single operating unit, although not before the Short Mountain lands had been added to the estate.[158] It is interesting to note that, aside from the general scramble for coal lands, a specific precipitating factor in the acquisition of the Summit Branch and Lykens Valley properties in 1869, appears to have been the fact that the Reading Railroad was grading a branch line to the collieries at the time, with a view to diverting a proportion of their tonnage eastwards.[159]

The second body of coal lands tributary to the Northern Central taken over by the Pennsylvania was that of the Shamokin Coal Company at the western end of the Western Middle Field. Early information states that the Railroad purchased 32,201 shares out of 40,000 on May 29, 1871. By 1873, the total had risen to 39,263 shares. However, after 1871, the 8 mines on the coal company's property were operated by the Mineral Railroad and Mining Company, whose stock was ⅓ owned by the Pennsylvania and ⅔ by the Northern Central. Some 7,808 acres of coal lands were included in the estate and the total purchase cost was $1,092,574, or $140 an acre.[160]

It is apparent that these early purchases at very low prices, in areas where the Reading was not yet a competitor for coal lands, were very shrewd and well timed. Although the prices paid in the Northern Field were twice as high, they were not expensive by the standards of that field. Only the Hazleton lands proved a speculation too far, but the railroad can perhaps be partially excused on the grounds of relying too heavily on expert geological opinion and too little on detailed proving of the property before purchase! Between 1869 and 1873, therefore, the railroad built up an estate of 27,950 acres, 3,000 of which were farmland and 2,119 of which were to be found effectively worthless, for a total consideration of $4,437,708.[161] The conservative investigating committee into the railroad's affairs in 1874, chastened by the panic of the previous year, took a rather less charitable view of the overall policy of the company during this period in relation to anthracite lands. While it applauded the early Shamokin and Lykens Valley purchases, it regarded both the Hazleton and Susquehanna Coal Company purchases as 'injudicious', and looked to the new management to shape policy 'into wiser and more conservative channels'.[162] They were undoubtedly correct in respect of the Hazleton lands, but the substantial and profitable shipments from the Susquehanna Coal Company's lands, over the following three decades, indicate that they were premature in their criticism of the commercial wisdom of acquiring this part of the railroad's anthracite estate.

THE ERIE

After the end of the Civil War, the position of the Erie in relation to the coal business was somewhat analogous to that of the Jersey Central. Each had important coal freight business, but both were dependent on other transportation companies to provide them with the coal to haul to market, as they did not have direct connections to the mines. The details of these relationships have already been examined in earlier parts of this study. However, by 1869, the Erie had set itself a clear strategic goal of becoming 'one of the most important coal carrying roads in the country'.[163] More specifically, its assessment of the road's geographical circumstances led it to believe that it should be carrying the majority of anthracite destined for the western part of New York state.[164] In part, the prospects looked bright, as it had by then a long-standing relationship with the Pennsylvania Coal Company, and the new connection with the Lehigh Valley's northern extension at Waverly was beginning to bring it an increasing business. However, the collapse of negotiations with the Lackawanna for its western coal traffic, after the latter gained control of the Syracuse, Binghamton and New York Railroad, was a forceful reminder that long-term reliance could not necessarily be placed on other participants in the industry. Following this loss of potential traffic, the Erie redoubled its efforts to complete the Jefferson Railroad connection to Carbondale, and make up its tonnage from the mines of the Delaware and Hudson. The new connecting railroad came into service in 1870.[165]

Prior to this, the Erie had little incentive to acquire coal lands itself, since it had no way of reaching them over its own tracks. Even after this, the urgency may have been limited, as very considerable quantities of coal began to reach its lines over its several connections. By 1872, for example, it collected 1.9 million tons of anthracite from Hawley and Honesdale, more than 400,000 tons at Carbondale for the Jefferson Railroad, and a further 439,000 tons at Waverly.[166] Nevertheless, the actions of the other companies could not be ignored, and steps were progressively taken to secure a foothold in the Northern Field. Details are unfortunately more sketchy than for the other railroads, perhaps as befitted a company described by Storrs of the Lackawanna as 'controlled by foreign stockholders and operated by American thieves',[167] but an outline of the overall process can still be given.

The Erie's developing interest in acquiring control of coal tonnage centred around two areas of the Northern Field and the operation of two companies that came under its influence. Of the two areas involved, the first was in proximity to the Jefferson branch and the upper portions of the Delaware and Hudson's railroad, i.e. the northeastern part of the coalfield. The second was below Scranton in the area between Old Forge and Moosic. The two companies in question were the Hillside Coal and Iron Company and the Glenwood Coal Company. The former was incorporated in April 1867, although it did not initially engage in any known mining operations or land acquisition.[168] The latter had apparently acquired control of the output of two properties in the Moosic area by the middle of 1869 in the interests of the Erie, because general reference is made to them in respect of an agreement, between the Erie and the Delaware and Hudson, dated May 1869. It is likely that the two properties in question were the Oak Hill and the Spring Brook mines. These were on coal leases of 220 and 253 acres respectively, the lessors being W.W. Winton and Sherwood and Company.[170] Prior to November 1870, the Glenwood Coal Company had also acquired the 'Whitmore' tract of about 782 acres near Carbondale, on which a new improvement was under development.[169] Other information indicates this improvement was the Erie Colliery.[171] Although the exact timing of all the Erie's investment in the Glenwood Coal Company is not known, by the end of 1872 it held $350,000 of its bonds, and the railroad's erstwhile president, Jay Gould, was said to be a large stockholder.[172] Available evidence points to a more exact indication of the date of acquisition of the charter of the Hillside Coal and Iron Company, which took place in the summer of 1870.[173]

At the beginning of 1870, the Erie also entered into a coal-shipping contract with E.A. Coray of the Lackawanna and Susquehanna Coal and Iron Company, whose property shared its western boundary with the Oak Hill property (Figure 10.4). Coray's previous contract, which was with the Lackawanna Railroad, expired on December 31, 1869, but there were still 63,000 tons of coal shipments outstanding under that agreement. The solution to this problem was sensible from the geographical point of view. An agreement was made for the Carbon Hill mine to make up this deficit. This was efficient because the Carbon Hill breaker connected directly to the Lackawanna and Bloomsburg Railroad of the Lackawanna on the western side of the Susquehanna, whereas Coray's Greenwood and Coray mines were on the eastern side.[174]

During the year 1870, it appears that the Glenwood Coal Company purchased the Carbon Hill, which had previously been leased to Malvern and Scott, and it is likely that the same was the case for the Oak Hill and Spring Brook mines. At the end of the year, the Glenwood Coal Company itself was formally bought out by the Erie. It is also likely, judging by the lease dates, that, in the first half of 1871, the Glenwood acquired leases on the Morris and Essex and the Massachusetts Coal Company tracts on behalf of the Erie. These lands were on either side of the Carbon Hill property. These acquisitions gave the Erie a substantial leasehold estate of 843 acres in a continuous block straddling the Susquehanna River, together with the control by contract of the Coray property, which adjoined it to the east.[175]

However satisfying this compact body of territory may have been from a geographical and management perspective, the same could not be said about its commercial prospects. Storrs of the Lackawanna, with his customary blend of conservatism and detailed local knowledge, asserted that there was a limited amount of coal on

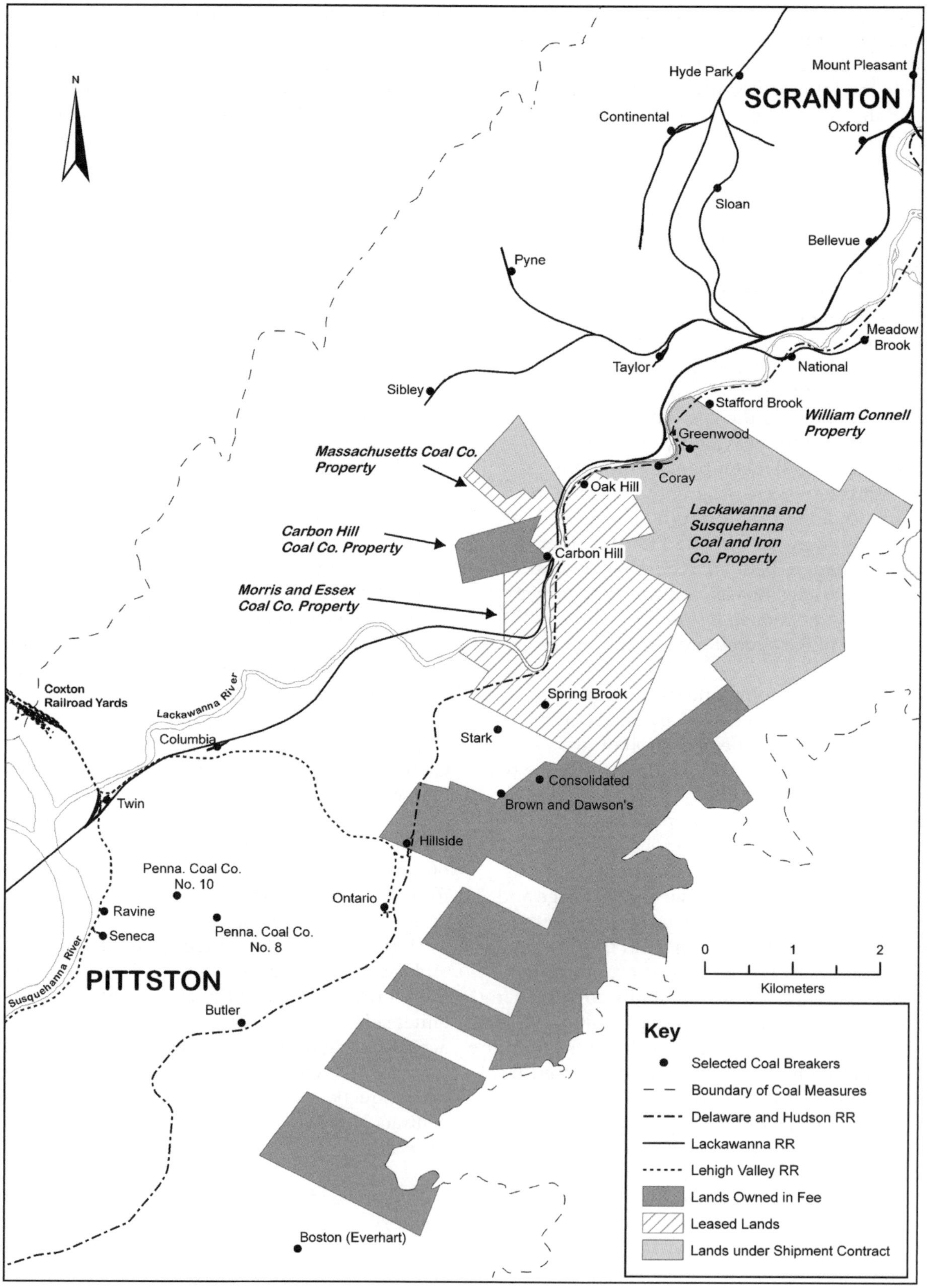

FIGURE 10.4
Map of the Coal Lands Controlled by the Erie Railway in the Northern Field Circa 1873.

the Oak Hill property, while Carbon Hill had already experienced problems with flooding and a squeeze in the coal around the shaft, which had necessitated a suspension of work in the latter part of 1870.[176] This was not good news in terms of the clauses in the various leases that required royalty to be paid on a minimum of 285,000 tons a year, to be shipped from the property as a whole. Such a figure was very demanding for a coal estate with question marks over its ability to produce sustained outputs.[177]

Unfortunately, these concerns were borne out in the latter part of 1872, as business conditions started to deteriorate prior to the 1873 panic. The Glenwood Coal Company went into bankruptcy with obligations of about $400,000. Later reports from mid-1873 indicate that the company was reorganized on an 'assured basis', but the disruption to the company's business was severe and extended. The shipment records indicate that Carbon Hill suspended shipments in August 1872 and did not restart until after the panic in November 1873. Oak Hill fared even worse, as there is no record of mining operations restarting there after the 1872 hiatus.[178]

Despite these setbacks, the Erie did not abandon its plans, but chose the Hillside Coal and Iron Company as the vehicle for future development of its coal estate. In the summer of 1873, the Erie purchased the Enterprise mine of J.H. Swoyer with 1,000 acres of coal land in Pleasant Valley. The mine was renamed the Hillside. It had previously been offered to the Lehigh Valley Coal Company, who had declined it.[179] These lands were adjacent to those purchased earlier. This purchase was to prove more satisfactory than the earlier ones, since the mine and its successor, the Consolidated, were to produce substantial tonnages for many years. Unfortunately, the Glenwood Coal Company abandoned the Carbon Hill and Spring Brook Drifts 1 and 2, in September 1876, because of the poor quality of the coal. The company itself, which only had the Erie Breaker left as an operating mine, was subsequently absorbed into the Hillside Coal and Iron Company.[180]

This rather inauspicious start to its period of coal land ownership was clearly not what the Erie originally had in mind. In 1871, it had tried to extend its control of coal properties further up the valley towards Scranton, by attempting first to lease, and then to buy, William Connell's substantial property that adjoined the Coray property on the east, but Connell preferred to continue his longstanding arrangement with the Lackawanna. They were more successful in securing a lease on property further north in Jessup that the Lackawanna had declined, but, as relative latecomers into the race, they were clearly not succeeding in securing the choicest properties in the coalfields.[181] Undeterred, in 1873 they began surveying a new railroad route down the valley from Carbondale to connect their lands with the Jefferson Railroad, but it appears that the aftermath of the panic put paid to such expansionary ambitions, as it was never built.[182] The ensuing depression also exposed the precarious finances of the Erie even more rapidly than those of some of the other anthracite railroads, and, by May 1875, it was in receivership. As part of a rescue plan, it was proposed that the Lackawanna and the Delaware and Hudson should take over the Erie's coal lands in exchange for a loan of $500,000, but the plan was eventually shelved because of legal problems.[183]

Hence the Erie's relative success as a coal shipper across New York state was not matched in the early years by equivalent success in the mining business. It had expended $1 million on several thousand acres of relatively poor quality coal lands, many of them encumbered with exacting lease demands, and the actual quality and quantity of output fell far below that which was anticipated. There was even local antipathy towards the company as a whole, at least from the Scranton press, which declared that it was a big mistake by the main anthracite companies to let the Erie get a foothold in the coalfield.[184]

THE PHILADELPHIA AND READING

The Philadelphia and Reading Railroad was the last of the major companies to join the race to acquire coal properties. Initially, it was unaffected by land purchases in the Northern Field, but increasingly it found itself faced with the prospect of the other companies obtaining land tributary to its lines and projecting new railroads into the Southern and Western Middle Fields, to divert the coal traffic eastwards.[185] Several of these initiatives have already been examined in earlier sections. At first, it tried legislative means to prevent encroachment on what it perceived to be its territory. Gowen lobbied through an amendment, excluding Schuylkill County from the 1869 act, which permitted railroads to buy stock in coal or iron companies. However, it is doubtful whether this was regarded as anything more than a holding strategy, since, at the same time, Gowen started buying coal lands in his own name. A good example of this is the Miner's Journal report of April 24, 1869, four days before he was elected president of the railroad, which stated that he had purchased a half-interest in the Trevorton lands for $310,000. As a humble legal counsel, he would not have had personal financial resources on this scale. It is likely that the funding originated from McCalmont Brothers, Merchant Bankers of London, who had a longstanding relationship with the railroad.[186] Other large stockholders of the railroad also became involved in acquiring coal lands, such as the Delaware Coal Company and North American Coal Company tracts near Pottsville which had 'been lying comparatively idle for the last 20–30 years'. These tracts covered nearly 2,300 acres.[187]

Hence the Reading began by utilizing the same charter circumvention mechanisms previously employed by its competitors, but it soon became apparent that progress using these methods was rather slow. Unfortunately, expediting the process of coal land acquisition required a volte-face in relation to the amendment to the 1869 act that had previously been passed in the company's interest at the time. A subsidiary company that could hold coal lands in Schuylkill County, while being controlled by the parent railroad, was now the first requirement. The legislative manipulation required to achieve this in 1871 is amusingly recounted by Schlegel in his classic biography of Gowen, but the outcome was the chartering of the Laurel Run Improvement Company, whose name was subsequently changed to the Philadelphia and Reading Coal and Iron Company. A second requirement was removal of the limit on the railroad's authority to borrow money, and this was also achieved. Armed with these pre-requisites for territorial expansion and the proceeds of a $25 million bond issue, Gowen was able to embark on an extraordinary orgy of coal land acquisition that was to change the face of the Southern and Western Middle Fields.[188]

Given the speed, scale, and secrecy with which many of the negotiations were conducted, it is not surprising that there are some minor differences between sources in the reported timing of expenditure and the final extent of the coal acreage that was acquired. Two sources indicate that about 70,000 acres were acquired in 1871, for a cost of $11.96 million dollars.[189] Another source gives a more precise figure of 65,605 acres and a claimed average cost per acre of $293, although it is difficult to determine how this average cost was calculated, based on the figures presented.[190] By early 1874, the Miner's Journal reported the coal estate had swelled to more than 92,000 acres, a figure that closely corresponds to the 91,149 acres given in Harris's later detailed report on the Reading's coal lands.[191] Of this total, only about half had veins below the Mammoth (46,627 acres), and, of the remainder, 26,253 also contained the Mammoth bed, while 18,269 acres were richly endowed with all the seams. A further 69,147 acres were timberland, believed devoid of any workable seams.[192]

Most of the lands were bought outright, often in large blocks such as the Manhattan Coal Company's Lands (19,000 acres), and the Swatara Lands (4,255 acres). However, some estates were acquired through purchase of a controlling interest in the stock of the companies that owned them. Examples here include the Tremont Coal Company (4,500 acres) and the Preston Coal and Improvement Company (2,346 acres).[193]

The relative paucity of surviving financial records for the railroad and the Coal and Iron Company going back to this time make it a little difficult to be precise about all the aspects of the Reading's coal land acquisitions. Fortunately, a slightly later compilation of the capital accounts and land ownership records of the Coal and Iron Company has survived and this enables a number of key points to be confirmed. Most importantly this record distinguishes between the actual cost of the lands themselves and other costs, such as commissions and interest payments. Up to 1880, these costs were as shown in Table 10.3.

TABLE 10.3
Costs of the Philadelphia and Reading Coal and Iron Company Property to 1880.[194]

Cost of Coal Lands	$26,148,408
'Interest and Commissions'	11,918,951
Timber Lands	590,308
Iron Ore Lands	655,820
Real Estate	1,365,909
Sundry Costs	263,964
Total	$40,943,360

The compilation also records that, of this amount, almost $26 million was spent by the end of 1872, and a further $4.5–5.5 million in each of the three following years.[195] The figures underscore the criticism of Gowen at the time for the capitalization of $7 million interest paid on money borrowed to purchase coal lands, and they confirm that this highly dubious accounting practice was stopped after 1875.[196] It should also be noted that the figures imply nearly $5 million was spent on commissions, indicating that the acquisition of so many mining and improvement companies in such a short time must have given a great deal of employment to bankers and stockbrokers of the period!

Taking simply the purchase cost of the lands, and dividing by the commonly accepted total of 91,149 acres from the Harris report, would give a cost of $287 per acre, very close to that quoted by Bogen.[197] Adding in the commissions and interest would raise the figure to $418 an acre. This accords with other information in the Coal and Iron Company records, which shows the majority of tracts were bought at figures in the range of $4–500 an acre, excluding mining improvements. However, there were a number of exceptions to this, as shown by the examples given in Table 10.4 These suggest a tendency for the lands in the Western Middle Field to attract higher prices. In part, this would have been because some of the lands were relatively recently opened and had large reserves; in part, also, they were more attractive to competitors, such as the Pennsylvania Railroad and the Lehigh Valley, who were already acquiring property at either end of the field.

Overall, the prices paid do not seem exorbitant when compared with the Northern Field, although it should be remembered that half the estate was not underlain by the valuable Mammoth seam. It is possible, however, that they were actually higher than they appeared at the time. A pencilled note in the Coal and Iron Company's records suggests that the actual area underlain by coal, instead of being 91,149 acres was later found to be only 81,552. If correct, this would have raised the cost per acre to $467 from $418.[199] Nevertheless, from a business-cycle viewpoint, the Reading's expansionary policy was well timed in securing the bulk of its lands in 1871, during and immediately following a period of recession, when companies and individuals would be more willing to sell at reasonable prices.c.f. Figure 10.1 It should be added that many companies had also been greatly weakened by the effects of the 1870 and 1871 strikes.[200]

Cost is one thing; value to the business is, of course, another. In typical fashion, in the 1873 Annual Report, Gowen claimed that the lands of the company, according to recent sales made to other companies would be 'worth at least double the amount paid for them'.[201] No doubt the figures he had in mind were realized just before the 1873 panic! By 1876, as the railroad's debts mounted to unsustainable heights, the value of the coal estate, in Gowen's imagination at least, had risen even further. 'The intrinsic value of the coal lands it is almost impossible to name', he asserted.[202] As is often the case, only receivership could reign in such flights of fancy. There quickly followed Harris's systematic and conservative assessment of the value of the property, taking account of likely future profit margins, the net present value of future income streams, and the time required to develop portions of the estate. His final valuation, based on a profit of 30 cents a ton on all coal mined, and some reduction in the interest rate on obligations connected with the lands, was $32,394,799. He also recommended disposal of lands costing over $5.2 million, which he valued at only $3.28 million, unless the bondholders involved could be persuaded to accept interest rate reductions.[203] As might be expected, Gowen railed against his assessment of the value of the property and the proposals to relinquish part of the estate.

TABLE 10.4
Sample High- and Low-Cost Coal Estates Purchased by the Philadelphia and Reading Coal and Iron Company.[198]

Tract	Date Purchased	Acreage	Cost/Acre ($/acre)
Black Diamond Coal Co.	Jan 4, 1871*	1,648	$255
Philadelphia and Mahanoy Coal Co.	May 18, 1871	2,545	$712
Carey, Hart et al. (St. Clair)	Jan 1, 1872	414	$726
Mt. Carmel and Locust Mountain Coal Co.	July 1, 1872	519	$715
Jas. McNeil Tract**	1872–1873, 1883–1884	569	$1,564

*Note that these lands were acquired in the Reading interest prior to the incorporation of the Laurel Run Improvement Company.
**There were a number of owners of this tract whose interests were bought up over a period of time.

'Those whose actions and policy he criticizes were wiser in their generation than himself and much more competent than he supposes to form a correct judgement upon such a question', he declared.[204] In the event, Harris underestimated the likely expansion of the anthracite industry in subsequent years, based on the stagnation of the 1870s. Likewise, Gowen underestimated both how long it would take to re-establish the fortunes of the Reading and the adverse effects of financial stringency on the ability of the company to develop its coal estate effectively.[205] At that time, the Coal and Iron Company did not relinquish those portions of its property identified by Harris. The overall estate continued to stand on its books at a valuation of about $41 million for a number of years, until re-valued upwards to $62.3 million, as a result of re-examination of the property in 1887–1888.[206]

A significant proportion of the lands were undeveloped when they were acquired, or, like the lands near Pottsville, had been essentially abandoned, because they required expensive deep shafts to reach the lower coals in deep geological basins. These raised long-term development and investment issues of the kind that Harris had examined in his report. However, in relation to the daily conduct of business, the most important issue in the short term was that control over a large estate, as early as 1872, had also brought with it control over approximately 100 collieries, 75 percent of which were in Schuylkill County, and the remainder in the adjoining counties of Columbia and Northumberland. Many of these collieries had previously been inefficiently worked by individual operators for short-term gain, to the detriment of the long-term value of many properties and their coal deposits. The challenge perceived by the Reading was to convert this old and uneconomical way of working into an efficient mining economy that matched the best practice being promulgated in the Northern Field, where the corporate approach had long dominated.[207] The nature of this rationalization process and the degree of success achieved by the end of the century is the subject of a later chapter.

In the space of a few short years, therefore, very largely between 1867 and 1874, the major railroad and mining corporations had increased their holdings of coal lands to upwards of 200,000 acres or nearly 70 percent of the entire area of the coalfields. These figures would increase even further, if account were taken of the indirect control over individual operators exercised by the railroads through shipment contracts, especially in the Northern Field. According to the tally in the Miner's Journal, by the beginning of 1874 there were only 35 collieries left out of 160 in the counties of Schuylkill, Columbia, and Northumberland that were situated on lands not owned by the railroads. Of these, 9 were on the Girard Estate, whose property could be leased, but not purchased.[208] Only in the Lehigh Region could individual operators still be regarded as a dominant force in the trade. These included long-established companies, such as those of the Pardees and Markles, as well as more recent entrants such as the Coxe Brothers. Even here, however, the interests of these operators were closely allied with those of the transportation companies, and their names were frequently to be found on the boards of directors of the latter concerns.[209] Concentration of economic power in a limited number of hands through control of coal territory was a necessary, but not sufficient, pre-requisite for the formation of industry-wide cartels. In addition, it would be necessary for previously competing interests to agree on a number of key issues. These included common business aims, appropriate means of dividing up market share, operating methods for the conduct of business, and enforcement of compliance with cartel agreements, including, where necessary, sanctions for non-compliance. All of these were subject to negotiation, on an annual basis, if not more frequently, and the ongoing ability of any individual member of the cartel to influence overall policy in its favor was initially dependent both on its market share and its financial standing. In later years, the background influence of financial institutions, and J.P. Morgan in particular, began to assume greater prominence. Hence the cartels are of interest, both in relation to the exercise of economic power and in respect of the evolving sociology of corporate behavior. At the outset, the cartel, as a mode of industrial organization, was a response to specific perceived economic and business problems. However, the subsequent operation of cartels, through their effects on the environment for subsequent decision-making by individual companies, had the potential to produce a range of unintended and unwelcome consequences that necessitated yet further adjustments in the conduct of business. The working out of these complex, temporally and locationally dependent effects, and their interaction with external business cycle changes, need to be traced across a number of scales. The next chapter will focus on the operation of the main cartels, while the range of impacts on the mining economy will be explored more fully thereafter.

Chapter 11

CORPORATIONS, COMPETITION, AND THE RISE OF THE CARTELS II: THE 1873 COMBINATION AND ITS SUCCESSORS

INTRODUCTION

The standard accounts of the combinations by Virtue and Jones,[1] to which reference has already been made, contain a wealth of chronological material about the early combinations from 1873 onwards, which therefore only needs to be summarized here. Jones's study is particularly detailed and perceptive in relation to combination activity after 1898. In terms of his time of writing (1914), this was as much contemporary economic analysis as historical investigation. Neither of the authors had access to the internal records of the anthracite railroads, nor did they apparently seek interviews with railroad officers at the time. Hence, their accounts are dependent on the sometimes-sanitized versions of events presented in railroad Annual Reports, or the judgement and knowledge of newspaper reporters and their ability to impute motives and intentions correctly. These authors also leave a number of important questions either partly addressed or completely unanswered. These concern such issues as the timing of the appearance of cartels, the nature of their operation in respect of the differential (and changing) economic power relations between the major transporting companies, the relationship of the cartel to the Pennsylvania Railroad, and the impact of cartel activity on mining development or decline in different parts of the coal regions. The detailed analysis of railroad financial performance, presented in Chapter 6, already provides important background to aid in answering these questions, but recourse must also be had to a wide range of newspaper and archival sources for further evidence. The latter sources were, in general, not available to the early students of the combination because the companies in question were still in operation at the time of their investigations. Hence their comments tend to reflect the public face presented by the railroads to the press of the time, or to government investigations, rather than the internal machinations that actually went on between the companies behind the scenes.

The above questions will now be examined in detail under two broad headings:

- the first industry-wide combination, and
- variations in subsequent combination activity with business cycle changes, looking at the major combination phases in turn.

In what follows, an attempt will also be made to restore, in some measure, the element of internal debate and negotiation between railroad presidents and their chief officers that characterized the combinations, supplemented by more quantitative charting and analysis of actual business behavior.

The basic (though not the only) mechanism employed by the combination to control the industry was the percentage allotment of shipments to individual railroad interests. A further distinction can then be made between years where formal percentage allotments were announced and those where a more informal arrangement was in place. Table 11.1 reproduces the information from a table in Jones that highlights the formal allotment years. Aside from the details of the figures themselves, several important points emerge. The first of these is the clustering of the formal allotment years into times of depression, the obvious initial supposition being that management of the trade was much easier during expansionary phases than during times of

TABLE 11.1

Percentage Division of Tonnage Between Railroads During Years of Formal Combination, According to Jones.[2]

Railroad	1873–5	1876	1878	1885	1886	1896
Phila. and Reading	25.85	25.57	28.625	}38.85	}37.82	20.50
Jersey Central	16.15	15.98	12.905			11.70
Lehigh Valley	15.98	15.80	19.75	19.60	19.07	15.65
Lackawanna	13.80	13.65	12.75	16.05	15.63	13.35
Delaware and Hudson	18.37	18.18	12.48	11.00	10.70	9.60
Pennsylvania Coal Co.	(9.85)	10.82	5.865	5.00	4.87	4.00
Pennsylvania			(7.625)	(8.00)	(10.45)	11.40
Erie				(1.50)	1.46	4.00
Ontario and Western						3.10
Del. Susq. and Schuylkill						3.50
Susquehanna and Western						3.20

N.B. Figures in parentheses represent probable tonnage of companies not members of the pool; the 1885–6 figures for the Reading and the Jersey Central are merged because of the Reading lease of the latter road; the Delaware, Susquehanna and Schuylkill did not accept its 'official' 1896 figure of 3.5 percent (see section below on the combination of 1896).

contraction, hence the need for formal agreements at the latter times. Secondly, it is apparent that in the later years it was necessary for the combination to recognize new entrants into the trade, such as the New York, Susquehanna and Western and the New York, Ontario and Western Railroad. Apart from the Delaware, Susquehanna and Schuylkill Railroad in the Lehigh Region, the other new entrants were in the Northern Field. This suggests that despite the initial and subsequent phases of coal land acquisition by the major corporations, sufficient coal property and actual/potential mining capacity remained outside their control for the establishment of new producing and transporting interests to be feasible. Finally, the table indicates that, at different times, several of the companies remained formally outside the combination. The most important of these was the Pennsylvania Railroad. Since such railroads continued to ship coal in quantity despite their non-participation, the effects of their likely shipments still had to be taken into account by those trying to manage the combination.

Agreements can be made, and agreements can also be broken or finessed, as the actual outworking of the combination would testify. Hence, agreed allotments at the beginning of the year and actual shipments at year-end were not necessarily the same thing. It is therefore important to chart, in summary form, the actual percentages of total anthracite shipments contributed by each of the main anthracite interests over the last three decades of the 19th century. These are shown in Figures 11.1–11.4. From the figures, it is apparent that over the medium term the fortunes of individual companies within the cartel varied quite widely. Prominent among the losers were the Philadelphia and Reading Railroad and the Delaware and Hudson, whose percentages entered a period of decline after they encountered financial problems in the 1870s. To a lesser degree, the same applied to the Pennsylvania Coal Company. Here the relative decline was attributable to a failure to increase its shipments over time as rapidly as some of the other companies, rather than to any specific financial difficulties. However, despite losing market share in the late 1870s and early 1880s, both the Philadelphia and Reading and the Pennsylvania Coal Company were able to stabilize their positions thereafter, even in the face of new entrants into the anthracite business in later years. The same could not be said of the Lehigh Valley and the Jersey Central, both of which lost ground in the 1890s, after a relatively strong performance during the 1880s. It should be noted that the Jersey Central figures have been kept separate in the figure from those of the Reading, despite its period of being leased to the latter road in the mid 1880s. The biggest gainers, at least in the period between 1880 and 1893, were the Lackawanna, which was a member of the combination, and the Pennsylvania, which was not, at least until the very end of the period. Both suffered sharp declines following the 1893 panic, although the change in the Lackawanna series was largely a function of the separation out of the tonnage of the New York, Susquehanna and Western Railroad in the early 1890s. These charts of relative business performance, in terms of market share, raise a number of questions, which will be examined further below, about the commercial benefits of participation versus

FIGURE 11.1
Philadelphia and Reading Railroad: Percentages of Total Anthracite Shipments 1868–1899.

non-participation in the combination. They also indicate that the combination was quite a pragmatic and dynamic organization, able to adjust, at least in some measure, to the realities of business life. However, whether all participants were equally satisfied about their allotments on an ongoing basis, remains to be seen. One final caveat, relating to comparison of Table 11.1 with the percentages on the graphs, is that the graphs include all shipments by the individual companies. In the initial stages, at least, the allotment percentages in the table only referred to coal shipped to what were termed 'competitive' points, roughly equivalent to the tidewater markets. Hence line shipments were excluded. This gave to the latter a special importance, because they remained essentially under the control of the individual companies themselves and tonnage shipped to them was not restricted.[3]

The first combination ran from January 22, 1873 until August 21, 1876. A period of free competition then followed until the combination was re-established for 1878. No agreement was reached for 1879, but the years between 1880 and 1884 were characterized, according to Jones, by an informal, 'friendly understanding' between the companies to restrict output when required to prevent or eliminate overstocking of the market.[4] A formal cartel was re-established on December 31, 1884 and it continued until March 31, 1887, although the intervention of J.P. Morgan had been required in 1886 to induce sufficient harmony between the railroad interests for the agreement to be extended for a further year. Thereafter, it was believed that formal combination, as in the past, had been rendered illegal by the passage of the Interstate Commerce Commission Act, which came into force on January 1, 1887. In consequence, the companies reverted to more informal arrangements, such as meetings between sales agents, as the means of managing the trade. Unfortunately, these arrangements gave only a limited degree of control and, in 1892, the Reading, in a bold expansionary move, leased both the Lehigh Valley Railroad and the Central Railroad of New Jersey, to give it control over nearly 70 percent of all anthracite shipments. Unfortunately, the latter lease was quickly voided on legal grounds and the former lease was abrogated in

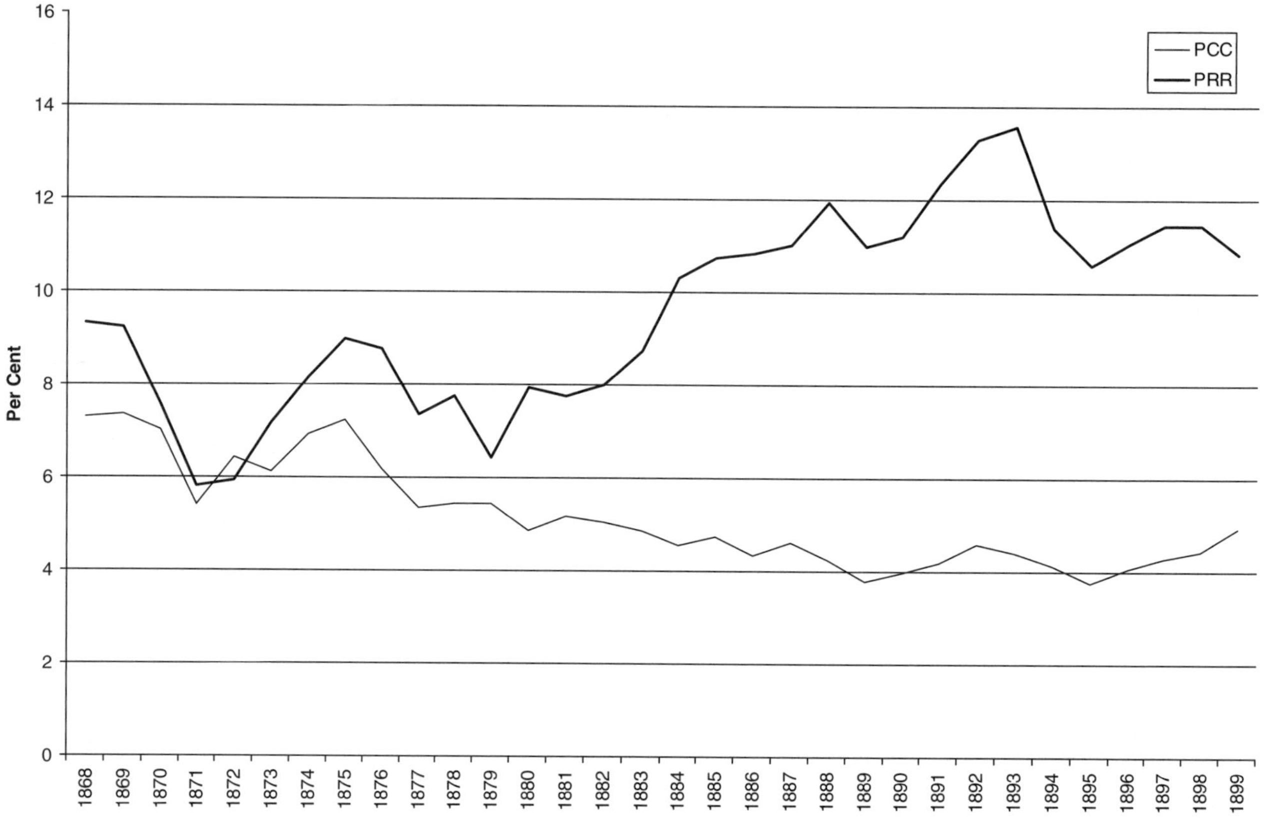

FIGURE 11.2
Pennsylvania Railroad and Pennsylvania Coal Company: Percentages of Total Anthracite Shipments 1868–1899.

February 1893, when the Reading entered its third receivership in thirteen years. Collapse of this attempt at a different method of formal control led to reversion to the use of sales agents meetings in 1895, but these once again proved relatively unsuccessful. A formal allotment system was re-introduced on February 1, 1896, to run until the end of the year. The agreement was not formally signed, however, in an attempt to evade the scrutiny of the ICC. Although it was relatively effective, it was not renewed in 1897, and the companies reverted to a policy of curtailment of production, with decreasing success into the middle of 1898. Thereafter, with the support of J.P. Morgan and others, the railroads began to develop a 'community of interest' through inter-ownership of stock and interlocking directorates. This proved highly effective as a means of co-ordinating the interests of the different companies, while putting the combination largely beyond the reach of the ICC. A number of years were to elapse before the legislators, using the Sherman anti-Trust Act, were able to bring the combination partly to account. The interested reader is referred to Jones's study for further details of this last phase, which lies outside the period of current interest.[5]

Each of the stages in the operation of the combination, up to the period of the 1902 Coal Strike, will now be examined in more detail, with particular emphasis on several facets of the problem of managing the industry. The first of these is the continuing process of negotiation between individual interests and the business factors that affected the stances adopted by each at different times. The second is the different mechanisms employed to manage the curtailment of production, and the third is the anticipated and unanticipated effects of these mechanisms on investment in the industry.

THE FIRST COMBINATION ERA 1873–6

Brief mention has already been made in an earlier section about the initial discussions, in 1872,

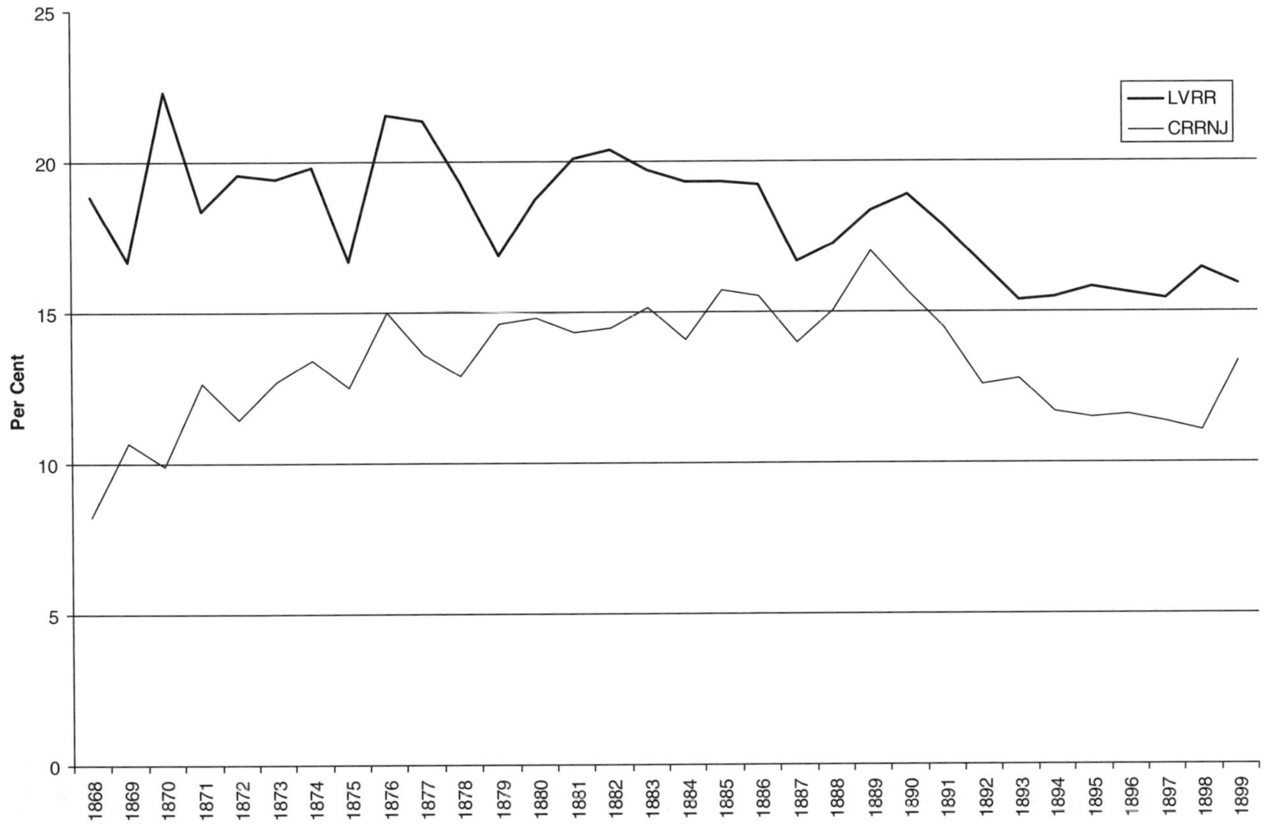

FIGURE 11.3
Lehigh Valley Railroad and Central Railroad of New Jersey: Percentages of Total Anthracite Shipments 1868–1899.

on the possible adoption of a scheme to allot tonnage to the different interests. It is significant that the proposal originated with Charles Parrish, although it was not supported by the Lackawanna or the Delaware and Hudson.[6] This indicates that the combination concept for the entire industry grew out of the experience of the Northern Field companies with operating the cartel for the western and northern trade. Specifically, the idea did not originate with F.B. Gowen and it is difficult to agree with Jones, who argues that the primary purpose of the combination was the need to meet interest charges on huge obligations incurred in buying coal lands. While this was undoubtedly a factor in their continuance, at the time Parrish was proposing his allotment scheme, Gowen was only part way through the process of acquiring coal lands. In words he doubtless later regretted, he also argued that the estate, to that time, had been built up 'without imposing any serious financial burden upon the Company'.[7] Similarly, as late as 1876, the Lehigh Valley Railroad was reporting that neither the lands owned by them nor by the Lehigh Valley Coal Company were subject to any bonded or other debt, with the exception of the guarantee on the bonds of the Delano Land Company. That said, they omitted to mention the substantial floating debt transferred to the Coal Company on its formation, but, even so, the interest associated with it would not have been crippling.[8]

One of the clearest rationales for the establishment of the combination can be found in a letter from Thomas Dickson to the Board of the Delaware and Hudson. He argued that, in early 1872, 'jealousies of producers prevented concert of action', but that by September 1872 it had become apparent, to both producers and transporters of coal, that they were making a net loss on every ton sent to tidewater.[9] A plan of action was then agreed by the major companies to set prices in October at a level that would leave a small margin of profit. He summarized the situation as follows:

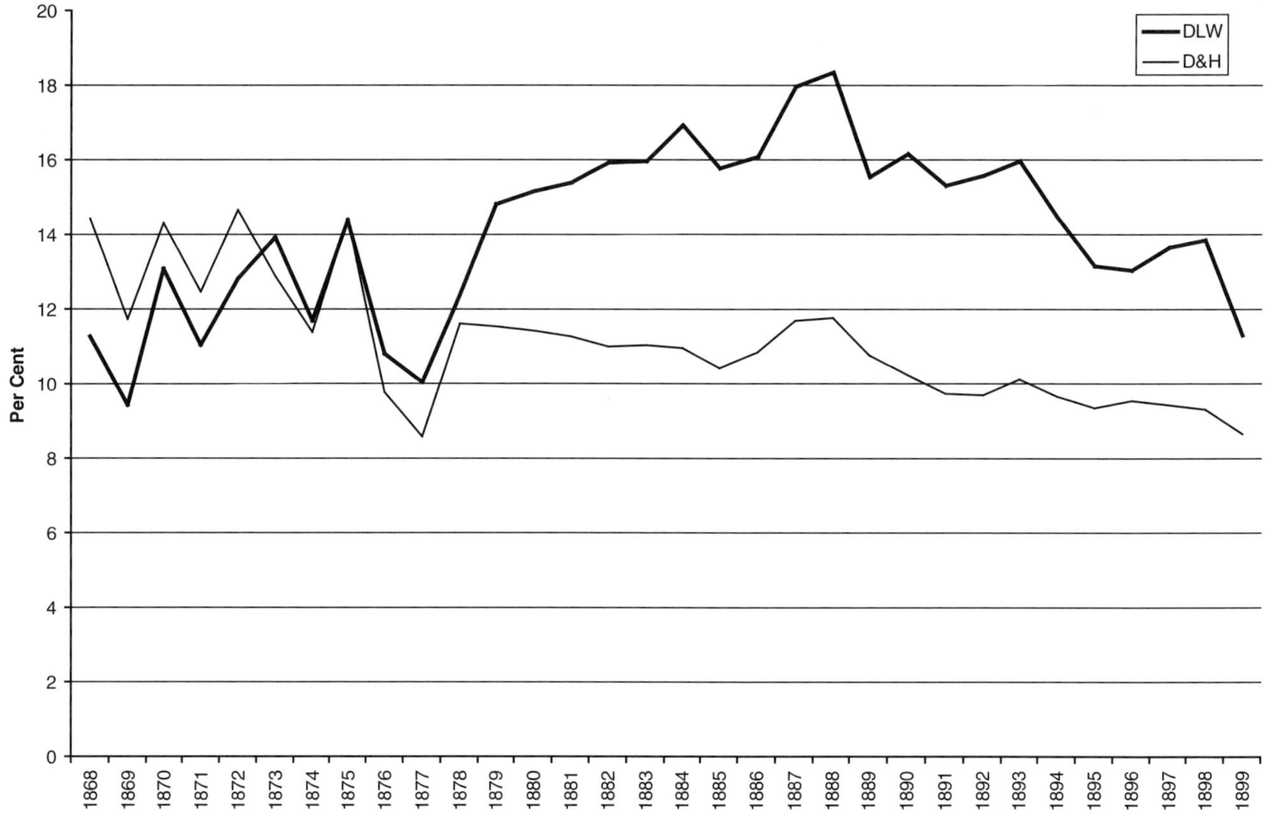

FIGURE 11.4
Delaware, Lackawanna and Western Railroad and Delaware and Hudson Canal Company: Percentages of Total Anthracite Shipments 1868–1899.

> 'The experience of the year has taught the parties who control this great and growing interest, the folly of exhausting their resources in ruinous competition and a movement is now in progress with every prospect of success that will regulate the supply to the demand and which without doing injustice to the consumer, will yield a reasonable profit to the producer'.[10]

Comparison of this account with that of Schlegel indicates that the driving force behind this price fixing plan was now Gowen, who had met with the presidents of the other companies in September 1872.[11] This was a sharp reversal of his 1871 views quoted earlier that increased demand at lower prices was the remedy for the industry's ills. His 1872 experience had shown him that the 'minimum wage' of the Schuylkill basis system had been set too high for the year, and recovering his costs from the consumer was the only short-term solution. The pressure was on him to organize an agreement with the other presidents, because it was mines in the Schuylkill Region tributary to his road that had been forced to suspend by the basis-wage problem and, hence, his future coal traffic was most at risk. Unfortunately also, this would put him in a position of needing to make concessions on his part to secure the support of other railroads, at the same time as his own roads' shipments were artificially depressed by the partial shutdown in the Schuylkill Region. Such a weak bargaining position was to have adverse consequences for the Reading and play into the hands of others, such as Sloan of the Lackawanna, who enjoyed a healthier financial position and had avoided losing financial flexibility by declining to implement the basis system.

In the generally positive business climate of the latter part of 1872 (c.f. Figure 10.1) this short-term price-raising agreement was relatively effective, and prices did climb upwards in the latter part of the year. Buoyed no doubt by this early success at concerted action, on January 16, 1873, Gowen again

PLATE 1: No. 3 Lee Drift and Slope *(courtesy of the Anthracite Heritage Museum)*

PLATE 2: Eaton and Company's Breaker at Archbald
(courtesy of the National Canal Museum and Wayne County Historical Society)

Plate 3: Gibson's of Rushdale *(courtesy of the National Canal Museum and Wayne County Historical Society)*

Plate 4: The Alexander Gray Breaker *(courtesy of the Luzerne County Historical Society)*

Plate 5: The Avondale Breaker circa 1888 *(courtesy of F.C. Petrillo)*

[illegible] (courtesy of the National Canal Museum and Wayne County Historical Society)

Plate 7: Landscape View of the Archbald, Hampton and Continental Collieries *(courtesy of Syracuse University Library)*

328-A
5-21-07
OLD DIAMOND BREAKER D.L.&W
BUNNELL
D.L.&W.
PHOTO

PLATE 9: Landscape View of the National Breaker and Washery *(courtesy of Syracuse University Library)*

COAL BREAKER
SCRANTON. PA.
BUNNELL
- PHOTO -
No. 159 - A

PLATE 11: The Manville Breaker (Formerly Jermyn No. 3) *(courtesy of Syracuse University Library)*

PLATE 12: The Hallstead Breaker (courtesy of Syracuse University Library)

Plate 13: The Holden Breaker *(courtesy of Syracuse University Library)*

"STORRS"
SEPT. 4 - '05
No 224 - A

Plate 15: The Woodward Breaker *(courtesy of Syracuse University Library)*

BUNNELL
D.L.&W.
D.L.&W. - PETTEBONE COLLIERY
LUZERNE, PA
#721 A
11-27-08

Plate 17: Woodward Trestling Under Construction circa 1886 *(courtesy of Hagley Library)*

Plate 18: The Bliss Breaker *(courtesy of Syracuse University Library)*

Plate 19: The Clinton Breaker *(courtesy of the Anthracite Heritage Museum)*

Plate 20: The Auchincloss Breaker *(courtesy of Syracuse University Library)*

met with the presidents of the other roads to agree on what has been termed America's first industry-wide price-fixing agreement.[12] The plan was to maintain prices at an average of $5 per ton wholesale in New York.

Early in the season, prices were to be kept low to encourage advance buying and they would be ratcheted upwards as the year progressed. In addition, the percentage allotments indicated in Table 11.1 were to be used, based on a target tonnage for the year of 10 million tons to tidewater.[13] This figure was almost exactly 50 percent of the total shipments to all points in 1872.

With the previous caveat that the allotments referred to tidewater tonnages while the graphs refer to total tonnages, some broad comparison between the two is still instructive. In Figure 11.1, it is apparent that in 1869 and 1871, the Reading shipped 34–35 percent of all tonnage. Although there were strikes in these years, all fields were affected, so the net impact was minimal. In 1870, by contrast, the extended strike, which did not affect the Northern Field companies, had a substantial depressing effect on Reading tonnage. The basis-related suspension problems of 1872 had a similar though less adverse effect than the 1870 strike. But importantly, the 1873 and 1874 combination agreements maintained the railroad's shipments at the depressed levels of 1872, well below what might have been considered its legitimate proportion of the trade, based on 1869 and 1871 proportions. The 1873 and 1874 proportions were, however, broadly in line with the agreed allotment percentages. The effect of the 1875 strike was to depress the proportion below the allotment, which seems to have resulted in a further slight reduction in its allowed proportion for 1876. Hence, the combined effects of the basis-related suspension of 1872 and the strike of 1875, when Gowen tried to institute wage reductions, led to a downward ratchet of the Reading's tonnage allotment during the period of the first combination. It is not difficult to see, therefore, why Gowen saw it as being in his business interest to destroy the miners' union and put control of wages back in his own hands.

In contrast, the Jersey Central, the Delaware and Hudson, and the Lehigh Valley seem to have done rather well out of the initial combination agreement. In the case of the first two roads, their allotments exceeded their apparent ability to ship coal, while in the case of the Lehigh Valley, the road does not seem to have felt constrained by the allotment limits. Its shipments exceeded them by several percentage points in 1873–1874, although, like the Reading, it was adversely affected by the 1875 strike. The Lehigh Valley certainly had a substantial line trade by this period, which would have had some impact on the figures, but the equivalent line trade on the Schuylkill had also increased since the late 1860s.

The Lackawanna performed broadly in line with its allotment and both it and the Delaware and Hudson benefited from the 1875 strike in the other regions. The Pennsylvania Coal Company was not formally part of the original combination, but it agreed to abide by a quota of 985,000 tons, hence it was given a notional allotment in the table. Its percentage of total shipments (Figure 11.2) appears rather small in comparison to its allotment, but this was because its line trade was limited compared to some of the other companies. In fact, it shipped a tidewater tonnage very close to the allotted amount in 1873, and 15–20 percent more in the following two years.[14] The Pennsylvania Railroad declined to participate and, so, did not receive a percentage of the allotment, nor were its shipments allowed for in the 10-million-ton estimated tidewater target. While it is correct, as Schlegel avers, that its tidewater tonnage in the direction of New York or Philadelphia, as opposed to Baltimore, was limited at this time, it is apparent from Figure 11.2 that the same could not be said of its shipments overall.[15] Since these are included in the total used to calculate the percentages on all the graphs, this also has the effect of slightly depressing all the percentage values in relation to a comparison with the allotment proportions. It is difficult to escape the conclusion that the Pennsylvania derived some significant benefit from remaining outside the combination in 1873–1875, as its proportionate shipments increased sharply during that period, only to decline again in the later years of the decade, when the combination was in operation just part of the time.

Reference back to Figures 9.6 and 9.7 in chapter 9 indicates the progress of the cartel from the preliminary price-fixing agreement of late 1872, which caused prices to embark again on an upward course, until they steadied at around the desired level of $5 per ton in New York. Since similar behavior was exhibited by the Scranton auction prices (Figure 9.6) as those of the Delaware and Hudson, it is reasonable to assume that the auctions were subject to manipulation (e.g. by agents acting for the combination buying lots themselves) to ensure the appropriate results were obtained. It was probably in tacit recognition of this fact that the auction sales were eventually suspended during 1875.[16] Running a cartel in the late stages of a bull market prior to the 1873 panic was a clear success story, since not only were prices raised significantly, but overall output also rose by more than 1.5 million tons. It even appeared that it enjoyed a measure of support from the media, if a mid-year report from the Philadelphia press is a reliable indicator. The report in question very much took the side of investors and the railroads, arguing that the coal business had been plagued by uncertainty for several years past. This uncertainty was very damaging in a business, which required large capital outlays over a period of years before reasonable returns could be obtained; hence it was necessary to establish the cartel to remove this uncertainty.[17] In terms of the present focus on business strategies and decision-making, this is an extremely important and profound observation, because it identifies the rationale for the construction of an industrial meta-organization such as a cartel, in terms of the need to reduce undesirable variability or unpredictability in the external business environment. For such a cartel to be effective, however, other sources of coal supply that could be substituted for anthracite had to be sufficiently limited and their output inelastic to prevent significant erosion of anthracite's share in the overall market for fuel. These conditions seem to have been met in the short term, but, as time went on, there was growing evidence of artificially high prices creating sales opportunities for competing sources of fuel supply. Some of these have been touched upon in chapter 9 and they will be examined further below.

In 1874, the cartel agreement was continued, with certain refinements to the rules concerning commissions, contracts, and allowable monthly price variations. Also, the Board of Control, as the railroad presidents came to be known, was given power to inspect the books of the different companies to ensure compliance.[18] An interesting report from the New York press in January of that year about a meeting of the Board of Control revealed both a harbinger of difficulties ahead and produced another eloquent apologist for the cartel, in the shape of Samuel Bonnell, a well-known figure in the trade, and a director of the newly formed Lehigh and Wilkes-Barre Coal Company. The indication of possible future problems is identifiable, with the benefit of hindsight, in that Judge Packer of the Lehigh Valley appeared not as the president of his own railroad, but as 'legal adviser [to] several of the largest coal mining companies in the state'.[19] As a high profile public figure and member of the judiciary, Packer was particularly sensitive to the charge that, since his railroad was a common carrier, membership of a cartel could be construed as acting in restraint of trade. On several occasions, and especially in later years, he sought to distance himself as an individual from the cartel for this reason. It is also interesting to note that the annual reports of the Lehigh Valley in the period 1873–1876 contain no reference to the cartel.

Bonnell, the apologist, stressed not the issue of uncertainty, but inexorably rising costs, as a reason for the cartel's requirement for more remunerative prices. These costs arose from the necessity for deep mining using shafts, greater distances of underground haulage, interest on the price of coal lands, higher wages, and consumer demand for broken and sized coal.[20]

Packer had good reason to fear the charge of acting in restraint of trade because the Lehigh Region still contained a number of substantial mine operators, who might insist on their rights to transportation from a common carrier railroad. The Reading was in a slightly better situation in terms of economic power, if not legal rights, because many of the individual operators tributary to its lines had either been bought out or now found themselves tenants on part of the Coal and Iron Company's estate. Although at the time, at least, there do not appear to have been any formal legal rulings on the matter, there seems to have been a general understanding of acceptable railroad behavior in relation to provision of transportation. This was based on the obvious point that there was not an infinite supply of empty railroad cars available for loading. Hence, if loaded cars were detained for any reason, such as blocked sidings, accidents, or shortage of vessels at the unloading docks, the supply of empty cars would, of necessity, be diminished for a period. Some kind of rationing system would then need to be employed, based on such criteria as equal proportions of the number of cars requested by different mines, or relative breaker capacity. Provided the restricted allocation was seen to be made on a fair and equitable basis, this would be regarded as acceptable practice. However, any suggestion that anyone was receiving preferential treatment, especially if the railroad was found to be sending proportionately more cars to its own mines than to those of its individual operators, would be greeted with immediate protests. A good example of the rationing process can be found at the end of 1874, when Frisbie of the Pittston and Elmira Coal Company complained about shortage of cars. Storrs confirmed to him that the cars to everyone were being reduced in the same proportions, adding 'I can stand the groans and howling when there is no discrimination'.[21] However, shortage of cars as a result of the vicissitudes of the trade was one thing, use of car allocation mechanisms to restrict overall levels of output for the purposes of the cartel was quite another. Schlegel attributes to Gowen's 'fertile mind', an earlier pre-cartel proposal in 1871 to control output by restricting car allotments to the mines.[22] However, the idea of car allotments was not new, even if they were not necessarily used for deliberate restriction of output. As early as 1863, the Lackawanna was using them, both for its own mines and those of individual operators. Further evidence from 1866 indicates that overall allotments were being made on the basis of breaker capacity, and further restriction, as a result of car shortages, was being arranged pro rata. With the possible exception of the leased Boston mine, the figures do not indicate any obvious discrimination by the company in favor of its own mines, as compared with those of its individual operators. By 1875, and doubtless before, the Reading was also distributing cars pro rata, based on breaker capacity.[23]

As the cartel continued, the role of car allocation mechanisms began to assume greater importance. There were two reasons for this. Firstly, if overall cartel tonnage allotments were to be adhered to, large and uncontrolled shipments by a minority of mine operators would adversely affect the output of everyone else shipping over the same road. Secondly, the cartel was effectively swimming against the tide by maintaining or increasing prices as demand began to slacken into the teeth of the 1870s depression. Maintenance of output while high prices dampened demand would result in available stocks of coal cars filling up with coal that could not be unloaded, because no one wished to buy it. The resulting lack of empty cars would result in increased rationing and fewer cars returning to the mines, until the latter eventually had to stop or suspend for short periods, because their storage pockets were full.[24]

Taken together, aspects of these two problems provided Gowen with an ingenious way of meeting the cartel requirements, while evading the legal strictures of his common carrier status and giving, to the untutored eye at least, the impression that he was operating in a fair manner. This was achieved as follows. When the railroad had reached the limit of its monthly tonnage allotment, in terms of shipments to Tidewater at Port Richmond, Gowen instructed the Coal and Iron Company to continue mining and loading cars at the mines it controlled. However, instead of sending these to market, the trains of loaded coal cars were switched on to sidings, thereby reducing the supply of empty cars. When the individual operators called for more empty cars, they were advised that there was a shortage and they could only be supplied with a proportion of what they requested. This proportion was as low as 50 percent in mid-1874.[25] This pro rata division would, of course, be scrupulously fair, leaving the operators no very obvious means of recourse, while Gowen achieved his aim of overall production restriction. He did not control the entire market, however, and slack demand in the summer of 1874 compounded the effect of his original intention, to the point where all the sidings as far north as Reading became blocked with loaded coal trains.[26]

The problem of restricting production was clearly more pressing for Gowen than for other railroads, as he had accepted a percentage allotment that was proportionally much lower in relation to the capacity of mines tributary to his road than was the case with the other companies. No explicit reference has been found to other companies manipulating the empty/loaded car balance in the way that Gowen did, but they were clearly troubled by the same concerns that led to his choice of tactics. For example, the Lackawanna expressed concern that they could not legitimately withhold cars requested by W.H. Meeker because of the type of shipment contract they had with him. At the time, they were only giving him 50 cars per day at the Winton breaker, two thirds of the number provided for in the contract.[27]

Despite these emerging issues and difficulties, the combination held up well throughout the year, prices were maintained at or above the $5 mark in New York, and total output only declined by about 1 million tons from 1873 levels. The only concession Gowen had been forced to make to market forces during the summer of 1874 was two short suspensions of shipments, each a fortnight long, in July and August. During the first of these the

Richmond St. bridge in Philadelphia was renewed, preventing shipments to Port Richmond, but it is not entirely clear whether this was a partial reason for the suspension or a convenient moment to undertake essential maintenance work with minimal disruption to normal traffic.[28] The Northern Field companies made no overall suspensions, although some of their shipments were curtailed during this period (see Figure 11.5). In the case of the Lackawanna, at least, this was achieved by having their own and their contractors' mines work full-time alternate weeks, rather than on a continuous part-time basis.[29] A close reading of earlier commentators on this period reveals some subtle but important differences of interpretation about the behavior of the cartel, and Gowen in particular, in the later part of the year. The differences are important because they reflect on the nature of the interaction between the railroads and the union, in the context of the breakdown of relations that led to the 1875 strike in the Schuylkill and Lehigh Regions. Aurand, writing from a labor history perspective, argues that in the fall of 1874 the Reading stockpiled coal in anticipation of a great battle with the union. It then forced a suspension of mining in conjunction with other companies, before announcing wage reductions, the response to which was a strike by the Workingmen's Benevolent Association (WBA) at the beginning of 1875. In June, at the end of this protracted strike, the men went back on the railroads' terms and this effectively signalled the death knell of the union.[30] Schlegel, in contrast, argues that there is no evidence that Gowen plotted the destruction of the WBA. In his view, it is likely that Gowen was not involved in drawing up the terms for wage reduction, although he was willing to enforce them with the operators behind them.[31]

Examination of the precise behavior of individual members of the combination in the second half of 1874 may shed some light on these differences in interpretation. The best way of doing this is by the laborious method of collating and tabulating weekly shipment figures for the period. These confirm that, for a very short period between

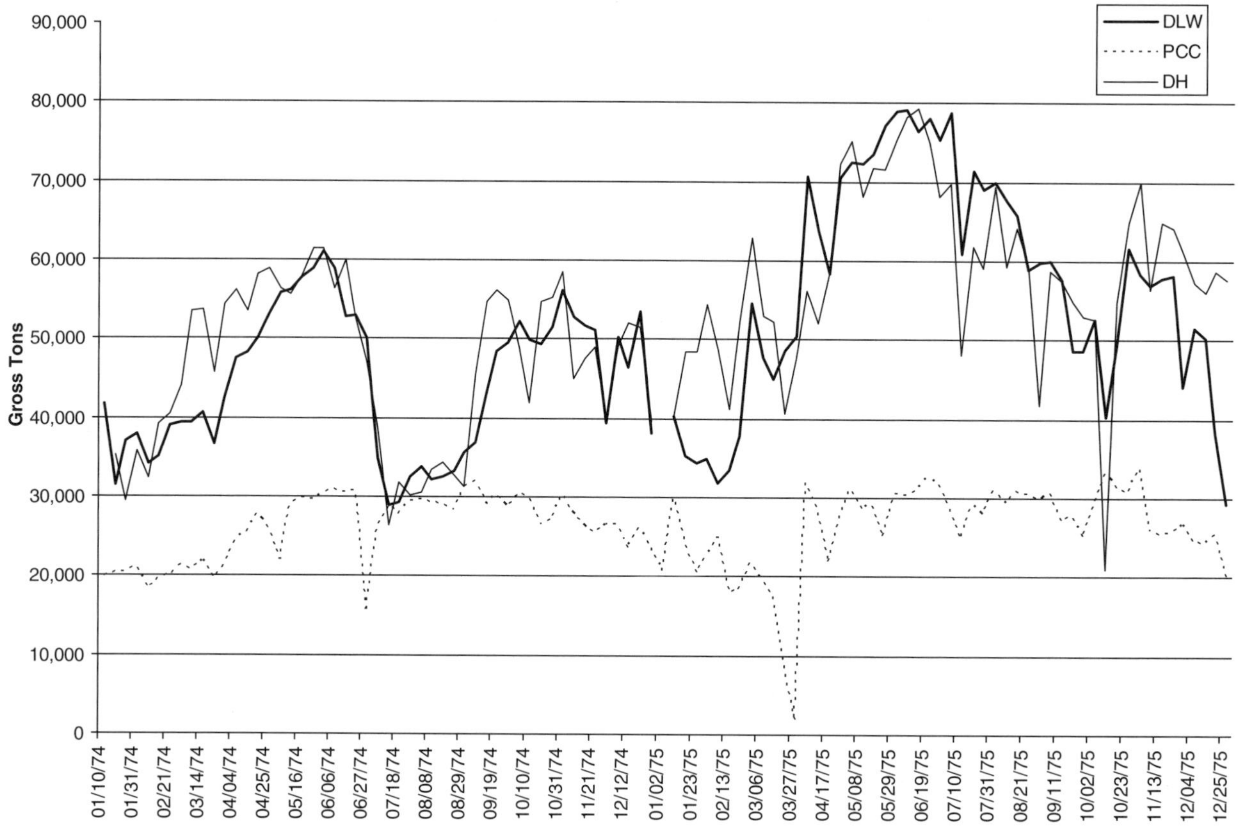

FIGURE 11.5
Weekly Shipments of Major Northern Field Companies 1874–5.

late September and late November 1874, the Reading did indeed push shipments up to the maximum possible levels of about 200,000 tons a week, from the average level of about 185,000 tons that had prevailed earlier in the year and also in the fall of 1873. Aurand unfortunately picks as his example the one week in the latter part of 1873, prior to the Christmas season, when shipments fell to around 80,000 tons, to make his comparison with 1874. This gives a misleading impression of the extent to which the Reading had increased shipments at the end of the latter year. While noting that having coal on hand would be useful in the event of a strike or suspension, Saward's Journal, which was usually well-informed, set even greater store by the fact that this burst of large shipments would allow the tonnage for the fiscal year ending November 30 to nearly equal that of 1873. The decision of the Reading, to apply the brakes in the latter part of November 1874 by shutting down 31 collieries, was interpreted as their realization that the trade really was as dull as that being experienced by other companies and they could not continue shipping, because the wharves were blocked up.[32] The notion that companies could stockpile coal in huge quantities to hold out against the miners was itself fanciful, since, as Gowen himself testified, the entire storage capacity of the anthracite production and distribution regions did not extend to more than 10-days coal supply,[33] and only a fraction of that storage was available at Port Richmond.

There is no evidence that the other companies increased shipments at the same time as the Reading. In the case of the Lehigh Valley, shipments were maintained at levels similar to those earlier in the year until the normal reduction of shipments into December. The shipments of the Northern Field companies can be seen in Figure 11.5 and, while there is one week of reduced shipments at the end of November, Aurand errs slightly in stating that the shipments of the Wyoming companies were halved at this time. Exact halving of shipments had undoubtedly taken place earlier in the year because of working the mines alternate weeks, but this was not related to the wages question. As an aside, it is also interesting to note that the Pennsylvania Coal Company, officially outside the combination in 1874, made a minimal effort to reduce its shipments in the summer of that year, almost all the restriction being borne by the larger companies.

Hence, there is no indication that the railroads were acting in concert to precipitate a strike. For Gowen to do this deliberately when others were not would have been foolishness, because of the selectively adverse impact on the Reading's tonnage alone and its future percentage allotments within the cartel. Nevertheless, none of the companies could ignore the clear signals coming from the market at the end of 1874 that coal could not be moved at the official cartel asking prices, although this was never publicly acknowledged. The secret was leaked in an anonymous letter to the New York Tribune, thought to be from a Lehigh operator. The letter confirmed that sales were being made in December for egg coal at $5.30 when the cartel price was $6.30. Ario Pardee also admitted that he had received a letter from the iron manufacturers requesting a $1 per ton reduction in the price of coal. He had replied saying that the transportation companies, operators and miners must all take their share of this reduction.[34]

In response to the weakening market, rumours began to circulate, doubtless started by the railroads themselves, that a 25 percent reduction in wages was to be sought. However, as late as the meeting of the combination in mid-December (at which the Pennsylvania Coal Company was also present), no definite conclusions had been reached, although it was apparent that wage levels in the Lehigh Region and possible curtailment of output there were major discussion topics.[35] Finally, announcements were made by the Lehigh and Schuylkill Coal Exchanges that wage reductions were to take effect from the beginning of 1875. In the Lehigh, there was a 10 percent reduction for outside labor and a 15 percent reduction for contract mining and yardage work. In Schuylkill, the equivalent figures were 10 and 20 percent.[36] A separate announcement by Gowen, following a meeting on December 30, 1874, ordered a reduction in railroad wages by 5–10 percent, depending on the category of staff, and significantly also, 50–60 cents per ton price reductions in the prices of the different sizes of coal.[37] The Northern Field companies reduced both inside and outside wages by only 10 percent.[38] As late as December 31, 1874, Storrs of the Lackawanna admitted that he had no indication of how the men would respond. The company was also unprepared for even a short strike, as they had little or no stock at some of their main interior shipping points, Syracuse, Ithaca, and Oswego.[39] In the event, most of the mines in the Lehigh and Schuylkill Regions came out on strike, as did those in the Plymouth area, but the big three Northern Field companies stayed at work, as did mine operators in Dauphin, Columbia, and Northumberland counties.[40] As previously noted, the protracted strike was ultimately unsuccessful, the WBA collapsed, and the men went back on the companies' terms.

With a substantial proportion of the productive capacity on strike, it might have been expected that the combination itself would collapse, but this was not the case. Meetings continued and a slightly more flexible framework for the year's business was progressively hammered out. Key aspects of this included suspension of the Scranton auction sales, prices were to be adjusted monthly to meet the market, coal prices at competing points were to be equalized, and the tonnage to tidewater points was to be adjusted on the previous year's basis.[41] The last point was of particular significance, because it implied that no advantage in tonnage terms was to accrue to the railroads still shipping versus those out on strike. An attempt to examine how this worked out in practice is shown in Figure 11.6, where the total shipments of the two groups of railroads are compared. In terms of the strike-bound roads, it is apparent that the strike itself was very effective. This is attributable not only to union solidarity, but to the efforts of Gowen in keeping the operators firm in their resolve to stand out against the men. For example, the president of the Philadelphia Coal Company later testified that Gowen had persuaded him not to start up during the strike, claiming it would 'break the combination'.[42] Once the strike had finished, however, tonnages soared and were maintained at close to maximum levels until near the end of the year. In contrast, the non-strike-bound companies barely managed to increase their output over 1874 levels in the early months of the strike. Reference back to Figure 11.5 indicates that only the Delaware and Hudson managed to increase shipments in January, while output for the other two companies actually tended to decline. Confirmation of the Lackawanna's unpreparedness for market

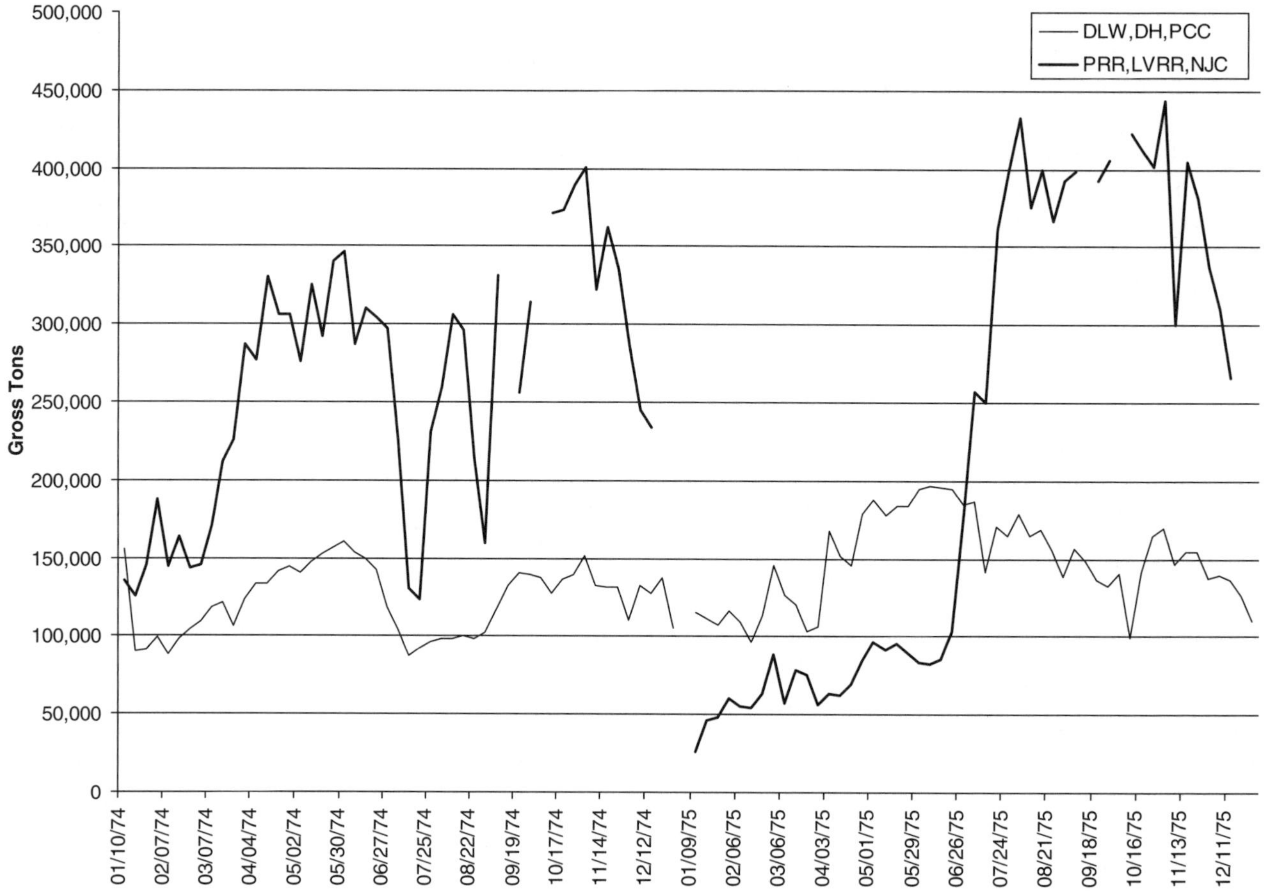

FIGURE 11.6
Weekly Tonnages of Strike-Bound and Non Strike-Bound Anthracite Railroads 1874–1875.

changes arising from the strike can be found in the fact that only 8 of its 18 collieries were in operation at the beginning of 1875, and several, such as Oxford, Hyde Park, and Archbald, had been effectively 'mothballed' for some time, owing to the dullness of the coal market and the cartel's output restrictions. Hence, it took some time to ramp up production again at several of their own and their contractors' mines.[43] In the period, April–June, the companies managed to increase production by about 30 percent over peak 1874 levels, before tailing off after the strike ended and the supply from the other regions came back on stream. It is interesting to note, however (Figure 11.5), that just as the Pennsylvania Coal Company failed to participate in the production restriction of mid-1874, it also failed to contribute to the shipment boom of mid-1875. It was, however, able to maintain a very stable level of shipments almost to year-end, which was very efficient in terms of the deployment of men and equipment. Whether the Northern Field companies would have increased their shipments still further, had they been able to with sufficient speed, is a moot point in terms of abiding by combination quotas. They would probably have increased the proportion of lump coal shipped still further, as this size avoided the cost and waste attendant on breaking coal. The company correspondence of the Lackawanna indicates that the demand for this size greatly outstripped supply at the beginning of 1875.[44] Even if tonnages were not as high as might have been expected under the circumstances, the companies certainly benefited in price terms. In April 1875, for example, although the cartel price for stove coal in New York Harbor was nominally set at $5.40, prior to this, actual sales had been made as high as $6.25, netting a very handsome profit for the producers in the early part of the year. Examination of the Delaware and Hudson price series (Figure 9.7) suggests that a brief period of price weakness followed, coinciding with the sharp increases in output by the Northern Field companies and optimism that the striking regions might go back to work. However, this was quickly reversed, and high prices ruled for the remainder of the year, despite the large shipments by the Schuylkill and Lehigh Regions from July onwards.[45]

It is also pertinent to consider whether other coal shippers were able to benefit from the strike, given the failure of the Northern Field companies to supply anything approaching the wants of the market. An obvious candidate was the Pennsylvania Railroad, but although its mines near Shamokin and in the Lykens Valley were unaffected by strike action, the Susquehanna Coal Company fell victim to the labor unrest in the Plymouth area and its men refused to work. Hence, gains and losses largely cancelled each other out. Total output by the latter coal company fell by 200,000 tons in 1875, while the Shamokin output increased by 180,000 tons. The Railroad, as a whole, only increased its anthracite tonnage by 130,000 tons for the year, largely as a result of greater output from the Lykens Valley area.[46]

The other mining interests that stood to gain both from the high prices of the cartel in general, and anthracite shortages during the strike in particular, were the bituminous shippers. However, as Figure 11.7 indicates, the Cumberland bituminous region appears to have derived minimal benefit from the 1875 strike, in terms of increased shipments. Its patterns of weekly output in that year differ only slightly from that of 1874. The reported Reading figures used in the graph, by way of clarification, include coal for company's use and a very small amount of bituminous coal. The other important source for bituminous coal moving eastwards was the Clearfield Region, served by the Pennsylvania Railroad. Although its 1873 and 1874 shipments were on the order of 600,000 tons per annum, these were not substantially larger than 1871. Unlike the Cumberland region, the following year saw a near 50 percent increase to 928,000 tons, and rapid expansion of shipments continued into 1876.[47] It is likely the 1875 tonnage would have been even larger if this region had not also fallen prey to strikes during the year.[48] As mentioned in a previous chapter, the fledgling coke industry of western Pennsylvania also benefited from the strike and the shortage of Lehigh lump coal in particular. Without the strike, the eastern ironmasters would undoubtedly have delayed making trials with coke for much longer, owing to the high cost of transportation. In the event, the forced introduction of this new fuel proved very satisfactory, and some coke continued to be used thereafter in varying proportions with anthracite, displacing a progressively greater proportion of the local fuel.[49]

It is very difficult to obtain a precise assessment of the combined impact of the 1875 strike and cartel on the overall pattern of the coal trade during the year. A summary table of shipments by carrier to both interior and competitive points was given in the trade press at the beginning of 1876, but the figures were not prepared by J.H. Jones, the cartel tonnage accountant, and, as printed, they contain

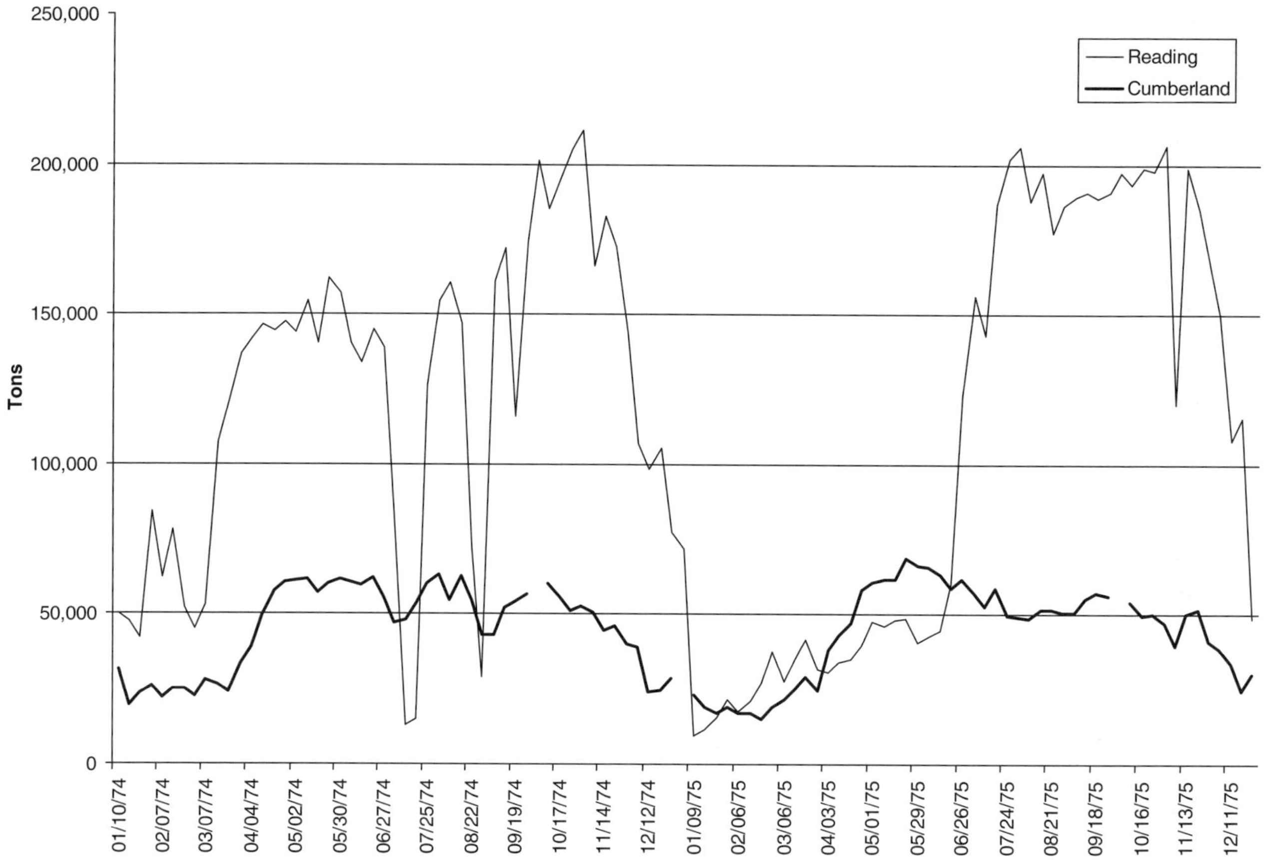

FIGURE 11.7
Weekly Shipments of all Reading Coal and Cumberland Bituminous 1874–1875.

inaccuracies and internal inconsistencies. Checking back against the 'official' cartel shipments and company Annual Reports, to correct the errors, suggests some broad conclusions. The greatest loss was suffered by the Lehigh Valley, which experienced a 17.7 percent fall in shipments compared to 1874, followed by the Reading with a 13.3 percent decline. The winners, as expected, were the Lackawanna and the Delaware and Hudson, with increases of 29.4 percent and 27.3 percent respectively. However, since the cartel controlled tonnage to competitive points, the latter two companies only modestly exceeded their shipments to these destinations as compared with the previous year, doubtless to avoid infringing on the goodwill of their cartel partners too severely. The corollary of this was that the bulk of their tonnage increases were concentrated on the line trade and interior points. This conveyed a double benefit, because prices in general were higher in these markets and the result was a substantial windfall for both companies. In contrast, the figures suggest that the Lehigh Valley suffered its greatest tonnage reductions in relation to these markets that offered the best prices, thereby compounding the financial penalty. There can be little doubt that these effects were attributable to the special circumstances of the 1875 strike, because in 1876 the Northern Field companies were not able to retain the temporary sales advantage they had exploited in the previous year.[50]

It was small wonder, therefore, that the Lehigh Valley was becoming increasingly unhappy about its participation in the cartel, since the majority of the benefits seemed to be accruing to other roads rather than its own. It was reported that Gowen had had extreme difficulty in persuading Packer to adhere to the combination, and the adverse business consequences of 1875 were clearly the last straw.[51] At the beginning of 1876, Packer announced that he would no longer be an active member of the cartel and would leave control of production up to the mine operators, thereby weakening the potential for concert of action.[52] Nevertheless, percentage quotas were agreed upon that did not differ markedly from previous years, although the Pennsylvania Coal Company was now officially party to the agreement. A novel aspect of the 1876 arrangement was the provision for fines of $1.50 per ton for over-shipping monthly quotas, although it was recognized that there was no legal means of enforcing such penalties.[53]

The year began badly for the combination. Dull and overstocked markets, following the surge of shipments in the latter part of 1875, necessitated suspension of mining for most of the first three months of the year, although prices were still largely maintained. During the first month back at work, in the absence of centralized co-ordination of output, the Lehigh operators over-shipped their quota, then, at the May meeting, declared they could not pay the fine! To quell the outcry from the other members, they promised to abide by the quota in future, but instead, further lack of discipline and tonnage overruns continued.[54] The markets also failed to improve, and further suspensions were arranged for June and July. The press charged that the combination was holding large quantities of unsold coal in 850 boats at a number of tidewater points, including Jersey City and Brooklyn, in an effort to prop up unreasonably high prices. They argued that, since there had been a general price reduction of 30 percent since 1873, the same should apply to coal.[55] Finally, in exasperation at the persistent non-compliance of the Lehigh interests during the year, the combination was dissolved on August 22. The Lackawanna immediately announced a return to its monthly auction sales, and Gowen began a sustained drive to increase output and claim market share, even at the expense of prices, which plummeted.[56] It also became clear that coal prices were not the only financial indicator that had been artificially held up by the combination. As the Wilkes-Barre press graphically remarked, as soon as investors on Wall Street heard the news of the collapse of the combination 'they at once commenced a general disgorgement of their long hoarded securities'. Lackawanna stock immediately dropped by 8½ percent, Delaware and Hudson by 7 percent and the Jersey Central by 13 percent.[57]

Before considering the later stages of the first combination phase, it is appropriate to consider whether the initial combination of 1873–1876 had, on balance, brought net benefits to the industry. According to Virtue, the operators were of the view that the restriction of production and maintenance of prices had mitigated the impact of the general depression on the anthracite business in comparison to its effects on other major industries.[58] The New York Daily Bulletin thought otherwise. It declared that 'this combination is today the great curse of trade and the worst of all obstructions to the recovery of business from its depression'.[59] Crucially, however, in the short term, the combination was able to succeed in charging high prices to consumers and manufacturers because it controlled more than 90 percent of supply, and the purchasers had no realistic alternative for their fuel supplies. The bituminous industry of the interior was insufficiently developed and railroad facilities for long-haul shipments lacked the capacity to transport large quantities of soft coal to seaboard, even had such quantities been available, which they were not. The only modest exception to this was the Clearfield district of central Pennsylvania, less than 250 miles from the coast, whose development was fostered whenever high anthracite prices offset its locational disadvantage, owing to higher transport costs to market.

The many and varied methods employed by the railroads for reporting their financial position at this time, and differences in their accounting relationships with leased lines or mining subsidiaries, make it very difficult to provide an accurate yardstick for comparing the net effects of the cartel upon each of them. With this caveat, a few general comments on some of the major players are in order. There can be little doubt that the overall effect of the first two years of the cartel on net earnings was overwhelmingly positive. The following year produced the expected divergence between those roads that were affected by the strike and those that were not. Despite this, both the Reading and Lehigh Valley paid 10 percent dividends in 1875. For the Lackawanna and the Delaware and Hudson, the period 1873–1875 was one of very sharp increases in profitability. For example, the Lackawanna doubled its net earnings between 1872 and 1873 from $4.16 to $8.41 million. An examination of overall accounting surpluses for the two companies, which allows for the effects of dividends, fixed charges, payments to leased lines, etc., in addition shows clearly that the Lackawanna derived the greatest benefit of all from the cartel, since it was able to increase its cumulative annual surplus, from only about $700,000 in 1872, to $5.8 million by the end of 1875. With the benefit of hindsight, better timing for the setting aside of a large accounting reserve would be hard to imagine.

Given the financial vicissitudes of the later part of the decade, the Delaware and Hudson was able to maintain a satisfactory surplus during the period, but it did not increase significantly. This implies that it needed the healthy trading conditions under the protection of the cartel and the windfall from the 1875 strike in other regions, just to service the financial requirements of its loss-making leased lines and other railroad construction activity to reach northern markets. This was enough to raise questions about future financial stability, as later events were to demonstrate.[60]

Alone among the cartel participants, the Reading, in its Annual Report for 1876, found major cause for complaint about the workings of the tonnage allotment system in general, and the behavior of the Lehigh Valley Railroad in particular. It insisted that, while it had faithfully complied with the various agreements, others had not. In consequence, over the period of the cartel, it had lost out to the tune of almost 800,000 tons, in relation to its agreed quotas. To reinforce its claim for injury, it noted that in the period June-August 1876, total profits of the Railroad and the Coal and Iron Company amounted to only $125,000, while in the following three months of free competition, after dissolution of the cartel, the profits had soared to $1.93 million.[61]

Despite its justifiable complaints, the Reading had derived one key benefit from the cartel that was different from the experience of the other companies. It had bought itself time to begin re-organizing much of the Schuylkill coal trade from an inefficient and high-cost enterprise into a more cost-effective operation, that could compete on more equal terms, especially with the Northern Field companies. Although the process was far from complete, between March 1873 and December 1874, Gowen had already managed to transform a business that previously incurred losses of 30 cents a ton into one where a profit of 50 cents a ton was being realized. This was substantially more than the usual margins of the middlemen that he was endeavouring to replace as part of his strategy.[62] These initial improvements in profitability were largely the result of higher prices. Further work was required to reduce costs significantly and rationalize output, topics that will be examined in more detail in a later chapter. However, the energetic prosecution of these policies over several years put Gowen in a better position to withstand the periods of intense competition that followed the collapse of the cartel. Without them, the finances of the Reading would have undoubtedly collapsed, sooner than they eventually did, under the weight of excessive debt and burdensome fixed charges, largely (but not wholly) incurred in the acquisition of its immense coal estate.

The attitude of the Pennsylvania Railroad towards the cartel was necessarily ambivalent, because it was the only one of the railroads that was heavily involved in the development of both the hard and soft coal trade at this time. Higher anthracite prices and freight tolls directly benefited the Northern Central and other lines which it controlled, as well as its own coal subsidiaries. At the same time, high anthracite prices also fostered the growth of competition from bituminous coal producers tributary to its lines, as noted earlier. However, the scope for future development of the eastbound bituminous trade was arguably greater, from its perspective, than that of anthracite, because of the vast area of the soft-coal deposits in western Pennsylvania. Although these would take many years to develop, it was obviously not in the railroad's long-term interest to favor its anthracite shippers over its bituminous shippers, but rather to maximize the volumes of both kinds of traffic over its own lines. This mix of long- and short-term strategies and balancing of commercial interests explains the tacit partial co-operation of the Pennsylvania in the anthracite combination. Yet it was also quite willing to exploit prevailing circumstances to its own business advantage, if it felt that it could do so without damaging relations with at least some of the cartel participants too severely. An excellent example of its approach can be found in the minutes of the Susquehanna Coal Company in 1876. In response to a letter from Gowen, proposing a suspension of anthracite mining in February and March, the board minuted the following response:

> 'It is not expedient for this company to concur in the suspension of mining proposed by Mr. Gowan (sic); but in consideration of the large stock of coal on hand in the company's depots, which is sufficient for the present demands of its customers, it is hereby ordered that mining operations at this company's collieries be suspended for thirty days from the 10th of next month'.[63]

Overall, it is difficult to argue that the Pennsylvania received any substantial advantage in the anthracite trade by staying nominally outside the cartel during this period. In the years 1873–1875 its share of total tonnage never exceeded that of 1869, although it did manage to restore some of the ground it had lost at the beginning of the 1870s.

Further twists to the overall relationship between the Pennsylvania and the cartel were a function of the very different kinds of interaction it enjoyed with the Reading, as opposed to the Lehigh Valley. Commercial relationships with the latter were of a close and mutually beneficial kind for many years, while those with the former were characterized by mutual suspicion, rivalry, and destructive competition, especially during the period when Gowen retained a measure of influence in the company. These themes would re-emerge at intervals during the combination era and will be referred to again below.

COMBINATION VERSUS COMPETITION 1876–1879

In the period of nearly three and a half years following the collapse of the first combination, a variety of tactics and strategies were followed by the different companies in response to both anticipated and unanticipated events and the actions of their competitors. While the basic outline of the main chronology of events is well known, a number of key aspects of corporate behavior at this time have neither been identified nor explained. In an attempt to understand the business interactions of this period and their wider implications more fully, recourse has again been had to the detailed weekly reports of coal shipments, provided by the companies to several trade and local newspapers during the period. Despite the gaps in the available runs of data in any single newspaper, taken together, almost complete shipment records have been reconstructed by cross-referencing the reports available in multiple sources.[64] Weekly data are essential for the actual operating behavior of the railroads to become apparent, since even monthly data introduce too great a degree of averaging of weeks of high and low shipments. The weekly shipment records for a number of the major companies 1876–1879 are plotted in Figures 11.8 and 11.9. Major points in the chronology of the period will be summarized in the context of these figures, before undertaking a comparative assessment of company behavior.

Immediately prior to the collapse of the combination in the late summer of 1876, the companies resorted to full or partial suspensions of shipments, on a 'week on, week off' basis, in an attempt to control output in the face of market weakness. This period can readily be identified on the graphs by the sharp output fluctuations visible at that time. Immediately after the combination ended. This behavior was replaced by a rapid expansion of shipments and a correspondingly precipitate fall in prices. Schlegel speaks of Gowen dumping large quantities of coal on the New York market at this time, but fails to point out that the other companies were doing the same.[65] Only the somewhat muted increase in shipments by the Lehigh Valley is surprising. As the main culprit for over-shipping its quota, it might have been imagined that it had substantial pent-up demand that it would aim to satisfy immediately, but this does not appear to have been the case. This would tend to support the conclusion that its motive was in fact purely to benefit from the combination, without submitting to its constraints, by getting more than its fair share of shipments at artificially inflated prices. Its high-minded insistence on the common carrier provisions in its charter appears therefore to have been largely a smokescreen for self-interest, a view doubtless shared by the combination members who voted for dissolution in August 1876.

While no one would have expected an instant return to co-operative behavior on the part of the railroad presidents, it is likely that there was a general expectation of a return to combination arrangements, after the salutary shock of a short, sharp spell of unfettered competition. This is important, because the rules as they had previously operated, based the quotas for the following year on the actual market shares of the preceding year. Hence it was very important, in the remaining months of 1876, to retain (if not enhance) market share for the next round of negotiations. This explains why the companies pressed their coal on dull markets, taking the price penalty in the process. Despite all their efforts, overall tonnage still fell by over 1 million tons compared to 1875. The Reading was only able to match its strike-restricted output of the previous year, while the Lackawanna and the Delaware and Hudson experienced large shortfalls as against the previous year, when they had profited from the Schuylkill strike.

After the surge in shipments in the latter part of 1876, the following year opened in a modest way despite the absence of any quotas. The Northern Field companies and the Lehigh Valley were shipping at about 50 percent of maximum weekly capacity, while the Reading was even more restrained. Prices

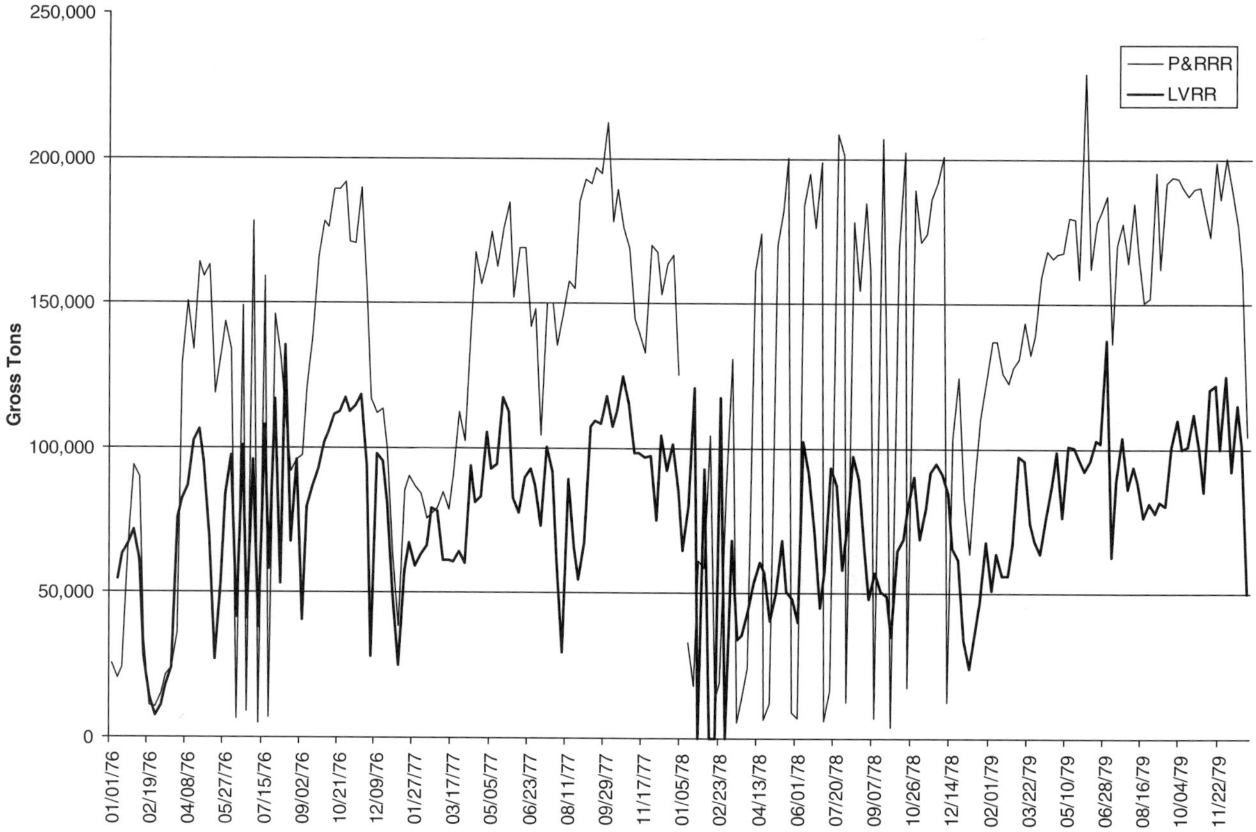

FIGURE 11.8
Weekly Shipments of Major Anthracite Railroads 1876–1879.

ruled very low, dipping under $3 a ton in New York, a far cry from the $5 a ton target under the old cartel. Meetings were held by the railroads and suitable postures adopted, but no agreement was forthcoming. Gowen was heard to remark that the trade would work itself out in a year or two, even without the combination. Yet, at the same time, he also indicated he would join any arrangement that met with mutual agreement.[66] Judge Packer felt constrained to explain again that his company could not participate in any arrangement that had 'for its object or purpose, directly or indirectly, the restraint of shipments over its road'.[67] However, in his personal capacity as a coal operator, he was willing to comply with any agreed arrangement to restrict production. As if to try and underscore the notional separation of interests linked to the Lehigh Valley, it was also reported that Robert H. Sayre had been delegated to represent the Lehigh Valley Coal Company (not the Railroad) at cartel discussion meetings, and the Lehigh coal operators had separately appointed G.B. Linderman, Charles Parrish, and John Leisenring to act as their representatives for the same purpose.[68] This was little more than window dressing, however, as individuals more closely associated with the interests of the Lehigh Valley and the Lehigh and Susquehanna Railroad would have been hard to find!

Gowen's relative detachment from the combination process at this time, as compared with earlier years, doubtless reflected in part his unwillingness to get his fingers burned twice by unreliable partners, in arrangements that were less advantageous to his own road than to any other. Yet the more significant motive for his behavior probably lay in the hard facts of cost accounting. By his earlier actions, he had successfully reduced the cost of coal delivered at Port Richmond to $1.59, putting the Reading in a very strong competitive position with respect to the other companies.[69] Hence, he could stand low prices, while also having the greatest capacity to make sustained large shipments from the 100 or more collieries tributary

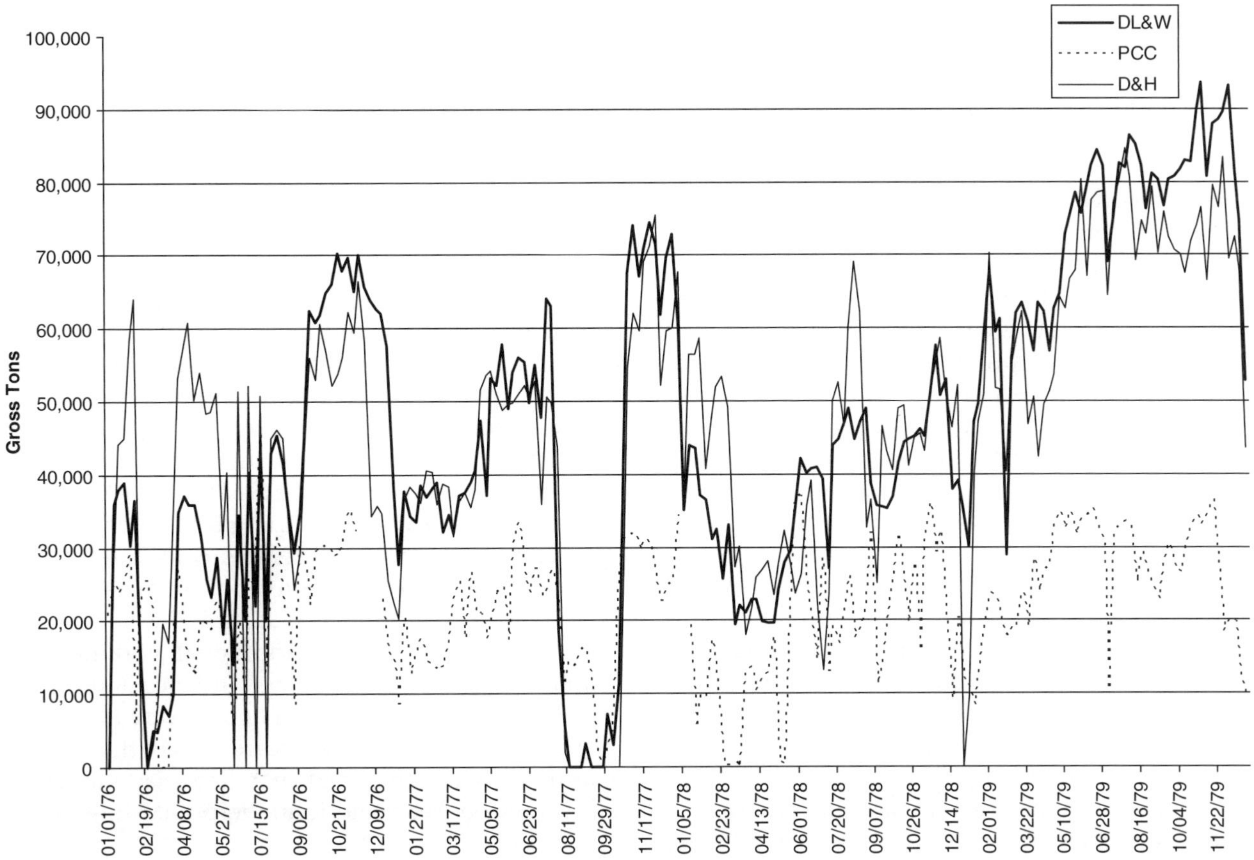

FIGURE 11.9
Weekly Shipments of Northern Field Companies 1876–1879.

to his road. The larger the shipments, the stronger his negotiating position for any future cartel agreement, if not in 1877, then the following year. This policy he managed to carry through. Shipments built up very sharply from March onwards and were sustained at high levels for much of the rest of the year. More than 1 million more tons of coal were delivered to Port Richmond for onward shipment than in 1876 and, by the beginning of September, overall the Reading had shipped 1.54 million tons more than in the equivalent period of the previous year. This was far ahead of the other companies, which had only managed very modest increases. The Northern Field companies were particularly adversely affected by the 1877 strike, which is clearly shown in Figure 11.9. The Lehigh Valley was partly affected, and only the Reading continued fully at work to take maximum advantage of the misfortunes of the other companies.[70] Gowen was also encouraged by a successful visit to London in the middle of the year, where he had been able to persuade the English stock and bond holders that the worst of the Reading's financial troubles were behind it.[71] Sadly, his optimism was eventually to prove unfounded, but, for the present at least, the Reading was a going concern, though buoyed up as much by Gowen's rhetoric as by the actual state of its finances.

Other companies were in a less fortunate position. The precipitate fall in coal prices in the latter part of 1876 immediately exposed financial weakness, inefficient or costly operation, and cash flow difficulties. The Lehigh and Wilkes-Barre Coal Company, owner and lessee of expensive coal lands, was first to fall, with debts greater than the value of its remaining property not subject to mortgage and without the means to meet its wages bill. It went into receivership on February 12, 1877, owing $2.5 million to the Jersey Central, which itself had been in a precarious financial position for some time. Unable to withstand this shock from its subsidiary coal company, the Jersey Central immediately followed it into receivership, and an extended struggle began to prevent the complete break-up of its mining and transportation system. Full details of the financial aspects of this process are given by Bogen and do not need to be repeated here.[72] Suffice it to say that the receivers were able to retain the leased Lehigh and Susquehanna Railroad, but in early 1878 had to relinquish the Lehigh and Delaware Division Canals, and the coal estate at the eastern end of the Southern Field, back to the Lehigh Coal and Navigation Company. The latter profited handsomely from this because the additional improvements on the lands in question made by the Lehigh and Wilkes-Barre, at a cost of over $1.1 million and an additional $700,000 paid by them in royalties for coal as yet unmined, were abandoned to the Navigation Company without payment.[73]

At the same time as the Jersey Central was losing the confidence of investors, the Delaware and Hudson also came under close financial scrutiny. This necessitated a spirited defence of its position by President Thomas Dickson, firstly to the Board of Managers and subsequently to the courts, the stockholders, and the press. In his report to the managers on the business of 1876, Dickson declared that 'the finances of the company are in a sound condition and the attacks that have been made upon its credit, are wholly unwarranted'.[74] At the same time, he admitted that the company's policy on leasing railroad lines in New York state had been severely criticized, but averred that the policy had been necessary to prevent these avenues to northern markets falling into the hands of competitors. His presentation of the 1876 accounts showed a loss of about $550,000 on these leased lines, leaving a profit of only $308,000 for the year.[75] The extended period of years required for these lines to become profitable has been the subject of discussion in an earlier chapter, but it is always in the times of deepest depression that current losses are the hardest to justify for the benefit of long-term gain. Certainly, a minority group of stockholders were unimpressed. A few days after Dickson's report, they brought legal action against the company, claiming it was bankrupt, had paid dividends out of unearned income, and had wasted large sums on unprofitable coal-land development. Although the motion for appointment of receivers was denied, Dickson felt it necessary to take further action to repair the company's damaged reputation, following the further fall in its share price as a result of the suit. As noted in Chapter 5, the company joined forces with none other than the Law Committee of the New York Stock Exchange, to offer a substantial reward for information leading to the conviction of any person found circulating damaging reports 'in regardto the credit of the company.'[76] At the annual stockholders meeting, a further committee of investigation was formed to double-check the valuation of the company's assets. It was noted in the subsequent year's Annual Report that this committee, 'after making full allowance for the shrinkage of values, and deducting all amounts considered doubtful, reported an aggregate of value which showed a surplus of over 4 million'.[77] Even this was not enough for elements of the trade press, who continued to cast aspersions on the company's valuation of its assets during the latter part of 1877.[78] In an attempt to allay remaining fears about its credit-worthiness and ability to pay its fixed charges, the company decided on a large issue of mortgage bonds to cover floating and other indebtedness and provide additional working capital. As if to express their confidence in the company, $5 million of these bonds were taken at 90 percent by a syndicate of directors of the company, which included such financial luminaries as John Jacob Astor and John Pierpoint Morgan.[79]

In comparison, the Lackawanna, by its standard approach of failing to provide stockholders with information, managed to avoid a similar mauling in the press or the courts. Nevertheless, Wall Street exacted just as great a toll on its share price, when the combination collapsed, as it did on the Delaware and Hudson. As can be seen from Figure 11.10, even the temporary restoration of the combination in 1878 did little to encourage investor sentiment. A sustained improvement in share prices had to wait until the final quarter of 1879, when evidence of recovery in the wider economy was already abundant.

By the end of 1877, aided by the strike in the Wyoming Region, Gowen had succeeded both in restoring his share of the trade to 32.82 percent, and in further reducing his costs per ton on his much increased shipments to only $1.24 delivered at Port Richmond. Since average prices received at tidewater had fallen to $2.41, this left a profit of $1.17. Although substantial, it was still not enough to cover the burgeoning fixed charges and floating debt interest with which the railroad was now saddled. In addition, Gowen, like Dickson, had faced mounting criticism during the year for his financial management, although he had been more

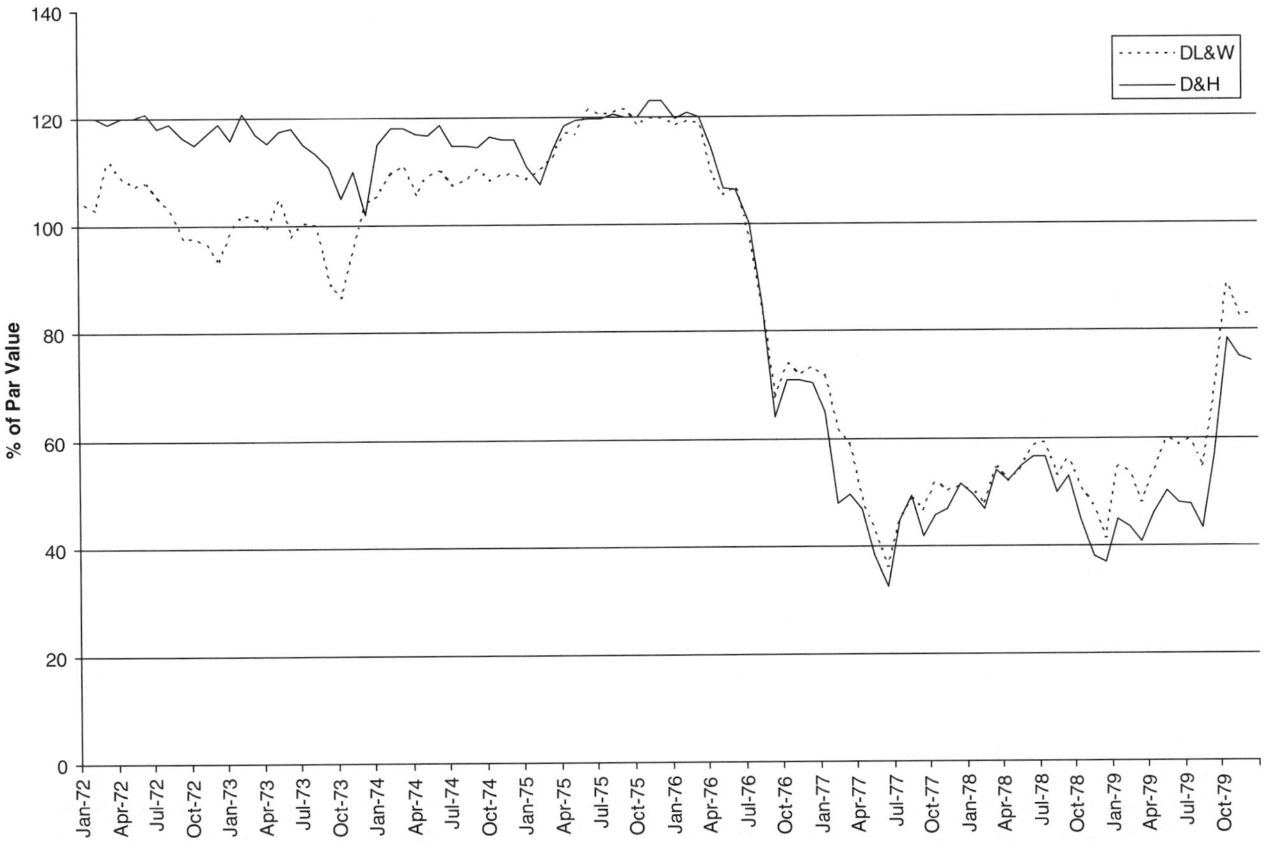

FIGURE 11.10
Monthly Wall Street Stock Prices 1872–9.

successful at quelling opposition to his policies, despite an accounting position that was much less favorable![80] Even lower prices than the Port Richmond average were recorded at the Scranton auction sales at the end of the year. In December, the New York prices received by the Delaware and Hudson fell to an all-time low of $1.92, and its shipments during the year had been nowhere near as buoyant as those of the Reading.[81]

Finally the railroads were ready to call a halt, to prevent a continuation of the 'ruinous competition' coupled with a continuing depression in the overall economy that was driving prices down inexorably.[82] Gowen's initial proposal for a new combination had some different features to previous agreements. Prominent among them were, firstly, that almost all tonnage was now to be included in any allotments, rather than just tidewater shipments as before. Secondly, voting rights were to be made proportional to agreed tonnage quotas and, thirdly, each company was to pay a certain amount (per ton shipped) into a fund that would be used to pay fines for over-shipment of quotas.[83] When the companies met together, they had before them a large array of tonnage statistics to compare performance as far back as 1868, but interestingly, at the meeting on January 2, 1878, when it came to a vote on the mechanism for assigning quotas, use of the last 5 years' data was voted down 2–5, and use of the previous 10 years was voted down 3–4.[84]
A compromise set of percentage allotments was then suggested, but the Lehigh contingent refused to accept their allocation, claiming it was too low, and the meeting dissolved without agreement. Subsequent behind-the-scenes negotiations involving E.C. Knight, past president of the Jersey Central, Gowen, and Scott of the Pennsylvania, resulted in a concession by Gowen that broke the deadlock. He offered to surrender a small portion of his quota, if others would do the same, to make up the Lehigh Valley quota to a level that they would find acceptable.[85] This pattern of the more financially strapped members of the combination

taking a leading role in securing final agreement was one that was to be repeated on future occasions.

Once agreement had been reached, the combination resumed business as usual. The primary restriction mechanism was the application of tonnage quotas, and Virtue argues that prices were left up to the individual companies. This latter assertion is contradicted by the Miner's Journal, which reported a meeting of the general agents of the companies to fix wholesale prices for the year.[86] In any event, the Delaware and Hudson quickly raised prices by 50–60 cents and the others followed. Unfortunately, the large shipments made by all the companies in the last months of 1877 had replenished stocks throughout the anthracite distribution region and the trade at the opening of 1878 was very dull. As can be seen from Figures 11.8 and 11.9, the companies responded in different ways to this. The Lehigh Valley operated shipment suspensions of 1–2 weeks for a brief period, while the Reading moved into a regular pattern of 2-week suspensions every month, with varying numbers of mines at work. The Lackawanna and Delaware and Hudson reduced shipments to only about half the level of the equivalent period in 1877, but avoided complete suspension, unlike the Pennsylvania Coal Company, which both suspended and reduced its levels of mining activity during periods when it was actually shipping. Only the mines of the Lehigh and Wilkes-Barre Coal Company set to work with a will, under the management of Charles Parrish on behalf of the receiver.[87]

A further interesting difference between the Reading and the Lehigh was that the Reading placed no quotas on individual mines, allowing them supposedly to ship to their regular capacity, provided, of course, they had the orders and the empty cars available for loading. In contrast, the Lehigh Valley, having finally agreed to participate, went forward with bureaucratic enthusiasm. A circular was issued notifying the operators that the mean of the largest single month's shipments in each of 1876 and 1877 was to be the basis of the percentage of the railroad's overall quota that was to be allocated to each mine. The Spring Mountain Colliery, for example, was allocated 4.905 percent of the total tonnage, or 29,430 tons to April 1st. Quota cards for each mine were printed on which the allowed tonnages were to be noted, so no one would be in any doubt about how the system was to be operated.[88]

Trade continued to be dull. By early March, 1.25 million tons less than in 1877 had been forwarded to market, with little immediate prospect of larger shipments, owing in part to the continued depression in the iron trade. The railroad presidents brushed off investigations into combination activity by the states of New York and New Jersey in a manner that was almost becoming habitual. They insisted that the combination was not in place to fleece consumers, but had been rendered necessary by the adverse state of the economy and the vicissitudes of the trade.[89]

As the year progressed, further price advances were made and, at their peak in October, the Delaware and Hudson was receiving $1.70 more per ton in New York than had been the case in January. As the prospects for a successful return to the gold standard at the beginning of 1879 grew more likely, business confidence began to show signs of improvement, and the Northern Field companies nudged output progressively higher. The first shoots of recovery were also being felt in the Lehigh Region, but this was greeted with discontent rather than pleasure. In characteristic style, the Lehigh operators complained that the new system of controlling all tonnage, not just that going to tidewater, was disadvantaging their interests. They were making increased line shipments, predominantly to the iron trade, because of improved business conditions, but the fixed quotas meant that this necessarily reduced their tidewater shipments, to a level below that to which they felt they were entitled. As a result, they refused to accede to the plan to extend the combination until April 1879, and further efforts to resolve the matter ended in failure.[90]

Gowen was now in serious difficulty. The increased profits from higher prices under the combination had been substantially offset by the restriction on shipments, and his tonnage had fallen back by 1.7 million tons compared to the previous year. As the burden of fixed charges and the costs of servicing the floating debt continued to rise, cash flow became an increasing problem, and delays in payment of wages were the result. A strike was narrowly averted early in 1879, when funds were scraped together to meet the payroll for disgruntled employees, but an overall upturn in the economy had been so long in coming that recovery prospects for the company were hanging in the balance.[91]

In the absence of a renewed combination agreement, the prop under prices was removed and they fell sharply by $1.50 a ton in New York to a figure only slightly above that of the dark days of the end of 1877. Wage cuts and strikes in various parts of the coal regions followed, for example at the Lehigh and Wilkes-Barre Coal Company and the Mineral Railroad and Mining Company in the Shamokin area.[92] Nevertheless, the successful return to the gold standard had finally begun to lift the economy out of the longest depression in living memory, and sustained growth in orders east and west, aided by very low prices, gave the first real stimulus to the trade for a number of years. Gowen had no option but to press for maximum output, aiming again to maximize both market share and cash flow, while paring costs to the bone. This he did quite effectively, as Figure 11.8 shows, increasing shipments by 900,000 tons compared to the previous year of unfettered competition in 1877. However, despite these exertions, this large tonnage gave him fractionally less than the 28.62 percent share of the trade to which the combination had entitled him in the previous year (and which he had slightly over-shipped). It was also significantly less than the figure of nearly 33 percent that he had achieved in 1877. The reason for this is not hard to find. Figure 11.9 indicates the stellar performance of the Lackawanna, and to a lesser extent the Delaware and Hudson, in increasing their shipments substantially during 1879. For the Lackawanna, the levels were almost double those in any of the three preceding years. A significant proportion of the increase by the latter company was sent north and west rather than to the eastern seaboard, reaching approximately 38 percent of the total. Low prices stimulated the trade, because they made Great Lakes grain carriers willing to take return cargoes of coal from Buffalo to Chicago at low rates. The fact that anthracite could then be sold in the latter city, at prices similar to those in Philadelphia, led to a marked increase in demand.[93] In contrast, the Lehigh Valley achieved a relatively indifferent performance, demonstrating for a second time an apparent inability to derive any benefit from the absence of quota restrictions, that it, and its tributary operators, had precipitated. In 1878, it had shipped 19.33 percent of the trade total, slightly less than its quota allocation, but in 1879, the proportion dropped to only 16.85 percent, based on calculations from the 'official' figures of J.H. Jones, the tonnage accountant to the combination.[94] An important contributing factor to this relatively poor performance was a series of wildcat strikes in the Lehigh Region in the summer of 1879. Again, this suggests that Gowen was more proactive and effective in dealing with labor grievance in the Schuylkill Region at this time than were the Lehigh Valley and the operators in that region.[95]

Aside from the differential willingness of the companies to abide by the rules and quotas of the combination, there are several other points relating to the methods adopted for restriction of shipments that merit attention. These are linked to the perceived efficiency of different approaches to operation of the mining economy when restriction was in force. As the shipment graphs indicate, the Reading and the Delaware and Hudson utilized the 'on-off' method of full-scale shipping for 1–2 weeks, followed by effectively complete shutdown for an equivalent period. The latter road only employed this approach in 1876, while the Reading used it in 1878 also. A similar pattern applied to the Lehigh Valley and the Lackawanna for the 'partial on-off method' where shipments were halved for short periods, instead of being entirely suspended. Again, the Lackawanna only used it in 1876. The Pennsylvania Coal Company seems to have employed a rather erratic mix of both methods and it is apparent that the years when combination was in force were a significant disturbance to its preferred manner of relatively modest but very regular shipments. There were further sub-variants of these methods, which involved some mines being told to suspend in one 'off' week, while others suspended in the next. Had the railroads all been common carriers only, i.e. owning no mines, such arrangements would have been extremely difficult to sustain, as they would have required all of the operators to agree with the railroads' timing of suspensions and tonnage allocations all of the time. This was part of the issue in Packer's continual insistence that the railroad would fall in line only if all the operators agreed to the combination.[96] There was also the problem, both from the railroad and mining perspective, of starting up and shutting down full-scale operations on a repetitive basis. This required large numbers of employees to remain on hand, yet be idle a significant proportion of the time, without becoming disaffected. In the case of the mines, there were also a number of costs that were incurred, regardless of whether the mine was actually shipping or not. Individual operators generally lacked the financial resources to stand these costs for extended periods. The fact that the railroads actually did own many mines allowed for the possibility of more drastic action to reduce capacity in the medium term, e.g. by removing pumps and letting mines fill with water. This reduced costs, but in the event of a sharp upturn, such as occurred in 1879, it required a period of months to bring such mines back into productive operation, reducing their flexibility to increase shipments above a certain level. Nevertheless, it would allow production to be concentrated on the most efficient low-cost operations. These issues will be examined in more detail, in later chapters, in relation to their impact on overall mining profits and losses.

Finally, a brief assessment of the wider impact of the cartel on the bituminous industry is required. Overall, the Cumberland coal industry performed badly against anthracite during the cartel era, despite the price protection the latter afforded it much of the time. Cumberland shipments had peaked at 2.67 million tons in 1873, but fell away sharply as the depression took hold, only recovering to 1.73 million by 1879.[97] Only the Clearfield Region, served by the Pennsylvania Railroad, made substantial progress over the period, albeit from a very low base. In 1872, shipments were a mere 430,000 tons, but this had nearly quadrupled to 1.63 million net tons by 1879. However, much of the overall increase was in the early cartel years, from 1873 to 1876, rather than in the later period.[98] Although detailed data are hard to locate prior to 1876, Figure 11.11 gives weekly shipments from 1876 to the end of 1879, with only a short break in the available data in the latter part of the first year. This reveals a slightly surprising conclusion. Namely, that even the most successful part of the bituminous trade, in respect of eastbound shipments, does not appear to have benefited from the anthracite cartel. The years of increased output are the non-combination years of 1877 and 1879. In the latter years, the same sustained increase in production can be observed, as was seen for the two major Northern Field anthracite railroads, although the soft coal tonnages involved were much smaller. Detailed investigation of the decision-making involved in the bituminous fields lies outside the scope of the present study, and there is little direct evidence in the major coal-trade publications. Nevertheless, it can be hypothesized that the Pennsylvania Railroad did not encourage a flood of soft coal to reach tidewater at times that would have weakened the cartel, although it was not averse to increased traffic during periods of free competition. The level of freight rates and rebates (drawbacks)

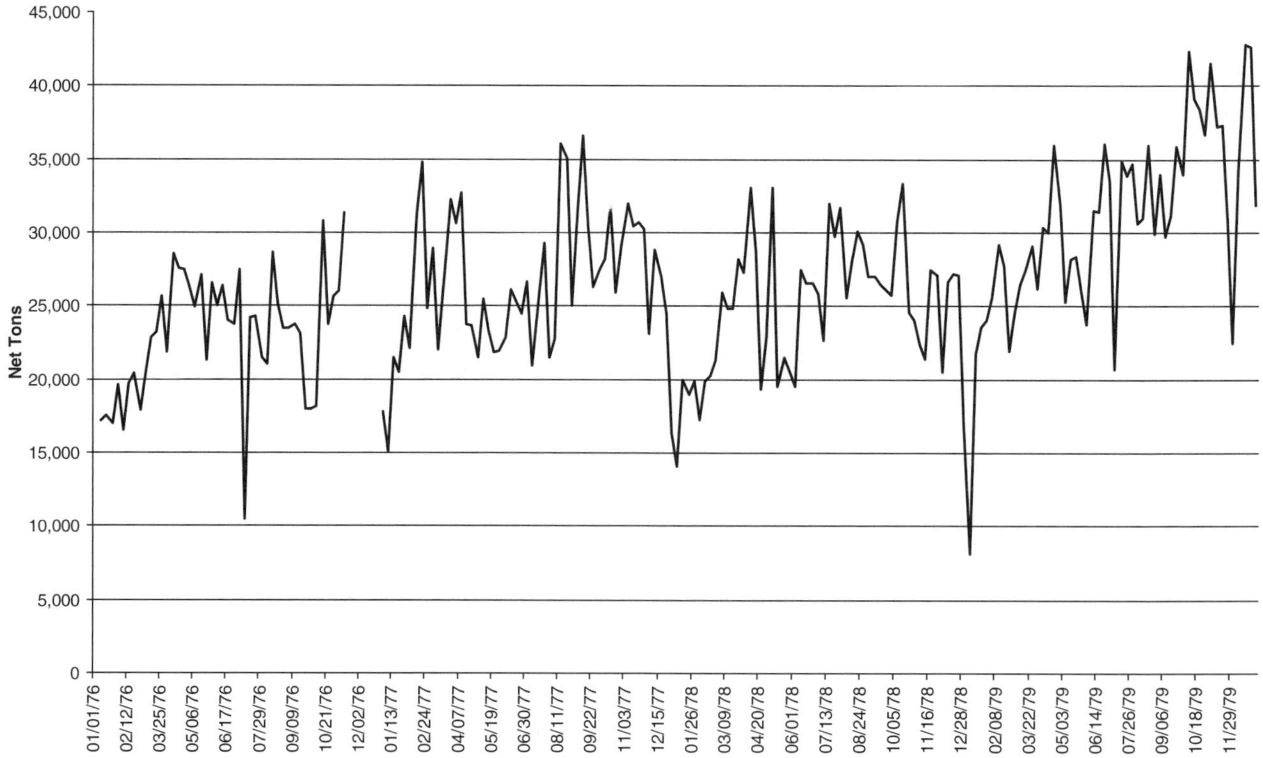

FIGURE 11.11
Weekly Clearfield Bituminous Coal Shipments 1876–1879.

were important in determining the competitiveness of the interior soft-coal operators in tidewater markets, and there is evidence that, by the beginning of 1878, the bituminous operators were in at least as bad a financial condition as those in the Anthracite Regions. They took the opportunity of the new cartel of that year not to try and undercut anthracite and increase their market share, but to meet, with a view to advancing their own prices in concert, because of the 'ruinous state' in which they found themselves. At the time, because of high freight rates, they were netting only 55–60 cents a ton at the mines, most of which was immediately paid out in miners' wages.[99]

By the end of 1879, there were some signs of a growing ability of the bituminous operators to fight for market share against anthracite, but the total shipments from Clearfield in that year were still less than 6 percent of the anthracite total. This was not enough to seriously challenge the hegemony of hard coal in eastern markets. The stage was therefore set for continued attempts by the anthracite industry to manipulate coal supplies and provide artificial support for prices.

THE 'INFORMAL' COMBINATION 1880–1884

The very low prices that accompanied the high volume of shipments in the latter part of 1879 again convinced the companies of the requirement for some harmony of action. A new agreement was drawn up at the beginning of October, but the Pennsylvania and the Lehigh Valley failed to sign it, so it became inoperative. Various attempts to raise prices by 25 cents then followed, and a further attempt was made, at the beginning of December, to persuade all the companies to suspend entirely from December 24 to January 5, 1880. This enjoyed a measure of success, although Sloan, who was willing to work ⅔ time throughout the month, would not commit himself to stopping on Christmas Eve, while Harry Packer, son of the late judge, took a contrary position for the Lehigh Valley on these points.[100] The less co-operative stance of the Lackawanna during the period correlates closely with increased 'clout', derived from its rapid growth of shipments during the year. Fortunately, a number of surviving letters written by Sloan during this period allow the development of this thinking to be followed quite closely. Early in 1879, Sloan had stressed to Storrs the need to maintain an independent stance and keep 'out of all alliances and connections'.[101] He felt that Dickson of the Delaware and Hudson had 'restriction on the brain'. At the same time, he could not understand why both the latter and the Pennsylvania Coal Company kept tending to reduce prices, rather than raise them.[102] As the year drew to a close, the outstanding success of the Lackawanna in increasing its output was becoming apparent to all, and to the Delaware and Hudson in particular. Sloan was moved to observe 'I am sorry D + H seem so jealous of everybody and disposed to work contrary to all the rest unless Mr. D[ickson]'s hobby of restricted production is carried out!'.[103] Despite this, he still favored working together whenever possible. He instructed Storrs to co-operate with his opposite numbers in the other two major Northern Field companies, with a view to working either ¾ or ⅔ time in December 1879.[104] The motivation for this was more than generosity of heart. He expressed on several occasions the need 'to keep our organization together' in the face of short-time working, in other words, to prevent the loss of good employees. The easiest way to do this was to try and ensure that other companies were not in a position to offer more days work than the Lackawanna itself.[105] However, even co-operation had its limits, and Sloan made quite clear that he intended to operate, first and foremost, in a manner that suited the Lackawanna's interests alone:

> 'We shall hold our tonnage relatively the same as we have the past season, and, I hope, widen still further the comparisons with our neighbors, it is high time the DL+W take and keep her position'.[106]

True to his word, Sloan began to exert his influence more forcefully, yet still diplomatically, in the subsequent years of the combination.

After the somewhat rocky start to renewed concert of action in the closing days of 1879, a new method of restriction was agreed on, for an initial experimental period in the first 6 months of 1880. This involved half-time working when restriction was required, for agreed 3-day periods in specified weeks. The standard arrangement was to work the first three days of the weeks in question and then suspend for the remaining three days. With but slight variation, this approach was utilized until the end of 1884.

Commentators as far back as Virtue in 1896 have argued that there was not a combination during this period similar to those of the 1870s, in the sense that there was no written agreement, no fixing of maximum tonnage, and no quotas. Despite this, in a slightly confusing caveat, he states 'there was perhaps as effective combination among the companies as ever, only on different lines'.[107] Unfortunately, the documentary evidence from company records does not entirely bear out these conclusions. There were several attempts to secure written agreements for specified periods of time, and the fact that co-ordinated restriction regularly took place indicates that at least all the major interests signed some of these. A good example is a surviving letter for a period of restriction from January to March 1881, which had already been signed by Hoyt, Packer, Dickson, and Gowen, and appeared to be awaiting Sloan's signature at the time. An interesting aspect of this particular agreement was that 'two thirds of the tonnage' could vote for discontinuance of the arrangement at any time, meaning that none of the signatories could take unilateral action.[108]

Arguably, the new combination was more effective at policing itself than in previous years. The indefatigable J.H. Jones, anthracite tonnage accountant, still kept a monthly score of relative company output, and the individual interests paid very close attention to the minutiae of each other's business. Two examples serve to highlight this very clearly. The first involved the Delaware and Hudson. In March 1880, employees at its Baltimore breaker were spotted loading cars and moving empty cars at 10:00 PM one Wednesday evening, when the arrangement stipulated that no coal was to be drawn from the breaker pockets or empties moved after the end of an ordinary day's work.[109] In the second case, the collieries tributary to the Lehigh Valley in the Beaver Meadow area were observed loading coal on a Thursday in May of 1882 when they should have been suspended. It was claimed the Lehigh Valley, ever willing to chance its arm, was putting in enough empty cars on the Wednesday to allow collieries to run on the Thursday. That said, the loaded cars were not actually transported until the suspension was lifted the following Monday, at which point a large daily shipment was unsurprisingly achieved! When this was brought to Mr. Packer's attention, it was reported that he was sorting the matter out and doing everything possible 'to line up to the letter of the agreement'.[110]

During much of this period, W.H. Tillinghast, initially receiver of the Lehigh and Wilkes-Barre Coal Company, and personal friend of Mr. Sloan, seems to have acted almost as lackey, and certainly facilitator, of the combination, and also as 'referee' or adjudicator when disputes arose. Since his company was the first to go into receivership in the late 1870s, it was clearly in his business interest, as the weakest member, to encourage, flatter, or simply nag the other combination participants to enter into agreements that would lift prices and improve his cash flow. In this, he clearly had a valuable part to play in the overall success of the combination arrangements. Prices rose rapidly after the initial 6-month experiment in 1880 was established, almost doubling to nearly $4 in New York within a matter of weeks. A period of remarkable price stability was then maintained for several years until serious price weakness began to set in at the beginning of 1884. In this context, it is difficult to believe Virtue's assertion that there was no agreement to maintain prices. Certainly, at the beginning of 1882, Sloan was quite specific in aiming to link curtailment with maintenance of prices, although he admitted that the Pennsylvania Coal Company and the Delaware and Hudson were not very good at doing this. In the case of the latter company, the fear seems to have been that upward pressure on prices would also cause the men to press for higher wages.[111]

Aside from the now customary bickering between presidents about the specific timing of particular restrictions, there was one other issue that arose on several occasions. This is of interest in the present context because it reflected on the broader management of the mining economy. The matter in question was the cost effectiveness of suspensions operated on a 3-day week basis. After the initial 6-month trial of the method in 1880, apparently S.B. Whiting of the Reading prepared a report for the combination, claiming it resulted in extra expense. However, there was no indication that it favored one company over another. Only Sloan opposed a change in the method, while all the others were willing to curtail in a manner that would keep the Reading receivers happy. Later evidence suggests this change would have meant suspensions of a full week of six days, or the variant of six successive working days at the end of one week and the beginning of the next, leaving only a Sunday break in between.[112] Not only was Sloan opposed to the 6-day suspension, he also annoyed his combination partners by consistently refusing to be bound to ship only half of what he could produce in 6 days, during the 3-day working week when restriction was in force. This stance appears obdurate and unnecessary at first sight, but it was based on several subtle aspects of the operating experience of the railroad that made it a reasonable view for the Lackawanna to adopt, in furtherance of its own interests. The first and most important of these was related to car supply. With a three-day week, a full car supply could be assured, thereby maximizing output and avoiding blockages of coal at the breakers. Over a 6-day period it was not possible to ensure a full car supply, thereby reducing working efficiency. Hence, more work could be done in 3 days than 50 percent of a six-day total. In addition, the railroad had no stocking grounds at local points along the line, and none for holding lump coal for the locomotives. A 6-day stoppage meant having to hold enough coal to meet these requirements in coal cars on the tracks, thereby significantly reducing the number of empty cars available to send to the mines. Some of the railroad's contract shippers also had their own versions of similar problems. For example, Connell and Co. could not produce and stock enough coal to keep one of their large customers, the Scranton Steel Company, fully supplied for the duration of 6-day stoppages.[113]

Another factor in the equation was the continued fear of losing employees, which has already been mentioned. This was particularly acute in the expansionary phase of the business cycle in 1881–1882, so any kind of curtailment would only worsen the situation. For example, in mid-1881, Storrs reported that as a result of a 3-week spell of half-time working, some mines had lost half their laborers. Shortly afterwards, he calculated that, overall, the Coal Department was 600 men and boys short of a full complement. This situation clearly persisted for some time, as he complained at the beginning of 1882 that labor shortages were creating forced curtailment, as the company could not make more than ¾ time at the breakers.[114] No explicit statement has been found arguing that alternate weeks suspension would aggravate the situation more than 3-day working weeks, but it is likely that this was feared, although it was not really put to the test during the period. There is no doubt, however, that the men intensely disliked very short-time and irregular work, such as could be produced by a mixture of curtailment and short car supply. They then had every incentive to seek more regular employment outside the mining industry, and many did so.[115]

Late in 1883, Gowen and Hoyt of the Pennsylvania Coal Company again attempted to change the suspension plan to alternate weeks. The former claimed that there was 'more economy and decided curtailment', i.e. better adherence to the rules, under his preferred plan, but Sloan persisted in his refusal.[116] By early the following year, under worsening business conditions, attention had begun to turn back to the possible re-introduction of tonnage quotas. Storrs was not averse to this under the favorable circumstances where a company shipped all its own coal, since this would allow some mines to be closed in a process of rationalization. However, the actual circumstances, where part of the overall shipments were provided by individual operators, he felt, were more difficult to regulate satisfactorily. A particular problem was the subdivision of the quota among the different individual concerns, which in his experience did not work smoothly.[117] Nevertheless, in the event, the companies did revert to an allotment scheme for the ensuing year, made more pressing by the business cycle downturn and the loss of confidence on Wall Street in the latter part of 1884.

Two other perturbing factors had to be addressed by the combination on a continuing basis. The first of these was the familiar problem of Pennsylvania Railroad anthracite tonnage; the second was the creeping threat of price competition from soft-coal producers, as their ability to deliver higher volume shipments to tidewater markets steadily increased. A very noticeable shift in the behavior of the Pennsylvania can be charted as the period progressed. As late as February 1879, the management of its mining subsidiary, the Susquehanna Coal Company, were sufficiently despondent about prospects for the anthracite industry that they seriously entertained the scenario of shutting down their operations entirely for a year. Their calculations suggested that it would cost them and the Pennsylvania Canal a total of $187,150 to do this, which was the same financial result as shipping 500,000 tons and losing 37½ cents on every ton, because of low prices.[118] Although they did not adopt this policy in the final analysis, postponement of investment in new mining capacity meant that they were relatively poorly placed to expand output in the short term when demand increased.[119] As a result, it was 1883 before their overall share of the anthracite trade returned to the levels of a decade earlier, but at that time, and subsequently, their proportion began to increase noticeably, at the expense of the other companies. This was attributed by the Lackawanna, at least, to the animosity between the Pennsylvania and the Reading and the bitter rivalry that developed out of the ill-fated South Penn Railroad project. This involved the Reading, and a syndicate headed by Vanderbilt.[120] In late 1883, the two companies were described as being 'at sword point', and Storrs subsequently attributed the refusal of the Pennsylvania to act with the rest of the coal trade as a deliberate device to injure, and, if possible, crush the Reading, now severely weakened by its second receivership in the space of four years. In controlling or crushing the Reading, the real object was, in suitable mining terminology, 'to sprag Vanderbilt'.[121] Despite this in-fighting, other members of the combination continued to include the Pennsylvania in talks about the future of shipment restrictions. The latter road wanted regular uniform work for economy, rather than repeated suspensions, and was expected to use the tried and tested device of running up its own shipments, then arguing for a future allotment based on the higher levels of output. In the case of the Pennsylvania, during 1883, not only had they worked full-time while others curtailed, but they gave bonuses to their men for increasing output.[122] Even they, however, were not entirely immune to market forces and, by the beginning of 1885, their collieries had ground to a halt for lack of orders. It was not felt that this would be sufficient for them to sign up to any new agreement, but at least there was now a strong likelihood that they would co-operate with any restriction.[123]

Their contribution to the expansion of the soft-coal trade was rather more impressive in this period than their performance in anthracite. The sources of increased bituminous shipments at this time have already been examined in chapter 9, and again it was the Clearfield Region, served by the Pennsylvania, that grew most substantially, as it had in the early 1870s (see previous section also). Shipments almost doubled from 1.63 million net tons in 1879 to 3.17 million in 1884. Although Cumberland coal shipments had still to return to their 1873 levels, other soft-coal areas, such as Westmoreland County, were also becoming significant contributors in the marketplace.[124] The overall effect was that soft coal had achieved a large enough market share by 1883–1884 to put sufficient downward pressure on anthracite prices, that the combination had to yield to it in limited measure.

An overview of the experience of one of the combination members during this period can be seen in Figure 11.12. Overlapping slightly with Figure 11.9 for purposes of temporal continuity, the unrestricted trade of 1879 contrasts markedly with the years that followed, where, in different ways, a managed control of output is clearly identifiable. The first noticeable phase covers the first eight months of 1880, a period of deliberate and precise restriction to about 60 percent of maximum weekly output. In the face of continued improvement in the wider economy, this restriction allowed a sustained rise in prices to be achieved, ushering in a period of relative prosperity for the anthracite carriers. A lesser degree of restriction was called for in the first half of 1881, followed by essentially unrestricted full work for the remainder of the year. Early 1882 was a period of labor shortages, but, in general, the year corresponded to the peak of business cycle activity, and there is evidence of capacity increases at this time in terms of maximum shipment levels, a trend that continued for the remainder of the period. Significant restriction of output by 50 percent or more in some weeks became apparent in the first half of 1883, when the wider outlook was still positive. This may have been slightly overdone, because very heavy shipments followed in the remaining months, to give an annual output for the railroad that exceeded 5 million tons for the first time. In the following year, however, a sharp business contraction set in. A shipment 'time signature', reminiscent of the dark days of 1876, reappeared, as output oscillated sharply, falling by two-thirds or more in some of the curtailment periods. Slightly ominously for its competitors, the Lackawanna was now demonstrating an ability to put out very large shipments at times when restriction was not in force, peaking at 160,000 tons per week. This was a figure double that which had been achieved a decade earlier. On this basis, it was able to increase its shipments, compared to 1883, despite the adverse business circumstances, but just as importantly, its market share rose to 16.9 percent, as compared with 11.7 percent back in 1874.

On the basis of both shipment and price data, it is difficult to escape the conclusion that the first four years of the new combination, from 1880 to 1883, were highly successful in regulating output and maintaining price levels. This is perhaps to be expected, since it is generally easier to manage expanding economic circumstances than those experienced when contraction sets in. And so it proved in 1884, when stop-go restriction methods became necessary to keep the combination from collapsing, and attention turned back to the possible

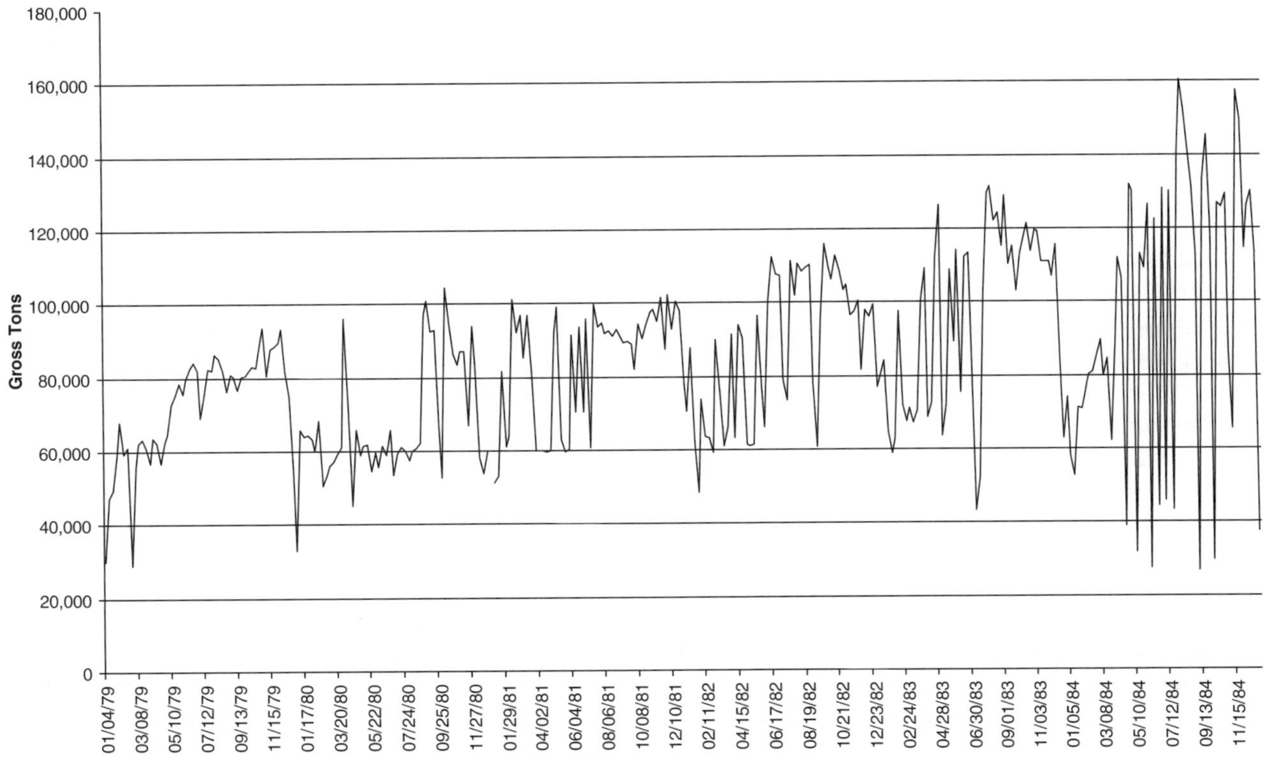

FIGURE 11.12
Delaware, Lackawanna and Western Railroad: Weekly Coal Shipments 1879–1884.

re-introduction of tonnage quotas and monthly allotments.

A final summary of the experience of these years can be found in Table 11.2, which compares the number of officially designated suspension days with the actual number of idle days in the Lackawanna Coal Department, taking 312 days (52 weeks x 6 days) as the maximum achievable. In looking at the figures, it should, however, be borne in mind that the number of operating mines varied year by year. Also, breakers were not necessarily able to work for all of the day every day, owing to minor blockages, machinery problems, and car shortages.

TABLE 11.2
Suspension Days and Idle Days During the 1880–4 Combination.[125]

Year	Suspension Days	Lackawanna Idle Days
1880	88	148
1881	48	87
1882	49	80
1883	60	98
1884	107	134

From the table, it is apparent that curtailment was not the only reason for lack of work at the mines. In general, about 10–20 percent of available time, or an average of 39 days a year, were lost through other causes. Company-wide averages may also conceal a wider disparity in days worked by individual mines. In the light of this information, it is easy to see why there was such concern among senior managers about loss of employees through short-time working, not to mention financial distress among lower-paid workers when the number of idle days became very large, as it did in 1880 and 1884. These figures also need to be borne in mind in relation to the wider questions of overcapacity and overproduction in the industry.

THE COMBINATION OF 1885–1887

As Table 11.1 indicates, the percentage allotments decided upon for 1885 were broadly in line with those of 1878, although the Lackawanna received more favorable treatment at the expense of the Delaware and Hudson and the combined Reading/Jersey Central interest. In the following year, slight downward adjustment of most of the quotas was made to accommodate the increasing shipment of the Pennsylvania, which continued its refusal to participate in the cartel. The assumed total tonnages for the trade in each of the two years were 30 million and 33.5 million tons respectively. In addition to the annual totals, a month-by-month sub-division was also agreed upon.[126]

The sub-committee appointed to determine the quotas at the end of 1884 consisted of F.B. Gowen, J.S. Harris, and G.A. Hoyt. They could not reach agreement on the length of time that the arrangement should last. Hoyt favored a short 15-month period, the others a 5-year period. The reason for seeking a long-term agreement was that they felt annual rounds of negotiation encouraged continued expansion of mining capacity, which was the major problem faced by the trade. Taking a longer-term perspective would allow better capacity planning and the abandonment of all surplus collieries.[127] In the event, the main committee appears to have overruled the recommendation of a majority of the sub-committee, although the rationale for this was not documented, and the termination of the arrangement was fixed for March 31, 1886.[128]

Despite the misgivings of Hoyt and Gowen about short-term agreements, there is evidence that, for the Lackawanna at least, the quota system was more effective than the curtailment system of 1880–1884 in encouraging rationalization of operations. Storrs observed, in December 1884, when details of the phased allotments became know, that scaling back of operations by the coal department would be necessary, as Sloan himself had suggested, because the productive capacity of their contractors and shippers had increased by 25–40 percent since discussions about allotments had begun earlier in the year. The implications of this was that 5 of the company's own mines would have to be idled, throwing 300 men and boys out of work at each. This was a considerable concern, primarily for the employees, but also because it increased the likelihood of unrest at the mines. After further consideration of the impact of the planned quota for January 1885, which was a mere 1.5 million tons for the entire industry, Storrs suggested a novel plan of working. Both he and his contract shippers felt the 2-day week resulting from the limited quota would be unsatisfactory. In an attempt to ameliorate the adverse consequences of this on employees with families, he proposed that work for all the single men in the operating collieries should be suspended, and their places taken by men with families from the idle collieries.[129]

The problem of very short time became less severe as the year wore on and monthly quotas were advanced in line with the requirements of the trade. Prices continued to slide, however, especially in the final quarter of the year, and early in 1886, as the initial agreement drew to a close. Campbell, who has examined surviving correspondence of the Lehigh and Wilkes-Barre Coal Company for this period, presents evidence of a breakdown in combination discipline at this time. Despite a tight monthly quota of only 2.1 million tons for the December 1885 trade, Tillinghast, of the Lehigh and Wilkes-Barre, issued an instruction to mine flat out so as to obtain a good tonnage for the year. Tillinghast's correspondence also suggests that the Northern Field companies more generally were taking a very relaxed view of the combination's directives on tonnage in the final quarter of 1885. In terms of renewal of the agreement for the following year, Sloan was again identified as the man least willing to compromise. He insisted on a tonnage allocation of 6.5 million tons, while the Pennsylvania Coal Company claimed he was only entitled to 5.5 million.[130]

Unfortunately, as the bickering continued, prices continued to fall and profits to slide. It was therefore not surprising that the New York bankers decided to take a hand. The railroad presidents would have known that their invitation to meet at the private residence of J.P. Morgan, on March 22, 1886, was not merely a social gathering. As the minutes of the meeting verbosely record, Morgan asked those present to

> 'take counsel as to the possibility of preventing further injury to the interests they represented by some concerted action looking to an arrest of the demoralization of business which resulted from the existing want of harmony'.[131]

Requests of this kind from J.P. Morgan were not easily set aside, and an agreement was rapidly forthcoming to renew the allotment system for a further year, to set an April quota of 2 million tons, and to make an immediate price increase of 25 cents per ton.[132]

The apparent ease of reaching agreement in the grand surroundings of Morgan's Madison Avenue residence belied the considerable effort made by J.S. Harris, in the preceding months, to secure information from the companies on their recent shipments and employment levels, as well as proposals that might hasten a negotiated settlement. Interestingly, within a month of the new agreement, Tillinghast, who himself had spent years cajoling and upbraiding other members of the combination, was caught by the cartel's tonnage accountant over-shipping his quota, probably by sending coal over the Lehigh Valley, that was not properly declared. His claimed excuses, based on uncertainty about how the combination was initially to work, seem lame, given that he was present at the meeting when the agreement was originally decided![133]

It is apparent from Figure 11.13, which is a continuation of the time series in the previous figure, that the heavy hand of the banking community provided some of the discipline that the companies themselves appeared to lack. More muted output levels replaced the comparatively vigorous shipments of the last quarter of 1885 in the following year, and some of the rapid oscillation characteristic of planned curtailment is readily apparent. Following the introduction of the Interstate Commerce Commission Act in 1887, the formal cartel agreement was not renewed, but restriction, coupled with improving business conditions, did much to promote price recovery back towards the $4 mark. This was further aided by the strikes in the Lehigh and Schuylkill Regions between September 1887 and the early part of 1888. These resulted in unusually large and sustained shipments by the Northern Field companies through the winter of 1887–8, when tonnage would normally have fallen off. Business was again brisk in the favorable business conditions of late 1888. In these two years, the Lackawanna enjoyed larger shipments than in the preceding or following years, largely at the expense of the Lehigh Valley, whose coal business was the most severely affected by industrial action. The more stagnant conditions of 1889 came as something of a jolt after two years of comparative out-performance, and the graph shows unmistakable evidence of managed restriction of output during the year. The first half suggests only ⅓ shipments

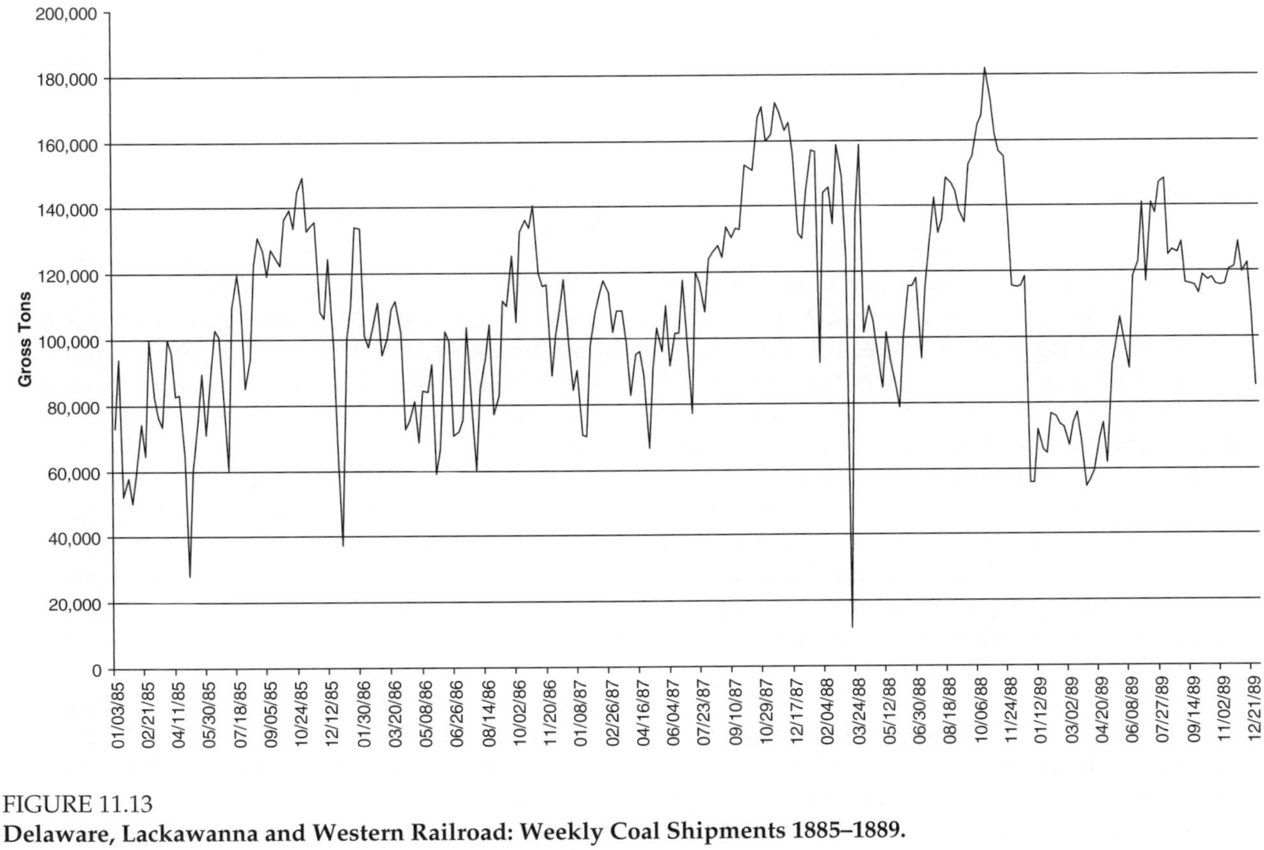

FIGURE 11.13
Delaware, Lackawanna and Western Railroad: Weekly Coal Shipments 1885–1889.

were being made, a pattern that was changed to roughly ⅔ time later in the year, based on an assumed maximum output of approximately 180,000 tons per week. The only caveat to this is that it is harder to be certain that the shipment figures for 1889 are reliable, than for any of the previous years. Virtue states that the companies withheld their normal reports from the public in 1889–1890. This seems to be confirmed by Storrs, who disclaimed any knowledge of where the Miner's Journal obtained its 1889 shipment figures from, implying they had not been provided by him, as heretofore.[134] Since there was little in the way of significant change in the overall proportions of tonnage shipped by the different concerns, including the Pennsylvania (which had also been represented at the Morgan meeting), at this time, it is reasonable to conclude that the mutual understanding was still in force, despite protestations that the recommendations from the regular meetings of sales agents were in no way binding.[135] Disturbingly, however, from the viewpoint of capacity planning, the Lackawanna had managed to achieve its restricted output of 5.23 million tons in 1889, while spending 173 days idle. Storrs also expressed surprise that President Corbin had complained in the Reading Annual Report of having to spend a mere 35 days idle, in the course of producing 7.13 million tons.[136] This latter point raises sufficient questions about the organization of production by the Reading as to merit further investigation in a later chapter.

Further confirmation of the existence of an understanding between the companies can be found in the trade press in early 1892. A brief tabulation simply lists the 1891 percentage quotas by company and compares these to actual shipment percentages, without giving any indication that this was either a new or an unusual development.[137] The quotas given are very similar to those published for 1886, with the exception of a modest increase in the Erie quota. This suggests that the earlier arrangements were simply continued informally, although there were no written agreements and virtually no reference to them in company correspondence, to evade the scrutiny of the Interstate Commerce Commission.

While the outward appearance of the coal trade, as represented by the various tonnage quotas, registered little change during this period, a series of shifting alliances and antagonisms between combination members, prominent capitalists, and bankers characterized early attempts to increase the degree of control over the industry. These precursors to the later 'community of interest', at the end of the century, tended to be partial, fragile, and subject to legal challenge, or collapse for financial reasons when business conditions became adverse. While some of these have been mentioned previously, they are listed again here for clarity. Early examples were the first lease of the Jersey Central by the Reading, which was strongly opposed by the Pennsylvania; and the alliance between the Reading and Vanderbilt on the South Penn Railroad project, specifically designed to antagonise the Pennsylvania, which responded by building a line down the Schuylkill Valley to parallel the Reading. While the Pennsylvania and the Lehigh Valley had a longstanding, close working relationship, there was no love lost between the Lehigh Valley and the Jersey Central, as a result of a disagreement over tidewater property in New York Harbor.[138] A syndicate headed by J.P. Morgan had acquired control of the Reading following its second financial collapse in 1884. Austin Corbin, the president they installed, then joined forces with G.F. Baker of the First National Bank of New York to acquire personal control of the New Jersey Central. When Corbin was ousted by pro-Gowen interests and replaced by A.A. MacLeod, he moved quickly, at the beginning of the 1890s, first to execute a formal lease with the Jersey Central, which Corbin had deemed unnecessary, and subsequently to do the same with the Lehigh Valley. Additional control over New England railroads, such as the Boston and Maine, was also secured. Unfortunately, this substantial corporate edifice to control the coal trade collapsed almost as quickly as it was constructed. The Jersey Central lease was declared illegal by the New Jersey courts in 1892, and the Lehigh Valley lease was abrogated the following year, after the Reading entered its third receivership.[139] Further to this, in 1893, there were reports of Vanderbilt acquiring substantial shareholdings in both the Lackawanna and the Delaware and Hudson. This was part of his policy of increasing his stranglehold over railroads in New York, especially those such as the Lackawanna, which were competitors for his trunk-line business.[140] Similar moves to acquire Lackawanna stock were made by President Maxwell of the Jersey Central and interests connected to him. This resulted in him and a second Jersey Central director joining the Lackawanna board, while President Sloan, in turn, became a Jersey Central director.[141]

MacLeod's ill-fated combine, of the Reading, Lehigh Valley, and Jersey Central, was a particular shock for the Pennsylvania Railroad, as the Lehigh Valley interests had previously fallen under its influence, but had now transferred allegiance to the Reading camp. As a contemporary trade commentator observed:

> 'The Pennsylvania is now *almost entirely,* if not so, shut out from the Lehigh region, her policy having formerly been to work in unison with the Lehigh Valley, and this time the latter has cleverly outwitted her. Her interests clearly do not seem to lie in a combination with the others now'.[142]

Ever combative, the Pennsylvania responded immediately by surveying routes between Wilkes-Barre and Scranton, with a view to competing with the new combination for a share of the tonnage of individual operators on the western side of the Wyoming Valley between Nanticoke and Coxton. Shortly afterwards, it secured control of the coal output of Van Wickle and Co., a large Lehigh operator at Milnesville. Further acquisitions were planned, including Myers and Co. at Yorktown, Audenreid, and Beaver Brook. Officials of the Reading Railroad claimed to be unperturbed by these aggressive actions, although similar moves to take control of the New Boston and Lawrence Collieries in the Schuylkill Region, and later the William Penn, may have rattled them slightly more, since they responded to the first of these by acquiring the Thomas and Coleraine Collieries themselves.[143]

These brief examples are sufficient to indicate that power struggles within the combination continued to be fought out at intervals on several fronts, including on Wall Street, in relation to trunk-line traffic, and to retain or enhance the coal tonnage tributary to their lines. The latter certainly involved new shipment contracts, but could also require construction of new railroad tracks to reach into the territory of competitors.

THE COMBINATION OF 1896

The events leading to the 1896 combination have received minimal attention in previous studies, yet they highlight very clearly many of the key problems in the management of the anthracite industry and the persistent difficulties in finding appropriate solutions. They also indicate the continued importance of the actions of the Reading in determining the degree of harmony in the trade.

The starting point must be taken as the establishment of the Reading Combine with the Lehigh Valley and the Jersey Central, which brought a majority of the anthracite industry under one management in early 1892.[144] The officers of the Reading were apparently taken aback by the level of public indignation against their actions, and state legislatures lost no time in placing the new organization under scrutiny. The New York Legislature, in particular, quickly established an investigating committee, which began its examination of the railroad presidents in March of that year. MacLeod of the Reading claimed before the committee that he did not have the power to arbitrarily raise prices without the agreement of other roads, and that the purpose of the new organization was not to raise prices, but to equalize them in different locations.[145] Such pronouncements, aimed at allaying public fears over possible price rises, were difficult to countenance, based on past experience, and scepticism was quickly justified. Substantial price rises followed in the second half of 1892, as the remaining companies found it in their interest to go along with the Reading (Figure 11.14). Indeed E.R. Holden, vice-president of the Lackawanna, was forced to admit to the investigating committee that the 80 cents price rise had been 'made without reference to the cost of production and carriage, but solely with a view to increase stockholders' profits'.[146]

The success of the new style combination was short-lived, however. It had withstood the annulment of the New Jersey lease, but the third Reading receivership, the consequent abrogation of the Lehigh Valley lease and the sharp collapse in business confidence as a result of the 1893 panic, sent prices on a downward trend that continued until August 1895. In 1891, the Reading had shipped 21.26 percent of all anthracite, reversing a weakening of its position at the end of the 1880s. It had achieved this by means of a program of substantial investment in its coal property, following the financial re-organization of 1887. However, in 1892 and 1893, mindful of its financial guarantees in relation to the Jersey Central and Lehigh Valley, it claimed to have preferentially moved traffic over these lines, to ensure as far as possible that they earned enough to pay their obligations. This reduced its own shipment percentages to 19.55 and 19.35 percent in the two following years, a strategy that quickly backfired in relation to other members of the combination.[147]

The problems began in early 1894, when the Anthracite Coal Operators' Association (ACOA) made proposals for the future conduct of the trade. This was a recently formed body of individual operators, such as William Connell, Clarence Simpson, and Charles Parrish, drawn largely, but not wholly, from the Northern Field.[148] The timing of the formation of this association was no doubt related to the disturbance in the trade caused by the rise and fall of the Reading Combine and the need to get shipping arrangements back onto some kind of normal footing. In addition, however, according to the trade press, actions of the railroads over the previous five years had left the individual operators no margin for existence, as the former had been mining 'for tonnage for their lines and not for profit'.[149]

The ACOA adopted the customary stance that, in common with every other large industry at the time, the mining industry suffered from an excess of productive capacity over demand. They argued that, although the 1893 anthracite output of over 43 million tons was the largest ever, it could have been increased by a further 50 percent, if all collieries had worked full-time throughout the year. Hence, as in earlier years, a plan was needed to work the collieries so output could be matched to demand. Their plan was deliberately couched in the form of suggestions, as they were adept enough to have sought legal opinion beforehand as to exactly what they could propose without themselves falling foul of anti-trust legislation.[150]

The key part of the plan was that it was based on determination of the capacity of collieries tributary to each railroad, to give a percentage of the overall capacity attributable to each interest. The monthly sales agents meetings would then estimate the likely demand for the following month and calculate what proportion of the maximum or capacity output that represented. The actual output of each mine and each railroad interest would then be scaled back from the relevant individual capacities, based on this proportion. Hence, if the total monthly capacity of the industry was 5 million tons and the market was estimated to require only 3 million, or 60 percent, then all would produce and ship to this percentage of their rated capacities.[151]

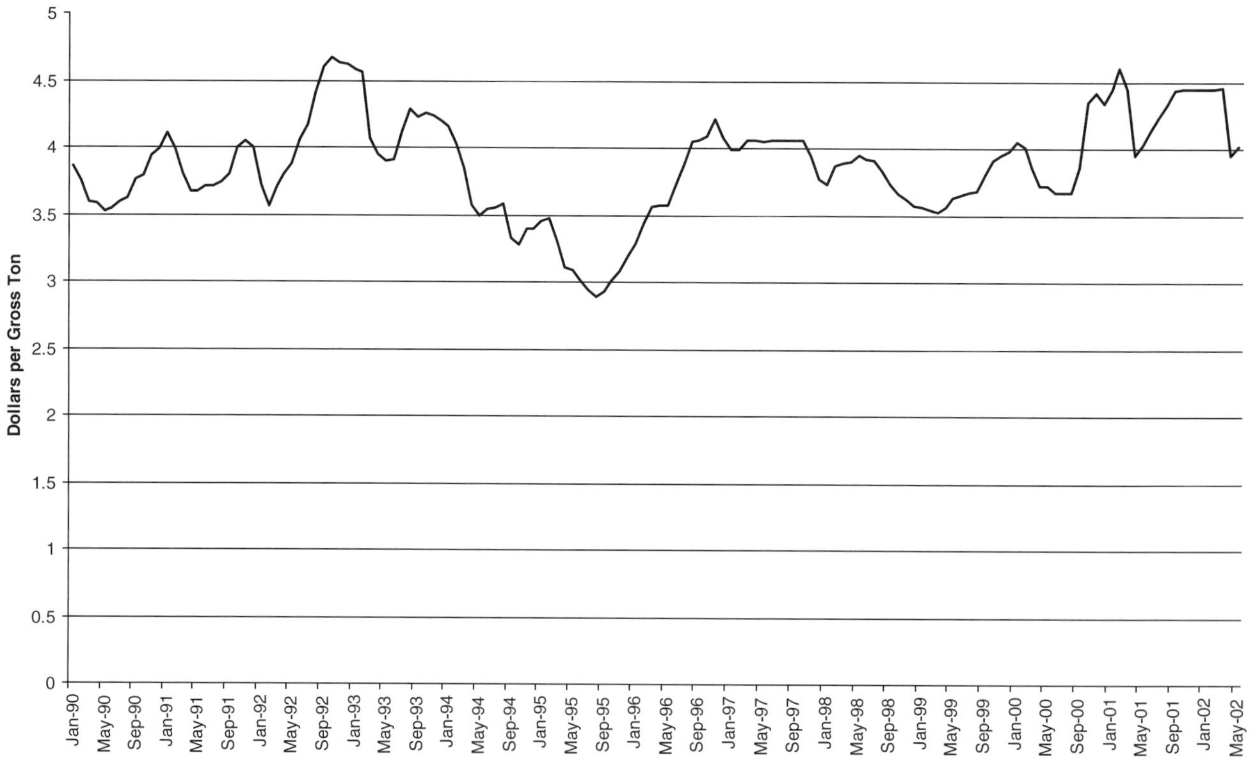

FIGURE 11.14
Average Wholesale Price of Stove Coal F.O.B. New York Harbor 1890–1902.

Although superficially attractive, because of its simplicity and apparent logic, the proposal initiated an argument between the Reading and other members of the combination that was to last the better part of two years before an approximation to harmony in the trade could be restored. In the process, several very important issues, about what constituted mining capacity and how it should be measured, were raised and debated almost *ad nauseam*, especially by J.S. Harris. He was a manager of great experience in the anthracite trade, but was also someone vitally interested in the outcome, on account of his recent appointment as one of the Reading receivers.

The difficulties began as soon as a representative of the ACOA visited the Reading to put to the receivers their proposal for the division of tonnage for 1894, based on the approach they had adopted. Unfortunately, their calculations indicated that the Reading was only entitled to 18.23 percent of the total production. This figure was less than its worst ever showing in 1888, following the labor troubles of that year, and much less than its percentage in 1891. Such a tactless opening gambit was hardly likely to endear itself to the receivers, charged with maximizing returns, not minimizing output! Further to this, however, evidence from the company's records, some of which is summarized in the 1895 Annual Report, spurred the receivers and the company's sales agent, Mr. Henderson, into action. His first task was to refute the proposals of the ACOA and his second was to develop a revised division of tonnage, which would naturally place the Reading in a more favorable light.

The first set of issues to address related to the means of calculation of the capacity of each colliery, since the capacity of each railroad interest was simply a summation of the relevant subset of colliery capacities. The ACOA had used the published Mine Inspectors' Reports for 1893 to calculate the capacity of each mine, by dividing its output by the number of breaker days worked. A breaker day was calculated by dividing breaker hours by 10, hence it was full days, not the number of calendar days on which some work was done, however little. This was a key concern for the Reading, because it claimed that Schuylkill coal came from the mine with a much greater percentage of impurities than coal in the Wyoming Region. Thus, in the latter area, the breakers often only had to work ¾ of a day to clear the output of a full day's mining, so conversion of these breaker hours to 10-hour days would overstate the feasible working capacity of these collieries, in comparison to those in the Schuylkill Region.[152] There may have been other more minor issues about whether breaker hours always represented full operation at maximum output and how idle mines were dealt with, but there is no specific record of these. However, a later calculation, by a committee chaired by Henderson of the Reading, but including other interests, reported the Reading's daily output capacity as 51,000 tons or about 21 percent of the total. This suggested that either the ACOA's calculations had not included all the relevant factors or that a correction, for the differential meaning of breaker days in different parts of the coalfields, resulted in a percentage more satisfactory to the view of the receivers.[153]

The second major complaint by the Reading about the ACOA's proposals was a more general one. Namely, that it was unfair to focus attention only on colliery capacity (although such an approach was perhaps to be expected from individual operators, who were miners but not shippers of coal). It was pointed out that investment in colliery improvements was less than one-tenth of that in coal lands and railroads, and the latter also needed to be taken into account. This was, of course, a matter of particular concern to the Reading, which controlled ⅓ of the entire coalfields but was being asked to accept a miserly allotment of only 18 percent of shipments. In the light of this, it was felt that the receivers' insistence on an allotment of 21 percent was more than reasonable.[154]

Having failed by tactless arithmetic to convince the Reading of its plan, the ACOA resorted to vague threats of large falls in the price of coal if the former did not accede to their plan or something similar to it. This gambit failed, since prices were already well below 1892 highs and the receivers stood their ground, despite further meetings and discussions on the topic. Later in the year, after negotiations had stalled, the Reading was asked to propose an alternative plan, given that the colliery capacity approach was unacceptable to them. This they did, calling their own method the plan of 'general capacity'.[155] True to their earlier arguments, they claimed this was based upon 'the joint capacity of the collieries to produce, the transportation lines to carry, and the sales departments to sell'.[156] An examination of monthly shipment data by company for previous years, suggested to them that 1891 had been free from disturbing elements in the trade, but these had been present in 1892–1893 and were still working their way out of the system in 1894. Hence, none of the years 1892–1894 by themselves could be used. Instead the proposal was to take the data for all of 1891–3, in addition to the 10 months of 1894 that had elapsed, and identify for each interest the three months of largest output, using the average of these as the definition of current capacity.[157] Having performed the relevant calculations, the resulting percentages of overall capacity attributable to each interest were compared with the figures derived by the Henderson Committee and revised by the sales agents, based on quite different calculations. The results are shown in Table 11.3.

The proposed plan and tabulations were distributed to the other interests, but elicited no response, doubtless because they supported the Reading's original contention that it was entitled to 21 percent of shipments. This was a point that the others were unwilling to concede, because its actual percentage in the two previous years had been well below these figures, and they had always worked on the basis of immediate past performance, not averaging of behavior over longer periods.

Trade was continuing to weaken as the depression of the Nineties took hold, so it was not felt that the matter could be left, despite the apparent stalemate. The railroad presidents met in December 1894 to appoint a further sub-committee, with representatives of all eleven interests, to investigate again the division of tonnage based on colliery capacity, but it reported that such a division was impractical. Yet another committee, set up with a looser remit, subsequently also failed to reach agreement and arbitration was then suggested. Unfortunately, the Reading receivers, as court officers, declared they were not empowered to hand over decisions about the conduct of their business to third parties, however impartial, and this avenue was also abandoned.[159] The final meeting of the presidents at this time was held in April 1895 and the Reading presented yet another paper, whose calculations, being slightly more favorable to its interest than on previous occasions, did not meet with acceptance. This led the company to bemoan the lack of consideration given to all its proposals over the period 'although no criticism of their justice or fairness was ever made'.[160]

Since all attempts at finding a new method of determining quotas had failed, the companies reverted to the old standby of the 3-day week for May 1895. True to its calculations, the Reading achieved 21.37 percent of the total output during the curtailment, which doubtless irritated the other interests. It also took to the June sales agents meeting both a proposed schedule of monthly tonnages and, rather riskily from a legal standpoint, a set of suggested monthly prices for stove coal in New York, rising from $3.50 in July to $4.00 by December.[161] Regrettably, no agreement was reached. Actual prices slipped briefly below $3.00, as Figure 11.14 indicates, before recovering slightly in the face of the brief and unexpected improvement in business conditions in the latter part of the year. Again, as if to prove its points, the Reading had achieved a market share of 21.05 percent on the whole business of 1895, when competitive conditions prevailed in all but the month

TABLE 11.3
Alternative Calculations of Possible Percentage Allotments for 1895.[158]

Railroad Interest	Henderson Committee Calculation (%)	'General Capacity' Calculation (%)
Philadelphia and Reading	20.98	21.36
Lehigh Valley	17.08	16.72
New Jersey Central	11.77	11.97
Lackawanna	12.64	13.22
Delaware and Hudson	9.61	9.29
Pennsylvania	8.30	9.77
Pennsylvania Coal Co.	4.29	4.44
Erie	3.95	3.65
New York, Ontario and Western	2.86	2.97
Delaware, Susquehanna and Schuylkill	4.93	3.79
New York, Susquehanna and Western	3.59	2.82
	100.0%	100.0%

of May. Indeed, in the final 6 months of the year, it achieved in excess of 22 percent, standing it in good stead for any future negotiations and laying to rest the adverse statistics of 1892–1893.[162]

In many ways, it was a pity that the Reading's proposals based on 'general capacity' did not meet with acceptance during 1895. Harris had worked hard to address the problems of the trade, arguing that any newly established quotas should run for 5 years to stabilize the industry. He also provided a mechanism for gradual shifts in output to accommodate evolution in the business position of the different companies. These were to be achieved by allowing companies who set a new production record in a specific month, when demand was large, to nominate that month's data for inclusion in the calculation of the average that was used to specify their general capacity for the following year's shipments.[163]

Continued price weakness at the beginning of 1896 brought the presidents back to the negotiating table, as so often before, and they were in no mood for prevarication. While reports that a letter of 'advice' had been received from J.P. Morgan were denied, a Reading internal memorandum noted that a 'friend' had strongly urged them to accept a quota of 20.5 percent, which they did for the 'sake of harmony'.[164] This information was taken to the meeting of presidents on January 30, 1896. After the necessary posturing and complaint about how the figure of 20.5 percent had been arrived at, they accepted the percentage so as 'not to continue the warfare longer'.[165] Unfortunately, after the Reading had finally broken the deadlock, the Delaware, Susquehanna and Schuylkill Railroad, a relatively new combination interest controlled by Coxe Brothers (a leading Lehigh independent operator), refused to accept the 3.5 percent allotted to it. All attempts to persuade them otherwise failed. In desperation, Roberts of the Pennsylvania proposed the unusual device of excusing them from voting on the allotments. Harris immediately protested that if Mr. Walter of the Delaware, Susquehanna and Schuylkill were allowed to duck out of his allotment this would have an immediate adverse impact on his own 20.5 percent quota, a state of affairs he 'absolutely refused to accept'. Finally, in a triumph of diplomacy over the constraints of percentage arithmetic, Roberts moved 'that Mr. Walter be excused from voting and that [the] Reading should be entitled to take 20.5 percent of whatever tonnage should be mined, and this motion prevailed'.[166]

The agreed quotas (see Table 11.1) took effect immediately for the remaining 11 months of 1896, and prices climbed steadily back to regain the ground lost over the two preceding years of strife, despite the large tonnage actually shipped during the year. This suggests that the degree of perceived harmony among producers and shippers was not as important a control on price behavior as actual supply and demand conditions. Although the agreement, or more properly, suggested allotments, were not formally restated at the beginning of 1897, the sales agents maintained a measure of discipline, by means of curtailments in the first part of the year. Many of these were quite severe, as the mines worked only 1.5–3 days a week.[167] Nevertheless, some interests began to take advantage of the situation, and an agreement was reached in July that those who had over-shipped their quotas in the first part of the year should 'refrain from pressing their shipments', while allowing the laggards to make up their deficiency over the following two months.[168] This adjustment proved difficult to organize when the market was somewhat depressed, and the idea of restriction was not adopted very enthusiastically, since in excess of 8 million tons were shipped in July and August, when it had originally been felt that the market would only take 7 million. The net result was a familiar story, namely, that all the interests, with the exception of the Reading, had over-shipped their quotas, the worst offenders being the Delaware, Susquehanna and Schuylkill, the Pennsylvania Coal Company and another of the relative newcomers, the New York, Ontario and Western. Harris called upon the Erie to assist him in preventing the trade from becoming demoralized again by lack of discipline and over-shipment.[169] However, by the end of the year, he was not very sanguine of success. In an internal memorandum he provided what, for the Reading at least, if not for all the interests, was a suitable epitaph for the previous 25 years of the old-style combination experience:

> 'It looks to me as if the anthracite business is in danger of working as the Joint Traffic Association business does; that those who stick to their agreement, get left behind in the race'.[170]

DEVELOPMENT OF A COMMUNITY OF INTEREST AFTER 1898

Although Morgan and Vanderbilt had been involved in the operation of the anthracite combination for a number of years, as the previous discussion has indicated, they had taken very much a background role, leaving negotiations as far as possible to the railroad presidents themselves. However, the continuing lack of harmony that made agreements difficult to obtain, the inability to enforce agreements even if they were made, and the increasing scrutiny of investigators under both state and federal anti-trust legislation made the search for a more profitable means of doing business increasingly imperative. It should also be mentioned that these issues were not confined to the anthracite coal trade. Evidence from the Reading Railroad indicates that similar quota arrangements were in place in the late 1890s for the bituminous coal tidewater shippers. Unsurprisingly, similar problems were also found in persuading individual railroads to stick to their allotments. Conversely to the anthracite case, the Pennsylvania tended to under-ship its quota, while the main offenders in terms of over-shipping were the Baltimore and Ohio and Chesapeake and Ohio.[171]

The solution to these problems was to develop a "community of interest" between the various railroads, based on extensions of methods that had already begun to be used earlier in the 1890s. These were inter-ownership of stock and interlocking directorates. Since, as noted earlier, the classic 1914 study by Jones deals with the details of these arrangements very fully, it is only necessary to highlight a small number of examples of the actions taken, to indicate how the community of interest developed.[172] It should also be pointed out that the process continued for several years beyond the period relevant to the present study. One important early step taken in the formation of the community of interest was the co-ordinated purchase by several roads of nearly 30 percent of the stock of the Lehigh Valley Railroad. The latter was perceived at the time (1901) to be financially weak, but receivership was regarded as an undesirable outcome for the trade in general. The Lake Shore and Michigan Southern, controlled by the New York Central, was financially best placed to take a large share of the stock, but the Reading, Jersey Central, Lackawanna, and Erie all acquired a significant holding. Inter-ownership of stock went hand in hand with interlocking directorates, and Mr. Thomas, who had been Erie president, took over at the head of the Lehigh Valley, and other notables from the Reading and Jersey Central, such as G.F. Baer and J.R. Maxwell, also joined the board of directors.

During the same period, the Pennsylvania secured large amounts of stock in the Baltimore and Ohio, the Chesapeake and Ohio, and the Norfolk and Western, all of which were part of the bituminous tidewater combination. In 1902, the Pennsylvania also bought over 43 percent of the stock of the Reading Company, before selling half of it to Vanderbilt interests. Although inter-ownership of stock declined somewhat in the years following 1902, the number of inter-locking directorates continued to increase.

At this time also, through the corporate device of the Temple Iron Company, the community of interest sought the progressive take-over of the remaining individual anthracite mine operators who were large enough to merit a voice in the conduct of the trade, to give it control of almost all of the productive capacity of the anthracite fields.[173] This topic will be explored in more detail when the relationships between the railroads and the individual or independent operators are examined in a later chapter.

CONCLUSIONS

In whole or in part, the experience of the anthracite industry from the Civil War era onwards, was closely bound up with attempts to avoid competition, which was always described as ruinous rather than healthy. The diagnosis of the industry's ills centred on either overproduction or its close cousins overcapacity/overinvestment, but strangely this claimed overcapacity never seemed to diminish, an observation curiously not made by any previous commentator on the industry. Other factors, such as the inflexibility of basic wage agreements with fixed minima, regardless of sale prices of coal, have also perhaps received less attention than they deserve. There is also no doubt that, prior to the 1880s the idea of retaining financial surpluses from good years as a buffer against future adverse business-cycle changes was not well developed in the railroad or mining industries. Furthermore, despite Juglar's early researches at the beginning of the 1860s, the notion that business conditions might have reasonably predictable cycles of different periodicities does not seem to have occurred to the managers of the time. Most actions were based on experience of the previous year with very limited attempts at future forecasting, an approach which combination members insisted on until the end of the century, despite the assiduous collection of shipment statistics by their own tonnage accountants over many years.

Although the detailed management of combination agreements operated over rather short timescales, other aspect of the anthracite economy required much more strategic, long-term thinking. Control over avenues to market and the development of the anthracite distribution region, which have already been examined, fell into this category. However, development of the coal estate of individual companies also required the same kind of approach at a more detailed geographical scale. Some of the constraints and time–dependencies involved interacted or conflicted with the short-term dictates of car allotments, tonnage quotas, and the like, resulting from combination activity. Effective management of the coal business across these different temporal and geographical scales simultaneously required skill of the highest order. On it depended the long-term financial health of the different companies, although very detailed analysis is required to identify and trace all the aspects involved.

Therefore, attention will now be turned to the detailed operation and management of the mining economy within the coalfields. It is necessary to examine the strategies and tactics employed in business decision-making, and to evaluate the importance of constraints on decision-making, derived from the timescales involved in mine construction and production processes, and from legal aspects of land ownership and leasing. In the course of this analysis, running over several chapters, the major factors shaping the evolving geography of the coal regions over the period will be identified and explained.

Chapter 12

DEVELOPING AND MANAGING THE COAL ESTATE

INTRODUCTION

Previous chapters have examined in detail the location and timing of investment in railroad and terminal infrastructure, which provided access to new geographical markets and led to the progressive articulation of the anthracite distribution region. Successive cartel agreements on prices and output quotas provided extended, if uneasy, truces against the pressures of free competition, and constrained the environment for both tactical and strategic decision-making by individual railroads and mining companies. However, in the final analysis, both transportation facilities and meta-organizational structures, such as the cartels, depended very heavily on long-term availability of coal resources, developed to provide a growing and reliable tonnage for shipment to market. A coal estate of ample proportions was, therefore, a necessary, though not sufficient condition, for maintenance (or, preferably, enhancement) of market share. Although, as the Reading discovered, acreage by itself was not enough to guarantee corresponding market share in the cartel, unless it was accompanied by a demonstrated ability to generate large-scale shipments from mines on its property.

The initial competitive acquisition of coal lands after the Civil War, and its impact on company finances, has already been examined, but ownership was one thing, development and management of the lands quite another. It is also important to realize that this first phase of securing control over coal property was only the first of several cycles of this type of investment activity. This was especially true in the Northern Field, although not all companies participated equally in each of the cycles.

No detailed examination of these issues has been undertaken in previous studies, with the limited exception of Jones's analysis of the contractual relationships between individual operators and the railroads at the end of the 19th century.[1] In consequence, two key types of processes involved in the development of the Coal Regions have not previously been understood. The first of these is the complex web of geographical constraints and legal contractual obligations surrounding the management and control of coal resources. The second is the relationship between investment in coal lands and business-cycle changes. Without a clear grasp of these issues, it is impossible to lay the foundation for a coherent explanation of the observed patterns of regional development and economic restructuring within the industry over the period of interest. Questions relating specifically to land resources will be examined in this chapter, while economic and geographical aspects of resource development for mining purposes will be addressed thereafter.

The discussion in the present chapter will be organized under several main headings. These are as follows:

- selection and valuation of coal lands for purchase or lease,
- contractual aspects of the coal leasing system,
- types of purchase and transportation contracts pre- and post-1880,
- the management of large coal property portfolios, and
- the spatio-temporal dynamics of land acquisition and leasing cycles.

Both for purposes of strategic planning by individual companies and for effective cartel participation, it was essential that the control of coal properties be as tight and risk-free as possible. However, this had to be tempered with a good measure of business pragmatism, since it was better to secure additions to coal tonnage, even for limited periods, than it was to have them fall into the hands of competitors. Also, while ownership of land might be the most desirable aim, in a number of cases landowners would not sell, they would only lease the mineral rights. Hence, the first three topics to be considered, land ownership, leasing, and the binding of the output of mines to specific railroads by contract for a period of years, represented a pattern of increasing risk in the long term. Leases could expire or be abrogated under certain limited circumstances. Tonnage or shipment contracts were often for a restricted period and might be terminated by bankruptcy or receivership, depending on the terms of the agreement. Large tonnage contracts involved purchase of the coal at the breaker by the railroad, while ad hoc transportation arrangements, whereby the railroad undertook to take coal on an agreed basis, might operate only on a 12-month renewable basis. Thus, W.R. Storrs was able to summarise his overall approach to the management of a coal estate very succinctly, as follows:

> 'Buying fee is the best way to control tonnage, then lease, then purchase of coal, then transportation'.[2]

A further legal point to note, in relation to this classification of degrees of control over territory, is that lands owned and leased counted against a company's authorized limit on landholding under its charter. In contrast, the lands owned or leased to individual operators, bound to the railroads by tonnage contracts, did not count against this limit for the railroads in question. This added a further dimension to the problem of managing large property portfolios, which is why this topic merits investigation in its own right. Although there were risks of sudden changes in the sources of coal supply, should short-term contracts not be renewed, it made business sense to encourage individual operators to make large shipments from their lands under the tonnage contracts, thereby exhausting their non-renewable resources, rather than those in the coal lands owned by the railroads themselves. Once again, Storrs had a suitable quote that expressed this intent very clearly:

> 'To the extent that we can make or get present tonnage from the operators, our own is held in reserve for the future and should appreciate in value'.[3]

This argument applied very clearly to coal lands held in fee, but for reasons related to the royalty payment mechanism that will be explained fully below, it did not apply with the same force to lands leased by the railroads.

From these introductory remarks, it is already apparent that railroad presidents and general coal agents responsible for the coal departments or coal subsidiary companies of the railroads faced a management task of some magnitude in relation to the coal estates in their charge. In the subsequent sections, an attempt will be made to indicate the full range of territorial, legal, and organizational issues faced by the senior officers of the companies during the period. A number of examples will be given of the practices that came to be used in the process of managing the often conflicting economic and business pressures that arose.

SELECTION AND VALUATION OF COAL LANDS FOR PURCHASE OR LEASE

Judging by surviving railroad and coal company correspondence, there were few issues that required such care and attention on a continuing basis as the selection and acquisition of coal lands by purchase or lease. Although the activity had a strong cyclical component, as will become apparent below, properties both large and small, varying considerably in quality and the degree of development of mining operations, were being offered to the railroads on a frequent basis. A wide variety of factors needed to be taken into account to determine whether any individual tract or tracts merited further consideration and detailed valuation. Several of these have already been highlighted in the course of examples of coal land acquisition given in an earlier chapter, but it is appropriate to examine these systematically here.

The first and most important question was whether the company wishing to purchase had spare capacity to hold additional coal lands under its charter, or could divest itself of low value timber lands to make room for high value coal tracts. This latter became an important issue for the Lackawanna during the mid 1880s, when, for example, it had to transfer 1,500 acres of non-coal-bearing land to its subsidiary, the Glen Alden Coal Company, for this reason.[4] Assuming charter restrictions were not a problem, the next set of questions was geographical and geological in nature. If the property was on the line of the company's existing railroad tracks or adjacent to property it currently owned or leased, it would immediately merit close attention, since development would probably not require significant outlay on new lateral lines. In the case of adjacent property, gangways could eventually be extended into it from current workings. If all the land involved was either held in fee, or was intended to be purchased rather than leased, there would also be no need to retain a large barrier pillar at the edge of each of the adjacent properties, thereby increasing the recovery rate of the coal. Tracts that were distant from the company's 'territory' presented more of a problem. New lateral construction might be expensive, or indeed infeasible, owing to distance. Careful attention to possible routes would also be necessary. If the lateral line would have to cross a competing company's lands or worse, its tracks, this could introduce considerable delay and difficulty into the lateral construction process. An alternative might be to make a trackage agreement with a rival company whose line was closer, but this might add 5 or 10 cents to the cost of every ton shipped. Indeed, the landowner wishing to sell might have approached both the original company and its rival, with the intent of bidding up the eventual sale price. Very often such issues were sufficient to result in the offer to purchase being declined, unless there appeared to be future potential for using the tract as a bargaining counter, in the process of negotiating an exchange of lands with another company, so each could consolidate their holdings to a greater extent. Such exchanges became quite important in the 1880s and 1890s. They involved companies such as the Susquehanna Coal Company, the Lackawanna, and the Lehigh and Wilkes-Barre, although they often required extended periods of negotiation before arrangements were successfully concluded.[5]

Assuming the proposed purchase or lease met the geographical tests of suitable location, proximity and accessibility, the next set of considerations was related to likely coal resources and ease of mining. Determination of likely resources was difficult in relatively undeveloped parts of the coalfields as late as the 1860s. This was due to the lack of a full understanding of the geology of all parts of the coal basins and, indeed, the actual extent of some of the smaller basins themselves. This knowledge obviously increased progressively as more shafts and boreholes were sunk over time, but it was rare for extensive proving of property to be possible prior to purchase, as the Pennsylvania Railroad found to its cost near Hazleton, recalling an example from chapter 10. Wherever possible, information from shafts or boreholes on neighboring property would be utilized, and it was accepted practice for mining engineers from different companies to be allowed to inspect any adjacent workings, to assist them with their decision-making.

The most important parameters were the number of seams present on all or part of the property and the estimated thickness and quality of each. Such estimates were, of necessity, rather inexact, since, in the apposite words of the noted mining expert, H.M. Chance, 'coal beds are not usually of the same thickness and quality over large areas'.[6] Seams might be pinched out on portions of the property or run into fault, necessitating expensive rock tunnelling to locate the seam again, but risks such as these were regarded as necessarily incident to the business of mining. Uncertainty would, of course, be greatly reduced if there was, or had previously been, a mine opening on the property, but, equally, the price per acre would then be correspondingly higher. Other aspects of the geology included the presence or absence of outcrops of specific seams, since such outcrops might permit the sinking of slopes, which were a cheaper and quicker method of opening the coal than sinking shafts. In certain locations, such as in parts of the Lehigh Region, the availability of outcrops of seams in shallow basins might also permit strip or surface mining, rather than underground workings, obviating the need for expensive pumping, hoisting and ventilating machinery. A good example of the period was the Hollywood stripping, four miles northwest of Hazleton.[7] The estimated maximum depth of the lowest coal seam was also important, since the deeper the shaft had to be sunk, the greater the initial capital outlay and the longer the time before coal production could begin. If the shaft cut a number of seams, however, it would be possible to open up several of these initially and sink to the lower veins at a later date, although there would be considerable disruption to production while the second stage of shaft sinking was in progress. This two-stage sinking process was used by the Lackawanna at the Sloan and Central shafts in Scranton, for example.[8]

Aside from the actual volume of coal believed to be on the property, other geological characteristics might affect the cost or even the feasibility of mining improvements in certain locations. The most important of these was the nature of the drift geology, i.e., surface deposits above the bedrock. Alluvial deposits of this kind covered the old glacial rock floor of the Susquehanna Valley to a considerable depth in parts of the Northern Field.[9]

They could be extremely treacherous deposits through which to sink shafts, and some early attempts, such as at one of the openings of the Maltby colliery, east of Kingston, ended in disaster, when the workings ran into quicksand and had to be abandoned.[10]

If a tract passed these basic criteria of suitability for purchase, a valuation of the property was then required. It was to be expected that the owner or lessor would err on the side of overstating the quantity and quality of the coal in the property. Conversely, it would be in the purchaser's interest to adopt very conservative estimates of these parameters.

The most detailed treatment of the issues involved in coal-land valuation during the period of interest can be found in the work of H.M. Chance, to whom reference has already been made, drawing on several decades of experience in both the anthracite and bituminous coalfields of Pennsylvania. According to this authority, a number of factors might need to be taken into account, depending on whether a sale or lease was contemplated and whether there were mining improvements already on the property or not. These included the following:

- quantity of recoverable coal,
- rate of interest,
- rate at which coal was to be mined,
- amortization and depreciation costs,
- royalty per ton (for leases), and
- taxes and miscellaneous expenses.

For anthracite, the total weight of coal per acre-foot of seam extent and thickness was generally taken to be about 1775 tons.[11] However, the recoverable coal was necessarily significantly less than this, because of the need to leave pillars in the mines to support the roof and the weight of super-incumbent strata.[12] Account also had to be taken of losses attendant on the mining and breaking of coal. The resultant rule of thumb, used by the Lackawanna at least, was that only ⅔ of the coal in the ground would eventually find its way to market.[13]

While the amount of coal in the ground was given by geological conditions, some of the other parameters involved in valuation were potentially more variable over time, depending on the assumptions made about the future conduct of business. The least contentious of these was the interest rate. Although it was subject to fluctuation as a result of macro-economic conditions at the international scale, in practice it was usually taken to be 6 percent in valuation calculations. In conjunction with the intended rate of mining, it was a crucial factor in determining the net present value of a property. The difficulty was that the price had to be paid now, but the planned revenue stream from profits per ton on the coal would be received over a period of years, even if production began immediately on the property. Hence, the value of any future year's increment of profit would have to be discounted to the present, by means of the interest rate, compounded over the number of years between the present and the time when the profit increment was realized. The best way of performing calculations of this degree of subtlety was still being debated by the American Institute of Mining Engineers on the eve of the Great War, as the discussion of Chance's paper testifies.[14] However, even if the calculations were performed correctly from the mathematical viewpoint, assumptions still had to be made about how quickly development of the property for mining would take place, and what rates of increase in output might be achieved over time periods possibly stretching into decades. Any such assumptions would be subject to wide error margins and, as will be seen below, business-cycle conditions could lead to unexpected changes in the timing of development of specific coal properties. Serious delays would significantly reduce the accounting valuation of properties, but the purchases had, of course, been made before these unanticipated events took place! This could tie up company assets in undeveloped coal lands that were effectively depreciating on the company's books. This depreciation was due to the opportunity costs of not being able to deploy the capital they represented more productively in other parts of the company's business.

It is appropriate to revisit the question of the coal land purchases of the Philadelphia and Reading Coal and Iron Company in this context. The company bought a mixture of developed and undeveloped lands and it used borrowed money to do it. Where mining was in progress, there was at least a cash flow to offset interest payments. On the undeveloped lands, however, the present value should have been discounted doubly—once, for the time expected before development was to begin and again, because of the annual interest payable on capital that was then sunk unproductively in the ground, at least for the medium term. Instead of a cold, hard accounting calculation, the Reading simply asserted that it was better to pay the interest on land not needed for present purposes than it was to have tonnage diverted to other railroads. As previously noted, in a Canute-like attempt to stand against the tide of arithmetic compounding, for a brief period it also capitalized the interest on funds used to buy coal lands, in support of its strategic view of its assets. This unacceptable accounting practice was fortunately not emulated by its competitors.[15] In this context, it is interesting to note that J.S. Harris, in his 1880 report on the same coal lands, disagreed with the company's earlier assessment and stated that there was no commercial justification for holding lands indefinitely simply to prevent other companies getting hold of them. Instead, he argued that the company should only hold a 100-year supply of coal and should divest itself of unnecessary holdings, or those which involved high interest costs. The latter proposal was precisely in line with the accounting considerations referred to above. The Harris Report is also of particular note, because he utilized both discounting to present values based on interest rate considerations, and also planned development scenarios given him by the company's engineers. This made his work a very early example of what might in contemporary parlance be termed 'best practice' in coal-land valuation.[16]

Some of his conclusions were quite stark and entirely justified on accounting grounds, although they were never entirely accepted by the company. For example, he argued very clearly that:

> 'The low values placed by the rules adopted on some of the properties whose development is distant, emphasise with great force the truth that it is ruinously costly to keep for many years properties against which there is a heavy interest charge constantly accruing, and from which no present revenue can be derived'.[17]

A further set of valuation questions arose if the property already had mining improvements on it. Even if these were newly built and of first class construction, it could not be assumed that they would last for more than 20 years. After this, at a minimum, extensive remodelling, or, more likely, rebuilding would be required. Hence, depreciation costs at 5 percent a year on plant also needed to be factored into the valuation calculations. Chance presents a table showing levels of net earnings per ton of output required to cover the cost of lands and improvements over several time spans, reflecting differences in the number of years to exhaustion of different properties. Moderately priced land, with medium-cost plant, would require a profit of 12–16 cents per ton just to cover the above costs, without regard to additional possible expenditure for fire, flood, or cave in the mines. Higher-priced land, with expensive plant, could require 23–31 cents per ton. This was more than the average profits of the business at the time Chance was writing. In accordance with the earlier argument, he also noted that enterprises funded on borrowed capital required still higher levels of earnings, if they were to be made to pay. A particularly clear feature of his tabulation was also that lands with large reserves, which required a number of decades to mine out the coal, needed a larger profit per ton to make them pay because of the interest rate discounting affect. His conclusions were as stark as those of Harris, since his figures showed

> 'how an operator may be handicapped by the purchase of a large acreage of relatively high-priced land requiring the more expensive types of plant, and how almost impossible it will be for him to meet the competition of those who have bought a smaller acreage, or cheaper land, or land that can be developed by a less costly plant'.[18]

Based on these accounting considerations, without regard for business strategy, it seems there is a strong case for arguing that the lack of charter constraints on land holding by the Coal and Iron Company, which allowed it to amass such a large estate, effectively guaranteed that it would have serious problems generating a positive net return over an extended period. This was a result of the impossibility of bringing the undeveloped portion of its holdings into production quick enough to make it pay. On this basis, the land-holding constraints in the charters of most of the other companies, that kept their holdings down to 20–30,000 acres, seem to have been a blessing in disguise. Had the Reading been constrained in the same way, it would have been able to select better paying lands, leaving the marginal or the undeveloped either to other companies or to individual landowners. Had the Lehigh Valley, for example, continued with its land purchases in the absence of competition from the Reading, it might well have fallen into the same trap of acquiring too large an estate, if funding or charter constraints had not previously curbed its expansionist aims. As it was, in the later period, the Lehigh Valley did endeavour to expand its land holding significantly from the levels of the 1870s.

Some of the adverse repercussions of this strategy will be examined further in a later part of the argument.

Extensive investigation of land-related records for a number of the other companies with charter-limited landholding has failed to reveal coal land utilization planning or valuation in quite as systematic a manner as that attempted by Harris in 1880. There is no doubt that conservative managers paid attention to the effect of interest rates on the determination of present values. With the smaller estates, however, it does not appear that a very rigorous and detailed long-term plan was deemed necessary for the entire property in the way that appears to have been undertaken by the mining engineers of the Coal and Iron.[19]

While over-optimism about coal-land valuation went hand in hand with imprecise accounting calculations, it was also possible to over-estimate the rate of exhaustion of properties or the rate of necessary accounting provision for depletion of the mineral resource, as the coal was mined out. Since every ton mined reduced the size of the remaining reserves on the property, and hence the overall value of the non-renewable resource, a sinking fund was required, to offset the progressive reduction in the book value of the asset over time. This normally took the form of an accounting charge, at a certain number of cents per ton, on all coal mined. However, in the 1850s and early 1860s, when knowledge of the geology was imperfect and only rough estimates of available coal could be made, it was possible for the mineral resources to be underestimated and sinking fund charges to be set too high in consequence. This could also result in the value of coal-land assets being given incorrectly on a company's books. Such a situation arose for the Lackawanna in 1869. It became apparent at that time that the sinking fund charges, which had been set in earlier years within the range 12.5–50 cents a ton for coal mined from the company's lands, were too high, and, therefore, the valuations were wrong. Simple arithmetic demonstrated that the sinking funds had risen to nearly the equivalent of the original book values of various properties, implying that they should be on the verge of exhaustion. However, inspection of the workings showed that this was not the case. The amount of coal in the ground had been under-estimated, which caused the sinking fund payments per ton to be set too high. As a result, these payments were reduced to only 10 cents per ton, which immediately increased the book profit per ton on coal from the lands in question.[20]

These sinking fund payments were often called royalties, even though the company was making provision in its own accounts, for the effects of mining from its own lands, on the value of its assets. The same principle applied when the company had leased the mining rights under lands it did not own in fee. In this case, the royalties paid under the lease amounted to recompense to the landowner for the privilege of removing coal from his or her land and reducing its future asset value. However, unlike an internal accounting charge, where the aim was correct identification of asset depreciation costs, royalties to external landowners were subject to market forces. It was in the interest of these landowners to have the royalties bid as high as possible, by companies competing to offer the best lease terms. A more detailed discussion of lease agreements and royalties will be found in the next section, but as an indication of royalty levels, these were as low as 10–12.5 cents in the 1860s. Later, 25 cents became the norm, and eventually, figures as high as 40–50 cents were paid for choice quality lands in favorable locations.

In the majority of cases, landowners were clear about the course they wished to pursue. Either they were willing to sell the property outright or they wished to retain the use of the surface and only lease the mining rights underground. In some instances, however, they were willing to entertain either possibility. Alternatively, land originally leased some years previously might be offered to the lessee for sale, if the landowners decided that they would now prefer a lump sum to an income stream over an extended period of years.

In either of these latter circumstances, it was necessary to determine the valuation equivalence of a purchase price and a specified level of royalties. Several examples of this kind of situation can be found in the Lackawanna Coal Department Records, and these enable the relevant equivalences to be indicated. For example, a lease with royalty payments of 12.5 cents per ton was deemed to correspond to a purchase at a rate of 2–2.5 cents per ton, of recoverable coal. Hence a 25-cent lease was equal to 5 cents in fee.[21] Determining equivalences, expressed in dollars per acre, was obviously much more difficult, because of variations in the thickness and quality of seams on individual properties. Nevertheless, it could be done on an average basis across all the lands leased by a company. In the early 1880s, the Lackawanna estimated that its leasehold estate was of similar value per acre to its lands held in fee, which would value lands leased for 12.5 cents per ton at $500 an acre and those at 25 cents at $1,000. The company believed its own fee lands to be worth the latter figure of $1000 an acre on average, but following the depressed conditions of the late 1870s, it had been unwilling to pay prices this high to acquire new properties in 1880. That said, Storrs also felt that the company had acted too conservatively, in not taking all the good lands offered to it for $1,000 or less, because once such lands fell into the hands of other corporations, they were unlikely to come on the market again.[22]

An interesting example of the deployment of these kinds of calculations in an actual example can be found in an 1881 case, where the company was offered the opportunity to purchase land it currently mined, under what was called the Fellows lease. The property was in the Scranton area and was offered initially for $200,000. The company estimated it would take 47 years to mine all the remaining coal covered by the lease and it would have to pay $10,000 a year in royalties during this period. Such a growing royalty total, with interest compounded at 6 percent over the requisite number of years, would produce, according to Storrs' calculations, a total figure of $2.4 million. In comparison, a purchase price of $150,000, the maximum figure favoured by Storrs, but well below the asking price of $200,000, would result in a total of nearly $2.27 million, when interest was compounded over 47 years.[23] These figures suggested there was not an overwhelming case for purchase. Conveniently, the surviving correspondence contains both Storrs' assessment of the situation and the responses of the Board of Managers committee that scrutinized the purchase proposal. This committee was likewise not entirely convinced of the business benefits of the proposal, given that the coal was already bound to the company by lease, but nor were they vehemently opposed to it. Clearly, the lower the purchase price, the more the interest account would run in favor of this option, but the sellers could not be shifted by any significant amount below the $150,000. Sloan noted also that the faster the coal was taken out the more it favored the purchase option, but he did not, in the correspondence at least, make the additional point that if the estimate of the coal in the ground proved very conservative, this would also accrue to the benefit of the purchaser. Under the lease, any additional coal that proved to be present would still have had to be paid for on the normal royalty basis, regardless of the original estimates. In the event, the property was finally purchased for $148,500.[24]

The Fellows lease was an established arrangement and the company was entirely familiar with conditions on the property, because it had already mined a substantial amount of coal from it by the time the purchase offer was made. On account of this, a factor that usually figured prominently in deciding whether to lease or purchase did not assume the normal degree of importance. This factor was the distribution of mining risk between the parties to the lease or purchase transaction. In the case of a lease, the risks attendant on mining, such as cave, fire, flood or fault fell on the lessor, whereas these obviously fell on the owner, if a purchase was made.[25] Although less of a concern (except for large properties), the same asymmetry of responsibility applied in relation to local taxes. In the early period, it was understood that these were borne by the lessor and by owners, thereby favoring a leasing approach. However, in later years, this was questioned in the courts and it was ruled that a lease of all the coal was a sale and hence the taxes would have to be paid by the lessees. This led to cases of disagreement, where lessors tried to use the ruling to force payment of taxes by lessees, even on leases made prior to the court ruling, and in which it was specifically stated that the lessors were responsible for the taxes. As Storrs ruefully remarked, 'the lessors want more money and they want it out of the companies'.[26]

In the light of the complexities involved in coal-land valuation and the uncertainties introduced by the necessity of estimating possible long-term income streams, it is appropriate to return to the second quote by Storrs, used in the introduction to this chapter, about holding the company's coal in reserve for the future, where it would appreciate in value. Based on the earlier analysis, such a statement would only be true if the net profit per ton on coal increased at a faster rate than compound interest deflated the present value of an asset to be mined in the future. However, evidence presented in earlier chapters indicates that the cartel was barely able to hold prices at similar levels in the last two decades of the 19th century, never mind increasing them progressively. Therefore, the only ways in which profits could be increased would be by continued reduction in the cost of mining and preparation and/or reduction in the share of the final selling price attributable to transportation. In terms of mining and preparation, incremental improvements in mining technology, such as the use of electric underground haulage, were offset by increased costs due to the extension of workings. Similarly, improvements in breaker technology and better utilization of fine coals were partly offset by greater waste in breaking coal, as the larger sizes went out of fashion in the market place. These changes in mining costs will be examined in detail in the next chapter, but they were not sufficient to change the fundamental pattern of costs in the industry during the period. Rather, it was a shift in the share of the selling price, away from transportation towards the mining economy, which became apparent as railroad competition for shipments intensified. This was to prove a key feature of the development of the tonnage contract system. By 1899, the process had progressed so far, that while formerly the Lackawanna had been unwilling to value lands for ownership at more than 7.5 cents per ton, individual operators were now offering their properties for sale at the considerably higher figure of 20 cents per ton of recoverable coal.[27] The changes incident to the growth of the tonnage contract system will be explored further in a later section, but enough has already been said to indicate the key role of the valuation of coal land resources within the overall mining economy.

CONTRACTUAL ASPECTS OF THE COAL LEASING SYSTEM

STANDARD LEASE PROVISIONS

Leases as a means of facilitating the development of coal lands had been a feature of the industry from the earliest days in the Schuylkill Region. Short-term leases in the region, of 5–10 years in duration, were often blamed for the proliferation of wasteful and inefficient mining operations that maximized immediate returns, to the long-term detriment of the properties in question.[28] Subsequently, leases were utilized in all the coalfields. They became less important in Schuylkill after the consolidation of landholding under the Coal and Iron Company, but they remained of considerable significance in the Wyoming Region throughout the period. In the latter area, leases were in general either long-term, for periods of 20 years, or perpetual, covering all the coal on the property. By 1907, for example, the Lehigh Valley Coal Company had executed nearly one hundred such leases, the majority of them in the Wyoming Region.[29] This company provided an interesting rationale for the prevalence of such leases, especially in the region named. In the first place, they argued that many landowners were unwilling to sell, because their property had been handed down to them by inheritance, and many could trace their roots back to the earliest white settlers in the Wyoming Valley. However, they were not averse to receiving the substantial remuneration that coal leases on the richer areas of anthracite deposits would bring. Secondly, owing to fragmentation of land ownership, many individual parcels of land were too small to merit interest in mining improvements. If the coal on several adjoining parcels could be secured by lease, mining investment would then become a commercial proposition. Thirdly, in many areas, the surface of the lands that contained coal deposits was laid out to city or town lots, in such a manner that it was infeasible to acquire the surface in fee. Hence leasing the mineral rights was the appropriate course to take.[30]

While the individual details of leases could differ widely, they all contained a number of standard provisions. These included the amount of royalty per ton mined payable to the landowner, which has been briefly discussed already, and the sizes of coal on which royalty had to be paid. In addition, there would be an annual minimum number or tons on which royalty was payable and a date of first payment. This royalty minimum was payable irrespective of the number of tons actually mined in any given year. Some leases also specified the term of years, if the lease did not cover all of the recoverable coal in the ground. Occasionally, an amount of coal to be mined was stipulated, rather than a period of years. These standard conditions were the main points of negotiation with potential lessors. The eventual outcome would depend on the intrinsic geological qualities of the property, the location, the number of competing companies interested in acquiring the lease, and the overall state of the market for coal properties.

For potential lessees, the amount of the royalty per ton and the annual minimum were of most immediate concern. Given the limited profit margins per ton shipped, in all but the most favorable of business conditions, high levels of royalty payments could be sufficient to deprive entrepreneurs of all their earnings, while the landowners enjoyed a substantial income stream, without any effort being required on their part!

As mentioned above, royalty levels tended to rise over time, as lessors, like the operators, tried to increase their share of the overall earnings from the coal business, despite an essentially static level of prices. In the early period of the 1860s, when significant areas of the Northern and Western Middle Fields were relatively undeveloped, landowners were, in a sense, competing with each other to attract the attention of corporations or entrepreneurs willing to invest in the development of their lands by means of new mines and connecting railroads. Hence, it was in their interest to offer attractively modest royalty rates. This would ensure an early start to their income streams from land that otherwise only had value for farming purposes or timber production. In the early 1860s, before competition for coal lands began in earnest, 12.5 cents per ton for coal of chestnut size or larger was a common figure in Northern Field leases. As competition for lands increased, the amounts quickly escalated to 16, 20 and then 25 cents per ton. The latter became the norm during the 1870s and the beginning of the 1880s, until further competition for the dwindling number of good quality leasehold properties still available, pushed royalties up to 40 cents and beyond. Unlike in the 1860s, increased marketing of the smaller sizes, below chestnut, resulted in these being included in the leases, albeit at lower royalty figures than the larger sizes.

Good examples of early leases from the Northern Field are the Fellows lease and the Luzerne Coal Company lease. The former, whose later purchase by the Lackawanna was described in the previous section, was originally made in 1860 to George and Selden Scranton, who were to become the operators of the Oxford mine, in the city that bears their name. The royalty was 12.5 cents per ton for all sizes above pea, and there was no charge for small coal or that used at the mines. In the first year, the royalty minimum was based on an output of 30,000 tons, i.e. the minimum payment was $3,750, rising to 60,000 tons in the second year and 80,000 tons in the third and subsequent years. The lease was for all the coal on two tracts, totalling approximately 337 acres and also all the coal in the 14-foot vein under the 40-acre 'Bridge' lot, but after 10 years the lease could be given up on 12-months notice. Royalty payments would be suspended in the event of strikes, breaker fires, or if the workings went into fault, until normal production could be restored. The lease was assigned to the Lackawanna in 1868 when it purchased the Oxford breaker and improvements for $170,000.[31]

The Luzerne Coal Company had owned land in Scranton since 1855, and in 1868, just before royalty rates began to rise, it leased its property of approximately 313 acres to the Lackawanna for 12.5 cents per ton. Since there were already mining improvements on the land, there was no escalating minimum royalty while these were brought into production. Instead, there was a fixed minimum of 70,000 tons per annum or $8,750. Provision was made for strikes, etc., as in the Fellows lease, except that any early shortfall in tonnage resulting from such circumstances had to be made up by the end of the first ten years, and subsequent shortfalls by the end of every five year period.[32] Many other examples could be quoted, but these are sufficient to indicate both the types of standard conditions and the variant clauses that were introduced in individual cases.

Another method of looking at leases is to examine what might be termed the 'royalty profile' of a company as a whole, bearing in mind that this would reflect a mixture of old established and recently acquired leases. Table 12.1 shows such a profile, calculated from the records of the Lehigh Valley Coal Company's agent, Israel W. Morris. Nearly all these leases are in the Northern Field, with the exception of a number of important leases allocated to the 40 cents band, which were on the Girard Estate in the Western Middle Field. These latter leases are difficult to represent in a static table, as the terms required them to start at 41 cents a ton when they began in the mid-1880s, but they escalated by 1 cent per ton for every year the lease was in operation.[34] It is noticeable that a majority of the leasehold property was held at royalties of 35 cents or more. Most of these leases dated from the late 1880s and would have been regarded as relatively expensive for the period. In general, these later leases also included royalty payments for the smaller coal sizes, which further increased their cost to the lessee. For example, the Shoemaker lease, in 1889, which leased property was tributary to the Dorrance breaker in Wilkes-Barre, had a high royalty of 50 cents per ton on larger sizes, 25 cents on pea coal and 12.5 cents on the sizes below pea.[35]

Although the structure of leases seemed straightforward as a means of encouraging the development of coal properties, the manner of their operation over time proved to have some unintended (and largely unanticipated) consequences, which in turn had significant effects on the development of the mining economy. These unintended consequences derived from the nature

TABLE 12.1
Classification of Lehigh Valley Coal Company Perpetual Leases by Royalty Rate 1868–1892.[33]

Royalty Band (cents/ton above pea)*	Acreage Inside Coal Measures	% of Total Leased Acreage	No. of Leases
10	145	2.0	1
12.5	0	0	0
15	604	8.4	2
17.5	0	0	0
20	301	4.2	3
22.5	412	5.7	1
25	1312	18.2	13
27.5	0	0	0
30	463	6.4	8
32.5	0	0	0
35	421	5.9	5
37.5	13	0.2	3
40	1714	23.8	6
42.5	0	0	0
45	1426	19.8	3
47.5	0	0	0
50	379	5.3	3
	7190	99.9**	48

*The lower bound of each band is given in the column, hence 10 refers to the band 10.0–12.49 etc.
**As a result of rounding.

of the royalty minimum payments. These had originally been conceived as part of the process of encouraging rapid mining development by the lessee and ensuring a regular income stream for the lessor. Small operators who failed to start mining quickly would rapidly go out of business, if they had no cash-flow with which to meet the minimum royalty payments, and the lease would pass to others, as frequently happened during the ante-bellum period in the Schuylkill Region.[36] However, in the Northern Field, when large corporations took leases, cash flow constraints were much less severe, in relation to the modest sums required for paying royalty minima in the short term, so the effect was not the same. Rather, the purpose of a lease initially was to take and hold control over a body of coal land, where purchase was not possible or was deemed inexpedient.

ADVANCE ROYALTIES

Regardless of the motivation for acquiring the lease, the financial mechanism was clear. Shortly after signature of the contract, royalty payments would begin, and they would continue for the duration of the lease. Unless the company leasing the tract could reach the new body of coal in a matter of weeks or months from its existing workings on adjoining properties, it would quickly begin to pay advance royalties, i.e., royalties on coal that it had not yet mined. In the early days, drift or slope mines from outcrops above water level could be opened quickly and relatively cheaply, so this did not present a major problem. In contrast, shaft sinking to a deep basin on newly leased, but undeveloped land could result in a 3- to 5-year time lag between the dates of the lease and the first regular shipments of coal from the property. Such a lag assumes the delay was only occasioned by the time required to complete the work of opening the mine. It does not take account of other possible delays for compelling business reasons. Hence, it is a minimum value rather than a maximum, although the delay would obviously be less if the shaft only needed to be sunk 2–300 feet, rather than 800–1000 feet.

These advance royalties would mount up quickly for an individual property, if the minimum royalty was large (either because the annual tonnage required was substantial, the royalty per ton was high, or both), if there was a substantial delay between the start of royalty payments and the onset of production, or if production was much lower than expected for an extended period of time. If advance royalties were accruing on a number of leased properties simultaneously, this would start to raise increasing cash-flow difficulties for the company in question. While in one sense, the advance royalties were business assets, because the coal on which royalty had been paid could be mined subsequently at no further charge, they were also liabilities, in a non-accounting sense. This was because they locked up capital in the ground, preventing funds from being deployed for other purposes.

The extent of the problem was already becoming apparent by the end of the 1870s. In 1880, the Lackawanna had paid advance royalties on more than 3.5 million tons of coal as yet unmined, and, by 1890, this figure had risen to over 7.8 million tons. At an overall average royalty of just under 20 cents per ton, this meant that the railroad had paid out more than $1.5 million for coal on which it had yet to get any return.[37] By way of comparison, in 1883, the Delaware and Hudson had paid $648,725 in advance royalties and the Lehigh Valley Coal Company $695,160. For the latter company, the figures rapidly climbed over the ensuing decade, to about $2 million by the end of 1891.[38] Sums of this magnitude were large enough to command the close attention of senior management. Also, the distribution of advance royalties, across the different properties comprising each company's coal estate, was to become a major constraint on the pattern of investment in new mine development, as well as the division of production between mines. These issues will be examined in detail in the next chapter. The advance royalty mechanism is, therefore, very important to an understanding of the regional development of the mining economy, especially in the Northern Field. Hence, it is very surprising that it has not been examined in any of the previous literature, with the exception of a brief discussion in the memoirs of a prominent mining engineer, compiled in the 1930s.[39]

A further twist to the problem of advance royalties arose in later years and it was to have an adverse impact on some older leases. It had originally been taken for granted by the companies that royalty payments would continue until all the workable coal had been mined, at which point they would cease. If the royalty account ran ahead of the mining account, it was further assumed that advance royalties would be payable until all the workable coal estimated to be present on the property had been paid for. The actual coal could then be mined at the company's leisure, with no further charge, assuming the lease was perpetual. Unfortunately, the terms of some early leases, especially those executed by predecessors of the Lehigh Valley Coal Company, were not very clearly drafted in this regard. When the standard interpretation was made the subject of legal challenge by landowners, the eventual ruling handed down by the Pennsylvania Supreme Court was decidedly unfavorable to the companies, if the leases did not contain an explicit clause for the cessation of minimum royalties. Under these circumstances, the Court decided that the minimum royalty was effectively a rental that was not controlled by the estimated amount of workable coal remaining. As a result, minimum royalties would have to be paid, as long as the company retained possession of the lease, regardless of the actual tonnage mined. A good example of this was the Everhart Lease of the Lehigh Valley Coal Company for lands near Pittston. The Court ruling meant that minimum royalties of $15,000 a year had to continue, even though the company maintained that it had already paid advance royalties for all the coal in the ground. The term of the lease ran until the exhaustion of all the coal by mining, but there was no clause for the cessation of minimum royalties.[40] At a stroke, such a ruling effectively converted all future advance royalties into additional mining costs, since all the coal in the ground had already been paid for, and these costs would continue to increase against the same reducing amount of remaining coal, as it was progressively mined out. Large increases in output could not be engineered overnight, forfeiture of the lease would result in all previously paid advance royalties being handed to the lessor, and it could scarcely be regarded as a commercial proposition to sell the lease to another company, given the obligations against the property. This left the Lehigh Valley Coal Company with no option but to continue mining as quickly as possible at a steadily increasing cost, until either the property was worked out, or it deemed the forfeiture of previously paid advance royalties less costly than continuing to pay the annual rental to retain the property. Fortunately, not all the Lehigh Valley Coal Company leases were as carelessly drafted as the Everhart lease, or the financial implications for the company could have been disastrous.

At the risk of trying the reader's patience on these technical and legal aspects of leases, it should also be noted that even more complex situations could arise, if leasehold properties were subsequently re-leased to other parties. A good example of this was the Forty Fort Colliery, opened in 1876 by J.H. Swoyer on land leased from the Shoemaker Estate. In 1882, the property was taken over by the Wyoming Valley Coal Company, which Swoyer had set up, before being leased, firstly to the Lehigh Valley Railroad in 1885, and subsequently to Simpson and Watkins, individual operators, in 1893. The original leases on the lands, totalling nearly 770 acres by 1882, did not contain clauses to stop payment of minimum royalties, if advance royalty payments had already covered all the coal estimated to remain in the ground, but the later sub-leases, in 1885 and 1893, did.[41] There were only two ways for the original lessees to protect themselves in such circumstances from having to continue royalty payments themselves, when their own sub-lessees had invoked clauses to stop such payments. The first was to charge higher royalties in the sub-leases to provide a financial cushion for such eventualities. The second was to set higher royalty minima, to encourage faster mining and claw back their own advance royalties on the back of the efforts of their sub-lessees.

A variant of the sub-leasing problem could also arise if a mining accident, such as a breaker fire, flood, or cave, prevented further mining from a property on which substantial advance royalties had been paid. There was no possibility of recovering **past** payments from the lessor, although lease clauses could be invoked to prevent further payments under these circumstances. A good example of this type of problem is provided by the chequered history of the Lackawanna's Hunt Breaker. The railroad stopped payments of advance royalties in January 1889, when they had reached a total of $44,000, claiming that all the coal in the ground had been paid for. Unfortunately, later that year, operations at the mine had to be suspended indefinitely because of flooding, 'trapping' all the advance royalties in the ground.[42] Some years later the company was fortunate to find a lessee, in the shape of the Wyoming Coal and Land Company, willing to try and re-open the property. Unfortunately, in 1897, after two years of operation, the breaker burned down and it was not deemed economic to re-build it, thus trapping once again the advance royalties in the ground. The only possible remaining option was to sub-lease the remaining coal to the Lehigh Valley Coal Company, which might eventually be able to reach it from an adjoining property, although such a proposal was deemed fraught with extraordinary risk, as a result of the past problems with flooding of the workings.[43]

MINING IMPROVEMENTS UNDER LEASES

The final significant aspect of the leasing system was that if there were not already mining improvements on the property, including coal breakers, it was nearly always the responsibility of the lessee to make them. In addition to the general inducement to early development provided by the advanced royalty mechanism, there might be specific clauses, requiring work on opening mines to begin by a certain date, or indicating the latest date by which shipments had to commence. Other related requirements might include stipulations about the capacity of the breaker, and the amount of insurance that had to be taken out on it. A good example was the lease by J.B. Schooley et al. to Cowan and Dininny in 1882, for land in Exeter Township near Pittston. The lessees were given 2 years to sink the shaft, put up the breaker, and begin coal shipments, but mine construction had to start within 3 months of the date of the lease. There was also an unusually specific clause allowing 9 months for breaker repairs/rebuilding in the event of a breaker fire. After expiry of the lease, the lessees had 90 days to remove their improvements from the property, unless the lessors had previously agreed to purchase them.[44]

A detailed examination of the leases of several of the major companies from the 1860s onwards has only turned up one early example of a lease dating from the Civil War period, where the onus to make improvements was placed on the lessor rather than the lessee. This indicates that it was a rare occurrence, in the Northern Field, at least. The lease in question was granted to the Delaware and Hudson on land in Blakely Township near Scranton, and it specified a series of precise requirements that the lessor had to fulfil. Among these were to sink a shaft and build a breaker at his expense. The shaft had to be at least 23' x 10' inside the timbers, the breaker had to be of 100,000 tons/year capacity and sufficient gangways had to be driven to allow the opening of at least 20 chambers ready for working. In return, the Delaware and Hudson had to maintain these improvements in good order and pay a minimum royalty of 25 cents per ton on 50,000 tons per annum.[45] Yearley gives other examples of ante-bellum leases in Schuylkill where the lessors also contributed to mining developments.[46]

In the early days of the industry when drift mining predominated and the initial cost of opening mines was relatively low, the general requirement for lessees to finance the improvements was not a serious impediment to the establishment of new enterprises. With the advent of deep mining and more elaborate breaker construction, the problem of access to start-up capital for new improvements by would-be individual operators became significant. The solutions to this difficulty, which usually involved borrowing funds from the railroads, will be examined further in a later chapter.

In summary, it is clear that the leasing system was initially well-adapted to the encouragement of small-scale individual mining enterprise, where costs of entry were low, although it did little to foster systematic and economical development of coal properties.[47] With the progressive shift to corporate control of the industry, longer time horizons, and higher initial investment costs, weaknesses in the system became increasingly apparent. Advance royalty problems, which began as unfortunate side effects of the contractual mechanism, came to dominate many aspects of the long-term management of the leasehold estate. They had marked effects on the evolving geography of the Coal Regions, especially in the Northern Field.

TYPES OF PURCHASE AND TRANSPORTATION CONTRACTS PRIOR TO 1880

Contracts for the purchase of additional tonnage from individual operators by larger corporations can be traced back at least as far as the simple fixed-price contracts negotiated in 1851 by the Pennsylvania Coal Company with Bowkley and Beyea, and Crawford, Fuller and Company of Pittston.[48] Reference to similar arrangements can also be found in the records of the Delaware and Hudson Canal Company in the latter part of the 1850s, pertaining to operators in the Carbondale area.[49] However, it seems to be to the Lackawanna that the industry owed the main development of the tonnage contract as a mechanism for formalizing the commercial ties between the individual operator and the shipping railroad. Fortunately, the company was quite assiduous in recording these contracts in the county courthouse deed books. By means of a systematic, though very time consuming, analysis of these documents, it has been possible to track the development of different types of contractual provision throughout the period of interest. Three main types of coal-purchase contract can be identified:

- sliding-scale price-basis contracts,
- labor-basis contracts, and
- sales-basis or percentage contracts.

The last of these was really a development of the first. Variants of the percentage contract were eventually to become the standard approach utilized by the major transporting companies in their efforts to control the tonnage of the individual operators. In addition to these purchase contracts, there were also transportation contracts. The provisions of these different types of agreement will now be examined.

SLIDING-SCALE PRICE-BASIS CONTRACTS

Sliding-scale price-basis purchase contracts were introduced prior to 1860 and were widely used before and during the Civil War. The contracts had several provisions. Firstly, they specified an amount of coal to be sold to the railroad, which would run into hundreds of thousands of tons. Secondly, there was a delivery schedule, usually divided into winter and summer requirements, the summer period being defined to run between April 30 and December 1 in any given year. Thirdly, the proportion of the different sizes to be shipped was to conform to the direction of the railroad's general coal agent. Finally, there was a schedule of payments per ton, which depended on the price received for the different sizes by the railroad at tidewater (Elizabethport). Details of the schedule are given in Table 12.2.

When the price rose or fell from the reference prices given in the table, the payment rose or fell by one-quarter of the associated price changes, subject to the minimum payment for each size given in the final column. There was no maximum constraint, which was a very useful attribute of the contracts in the inflationary Civil War period! A flat rate payment of 50 cents per ton was made for any coal of chestnut size, which was of limited importance at the time. It should also be noted that all the payment figures assumed Scranton to be the shipping point. If the railroad had to transport the coal from the operator's mines to Scranton, deductions for lateral tolls were made in addition. It was also common for

TABLE 12.2
Reference Payment Schedule by Size for Sliding Scale Price Basis Contracts.[50]

Coal Size	Elizabethport Price Received Per Gross Ton	Resulting Payment to Operator Per Gross Ton	Minimum Payment by Size
Lump	$3.90	$0.875	$0.80
Steamboat	$3.95	$0.90	$0.8375
Furnace	$3.95	$0.95	$0.9125
Egg	$4.20	$1.125	$0.105
Stove	$4.35	$1.35	$1.2625

10 percent of the payments due on past shipments to be retained for a period of 4 months, as a precaution against non-compliance with the contract.[51]

A number of contracts of this kind were executed over the period 1859–1863, and these are shown in Table 12.3. There are several striking features about the information given in the table. The first of these is the large quantity of coal being bound to the railroad in these early agreements, which totalled almost 3 million tons, excluding the very substantial shipments made by the Nay Aug mine over the following two decades. Secondly, as a direct corollary, the contracts had the effect of extending the coal estate of the railroad by several thousand acres for the duration of the agreements. Yet, as the lands were neither owned nor leased by the railroad itself, they did not count against any charter restrictions on land holding. Thirdly, the total tonnages involved for each mine were large enough, in relation to the daily rate of shipments, that the agreements would last for several years at least, lending stability to the business arrangements between the railroad and the operators. Related to this, the agreements also stressed the need to make the shipments as regular as possible, thereby facilitating steady work at the mines and efficient deployment of rolling stock. This situation contrasts with the later period of frequent suspensions and fluctuating traffic during the 1870s cartel.

LABOR-BASIS CONTRACTS

After the Civil War, a different mechanism was introduced for controlling the payments made to individual operators, in the form of labor-basis purchase contracts. These contracts used, as their reference point for calculation, the average amount paid for mining at the Diamond (Plate 8), Hampton,

TABLE 12.3
Lackawanna Railroad Sliding Scale Basis Contracts 1859–1863.[52]

Operator	Mine(s)	Total Tonnage Purchased	Tons/day Winter	Tons/day Summer	Contract Date
Judson Clarke	Clark	500,000	200	400	Mar. 23, 1859
Old Forge Coal Co.	Old Forge	330,000	150	300	Mar. 28, 1859
Phinney and Schott	Greenwood	633,000	162	325	Feb. 15, 1860
Hyde Park Coal Co.	Hyde Park	300,000	-	-	Apr. 11, 1860
Susquehanna and Wyoming Valley Coal Co.	Stafford Brook	375,000	150	300	Jan. 1, 1861
Hurd and Flagg	Seneca	200,000	100	150	Jan. 11, 1861
Thomas Oram	National	210,000	150	300	July 15, 1861
T.F. Hunt et al.*	Nay Aug (Roaring Brook)	All their Coal			Nov. 28,1863
D. Howell	Mount Pleasant	687,000	150	300	Dec. 1, 1863

* The Nay Aug agreement specified delivery of 70,000 tons in the first year of the agreement and 100,000 tons each year thereafter.

Bellevue, and Continental mines of the Lackawanna in Scranton. For every rise and fall of the mining amount, there was a corresponding change in the size of the payment made to the operator. For example, a contract with William Connell, dating from 1870, was for the sale of 1 million tons of coal from the National (Plate 9) and Stafford Brook mines, in and near Scranton, at a rate of 100,000 tons per year. When the price of mining was 70.94 cents/ton at the four Lackawanna mines, Connell received $1.7723 per ton for his shipments. This latter figure would rise or fall by 8.5 cents for every 5 cents change in the price of mining. Importantly, there was no specified minimum, as there had been on the earlier sliding-scale price-basis contracts. This particular contract also contained a clause about delivery of railroad cars for shipment. The railroad was not obligated to ship coal at Connell's convenience, but only to deliver cars 'with as much regularity as their business at other collieries and their sales of coal [would] admit of'.[53] A number of contracts were made on this basis. Some, such as that with the Mount Pleasant mine (plate 10), adjacent to the Diamond, were renewals of earlier contracts, which had been on the Elizabethport price-basis. Others, such as the contract with Charles Hutchison for the output of the Black Diamond mine above Wilkes-Barre, were new arrangements.[54] Wherever possible, the railroad tried to use contracts of this form, but it was only able to require them where it had a strong negotiating hand. This was the situation when, at the time, it was effectively the only railroad that could transport the coal from the mine in question, because other lines were too distant for construction of branch lines to be economic, or they simply did not pass through the relevant portion of the coalfield.

SALES-BASIS OR PERCENTAGE CONTRACTS

By the early 1870s, however, some properties were already potentially accessible to more than one competing railroad, and this put the individual operator in a position of negotiating strength. Under these circumstances, the Lackawanna might have to go along with a new variant of the old sliding-scale price-basis contracts. The original version of these shown in Table 12.2, represented a payment of 22–31 percent of the tidewater price to the operator, depending on the size. So the great majority of the sales revenues accrued to the transportation, rather than the mining interest. The new 1870s variant, called either a percentage contract or a sales-basis purchase contract, began to redress the balance of remuneration in the direction of the operator, by giving the latter at least 40 percent of the tidewater price. A good example is given by the property of the Lackawanna and Susquehanna Coal and Iron Company, which had a large estate of 1,900 acres below Scranton. This company signed a sales-basis purchase contract with the Lackawanna in July 1872, although it already had an agreement with the Erie, as mentioned in chapter 10. The total tonnage to be sold was large, at 1.425 million tons, and was to be shipped at a rate of 75,000 tons in the first year, doubling to 150,000 tons for the second and subsequent years. When the price delivered at Elizabethport or Hoboken was $4 or less, the operator would receive 41 percent of the price, between $4 and $5 the percentage increased to 42 percent and above $5, a figure of 43 percent was payable.[55] In the light of subsequent events, it is interesting to question the business rationale for this contract from the perspective of the Coal and Iron Company. In 1873, it only had one breaker, the Greenwood, in operation, and all the tonnage from this mine must have gone to filling the Erie contract, since none of it appears in the Lackawanna Railroad shipment records. This left the Coal and Iron Company with no capacity to fill its contract with the Lackawanna, and its planned new breaker, the Phinney, was not completed until mid-1876.[56] To meet the contract shortfall, it apparently made an arrangement with Elliott, Koerner and Company, proprietors of the new Sibley breaker, on lands leased from E.A. Coray, one of the principals in the Coal and Iron Company, to ship coal on its contract. This was not dissimilar to the arrangement it had made after 1869, for the Carbon Hill breaker to make up the deficit on its previous contract with the Lackawanna! A further twist to the involvement of Elliott, Koerner and Company was that, prior to the Lackawanna and Susquehanna agreement being negotiated, they had independently approached Storrs of the Lackawanna Railroad, offering to sell 500,000 tons of coal for 45 percent of the tidewater price from their mine, which was still under construction. This proposal had been turned down because the price was too high. The Phinney breaker belatedly joined in shipping on the contract, but, by the end of 1878, when more than 800,000 tons should have been shipped, the total had only reached 514,817 tons.[57] Falling behind its contract to this extent was suggestive of over-optimism in the company's earlier business arrangements, or poor management, or both. Unfortunately, its creditors seem to have reached the same conclusion early in 1879, when the company went into receivership, 'unable longer to sustain itself at the present low price of coal'.[58]

COMPARISON OF LABOR- AND SALES-BASIS CONTRACTS

Given that two different kinds of contracts were being used at the same time, it is appropriate to consider the relative merits of each from the different standpoints of the railroad and the individual operator. Storrs argued that the rationale for labor-basis contracts was that they were intended to ensure that the contractor always received a price that bore the same relationship to the cost of mining, as had pertained at the time the contracts were first established after the Civil War.[59] However, wages and prices did not always change exactly in tandem in the Northern Field, because the Schuylkill wages basis system was not in general use. This makes it very difficult to draw any simple conclusions about the benefits of the two types of contracts, as the relationship between wages and prices fluctuated over time. Fortunately, the Lackawanna's own analysis of this problem has come down to us and the results are shown in Figure 12.1. In the light of the discussion of price behavior and cartel activity during the 1870s, presented in earlier chapters, several clear conclusions emerge. It is clear that the artificially high prices created by the 1873–1876 cartel favored the sales (percentage) contracts over those operating on a labor basis. However, in the competitive periods of 1876–1877 and 1879, the position was reversed. During 1878, when the re-established cartel struggled to hold prices near the $4 mark in New York, there was little to choose between the types of contract. This balance point, at a price approaching $4, is also of interest for an extrapolation of the analysis beyond the period of the Lackawanna's own calculations. Prices were relatively stable at approximately $4 from 1880 to 1884, while there were only limited wage increases

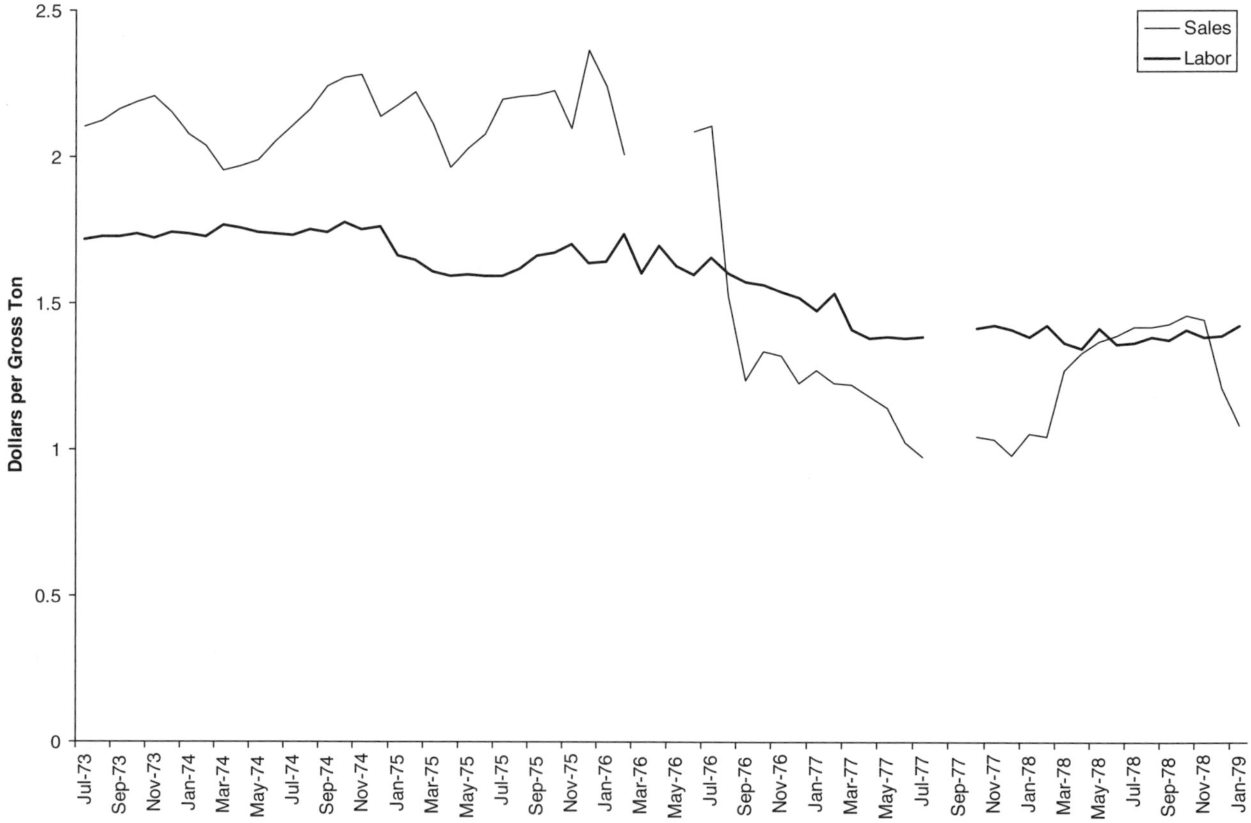

FIGURE 12.1
Delaware, Lackawanna and Western Railroad Coal Department: Prices Paid to Contract Shippers on Sales (Percentage) and Labor Basis 1873–1879.

as business conditions improved. This suggests that there was again not a great deal to choose between the types of contract during this later period. Since the mining wages, on which the calculation of labor-basis contracts depended, were less volatile than prices, it is also apparent that, while the labor-basis contractors under-performed financially in periods of high sales prices, and outperformed when prices were low, they always received a reasonable level of renumeration for their efforts. In contrast, the sales-basis contractors enjoyed excellent returns during the good times and miserable returns at times of low prices. Indeed, the figure fell so low in 1877, for example, that smaller high-cost operations might be expected to have reduced output or suspended altogether, rather than continue shipments at a loss. Inspection of the Lackawanna shipment records suggests that such behavior was indeed observed. Labor-basis contractors tended to increase their output in years of free competition and low prices, such as 1877 and 1879. Conversely, sales-basis contractors increased their shipments in the combination years of 1878 and 1880 when prices were higher, even though output restrictions were in force at these times.

Although labor-basis contractors were shielded from sudden coal-price falls by the damped response of wage rates to such price fluctuations, they chafed at the effects of cartel output curtailment on the operation of their contracts. William Connell, operator of the Meadow Brook Colliery in Scranton, summed up the problem very concisely:

> 'As our contract is not based on the price of coal and we gain nothing from any advance that may occur, we do not consider it equitable to be called on to contribute to such advance by a curtailment of our contract, and we would respectfully request that we either receive a full supply of cars in accordance with the contract or else so change the terms of our contract that we may have an interest in the price of coal'.[60]

In fairness to the Lackawanna, Connell was putting the best possible interpretation on his contract, in claiming a full supply of cars. The Railroad was only actually obligated to treat him fairly with regard to the overall level of business at the time.[61]

Tensions of this kind between holders of different kinds of contracts and the railroads were to become a recurrent feature of the industry from the 1870s onwards. However, it was not in the interest of the transporting companies to antagonize their business partners, if it could be avoided, and a variety of remedies were adopted over the years, in striving to maintain friendly relations. These will be examined further below.

TRANSPORTATION CONTRACTS

The final type of contract that became significant prior to 1880 was the transportation contract. At first sight this was hard to distinguish from the two types of purchase contract, but in fact there was a key legal difference. Under the former type of contract, the railroad actually agreed to purchase a defined amount of coal at the breaker and pay for it in a designated manner. With a transportation contract, the railroad did not purchase the coal, instead it agreed to transport it when delivered in cars at the breaker, either to tidewater or to western markets. The operator or his agent then paid the railroad for its services. This payment was usually a percentage of the tidewater price received by the operator, lending the contract a superficial similarity to the sales-basis contract. Transportation contracts could be of long or short duration, but the agreements were not couched in terms of a specified total number of tons involved, as was the case with purchase contracts. Instead, tonnages to be shipped per annum were indicated. Since transportation and delivery also implied an ability by the operator or his agent to deal with the coal at the destination point, the contracts also tended to contain clauses concerned with stocking grounds or loading of vessels, to ensure that the railroads' cars could be emptied and returned in a business-like manner.

The following example of an early transportation contract from 1873 indicates the complex nature of the provisions they could contain, and also how they could create both long-term business and geographical relationships between the railroad and the operators/coal merchants involved. The contract in question was between W.H. Meeker, a New York coal merchant and his associates, and the Lackawanna Railroad, but it also included unrecorded agreements between Meeker et al., members of the Winton family, Edward Dolph, George Filer, and Thomas Livey. Filer and Livey, in particular, had secured from the Wintons and Dolph the rights to develop several tracts of coal land. Meeker agreed to secure the right of way for a new railroad connecting with the Lackawanna that was to link up the new mines to be opened by Filer and Livey.

The Lackawanna was to fund the construction of the new road, which became known as the Winton branch, by means of $180,000 of construction mortgage bonds of the Morris and Essex Railroad. Such a major railroad construction project as part of a contract implied a long-term relationship, and the agreement was to last for 36 years from 1874. It covered all the coal to be mined from the new collieries, beginning with modest deliveries of 60,000 tons in the first year, rising to 300,000 tons in 1876 and subsequent years. Since Meeker was a large New York coal merchant, the contract terms provided that 85 percent of shipments to Hoboken had to be loaded into vessels at that destination within 24 hours. If Meeker failed to provide these vessels for loading, then the Lackawanna could decline to furnish him with further cars for shipment until the problem was rectified.

In return for these transportation services, Meeker was to pay the Lackawanna 52½ percent of the tidewater price. For purchase contracts, at this time, the railroad was retaining 57–59 percent. It appears that the large annual tonnage and the length of time the agreement was to run were factors in the favorable terms for Meeker, but he would also have had to shoulder selling costs that fell on the Lackawanna in the case of purchase contracts.

Enforcement of these agreements by the railroad, in the sense of being able to require the operator to continue shipping at the specified level for the specified time period, was effectively impossible. Instead, there were rather lame clauses allowing the railroad to abrogate the contract if shipments were less than ⅔ of the required level for the preceding two-year period.[62] This, however, scarcely constituted tight long-term control of the tonnage, which the railroads desired (and the operators desired to avoid), in case future circumstances might lead to more satisfactory business arrangements with other parties. While the status of transportation contracts was regularly checked, they were not monitored as closely as the purchase contracts, where the exact number of tons still due to reach the contract total were calculated and reported each year in the general coal agent's annual statement.[63] However, enforcement clauses in the purchase contracts were usually not much more effectual than in the case of transportation contracts, although, as noted earlier, some allowed for retention of 10 percent of the proceeds of sales payable to the operator for up to 4 months at a time, as an insurance against contract non-performance.[64] In consequence, some contracts with competent businessmen such as William Connell and William T. Smith of the Mount Pleasant Colliery were well kept. Others, by contrast, such as those with Charles Hutchison of the Black Diamond near Wilkes-Barre and the Pennsylvania Anthracite Coal Company, dragged on for years, with shipments at a level well below that originally anticipated.

One interesting example from the later part of the period had a number of novel clauses, that both facilitated contract enforcement and gave the railroads involved a much greater degree of control over the tonnage than had been the case in earlier contracts. The contract was an unusual joint arrangement between the Delaware and Hudson and the Delaware, Lackawanna and Western companies, and John Jermyn, for coal from the eponymous Jermyn No. 3 mine in Scranton. It was a labor-basis contract for 150,000 tons per annum and the price paid per ton was higher than normal. However, there were three additional clauses not usually found in such contracts:

- The railroads had the right to take over the property if Jermyn failed to supply coal as required, when there were no extenuating circumstances such as strikes.
- Jermyn was not permitted to assign the contract to other parties without the agreement of the two companies.
- In return for purchasing his coal at a premium rate, the companies had the right to take over his rights and all the mining developments, improvements, and personal property associated with this particular colliery. These could be bought out by them for $200,000 between January 1, 1880 and January 1, 1882.[67]

In the event, the third of these provisions was indeed invoked and the companies took over the mine under a curious joint arrangement, whereby the accounts were charged to each in alternate months. The mine was renamed Manville (Plate 11) and it made its first shipments under the new joint management in 1882.[68]

One final legal quirk of the transportation contracts made them potentially less attractive to the shipping railroads, but more attractive to the operators. Like a number of other aspects of the coal business, this centred on rights and responsibilities in the provision of cars for shipment. The Meeker contract, for instance, contained an explicit clause that required the Lackawanna to supply cars, to the mines covered by the agreement, with the same degree of regularity and promptness as the railroad supplied cars to its own mines. This was sufficient to prevent the railroad discriminating against Meeker in the distribution of empty cars to the mines. However, the contracts related specifically to the provision of transportation services, and railroads, despite the cartels, were still, in law, common carriers. As a result, it became a moot point for the legal advisers, as to whether transportation contractors could be denied all the cars they asked for, regardless of whether this meant discriminating in their favor against the railroad itself or its purchase contractors.[66] The latter, of course, had not contracted for transportation services, they had simply sold their coal, so the legal arguments did not apply with the same force to them. Prior to 1880, this legal question about transportation contracts had not developed into a major issue, but as the number of such contractors grew in later years, it was to assume a more serious aspect.

From these examples, it is very clear that, by the end of the 1870s, the Lackawanna, in particular, had developed and refined a variety of contractual mechanisms, for securing and holding coal tonnage, that did not adversely affect its charter restrictions on coal-land ownership or leasing. The Delaware and Hudson was also quite an active participant in securing the tonnage of individual operators. The two companies compared notes on types of contracts and indeed entered into them jointly, as has been seen.[69] Other companies, such as the Pennsylvania Coal Company and the Lehigh Valley Coal Company were much less involved in this type of activity during this period. Indeed, in the Lehigh Region itself, such contracts were effectively unknown, and the independent operators there could not understand why some of the Northern Field companies insisted on them.[70] Jones made a related point in his study of the anthracite combination. He argued that the percentage or sales-basis contract system was not the customary mode of marketing the output of the independent operators until it was popularized by President McLeod of the Reading in the 1890s.[71] However, on closer examination, this is a slightly odd statement, since the term 'independent operator' was usually taken to mean a firm that was not bound by any form of shipment contract. Leaving that aside, it would seem that the importance of such contracts grew with the increasing dominance of the Northern Field and other transporting companies progressively emulated the methods for controlling tonnage pioneered by the Lackawanna and the Delaware and Hudson. The later stages in this process, after 1880, will now be examined.

PURCHASE AND TRANSPORTATION CONTRACTS AFTER 1880

PURCHASE CONTRACTS AND NEW RAILROAD COMPETITION

The contract system continued to develop in the final two decades of the 19th century and the percentage-basis purchase contracts, in particular, became progressively more important over time. The major reason for this was that they provided an easy method for new railroad competitors entering the coal business to secure coal tonnage for their lines. Offering a higher percentage of the New York price than currently prevailed would attract operators, with expiring contracts, who could be reached by the new railroads, and it might even tempt those with short-term transportation contracts to switch to a longer-term purchase contract. Extension of the tracks of existing railroads, such as the Lehigh Valley, within the Northern Field, also allowed the mechanism to be used by current members of the cartel to divert tonnage from their collaborators/competitors.

The first railroad scheme to disturb the prevailing arrangements in the trade was the New York, Susquehanna and Western Railroad at the beginning of the 1880s. This was followed nearly a decade later by the New York, Ontario and Western and other projects such as the Wilkes-Barre and Eastern, which was eventually built, and the New York, Wyoming and Western Railroad, which was not. An important aspect of many of these schemes was the building of alliances between individual operators dissatisfied with their present shipping arrangements and potential backers with the necessary finance. In the case of the Ontario and Western, many of these investors were from Britain.[72]

Conceived and built during the boom period of the early 1880s, the original main line of the New York, Susquehanna and Western from Jersey City did not actually reach the anthracite fields. Instead, as a result of a traffic interchange agreement with the Lackawanna, it made its terminus at Gravel Place near Stroudsburg. The two railroads shared the freight receipts from the Coal Regions to tidewater on a pro rata basis.[73] Even before it was completed, the new railroad began offering to buy coal at 50 percent of the tidewater price, a significant increase over the prevailing rates at the time. It also managed to secure control of several thousand acres of coal land, including the property of the ill-fated Pennsylvania Anthracite Coal Company, which was still in the hands of the receiver after its earlier bankruptcy. Part of this estate was already largely worked out, and most of the other lands were not regarded as being of high quality.[74]

From this rather inauspicious start, the Susquehanna and Western managed to build up its tonnage in only a modest way and it required a decade of operation before its shipments exceeded 1 million tons.[75] The high prices it was willing to pay for shipping contracts had a greater impact on the trade than the shipments it was able to make, as the prospect of higher returns encouraged new mining ventures and created dissatisfaction among those bound into existing contracts at lower figures. For example, in 1881 the Roaring Brook Coal Company in Scranton, one of the Lackawanna's longest-standing contractors, tried to lease coal lands adjoining its current property, which was on the verge of exhaustion. It also sought to purchase other lands for $1,200 an acre. However, to make its continued operation on these properties financially viable, it claimed it needed 50 percent of the New York price, which was 28–30 cents a ton more than the Lackawanna was paying under their existing contract. Sanderson and Simpson of the Pancoast Coal Company, who had recently begun developing a mine east of Scranton, made a similar proposition to the Lackawanna. They also wanted provision for a further increase in the percentage if that was secured by others, claiming the Erie would agree to such terms, not to mention clauses giving them unfettered access to western markets served by the Lackawanna. In reporting this information to Samuel Sloan, Storrs could only observe 'it is Frisbie [of the Erie] and the liberal offers of the New York and Susquehanna Railroad (sic) that suggest these points to the parties'.[77]

Potential or actual increases in prices paid under purchase contracts naturally attracted the attention of the long-suffering labor-basis contractors, who pressed their case of relative penury. First in the queue was William Connell, who had a very large contract, with 1.75 million tons still to be delivered. He argued for an increase in payments to him as a result of increases in the percentage of pea coal being shipped. The reasoning was that the labor basis price was calculated by dividing total miners earnings at the four benchmark collieries by total tonnage shipped. Increasing the percentage of pea being shipped from the same amount of coal tipped into the breaker, increased the division in the payment calculation, thereby reducing the payment to the operator. Agreement was finally reached for the Lackawanna to pay him an additional 6 cents a ton from February 1, 1882, but on a separate accounting voucher, in the vain hope of preventing other contractors finding out about the extra payment!.[78] This ruse worked for a time, but, later in 1883, the secret leaked out and all the other labor-basis contractors demanded the same financial accommodation, backdated to early 1882, at a cost to the railroad of $23,118.[79]

A further unwelcome aspect, at least to its competitors, of the Susquehanna and Western's attempts to secure tonnage under percentage contracts, was that they also offered to furnish money to build mining improvements. This was an attractive proposition for potential entrepreneurs with access to coal leases, since it removed the main barrier to entry into the coal business for those of limited means.[80] Such loans were not unheard of, but nor were they commonplace prior to this period.

Following the largesse proffered by the Susquehanna and Western, although they themselves had few takers, it quickly became the order of the day for all types of contractors to come to the railroads, cap in hand, seeking loans. Storrs of the Lackawanna, ever the conservative financial manager, could hardly be described as enthusiastic about this new development in the business. This was because he was well aware of the risks involved and had little faith in the budding entrepreneurs, whose optimism was boundless, but in those financial resources were very limited. Two good examples were the labor-basis contractors Samuel Stetler and Messrs. DeWitt and Cake. Stetler will be the subject of more detailed investigation in the next chapter. He required $70,000 for his improvement.[81] DeWitt and Cake were the directors of the Clear Spring Coal Company in West Pittston and they eventually needed $100,000 to complete their improvement, having originally only requested $65,000.[82] Even the transportation contractors felt the need for money. Haddock and Steel, who took over the Black Diamond and Dodson breakers on the Lackawanna and Bloomsburg Division of the

Lackawanna Railroad in the lower part of the coalfield, provide a good example. Unfortunately, to pay off outstanding liens on the property, they had an immediate requirement for $150,000 before they could begin shipping. The loan was secured by mortgage on the leasehold rights and was to be paid off at 10 cents a ton on Black Diamond shipments and 5 cents a ton on those from Dodson.[83] In the event of non-performance under the contract, the Lackawanna could take possession. Some of these entrepreneurs, such as C.F. Steel, had financial resources.[84] Others had links to wealthy local landowners or capitalists in New York, Philadelphia, or Boston. Many, however, had very limited means, and local banks were not in a position to lend sums of the large magnitude required. Nor could the banks have readily recovered their loans from risky mining investments had they been able to make them. Hence it was left to the railroads to finance the development of the mining economy. They had a large enough cash flow to make the loans in the first place, they were best placed to assess the risks involved, and they had the expertise to take over the mining properties, or lands that provided the security, in the event of defaults. This important topic will be examined in more detail in Chapter 13.

TONNAGE CONTRACTS AND COMPETITION FOR TONNAGE: CASE STUDIES OF MALTBY AND THE WYOMING VALLEY COAL COMPANY

The New York, Susquehanna and Western proved to be only the first in a series of disturbances to the settled pattern of cartel interrelationships during the 1880s. Although the Lackawanna and the Delaware and Hudson were at pains to avoid treading on each other's toes in respect of new tonnage acquisition, the Lehigh Valley had few such qualms, especially in the Northern, but also in the Western Middle Field. Storrs, who observed that their bridge over the Susquehanna at Coxton near Pittston gave them the potential to extend their line down the west bank of the Susquehanna, quickly recognized this. This would give them the opportunity to be an 'ugly competitor' for the tonnage of individual operators down the valley when the existing contracts of these concerns with the Lackawanna expired.[85] Such a threat was very significant, as the Lackawanna and Bloomsburg division of the railroad had enjoyed an important element of spatial monopoly power in relation to the development of this part of the coalfield for more than two decades. The threat quickly became a reality, when, in June 1882, the Lehigh Valley purchased the Maltby breaker, leasing lands held in fee by the previous owner and taking an assignment of the leasehold on other tributary lands. Taken together with another adjoining lease it had acquired earlier in the year, this gave it a substantial property of more than 760 acres of coal. This was a singular blow to the Lackawanna, which had a transportation contract on all the coal from the breaker, dating back to 1879. Despite this, the railroad did not even hear about the purchase until after it had taken place.[86] Although specific confirmation is lacking from available sources, it would appear that, in this case, the transportation contract was abrogated by the breaker purchase. This had the effect of withdrawing almost 100,000 tons a year from the Lackawanna's shipments, and reducing the rated mining capacity tributary to its lines in terms of cartel calculations. In fact, some tonnage was still shipped from the mine over the following two years, but at a rapidly decreasing rate. This must still have gone over the Lackawanna on a non-contract transportation basis, since the Lehigh Valley had not yet extended its line down the valley to the breaker. In 1885, the latter company shut the breaker down altogether for a period of years.[87] In business-cycle terms, this was perhaps not surprising, since business conditions were very adverse at the time. From the viewpoint of a large leasehold property, which continued to generate substantial advance royalties during periods of idleness, it is more difficult to comprehend, especially as the Lehigh Valley had begun construction of its branch road from Coxton in the direction of Maltby by the early part of 1885.[88]

Once Maltby had changed hands, there was the prospect of other mines following suit in a 'domino effect' down the valley, in the expectation of the arrival of the new Lehigh Valley line. Two of the closest candidates were the Ellenwold, operated by Albright and Company, and the Forty Fort, which J.H. Swoyer had just transferred into his new concern, the Wyoming Valley Coal Company, together with the Wyoming colliery on the other side of the Susquehanna. Swoyer already had a substantial 10-year transportation contract with the Lackawanna, dating from 1880, for 200,000 tons per annum from the Wyoming colliery. He also had a shorter five-year transportation contract with the same railroad from the beginning of 1880 for 100,000 tons a year from Forty Fort.[89] In a deft move, Swoyer acquired the Ellenwold in June 1882, before the Lehigh Valley could get to it, and renamed the breaker Harry E. This was a sensible step, as the mine was adjacent to Forty Fort and the two properties could be worked together more efficiently than if they were under separate management. Given that Swoyer had the two large contracts with the Lackawanna, it was no surprise that he declared his interest to be aligned with that railroad. At the same time, he charged that he had been the subject of discrimination, in relation to the payment of lateral tolls, and that his coal was going astray on its way to Hoboken. As a result, he claimed that he did not intend to stay with the Lackawanna beyond his current contractual obligations, which, in the case of Forty Fort, meant the end of December 1884. Following an investigation of the matter, and with the loss of the Maltby tonnage fresh in his mind, Storrs was quick to identify that there *was* substance in the allegations and the lateral tolls problem was rapidly adjusted in Swoyer's favor.[90] This prompt action improved business relations. The subsequent willingness of the Lackawanna to loan Swoyer $75,000, to fund development of the Harry E property, resulted in a further substantial transportation contract for 750,000 tons from this breaker and Forty Fort in July 1883. As was now customary under the cartel, the contract contained a clause stating that Swoyer was only entitled to the same proportion of cars in relation to rated capacity as were other shippers.[91] As the business-cycle downturn began to gather pace, Swoyer would have been aware that this implied the mine would not be running anywhere near its productive capacity for the foreseeable future, but at least he was guaranteed a limited share in the Lackawanna's overall tonnage allocation from the cartel.

By the following Spring, as business conditions worsened, Swoyer began offering his property back to the railroads, either at a lease royalty of 60 cents a ton, or as a sale for $2 million, both very high figures. Storrs took the view that Swoyer had made money, and exhausted most of the choice and cheaply mined coal on the Wyoming property, and was now seeking to unload his obligations at a time when the railroads still wanted tonnage. Storrs must have learned of the high asking price with a wry smile, since the Lackawanna had refused the lands under the Harry E breaker, a dozen years previously, on account of the relative lack of coal and the expense of mining.[92]

Since the Lackawanna already controlled most of the tonnage by contract, the railroad was not in a rush to take over the property when Swoyer was carrying all the risks of mining. Again, however, they had to contend with the Lehigh Valley, whom Storrs believed had been upset by the loss of the Wyoming breaker tonnage in 1880 and were still trying to get even. The Lehigh Valley entered negotiations with Swoyer in August 1884 to lease the Wyoming breaker and 285 acres of tributary land. The terms included a large $75,000 minimum annual royalty at 50 cents a ton above pea, but the arrangement also included a requirement to fulfil the original transportation contract with the Lackawanna, rather than abrogating it. In a further clever manoeuvre on his part, Swoyer also negotiated a second transportation contract on Harry E and Forty Fort with the Lehigh Valley, equal in amount to the tonnage on the Wyoming contract with the Lackawanna, which obligated the Lehigh to provide him with transportation facilities. In this way he was able to secure for his Wyoming Valley Coal Company a guaranteed profit of $75,000 a year on the Wyoming breaker, despite the depressed state of the coal trade. In addition, by stacking up multiple transportation contracts with two railroads rather than one, he was able to secure a proportionate share of each of their cartel tonnage quotas, thereby maximizing the number of empty cars to be delivered to his mines. Even the disastrous fire in November 1884, which destroyed the Forty Fort breaker, was not enough to demolish his scheme. He quickly built a connecting railroad back to the Harry E breaker, using the money from the fire insurance claim, so the output of both mines could be prepared there. Swoyer claimed the end result would be the ability to run the Harry E breaker at nearly 1,000 tons per day for 1885 and at a profit of $75,000 on the joint property. Given the adverse state of the coal trade in 1885, as the depression bottomed out, this was a rather bold prediction. In the event, the company managed to run for about 250 days, when the district average was 176 days in that year. It mined a total of 208,750 tons as compared to 186,933 in the previous year, when both breakers were operational. This generated a net profit of $62,364, before accounting for bond interest and sinking fund payments. A profit of this size was a signal achievement under the circumstances. It indicates the skill with which Swoyer was able to manipulate the railroads' appetite for tonnage to further his own interests as an individual operator, despite the constraints of the cartel quotas and his misfortune in losing a valuable breaker at a critical moment.[93] As a postscript to this intricate contractual episode, correspondence from many years later reported on the final outcome of the original Wyoming breaker transportation contract with the Lackawanna. Of the 2 million tons that should have been shipped, the Wyoming itself managed to produce 1,044,045 tons. A further 414,765 tons were transported by re-directing the empty cars to Forty Fort and Harry E to be loaded, as instructed by Swoyer himself. This left a shortfall of over 500,000 tons, but the writer concluded:

> 'In view of the saving and qualifying clauses in said contract, coming from restricted production, obstructions at mines causing idleness, a short supply of cars, which were frequent, and other causes, I would consider that said contract had morally at least, if not legally been fully performed'.[94]

This example of the Wyoming Valley Coal Company has been discussed in some detail to show the complexities of possible interactions between individual operators and the railroads. On an industry-wide scale, the latter were broadly co-operating within the cartel framework, but at a more local geographical scale, they were often competing vigorously for territory and tonnage. Nor is the example particularly unusual, as many of the same issues arose in relation to other operators in the Northern Field, although often the eventual outcome was less favorable than in the present case.

Continued extension of railroad branch lines enhanced the negotiating position of individual operators who were fortunate enough for their mines to be accessible from two or more lines, actual or proposed. In consequence, new agreements on the old-style labor basis became increasingly rare from the mid 1880s onwards, and percentage-basis purchase contracts or transportation contracts were employed instead.

As would be expected, those still working all or part of their available tonnage under an existing labor-basis contract continued their loud complaints wherever cartel restrictions curtailed shipments significantly. The years 1885–6 were particularly bad in this context, with 12–13 days idleness per month being a regular occurrence. William Connell argued forcibly that he was receiving 30–40 cents a ton less on that portion of his tonnage sold under the labor basis than on that sold at 50 percent of the Hoboken price. Storrs insisted that everyone, including the coal department's mines, suffered the same degree of restriction, and Connell admitted that the labor-basis contract gave them work that was as regular as possible under the circumstances. A compromise of 15 cents a ton additional to the standard labor-basis price was finally agreed, as a means of maintaining friendly relations and preventing Connell looking to competing railroads for transportation.[95]

NEW CONTRACTS AND NEW RAILROAD SCHEMES OF THE EARLY 1890S

Such palliative measures to maintain stability in contractual relations provided only temporary relief from competitive pressures. Late in 1888, rumours began circulating that a new railroad project would offer to transport coal to tidewater for 40 percent of the New York price. This railroad, originally known as the Scranton and Forest City Railroad, was to become part of the Ontario, Carbondale and Scranton Railway or the Scranton branch of the New York, Ontario and Western Railway. The latter, in turn, had originally been called the New York and Oswego Midland Railroad when it was chartered back in 1866. The 54-mile branch to Scranton from its main line at Hancock Junction was opened in June 1890. By 1893, approximately $6 million had been spent on this line, including $157,000 on branch laterals to coal breakers.[96] The line was a particular threat to the Delaware and Hudson in the upper part of the coalfield, to the point that the latter attempted, albeit unsuccessfully, to serve an injunction preventing the Ontario and Western laying tracks across their land near Carbondale. It was also a threat to the Lackawanna in Scranton itself. Indeed, Storrs went as far as to claim it would strike the Lackawanna more than any other interest.[97]

The very low transportation rates offered by the Ontario and Western were not quite as favorable as first thought, since selling expenses had to be borne by the operators. Slightly later, they began offering genuine sales-basis purchase contracts for 55 percent of the New York price, a clear advance on the 50 percent rate that had been introduced by the Susquehanna and Western some years before.[98] The Ontario and Western was more successful in attracting tonnage to its line than the Susquehanna and Western had been. Shipments exceeded 1.1 million tons in 1893, its third full year of operation. A total of 14 breakers contracted to ship over its lines, including those of the Northwest Coal Company, the Elk Hill Coal and Iron Company, and the Mount Jessup Coal Company. Two new breakers, the Richmondale on Elk Creek and the West Ridge in

Scranton, were added in 1893. Substantial loans were made to tempt individual operators to join them or to facilitate new breaker construction. By mid-1893, these loans totalled over $680,000. The West Ridge Coal Company secured a large loan of $100,000, repayable at 20 cents a ton on coal shipped, in return for guaranteeing, to the railroad, shipments of at least 100,000 tons a years.[99] It is interesting to note that John Jermyn, a former superintendent of the Susquehanna and Western Railroad's coal properties and long-term shipper over that road, became so disgruntled with the shortage of cars and the poor transportation service offered by them during the 1891 depression, that he sought release from his contractual obligations so he could ship over the Ontario and Western instead. This request was partly acceded to, at least on a temporary basis.[100]

The rapid resurgence in business confidence that began in late 1891 encouraged yet further railroad schemes. The first of these was the Wilkes-Barre and Eastern Railroad, chartered in March 1892 under the presidency of William Ryman, a Wilkes-Barre entrepreneur with financial interests in the Susquehanna and Western. John Jermyn was also involved. The new line connected with the latter road at Gravel Place and ran down to Wilkes-Barre, linking in the Susquehanna and Western's coal properties and contractors. An important aim of the project was to end the arrangement with the Lackawanna for shipping Susquehanna and Western coal tonnage out of the Northern Field.[101] The building of the road initially secured only a limited amount of extra coal, in the shape of the Langcliffe breaker, on which the very large sum of $350,000 was loaned, repayable at 15 cents a ton on future shipments. On its completion in October 1893, the trade press remarked that it was 'another blow at DL&W supremacy'.[102]

The second new scheme was, of course, the ill-fated leasing by the Reading Railroad of both the Lehigh Valley and the Central Railroad of New Jersey. Since both the latter roads had extensive connections into the Eastern Middle and Northern Fields, this briefly gave the Reading a very extensive hold on the majority of the industry. A key part of the plan of the 1892 Combination was for President McLeod to tie up the shipments of the individual operators by means of long-term tonnage contracts. To achieve this, even more in the way of incentive was offered, the Reading Coal and Iron Company agreeing to pay 60 percent of the New York price. The same company also agreed to purchase the output of the Lehigh Valley Coal Company and the Lehigh and Wilkes-Barre Coal Company on the same terms. Most of these contracts were for a period of seven years, though some were of shorter duration. A good example was the contract with the Babylon Coal Company in the Northern Field. The contract remained in force for the duration of the lease of the Lehigh Valley Railroad to the Reading, and the 'New York price' used was a complex average of prices received by the Coal and Iron, the Lehigh Valley Coal Company, the Lehigh and Wilkes-Barre, and the Lackawanna, at tidewater points between Hoboken and Perth Amboy. The 60 percent applied to all sizes above pea, until the price exceeded $4, when only 40 percent of the additional price was payable. Above $4.50 the percentage dropped to 30 percent. For pea and buckwheat No. 1 and No. 2, the starting percentages were 45, 32 and 20 respectively and various increments were payable as prices rose. In an unusual departure from many earlier contracts, the operators had the option to cease shipments entirely if prices fell below $3.25 for sizes above pea. A strict tonnage allotment was to be made to the colliery, based on the ratio of its rated capacity to the overall rated capacity of mines on the Lehigh Valley in the Wyoming Region. The rated capacity was to be determined by a 'standing committee' of three mining engineers, and was subject to revision every six months.[103]

These agreements met with favor from many individual operators, since they were now receiving the majority of the selling price of coal, unlike in earlier times. Hence, it was a problem for them when the collapse of the Reading leases led to abrogation of their contracts. Fortunately, the Lehigh Valley Railroad, once freed from its lease constraints, agreed to reinstate the agreements in large measure, paying the 60 percent (less a 15 cents a ton commission for selling the coal). However, the initial term was only one year rather than seven.[104]

Responding to this new competitive challenge from the enlarged Reading combine, Storrs recommended to President Sloan that the Lackawanna should make similar long-term contracts, before the terms became even more favorable to the operators! Further to that, two of their long-standing contractors, William Connell and William T. Smith of the Mount Pleasant Breaker, began asking for the new rates to be paid on their old contracts. On this occasion the Lackawanna was willing to indulge them, since the former had 1.7 million additional tons of coal that he wanted to put under contract, and Smith still had a large body of coal left that other railroads were keen to get their hands on. Since his existing contract would shortly expire, it was essential to persuade him to renew it for a further term. Having successfully negotiated these new contracts and committed the Lackawanna for the rest of the decade to these high prices, it must have been somewhat galling to find the Lehigh Valley shaving the 15 cents a ton commission from its contractors after the Reading lease collapsed![105]

Unlike the Northern Field, where contracts had long been an accepted part of business practice, some of the larger Lehigh operators continued to hold out against them. The Pardees of Hazleton refused to commit themselves to them and the Coxe Brothers set about building their own railroad, the Delaware, Susquehanna and Schuylkill in 1890–2, as mentioned in the previous chapter. This was the third major railroad scheme to date from this period. The Coxes' decision to invest on this scale followed their legal battle with the Lehigh Valley Railroad over freight rates on anthracite that they considered excessive in comparison to those on soft coal. Coxe Brothers had also had a related and serious disagreement with the Pennsylvania Railroad, because the latter had set freight rates from Drifton westwards to Chicago and Buffalo at a level 21 cents higher than from any other point in the coal regions. The avowed purpose of this was to force more of the Coxe Brothers tonnage on to the Lehigh Valley Railroad, with whom the Pennsylvania were working closely at the time.[106]

Despite Jones's assertion, in 1914, that one of the main aims of percentage contracts was elimination of the independent operators, it is difficult to find compelling evidence that this was actually the case prior to the establishment of the railroads' 'community of interest' at the end of the 1890s. No such statement has been found in the voluminous correspondence of the Lackawanna Railroad coal department or in the papers of the other railroads consulted in the course of this study. Indeed, as a counter-example, it was reported as late as 1893 that the Reading was actively considering giving up mining and farming their property out to small operators, to be worked on a royalty basis.[107] Rather, as this examination of the structure and evolution of the various contractual mechanisms has already shown, such arrangements were of very long standing. The symbiotic relationships between railroads and operators, lubricated by loans, special payments, and suitable contractual adjustments when necessary, provided important elements of stability in the industry, especially in times of recession. The combination of contracts and associated loans also facilitated the early development of more marginal or risky properties with uncertain coal reserves. Had the aim of the contracts been to drive the smaller operators out of business, it might have been expected that the proportion of overall coal revenues per ton accruing to the transportation interest, rather than the mining interest, would have increased progressively. In fact, competitive pressures and the constant drive for tonnage produced the reverse effect, with the benefit increasingly being felt by the mine operators rather than the shippers of coal.

Equally, however, it could not be denied that the individual and independent operators had been important forces behind the series of new railroad schemes from the 1880s onwards that had been particularly prominent in the Northern Field. These schemes had been pressed forward with every major surge in business confidence and had lain dormant in every recession. And so it proved with the final turn of the competitive wheel, until the Morgan-inspired community of interest between the railroads effectively ended the role of smaller operators as a competitive force.

FINAL COMPETITIVE CHALLENGES AND THE COMING OF THE 65 PERCENT CONTRACTS

The first stirrings of competitive activity, after several years of depression in the mid 1890s, began with the Susquehanna and Western. In 1896–1897, it chartered and built the Susquehanna Connecting Railroad from Suscon on the Wilkes-Barre and Eastern Railroad below Moosic to the Greenwood breaker in Minooka, with a branch to Jermyn's mines at Old Forge. Its charter provision to extend as far as Winton, above Scranton, was never fulfilled. The purpose of the road was to reach its mines and contractors directly over its own tracks, without needing to rely on trackage over the Lehigh Valley and Delaware and Hudson.[108] In the process, it also put itself in a position to divert Delaware and Hudson and Pennsylvania Coal Company tonnage away from the Erie, a situation that the latter could not countenance for long. In January 1898, J.P. Morgan began to buy up Susquehanna and Western stock for the Erie interest and a formal lease was executed in the following month. At the same time, the Hillside Coal and Iron Company, the Erie's coal subsidiary, took over the equivalent subsidiary of the Susquehanna, and Western, namely the New York, Susquehanna and Western Coal Company. Among its assets, the latter had a 60 percent perpetual contract for all the coal from Jermyn's four mines above and below Scranton.[109]

The second railroad initiative originated, as was now customary, with dissatisfaction among the smaller shippers over the operation of the 60 percent contracts. They sought a still higher proportion of the tidewater price, as the old contracts dating from 1892–1893 were coming to an end, but members of the cartel naturally resisted this. Hence, yet again the operators organized a new company to build a further independent road to tidewater, which would purchase coal for 65 percent of the New York, price. The new road, called the New York, Wyoming and Western, was organized in May 1898. E.B. Sturges, a prominent coal operator who had previously promoted both the Susquehanna and Western and Ontario and Western projects, was elected president. Again, in the usual way, the new railroad was promised several million tons of coal reserves not then bound to existing contracts. Most importantly, Simpson and Watkins, major shareholders in the new project, offered 500,000 tons of shipments. Simpson and Watkins was one of the limited number of independent (as opposed to individual) operators in the Northern Field, i.e., their entire tonnage was not bound into long-term tonnage contracts. They were associated with a number of mining operations, such as the Edgerton Coal Company and the Mount Lookout, and they took over the lease of the Forty Fort and Harry E collieries in 1893. In 1899 they purchased the Sterrick Creek breaker near Peckville, from under the nose of the Erie.[110]

It appears that it was at this point that the members of the cartel, and the bankers who backed them, finally decided that further competition and the associated downward pressure on freight tariffs could no longer be countenanced, and the activities of the individual and independent operators had to be brought under control, once and for all. The plan to achieve this began with the buy-out of Simpson and Watkins, which withdrew so much tonnage from the proposed new road that the project had to be abandoned. The buy-out was achieved by means of a corporate vehicle known as the Temple Iron Company. An extended discussion of the background to this company is given in Jones, based on testimony from a Sherman anti-trust case, so only the main points need to be re-iterated here.[111] The idea for using the Temple Iron Company came from President Baer of the Reading. The company had a very broad charter and he reorganized and refinanced it, arranging for its shares to be held by a majority of cartel members in proportion to their anthracite shipments in 1898. It is interesting to note that neither the Delaware and Hudson, nor the Ontario and Western were listed as shareholders in the company.[112] In other words, he created a device for joint ownership of coal properties by a group of the major cartel players, rather than any individual railroad, and it would be some years before anti-trust lawyers would finally catch up with this masterstroke of corporate engineering!

Now duly configured to do the bidding of the cartel, the Temple Iron Company purchased the collieries of Simpson and Watkins in a complex deal brokered by J.P. Morgan and Company, for the substantial figure of $5 million. Given that the Ontario and Western was not among the Temple Iron Company stockholders, it is also interesting to note that a matter of days before Simpson and Watkins sold out, it was reported that the Ontario and Western were also trying to buy their mines. This suggests that the cartel was not operating entirely harmoniously, even at this time. Although they failed to acquire the Simpson and Watkins properties, the Ontario and Western were more successful in acquiring the coal estate of the Lackawanna Iron and Steel Company in Scranton for $4 million, when the latter decamped with minimal notice to Buffalo. This purchase greatly surprised the Lackawanna, unaware it was about to happen. This provides further evidence that the Ontario and Western was pursuing a somewhat independent course from the rest of the combination.[113]

Having successfully headed off a second challenge to their control of the industry, the major partners in the combination were quickly faced with a third and final challenge from an unexpected quarter. The Pennsylvania Coal Company, which had operated the Erie and Wyoming Valley Railroad from the Northern Field to Hawley, decided to extend its line eastwards to Rondout on the Hudson River, to give itself an independent outlet to tidewater and remove its reliance on the Erie. In an ironic move, considering the extended legal battle between the two companies several decades earlier, it purchased the abandoned roadbed of the Delaware and Hudson Canal to provide the ready-made route to tidewater. Although its own coal shipments were substantial, of the order of 2 million tons in 1899, it sought to attract other coal operators to use its new line by offering the new 65 percent purchase contracts as an inducement. Naturally, such a move attracted considerable support from the dwindling numbers of individual and independent operators and caused panic in the Boardroom of the Erie, which stood to lose a huge amount of coal tonnage. President Thomas of the Erie persuaded J.P. Morgan and Company to acquire a controlling interest in the Pennsylvania Coal Company, in a successful attempt to halt the construction of the new road. Purchase of this controlling interest was a costly and difficult undertaking, owing to the prosperous condition of the Coal Company. The shares in the latter, together with all the stock of the Erie and Wyoming Valley Railroad and the company chartered to build the new tidewater road, the Delaware Valley and Kingston Railroad, were finally turned over to the Erie in January 1901 for the handsome consideration of $32 million.[114]

The rapid closing-off of these potential avenues for future competition greatly weakened the position of the remaining individual and independent coal operators as a bargaining force within the industry. In the final stage in the evolution of the purchase contract system within the period of interest, they were unwittingly assisted by the resurgent power of trade unionism, under the guiding hand of the great John Mitchell. As part of the settlement of the 1900 coal strike, the miners demanded that all parties in the industry agree to their wage advance, including the individual operators. The latter immediately sought freight-rate concessions from the cartel to make up for the wage advances. After various meetings, a solution was found in a standard 65 percent purchase contract. Most of the provisions were similar to those of the earlier 60 percent contracts described above, but, crucially, the higher price received by the operator was traded off against future independence of action, since these contracts were not for a limited period of years, but for the life of the mine, thereby tying the output in perpetuity to the shipping railroad. The converse of this for the transporting company was that, although it obtained lower revenue per ton on shipments under these contracts than from its published freight rates, this was offset by the guaranteed tonnage and the removal of future competition. Jones provides a very interesting table, which analyses Lehigh Valley Coal Company freight rate data in the period 1900–1901. This shows that the 65 percent purchase contract tonnage paid 12–36 cents a ton less for transportation than the published rates for non-contract tonnage, a measure of the value of guaranteed tonnage to the company.[115]

As the 60 percent contracts expired, operators progressively switched to the 65 percent agreements or were bought out in the years after 1900. The actions of the Erie, in the years 1898–1901, had substantially increased its coal estate and raised its standing within the cartel. Likewise, after 1900, the Lehigh Valley Railroad and its coal company were active in buying out operators tributary to its lines, culminating in the takeover of Coxe Brothers and the Delaware, Susquehanna and Schuylkill Railroad in 1905 for $18.4 million. This further rationalized the cartel and finally silenced one of the most influential independent voices in the industry. More pragmatically, it also stopped the Pennsylvania Railroad buying the property and diverting its tonnage westwards![116]

Summarizing the evolution of contractual ties between the railroads and the individual operators over the period, several points can be made. In the early years, the operators were beholden to the railroads because of a substantial element of spatial monopoly power on the part of the latter. Few operators enjoyed the luxury of being linked to more than one railroad (thereby engendering competition for their business) because of the limited number of lines that had been built. As a result, approximately ⅔ of the tidewater price accrued to the transporting companies. Over time, as more lines were constructed, the element of spatial monopoly was progressively whittled away, until by the turn of the century, the proportions were reversed and the 65 percent contracts meant that almost ⅔ of the tidewater price was now paid to the operator. However, in the course of this process, the freedom of action of all but a few large mining concerns to ship over roads of their choosing was also steadily eroded, until the majority of tonnage not actually owned by the railroads was bound to them in perpetuity by the same 65 percent contracts.

At the same time, as the operators found themselves progressively ceding economic power to the transporting companies, their relative importance in the trade was also declining. However, this decline only became rapid at the very end of the period of interest. Prior to that they remained of considerable importance, not least because some railroads, such as the Lehigh Valley, continued to have a policy of using tenant operators to develop some of their lands.[117] As late as 1890, in terms of actual tonnage mined, the Anthracite Coal Operators Association calculated that 42 percent of all coal was produced by individual and independent operators. By 1900, however, the figure had fallen to 29.7 percent and there was a further sharp reduction to 17 percent in 1901, following the wave of takeovers of mining operations by the railroads at that time.[118] Based on ownership of coal resources, rather than annual output, John Mitchell, president of the United Mineworkers Union asserted, in 1901, that 90 percent of all anthracite was owned by the railroads. Again this was a sharp increase from his figure of 75 percent for the period immediately prior to the coal strike of 1900. These statements were accompanied by a prediction that the Morgan-backed community of interest among the railroads would soon own 'absolutely all anthracite coal'.[119]

The trade press provided a rather sad epitaph for the loss of any control within the industry by the mine operators in 1900. Reporting on the sale of yet another concern to one of the railroads, the Coal Trade Journal declared in relation to the operators that 'their force consists of wind and the supply of that is inexhaustible'.[120]

While their actions and decisions may have become of limited consequence by the very end of the 19th century, it should not be assumed that this had previously been the case. Indeed, in the Northern Field, where much of the growth during the period had been concentrated, management of the relationships and contracts with the operators was of such importance to the railroads, in the maintenance of their share of overall output, that they devoted as much, or more time to it than they did to the management of their own lands held in fee or by lease. In addition to this, the development of their own lands was undertaken with close regard to the commercial requirements of their contractors. The changing nature of these broad management strategies will now be examined in general terms, because of the light they shed on the evolving economic relationships with the industry. More detailed points arising from this discussion will be taken up in the following chapters.

THE MANAGEMENT OF LARGE COAL PROPERTY PORTFOLIOS

The discussion in the previous section has already highlighted a number of aspects of the complex 'decision space' faced by managers of large coal properties. However, the general coal agents were left in no doubt by the presidents of their parent railroads as to the fundamental principle that was to guide their actions. As President Sloan observed in July 1879:

> 'Our investment capital in transportation is large and heavy and we cannot make it pay unless we control tonnage'.[121]

Or again, when Storrs had the temerity to suggest to Sloan that the company might divest itself of a recently acquired mine, because the tract on which it was located was limited in size:

> 'We must not under any circumstances part with a ton or acre of our coal lands and properties . . . for we have too much money invested in the transportation department to risk the loss of our coal lands or *any portion of them*!' (Sloan's emphasis).[122]

Thus, for the major transportation companies, the primary aim was not to maximise profits in the coal business per se, but to safeguard long-term tonnage, and hence cashflow from tolls, for the railroads. Against this background, however, cyclical business fluctuations and the normal vicissitudes of the trade left a fair degree of scope for tactical and strategic decision-making by senior managers. At the same time, the workings of the cartel introduced important constraints on the volume and timing of shipments, and significant increments to mining capacity required extended periods of time to bring into production. Investment in undeveloped lands held in fee carried the opportunity costs of potentially lucrative investment elsewhere that was foregone, not to mention interest payments on mortgages and the burden of taxes. Similarly, undeveloped leases accrued growing advance royalty payments that depressed short-term earnings and required considerable investment of time and money to recover.

In the light of these and other detailed considerations that will be examined in the next chapter, it is instructive to compare the managerial responses of several of the companies over lengthy timescales. These might be measured in decades, longer even than the extraordinary partnership between Storrs and Sloan, which lasted more than thirty years prior to 1899. To do this, a main 'policy variable' has been calculated, based on data from newspaper reports and a variety of company Annual

Reports in both printed and manuscript form. The policy variable chosen is the proportion of annual tonnage shipped from the company's mines (on land owned in fee or leased), as compared with the proportion from contractors' mines (purchase or transportation contracts). It is interesting to note that no record has been found that unambiguously states that such a variable was actually used in decision-making. Despite this, the fact that the requisite data to calculate it can be identified over quite long periods, in some cases coupled with the results of the calculation themselves, imply very strongly that it must have been a key 'rule of thumb' in setting medium-term policy for the management of coal estates.

The clearest example of this can be seen in Figure 12.2, which compares the percentage of total shipments by the Lackawanna derived from its own mines, with the equivalent figure for its contractors. The lines on the graph are necessarily mirror images of each other, as the total in each year has to sum to 100 percent. Despite some early gaps in the record, the overall picture, stretching back across almost the entire second half of the 19th century, is very clear. Most striking of all is that a number of major policy shifts can be identified. The triggers for these shifts correspond to major financial events and upheavals of the period, and the banking panics of 1857, 1873 and 1893, in particular. Looking at the graph in more detail, the phases can be described as follows. From 1854–1857, the company saw a rapid shift from a 30 percent shipment contribution by contractors to a figure closer to 70 percent in a very short time. Thereafter, a long decline in the proportionate shipments by contractors took place that lasted until 1873, although it was punctuated by brief periods of a 50:50 split and then a 40:60 split with company output. Thereafter, the contractors' share was allowed to increased again until it reached 50 percent, an approximate figure that was broadly maintained for an extended period of years until 1893, when again

FIGURE 12.2
Delaware, Lackawanna and Western Railroad Coal Department: Percentage of Total Shipments from Company's Mines and Contractors' Mines 1853–1900.

the contractors' contribution went into a sharp decline that persisted until the end of the period. These policy shifts cannot be attributed to any single cause or mechanism, neither are they random fluctuations, which makes the association with the Kuznets or Juglar cycle peaks all the more remarkable (cf. Table 2.1, chapter 2). For example, a different representation of the same data, in terms of absolute tonnages (Figure 12.3), suggests the output of the company and its contractors grew at broadly the same pace until the end of the Civil War. Thereafter the company outperformed relative to its contractors, largely because of a substantial amount of mine construction activity, which will be examined more closely in the next chapter. This expansion was deliberately reversed after 1873 as the longer depression set in. The Lackawanna 'mothballed' or suspended operations at a number of its mines, such as Archbald and Oxford, to leave a greater share of total tonnage to its contractors.[123] This reduction in output was sharply reversed in 1879, with the return of business confidence. Further increases in company output matched the contractors initially, although company shipments lagged a little behind as the 1880s progressed.

Unlike the period prior to the 1873 panic, when rising absolute shipments by the contractors had corresponded to a relative decline in their proportion of shipments, the period after the 1893 panic was characterized by both relative and absolute decline in the importance of the contractors. The main trigger for this was undoubtedly the diversion of the tonnage of the New York, Susquehanna and Western to the newly completed Wilkes-Barre and Eastern Railroad, which has already been described earlier in this chapter. However, as Figure 12.4 indicates, this was not the whole explanation. The Susquehanna and Western tonnage was classified under the transportation contract heading. While the loss of that tonnage is very clear in this more detailed breakdown of

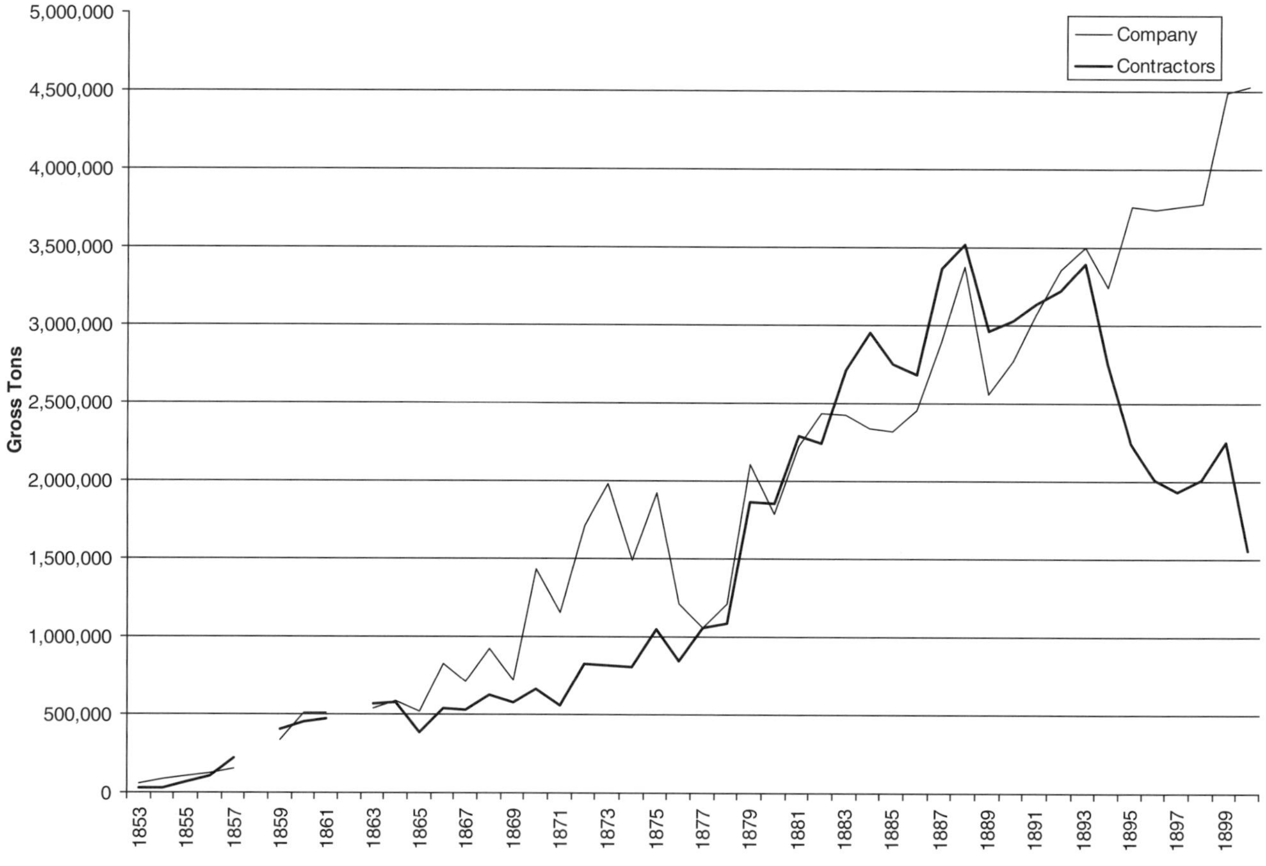

FIGURE 12.3
Delaware, Lackawanna and Western Railroad Coal Department: Tonnages Shipped from Company's Mines and Contractors' Mines 1853–1900.

FIGURE 12.4
Delaware, Lackawanna and Western Railroad: Shipments by Type of Contractor 1876–1900.

shipments by type of contractor, it is also apparent that, after 1895, the tonnage supplied by purchase contractors also went into decline, falling more than 40 percent from the 1895 peak by 1900. These declines were compensated by sharp increases in the output of the company's own mines during the same period, as can be seen in Figure 12.3.

The combined interpretation of these three graphs indicates that not only were the policy changes deliberately initiated either by the company or, in the case of the Susquehanna and Western, by one of its contractors, but that the prevalence of percentage splits, such as 70:30 or 50:50, shows that a high degree of control over quite specific tonnage output targets was maintained over long periods of time. Major shifts in new capacity construction, or in suspension of operations, also had significant impacts on the local patterns of employment during business-cycle fluctuations, e.g., in the Scranton area, where many of the Lackawanna mine closures/suspensions were concentrated during the 1870s depression.

Although the period of available data is very short, the same ability to establish precise proportions of shipments between company and contractors' mines is apparent for the Delaware and Hudson. As Figure 12.5 shows, it appears that the 70:30 model predominated, as the company moved to establish a policy of producing the majority of its output from its own mines during the Civil War period.

A longer period of record is extant for the Lehigh Valley Railroad, although the 'policy variable' is measured somewhat differently. The Lehigh Valley shipped significant amounts of coal from the Eastern and Western Middle Fields, where the tonnage contract system had not been widely

FIGURE 12.5
Delaware & Hudson Canal Company: Percentage of Total Shipments from Company's Mines and Contractors' Mines 1860–1868.

used, in contrast to the Northern Field. Hence, a simple division into company and contractors' mines was not appropriate. Instead, the company reported three subdivisions of its output, namely coal from its own mines, from tenants on its lands, and from other shippers. Mines in the last of these three categories from the Northern Field may well have had shipping contracts, but mines in the other fields in general did not, prior to the 1890s. The percentage contribution to overall tonnage by the three different types of shippers is shown in Figure 12.6 for the last two decades of the 19th century. At the start of the period, only 20 percent of shipments came from tenant mining companies on property owned or leased by the railroad or its subsidiary, the Lehigh Valley Coal Company (LVCC on the graph). This percentage rose throughout the 1880s, peaking at nearly 50 percent in 1893. This increase was matched by a corresponding decline in the tonnage of 'other shippers' during the same period. However, the sharp decline in the output of the tenant mining companies during the mid 1890s did not result in a compensating increase from the 'other shippers'. Instead, much of the shortfall in tonnage was made up by expansion of shipments from the mines operated by the Lehigh Valley Coal Company, with some recovery in shipments by the tenants in 1901. The absolute tonnages shipped by the different categories (not shown on a graph) indicate a very similar pattern to the percentages. They also show that the Lehigh Valley Coal Company pursued a policy of shipping a target output of approximately 1.5 million tons from its own mines throughout the 1880s, as the tonnages were all very close to this figure. Fluctuations in demand were therefore absorbed either by the tenants or the 'other shippers'. These findings lend support to the contention, earlier in the chapter, that the railroads did not pursue an active policy of trying to eliminate independent operators

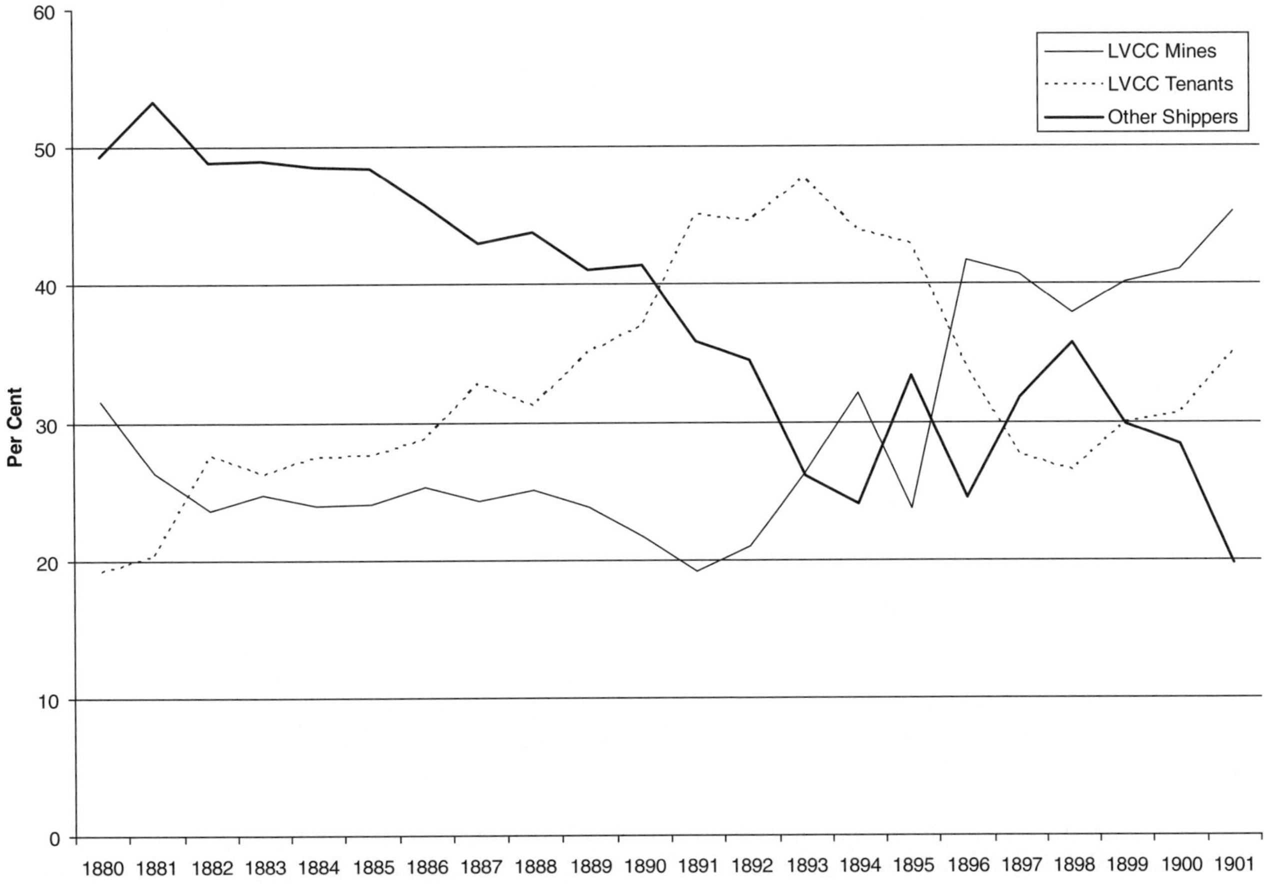

FIGURE 12.6
Lehigh Valley Railroad: Percentage of Total Shipments by Category of Shipper 1880–1901.

prior to 1893. Confirmation of this strategy in the case of the Lehigh Valley, as late as 1890, can be found in the Annual Report for that year, which in turn referred to a long-standing policy:

> 'It will be noted that the coal shipped by our tenants has been steadily increasing. . . . This accords with the statement made in the annual report for the year 1873, as follows:
>
> "The policy of the company has always been to encourage development by private partners, and in a great degree, reserve our own lands for future workings"'.[124]

Thereafter, in the case of the Lehigh Valley, a combination of takeover of other mining concerns, acquisition of new lands, and development of its existing estate led to an increasing proportion of its shipments being under its own direct control. The continued expansion of the Lehigh Valley's coal estate during this period will be examined further in the next section.

In conclusion, there is clear evidence that each of the companies examined utilized well-defined policies, in terms of shipment percentages or absolute tonnage quotas, to control the broad allocation of shipments between their own mines and those of individual or independent operators. In the case of the Lackawanna and the Lehigh Valley, there is no compelling evidence that these companies consistently favored their own mining interests over those of their tenants or independent operators prior to the 1890s. Indeed, for the Lackawanna during the 1880s, the evidence would point in the reverse direction.

Similarly, the impact of major cyclical events, such as the Juglar- and Kuznets-cycle peaks and the financial revulsions that followed them in the shape of the various banking panics, has been shown to have an important influence on the timing of policy changes or re-assessments. The non-recurrent event of the Civil War also had this effect on the Delaware and Hudson, although not on the Lackawanna. Since these policy changes affected the relative pace of mining development both within and between coalfields, and, by extension, employment opportunities, population growth, and urban development, they provide a key means of linking economic change at the national level with corresponding adjustments at the regional and local levels.

THE SPATIO-TEMPORAL DYNAMICS OF LAND ACQUISITION AND LEASING CYCLES

Although a number of key aspects of the initial rush to acquire coal lands have already been documented in chapter 10, the comparative experience of some of the different companies in securing their coal estates now needs to be set in the wider context of changing business-cycle conditions over an extended period of time.

Given the extremely large number of land transactions involving the anthracite-mining and -carrying companies over the period, any comprehensive analysis would be infeasible. Fortunately, for two companies, the Delaware and Hudson and the Lehigh Valley Coal Company, records summarizing land transactions have survived, although in the case of the latter company, substantial additional effort was needed to process the data into a usable form for analytical purposes. To provide a further comparison, similar data were also collected for the Lackawanna Railroad coal department. This proved to be a very complex and onerous task, requiring extensive recourse to the original deed-book records in the Luzerne and Lackawanna County courthouses. Nevertheless, it has been possible to generate broadly comparable data series for these companies that shed important light on the relative timing of land and lease acquisition from the Civil War era onwards.

The temporal patterns of land purchases by the different companies are shown in Figures 12.7–12.9. A number of issues had to be addressed in preparing these data series, and the details of these can be obtained by consulting the notes to the figures. Hence, the discussion here will focus on the interpretation of the data rather than on how they were obtained. As the title to Figure 12.7 indicates, up to four elements could be involved in land purchase. The first of these was land purchase in fee, as normally understood. However, it might be restricted to underground mining rights without surface ownership, or vice versa. More unusually, both mining rights and surface rights were included, although the acreage of coal and the acreage of surface might differ. The second element was lands held in trust. These were lands purchased in the interest of the company, but held in the name of a company director or third party. The aim was either to circumvent charter limits on landholding, as noted earlier, or to conceal from the casual observer the extent of a policy of land acquisition. The Lackawanna Railroad extensively used this method over quite a lengthy period of time, so it is specifically highlighted in the title to the figure. Similar methods were employed by the other companies represented, but to a much lesser degree and for shorter periods. Thirdly, the cost of leaseholds has been included. In general, coal leases formed a distinct method of controlling coal lands and there was no initial payment, only a schedule of royalty payments on tonnages mined or to be mined, as earlier sections of this chapter have shown. However, on occasion, lessees were paid initial sums to secure the agreement, or payments were made to acquire leases originally executed between other parties.[125] Such sums were traditionally distinguished as capital items in company accounts, unlike normal lease payments, which were classed as expenses. This approach is continued here. Finally, if a company acquired lands that were already partly developed by one or more mining improvements, the cost of these improvements, though not the personal property (i.e. mules, mine supplies) was usually included in the price, without any breakdown of the overall cost being available.

Taking these different categories of expenditure together, in Figure 12.7, gives a good indication of the pattern of expenditure on land-related resources. The striking peak on the graph in 1870 is the result of the $3 million merger of the Nanticoke Coal and Iron Company into the Lackawanna Railroad, which brought with it a variety of coal properties and several mines, including the Avondale and Taylor breakers (see Chapter 10). However, as the notes to the figure explain in more detail, many of the Nanticoke Company's lands had originally been bought in earlier years by Lackawanna directors and transferred into the subsidiary holding company at various times. Using the original deed records, these

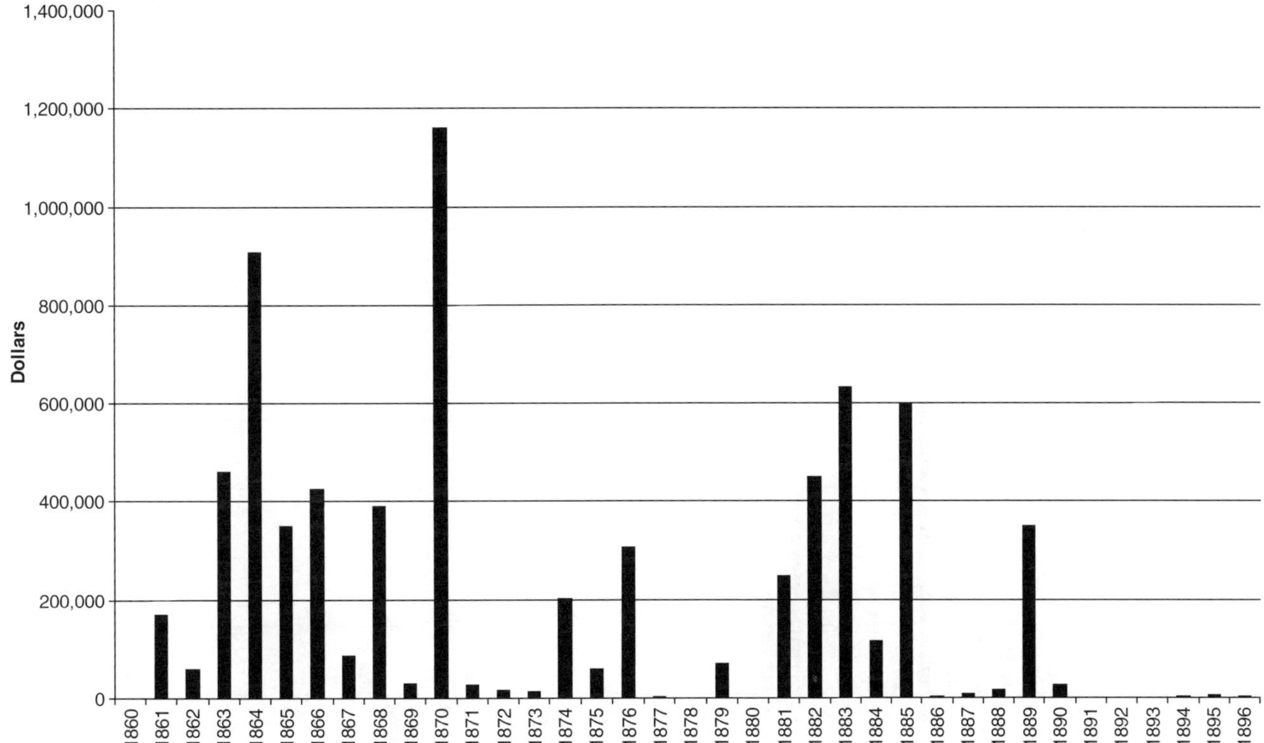

FIGURE 12.7
Delaware, Lackawanna and Western Railroad Coal Department and Subsidiary Coal Companies: Estimated Purchase Costs of Fee Lands, Lands Held in Trust, Leaseholds, and Mining Improvements 1860–1896.

purchases have been placed on the graph in the original year when they came under the control of the railroad. This has the effect of spreading the costs back as far as 1863 and correspondingly reducing the peak in 1870 to its correct level in investment terms, rather than accounting terms.

Based on these adjusted figures, it is apparent that the Lackawanna began a first major cycle of coal-land acquisition in the early 1860s, which ended with the Nanticoke Coal and Iron Company merger in 1870. The absence of purchases at high prices in the phase of speculation prior to the 1873 panic is noteworthy, as is the moderate level of land acquisition after the demand for new lands had collapsed during the aftermath of the panic. A second cycle of purchases coincided with the railroad boom of the early 1880s. The unexpectedly high level of expenditure in the depression year of 1885 is explained by the acquisition of several ailing or bankrupt individual operators, following the loss of confidence on Wall Street in the latter part of 1884. The mines in question were those of Stetler and Company, the Amity Coal Company, and the Fuller Coal Company, which were to become respectively the Hallstead, Holden, and Hunt breakers of the Lackawanna Railroad coal department.[126] Further discussion of this period will be found in the next chapter. Thereafter, acquisition remained at a low level, with the exception of 1889, where the figure is explained by the large payment made to secure the lease on the prime Nesbitt lands near Plymouth.[127]

The Lackawanna graph contrasts sharply with that for the Delaware and Hudson (Figure 12.8). In this case, almost all the significant land acquisition activity is concentrated in the period 1864–1874, including the key purchase of the Baltimore Coal Company's lands in 1867, which has been described in chapter 10. It is apparent that the Delaware and Hudson activity followed on from that of the Lackawanna in the 1860s, but the pattern was not repeated in the early 1880s, when the former company scarcely entered the land market.

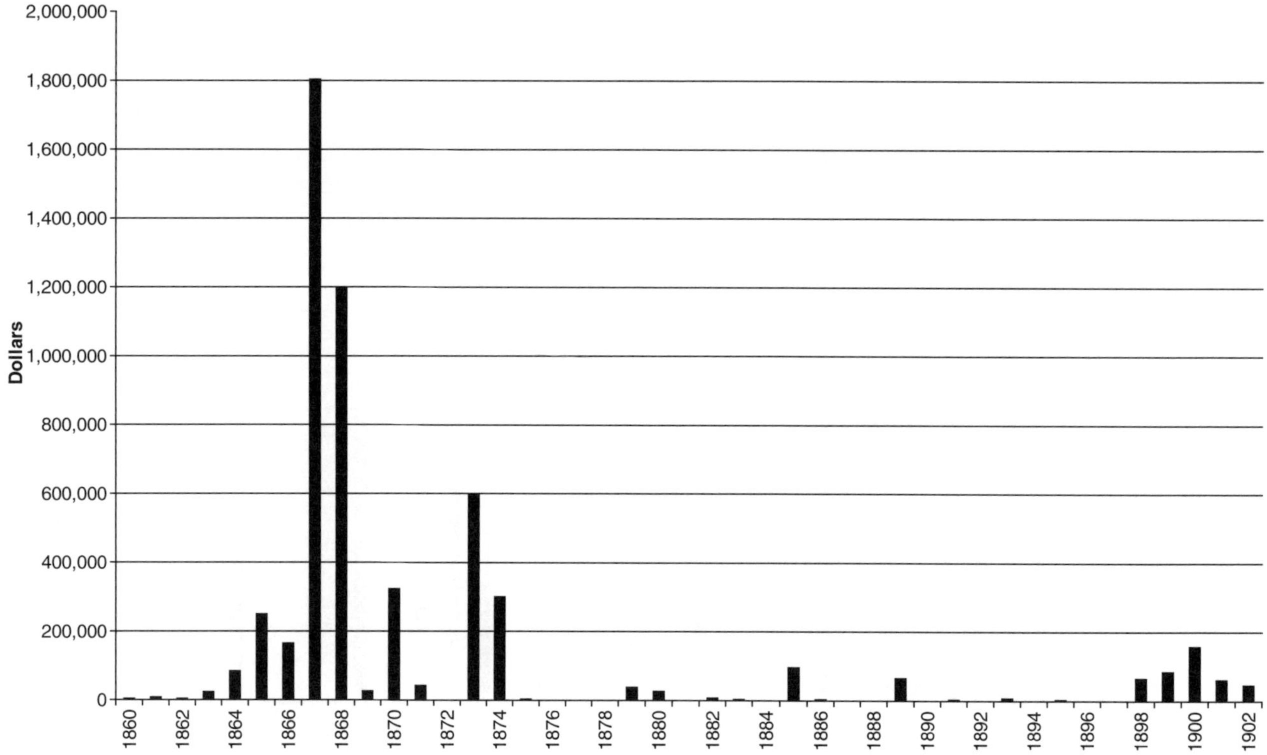

FIGURE 12.8
Delaware and Hudson Canal Company and Northern Coal and Iron Company: Purchase Costs of Fee Lands and Mining Improvements 1860–1902.

The data in Figure 12.9 show that the experience of the Lehigh Valley Coal Company, which acquired lands during the period right across the anthracite fields, differed again from that of the two Northern Field companies already examined. A comparative latecomer into the business of coal-land acquisition, the Lehigh Valley made substantial investments between 1868 and 1874, many of them in the period of higher prices prior to the panic, when the Lackawanna, at least, had largely withdrawn as a buyer. Some further acquisitions were made in the boom years following 1879, but the most noticeable difference was the continuation of investment from the late 1880s through into the depressed years of the mid-1890s. Several of the investments in this period were made in the Schuylkill Region.[128] It should, of course, be remembered that even the substantial purchases of the Lehigh Valley Coal Company were dwarfed by the burst of land acquisitions by the Reading Coal and Iron Company, the last entrant into the race. These acquisitions were, however, concentrated in a very short period of time in the early to mid-1870s, as chapter 10 has indicated.

The temporal pattern of lease acquisition reinforces the trends already evident from the land-purchase data, although the timing differs somewhat at a detailed level. Unfortunately, it has not been possible to analyse the Delaware and Hudson lease records, but the data for the Lackawanna and the Lehigh Valley are shown in Figures 12.10 and 12.11 respectively. As before, where subsidiary companies were established in the interest of the parent railroad from the outset, or where leases were held in trust, the original date of the lease is used. Where previously existing mining companies were taken over, the date used is the date of the takeover, not the date when the original leases were executed by the companies in question.

In the case of both railroads, it is very apparent that the major depressions of the 1870s and 1890s brought an almost complete cessation of leasing

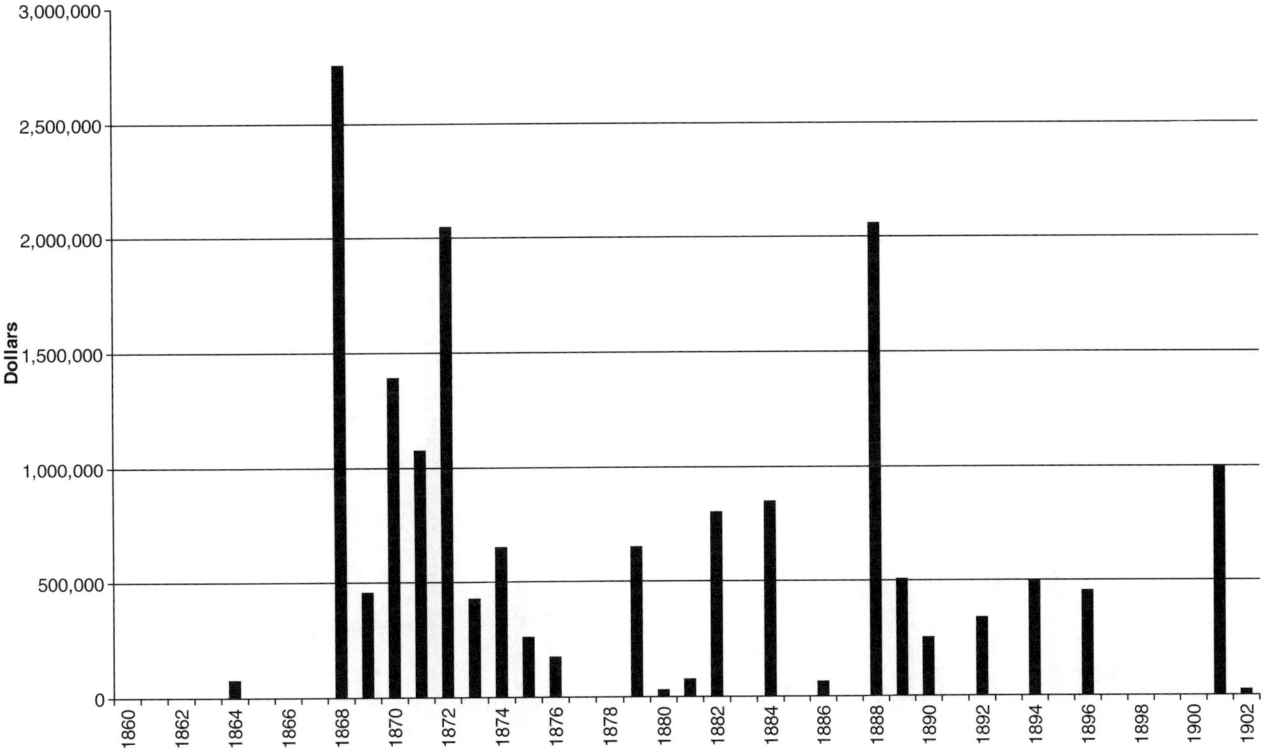

FIGURE 12.9
Lehigh Valley Railroad and Coal Company: Purchase Costs of Fee Lands, Leaseholds, and Mining Improvements 1860–1902.

activity. In periods where leases were being negotiated, it is apparent that the Lackawanna was most active in the earliest part of the period from 1860 to 1872, although the early and late 1880s are also quite well represented in the time series. In the same way as for land purchases, the activity attributed to 1885 is a result of the takeover of the several small companies listed above.

In contrast to the Lackawanna, the most vigorous period of expansion of the leasehold estate by the Lehigh Valley was in the years prior to the 1893 panic, although 1868–1874, the early 1880s, and the turn of the century are also well represented. This confirms the earlier impression that the Lackawanna was the prime mover in securing its coal estate by whatever means possible and the Lehigh Valley was a laggard across the board.

Another less obvious, but very important, aspect of the decision–making that lay behind the timing and extent of land and lease acquisition was the nature of the charter restrictions on landholding that applied to the different companies. In the course of creating or acquiring subsidiary mining and/or railroad companies, each of the large anthracite carriers obtained rights to hold varying proportions of coal lands, as discussed in Chapter 10, and the details do not need to be repeated here. However, in the case of the Lackawanna and the Delaware and Hudson, all the individual charter rights from the subsidiary companies pertaining to coal-land holding were themselves limited. Thus, for example, the Northern Coal and Iron Company could hold 3,000 acres of land and the Union Coal Company 2,000 acres, as part of the 25,000-acre total owned by the Delaware and Hudson as a whole.[129] Similarly, the overall limit for the Lackawanna was 20,000 acres after all its subsidiary companies had been absorbed and their land-holding rights totalled up.[130]

In contrast, the Green Land Company, one of the predecessor companies to the Lehigh Valley Coal Company, had originally had a limit of 5,000 acres of land ownership written into its 1871 charter, but this limit had been removed by a charter supplement in

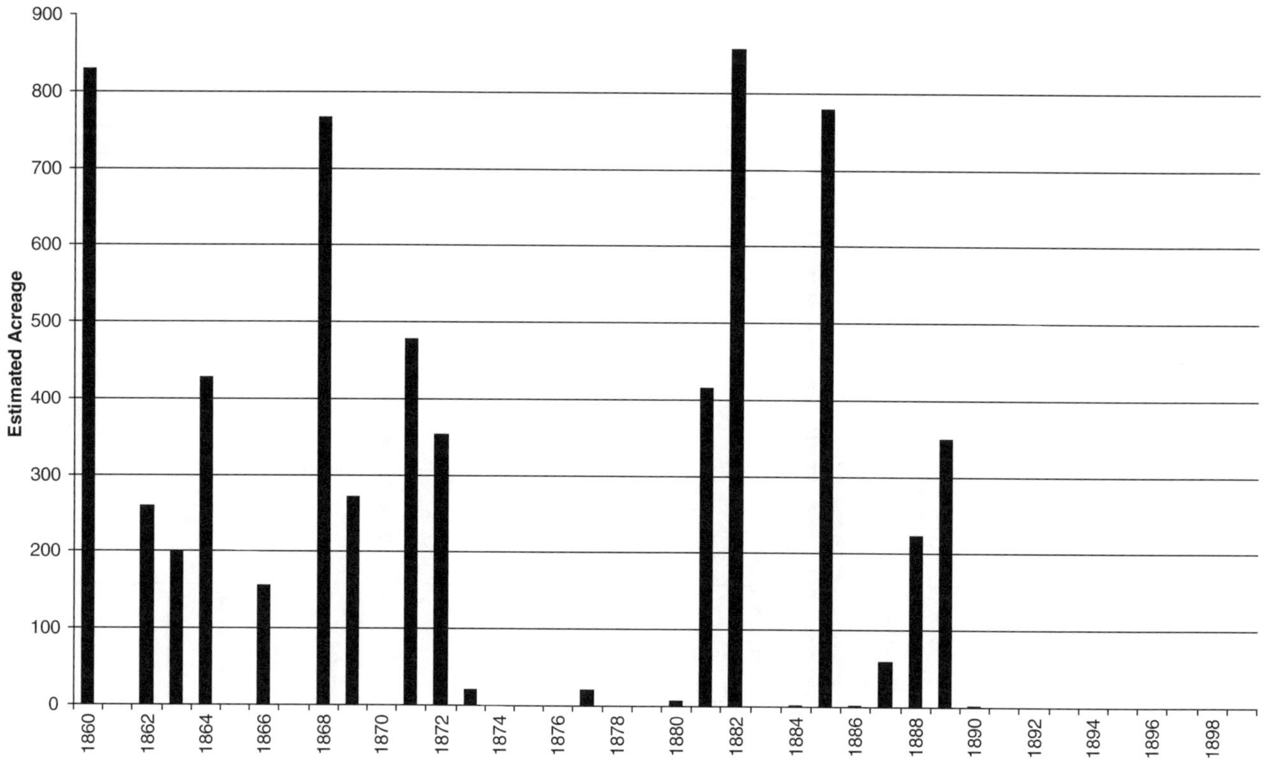

FIGURE 12.10
Delaware, Lackawanna and Western Railroad Coal Department: Lease Acquisition and Renewal 1860–1899.

April 1872, two months before the charter was purchased by the Lehigh Valley.[131] With equal deftness, the charter of the Laurel Run Improvement Company, precursor to the Reading Coal and Iron Company, also avoided placing any limits on the acreage of coal land that the company could hold.[132] Although mounting debt and financial stringency rapidly ended the late-stage orgy of land acquisition by the Reading, the Lehigh Valley was able to continue to expand its land portfolio at opportune times throughout the remainder of the century, as Figures 12.9 and 12.11 have shown. A summary tabulation, from 1906, states that the Lehigh Valley Coal Company had almost doubled its coal-land holdings to 37,378 acres from the 18,702 acres held at the time of the 1873 panic (see Table 10.2). Of the 1906 total, approximately two-thirds were lands held in fee and the remainder by lease.[133] It is interesting to speculate on the possible effect of this relative lack of legal constraint on management thinking within the Lehigh Valley Coal Company. It meant there was not a need for regular re-evaluation of the commercial value of different parts of the estate, to ensure that only potentially productive lands were being held and only the best lands were being acquired for future use. This contrasts strongly with the acreage-limited Lackawanna, which had to monitor its land holdings closely and ensure that it divested itself of any unproductive property, while maintaining a small margin of safety for new acquisitions, without exceeding its 20,000-acre total.[134] The issue for the Reading was different again. Once the decision had been taken not to break up the property at the beginning of the 1880s, the necessary management focus became quite clear. Primary consideration had to be given to the most efficient and cost-effective means of developing large and contiguous blocks of coal territory, both in the short and long term, on the basis of existing mining improvements and prospective sites for new operation. A detailed examination of how successfully this was achieved will be undertaken in Chapter 15.

These findings also raise interesting questions about the relative merits of different strategies in general terms. Lands and leases acquired early in the period were relatively low cost, while prices per acre and royalty payments per ton rose progressively through the period. First-mover advantage allowed

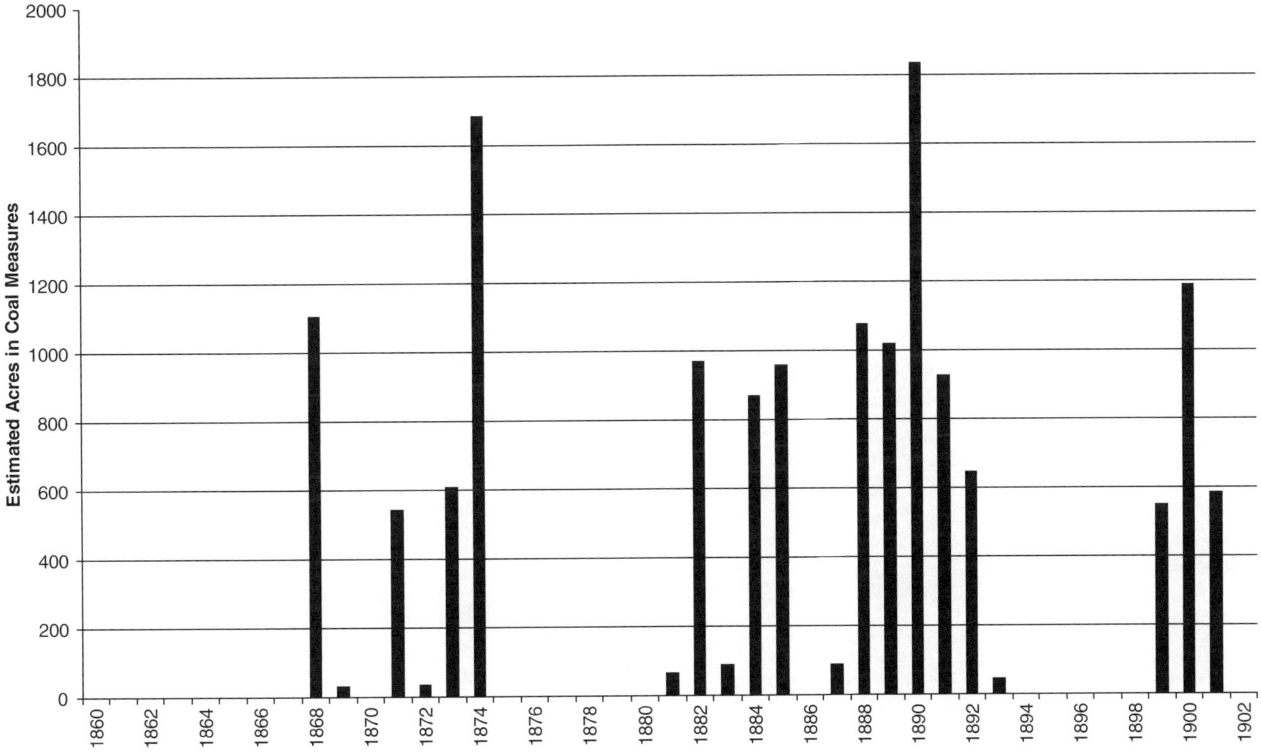

FIGURE 12.11
Lehigh Valley Railroad and Coal Company: Lease Acquisition and Renewal 1860–1902.

a company the greatest range of choice over premium lands, assuming that the landowner was willing to sell or lease, which was not always the case. However, one of the reasons that land was initially cheap was that mining improvements and transportation networks were relatively undeveloped and considerable investment would be needed to realize a return on the coal resource in the ground. Purchasing land that could not provide this return for many years therefore had significant opportunity costs associated with it, as well as real costs associated with the payment of annual taxes.

In the case of coal leases, the inexorable increase of advance royalty payments, based on the annual minimum required tonnages, raised similar questions about investments that were unproductive in the short term. Leases executed in the early period were generally at favorably low rates, but those dating from the 1880s onwards tended to have increasingly higher rates and comparatively large minima. Under these circumstances, advance royalty payments could quickly become a costly burden on a company's finances. However, delaying the acquisition of leases to avoid such costs was a very risky undertaking, as they would likely be snapped up by competitors and removed from the marketplace, almost in perpetuity. Long-term loss of tributary lands and tonnage was to be feared more than short-term cash-flow problems, since it undermined the very basis on which the growth and expansion of the anthracite railroads was predicated. Nevertheless, as the estates of the different companies expanded both in size and the level of associated financial obligations, effective management of the development of coal lands was to become one of the most pressing issues faced by the railroads in the latter part of the 19th century. The different approaches to the management challenge and the varying extent to which individual companies met the challenge, form the subject matter of the next three chapters. The detailed outworking of development decisions on the ground was to have a profound impact on the industrial geography of the region and the distribution of settlements and employment opportunities.

Chapter 13

REGION BUILDING I: FINANCING DEVELOPMENT IN THE MINING ECONOMY

The aim of the next two chapters is to investigate the adjustment processes involved in developing new productive capacity within the coalfields to meet increasing demand. Particular attention will be given to the Northern Field here, as the domain of the Reading Coal and Iron Company in the Southern and Western Middle Fields raised different issues, which will be examined in chapter 15.

These processes were of two main types. The first were financial in nature, the second were related to the operational decision-making and productive processes involved in making physical improvements to the coal estate in terms of new mine openings, coal breakers, and lateral railroads. Financial processes are considered in this chapter and operational processes in the next.

The discussion of contractual arrangements above has begun to elucidate the complex web of legal and financial interrelationships that increasingly bound together the anthracite industry in the second half of the 19th century. However, the financial aspects now need to be explored in more detail if the interrelationships are to be fully understood.

At its simplest, development of the mining economy was dependent on the availability of capital for investment in new productive capacity. Several potential sources of capital can be identified. Among these were retained earnings from ongoing operations; cross-subsidies from parent companies, usually railroads in the present context; and borrowings. Retained earnings implied the ability to generate profits. These in turn were dependent not just on variations in selling prices, which have already been discussed, but also mining costs, which have not previously been examined in any systematic way for the anthracite industry. This side of the equation must therefore also be taken into account, in relation to the ability or inability of concerns of different sizes to generate operating surpluses. Loan finance could be derived from several sources, depending on the nature of the concern involved. Larger incorporated companies had access to the financial markets via bond and stock issues. Small companies, partnerships, and individual entrepreneurs might borrow modest sums from wealthy backers or local banks, although the latter did not have the resources to fully fund mining developments on the scale required by the 1880s. Alternatively, they could turn to the railroads for loans, secured on future shipments, their mining assets, or both.

These different aspects of the financing of the mining economy will now be investigated and the interrelationships between them explained, using a variety of case-study examples drawn from different segments of the industry.

SOURCES OF DEVELOPMENT FINANCE: INTRODUCTION

While a number of potential sources of finance can be identified, it is important at the outset to make a distinction between the small individual mine operators and the mining subsidiaries of the large railroad corporations. The mining subsidiaries were legally of two kinds. The first group were companies with separate charters from the parent railroads, which issued their own shares and/or bonds, such as the Pennsylvania Railroad subsidiary, the Susquehanna Coal Company or the Lehigh Valley Coal Company. Most railroads adopted this approach. A second type was exemplified by the Lackawanna Railroad, which merged its earlier subsidiary coal companies into its own coal department, which did not have a separate corporate existence, although it did have separate accounts. Nominally, at least, subsidiary companies in the first group could go to the financial markets to raise capital for expansion by means of share or bond issues. In practice, this was rarely done. The shares, and often large numbers of bonds, were tightly held by the parent company or individuals associated with it, and new investment was financed by retained earnings, if indeed book profits were made at all. Failing that, necessary investment might continue regardless, funded by temporary loans or cross-subsidies from the parent. One good example of this type of behavior was the loan made to the Susquehanna Coal Company by the Pennsylvania Railroad, to facilitate its program of new mine construction at the beginning of the 1870s.[1] A second example was the substantial financial lifelines extended to the Lehigh Valley Coal Company in the 1890s, by the parent railroad, so it could meet its obligations and continue with investment in mining improvements. The mechanism involved payment of bond interest as it came due, amounting to the substantial sum of $5.4million in the years 1896–1902.[2] The relationship between the Lehigh Valley Railroad and its subsidiary, from 1875 onwards, was aptly summed up in the coal company's corporate history in the following words:

> 'The Railroad Company continued to advance the necessary funds for the acquisition of property, the development of the same, and the making of all necessary improvements preparatory to the mining of coal'.[3]

Hence the situation with the first group of subsidiary companies was in practice very little different from the second group, exemplified by the Lackawanna Coal Department which was also financially dependent on its parent railroad. However, the coal department did not have the option of issuing its own stock or bonds. In consequence, the subsidiary coal companies of both types were reliant on either their own retained earnings, if any, or those of their parent railroads to provide the financial wherewithal for mining improvements. The extent to which the mining operations of the different companies were indeed self-financing will be examined in the next section.

Among the individual and independent operators, it is also important to make a distinction between those that were incorporated companies and those that were not. While the majority fell into the former category, there were also a number of unincorporated smaller concerns, run by individuals or co-partnerships. Under the general incorporation laws of Pennsylvania of April, 1849 and its supplements, as they applied to coal companies in Luzerne County for example, any five persons who subscribed for $20,000 of stock could incorporate a company for mining purposes. Such companies could own up to 2,000 acres of coal land and issue up to $500,000 of stock without further recourse to the Legislature. An early example of one such incorporated company was the Hartford Coal Company, established in 1855 to conduct mining operations near Wilkes-Barre. The general legal provisions also allowed such mining companies to issue bonds to an amount not exceeding the amount of capital stock paid in.[4] Another contrasting example was the Honey Brook Coal Company, set up in 1864 to conduct mining operations in the Eastern Middle Field. This company required a special act of incorporation because it wanted to issue $3 million in stock, own 3,000 acres of coal land, and have less restrictive conditions on its ability to sell bonds, secured by mortgage on its property.[5] Such companies were therefore able to raise substantial amounts of capital on the financial markets to fund mining improvements. It is also no accident that the timing of incorporation, in both of the above examples, corresponded to peak periods in business activity in general, and mining industry speculation in particular, a point that has already been noted.

Apart from the financial markets, the other two sources of funds for investment were potentially open to both incorporated and unincorporated mining concerns. These were banks and the railroads. The former were doubtless important in many cases, but it is extremely difficult to find reliable evidence of the levels of their exposure to the coal industry at different times. One valuable insight can, however, be gained from the collapse of the Wilkes-Barre banking house of Bennett, Phelps and Company at the beginning of 1879. Ironically, this failure came the day after the 1870s depression ended on January 1, with the return to the gold standard. The bank collapse was directly attributable to the 'present inability of the Forty Fort Coal Co., Riverside Coal Co., J.H. Swoyer and others to meet their obligations'.[6] The quote is perhaps slightly misleading because Swoyer was involved in the Forty Fort and Riverside Coal Companies, so the failure was quite specifically generated by the cash-flow problems experienced by him and his business associates. In addition to carrying Swoyer's notes, a later inventory of the assets of the banking house revealed that it held 70 Riverside Coal Company bonds and 250 shares of the Wyoming Coal and Transportation Company. Mr Phelps, one of the proprietors of the bank, also personally owned 238 shares in the latter company (par value $20), whose president, coincidentally, was again J.H. Swoyer![7]

However, as these figures show, assuming a par value of $100 for the Riverside Coal Company bonds, the exposure of the bank to these and mining company shares was not large. While Swoyer was clearly using the bank for his working capital requirements, it is not apparent that the local banks had the resources to provide loans large enough for acquisition of coal lands or the construction of new mines. For example, in 1877, each of the First, Second, and Third National Banks of Scranton was only capitalized at $200,000, but Swoyer, on his Wyoming (or Riverside) colliery property alone, had mortgages of $175,000 on the land and $70,000 on the mining improvements. Sums of this kind would have severely taxed even the largest of the Philadelphia banks of the period.[8]

Aside, therefore, from the financial markets, the only remaining potential sources of funding for mining development were the railroads themselves, since only they had the financial strength to make loans of the size required to fund mining improvements. In addition to having a commercial rationale for making loans, in terms of the increase in productive capacity tributary to their lines, they also had a ready-made means of recouping the loan, by retaining a certain amount of the selling price per ton on all the coal they sold on behalf of the producer, in addition to the standard shipping tolls. As a result, the role of the railroad as 'venture capitalists' and bankers of last resort became extremely important, especially in the development of the industry in the Northern Field. Both the anticipated and unanticipated consequences of mining improvement loans made by the railroads will be examined in detail later in this chapter.

SOURCES OF FINANCE: PROFITS, LOSSES, AND THE AVAILABILITY OF RETAINED EARNINGS FOR INVESTMENT I: THE CORPORATIONS

Investigation of the financial performance of the different types of coal companies is made difficult by the relative paucity of accessible information for the major concerns and the general lack of archival materials pertaining to smaller companies. For example, the Annual Reports of the Lehigh Valley Railroad and the Delaware and Hudson contain very little financial information about their mining subsidiaries. The failure of the Lackawanna to publish Annual Reports during much of the period has already been noted in an earlier chapter. Only for the Susquehanna Coal Company, and to a certain extent the Reading Coal and Iron Company, do the printed reports contain relevant information for a worthwhile sequence of years. Fortunately, the main account books have survived for some companies, such as the Lackawanna and the Susquehanna Coal Company, and other archival sources can be brought to bear in the case of the Lehigh Valley Coal Company. Correct interpretation of the figures remains a challenge, however, as will be seen below. For individual operators, even less information is available, and that which survives will tend to favor the successful rather than the unsuccessful companies, for obvious reasons. Nevertheless, some indicative data can be provided by way of comparison with the larger concerns.

The first stage in this investigation is to determine whether the mining business of the railroad coal subsidiaries was in general profitable and, if so, whether that profitability tended to increase or decrease over time. Unfortunately, the lack of standardization in 19th-century accounting practices, both between firms and, indeed, within a single firm over a period of years, makes the apparently straightforward question difficult to answer with complete confidence. As a result, considerable effort has had to be expended to reduce the surviving figures to a broadly comparable basis. This should be borne in mind in interpreting the evidence now to be presented.

The simplest approach to the problem is to attempt to identify the annual net surplus or deficit on the operation of the business, after all charges have been taken into account. Put simply, subtraction of gross expenses from gross earnings yields the net earnings. From these all dividends, bond interest, and sinking fund payments need to be deducted, to give the final surplus or deficit figure. Short of bankruptcy, bond interest had to be paid regardless of financial circumstances, and dividends might continue to be paid, even in deficit years, if the company had an accumulated surplus from previous years. This immediately highlights a difference between a coal department of a railroad and a coal subsidiary with its own bonds and shares.

The latter would obviously have potentially larger deductions than the former. It should also be noted that the annual surplus calculations will already include the cost of construction charges for new mining improvements, so in that sense they record the financial results after retained earnings had already been re-invested in the property.

To give an indication of how these calculations work out in practice, Figure 13.1 shows the time pattern of earnings and expenses for the Susquehanna Coal Company over the last three decades of the 19th century. The graph reflects the surge in output during the 1880s as its parent, the Pennsylvania, increased its market share of the anthracite trade. Figure 13.2 takes the data from the previous figure and presents the summary of company financial performance after dividends and bond-related fixed charges have been subtracted. The period of accumulating reserves during the 1880s stands out very clearly, but the healthy situation is progressively reversed during the depression of the 1890s. As an interesting aside to these findings, between 1875 and 1900, the bonded debt was reduced from $2 million to about $1.5 million, and there was no change in the capital stock of $2.14 million. The latter is perhaps surprising as the company had invested in a number of major mining improvements during the period and its coal property on the Nanticoke side of the Susquehanna River was far from exhausted at the beginning of the 20th century.[9]

Further examples are given in Figures 13.3 and 13.4. The financial record of the Reading Coal and Iron Company is perhaps as dismal as the adverse fortunes of its parent railroad during the period would lead one to expect. The tracking of the 10-year Juglar cycle by the worsening losses is particularly striking. It should be noted that the accumulating

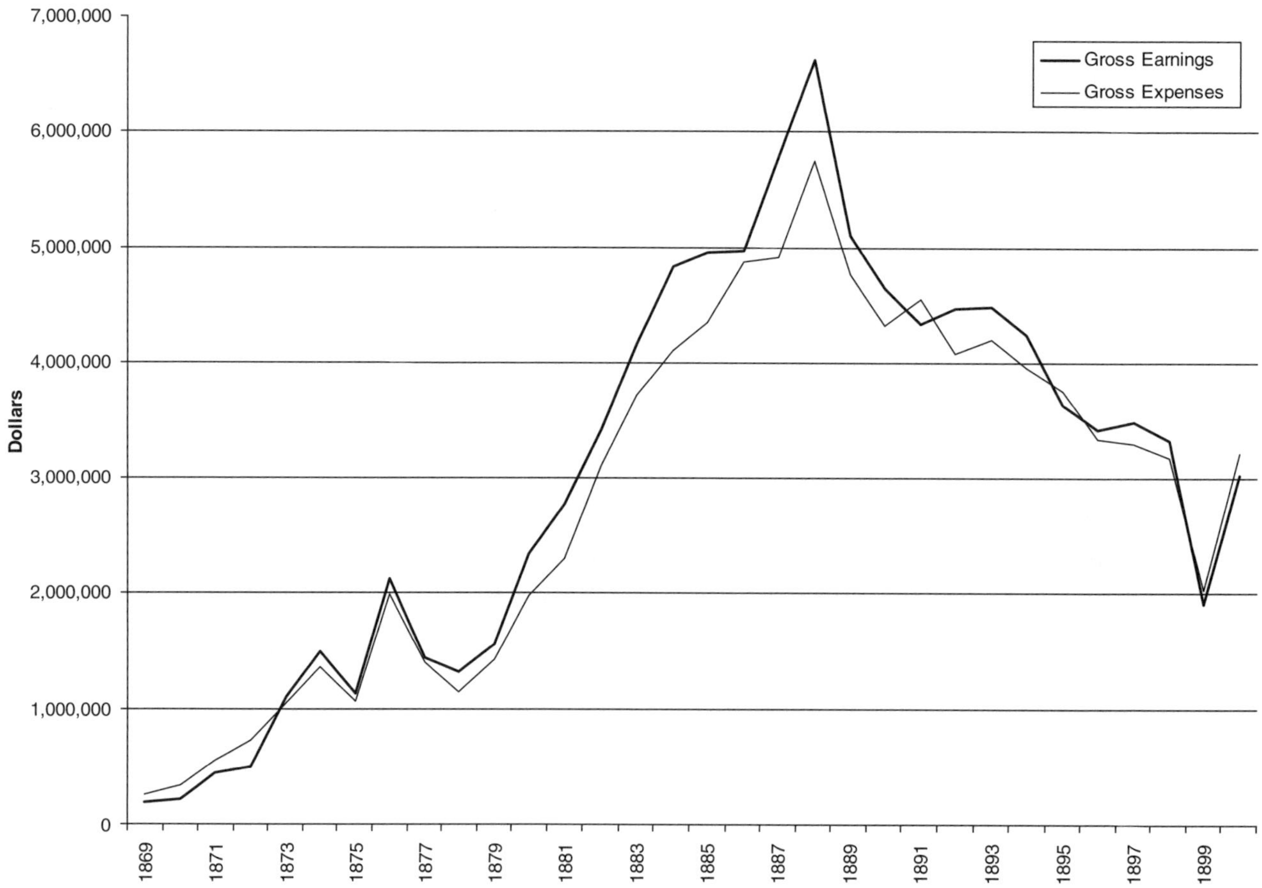

FIGURE 13.1
Susquehanna Coal Company: Gross Earnings and Expenses 1869–1900.

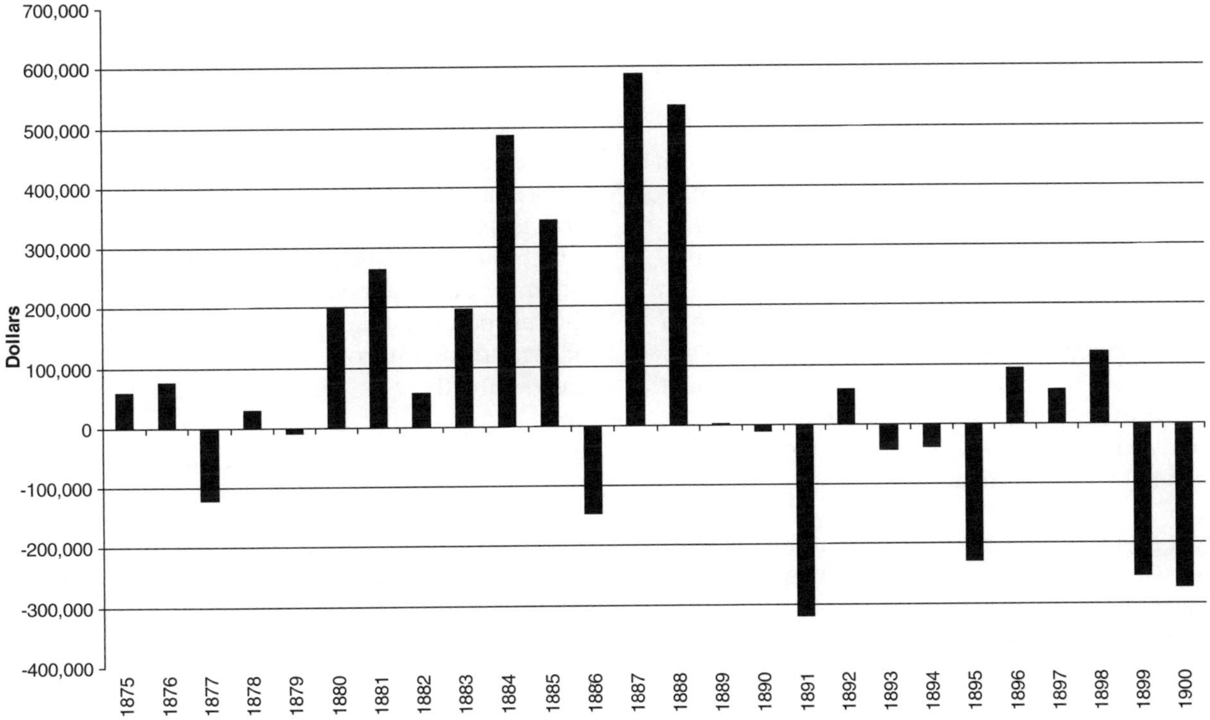

FIGURE 13.2
Susquehanna Coal Company: Annual Net Surplus/Deficit after all Charges 1875–1900.

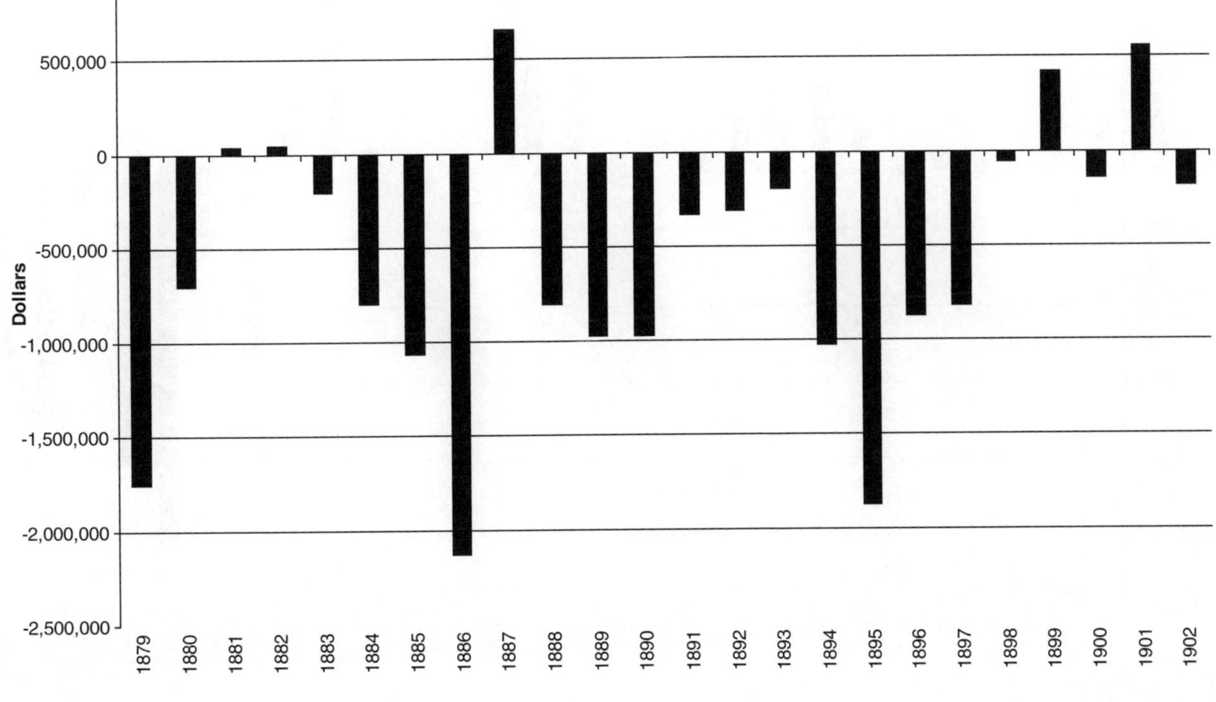

FIGURE 13.3
Philadelphia and Reading Coal and Iron Company: Credit and Debit Balances of Profit and Loss Account (Including Provision for Reading Company Loan Interest) 1879–1902.

floating debt of the company which emerged in the 1880s, after the cavalier accounting of the 1870s came under the scrutiny of successive receivers, is not included in the data on the graph. However, a formal loan from the parent railroad required interest payments in excess of $800,000 per annum to be made as soon as the Coal and Iron finally moved into the black, and these charges are reflected in the data for the years 1900–1902.[10] Although the accounting terminology used in the Reading Annual Reports is different from that of the Susquehanna Coal Company, examination of the respective accounts and balance sheets indicate clearly that the same types of charges have entered into the calculation in both cases, and the financial results are therefore broadly comparable.

At first sight, the performance of the Lehigh Valley Coal Company in Figure 13.4, based apparently on comparable data, looks markedly superior to that of the Coal and Iron, except at the end of the 1890s, when the Railroad had to bail out its subsidiary by meeting its bond interest payments. However, this proves not to be the case. While the Coal and Iron appears to have moved to more transparent accounting during the 1880s and clearer separation of its business from that of its parent, the same could not be said of the Lehigh Valley Coal Company until a much later date, according to its own corporate history. In a remarkably candid assessment of accounting policies, the author of the history identified ways in which the Coal Company had understated its actual expenses and had been extensively subsidized by the railroad in a relatively hidden manner. In terms of accounting impropriety, it is best to quote his views directly:

> 'There may have been good reason for capitalizing the cost of maintaining a property during idle periods, such cost being for the keeping of buildings and machinery in repair, the openings and gangways in order, and the mine free from water, but at the time of this writing such a policy would not obtain'.[11]

He went on to mention that the cost of extinguishing mine fires, amounting in two cases

FIGURE 13.4
Lehigh Valley Coal Company: Annual Surplus/Deficit Allowing for Bond Interest Payments by Railroad 1874–1901.

to over $180,000, was also capitalized.[12] Finally, he divulged the extent to which the parent railroad had provided working capital to its subsidiary over the three decades from 1873. In the 1870s, the outstanding amount generally exceeded $5 million, and it climbed to a peak of over $11 million, in 1892, before falling back sharply when the railroad was leased by the Reading. This revolving loan amount was not apparently reflected in the coal company's reported annual surpluses and deficits.[13]

Access to a full set of surviving account books has allowed the fullest picture to be derived for the Lackawanna Coal Department from 1860 to 1902. This is shown in Figure 13.5. Some of the main cyclical peaks and troughs are apparent in the usual way, such as for the Civil War period, the boom years of the early 1880s and the depressions of the mid-1870 and 1890s. These match most closely to the Juglar periodicities (cf. Table 2.1), although the Juglar peak of 1889 is not very strongly represented. Years when the cartel was not in operation, such as 1877 and 1879 also register substantial losses. The very strong financial performance during the Civil War stands out, although inflationary pressures amplified the actual scale of the earnings. Without taking account of this inflation, the coal department had net total profits of $9.05 million over the entire period. While substantial in sum, on average it amounted to little more than $200,000 a year. This is less than impressive, given the scale of the company's mining activities, the risks and cost inherent in developing large coal estates, and the absence of any financial obligations in the form of dividend payments or fixed charges, all of which fell upon the parent railroad. Compared to the Susquehanna Coal Company, for example, which paid both dividends and bond interest, the organizational structure of the coal department allowed an effective though hidden subsidy to be provided to the mining business. The cash-generating power of the Lackawanna Railroad and the rapid accumulation of a large reserve fund, especially in the 1880s, as explained in an earlier chapter, also provided an important financial buffer for the coal business. This allowed decision-making for the optimum

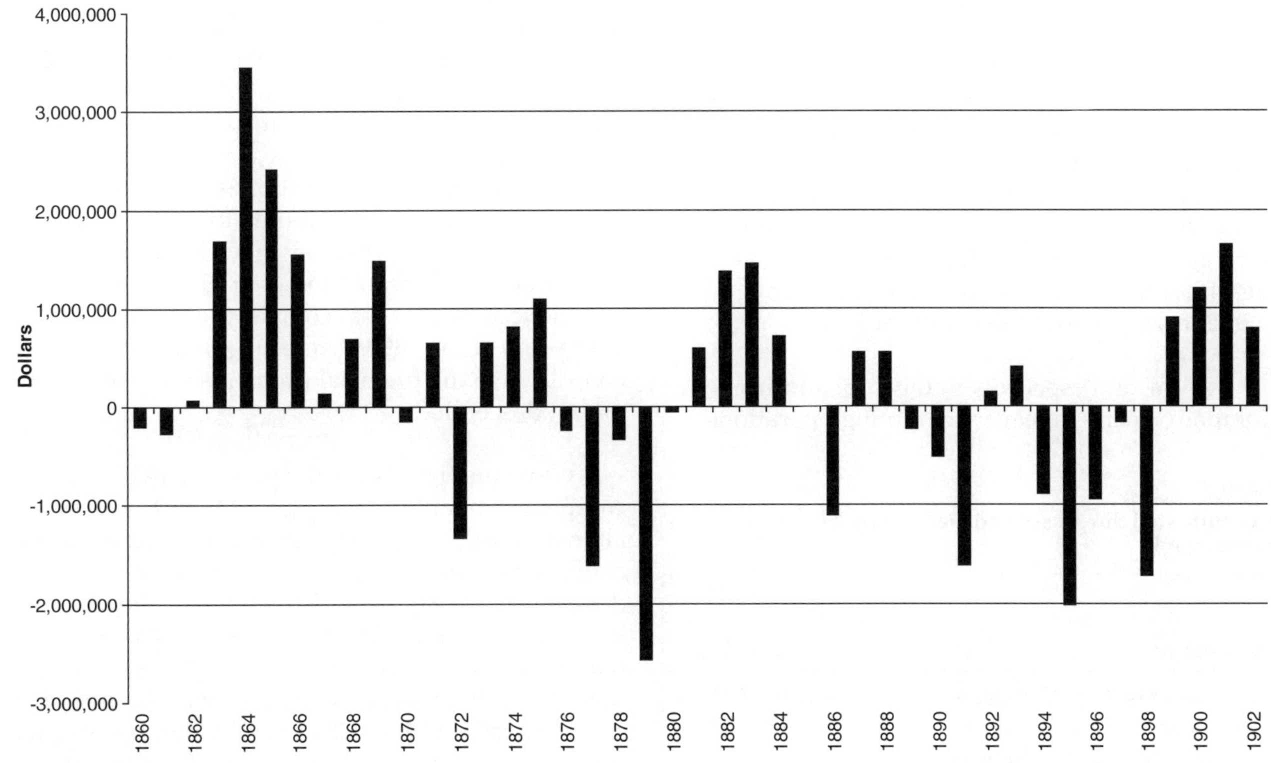

FIGURE 13.5
Delaware, Lackawanna and Western Railroad Coal Department: Profits and Losses 1860–1902.

development of its property to proceed without the serious constraints experienced by companies such as the Coal and Iron, whose financial condition was impaired over several decades.

By way of stricter comparison, it is appropriate to note the cumulative surpluses or losses experienced by all the companies so far mentioned, during the period 1879–1900. These figures are shown in Table 13.1.

A variety of reasons can be given to explain the rather sorry state of financial affairs that the table seems to represent for the mining business in the last two decades of the 19th century. A first observation would be that losses seemed to multiply roughly in proportion to the size of the coal estate. Secondly, it would seem that the cartel was at least partially effective in reducing losses during periods of depression, although the Coal and Iron benefited from this to a lesser degree than the Lackawanna, for example. Thirdly, the calculated distancing of the Susquehanna Coal Company from the cartel by the Pennsylvania Railroad, during the 1870s and early 1880s, produced favorable financial results, not least because the company was able to work for a greater number of days per year than would have been possible under cartel quotas.[15] Such benefits could only accrue, of course, provided one single player failed to join the cartel game, as proved to be the case. Finally, there may be additional aspects of the overall financial equation, specifically, those concerned with variations in the cost of mining in different parts of the coalfield, that cannot be identified from these aggregate figures. This question will be examined further below.

In sum, and especially in the Coal and Iron Company, many managers of mining operations must have wished they had paid more heed to Franklin B. Gowen's famous dictum before the 1875 Investigating Committee:

> 'I practiced law for seven years in the County of Schuylkill, and in all that time, and up to it, there were but three men who had ever retired from the business of mining coal with any money, and one of those died in an insane asylum and another had softening of the brain, the shock too great to their systems.'[16]

TABLE 13.1
Accumulated Surpluses and Deficits by Major Company 1879–1900.[14]

Company	Accumulated surplus/deficit (–) 1879–1900
Susquehanna Coal Company	+ $1,686,524
Lackawanna Coal Department	– $3,897,483
Lehigh Valley Coal Company	– $6,778,690 (excluding LVRR loans)
Reading Coal and Iron Company	– $13,871,851

It is clear that, within the first thirty years of its existence, the Coal and Iron did not manage to earn anything remotely resembling the profit it would have required to fund the proper rationalization and development of its estate out of retained earnings. All its investment was necessarily derived from borrowed money, channelled to it by an equally needy railroad parent. An analysis of its achievements in the light of these straitened financial circumstances will be undertaken in chapter 15.

Much the same can be said of the Lehigh Valley Coal Company, which fared little better than the Coal and Iron, and at times possibly worse—when the extent of its loans from the parent railroad are taken into account. However, this company lacked the excuse of a single catastrophic episode of over-expansionary zeal in the territorial domain. Instead, there is evidence that strategic errors in the acquisition and management of its coal estate over long periods of time contributed to its worsening financial condition. These events will be examined further below. Repeated losses on the scale experienced by this coal company, especially in the later years of the 1890s, were large enough to contribute to the financial problems into which the parent descended.

While the figures for the previous two companies in table 13.1 represent their losses over almost the entire period of their existence, the same could not be said of the Lackawanna coal department, as Figure 13.5 has already shown. The loss of nearly $4 million, from 1879 to 1900 was more than offset by the $10.5 million surplus on mining operations that had accrued between 1860 and 1878, and the profits of 1901–2 also cut in half the accrued deficit of the previous two decades. During the 1860s cumulative mining profits of nearly $11 million were nearly as great as the $13million profits earned from transportation (before fixed charges and dividends). This greatly bolstered the financial strength of a company that had been severely weakened at the time of the 1857 panic.[17] Strong earnings from the mining business enabled the coal department to embark on its first phase of acquisition of existing mines and construction of new ones during the 1860s and early 1870s. Later phases in the 1880s and 1890s, when mining profits were on average much lower, were made possible by the continued financial strength of the railroad, as a result of its profitable transportation business. This enabled it to stand short-term losses in the mining business, provided the flow of coal traffic on which its prosperity depended could continue unabated.

Lastly, the accounts of the Susquehanna Coal Company have confirmed that it was the most successful of the available-example companies during the period. Initial investment in improvements was partly financed by the Pennsylvania Railroad, as noted above. The subsequent phase of protection from the restrictions of the cartel was then a major factor in its profitable operation until the mid-1880s. During this time, profits were large enough for the company to fund its own program of mining development from retained earnings, without recourse to further railroad loans or stock/bond issues on the financial markets. However, as soon as the Pennsylvania Railroad began to adopt a more co-operative approach to the cartel in the latter part of the 1880s, additional profits essentially disappeared, as Figure 13.2 testifies. This period was also marked by a sharp reduction in the average number of days worked by the company's breakers.[18]

In summary, the ranges of examples presented indicate a very varied set of financial experiences among the different subsidiary coal operations of the railroads. Financial success from mining alone rarely continued for any extended period of time, so financial risks were carried by the railroads to varying degrees. In most cases, acquisition of coal territory and development of new mining capacity was able to continue, even when mining earnings were insufficient to fund it, simply because it was in the railroads' interest for shipments to be maintained or expanded. However, in extreme cases of financial loss and repeated receivership, such as that experienced by the Reading from the 1880s onwards, cost controls undoubtedly impeded development, as will be seen below.

SOURCES OF FINANCE: PROFITS, AND THE AVAILABILITY OF RETAINED EARNINGS FOR INVESTMENT II: THE INDIVIDUAL AND INDEPENDENT OPERATORS

In contrast, the individual and independent operators were answerable only to themselves, or if incorporated, to their stock- and bondholders. (Hereafter, in this section, 'individual' will be used as shorthand terminology for both 'individual and independent' operators). In the context of the foregoing discussion, it is pertinent to investigate whether individual operators were able to mine profitably over extended periods, especially at times when the railroad coal subsidiaries were not able to achieve this. A subsequent question would then be whether, if substantial profits were earned, they were ploughed back into development of the property or largely used to remunerate the original investors.

Given the large profits earned by the major companies during the Civil War, it would not be surprising to find that individual operators were also able to share in the financial bonanza of the time. Although the actual earnings of individual entrepreneurs are extremely difficult to determine, a good example of an established mine owner who derived a handsome return during the period was William H. Johns, who had operated the Eagle Shaft at St. Clair near Pottsville since 1846. Johns died in 1865 at the age of 60 and his obituary in the Miner's Journal conveniently recorded his taxable income for the years 1862–1864. The sums in question were $118,600, $236,151 and $260,113.[19] While inflation greatly eroded purchasing power during the war period, these figures still suggest substantial mining profits, although he doubtless derived part of his income from other investments. Daddow and Bannan further asserted that the colliery had generated for its 'prudent and fortunate possessors' more than $1million in profits over nearly 20 years of operation, and was one of the most successful collieries in the coalfield.[20]

A useful comparison can be made with a more newly developed mine in the Northern Field during the same period, which had a broadly similar capacity. In 1863, the Stafford Brook mine of the Susquehanna and Wyoming Valley Railroad and Coal Company, situated below Scranton, produced 75,315 tons, to generate for its owners profits in excess of $36,000. This constituted a useful return on $320,000 of capital stock, but probably nothing on the scale of the Eagle Colliery profits.[21] In terms of tonnage, the latter mine produced 84,558 tons in 1864.[22] It is interesting to note that the operators of Stafford Brook were wasting no time in re-investing their surpluses after payments of a 10 percent dividend. It is recorded in the Annual Report for 1863 that they had already contracted to purchase both additional land and new railroad cars—for the transportation of coal to New York over the tracks of the Lackawanna.[23]

In the later years, a paucity of surviving archival accounting records for individual operators makes detailed examination of profits a rather difficult undertaking. Nevertheless, a good example from the Northern Field is available and it sheds some interesting light on this segment of the coal industry.

The company in question is the Florence Coal Company, whose breaker in Pittston Township commenced operation in 1883.[24] The six directors of the company held the entire capital stock of $48,000, half of which was subscribed by the treasurer, C.P. Matthews. Unfortunately, after a total of only about 35,500 tons of coal had been shipped, the breaker was burned to the ground on August 1, 1884. The fire was caused by a pump-runner, who had attempted to fill a lighted lamp from a 2-gallon can of kerosene, with disastrous results. The financial loss of the breaker was put at $60,000, against an insured value of $35,800. After an acrimonious battle with the insurers, the final settlement was agreed at $28,511. The breaker was promptly rebuilt, given the name of 'Elmwood', and shipments were restarted on May 1, 1885. This exquisitely bad timing, of course, coincided with the bottom of the business depression of that time![25] This unfortunate experience serves to indicate the vicissitudes of the mining business and the unexpected problems and expenses that could arise without warning, and frequently did.

The company apparently began business with a loan from the Pennsylvania Coal Company of more than $50,000, but very shortly afterwards, in early 1884, for reasons that are not documented, they decided to ship via the Lehigh Valley Railroad instead. This required a second loan of $70,000 from the latter railroad, to repay the first loan with interest, and cancel the original contract. Incidentally, it also required a change in the gauge of the railroad branch line to the mine. Immediately after completion of the breaker rebuilding in 1885, the directors of the company must have realized that these were inauspicious times for the coal business, as they offered to sell the colliery to the Lehigh Valley Railroad, and assign their coal lease to them, for $125,000. They claimed this sum was $15,000 less than they had invested in the property to that time. The fall-back position was to ask the railroad for a further loan of $25,000 to see them through their cash-flow difficulties.[26] However, the Railroad would have none of it. They replied briskly:

> 'The well-known financial ability of the gentlemen associated with you in this operation should enable you without further assistance from us to carry this work to a successful conclusion'.[27]

While the Railroad may have been keen to see the directors dip still further into their personal financial reserves, the directors themselves were apparently less keen! By July 1885, the company had a floating debt of over $5,000 and a note from the Lackawanna Valley Bank due in August for a similar amount. Funds to cover these obligations came not from the existing directors, but from the established coal merchants W.B. Whitney and M.S. Kemmerer, who replaced other directors on the board. In return for their injection of capital, they also secured the right to have all the company's future production consigned to them on 20 cents sales commission.[28]

After the extremely shaky and eventful beginning to their mining operations, with no financial reserves but, at least, all past debts covered, the company looked to the future with confidence. Fortunately, the company secretary was kind enough to leave, (spread on the minutes), very useful summary statistics of the company's fortunes on a monthly basis for the next seven years. Charts of these, and calculations based on them, are shown in Figures 13.6 to 13.8. Despite low output, high costs, and initial losses, the company persevered until external circumstances brought it a measure of good fortune, which finally recompensed it for its inauspicious beginnings and years of financial tribulation. These circumstances came in the form of the labor troubles of 1887–1888, which particularly affected the Lehigh and Schuylkill Regions. For a limited period, as the graphs indicate, these stoppages in other regions provided ready markets for a much larger tonnage than normal, and at much higher prices than prevailed in the preceding depression years. Such a felicitous set of business circumstances produced extraordinary profits, of a kind that were not to be repeated in the remainder of the period for which data are available (Figure 13.8).

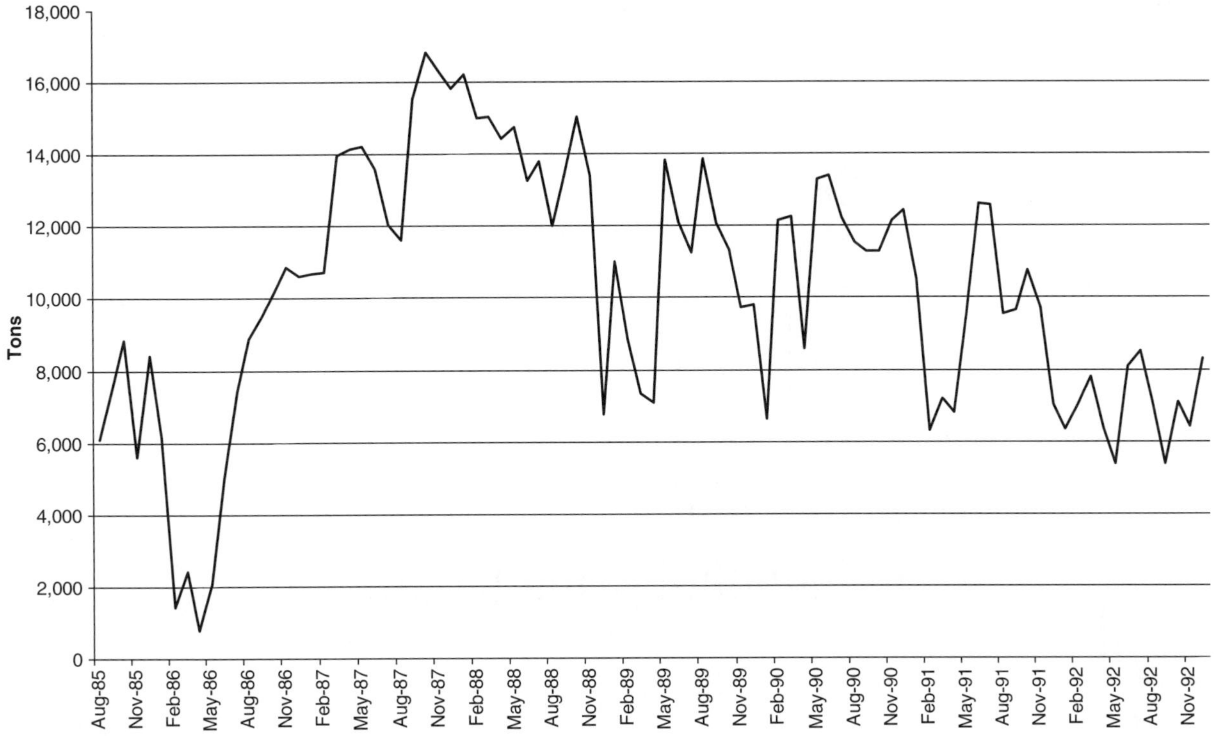

FIGURE 13.6
Florence Coal Company: Monthly Tonnage Mined 1885–1892.

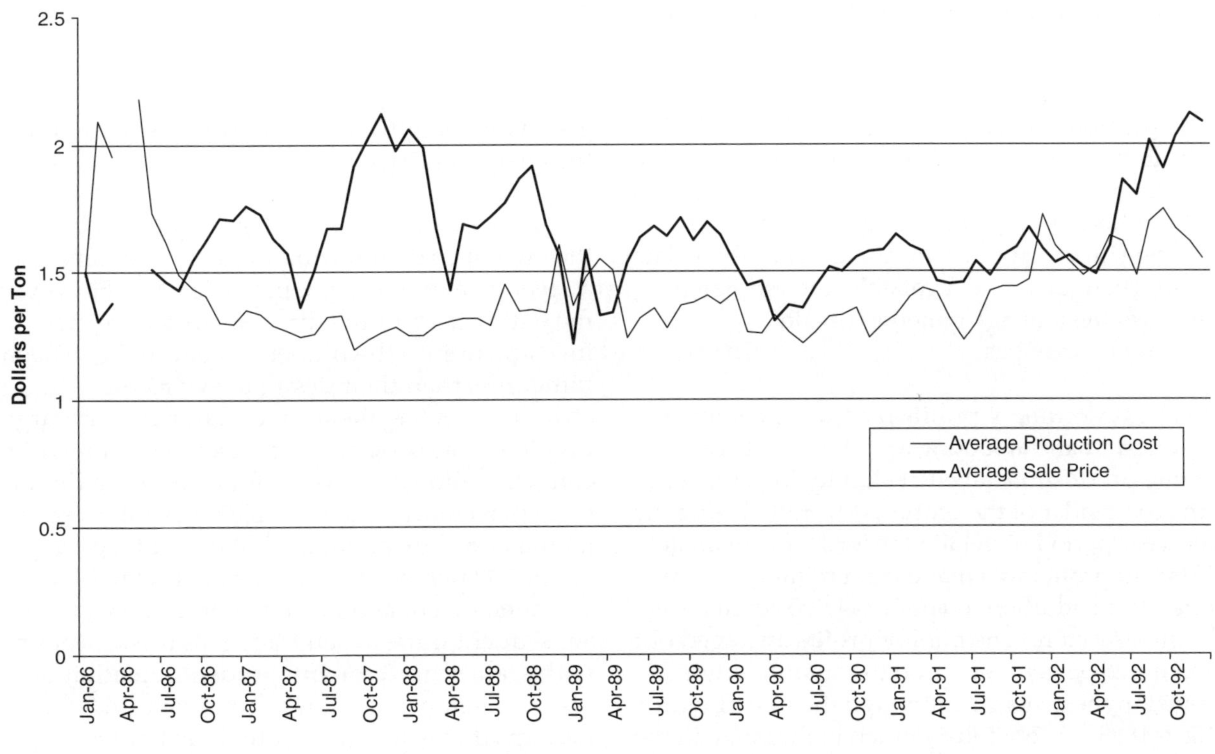

FIGURE 13.7
Florence Coal Company: Average Costs and Prices Received 1886–1892.

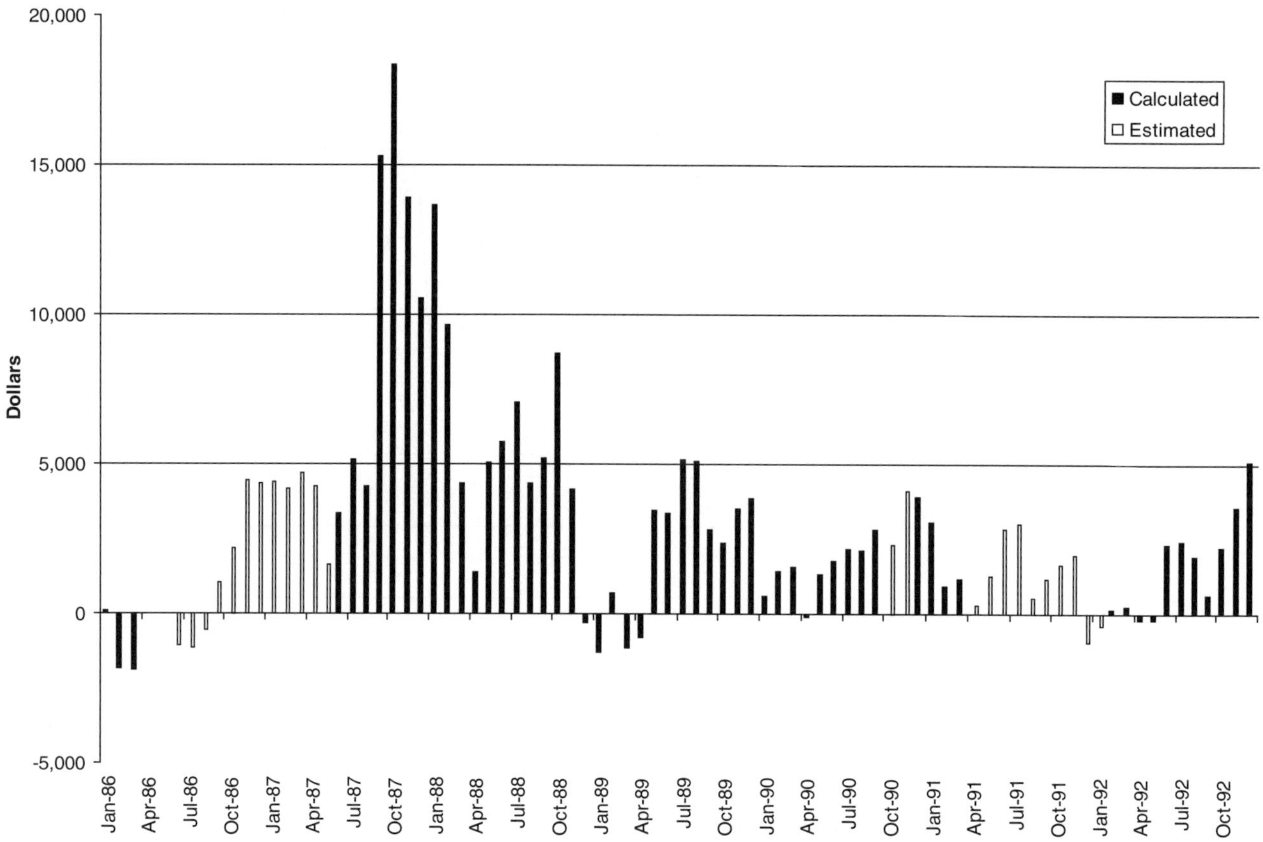

FIGURE 13.8
Florence Coal Company: Monthly Net Profits/Losses 1886–1892.

The distinction between calculated and estimated profits in the figure arises because it is only possible to calculate sales revenues precisely from the data in months where changes in the stock of coal at the tidewater shipping points are also given. Where data on stock changes are not available, the estimated figures use the tonnage mined to obtain approximate revenues.

The extraordinary profits placed the company on a sound financial footing and it was able to continue production, albeit at slightly lower levels, for the remainder of the century. Overall, during the seven-year period of detailed record, approximately $250,000 in profits was made from mining, before dividends, in addition to about $44,000 profits from the company store. The mining profits averaged out at about a 25 percent return on the amount originally invested in the property, although the rate of return in 1887–1888 was obviously much higher than in the later years. Given the potential for loss in the early period of the company's existence, such a return was gratifyingly high, and arguably necessary to offset the attendant risks.

There is evidence of later attempts to increase the area of coal land tributary to the colliery and a second shaft was sunk in the 1890s.[29] However, the enthusiasm of the directors to put further investment into the property seems to have been rather less than their desire to award the shareholders (i.e. themselves) large or very large dividends, as soon as the financial condition of the company allowed it. Thus, for example, a dividend of 20 percent in December 1887 was followed in January 1888 by another of 50 percent, and a further 30 percent four weeks later. Other large dividends were announced later in 1888, and at subsequent times when trade was brisk, seemingly with scant regard for any prudent retention of reserves against any future contingency.[30] Or perhaps it was fear of eventual further loss that encouraged the directors to enjoy the good times while they lasted!

Since the Florence Coal Company was effectively still a new concern at the time the detailed monthly records began, it is instructive to compare the pattern of earnings with the equivalent data for the early years of a railroad subsidiary, the Susquehanna Coal Company (Figure 13.9). In 1869, this was also a new operation, but its initial investments included both refitting existing mine improvements and developing new ones. This comparison further highlights the benefits provided by the 1887–1888 labor troubles to the Florence Coal Company, as they enabled it to be strongly cash-generative at a critical time in its early development. The worsening losses of the Susquehanna Coal Company in its first four years of operation, that only seem to have ended with the onset of the 1873 cartel, also underline the importance of financial support from the parent railroad in the critical early years.

A further comparison that sets the experience of the Florence Coal Company in context is shown in Figure 13.10. Although the Wyoming Valley Coal Company was a new corporate entity in 1882, it took over the operation of Swoyer's existing collieries, as has already been discussed in the previous chapter. Hence there are no 'start up' costs involved, but the impact of wider business conditions can be seen very clearly. The effect of depression periods on earnings in 1885–1886, 1890–1891 and the mid-1890s is very apparent, while the 'Labor Troubles' effect stands out in a positive fashion in 1887–1888. It should also be remembered that the company's collieries, Harry E and Forty Fort were leased to Simpson and Watkins in 1893, so its earnings became somewhat insulated from market fluctuations thereafter.[31] One final caveat about the financial performance of individual operators needs to be mentioned, however, at this juncture. Although the earnings in Figure 13.10 include bond interest, apparently they did not include any accounting provision for depreciation of coal lands as a result of progressive exhaustion of the coal seams. To correct this, a separate charge of more than $528,000, not included in the earnings calculation, was made in the years 1896–1901. Had this been applied as a regular monthly debit to the expenses account over the period, it would have reduced net earnings by approximately 50 percent.[32] Taking account of taxes and improvement charges, in addition, reduced the total earnings of the

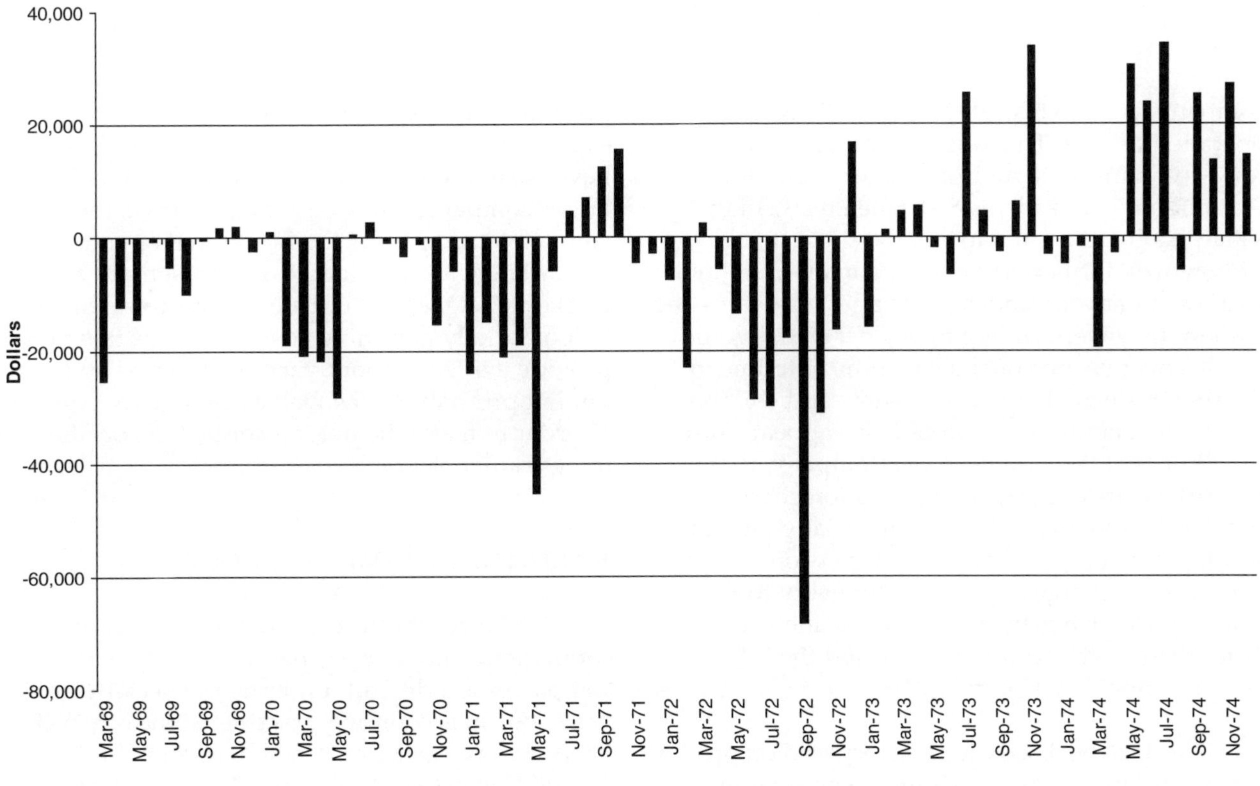

FIGURE 13.9
Susquehanna Coal Company: Monthly Net Earnings and Losses 1869–1874.

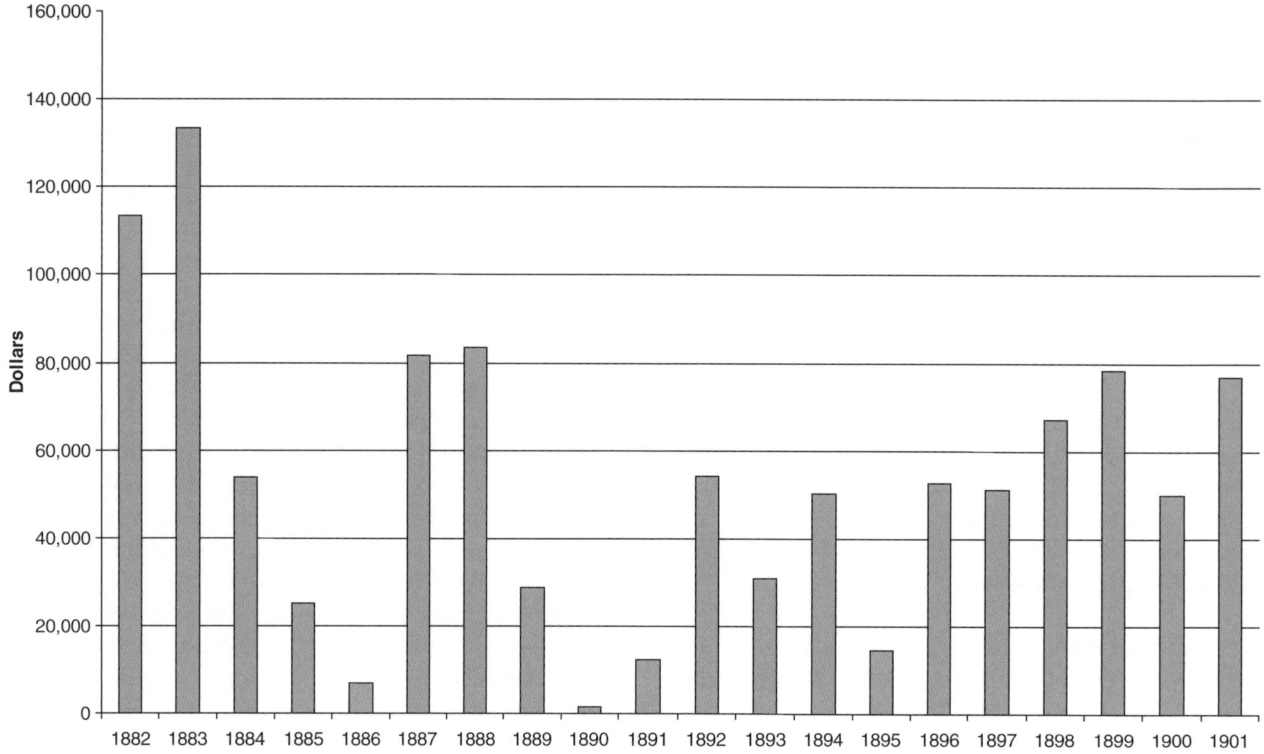

FIGURE 13.10
Wyoming Valley Coal Company: Net Earnings Before Taxes 1882–1901.

company to about $363,000 over its 20 years of operation to 1901. This was a very meagre return for a company with an initial share capital of $1.2million.[33] Indeed, in 1890, when the Lehigh Valley Railroad offered to buy the shares of J.H. Swoyer, former president and main shareholder in the company, (from his estate after his death), the amount they were willing to pay per share would barely cover his personal debts from a lifetime in the coal business, leaving his widow and children effectively penniless.[34] It would also appear, from the records of the Florence Coal Company, that in the early years at least, no depreciation charges on coal lands were made, thereby artificially inflating earnings. In comparison, a 10-cent-per-ton depreciation charge to expenses for every ton shipped from lands held in fee was standard practice for both the Lackawanna and the Susquehanna Coal Company.[35]

Thus far, the discussion has focussed on profits, since without them there would be no retained earnings to invest in new mining improvements. The experience of several companies has been presented, and a variety of contributing explanations for variations in profits and losses have been offered. While the available data for the larger companies may be representative, the equivalent sample for the individual operators is far too small for any degree of representativeness to be claimed. One way of trying to address this is to look briefly at mining costs per se, as this may provide evidence about variations across space in mining profitability, as well as changes over time, in order to make the overall context of specific mines rather clearer.

VARIATIONS IN MINING COSTS

Unfortunately, accurate determination of comparable mining costs per ton by different companies is a difficult undertaking because of variations in accounting practices, some of which have already been mentioned in relation to the Lehigh Valley Coal Company. A standard problem was that some companies charged the cost of new improvements to mining expenses, others maintained separate construction accounts. The Susquehanna Coal Company was an example of the former; the Lackawanna adopted the latter practice.[36] There were also variations in the extent to which royalties on leases, taxes, and depreciation of fee lands were taken into account, as the Wyoming Valley Coal Company example has indicated. These differences were as much a source of confusion to the industry of the time, as they have been a source of exasperation to the present author, when seeking comparability between different sets of accounting data! This is apparent from Lehigh Valley Coal Company correspondence in 1898. Israel Morris, general land agent, commenting on a transcript of a recent legislative investigating committee looking into the coal industry, observed:

> 'You will notice that various parties estimate the cost of mining coal, anywheres (sic) from $1.50 a ton up to $2.75 a ton, the difference being occasioned mainly by the different items which the respective parties charge to the cost!'.[37]

The costs claimed by mining company officials in this transcript, and the stated items of expense included or excluded from the calculations, will provide a useful check on the figures to be presented below.

To begin the analysis with the familiar Susquehanna Coal Company, Campbell has identified a contemporary assertion that it made more money than any other large concern in the anthracite regions.[38] Unfortunately, the information comes from an undated newspaper clipping, although it is supported by the earnings figures given earlier in this chapter. Since there is no evidence that the company's coal enjoyed a price premium over other Northern Field coals, this would imply its costs were at the low end of the spectrum. A high degree of plant utilization in the period of unconstrained operation, during its period of protection from the cartel by the Pennsylvania, was undoubtedly a factor in this, as already mentioned above. Equally, favorable geological conditions or low cost operations from mines above water level could also have been factors. To examine this question, data from the company's accounts and Annual Reports have been re-calculated to distinguish improvement or construction costs from production mining costs, as shown in Figure 13.11. The only caveat is that this will absolutely minimise apparent production costs, because it includes improvements at existing as well as new mines. Other companies may well have charged the former (or some of them) to expense rather than construction accounts. Nevertheless, it is clear that, in the first part of the period, although there is significant variation from year to year, costs at times were extremely low by the standards of the time. By way of comparison, A.J. Davis of the Warrior Run Coal Company, several miles away, estimated his costs at $2.50 a ton in 1875, exclusive of royalty, while the Susquehanna figure of $1.07 includes royalty, albeit at a very low figure.[39] In later years, when the benefits of operating outside the cartel had disappeared, mining costs stabilized at around the $1.40 mark, inclusive of royalty and depreciation. This itself is of interest, because the company had argued in its 1885 report that costs would increase over time, as follows:

> 'Mining will, as elsewhere, become gradually more dangerous and expensive, proportionately with greater depth, larger areas of ventilation and longer hauls for coal and materials'.[40]

Fortunately, the Annual Reports give a detailed breakdown of costs, both inside and outside the mine, for the years 1876–1893, and details of trends in some of the major cost headings can be seen in Figures 13.12 and 13.13. From these, it is clear that increases in inside expenses were offset by a reduction in outside expenses from 58 cents per ton to 28 cents. Within the two main categories, significant reductions in repair costs/general expenses were apparently made. This may suggest the use of high-quality equipment and rigorous maintenance schedules. However, despite building some of the largest breakers in the entire coalfields during the period, all but one of which came into operation prior to participation in the cartel, apparently no economies of scale were realized in the preparation of coal (Figure 13.12). This suggests that economies that might have been achieved were lost as a result of the reduction in days in operation under the cartel. In Figure 13.13, the matching divergence of mining-coal costs and repair costs, after 1886, may reflect a change in accounting practice more than the explanation advanced above. Nevertheless, there was a clear upward trend in the costs of timbering throughout the 1880s, which was enough in itself to impart an upward drift to inside costs as a whole. It should be noted that, until 1887, no charges for hoisting and pumping were made and, even by 1893, these had only reached 1.9 cents a ton, which was a low figure for the industry. Royalty payments were also at a similarly low level, because

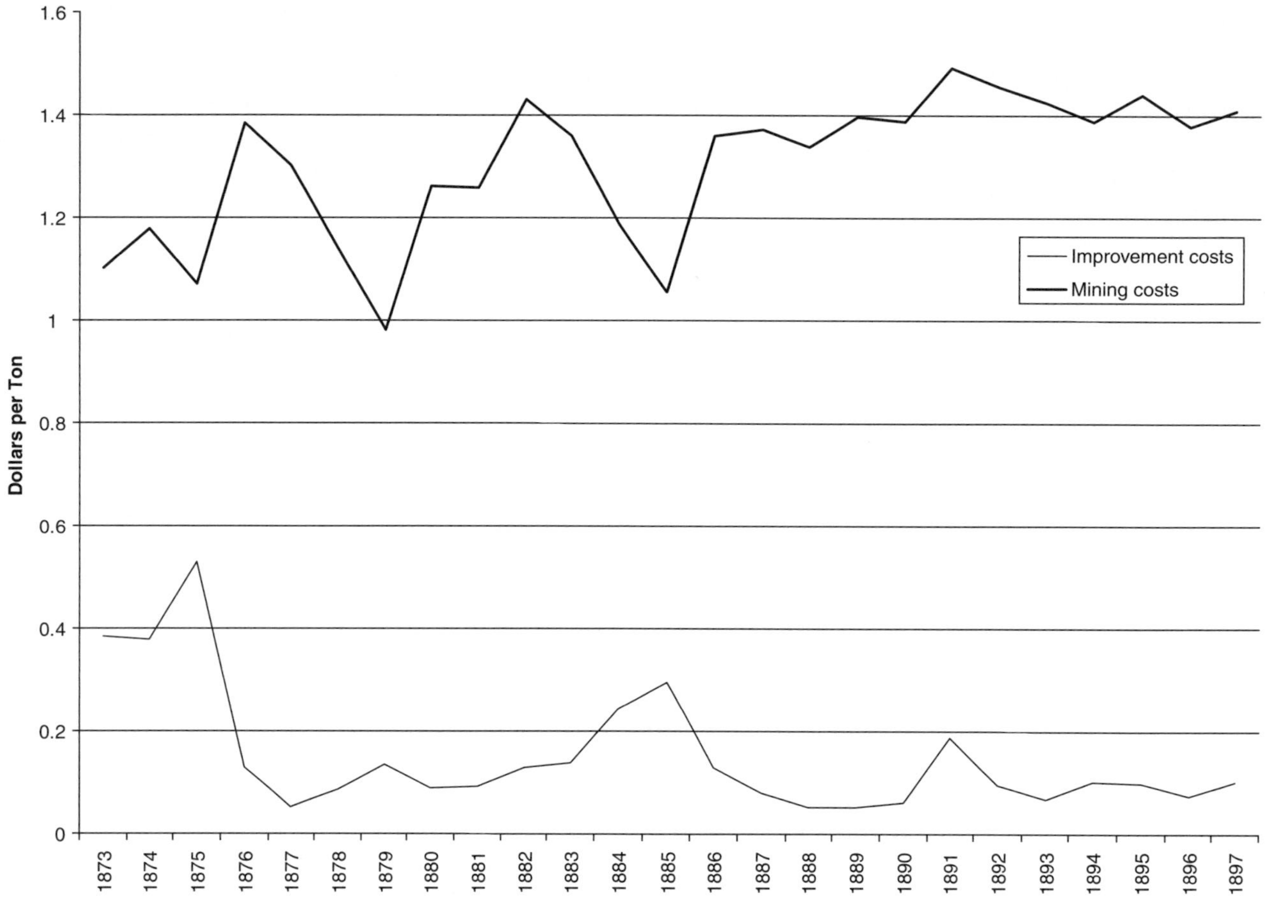

FIGURE 13.11
Susquehanna Coal Company: Improvement and Mining Costs 1873–1897.

nearly all the mining was done from fee lands, for which the customary 10 cents per ton mined was charged to expenses, in the form of depreciation.[41]

By way of direct comparison, figures can also be given for mines in other coalfields, but under the same overall management, so the accounting conventions should be very similar. For example, in 1885, the Mineral Railroad and Mining Company had outside expenses of 42 cents per ton. This was a function of higher repair costs than the Susquehanna, while coal preparation costs were similar. Overall, inside expenses were about 20 cents higher than the Susquehanna, largely due to higher deadwork costs and 5 cents per ton required for hoisting and pumping. While outside expenses at the Summit Branch Railroad Company's mines were broadly the same as in the previous example, inside expenses were nearly 58 cents a ton higher than the Susquehanna, due to high deadwork and repairs costs, and 8.7 cents per ton for hoisting and pumping.[42]

At a later date in the 1890s, after the Pennsylvania Railroad had acquired the William Penn Colliery in the Schuylkill Region, a further opportunity to examine trends in costs becomes available from surviving records. In 1897, in terms of outside expenses, coal preparation costs at this mine were double those of the Susquehanna in 1893, and repair costs were higher, to give an overall outside cost of nearly 44 cents a ton, excluding royalty. This figure was comparable to the Mineral Railroad in 1885. Inside the mine, deadwork and other costs were low and pumping was as cheap as at the Susquehanna Coal Company's mines, for an economical overall inside figure of only $1.10 per ton. Unfortunately, the low-cost underground

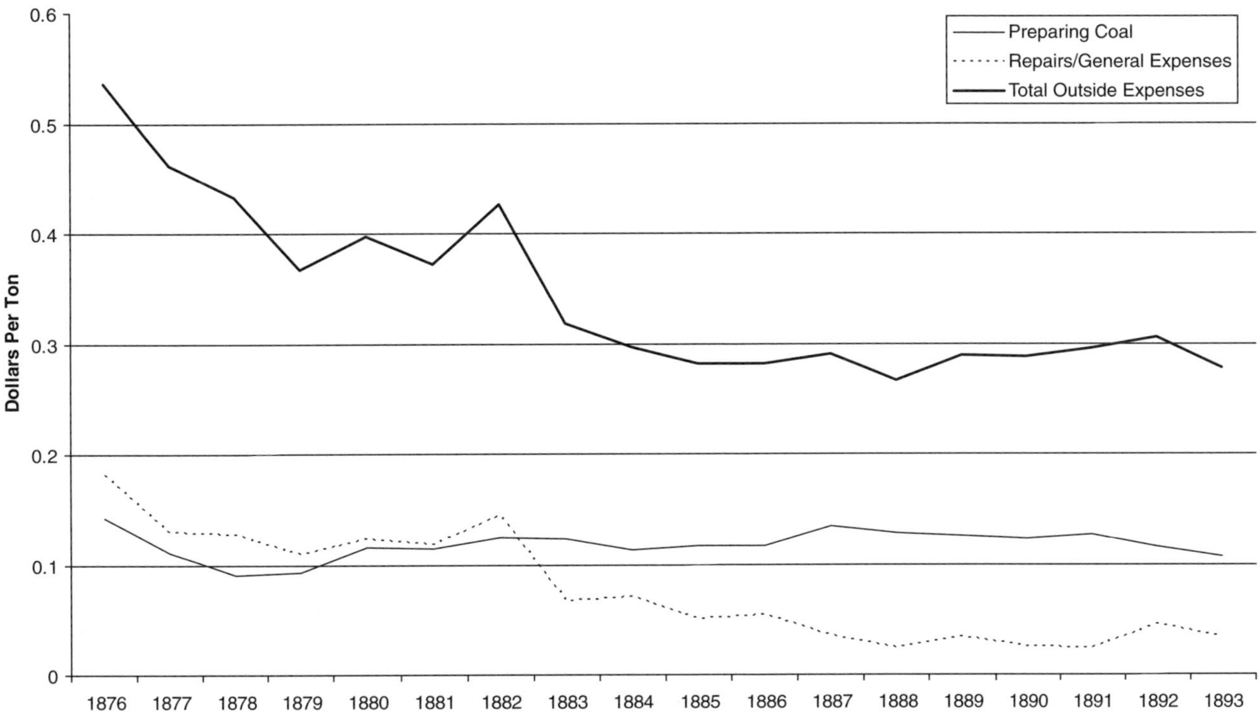

FIGURE 13.12
Susquehanna Coal Company: Trends in Outside Expenses 1876–1893.

mining was negated by high royalty costs, which averaged nearly 40 cents a ton during the period. The overall costs of mining are given in Table 13.2 for the years 1893–1898.

In the light of these figures from the Pennsylvania Railroad's records, it is gratifying that, in his testimony before a Legislative Committee, W.H. Lewis from the William Penn Colliery declared that '$2 to put a ton of coal on the cars at the mines would be a fair price'.[44]

TABLE 13.2
Total Cost of Mining (including Royalty) at the William Penn Colliery 1893–1898.[43]

Year	Cost ($/ton)
1893	1.74
1894	1.70
1895	1.69
1896	1.80
1897	1.93
1898	2.02

A further comparison can be made with the coal department of the Lackawanna Railroad, as equivalent figures can be calculated for 1892–1900 on a monthly basis, using surviving expense account records.[45] The results of these calculations can be seen in Figure 13.14 under the heading of 'cost of company-mined coal'. This is the cost of all the coal taken from the company's mines, both from leased lands and lands held in fee. The second series on the graph is the amount paid by the coal department for coal purchased from contractors' mines, excluding freight charges. Apart from the low cost 'blip' in December 1895, which is an artefact of a year-end accounting adjustment, it is apparent that the monthly expenses figures generally fluctuate within a band between $1.10 and $1.60. The higher values generally occurred in the winter months when production was reduced. On an averaged annual basis, in the years 1892–1898 the values fell within a narrow range of $1.22–$1.37. Costs were visibly reduced in 1899 and 1900, the averages falling in these years to $1.19 and $1.16 respectively. It should be noted that all these figures exclude 'royalty', which in Lackawanna accounting terms was a mixture of the standard 10-cent

FIGURE 13.13
Susquehanna Coal Company: Trends in Inside Expenses 1876–1893.

depreciation charge for coal from fee lands and any actual royalty attracted by tonnage mined from leased lands. In 1896–1897, this royalty averaged 14 cents per ton, as can be seen from Table 13.3. Hence for comparison with the William Penn figures in Table 13.2, the 1892–1898 average fell within the range $1.36–$1.51, and in 1899–1900 the figures were $1.33 and $1.30. This is of particular interest, as costs in the Lackawanna Coal Department were clearly not drifting upwards during the 1890s. The graph also demonstrates that purchased coal was generally higher cost than company coal by a margin of 30–50 cents or more. The margin tended to increase when coal demand was high and decrease in more depressed times. Part of the explanation for this would have been attributable to the higher royalty costs paid by individual operators, the majority of whom were working on leased lands with royalty rates of 25–30 cents per ton. Nevertheless, such a margin of difference indicated that a balance had to be struck, in management terms, between maintaining mining capacity tributary to the railroad for cartel purposes and paying over the odds for high-priced contractors' coal when company coal could be mined more cheaply. The greater the margin of difference, either because of high prices of contractors' coal or because of reduction in the cost of company coal, the greater the incentive to reduce reliance on the contractors and divert a greater proportion of production quotas to be filled by the company's own breakers.

Since the Lackawanna Coal Department operated up to twenty of its own mines at any one time, in addition to receiving the product of a number of contractors' mines, the above averages also concealed a significant degree of variation in the economic performance of individual mines. It is

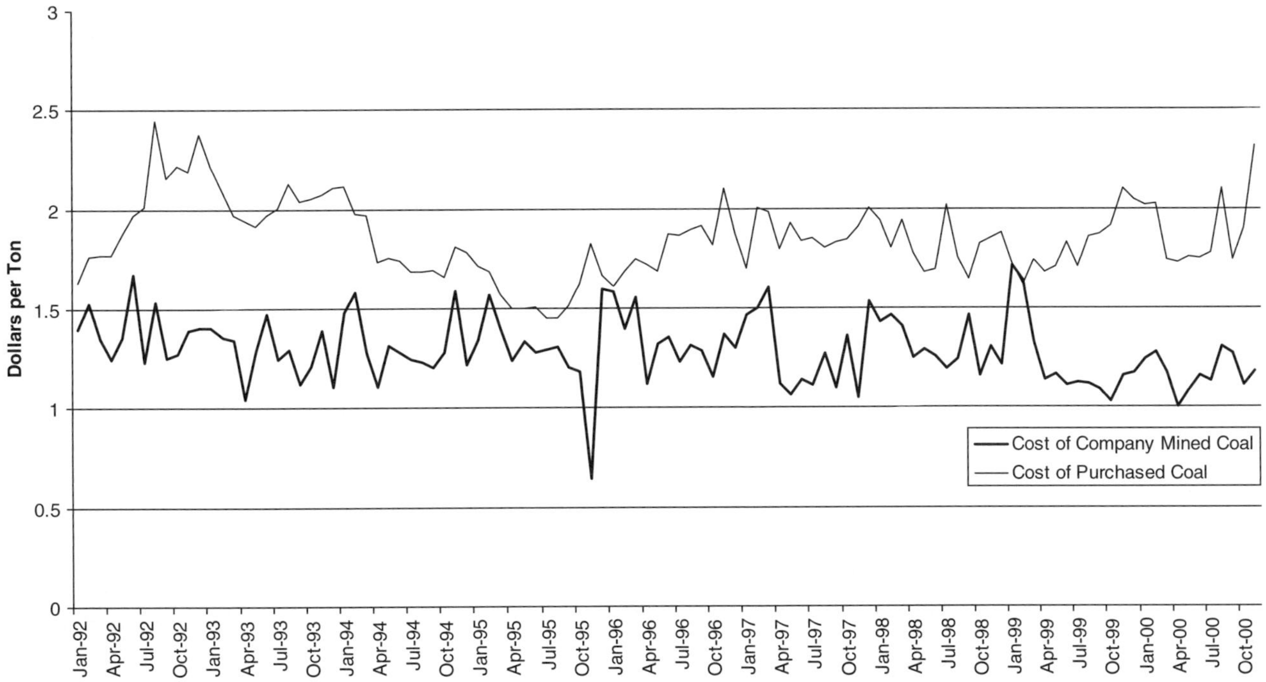

FIGURE 13.14
Delaware, Lackawanna and Western Railroad Coal Department: Coal Mining and Purchase Costs 1892–1900.

highly unusual to find consistent data down to this level for a large group of mines but, fortunately, over a 16-month period, between November 1896 and February 1898, the letter books of the coal department record just this information for eight different months, for all the company's mines. These data have been analysed and are presented in Table 13.3 as averages by mine for the eight recorded months, with and without royalty added. The mines appear in descending order of output for the 12–month span that most closely matches the months when costs are recorded. This is to facilitate identification of possible economies of scale in mining. A weighted average of monthly costs per ton was used because it was apparent, from inspection of the raw data, that costs were higher in months where the breaker worked very few days. Since the tonnage would then be small, the high cost month will only carry a small weighting in the overall average.

From the table, it is apparent that costs varied across a wide range, from $1.09 at Pyne to $2.09 and $2.10 at Pettebone and Holden respectively. The last of these was originally owned by a contractor, but was taken over by the Lackawanna in the mid-1880s after the original owners, the Amity Coal Company, ran into financial difficulties, as will be explained further below.[47] Two other points are worthy of note. Older mines such as Continental and Diamond in Scranton were still operating in a very cost-effective manner, but some newer mines, such as Woodward, Pettebone, and Bliss, were proving expensive to operate. The first two of these three mines also attracted quite a high royalty, as they were on leased lands.

Since the Lackawanna mines were spread across the Northern Field, from Scranton to below Nanticoke, and were at widely differing stages of development of the underground workings, the figures in the table can be taken to give a good idea of the range of costs found across large parts of the industry. Returning to the evidence of the legislative committee from the late 1890s, it is therefore useful to mention production costs given for other mines and companies in the course of the hearings. For example, the Kingston Coal Company, operating near Wilkes-Barre, claimed a figure of about $1.80, excluding royalty; the Lehigh Valley Coal Company $2.12, including royalty; and the Lehigh and Wilkes-Barre Coal Company $2.50, including royalty and

TABLE 13.3

Comparative Costs of Mining at the Mines of the Lackawanna Coal Department 1896–1898.[46]

	Tonnage Dec 1896–Nov 1897	Weighted Average Cost/Ton 8 Months Data Nov 1896–Feb 1898	Weighted Average Royalty/Ton	Weighted Average Cost Without Royalty
Storrs	428,140	1.29	0.12	1.17
Pyne	259,900	1.09	0.11	0.98
Sloan	250,743	1.28*	0.12	1.16
Diamond	233,381	1.38	0.13	1.25
Bellevue	218,595	1.40	0.11	1.29
Woodward	211,691+	1.95**	0.29	1.66
Cayuga	204,910	1.19	0.15	1.04
Archbald	200,326	1.32	0.10	1.22
Continental	184,189	1.31	0.10	1.21
Taylor	183,656	1.25	0.10	1.15
Brisbin	179,876	1.57	0.12	1.45
Dodge	176,991	1.25	0.10	1.15
Hampton	153,367	1.33	0.10	1.23
Hyde Park	153,120	1.41	0.11	1.30
Oxford	137,423	1.53	0.17	1.36
Pettebone	132,340	2.09	0.25	1.84
Bliss	127,701	2.03	0.10	1.93
Hallstead	87,646	1.82	0.15	1.67
Holden	51,589	2.10	0.31	1.79
Total (all mines)++		1.54	0.14	1.40

*7 months data only.
**6 months data only.
+ 3 months idle.
++ based on 14 months data within the 16 month period.
(Avondale is omitted because of a serious squeeze and flood in 1897 that prevented normal mining operations).

provision for extraordinary expenses attributable to gaseous and dangerous mines. A further comparison is provided by Thomas Righter, superintendent of the Mount Carmel Breaker in the Western Middle Field, who reported a high figure of $2.70, all inclusive. As an addendum to this list, the Lehigh Valley Coal Company was conveniently asked to calculate costs again, to match the basis used by the Reading Coal and Iron Company. This method excluded taxes, insurance, and various other accounting provisions. The directly comparable figures for the two companies then emerged as $1.786 and $1.956 respectively, indicating an average advantage of 17 cents a ton for the Lehigh Valley Coal Company. This was a significant differential, given the size of both mining estates at the time.[48]

A final source of information for a slightly earlier period is the 'industrial espionage' reports on Northern Field mines, commissioned by the Reading Railroad to gather information about industry competitors. These reports contain often quite detailed assessments of production costs at different mines, including both large operations and small concerns. Samples of these from different parts of the field, especially from areas not covered by the Lackawanna Coal Department's mines, are given in Table 13.4.

It is apparent that a significant proportion of smaller mines were able to match the returns achievable by the larger companies, while others suffered during periods of trade depression or from high royalty charges. The figures for the Simpson Breaker, for example, indicate the vulnerability of concerns in the early stages of development, although further comments in the report on the same mine indicate the expectation that mining costs would drop significantly as production became more established. Mines in urban areas, such as the Green Ridge in Scranton, experienced additional costs because of the extra care in mining required to

TABLE 13.4
Estimated Costs of Mining in the Northern Field 1888–9.[49]

Mine or Company	General Location	Cost Per Ton Excl. Royalty ($ Per Ton)	Royalty on Size Above Pea ($ Per Ton)	Comments
Pennsylvania Coal Company	Pittston	1.15	0.20	Expected profit >50 cent per ton
Dolph	Peckville	1.25	?	Store profits will reduce costs
East Boston	Kingston	1.50	0.35	Proprietors statement under oath
East End	Mocanaqua	?	?	Suspended 1886–1888, unprofitable
Eaton Colliery	Archbald	1.15–1.20	?	
Edgerton	Carbondale	0.95–1.15	0.25	Estimated royalty
Simpson	Carbondale	1.40–1.60	0.575	Costs high owing to recent construction
Grassy Island	Winton	1.05–1.15	0.25	Low-cost mining
Green Ridge	Scranton	1.20–1.25	?	Not very profitable in earlier years

? No data available

support the surface and avoid mine caves, for obvious reasons. A final very informative case from the Northern Field was the Red Ash Coal Company, which operated two breakers near Wilkes-Barre and paid a royalty per ton of 35 cents on the larger sizes of coal. According to the report, the mines had initially been very unprofitable and had been put up for sale during the 1885 depression. However, since then, they had cleared $200,000 in profits after paying for all new improvements. These profits were attributed to the extraordinary gains during the Labor Troubles of 1887–1888, when profits of $1.00–$1.50 a ton were made.[50] This confirms the experience of the Florence Coal Company in the earlier example—namely, that the benefits of the strikes in the Lehigh and Schuylkill Regions were very important to the success of individual operators, especially those which were relatively new to the business and had managed to avoid bankruptcy during the 1885 depression. Seemingly, however, such windfall profits were not to recur during the 19th century. So, in the ensuing years, good management and careful cost control would be essential for continued debt-free operation.

To complete the geographical coverage of data on indicative mining costs, recourse can be had to Eckley B. Coxe's testimony during the investigation into the selfsame Labor Troubles of 1887–1888. He had looked into the matter closely, as a result of questions put to him by the investigators, and had concluded that, for Coxe Brothers, a figure 'not very far from $1.50 per ton' for coal in the cars at the breaker was appropriate.[51] He also felt there was a tendency for costs to rise slowly from year to year, on account of the increasing difficulty of working older mines. Judging by an accompanying table he provided, this did not include royalty, which averaged more than 40 cents a ton in the Eastern Middle Field, although some leases as low as 25 cents could be found.[52] Over the long term, he argued, the company had averaged a profit of 12 cents a ton at most, although much of this would disappear if depreciation were fully accounted for.[53] In years of depressed trade, profits would naturally be lower. The company had only made 4.5 cents per ton in 1886, but Coxe also commented that the Reading Coal and Iron Company had reported a loss of 40 cents a ton in the same year. By way of explanation of the differential, he declared, 'I do not think they manage their business as well as we do'.[54] Overall, he was not very sanguine about the profitability of mining as an occupation: 'There is no business on the face of the earth subject to as many casualties'.[55] His final observation on the previous quarter century in the coal business was that most of the money was made during the Civil War, and

> 'a majority of the men who made fortunes during the war would have been better off if they had gone out of the business after the war and invested their money in Government bonds'.[56]...

Or, one might add, they had waited until 1886 before re-investing in the Northern Field, so the

Lehigh and Schuylkill mines and operators could again provide them with windfall profits!

To conclude, the question of mining costs and their variation over space and time has merited close attention because of the wide variety of mining circumstances and geological conditions experienced. These cost variations will now be summarized. Conditions were such, in some parts of the Northern Field, that individual operators could earn sufficient profits to remain in business for extended periods of time and have enough funds to develop their properties on a modest scale. A favored few were blessed with a very low cost of mining that put them on equal terms with the most efficient of the large corporations. Others, in less accessible locations, or when the qualities of the seams were poor, struggled to survive, either by frequent change of ownership, or extended suspensions of operations during depression periods. The timing of initial investment in relation to cyclical business conditions or unforeseeable events, such as the Labor Troubles of 1887–1888, could also make or break mining operations, which were particularly susceptible to financial problems in the early stages of their development. Although the supporting evidence is limited, owing possibly to the restricted number of individual operators left in the Schuylkill Region after the formation of the Reading Coal and Iron Company, it would appear that mining costs continued to be significantly higher than in the Northern Field. This finding is consistent with the generally less favorable mining conditions, characterized by deep mining on steeply inclined seams. Much of the cheaper, more accessible coal above water level had been mined out by the time of the Civil War. Individual operators were usually at a disadvantage to the larger companies, because they rarely owned lands in fee and their leases were often at relatively high rates for less than choice tracts of coal lands. A 15–20 cent higher level of royalty payment per ton could be the difference between accumulation of profits and a knife-edge existence, where every business downturn threatened financial collapse. Only a small number of independent operators, such a Coxe Brothers in the Eastern Middle Field and, later, Simpson and Watkins in the Northern Field, were able to build up their mining operations, acquiring additional lands or leases, and building new breakers, by the re-investment of retained earnings. However, to assume that all individual operators were entirely at the mercy of market forces, would be to overlook the crucial vested interest of the railroads in maximizing the tonnage available for transportation over their lines. This had important implications both for potential new entrants into the mining business and for the cash-flow of established concerns. These issues will be examined further in the next section.

Finally, several of the major Northern Field companies, such as the Susquehanna and Pennsylvania coal companies, together with the coal department of the Lackawanna Railroad, constituted a group of efficient, low-cost producers. They had sufficient earning capacity and, in most cases, railroad backing to enable them to pursue a well-planned approach to the optimal development of their properties over the medium and long-term, largely unfettered by any serious financial restrictions. A very clear example of this was the construction of the Susquehanna Coal Company's Breaker No. 5. This investment had commenced in 1875, but was then postponed during the ensuing depression years. With improving business conditions in 1879, $36,180 was appropriated from retained earnings to press the project forward, so shipments could begin from the new breaker in 1880.[57] The Lehigh Valley Coal Company was in a somewhat less favorable position, since it had a mixture of good-quality lands that were economical to mine, and others, such as the Girard leases, where royalties were high enough to have an adverse impact on levels of profitability. As the century wore on, the substantial financial dependence of the coal company on an increasingly ailing Lehigh Valley Railroad resulted in financial difficulties and cash flow problems, especially after the termination of the Reading lease of its parent in 1893.[58] Among the large concerns, the Reading Coal and Iron Company undoubtedly had the greatest management problems to solve and had considerable difficulty keeping costs under control. However, the Lehigh and Wilkes-Barre Coal Company should not escape mention also. Although its output included premium-quality coals, both from the Lehigh Region and the Plymouth area of the Northern Field, it had also paid very full prices for its estate, its royalty payments were high, and its mines were costly to run, as previously noted. Such an adverse combination of circumstances placed it among the more financially precarious of Northern Field companies, as its bankruptcy during the late 1870s had already demonstrated. This financial reverse limited its ability to develop its estate and expand its output in a well-planned and effective manner.

Understanding of these financial differences between the major players in the mining business will facilitate final interpretation of the evolution of the industry in the concluding sections of this study, as well as the subsequent examination of approaches to investment in new mining capacity during the period.

SOURCES OF FINANCE: THE DUAL ROLE OF THE RAILROADS AS 'VENTURE CAPITALISTS' AND BANKERS OF LAST RESORT

The background to this topic has been introduced briefly, in the previous chapter, in the context of purchase and transportation contracts, and further anecdotal evidence relating to it has emerged in the course of examples used above. However, it is of such importance, in understanding the symbiotic relationship between railroads and their individual operators, that it needs to be examined more systematically here.

Put at its simplest, the individual operators were, in general, men of limited means who required substantial injections of capital to enter the coal business, because of the high start-up costs of deep mining by shaft. This method of mining became increasingly necessary as the more accessible seams above water level became exhausted. A very small number of individuals who had entered the coal business prior to the Civil War, such as William Connell and John Jermyn in the Northern Field, were more independent financially, but they were the exception rather than the rule. The railroads, in turn, needed the individual operators for a variety of reasons. These were as follows: firstly, provision of traffic in the early period of a railroad's existence, perhaps before its own mines were fully operational; and secondly, to preserve the coal resources of the railroads' own coal lands held in fee. Retention of reserves was certainly a reasonable strategy, although the original expectation of higher levels of future profit was to prove somewhat illusory. Finally, individual operators made an important contribution both to the total of a given railroad's shipments and the rated productive capacity tributary to their lines. These two measures were important in the calculation of market share for cartel allocation purposes at different times, as a previous chapter has shown. Hence, both the encouragement of individual operators, and retention of these shipments once established were key aspects of strategic behavior by the railroads within the coalfields. Fear of losing potential mining developments to competing roads or of failure to bind new companies into long-term tonnage contracts, were also important motivating factors for those charged with the management of large coal estates. However, as will become apparent below, such competitive behavior, especially during business-cycle upswings, could prove to have unanticipated and unwelcome consequences for the industry as a whole.

Examples of loans to individual operators can be traced back as far as 1860 when William Brennan, mining near Carbondale, asked the Delaware and Hudson for a further loan of $5,000 to aid him in putting up his breaker.[59] However, the paucity of further references to such arrangements, in surviving records over the next two decades, suggests they were not very prevalent until after the end of the 1870s depression. They appear to have been most widely used in the Northern Field, although examples can be found from the Western Middle Field also.[60] The companies most heavily involved were those competing most vigorously for tonnage, namely, the Lehigh Valley, the Lackawanna, and the Erie, as subsequent examples will show. The Jersey Central certainly participated, at least to a certain extent,[61] and the Reading probably entertained such propositions also, although specific evidence is lacking, and it faced less in the way of competitive pressures from other roads, in many of the areas tributary to its lines, than was the case in the Northern Field.

The letter books of the Lackawanna coal department, in particular, bear ample testimony to the importance of the loan mechanism. Although there is a risk of over-reliance in the discussion on one company's perspective, the competitive nature of many of the negotiations with the operators means that the actions of other railroads were also reported on a regular basis. While numerous examples could be cited, one has been selected to elaborate in detail, because of the range of aspects of the complex, evolving relationship between the operator and the railroad that it illustrates.

Prior to the example, however, it is useful to make some general observations about the Lackawanna Railroad's perspective on two issues. The first of these was the process of lending to individual operators, in terms of financial risk. The second was the railroad's assessment of the business character of the operators themselves. Fortunately, some checks on the validity of these subjective estimates can be found in the labyrinthine entries of the surviving 19th-century R.G. Dun and Company credit ledgers.[62] These ledgers have the signal merit that they have as much, (or more), to say, about the untrustworthy character and the bankrupt, as they do about the upstanding businessman of the highest character. Such credentials make them entirely suitable as a commentary on the coal trade!

The return of business confidence in 1879–80 re-awakened the interest of all participants in the coal trade in possible investment opportunities. As Storrs remarked, 'there is a buzzing of all the parties who have coal property to have them improved'.[63] However, he was aware that President Sloan had been reluctant for some time to negotiate any new contracts with operators on a labor basis, although requests of this type were starting to reach his office.[64] He also took the view when writing to Sloan that 'inferior and ordinary properties it is better that others work', but it was still important to secure the tonnage from such properties, because 'transportation arrangements decide losses and profits'.[65] When speaking to an operator, however, his emphasis was somewhat different.
He reported having informed one individual that the railroad would not give him money to make improvements on land that the railroad would not itself undertake to improve and work.[66] President Sloan and his faithful coal agent were very much of one mind, in terms of their cautious stance at this stage, in relation to mining and financial risk, yet they kept one eye very clearly focussed on cash flow from coal transportation. As Sloan remarked at the end of 1880:

> 'Of course we must not encourage any weak enterprises to our injury but we must not turn off any tonnage as they will go elsewhere or encourage new enterprises'![67]

By 'new enterprises' Sloan was referring to the construction of new competing railroad lines, the only kind of development that was more to be feared than loss of a specific mine's tonnage to an existing competitor.

Such caution was understandable after so many years of depression in the trade and unremunerative prices. It also partly explains the relative lack of enthusiasm for making financial advances to operators in the early stages of the recovery. Even when such advances were made, the company was very careful to make detailed calculations of the total tonnage in the property, the likely cost per ton, and, hence, the amount per ton shipped that could be applied to pay off the loan, without eating up mining profits entirely. This amount might normally range from 10 to 25 cents a ton. Further to this, and related to the comment made to the operator above, wherever possible, the railroad would seek additional security for the loan. One method of doing this was to insert a clause into the contract, giving the railroad power to step in and take over the mining operation in the event of failure to ship coal or make loan payments. Hence, after allowing for the distinct possibility of failure by the individual operator, the railroad would still prefer to be saddled with a mine of reasonable promise and productive potential than one that had no hope of ever generating a good return on investment. A second approach might involve a mortgage on lands held in fee, if the potential entrant to the coal business owned title to some tracts in addition to his lease.

Interestingly, as business conditions improved and the competition to secure the tonnage of remaining coal tracts intensified, a discernible shift in the company's attitude towards financial risk became apparent, although it was still tempered by financial prudence and sound judgement. For example, in the spring of 1883, Sloan declared, 'We must try and be secure in all our advances and yet keep in mind both LV [Lehigh Valley] and Erie are watching us very closely'.[68] Later, in the summer of the same year, after Storrs had set before him his assessment of a particularly dubious proposal for a loan, he was moved to express his assessment of the prevailing situation at greater length:

> 'While transportation of coal will urge me to make advances and take risks, and large ones, I fully agree with you in your conclusions—we must not take such risks as is (sic) proposed in this case. The Reading and others will no doubt take large risks and do almost anything to thwart us in any of our territory, they all feel jealous of our success in coal transportation'.[69]

The reference to the Reading in this quotation related to their lease of the Jersey Central earlier in the year.

In general, the very full surviving correspondence of both Storrs and Sloan for this period gives clear evidence of careful appraisal of all loan proposals on their merits or otherwise, and relatively conservative financial assessment, tempered with sensitivity to external competitive pressures. Above all, it shows a clear strategic sense of which actions were essential to safeguard the company's long-term transportation interests and which were of more marginal importance, only to be pursued if the commercial terms were sufficiently favorable.

Aside from the specifically financial and contractual aspects of the railroad's interaction with the operators, Storrs also took account of the personal characteristics of the individuals involved, and, if appropriate, of the investors who might be backing them. While more entertaining for both writer and reader than dry contractual matters, these assessments were also to be taken very seriously, given the importance of men such as Storrs and Sloan in the coal trade more widely. They can also usefully be compared with the more detached observations of the credit agents of R.G. Dun and Company.

Judging by his letters, Storrs' classification ran from the 'visionary', whose imagined schemes bore little connection to commercial reality, through to the 'practical man', the highest form of understated praise that he could muster. Other intermediate categories included the 'restless speculator' and his ilk, who specialized in taking risks with other people's money. For example, Messrs. DeWitt and Cake of the Clear Spring Coal Company slipped down the rankings somewhat because one of them (never identified) was both 'visionary and impractical'.[70] Fortunately for them, since they had the best coal between Pittston and Kingston, Storrs was willing to negotiate with them despite such telling character flaws![71] C.M. Sanderson, operator of the Pancoast mine near Scranton, fell into the 'restless speculator' category, no doubt in part because of his seemingly endless pecuniary requests for a variety of planned, as well as current projects.[72] Finally, there was William Connell, 'practical man' *par excellence* and leading citizen of Scranton.[73] Almost by definition, his exalted status among entrepreneurs meant he had very little need to call on the railroad for loans. Nonetheless, as we have seen, he was first in the queue to request additional payments on his contract when he felt market circumstances were moving against him.[74]

Having established the basic outlines of decision-making criteria employed by the Lackawanna Coal Department, it is now appropriate to consider in detail the example of the entrepreneur Samuel Stetler and his business relationship with the coal department over a number of years. Stetler was originally an outside foreman at the coal department's central breaker in Scranton, and apparently one of the company's best employees, before he decided to go into business for himself at the beginning of the 1880s.[75] Unfortunately, his attributes as employee did not seem to carry over into his entrepreneurial role. In the latter, Storrs judged him to be neither safe nor judicious as a manager, and worst of all to be 'wild and visionary in his schemes'.[76] In consequence, the Lackawanna was initially very reluctant to give him any financial support, but eventually decided to do so, to a limited extent. One unintended consequence of the latter action, that they were probably unaware of, was to raise the business credibility of Stetler in the eyes of the credit-rating agency. In August 1881, the R.G. Dun report on Stetler read as follows:

> 'He is a hardworking…man who has the confidence of the concern backing him [i.e. the Lackawanna] and in whose employ he has so long been and who should know his traits'.[77]

This slightly misplaced confidence was also to lead the credit-rating agent to make a more positive assessment of Stetler's prospects than was perhaps wise, as will be seen below.

Stetler's first approach to the Lackawanna came as early as the spring of 1880 when he notified them that he had acquired an option to lease nearly 200 acres of land east of Pittston. Storrs felt that the lands were not of the choicest kind and expensive to mine, so he offered Stetler a transportation arrangement. The latter declined, requesting a price- or labor-basis contract. Storrs's initial judgement was that this was a mistake, declaring 'I don't think Stetler or anyone can save themselves on this property…but he insists on carrying it through'.[78] Apparently, on this occasion, Stetler realized or was persuaded of the dangers attendant on investing in this particular tract, so he let the option lapse. Subsequently, he leased another tract slightly lower down the valley, agreeing to sell to the Lackawanna on a labor basis. He claimed he could get money for improvements, but Storrs felt there was a distinct possibility of failure, so he insisted on a clause in the contract giving the coal department the right to take over the property on non-performance under the agreement.[79] Such a clause was to be extensively used in dealing with a range of budding entrepreneurs in the ensuing years. Again, the credit agent, unaware in turn of Storrs' sober assessment of the situation, recorded the final contract provision in a positive light as follows:

> 'The company contract is so constructed that if Mr Stetler does not turn out a given amount of coal in a given time they have the right to take the Breaker and run it themselves which there is no danger of.... We believe his prospects are good and that he will make money'.[80]

However, in the early part of 1881, justification for Storrs's early doubts began to surface. Stetler was clearly having more difficulty raising the money for improvements than he himself had expected. His first ploy was to ask the Lackawanna for the money, saying otherwise he would have to give up the contract. Storrs initially declined, but offered to change the contract to a transportation basis if that would assist Stetler in raising funds. Despite his obvious lack of capital, Stetler then tried to acquire the nearby Phoenix Colliery, apparently with a view to switching its tonnage contract away from the Lackawanna towards the Lehigh Valley and the Erie. This seems to have been partly a ruse to try and put pressure on Storrs. There followed in quick succession a highly complex series of negotiations, as Stetler tried to raise money for his improvements. These involved parties such as Langdon and Company, who were interested in the Phoenix; Packer and Knowlton, coal merchants; and Haddock and Steel, mining entrepreneurs from further down the valley. At one point, Stetler's lessor also threatened to sell the property from under his feet! Finally, Storrs decided to put an end to the merry-go-round of negotiations to avoid potential loss of tonnage. He reversed his original decision and agreed to loan $70,000 to enable Stetler to commence his improvements, although the contract was to remain on the originally agreed labor basis.[81] Once the funding was placed on a sure footing, shaft sinking commenced in August 1881 and continued into 1882.[82]

However, by May 1882, Stetler was back in Storrs's office, looking for further funds to finish his work and start mining. The property had proved to be better than expected now that the shafts were sunk, so Storrs took the view that Stetler was good for a further loan of $30,000.[83] Not content with this, Stetler very soon found himself in need of a further $50,000 for mine cars, mules, houses, and to pay running expenses until the mine was self-supporting.[84] Fortunately, he was able to secure this very quickly from one of his wife's relatives, the wealthy John Duryea, associated with the Glen Cove Starch Works on Long Island, in exchange for a half share of the mine.[85] By the end of 1882, the works were completed and 35,125 tons had been shipped from the new Stetler breaker.[86]

Unfortunately, it appears that Stetler was not aware of his own personal and financial limitations. Within weeks of securing the funding for his first enterprise, he embarked on another, joining with E.L. Fuller and Attorney R.N. Archbald in the Fuller Coal Company. Stetler himself subscribed $6,000 of the capital stock and his partners $14,000 and $12,000 respectively. Dun and Company regarded Archbald as the most financially secure of the three.[87] Storrs had encountered Fuller before in relation to an earlier proposed scheme, and he was quickly assigned a position on the general coal agents' continuum of (in)competence: 'a young man said to have no money with more cheek than solid merit and no experience in the coal business'.[88] His cheek had clearly not deserted him—his opening request, to develop a tract of 425 acres in Wyoming, below Pittston, was for a loan of $100,000. This amount was insufficient for a first class improvement, in Storrs's judgement. Fuller's partner, Stetler, followed up with a further letter to President Sloan a fortnight later, requesting a more substantial loan of $150,000, to be paid back at 10 cents per ton, in return for a labor-basis contract.[89] Storrs initially felt the risk was too great to contemplate backing them at all, but unusually, Sloan suggested that the Lackawanna was perhaps being too risk-averse in the light of growing coal demand and escalating prices for coal lands. As a result, they agreed on a compromise proposal. Provided that Fuller et al. found the first $30,000 for the improvement themselves, the Lackawanna would loan $100,000, repayable at 15 cents a ton. The property in question was near the Maltby breaker, which had just been snatched from the Lackawanna by the Lehigh Valley. So there was also a fear that if they did not secure the tonnage, the Lehigh Valley would extend their influence further down towards Wilkes-Barre and siphon off more and more traffic from the Lackawanna and Bloomsburg division.[90] These negotiations proceeded more smoothly than with the Stetler mine and the price-basis tonnage contract was signed in February 1883, shortly after work on the shaft had commenced.

Naturally, it contained a clause allowing the Lackawanna to assume control in the event of non-performance.[91]

Initially, their prospects had seemed bright, according to the Dun and Company ledgers, and Storrs' early pessimistic reaction to this second enterprise that 'Stetler should have stuck to one thing' seemed overdone.[92] Unfortunately, even before the mine had begun to ship in the summer of 1883, it was becoming clear that the coal deposits on the property were not of the quality or quantity originally expected (at least by the Fuller Coal Company), since only one vein 4.5 feet in thickness had been located, instead of the three anticipated. Further investigation revealed that the vein covered more of the property than first thought, but there was a growing concern that it could not be mined profitably.

Astonishingly, when this enterprise, named the Fuller mine, was barely off the ground and its future was looking dubious, Stetler, Fuller, and their compatriot embarked on yet another scheme, this time near Taylor, below Scranton, under the different corporate guise of the Amity Coal Company. Not only did they want $150,000 as an improvement loan, they also wanted the railroad to build the necessary branch line. Some of the lease terms were disadvantageous because of an unreasonably high annual maximum tonnage, and another part of the property, owned by a Mr. Felts, was risky because of large mortgage liens against it.[93] An exasperated Storrs could only exclaim:

> 'Stetler and Fuller, presuming on what we had done for them at the Stetler colliery and the Fuller in Wyoming have lost their heads, and without judgement or reason are ready for the wildest schemes'.[94]

The Delaware and Hudson, the Pennsylvania Coal Company, and the New York, Susquehanna and Western had on previous occasions declined to have anything to do with the Felts property. However, detailed local geography was significant in this instance, as Stetler and Fuller doubtless appreciated, because all the land surrounding the Felts property was already owned by the Lackawanna.

Once again, despite the initially adverse reaction to the Amity proposal, the tension between financial risk and the danger of strategic loss of tonnage began to weigh on the minds of Storrs and Sloan. In addition, the main property tract proved to be larger than had been thought, when it was properly surveyed. This meant that it contained about 1 million tons of coal, which was better security for a loan. President Sloan, in particular, was struck by Storrs's observation that there were not many 1 million ton lots left outside the hands of the major companies. Hence, he was keen that control of them should pass to the Lackawanna rather than its competitors. On the strength of the detailed survey, Storrs had decided to make an initial offer of a $100,000 loan and a railroad branch line, but Fuller, in response, had declared himself committed to other parties, i.e. the Lehigh Valley or the Erie, with whom he had also been negotiating. Fuller then set off for New York, hoping to improve on Sloan's offer, but in an amusing twist, Stetler, who apparently had a little more loyalty to his major paymaster, set off after Fuller to try and persuade him to give the transportation contract to the Lackawanna![95] Finally, an improved offer of a loan of $125,000 brought the tonnage to the Lackawanna, by a contract dated March 5, 1884,—although they had to snatch it away from the Erie, when a last-minute dispute over the terms of a draft contract with the latter gave them opportunity.[96] After two previous successes in attracting financial support, and with a third imminent, Stetler's thirst for other people's capital seemed to be reaching unquenchable proportions. In December 1883, he was reported to be hatching a scheme to sell bonds on his existing property. This was to raise funds to buy lands, considered by Storrs to be grossly overpriced, from Groves Brothers of Danville.[97]

While fortunately nothing more was heard of this particular scheme, murmurings began to be heard in the business community, as the costs attendant on starting up multiple new mining enterprises began to tell on the credit of Stetler and his partners. They had leveraged the optimism in the coal trade to the maximum during the expansionary phase of the early 1880s. However, the storm clouds were beginning to gather over Wall Street, just as the Hunt and Amity Coal Company's mines were entering that crucial early phase of development when large investments had been made, but cash-flow was still small. Although Stetler had had some success with his first concern, the Hunt proved to be the Achilles heel of the overall operation. As a consequence of the disappointing yield of coal, the mine was already being regarded as a failure by the spring of 1884. Then, to add insult to injury, a serious problem with flooding and quicksand breaking into the shaft suddenly occurred in April of the same year. Storrs estimated this would require $50,000 to resolve properly.[98] Within weeks, the Fuller Coal Company were in difficulties, looking unsuccessfully for short-term financing to cover their costs, after Storrs assisted them by buying their coal for a month to tide them over. Realizing the precariousness of their position, they subsequently made unsuccessful attempts to sell the colliery to the Lehigh Valley Railroad and other parties. By this time also, Dun and Company had caught up with them and declared their credit to be seriously impaired. They were behind with their payments and suits were being instituted against them. It was also reported that they were in the process of abandoning operations at Hunt and moving some of their property to the Amity shaft.[99]

Undaunted by looming failure at the Fuller mine, Stetler approached the Lackawanna again, to ask for a further $60–75,000 to buy out Duryea's interest in his first concern, and to offer them a majority stake in the Amity mine, but Storrs was not in any mood for hasty decisions.[100] Problems continued to worsen at Fuller until they finally abandoned the property to the Lackawanna under their contract, with debts of about $100,000, not to mention back royalties owing on the lease.[101] Unfortunately, the latter had to be paid immediately for the Lackawanna to gain possession, or the lease would be forfeit. This left the railroad no option but to stump up the necessary funds. Temporary possession was one thing, however; satisfying creditors to the tune of $100,000 quite another. A decision therefore had to be made as to how to proceed. The cost of mining was 25 to 40 cents above average, and, with existing large debts, it was scarcely an attractive proposition for any new investor during an accelerating business-cycle downturn. The decision rested on the perceived benefit to the railroad of future coal tonnage from the mine, taking account also of their prior investment by way of the earlier loan to the Fuller Coal Company.[102] While President Sloan's views on the matter have not come down to us, the county courthouse deed book recorded that the mine was purchased by the Lackawanna at sheriffs' sale, by deed of March 13, 1885, for the sum of $100 and settlement of outstanding debts, with interest from January 1, 1885.[103]

Growing adversity seemed to stimulate Stetler to even greater exertions. He generated a plethora of further proposals for offloading the Stetler mine onto Lackawanna, to free up funds for re-investment elsewhere. Compared to the Hunt, it has to be said that some of these proposals were quite attractive. The Stetler mine had a very respectable mining cost of $1.40 a ton, including royalty, and was generating profits of 33 cents a ton. The possibility of Stetler delivering additional lands, and the Feeder Dam property in particular, was also present.[104] Finally, with Stetler very keen to sell, Storrs purchased the mine for $200,000, by deed of January 1, 1885, thereby securing its tonnage and the possibility of an additional $70–80,000 profits during the lifetime of its operation. Unsurprisingly, a last minute ploy by Storrs to postpone payment to Stetler by a year was not successful![105] Estimates reported by Dun and Company of how much Stetler himself had netted from the sale ranged from $75–100,000 down to $30–40,000. The latter figures were felt to be more plausible. Stetler was reported to be engaged in real estate deals in Scranton, but ominously, was also slow in paying his bills.[106] Further reports of arguments between Stetler, Fuller, and Archbald began to circulate, related to developing problems at the Amity shaft, as business conditions progressively worsened in the spring of 1885. Storrs became alarmed, because, unlike the other properties, their transportation contract with Amity contained no forfeiture clause allowing them to take possession of the mine, although their loan *was* secured by a mortgage against the property.[107] By the summer of 1885, the Amity Coal Company had collapsed, once again owing back royalties on its leased lands. During and after the resulting sheriff's sale, Storrs displayed quite extraordinary business acumen in finally securing the mine at a reasonable price for the Lackawanna. Initially, at the sale, Storrs was outbid by Fuller himself, who topped Storrs's best offer of $182,100. However, Storrs correctly judged that Fuller would not actually be able to raise the money in the time allowed. Another bidder stepped in, claiming to have $175,000, then asked Storrs if he would bid the same, but he declined, and they backed out. Having cleared the field of competing bids, Storrs finally stepped in to secure the mine with a bid of about $167,000, some $15,000 less than his original unsuccessful bid at the sheriff's sale itself![108] Storrs's overall rationale for the purchase was simple. He did not expect any profit from the mining itself, unlike at the Stetler mine, and he anticipated losing money on the purchase, in relation to his assessment of the mine's value as an improvement. However, the loss would be made up by 25 cents a ton profit on 100,000 tons of shipments and there were at least 8–900,000 tons in the property.[109]

Thus, in the space of a few months, the optimistic aspirations of several budding entrepreneurs led by Samuel Stetler, the only one with genuine mining experience, had come to almost nothing. They had ridden the rising limb of the business cycle and the mood of the times to extract hundreds of thousands of dollars from the Lackawanna for enterprises on coal property whose value Storrs had questioned from the outset. They began to falter as the business expansion of the wider economy stalled and Dun and Company reported that 'their capital [did] not appear to have been hardly adequate for their wants'.[110] Finally, they crashed or bailed out, as the business depression hit bottom in late 1884–1885, leaving the Lackawanna with partly repaid loans of over $300,000 and further bills of more than $470,000 for the privilege of acquiring three new improvements. Of these, only one, Stetler's original mine, appeared to have commercial promise for the future.

With the benefit of hindsight and statistics from the Mine Inspectors' Reports, it is possible to assess the eventual benefit or otherwise to the Lackawanna of its involvement in this extraordinary saga of entrepreneurial ambition and eventual collapse. The Stetler, renamed the Hallstead after its purchase, provided a handsome return on the amount invested, shipping a total of almost 2.3 million tons of coal by 1902. The Amity, renamed the Holden, produced a solid level of output into the early 1890s, but tailed off thereafter, for a respectable total of 1.38 million tons by 1902. This was rather more coal than Storrs originally anticipated mining from the property. Photographs of these two mines under Lackawanna management can be seen in Plates 12 and 13. The Fuller, renamed the Hunt, proved to be an unmitigated disaster. As a result of further flooding and a disastrous fire, its operations were suspended for long periods. Even including mining by a later lessee of the property, less than 280,000 tons had been shipped in the first 20 years of its existence up to 1902.[111] Allowing for an optimistic 30 cents a ton profit on mining at Hallstead, no profit at the other mines, and 25 cents a ton transportation profit, the Lackawanna would have made about $1.68 million over the period. After deducting the cost of their original outlays, however, it is clear that had these been their only mining properties, they would have done as well to heed Mr Coxe's advice and invest in risk-free Treasury bonds instead.

While Storrs and Sloan, already inured to the financial risks of mining, emerged largely unscathed, if not resplendent, from this particular episode, the same could not be said of Samuel Stetler. His final entry in the Dun and Company ledgers in August 1885 exclaimed 'it is now rumoured on the streets that he has absconded!' This was not particularly surprising, since on the same day two lawsuits for recovery of debts were instituted against him by local commercial houses and a judgement for $7,000 was entered against him in favor of one of Scranton's prominent citizens, B.H. Throop.[112] Nothing was heard of him for some years thereafter, until he surfaced again in 1892, penniless and paralysed by a stroke. Ever generous of spirit, Storrs recalled to President Sloan the assistance rendered to the company by Stetler, in relation to the Amity foreclosure, and subsequent assistance with a problem lessor. He suggested giving Stetler a small pension of $50 a month for a time, although Storrs also felt it wise to keep the vouchers private from the other employees in his office. Stetler was grateful for this kind consideration, which enabled him to go to Buffalo for treatment.[113] While he never fully recovered, Stetler eventually bounced back. In 1899, he was once again knocking on the coal department door. This time it was W.H. Storrs's turn to take up the familiar refrain: 'their financial standing is not strong. Mr. Stetler himself has no money.' In the light of past experience, Storrs Junior was unable to recommend the latest proposition 'if it means we have to advance them money to make improvements'![114]

The case of Samuel Stetler has been recounted in some detail, not just because of its obvious human interest, but because it epitomises the interplay of so many of the factors that shaped the development of the coalfields in the latter part of the 19th century. Among these were, firstly, the mixture of symbiosis and unequal financial power relations between the entrepreneur and the railroad; secondly, the way in which competition for lands and future tonnage fed off a climate of business optimism; thirdly, the speed at which entrepreneurs would try and enter the trade whenever a margin appeared on the upswing of the business cycle, not realizing that the time required to develop the mine to full capacity would leave them with cash-flow difficulties, just as the ensuing downturn became most severe; fourthly, the significance of personal and friendly business relationships, which nonetheless were subservient to the demands of prudent accounting; fifthly, the voracious capital requirements of deep mining of anthracite and the considerable risks that were attendant on any new mine-improvement project; and finally, despite the seemingly inexhaustible wellspring of unwarranted optimism displayed by successive generations of entrepreneurs, the economics of mining on its own barely justified investment, because the attendant risks and costs were so high. Only the promise of reliable profits on coal transportation provided a reasonable rationale to explain the willingness of railroads to lend such large sums to individuals, only a proportion of whom had experience or knowledge of the mining business.

It might be imagined also that Stetler and his business partners would have been sufficient by themselves to exhaust the patience (and the resources) of any railroad with which they come into contact. However, this was far from being the case. In the space of a few short weeks in 1881, for example, Swoyer had approached the Lackawanna looking for funds for his Forty Fort mine. He was closely followed by Sanderson in need of $30,000 for the new Pancoast mine near Scranton, and Waddell looking for an additional $15,000 on top of the $10,000 he had already been given. A week later, Hutchison, who owed creditors $200–225,000 was looking for at least $75,000 for improvements, while Haddock and Steel wanted to borrow $150,000 so they could buy out Hutchison and start mining themselves! Last, but not least, were DeWitt and Cake, mentioned earlier, who had now realized that their previous request was not large enough, and $85–100,00 was actually required.[115] Some of these operators, in favored locations, were able to establish themselves reasonably effectively after an initial injection of loan finances. However, in other instances, the financial relationships with the railroad turned into long running sagas that made the Stetler episode appear brief by comparison. The best examples of these were Haddock and Steel (later just Haddock) and Sanderson at Pancoast. They made numerous requests for further loans during the 1880s and 1890s, and, at times of acute cash-flow problems, the railroad had to institute a monthly revolving credit arrangement for them. This enabled the operator to meet his payrolls and pay his most pressing bills. Each month's loan was repayable at an agreed number of cents per ton on all shipments during the month. Space precludes detailed discussion of these individual cases, but a good illustration of Sanderson's behavior can be found in 1891–1892. He was experiencing cash-flow problems because of depressed business conditions and short-time working and needed a revolving loan of about $30,000 a month to keep him afloat. Later in the year, he persisted in requesting a much bigger loan of $300,000, which the Lackawanna board declined. Apparently, he could not understand why the railroad would not give him all the money he requested, to which Storrs's straightforward answer was, because his indebtedness increased year by year.[116] Indeed, Haddock and Sanderson were so persistent and so importunate that an exasperated Storrs could only declare them to be 'two hard problems', men who acted 'as though the world owed them a good living'.[117] As ever, the tiresome dealings with such operators had to be endured in the interests of retaining tonnage for the railroad.

CONCLUDING COMMENTS

While the invaluable company correspondence of the Lackawanna has enabled these financial and contractual inter-relationships to be followed closely over a number of years, the railroad was by no means alone in pursuing such a policy of courting the tonnage of operators tributary to its lines. As the Florence Coal Company example demonstrated and other evidence incidental to the Stetler case also showed, the Pennsylvania Coal Company, the Lehigh Valley, and the Erie were making, or trying to make loans of similar magnitude to the Lackawanna, and at the same moments in time. Both the New York, Susquehanna and Western and the Ontario and Western were also very active in this regard, as was mentioned in the previous chapter, using loans as a vehicle to attract traffic to their new lines, generating market share in the process. However, once such a mechanism had been introduced, it was hard to stop, and any company not participating during periods of business optimism would undoubtedly have lost out in relation to its future positioning within the trade. Adverse side effects of such a policy were perhaps difficult to anticipate at the time, but no doubt they would start to emerge during periods of business contraction rather than expansion, and so it proved. Storrs perhaps spoke for all the major Northern Field companies when he asserted in 1884:

> 'Our action and that of others in giving money to parties who have little of their own to go into the

> coal business has stimulated competition and inflated values beyond the point of safety'.[118]

And he might have added, had he known, this would shortly lead to the unexpected acquisition of a number of contractors' mines. Some of these were expensive to run, on lands of doubtful quality prone to flooding, and their tonnage had come on-stream just when it was no longer required, as a result of the worsening depression in the trade!

Chapter 14

REGION BUILDING II: INVESTMENT IN NEW MINING AND RAILROAD CAPACITY

Once the availability of investment finance was established, a further series of operational processes were called into play to enable physical development 'on the ground', and, indeed, underground, to take place. These processes were of three interrelated types, namely, decision-making, construction, and management activities. Since they took place within the context of fluctuating economic conditions, and a range of financial, contractual, and geological constraints, they were necessarily adjustment processes of quite a complex kind. Also, given the size of construction expenditures and the workforces involved, these adjustment processes had a significant impact on the developing pattern of economic activity in the coal regions.

The initial group of decision-making processes centred on two main issues. These were determination of the timing of new investment and choice of the geographical locations for new mines and associated lateral railroads. In addition to the broader economic and territorial context, decision-making at a detailed level was also closely tied to strategies for balancing development of different categories of coal lands, and especially to the management of the royalty system on coal leases, which has been explained in an earlier chapter.

The second group of processes were concerned with the physical construction of new mining and railroad capacity. It might be imagined, at first sight, that such 'nuts and bolts' activities would be straightforward and undeserving of further attention. However, this proved not to be the case. The timescales involved to bring some of the larger mine developments into operation were sufficiently long and the expenditure involved so large, that there were necessarily important interactions with other aspects of company decision-making, and these interactions need to be explored.

The third group of management-related activities obviously ran as a thread through the entire construction process. However, they were of particular importance for larger companies, when multiple construction projects in different parts of the coalfield were in progress at the same time. An additional set of management issues surrounded the integration of new capacity into the production system. While, again, these might appear to be unproblematic, the market conditions prevailing at the time a mine first started shipping might be very different from those envisaged when investment was initiated, and mines could also require extended periods of time, after start-up, before they reached full production. While larger railroad coal subsidiaries might be able to absorb some of these uncertainties and risks, they could have very adverse effects on the cash flow and financial fortunes of smaller concerns. The wider ramifications of such effects on the production system as a whole, therefore, need to be investigated.

These distinct but interacting components of the region-building process will now be examined in an attempt to convey the richness of the economic and spatial dynamics displayed by a major segment of the industry.

FACTORS AFFECTING THE TIMING AND LOCATION OF NEW INVESTMENT

In this section, specific consideration is required, firstly, of the factors affecting the timing of investment; and, secondly, of the factors that helped determine the geographical locations where capital would quite literally be sunk into the ground. In both cases, but especially in relation to locational factors, the implications of the advance-royalty payment system will prove to be of major importance.

Beginning with the timing of investment, the aggregate nature of investment patterns by different companies in the light of changing economic circumstances will be examined. This will be followed by an overview of the more detailed temporal and locational factors specific to investment decisions on individual new-mine improvement projects. Finally, an extended worked example of the interplay of these factors in the development strategy of the Lackawanna Coal Department during the 1880s will be given. This example will also be elaborated in later sections of the chapter, because the quality and detail of the available data are such as to allow the complex processes involved to be illustrated very clearly.

BUSINESS-CYCLE-RELATED FLUCTUATIONS IN INVESTMENT

As would be expected, and as the case study of Samuel Stetler has already indicated, judgements about the appropriate timing of mining investment were greatly influenced by the prevailing mood of business optimism or pessimism. In turn, this subjective assessment was closely linked to more objective external economic circumstances and the stage of the business cycle, although explicit discussion of 'cycles', as such, did not feature in the business correspondence of the period. For three companies, the Susquehanna, the Lehigh Valley and the Lackawanna, the surviving records allow these individual responses to the general business climate to be examined. Fortunately, the accounts have come down to us in sufficient detail that it has been possible to generate annual time series of expenditure on mining improvements that are as closely comparable as the rather variable accounting standards of the period will allow. More specifically, in the cases of all three companies, charges relating to depreciation of either coal lands or the improvements themselves have been excluded, so only the original investment expenditure is represented. However, it is not possible to ensure that the attribution of costs between running expenses and construction or improvement accounts, particularly for new investment at an existing mine, e.g., sinking of a new slope, is exactly comparable in each case.[1] Some of the sharp peaks in the graphs are also the result of mine purchases. This effectively concentrated investment into a single year that would have been spread over a longer period, if the company had originally developed the mine themselves.

The longest period of available record is for the Lackawanna (Figure 14.1). It stretches from the inception of its coal business almost to the end of the century. Essentially, four main phases of investment can be identified. The first was in the mid-1850s, but came to an end after the 1857 panic. The second began after the Civil War and drew to a rapid close after the 1873 panic. The large peak in 1870 is an artefact of the merger with the Nanticoke Coal and Iron Company in that year, as mentioned previously, and the transfer of mining properties, such as the Avondale, to the Lackawanna itself. Although it was possible in chapter 12 to distribute the land component of this merger back to the years when the original properties were acquired, the accounting records are not sufficiently complete or consistent to allow this to be done for the improvement expenditures made by the Nanticoke Coal and Iron Company. As a rough guide only, the bulk of the 1870 'expenditure' should be spread back over the previous years to 1866, to reflect the actual time pattern of improvement expenditure in the Lackawanna's interest. Mines completed in the period 1866–1875 included major new developments such as the Sloan, the Pyne, the Cayuga, and the Brisbin. The third phase began with a surge of new investment in the upswing of the 1880s railroad boom, which initiated construction of the Woodward, Pettebone, and Storrs breakers.[2] Expenditure was sharply curtailed for a time following the 1885 depression, but continued thereafter. This period will be examined in more detail below. The fourth and final phase seemed to be an appropriate response to the improving business environment in 1892 after depressed conditions in the previous year. However, this was to prove a false dawn when the 1893 panic struck the following year to usher in the depression of the 1890s. Unfortunately, by this time, the Lackawanna had already committed themselves to the development of the Bliss and Auchincloss mines on their Hanover lands below Nanticoke. A map of the locations of the Lackawanna breakers across the Northern Field can be seen in Figure 14.2.

The pattern of expenditure by the Lehigh Valley Coal Company shown in Figure 14.3, has some similarities to that of the Lackawanna, but also a number of marked differences. The 1870s depression produced the same effect of ending any thought of new investment for several years, but, with the exception of 1882, new investment was much more subdued in the first part of the 1880s than in the case of the Lackawanna. The 1882 peak is explained by the

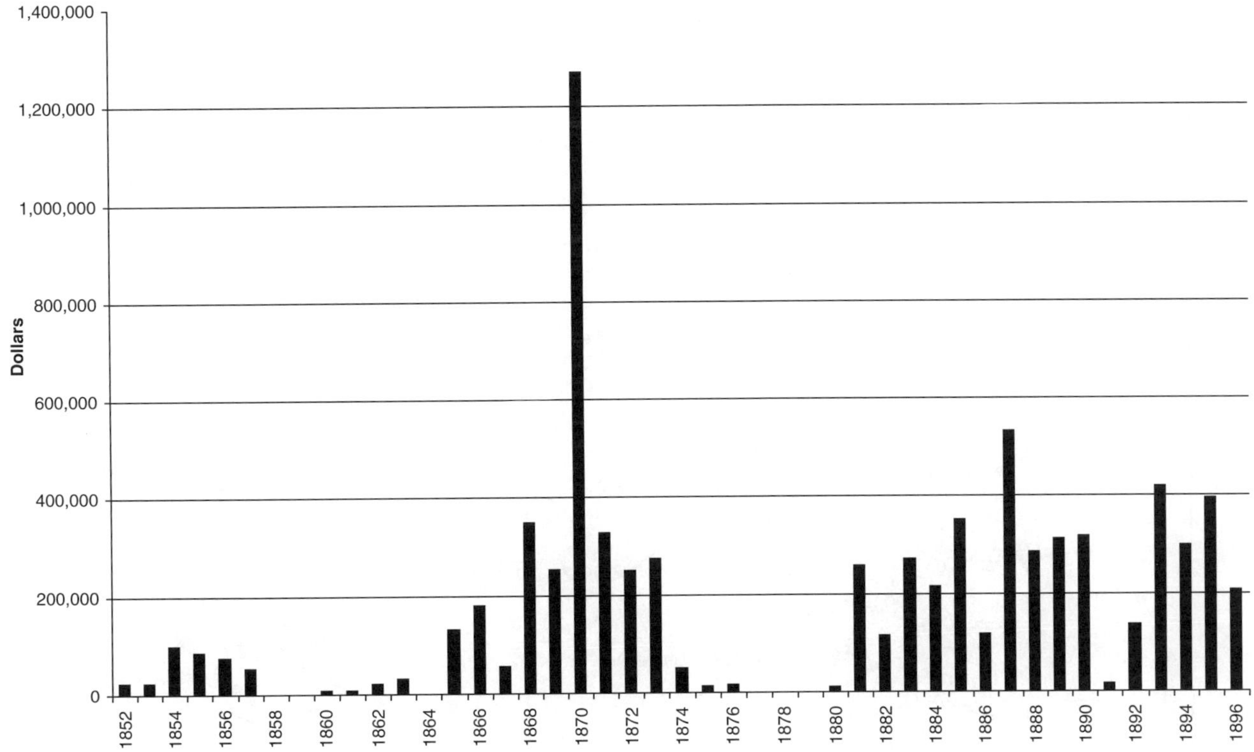

FIGURE 14.1
Delaware, Lackawanna and Western Railroad Coal Department: Annual Improvement Expenditure including Mine Purchases 1852–1896.

acquisition of the Packer improvements on the Girard Estate coal lands in the Western Middle Field.[3] These were major additions to the company's productive capacity; so the only new improvement it actually initiated during this period was the Dorrance in Wilkes-Barre.[4] In general, the years of particularly large expenditure, such as 1871, 1882, and 1892, tend to reflect major purchases of mines, which the company undertook to a greater extent than did the Lackawanna. This does not apply in 1887 however, when large new improvement expenditures were made at existing mines belonging to the company, namely Prospect and Packer No. 5. Similarly, the period of 1889–1891 was characterized by large investment in the recently leased Franklin mine near Wilkes-Barre and the new Blackwood and York Farm developments in the Southern Field.[5] Arguably, one advantage of buying existing improvements, rather than the company constructing its own, was that the capacity increment was immediately available for deployment. A second advantage was that if business conditions suddenly became adverse, as they did in 1893, there were fewer ongoing improvement commitments, so new investment expenditure could be curtailed very quickly. This is what appears to have happened in 1893–1894. Lastly, the Susquehanna Coal Company maintained a substantial level of improvement investment in the early period of the development of its property from 1869 to 1875, barely pausing for the 1873 panic, as indicated in Figure 14.4. Major new works during this time included Breaker No. 3 at West Nanticoke to replace the demolished Harvey Breaker and Shafts Nos. 1 and 2, on which considerable sums were expended in 1875.[6] As with the other companies, further expenditure was sharply curtailed in the late 1870s and did not accelerate again substantially until the upswing of the 1880s was well under way. The major new development at this time was Breaker No. 6 and its associated shaft and slope, situated 4 miles west of Nanticoke.[7] First shipments from the breaker were consummately poorly timed to coincide with the bottom of the 1885 depression. The final peak in the graph in 1891 coincides with the construction of Breaker No. 7, which shipped a substantial 428,000 tons in its first year of operation in 1892.[8]

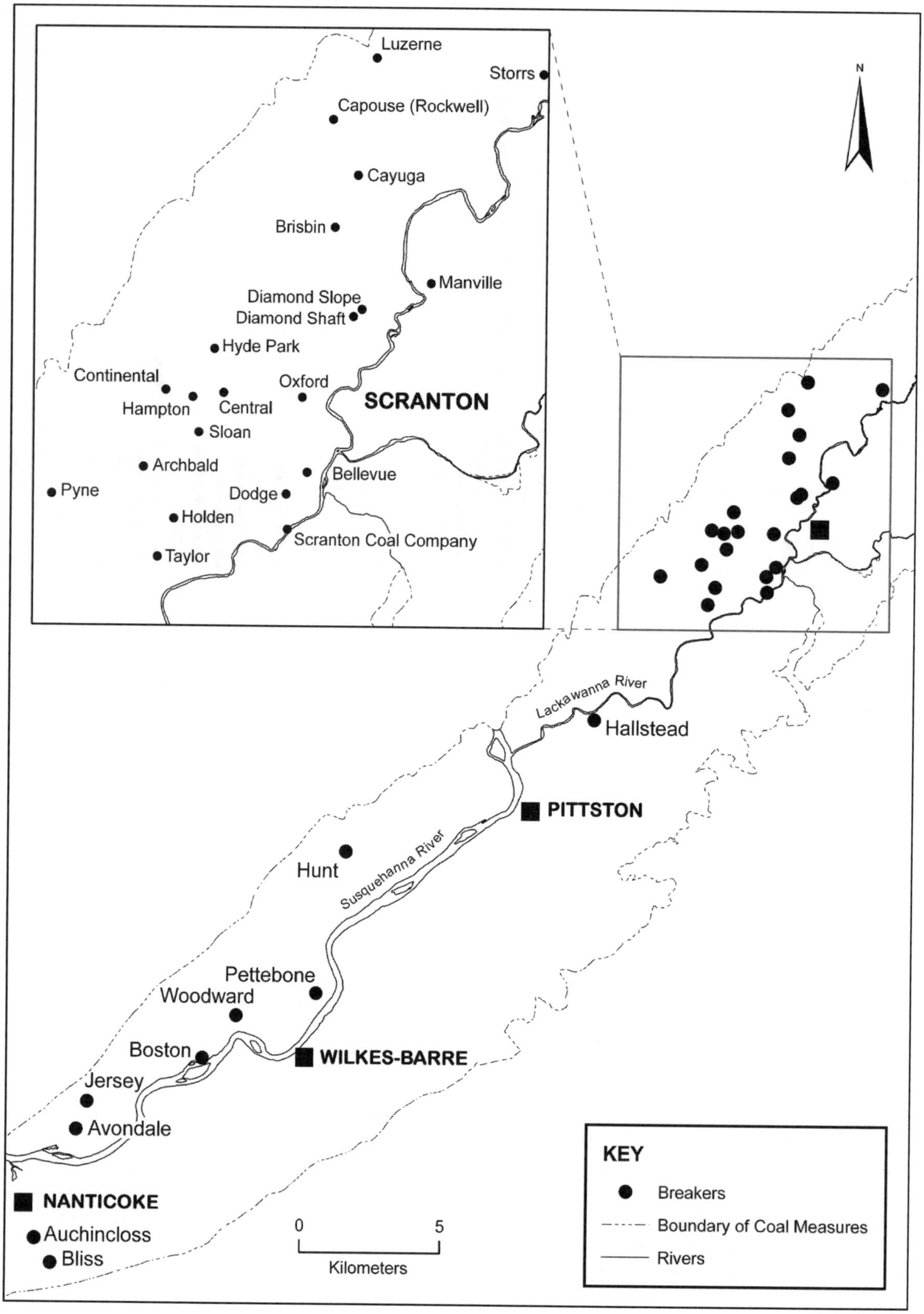

FIGURE 14.2
Delaware, Lackawanna and Western Railroad Coal Department: Location Map of 19th-Century Coal Breakers.

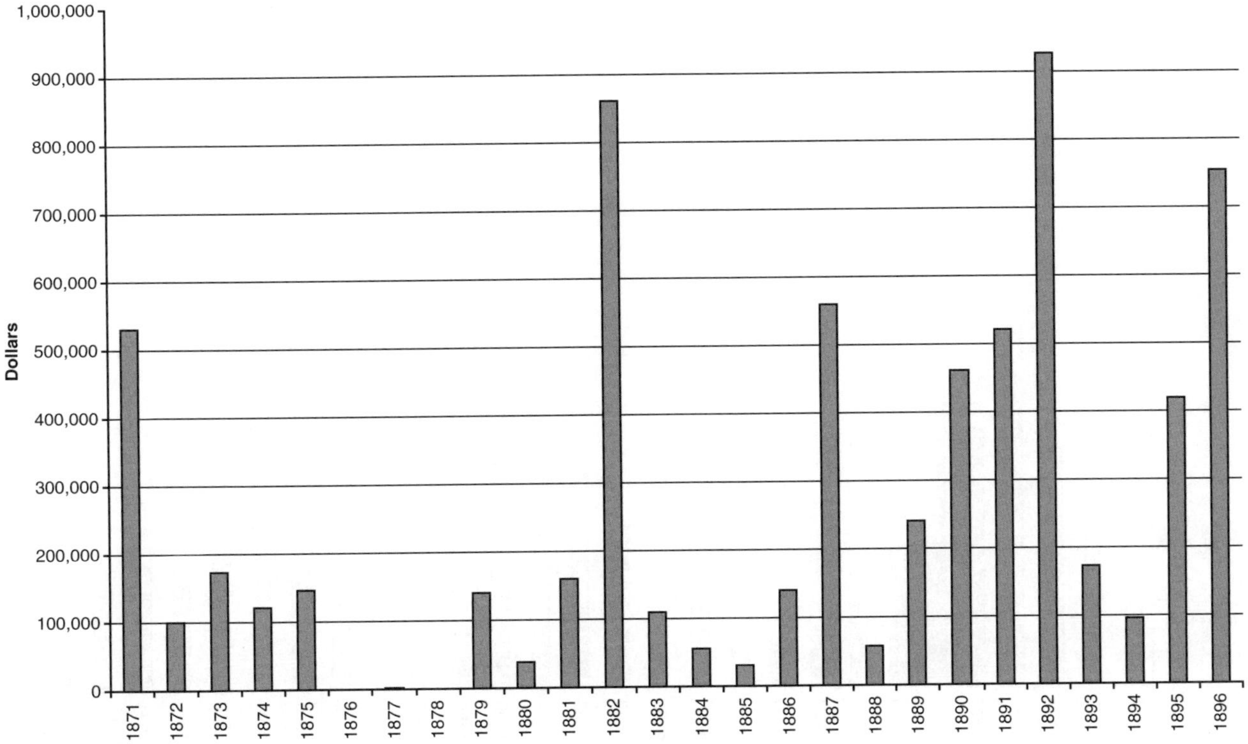

FIGURE 14.3
Lehigh Valley Coal Company and Predecessor Companies: Annual Improvement Expenditure Including Mine Purchases 1871–1896.

Otherwise, in the preceding years of 1889–90, expenditure trends were the reverse of the Lehigh Valley. This probably reflected the more subdued outlook for business as a result of a greater measure of tacit co-operation with the other companies. Since the three graphs necessarily have different scales, this makes visual assessment of magnitudes more difficult than identification of temporal patterns. To summarize, therefore, in the period 1871 to 1896, for which the data are present for each company, the Lackawanna invested about $5.2 million, the Lehigh Valley $6.8 million, and the Susquehanna Coal Company $3.9 million. The last of these, in particular, was a substantial sum, in the light of the moderate tonnages being shipped in the late 1890s.

Also, to gain a sense of the overall effect of these time-varying investment strategies, Figure 14.5 combines the annual totals of the three companies into a single graph. The most striking feature is the extreme investment 'trough' that followed the 1873 panic and the sharpness of the recovery in 1881–1882. Thereafter, differences in expenditure pattern partly compensate for one another, but the depression of the mid-1880s stands out, as does the impact of the Labor Troubles in 1888 on the Lehigh Valley. Of particular interest, however, is the fact that the impact of the 1890s depression seems to have been far less severe on investment by these major companies than was the 1870s depression. This, in turn, may well have been an unexpected benefit of the reduced dependence of anthracite on industrial, (as opposed to domestic) markets for coal. Overall, throughout the period in question, the three companies alone invested a total of $15.9 million in new capital projects in the coalfields.

CAPACITY-RELATED FACTORS AFFECTING THE TIMING OF INVESTMENT

A range of possible factors could affect available levels of productive capacity and the perceived extent to which it matched expected future growth or reduction in shipments. For example, during the 1870s depression, when trade was very dull and prices unremunerative, the Lackawanna and other major companies idled a number of their mines to concentrate production at the remainder for more economical operation. By the beginning of 1879,

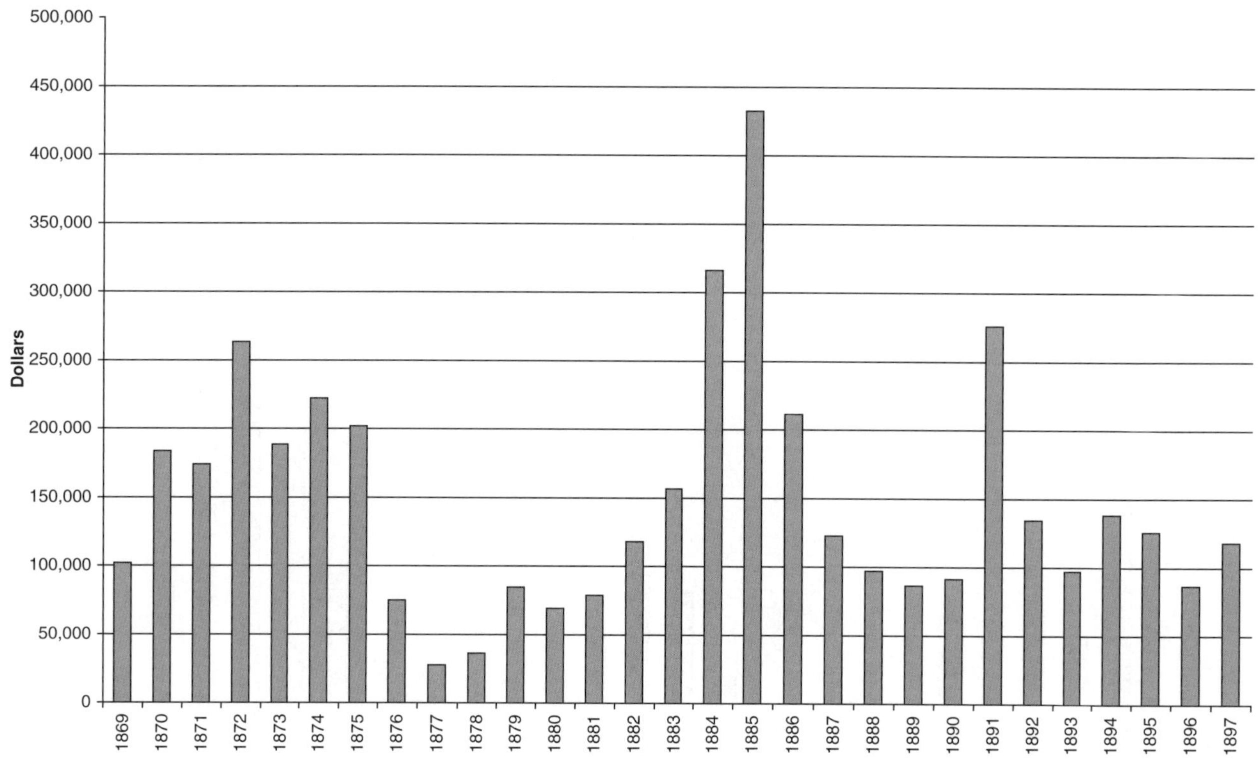

FIGURE 14.4
Susquehanna Coal Company: Annual Improvement Expenditure 1869–1897.

the Lackawanna had only nine mines still in operation. Four more came back into operation during the year, as the upturn accelerated, and, by 1881, fifteen were in production.[9] Clearly, under these circumstances, the collapse of investment in the late 1870s, shown in Figure 14.5, is easy to understand. The companies could meet all anticipated increases in demand for the immediately foreseeable future simply by restarting existing capacity that was not being utilized.

Over a slightly longer timespan, however, other factors would enter into the equation. Among these were possible exhaustion of older mines; termination of fixed-term leases, made either to the company or one of its contract operators; and progressive dilapidation of breakers that would eventually need to be stopped for overhaul or rebuilding. Account would also have to be taken of possible loss of output whose timing could not readily be predicted. This could be from causes such as flooding, mine caves, and fires underground or in breakers. Looking across the coalfields, these were all too common occurrences for which prudent provision needed to be made. A related point to emphasise is that the necessity of having coal breakers meant that output capacity was a rather 'lumpy' investment. In general, once a mine and its accompanying breaker were fully operational, the level of daily output could not be varied upwards across a very wide range. Nor would any purpose be served by increasing employment levels underground, since there was a limit to the number of mine cars that could be hoisted and put through the breaker per unit of time. Hence, to expand output significantly, either existing breakers would have to be rebuilt with much larger capacity, or new mines would have to be acquired or constructed.

A final concern for senior managers, that became more pressing over time, was the potential loss of contract shippers, when their contracts expired, because of more favorable terms offered by rival railroads. In earlier years, this had not been a serious issue because most mines were only accessible from a single line. As the density of lines increased over time, however, the likelihood of sudden loss of tonnage grew sharply, as has already been documented.

FIGURE 14.5
Lackawanna, Lehigh Valley, and Susquehanna Coal Company: Combined Annual Expenditure on Improvements and Mine Purchases 1871–1896.

ADVANCE ROYALTY PAYMENTS AND THE TIMING AND LOCATION OF MINING INVESTMENT

The mechanism by which apparently innocuous lease provisions could result in the accrual of large advance royalty payments on land tracts was explained in chapter 12. It is now necessary to consider the implications of this mechanism for the timing and location of mining investment.

To recapitulate, where leases were at a very low rate of royalty, such as 12.5 cents per ton, and the annual minimum tonnage was small, even if mining did not begin on the property for some time, the amount of advance royalty owing would not increase sufficiently rapidly for the matter to become a concern for senior managers. Clearly, however, if the royalty per ton was high or the annual minimum was large, or both, the advance-royalty account against the property could climb to a high figure quite quickly.

Since leases were executed at different times on different tracts, with widely varying royalty clauses, the royalty accounts for each property had to be maintained separately. A statement of the condition of coal leases thus became a standard part of the regular reporting mechanism by general coal agents to the boards of managers of their respective railroads. As Sloan remarked to Storrs, following the receipt of his Annual Report for 1880:

> 'We were deeply impressed with your tables—the large deficit in our leases and contracts! What can we do to improve this account; it is so very large!'[10]

By 'deficit' Sloan was referring to the great disparity between the tonnage on which royalty had been paid and the tonnage that had actually been mined from the property. In some cases the tonnage mined was nil, in others, especially if the property was large, some parts of it might have been mined from existing breakers, but other parts were essentially virgin territory. This, of course, was of no concern to the lessor, who based the annual minimum on the potential yield from the entire property, and it was up to the lessee to make arrangements for mining the coal in an efficient and timely manner.

Hence, the level of advance royalties on particular properties had to be monitored closely, and the information was fed into company decision-making at the highest level. It is apparent therefore, with the benefit of hindsight, that the issue was almost bound to become a key determinant of investment decision-making concerned with the improvement of a company's leasehold estate, and so it proved. A clear priority order for development of different leased tracts would emerge as the advance royalty accounts built up over time. Some leases might increasingly engage the attention of managers because they had not been developed for a long period after the lease was executed. Others might come quickly to the forefront because of the rate of increase of the royalty account. This would be a result of a large minimum tonnage and a high or escalating rate of royalty per ton. Also, new mining improvements, especially where shafts as deep as 700–1,000 feet were required, would take several years to complete and still more time to bring into full production. Hence, not only the current but also the projected status of the royalty account had to be taken into consideration.

A clear implication of the foregoing is that the advance royalty system would come to control, in large measure, not only the temporal sequence of new mining developments on the leasehold part of a company's estate, but also the geographical locations where the mines would be sited, *because the royalty accounts were tract-specific*. Since the mines constructed by the major companies in the later part of the period were generally very large and employed a substantial workforce, each new investment decision would have a significant impact on the course of economic development at the sub-regional level and also on the associated pattern of mining settlements that would follow in its wake.

The interplay of factors just described is best illustrated by a detailed case study of the Lackawanna Coal Department's approach to the improvement of its leasehold estate during the 1880s, although some of the decisions to acquire the leases in question had been made much earlier. Initially, the case study will be confined to royalty accounts and investment decisions. However, in a subsequent section, further details of the actual construction processes during the same period will be given. This will provide a comprehensive analysis of the multiplicity of issues involved in developing a coal estate in the medium to long term.

CASE STUDY: LEASEHOLD-ESTATE DEVELOPMENTS BY THE LACKAWANNA COAL DEPARTMENT IN THE 1880S

During the 1880s, the Lackawanna Coal Department decided to invest in the development of three new substantial mine properties. The three mines were named the Storrs, the Woodward, and the Pettebone, and their construction began in the years 1880, 1881, and 1882 respectively.[11] Photographs of the mines, commissioned by the railroad, dating back to 1905, can be seen in Plates 14 to 16.

All three mines were developed on leased lands, but the original dates of the leases involved varied quite widely. The Storrs was located in Blakely, near Scranton, with a view to mining from the Central and Luzene Coal Company leases. A portion of the estate had originally been leased to the Lackawanna in 1860, and further tracts had been leased to J.J. Phelps in trust for the company in 1864. The final components of a valuable mining property of over 1,000 acres were added by means of a lease to the Nanticoke Coal and Iron Company in 1868.[12] The Woodward was located in Edwardsville near Wilkes-Barre. The original lease was executed in 1872, during the period of competitive acquisition of lands and leases prior to the 1873 panic. Other leases were added to the property in later years. Finally, the Pettebone was located further up the valley, on a property leased in 1881 during the next cycle of lease acquisition in the coalfields.[13]

Given the time that elapsed, in the case of both Storrs and Woodward, between lease execution and the onset of mine construction, a brief explanation is required of the tactical and strategic decision-making involved in these extended delays, based on the limited information that has survived. According to Storrs's 1869 Annual Report, borings on the Luzerne Coal Company property commenced, in the summer of 1869, with a view to siting one or more improvements there. Unfortunately, the initial results were not promising, although further exploration was planned to prove the lands more fully. There was clear intention at that time to commence improvements as soon as the final boring results were obtained.[14] At the same time as these lands were being proved, examination of the Hanover lands of the company, held in fee near Nanticoke, was also in progress. It was the intention to site new improvements there, in addition.[15] By the following year, apparently no better results from exploratory boring at the Luzerne Coal Company property had been obtained and an alternative approach was being pursued. This involved contracting with W.H. Richmond for him to mine from the adjacent Central Coal Company lands and sell the coal to the railroad. Based on the borings on the Nanticoke lands, sites for two improvements had been proposed, but Sloan had decided to defer decisions on these.[16] Little apparent progress was made in the negotiations with Richmond over the following year, but Storrs' next report indicates a certain nervousness over the future reliability of local shipments from some of their contractors. For example, the previous contract for the product of the Stafford Brook and National mines, below Scranton, had expired in December 1871, and there was also a risk of losing the tonnage of the Pittston and Elmira Coal Company's breakers.[17] At the beginning of 1872, Storrs offered the following observations to Sloan on the outlook for the trade, in relation to decisions about embarking on new improvements:

> 'The present demand and price of coal do not offer much encouragement for increasing product but of the policy of further development so as to be ready for future demands you can better judge than myself'.[18]

Having neatly side-stepped any responsibility for major investment decisions that might rebound adversely in the future, Storrs went on to advise Sloan that Richmond was willing to proceed at once, but the matter was delayed because the relevant lease paperwork was still in the name of Mr. Phelps rather than the railroad itself.[19]

Not long afterwards, circumstances took an unexpected turn. As a result of the short-lived union with the New Jersey Central, negotiations with Richmond were suspended and, in fact, never resumed. The reason for this was that to reach him would have required expenditure of $125–150,000 on new branch railroad-line construction, and it was thought that this expense could be avoided if improvements were opened on the Nanticoke lands instead. The sites chosen were very close to existing tracks of the Jersey Central's Lehigh and Susquehanna branch, so the requirements to build new connecting stretches of road would be minimized.[20]

By the beginning of 1873, Storrs was also able to report that the lease on the Woodward property had been negotiated. The proposed site for the improvement, being adjacent to the Lackawanna and Bloomsburg division, would not require any great expenditure on new branch-line construction. Boring was also to commence in the spring of 1873 on these lands to prove them more fully.[21]

It should also be noted that, while these investigations and borings were taking place, major breaker construction projects were already in progress on other properties. The most recent of these projects included the Sloan, Pyne, and Brisbin breakers in the Scranton area, which were mentioned earlier. Thus, by the beginning of 1874, Storrs could advise that the company now had 18 breakers, taking the coal from 16 shafts, five slopes, and one tunnel or drift. Since the cartel was in operation, and the 1873 panic had adversely affected business confidence, it was therefore not surprising that Storrs reported no planned improvements for 1874. Rather, as he observed, mine construction was shelved because 'we had too much coal', although Sloan had also issued a presidential edict, in late 1873, stopping construction work of all kinds and demanding 'the most rigid economy in every department of the Company's service' because of the adverse financial climate.[22] The company's own productive capacity, in conjunction with new agreements with its contract shippers was now more than adequate for the wants of the market and would fully occupy the railroad's transportation capabilities for the next 2–3 years.[23]

The deepening depression that followed meant that all improvement plans were shelved for a further five years, and it was not until 1879 that new factors led to the question of potential new investments being revisited. An excellent quotation from Storrs summarized very clearly the tactical and strategic choices facing the company at that time:

> 'Our purchase contract with Kingston Coal Co. will be full in about a year and that with Hutchison in 2 years. The Boston lease will expire in about 2 years by exhausting all we have the right to mine and the Delaware and Hudson, the owners, will hardly extend. Our present idle collieries can make up some of the deficit, but we must anticipate the future and its demand. My suggestion would be an improvement on the Woodward lease at Kingston and on the Central lease at Blakely, while we secure all the outside coal that can be had for transportation or on reasonable terms for purchase.'[24]

The role of advance royalties on the Woodward and Central leases will be discussed in detail below, but it is apparent that a long view of capacity requirements was being taken, and that new mine construction was an important part of the planned solution for the 1880s. It was intended as a direct replacement for both existing company-controlled tonnage, in the form of the Boston mine, and the loss of substantial purchase contracts that might be less easy or more expensive to replace in future. It is also worth noting that the takeover of the Hunt, Holden and Hallstead mines from individual operators in 1885, which has been discussed in Chapter 13, was not part of any original plan to increase the company's own number of mines. Instead, they were part of the aim of securing additional transportation or purchase contracts, but the acquisitions became necessary to avoid losing the tonnage. By the time they happened, too much had been invested in construction, at Woodward, Pettebone, and Storrs, for any thought to be given to shelving all or any of these projects.

LANDS AND LEASES FOR NEW MINE CONSTRUCTION

By the end of 1878, four coal leases stood out on the Lackawanna Coal Department's books as having large outstanding royalty payments against them. These were the Luzerne Coal Company, the Central Coal Company, the Pringle, and the Woodward and Williamson leases. The former two were located in Blakely township, east of Scranton, the latter two near Kingston, to the west of the city. Details of these leases are given in Table 14.1 and, between them, they accounted for the majority of the advance royalties paid by the company up to that date. The Pringle was the first to be dealt with, by means of a large contract with the Kingston Coal Company. The aim was to produce up to 150,000 tons a year from the lease, thereby rapidly reducing the tonnage paid for, but not mined. In 1879, over 100,000 tons were shipped under this agreement.[25] To address the problem on the other leases, two improvements were proposed. The first mine, Woodward, was to take the name of the lease; the second became known as the Storrs mine. This improvement was to mine from the Central lease and reach the Luzerne lease in due course. As the table indicates, some coal on these latter two leases could also be reached from the existing Cayuga shaft.

At the end of 1879, the company's executive committee approved Storrs's proposals for construction of the Woodward and Storrs mines,[26] but it was subsequently discovered that there were legal problems with the Woodward lease, in the shape of a mortgage that had a prior claim on the land. The implications of this had not been appreciated when the lease had originally been signed, but it was now feared that, as a result, the advance royalty payments already accrued might be forfeit. A further difficulty was that Charles Parrish, a well-known but mistrusted coal operator, was involved, seemingly with a view to fleecing the railroad for his own benefit. The matter was, however, eventually resolved in the railroad's interest, although not without cost. Parrish had only been thwarted by his inability to raise the necessary cash to buy one of the lessors' interest in the property and release the mortgage.[27] Within days of the legal problem being sorted out, construction activity began, in May 1881. It is clear, therefore, that the major delay from 1874 to 1879 was deliberate policy, while the additional relatively short delay, during the latter part of 1880 and early 1881, was due

TABLE 14.1
Condition of Coal Leases with Large Outstanding Advance Royalties 1878.[28]

Name of Lease	Luzerne Coal Co.	Central Coal Co.	Pringle	Woodward & Williamson
Intended Means of Development	Cayuga [Storrs]	Cayuga [Storrs]	Kingston Coal Co.	[Woodward]
Annual Minimum (tons)	70,000	90,000	51,000	Year 1 50,000 Year 2 50,000 Year 3+ 75,000
Royalty Per Ton (cents)	12.5	12.5	10	25
Accumulated Total Tonnage Required	717,500	1,162,500*	765,000	325,000
Total Tonnage Actually Shipped to Date	30,255	511,621	–	–
Tonnage Shipped 1878	16,500	–	–	–
Deficit on Lease Shipments (tons)	687,245	650,879	765,000	325,000
Advance Royalty Paid ($)	85,906	81,360	76,500	81,250

*Takes account of 1877 allowance for effects of strike (7,500 tons reduced).

to unexpected factors. These latter were countered with the Lackawanna's standard methods, namely, good legal counsel and the immediate availability of large amounts of funds, to ensure that any minor obstacle to its strategic aims (and legitimate claims) could be overcome.

While there was ample coal in the Central and Luzerne leases for future mining, as well as a further 942 acres owned by the company in Blakely township,[29] the Woodward property was smaller, comprising only 239 acres. A policy of purchasing or leasing adjoining lands was therefore set in motion very quickly to consolidate the estate. Some lands were specifically sought, others were rapidly offered through the well-established 'neighborhood effect', resulting from interest by the large corporations in specific areas of the coalfield. Individual landowners would see revenue opportunities, if the companies began construction activities on lands adjoining or close to their own, and would seek the best possible terms, particularly when the coal business was brisk.[30] By the end of 1881, a further 11 leases had been added to the original one and, in 1889, an agreement was finally reached with A. Nesbitt for a combined cash payment and lease on 200 acres of coal land and 75–80 acres of river warrant (right to mine under the Susquehanna).[31] The Company had been interested in this property since 1882, because it was the best land remaining in private hands in the Northern Field and was also 'a bait for other companies to build [railroad lines] on the Bloomsburg side of the river',[32] i.e., in direct competition with the Lackawanna's own tracks. The company had, however, declined to take a lease at 40 cents royalty in 1882, preferring to 'secure others at lower price before setting such a precedent', while trying to persuade Nesbitt to sell outright.[33] This he consistently refused to do. The eventual compromise, in 1889, was a one-off payment of $350,000 in cash, followed by a royalty and minimum, on a rising scale, to 50 cents and 100,000 tons by the sixth year of the lease. This was instead of the original 1888 proposal for a lease at 75 cents on all the coal.[34] The $350,000 was in cash and 'not to be on record'. The company also deliberately delayed recording the lease in the county courthouse for nearly eight months, to prevent, for as long as possible, information about the arrangement being available to others.[35]

Details of all the leases comprising the coal estates for both the Woodward and the Pettebone mines are given in Table 14.2. For Woodward, this table updates to 1890 the information about the initial Woodward and Williamson lease, originally given in Table 14.1. It is apparent that the original lease, the subsequent batch of smaller leases, and the Nesbitt arrangement formed three major sets of contributions to the costs of establishing the Woodward property, quite apart from the costs of mining development. By the end of 1890, the smaller leases had accrued a total of $293,816 in advance royalties, representing 1,175,268 tons paid for but not mined. This was more than the amount outstanding on the original lease, but less than that payable immediately on the Nesbitt property, which can also be viewed as a kind of 'lump-sum' advance royalty. In total, this meant that over $920,000 had been paid in advance on more than 2.25 million tons of coal, while only 187,561 tons of coal had been produced since Woodward had begun shipping in 1888. In 1890, the total annual minimum across all the leases was 223,000 tons, rising to 323,000, six years later, under the provisions of the Nesbitt lease. Although production rose steadily, by 1892 it had only reached 186,530 tons,[36] so it was to be a number of years before the minimum was to be covered, not to mention a reduction effected in the outstanding advance royalties.

In comparison, the financial pressures were less severe for developments to take place at the Pettebone mine, since advance royalties had only reached $155,000 by the end of 1890. While the minimum was high at 80,000 tons, the lease had only been in operation since April 1882, unlike the main Storrs and Woodward leases, which dated from the 1860s and 1870s respectively. At the same time, the construction of mining improvements could not be delayed indefinitely because the lease contained a provision that these had to be started by April 1, 1890 or the agreement would be forfeit.[37] The coal department therefore decided to proceed with mining developments very soon after the lease came into operation. Figure 14.6 shows the detailed land-tract geography of part of the Northern Field, indicating the relative position of the leases for the Woodward and Pettebone mines.

The Pettebone lease was not part of the company's long-term development strategy, as Woodward and Storrs had been, because the first suggestion that the property might become available was not made until late in 1880. Initially, W. R. Storrs was unimpressed at the proposal because of the high minimum. Some of the lands were also 'of very doubtful character', and there was a likelihood of having to drive the shaft down through up to 200 feet of alluvial deposits on the Susquehanna River

TABLE 14.2
Condition of Coal Leases for Woodward and Pettebone Mines 1890.[38]

1	2	3	4	5	6	7	8	9	10
1	W	1/74	1 50 2 50 3+ 75	25	1,225	119	47	1,106	277
2	W	1/82	1+ 6	25	57	–	–	57	14
3	W	1/82	1+ 17	25	156	–	–	156	39
4	W	1/82	1+ 13	25	117	2	2	115	29
5	W	1/82	1 1 2+ 2	25	20	–	–	20	5
6	W	1/82	1 8 2 10 3+ 12	25	101	–	–	101	25
7	W	1/82	1 2 2 3 3 5 4+ 6	25	47	–	–	47	12
8	W	1/82	1+ 2	25	17	–	–	17	4
9	W	1/82	1+ 24	25	217	64	32	153	38
10	W	1/82	1 2 2+ 3	25	27	–	–	27	7
11	W	1/82	1 12 2 23 3 35 4+ 46	25	348	–	–	348	87
12	W	1/82	1+ 15	25	137	3	3	134	34
13	W	4/89	1 350* 2 25 3 50 4 65 5 85 6+ 100	– 30 35 40 45 50	–	–	–	–	350
14	P	4/82	1 30 2 50 3+ 80	25	620	–	–	620	155

Column Headings:

1. Lease No.
2. Mine from which lease was to be developed, W - Woodward, P - Pettebone.
3. Date of first royalty payment on lease
4. Annual minimum (year of lease operation followed by minimum for the year in question in '000 tons)
5. Royalty per ton (cents)
6. Accumulated total tonnage required under the lease ('000 tons)
7. Total tonnage actually shipped to date ('000 tons)
8. Tonnage shipped 1890 ('000 tons)
9. Deficit on lease shipments ('000 tons)
10. Advance royalty paid ('000 dollars)

TABLE 14.2 *(Continued)*

Lease names corresponding to lease numbers:

1. Woodward and Williamson
2. Horton
3. Loveland
4. Hoyt
5. Strong
6. Shonk
7. Welles
8. Bailey
9. Goodwin
10. McClintock
11. Mitchell
12. Parke
13. Nesbitt
14. Vaughn

*This figure in the table corresponds to the $350,000 initial payment made in cash for the Nesbitt lease.

floodplain. This would introduce a number of engineering problems that might be costly to overcome. At the same time, there was known to be good coal on the adjoining property worked from Hutchison's mine.[39]

Fears that the Lehigh Valley Railroad, who were bridging the Susquehanna at the time to bring their tracks into the area, would acquire the lands, encouraged the Lackawanna to make a move. This was reinforced by the general competition for remaining coal properties at the time, resulting from the upturn in business conditions, which had also been a factor in consolidation of the estate at Woodward. With customary caution, however, the lease for 235 acres was entered into, subject to satisfactory results from exploratory drilling. In the event that these tests showed the lands to be of uniformly poor quality, the lease could be relinquished before any payments had to be made on it.[40]

While these lands were undeveloped, the same conservatism attached to a valuation of the adjoining Hutchison mine property, which also became available for sale in 1881. This comprised ownership of 15 acres, 63 acres whose lease had expired in 1889, and a further 200 acres with a lease that had a 72,000-ton minimum. The asking price of $185,000 was deemed too high for a property from which most of the best coal had been mined, but the Lackawanna's offer of $140,000 was refused. The company retained, however, a considerable interest in the property, because Hutchison, the operator, still had 225,000 tons to provide under his existing purchase contract with them, and this would lapse if the land went to a competitor.[41] The solution was for Haddock and Steel, the individual operators, to purchase the property, (changing its name to Black Diamond), helped by a loan of $150,000 from the Lackawanna. In return, the Company was to get an extended transportation contract on the coal from this and the Dodson mines, which they already worked. The loan was to be paid back at 10 cents a ton on coal shipped from both mines.[42] The combination of the Pettebone lease and the Haddock and Steel transportation contract gave the company at least a measure of control over a block of approximately 500 acres of coal land. The Pettebone lease, on its own, was relatively limited in size compared to the Woodward estate, but there were no further additions of adjoining property during the 1880s.

In total, by the end of 1890, 5,535,200 tons worth of advance royalty, amounting to $1,406,756, had been paid for on the Woodward, Pettebone, and Storrs leases, out of a total of 7,863,066 tons and $1,901,136 for the company as a whole. Although Woodward had begun shipments in 1888, and Storrs at the end of 1889, Pettebone had still to commence volume output, having a mere 444 tons to its credit.[43] The leases on which these mines stood had been acquired during three different periods of competitive acquisition of coal lands, during the late 1860s, early 1870s, and the early 1880s. The price, in terms of cashflow and deferred income, was high, (but not prohibitive, given the Lackawanna's large surplus), and it secured assets for the company that were to underpin developments stretching well into the 20th century.

As a postscript to these detailed advance royalty calculations, it is instructive to compare the experience of the three mines during the remainder of the period to 1902. Taking the figures forward for the lease requirements and comparing them with total shipments to that date, the following is clear in very broad terms. Firstly, by 1902, Storrs

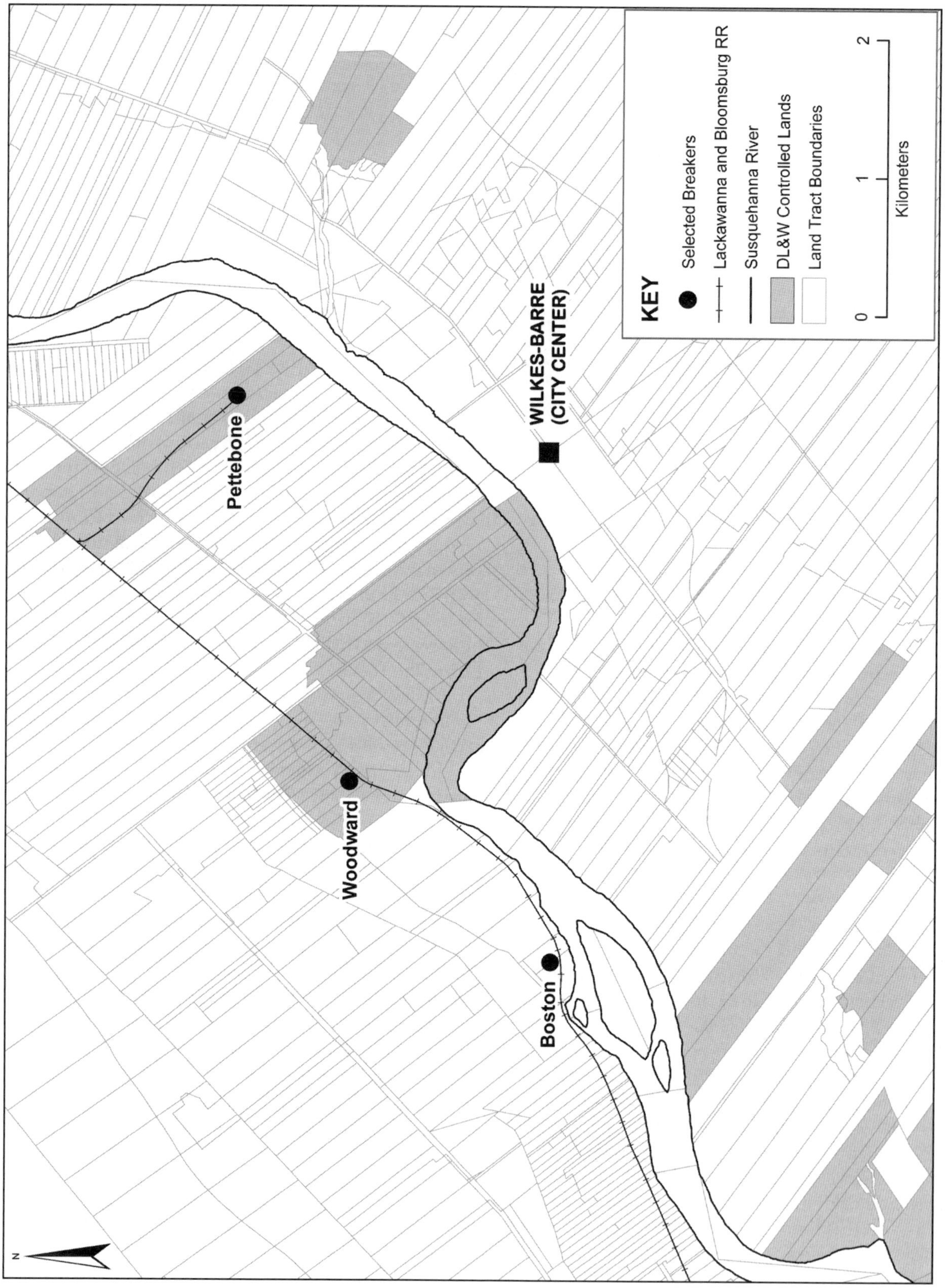

FIGURE 14.6
Map of the Coal Leases Tributary to the Woodward and Pettebone Mines 1887.

had eradicated the deficit on its leases and was mining more than double the required minimum tonnage. Despite a rather chequered history of output in the mid-1890s, Pettebone had also squared its account. Woodward, by contrast, had to contend with a very large minimum, in excess of 300,000 tons, and its shipments had not been averaging much more than that, so the deficit on its leases remained broadly as it had been in 1890. Very shortly afterwards, however, its tonnage soared to set a then world coal-output record of more than 1 million tons in 1905, enough to eradicate even the largest of lease deficits very quickly.[44]

Hence, although it required several decades to achieve, the Lackawanna management were ultimately successful in releasing the capital tied up in the ground in the form of advance royalty, so it could be used for other company purposes. In doing this, they had been able to compensate for the competitive excesses of earlier speculative cycles. These had essentially forced them to acquire control of coal leases well in advance of actual requirements, and the burden of those decisions had to be carried on their balance sheets for many years.

Unfortunately, all the other companies in the trade did not enjoy equivalent management success in the long term. As late as 1907, the Lehigh Valley Coal Company reported an advance royalty account that still stood at the very high figure of $5.4 million, so they had clearly failed to ensure that mining operations had kept pace with their lease obligations.[45] Part of this was doubtless that they had overstretched themselves by acquiring an estate larger than that required for present or medium-term needs, given their average share in the cartel/community of interest. A second very important factor, however, was the acknowledged strategic errors, in the period 1873–1892, of holding back tonnage from the company's own properties to make room for the output of individual operators. This approach was discontinued subsequently, but it left the Lehigh Valley with an immense capital sunk in the ground that it could not release in a timely manner.[46]

PHYSICAL CONSTRUCTION OF NEW MINING IMPROVEMENTS

In the early days of mining above water level, very little capital had been required to enter the industry. Over time, costs increased rapidly as the deeper basins were tapped by shafts, and new mine safety legislation, following the Avondale disaster, required two openings for each mine.[47] Breakers also became progressively larger and incorporated new types of machinery to separate coal from rock and dirt, until, by the first decade of the twentieth century, the coal regions boasted some of the world's largest coal-preparation plants.

In consequence, the scale of investment required was so large, especially if several new improvements were being prosecuted simultaneously, that the expenditure had to be managed very carefully. While construction was still in progress, the money expended was, obviously, not providing any return. Coupled with advance royalty payments, these construction accounts could therefore have a very adverse impact on cashflow, especially at times when trade was depressed. Also, the time required to complete a construction project tended to lengthen as the projects themselves became larger. This meant that the timescales involved were measured in years, not months. It also meant that the business conditions anticipated, when the original strategic investment was made, might not correspond with the actual circumstances that pertained when the new mine entered production some years later.

To illustrate the range of decision-making issues involved, stages in the mine construction process will now be outlined in general terms. This will be followed by a detailed case study of the construction of the Woodward, Pettebone, and Storrs mines, in order to highlight the different management strategies that could be adopted in particular circumstances.

STAGES IN MINE CONSTRUCTION

In general, several distinct stages in the mine construction process can be identified as follows:

- exploratory borehole drilling to prove the lands,
- shaft/slope sinking by company men and contractors,
- breaker construction, and
- initial opening-out of the mine underground.

Some of these stages were necessarily sequential, in that one had to be completed before the next stage could begin, but others could be overlapped, if desired. This would increase the monthly rate of expenditure but shorten the time to completion. So, the phasing could be varied by management decision, either for financial or other reasons.

Mention has already been made of exploratory drilling in earlier sections. This was primarily used by the larger companies to assess the likely yield of properties, in terms of the number of veins present and in what thickness. It also provided some information on local variations in geological conditions. Since drilling was quite costly and produced no immediate return on expenditure, it was not generally undertaken by individual operators, who preferred to rely on optimism and energy, rather than evidence, to underpin their success (or lack of it)! The normal practice would be for drilling to precede any shaft-sinking activity, but additional drilling might also be undertaken under other circumstances. For example, if a first shaft, once sunk, failed to tap the expected veins, further drilling might be used in an attempt to identify a more suitable location for the second shaft. Boreholes might also be drilled in the base of a partly sunk shaft to determine the feasibility of sinking down to lower veins on the property.

Once the precise location for a shaft or slope had been determined, sinking could begin. Initial groundbreaking was usually undertaken by company men who were paid a normal daily wage. In the case of a slope, this might involve a limited amount of preparatory work and timbering to secure the slope entrance against loose superficial ground material. In the case of shaft sinking, the initial work by company men would usually be of a more substantial kind. It would involve the first stage of shaft sinking, through sand, gravel, and overburden, until solid rock was reached. At this point, it was normal practice to hand the work over to specialist shaft-sinking contractors, who had a formal written contract to sink the shaft a specified number of yards, at a stated rate per yard.[48] These rates would vary with the size of the shafts and the nature of the geological strata through which it had to be sunk. A similar situation applied with slope sinking, although the rates were lower, since the work was not as difficult and the opening was driven in the coal, rather than through solid rock.

There was scope to vary the monthly level of expenditure and the timescales involved, because the first and second openings could be sunk simultaneously or sequentially. There might even be a gap between the sinking operations, if the first opening did not look promising and additional boring was required. However, for an individual opening, costs could not be varied at will because of the formal legal nature of the shaft-sinking contract. So a company could not stop sinking a shaft part-way through the contract simply because it wished to save money, nor could the contractor walk away from the job because the going proved harder than expected. Only bankruptcy of either party could terminate the contract. Unfortunately, this was a relatively frequent occurrence. For example, Storrs reported that two sets of contractors had failed while sinking the shaft at the new Pyne improvement in 1872.[49]

Depending on the size of the breaker, it could be constructed on a timescale ranging from a number of months to a year or more for the larger edifices. The precise timing of breaker construction was dependent not only on the size of the breaker itself but also on the depth of shafts and slopes being sunk and on prevailing business conditions. If the required shaft depth was limited, say 2–300 feet, breaker construction would have to take place while shaft sinking was in progress, otherwise it would not be ready when the shafts had reached the required depth. Conversely, if the shaft were very deep, perhaps 900–1000 feet, this might take several years to accomplish. If breaker construction were to be started prematurely, it might be finished well before shaft sinking was completed, and a large investment would have to stand idle for an extended period. Under these circumstances, shaft sinking might be well advanced before breaker construction was initiated.

To return to the problem of investment timing, mentioned at the beginning of the section, unwelcome circumstances could also arise where a major business downturn occurred some time after the shaft sinking had commenced. In the judgement of managers, it might then be prudent to delay breaker construction until business conditions improved. Such a situation was very unsatisfactory from every viewpoint, because large investment would have been made in shaft sinking, but no coal could be mined to provide any return on the capital. At the same time, proceeding with breaker construction would incur further large outlays and the coal, once mined, would have no market at remunerative prices. This latter was the classic circumstance under which individual operators, such as Stetler and his ilk, would find themselves on the brink of bankruptcy.

Finally, a certain amount of opening of gangways and airways within the mine was necessary after completion of shaft sinking, or there would be very little coal available to put through the breaker when it commenced full operation. If the veins did not open well in the first shaft, gangways or rock tunnels might also be driven underground in search of better coal. Connections would also have to be driven to connect up the first and second openings to comply with the mine safety legislation after 1870. The standard procedure was to charge this initial deadwork to the construction account until the breaker was in regular operation. Thereafter, it would form part of the normal running expenses of the mine.

An excellent example of unusually rapid mine construction is provided by the Jermyn No. 3 or Centennial shaft and breaker (later renamed the Manville), which has already been mentioned in another context in Chapter 12. The mine, situated in Green Ridge, Scranton, was developed by John Jermyn, a well-established individual operator. Sinking of the shaft commenced on June 21, 1875. The main shaft was 32 feet by 17 feet in cross-section, and solid rock was reached after five weeks had been spent sinking through 46 feet of quicksand below the surface. A further 255 feet of sinking through rock was required for a total shaft depth of 301 feet. The sinking was completed in 6 months and 11 days, which was regarded as extremely rapid progress at the time. A second smaller shaft, 15 feet x 10 feet, was also sunk. At the same time a substantial breaker of 800–1000-tons-per-day capacity was constructed that utilized about one million feet of lumber (see Plate 11). The entire construction project was completed in just 11 months and 2 days. The operator was applauded for his untiring energy and enterprise, and more importantly, 'not the least accident happened to anyone from beginning to end'.[50] This brief description of a very rapid and well-executed project, for a mine of limited depth, provides a useful contrast to the much deeper and more expensive shaft-sinking operations required on the Lackawanna Coal Department properties. These larger projects showed the necessity to utilize a much wider range of management strategies to ensure successful completion.

CASE STUDY: MINE CONSTRUCTION BY THE LACKAWANNA COAL DEPARTMENT IN THE 1880S

The Lackawanna has been chosen again for this second case study for several reasons. Firstly, it provides continuity in the overall argument in relation to the interplay of decision-making factors, costs, and business-cycle conditions that affected the processes of developing a coal estate. Secondly, the Lackawanna made major investments in new construction throughout the period in question on tracts with contrasting geological conditions. Thirdly, and most importantly in the present context, not only is a commentary on the process available from the company managers through their correspondence, but also a very comprehensive set of accounting records have survived for the construction. Although it proved to be a substantial and difficult undertaking to analyse these, the resulting cost estimates give a monthly breakdown of expenditure, with relatively few gaps, for each of the three mines—Woodward, Pettebone, and Storrs—constructed during the period.[51] The detail of these cost series also enables certain decision-making strategies to be inferred directly from the accounting evidence, even though there is no direct record of them in the text of surviving correspondence. This underlines the importance of utilizing both quantitative and qualitative (textual) data sources in investigations of this kind. The different stages in construction will now be examined in relation to the three mines of interest.

EXPLORATORY DRILLING OPERATIONS

The different timing of drilling operations at Woodward, Pettebone, and Storrs indicates some of the varying circumstances that could arise in the opening up of new coal properties. At Woodward, drilling began at the first hole in October 1873 and continued for six months to a depth of 948 feet, the first 700 feet at a rate of $3.50 per foot, the remainder at $5.00.[52] A second hole was drilled between July and November 1874 to a depth of 930 feet, with costs per foot the same as the first hole. It is possible that a third hole was begun but then abandoned in 1875, but there is insufficient voucher information to confirm this.

Such a measured approach was not possible at Pettebone, because the lease was signed on October 7th, 1881 and the drilling was supposed to take place before the date of first payment, April 1, 1882, so the lease could be relinquished if necessary.[53] In the light of the time required for drilling at Woodward, this was clearly rather optimistic, and so it proved. By March 30, 1882, the first hole was only down 470 feet, without satisfactory result, because of delays in getting the Pennsylvania Diamond Drill Company men on site and the slow progress of the work. Therefore, a decision in favor of retaining the lease had to be made in the absence of proper drilling information. This was a matter over which Storrs doubtless had some second thoughts, as he reported later that the results of the borings were 'not as good as expected'.[54] A total of $12,618 was expended on drilling between January and November 1882, but unfortunately no details of the cost per foot have been found. The voucher record indicates that the machinery was moved to Storrs (in January 1883), where drilling was undertaken and pursued until the end of July of that year, at a total cost of $12,757.[55] This operation is noteworthy because it took place *after* initial sinking of the first shaft in 1881–1882, rather than before. This might explain some of the difficulties experienced early on at Storrs in finding a sufficient quantity of mineable coal![56] The much earlier borings, mentioned in a previous section, which had been undertaken in 1869, were on a different part of the Lackawanna's lands in this area.

From these examples, it would appear that the benefits that should have accrued from prospecting and drilling operations were not always realized. This was either because commercial pressures forced decisions to be made before drilling yielded up the necessary information, as in the case of Pettebone, or, perhaps, because of haste to begin developing the property, as at Storrs, when drilling may only have been initiated after problems finding suitable coal were encountered. Nevertheless, drilling was undertaken in each case and, at least for Woodward and Pettebone, the results were available before shaft sinking commenced.

SHAFT SINKING

At the outset, it is appropriate to summarise several of the key parameters associated with this part of the construction process in each case. The relevant figures are given in Table 14.3. These reveal an important difference between Woodward and Storrs, on the one hand, and Pettebone on the other. The former two mines were both designed for a large productive capacity. As a result, both the main No. 1 shaft and the No. 2 shaft were fitted out for hoisting coal, whereas, in smaller mines, the No. 2 shaft was only used for ventilation, as was the case at Pettebone. Voucher information for Pettebone indicates that Pettebone No. 2 must have been approximately 18′ x 12′, a standard size for a ventilation shaft, but, strangely, no precise information on its dimensions is given in any of the available sources. Woodward No. 1 shaft was also a 'double' shaft, which had four hoisting cages rather than the usual two. Of these, half were to be used for working the Bennett seam and the rest for the lowest red ash seam at the base of the shaft.[57] The Storrs No. 1 shaft was slightly larger than normal to reduce friction in the upcast for ventilation purposes, but otherwise it was of standard construction.[58]

In terms of depth, Woodward and Pettebone were deep shafts, sunk to more than 1,000 feet, while Storrs No. 1 reached a depth of only 416 feet, having cut the Big (or G) Vein at 285 feet. The depth of Storrs No. 2 is only given down to the Big Vein in the available sources, so it is uncertain whether the shaft was sunk below this at that time.[59] The combination of large dimensions and great depth made the contract sinking of Woodward No. 1 a very costly enterprise, on which nearly $94,000 was spent. The figures for Pettebone are disproportionately lower because company men undertook more of the sinking, as explained further below, while Storrs was an altogether cheaper operation at this stage.

As the table indicates, quite different arrangements pertained at the three mines in relation

TABLE 14.3
Shaft Sinking Data: Lackawanna Coal Department Mines.[61]

	Woodward		Pettebone		Storrs	
	No. 1	No. 2	No. 1	No. 2	No. 1	No. 2
Dimensions (ft)	53 x 12*	35 x 12*	37 x 12	18 x 12 (approx)	40 x 10	35 x 10
Final Depth (ft)	1,039	1,003	1,126	1,143	416	302 (to Big Vein)
Contract Sinking Cost (nearest '000)	94	53	42	31	16	13
Period(s) of Main Sinking	1881–1885	1882–1884	1883–1884 1888–1889	1885–1888	1881–1882	1887–1888

*While 12′ was the design dimension, after the shaft was completed, the size was described as 10′ because of the extra timbering that had to be put in which reduced the available space.

to the timing of the shaft-sinking contracts. The sinking of Woodward No. 2 commenced nearly a year after that of No. 1, but finished before it, since its smaller dimensions meant more rapid progress could be made. So, for much of the time, the two shafts were being sunk simultaneously. Only 60–90 feet of superficial material had to be cut through by company men before the contract sinkers could begin. At Storrs, the thickness of overburden was only 40 feet[60] and contract sinking of Shaft No. 1 was able to commence in September 1881. It continued until October 1882. Unfortunately, at the end of 1882, insufficient coal had been found to warrant the construction of a breaker or, indeed, the sinking of a second shaft, so gangway driving, out from the shaft, was initiated in search of good coal. At the end of 1883, the situation had not improved. By the close of 1884, it was reported that it was necessary to deepen the shaft, but work on this had been delayed. Finally, in 1885, enough coal was found in the west gangway to make further development economical.

Preliminary sinking of a second shaft began in 1886, but contract sinking was not initiated until mid-1887.[62] Hence, in this case, it was largely problems of a geological nature that resulted in a gap between the sinking of the first and second shafts. Only a skeleton workforce was employed at the site in the interval. However, it is likely that the gap was extended because the eventual discovery of good coal in the main shaft gangway happened to coincide with the 1885 depression.

The sinking arrangements at Pettebone proved to be somewhat unusual, although both geological and economic conditions were again implicated. This mine was sunk on the alluvial flats of the Susquehanna River 'down 78 feet to the rock, through gravel, clay and quicksand'.[63] Apparently, the sinking of a shaft on such treacherous drift geology had not previously been attempted, and competitors, with smug pessimism, were 'predicting certain failure'![64] Nevertheless, with considerable expense in additional timbering, masonry, and caulking to make the shaft watertight, it successfully reached the rock base at the beginning of 1884.[65] The difficulty of the task can be gauged from the fact that miners were first employed at the shaft in June 1883, according to the payroll records, and it took the remainder of the year to sink the 78 feet. Jermyn's men, in the example from the previous section, had sunk a slightly smaller shaft 301 feet in a similar time, where the geology was much more favorable.

Unlike Woodward and Storrs, the next phase of sinking through the rock was not undertaken by contract sinkers, nor was it continued to the eventual depth of the shaft. Instead a number of company men were employed to sink the shaft to a depth of 300 feet by the end of 1884.[66] This further sinking through rock by the company men was presumably, in part, to ensure that everything was in good order before the task was eventually turned over to the contractors. A separate report at a later date implied that the depth reached was close to the final intended depth of this stage of the sinking, because the first lift, (or weight of column of water), for pumping out the shaft was 310 feet, the depth of the first workable coal seam.[67] Continued employment of company sinkers in the first few months of 1885 was presumably to finish off this part of the work. However, there was probably more to these circumstances than initially meets the eye. In November 1884, in the wake of financial disarray on Wall Street and a worsening business outlook, President Sloan had directed Storrs to reduce expenditure as far as possible. The latter had responded by trying to place as much of the company workforce as possible on short-time, including some construction workers. At the same time, he pointed out that extant sinking contracts needed to proceed to completion and should be 'pushed rigorously'.[68] However, warning signals on Wall Street had appeared as early as the summer of 1884, and, by September, Storrs and Sloan had already begun discussing cost-control measures.[69] So it is possible that a decision had been taken to sink Pettebone only to the first workable vein, and hence dispense with sinking contractors, whose work could not easily be stopped, as an early step in the direction of cost reduction. Subsequent developments at the mine would seem to bear this out. In the early summer of 1885, the payrolls indicated that the company sinkers were transferred to the No. 2 shaft and they were reported to be sinking through '90 feet of drift' in August of that year.[70] Company sinkers continued the work through rock, to a total depth of 164 feet, before sinking contract work was let, which commenced in July 1886. By this time, the worst of the 1885 depression had already passed, so this contract work continued, not to the first workable vein, but to the bottom of the coal basin at 1,143 feet (Table 14.3). This depth was reached in January 1888, leaving a strangely lop-sided improvement, whose main shaft was only 300 feet deep, but whose second shaft cut all the veins on the property. Fortunately, the ending of the contract on the second shaft coincided with the period of windfall profits from the labor troubles in the other regions, so the company let a new sinking contract back at the main shaft. This commenced in February 1888 and continued until mid-1889, when the shaft had been sunk a further 826 feet below the level of the first workable vein to reach its maximum depth of 1,126 feet. The effect of this convoluted sequence of construction activity at the two shafts meant that the overall process of shaft-sinking had been spread over six years, but the bulk of the work, and hence the costs, had been concentrated in the years 1883–1884 and 1887–1889, thereby minimizing expenditure in the depression period of 1885–1886.

BREAKER CONSTRUCTION

The timing of breaker construction activity, especially at Woodward, also confirms the direct effect of wider business-cycle conditions on investment, and resultant employment, at the local level. Although contract sinking of both Woodward shafts had been pushed to completion at the end of 1884 and the beginning of 1885, there was no pressing need for the coal in the market at that time. The company therefore had to decide either to proceed with breaker construction and fitting out the mine, at an additional cost of several hundred thousand dollars, which would obviously affect cash flow, (with no immediate return), or stop the construction process until business conditions improved. In the event, they chose the latter course. A local newspaper reported in July 1885 that all work except essential maintenance had been 'discontinued for an indefinite period' and the company officials had decided 'not to build the great breaker for many months to come'.[71]

As business conditions began to improve in the spring of 1886, work resumed on the headhouse for the shaft, and, in April, preparations began for laying the breaker foundations. Perhaps surprisingly, the two pristine shafts had been allowed to fill with columns of water 800–900 feet deep, all of which had to be hoisted out in tanks.[72] Although the main shaft was cleared of water quite quickly, the airshaft remained flooded for a considerable period of time 'pending surface improvements'.[73] The voucher records confirm that bulk deliveries of hemlock, oak, pine, and spruce timber commenced at this time and

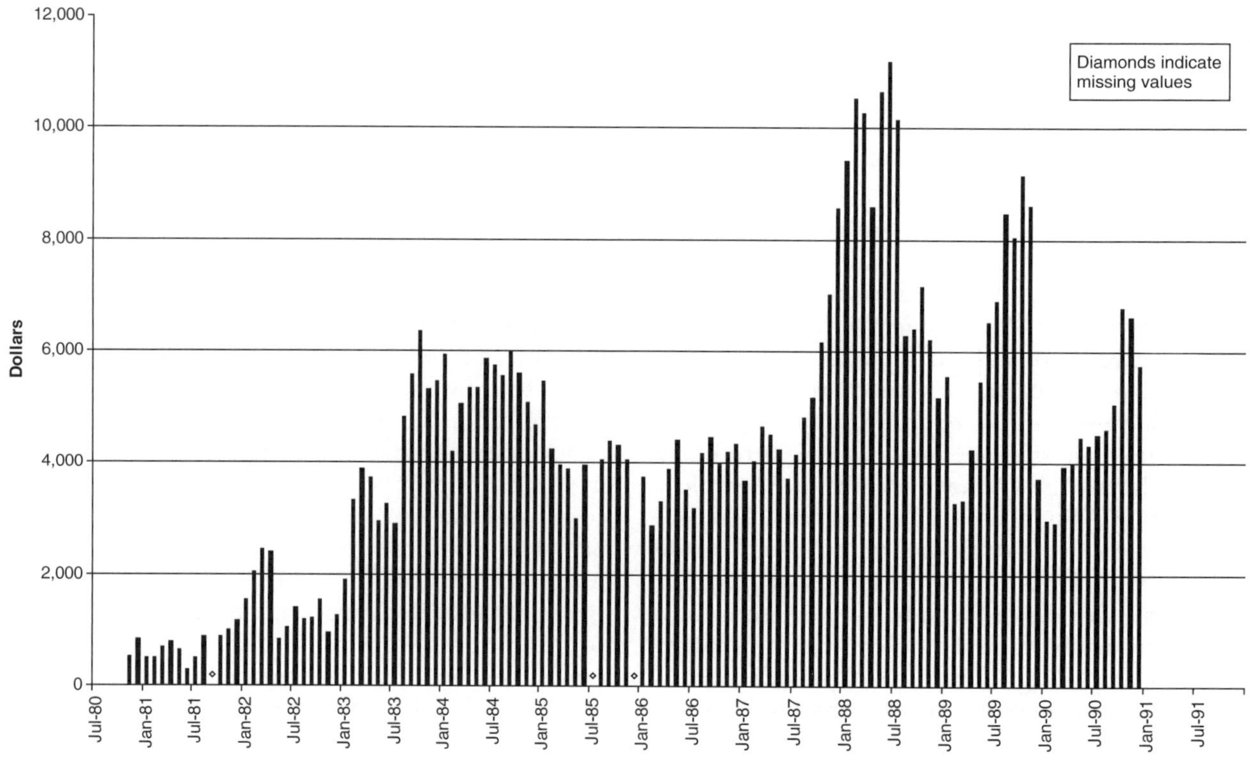

FIGURE 14.7
Company Men Labor Costs: Woodward, Pettebone, and Storrs Mines.

continued until the breaker came into full operation in the summer of 1888. Even before the 3,000-ton-per-day-capacity breaker itself was built, an immense trestlework structure was erected to connect the main shaft to the breaker. This was nearly 600 feet long and over 130 feet tall at its highest point. A rare photograph of this trestling under construction can be seen in Plate 17.[74]

Despite the deliberate postponement of investment in 1885–1886, it was no accident that Woodward was the first breaker to be brought into production, because the rate of increase of the advance royalty account was greatest at this location. Although the labor troubles of 1887–1888 brought a welcome return to prosperity in the Northern Field, the memory of 1885–1886 was still sufficiently fresh for the company to proceed cautiously with its other investments at Storrs and Pettebone. As soon as the Woodward breaker was operational, construction effort switched to the Storrs, which had the second most pressing problem of advance royalty. The same broad pattern of expenditure was observed as had been seen at Woodward, accompanied by large deliveries of different kinds of lumber. The main surface improvements were built more rapidly than at Woodward, since no large trestling was required, and the breaker became fully operational in the latter part of 1889. Finally, attention turned to Pettebone to complete the trio of new improvements, although construction here continued until the end of 1890. The deliberate sequencing of investments in new breakers, which also took account of the timing of contract sinking in the latter part of the 1880s (Table 14.3), can be seen very clearly in Figure 14.7, which shows payroll expenditure on company men employed in the construction workforce.

OPENING-OUT OF MINES

Compared to the other stages of mine construction, opening-out was much less costly, since neither large supplies of timber nor large engines required. Nevertheless, it was essential preparation for full production, although the nature and extent of the initial developments could vary widely. The timing was equally variable, since gangway driving could take place in one shaft before the second was sunk. It might also precede breaker construction, although a large amount of driving in the coal would be problematic under these circumstances, because the resulting cars of coal could not be processed and shipped out. In the early stages, gangway or airway driving might be undertaken—either by contract miners working by the yard, or by miners employed by the day on the company payroll. As completion of the mine approached, work was normally done by contract. Often, the miners would continue blasting and loading coal from the places they had opened during the construction phase, when normal production began.[75]

Reference has already been made to the exploratory gangway driving at Storrs in the mid-1880s. Between December 1884 and April 1887, the payrolls record more than 800 yards of airway and over 400 yards of heading driven. Apart from this period, only one additional month of development work, in November 1889, was charged to construction, during which a further 114 yards of airway and 69 yards of rock were cut, and seven chambers were opened for mining. The underground development period at Woodward charged to construction began in November 1887 and continued until July 1888. During this time, more than 1,400 yards of gangway, 280 yards of heading, and 90 yards of airway were cut, but there is no evidence of any chambers being opened. It is interesting to note that, by the beginning of 1890, still only a few chambers had been opened at Woodward in its first 18 months of production. In contrast, more than 80 were opened at Storrs in its first 13 months of operation. No contract-mining charges of this kind were made at Pettebone during its construction, but more than 20 chambers were opened in its first full year of operation.[76]

SUMMARY OF THE CONSTRUCTION PROCESS

The sum total monthly construction expenditures for the three mines are shown individually in Figures 14.8–14.10. These include company and contract sinking costs, outside labor, lumber, and engines and machinery. Sharp 'spikes' in the graphs tend to be associated with the delivery of expensive capital machinery, such as pumps and hoisting engines.

In terms of labor costs alone, approximately $339,000 was incurred at Woodward, $320,000 at Pettebone, and $130–140 thousand at Storrs. Figures for Storrs are less precise, because of the lack of accounting records right at the start of the period. Total costs of all kinds, in each case, give estimates of $734,000 for Woodward, $602,000 for Pettebone and $353–362 thousand for Storrs. In relation to its size and capacity, Pettebone was a particularly

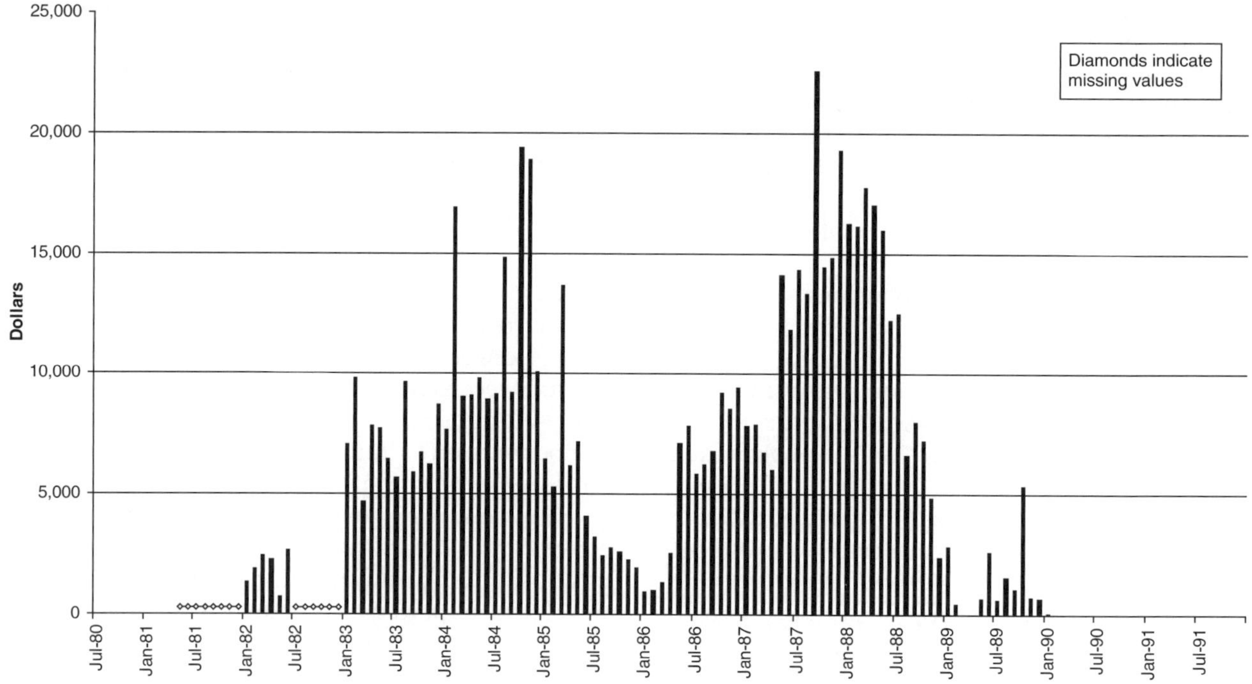

FIGURE 14.8
Total Construction Costs: Woodward Mine.

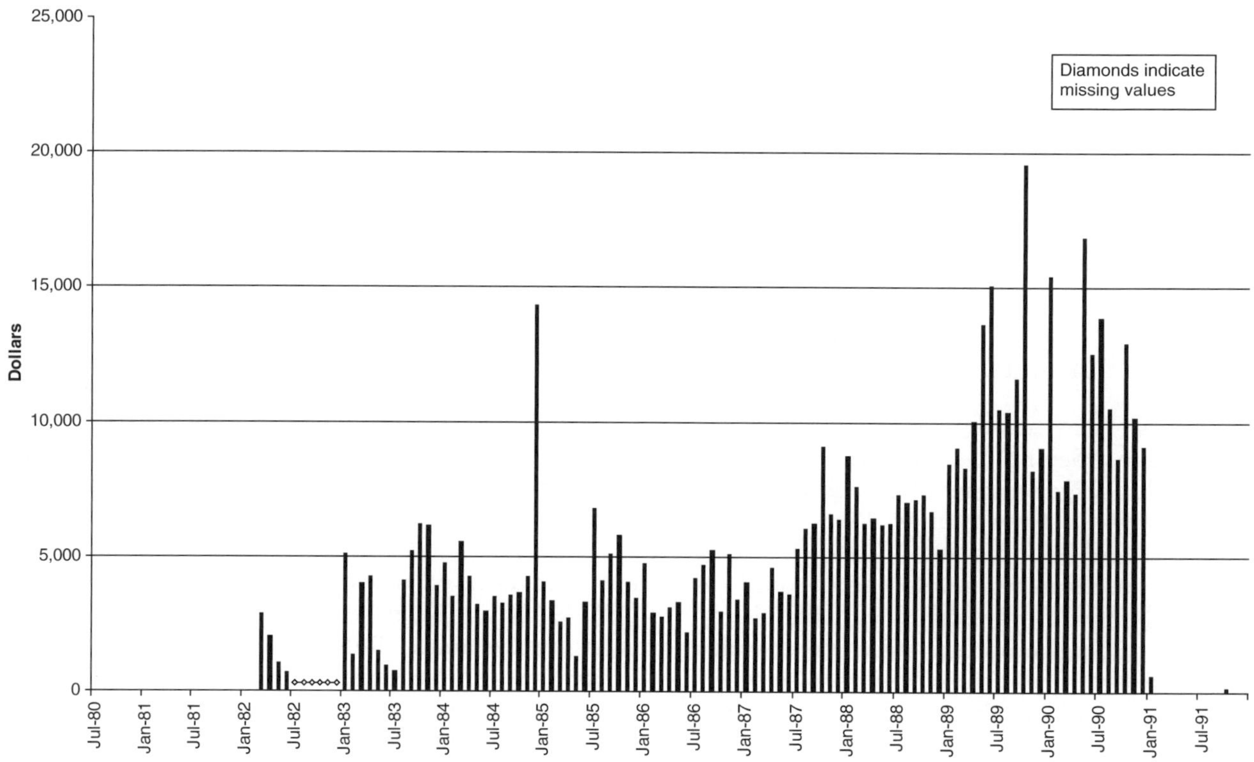

FIGURE 14.9
Total Construction Costs: Pettebone Mine.

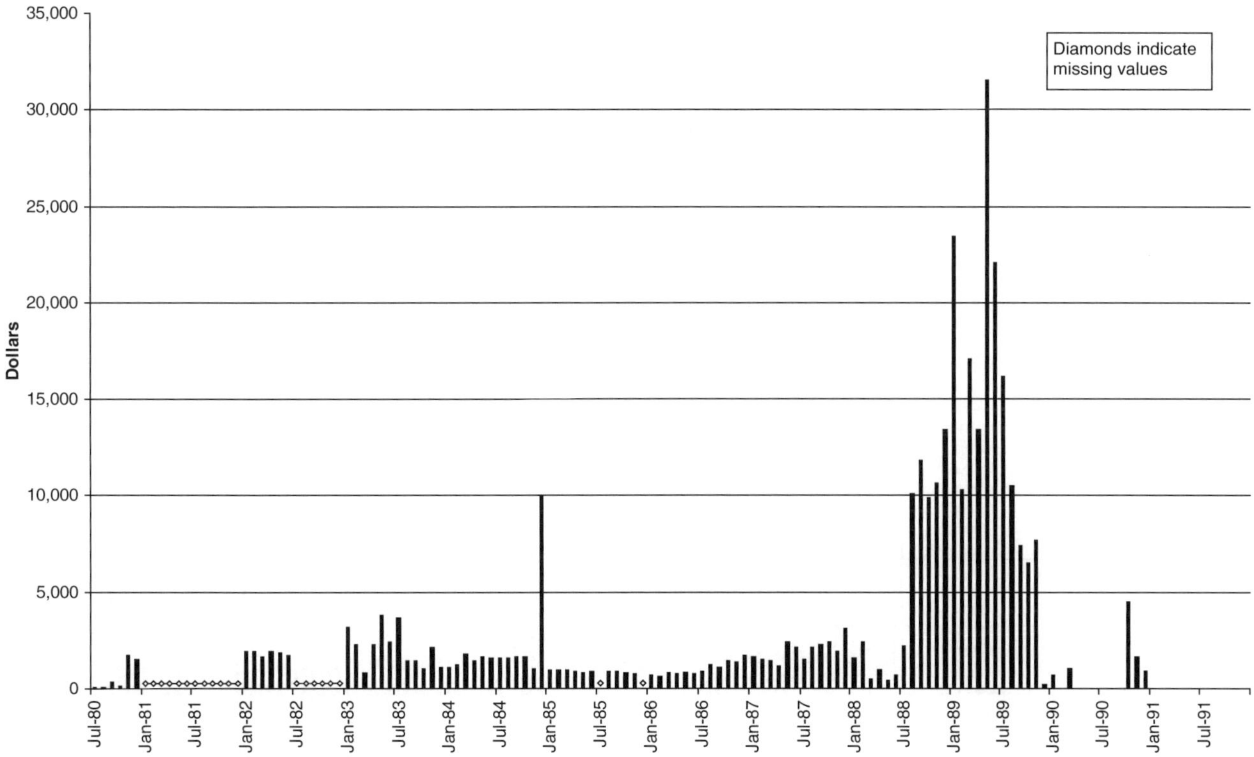

FIGURE 14.10
Total Construction Costs: Storrs Mine.

expensive mine to construct, largely due to the problems of protecting the shafts against quicksand and caving-in.[77] An indication of this comes from the high timber costs, which at nearly $51,000 were almost as large as at Woodward, where the extensive additional trestling had been built.

The overall monthly construction costs at all three mines taken together are shown in Figure 14.11. Although the 'lumpy' purchases of capital equipment make the overall pattern slightly less clear, several key features of the pattern of expenditure can still be identified. Firstly, Mr. Storrs and his superintendents were successful in reducing expenditure significantly during the 1885–1886 depression, despite the expensive Woodward sinking contracts that had to be carried to completion. The wages component of overall costs was reduced by 35 percent in the period from the end of 1884 to mid 1887. The construction workforce was reduced by 20 percent, short-time working reduced hours by 23 percent, and there was also a 14 percent pay cut in November 1884, based on the evidence in the payroll records. Secondly, the delays and postponements of new construction work, which formed a major component of the overall adjustment process, eventually extended the timescale, over which new capacity became available, by several years. Certainly, Woodward and Storrs could have been completed by 1886 and 1887, if not slightly sooner, if the company had been willing to overlap breaker construction with other work and to invest in more than one breaker at once, as had been the case in the late 1860s. With the benefit of hindsight, this would have provided more output capacity to benefit from the 'labor troubles bonanza' of 1887–1888 and would have allowed the advance-royalty accounts to be reduced more quickly. This reduction would have been particularly beneficial in the Woodward case. Thirdly, despite the delays and difficulties, three major mines were eventually brought into production, at a total cost of $1.7 million. Approximately 2,500 different individuals were employed at some time during the construction period at these mines. When they reached substantial production levels by the turn of the century, their total combined workforce exceeded 3,000 men. Allowing for dependent family

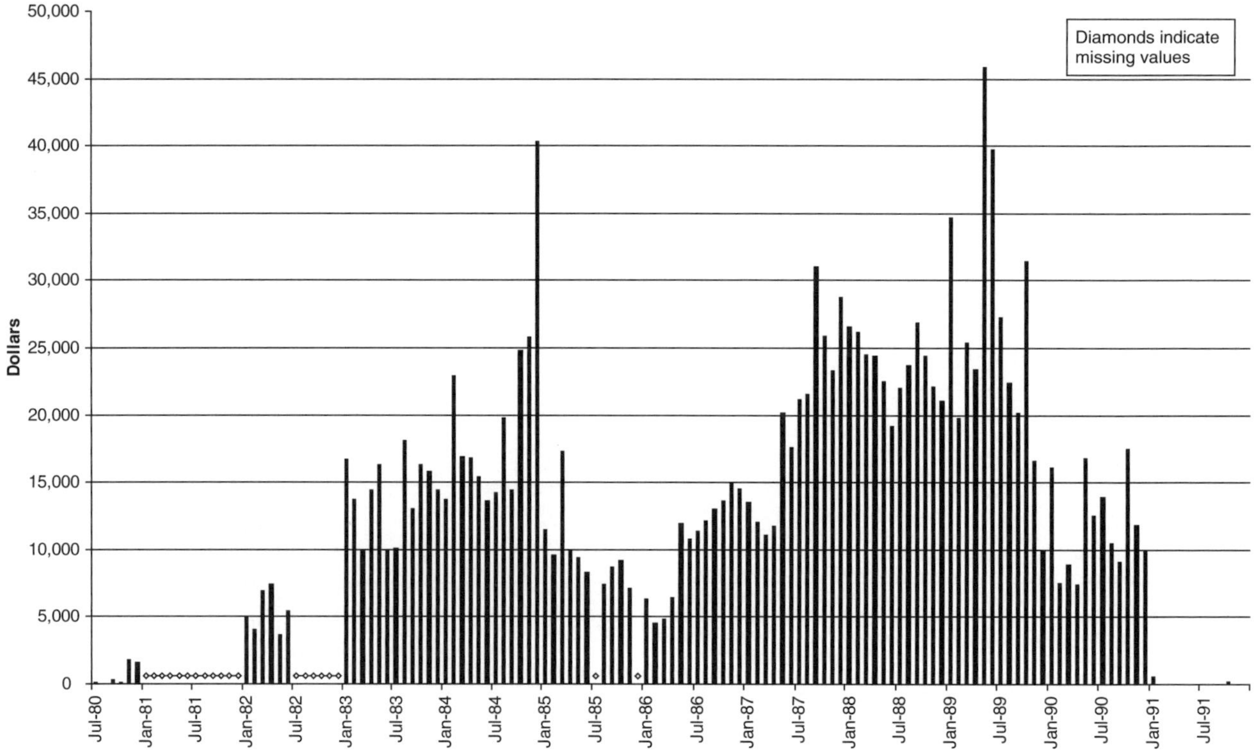

FIGURE 14.11
Total Construction Costs: All Three Mines Combined.

members, this was the equivalent of a good-sized town.

Finally, such tactical adjustment of long-term strategic plans, in the light of fluctuating economic circumstances or unexpected geological conditions, was not unique to the Lackawanna. However, the railroad's solid financial position made it much easier to ride out temporary reversals in trade in order to ensure efficient and economical development of their coal properties over the long-term. The overall impact of these changes in the timing of investment, both by the larger companies and the smaller individual operators, on the paths of development in different parts of the coal regions, will be examined in Chapter 16.

INTEGRATION OF NEW MINING CAPACITY INTO THE PRODUCTION SYSTEM

It might be assumed that once mines entered the normal production system there would be no further issues to account for in relation to capacity planning for the future. However, production statistics do not bear this out. There are two specific questions that merit brief examination here. The first is the time required to reach capacity output in tons per day; the second is the time required to reach efficient productivity levels, measured in terms of tons per man per breaker day. This last measure has not previously been computed for mines within the industry, and it can be quite revealing in the present context. Saward's Coal Trade Annual drew attention to the apparent decline in productivity per man in the anthracite industry from 535 tons per man per year in 1870 to 350 by 1900, but these data are not standardized by average days worked per year, which substantially reduces their usefulness.[78] Further difficulty, in determining representative overall industry averages, arises from the practice of reporting of shipments by breaker. Some companies sent tonnage from several mines to the same breaker, but the mines involved could change over time, while other companies had one breaker for each of their main openings. Inspection of available data, over the period 1876–1902, suggests that a rough figure of 2 tons shipped per man per day represented a good indication of an efficient working operation. For example, the Pyne breaker, whose costs were very low, ranged between 2.1 and 2.9 in different years during that period, while the Prospect breaker of the Lehigh Valley Coal Company ranged from 1.52 to 2.38.[79]

Examination of the shipment time series for Woodward, Pettebone, and Storrs shows that it could take some time to ramp up to full-scale production. In the case of Woodward, output did not surpass 300,000 tons until 1894, six years after production began, and it did not manage to exceed about 400,000 tons prior to 1902.

In contrast, Storrs passed the 300,000-ton mark in 1892, and managed to double this again to about 600,000 tons by 1899. It is difficult to include Pettebone in this comparison because of operational problems that affected its output substantially between 1894–1897. Instead, the Bliss breaker, completed on the Hanover lands of the Lackawanna in 1895, can be used (Plate 18). Here again, a substantial output of more than 300,000 tons was not reached until 1901, although, in the absence of the 1900 coal strike, it might have achieved this figure one year earlier. Nevertheless, for long-term planning purposes, it is clear that not only was it necessary to allow time for new mine construction, but then a further period of several years could be required before the full benefits of the investment would be realized.

A somewhat similar point applies in relation to efficient operation based on the admittedly crude statistic of 2 tons per man per day. Storrs only managed a figure of 1.37 in its first year, rising to 2 in its fourth year. Thereafter, it maintained values in the range of 2.1 to 2.4 until the end of the century. A similar situation pertained at Bliss. In 1896, its first full year of operation, it achieved a figure of 1.24, reaching 2.05 in 1900. Woodward made a better start with 1.8, but slipped back to 1.6 before recovering to 2.0 in 1900. These values can be compared with the Maxwell No. 20 of the Lehigh and Wilkes-Barre Coal Company, completed in the same year as Bliss. It made a positive start with a figure of 2.29 in its first full year of operation, and managed to sustain it with values ranging between 2.26 and 2.87 up to 1902.[80]

The implications to be drawn from these simple statistical measures are clear. Immediate profitable and efficient operation could not be guaranteed. In some cases it was readily achieved, but in others it had to be attained by diligent effort over several years of production. During this time, a new mine could be a drain on company finances, rather than an asset, because of inefficient operation and, hence, relatively high production costs. Such matters would also need to be taken into account in the course of long-term capacity planning.

CONCLUSIONS

The foregoing detailed analysis of mining costs, royalty accounts, and construction processes has been necessary to demonstrate the complexity of the management issues facing companies wishing to develop coal properties, both large and small. Small individual operators did not have the luxury of spreading production and mining risk across tracts in different parts of the coalfields. Nor did they possess the reserves of capital necessary to carry them through the initial start-up phase of production, especially if business conditions became adverse or geological conditions proved unfavorable. So they became very dependent on the railroads to advance funds at times of necessity. Otherwise they would have to abandon their properties to the shippers, or be left at the mercy of the sheriff's sale.

Even for the coal departments or coal subsidiaries of the railroads the problems were far from simple, despite their relative immunity from daily cashflow problems. Exhaustion of lands, loss of contract shippers, and accidents at the mines—all could reduce the flow of coal to market over their roads. These situations had to be foreseen and hedged against, as far as possible, by a carefully planned program of investment in their coal estate. A classic case of how unforeseen circumstances could conspire to reduce production capacity tributary to a railroad is provided by the correspondence of Israel Morris of the Lehigh Valley Coal Company, in 1897. He observed that, in 1894, 18.86 percent of the productive capacity of the industry was tributary to the Lehigh Valley Railroad. However, he wished to sound a note of alarm and urge corrective action, because this figure had now dropped to 16.54 percent. The explanation given, in terms of loss of tonnage at the different mines, was as follows:

Seneca	shut down by last June's disaster
Babylon and Lawrence	reduction in rated capacity because of faulty ground
Linderman and Skeer Mines	abandoned

Logan Colliery	destroyed by fire
Blackwood Colliery	unsafe condition of breaker
Maltby	destroyed by fire
Centralia Breaker	torn down for rebuilding
Spring Mountain No. 4	destroyed by fire
Bellmore Colliery	shut down indefinitely

By his calculation, this reduced the railroad's tributary capacity by nearly 9,000 tons per day. Urgent action was needed to rebuild breakers and prosecute other new improvements, such as the Hazleton Shaft, which had previously been suspended.[81]

At the same time, multiple cycles of speculative fever over coal lands and leases had pushed all the major companies to acquire lands and mining rights well in excess of their actual requirements in the short and medium term. Unfortunately, they could not escape the financial consequences of this competitive acquisition of tracts under lease, and advance royalty accounts had to be managed and controlled. Otherwise, cash-flow problems might develop in an inexorable fashion, until they spilled over to burden the parent railroad. Nor could these problems necessarily find rapid resolution. Mines 1,000 feet or more in depth were not built in a day, and business-cycle depressions could turn a substantial investment undertaking into a construction enterprise lasting a decade or more.

This last point is very clearly exemplified, if the three Lackawanna mines in the earlier example are revisited, now that all aspects of the investment decisions associated with them have been examined. The overall timescales of investment, between initial acquisition of the leases and first output from the new improvements, can be summarized as follows. For Pettebone, the lease was signed in 1881, first production took place in 1891, and a further decade was required to clear the accrued advance royalty account. For Woodward, the corresponding dates were 1872 and 1888; a total of 16 years elapsed, and the advance royalty account was still in major deficit in 1902, 30 years after the original lease was signed. At Storrs, leases were executed at various times during the 1860s, but mining only began in 1889, and a further decade of substantial shipments was required to square the royalty account. This was an extraordinarily long timescale over which to execute an overall plan of investment in mining property for the benefit of the railroad tonnage it could provide. Nevertheless, major financial benefits were to be reaped in the first two decades of the 20th century because of the prior expenditure on high-quality plant, situated on productive and valuable lands that had many years of coal reserves. It is also a further testimony to the skill of Storrs and Sloan, whose conservative financial management, coupled with strategic foresight and tactical judgement, found few equals in the 19th-century anthracite industry.

Chapter 15

REGIONAL RETRENCHMENT: RATIONALIZATION AND REORGANIZATION IN THE SCHUYLKILL REGION 1872–1902

INTRODUCTION

In keeping with the general thrust of the argument of this study, it is appropriate to single out the Philadelphia and Reading Coal and Iron Company (hereafter called the 'Coal and Iron') for separate examination, because the business strategies it was required to adopt differed in a number of ways from those of the companies in the other fields, which have been discussed in previous chapters.

There were a number of reasons for this. First and foremost, there was the large size of the estate, compared to that of its competitors in other fields. Secondly, a substantial proportion of this estate was at least partially developed by a wide range of independent operators. Some of these were long established and experienced in the coal business, while many others shared with their Northern Field brethren the multiple deficiencies of under-capitalization, inexperience and over-optimism. Thirdly, the Coal and Iron was doomed to an extended period of penury, almost from its inception, because of the excessive borrowing by its parent railroad, just to establish the estate it was supposed to manage.

No simple 'quick fix' solution could resolve the range of problems the company faced at the beginning of its existence, but some required more immediate attention than others. It would be difficult to summarize the situation in the early 1870s more succinctly than in the words of General Henry Pleasants, the company's chief engineer. Writing in the 1876 Annual Report, he observed that, at the time the company was formed:

> 'the coal business in the Schuylkill region, owing to overproduction and high wages, was in a deplorable state. Coal had fallen in price, but labor continued high; a powerful union of miners and laborers, controlled and terrorized by the murderous secret society of Molly Maguires, prevented a reduction in wages corresponding to the decline in the price of coal; the mines and mining machinery could not be kept in order by financially embarrassed operators; and the collieries of the Schuylkill region became less and less able to compete with those of Lehigh and Wyoming'.[1]

Setting aside the contentious nature of the claimed link between the Molly Maguires and the Miners' Union, although such assertions were doubtless company policy at the time, the quotation clearly identifies the loss of competitiveness of the Schuylkill Region, because of high and inflexible costs in a period of falling prices.[2] The 1872 coal season had provided an object lesson in this, as a previous chapter has noted. At that time the lower bound of the wages basis system had been set too high and Schuylkill coal could not be produced at a profit. While wage levels provided a readily identifiable scapegoat, General Pleasant's observation about impecunious operators also points up two other types of problems. One of these was lack of working capital; the other was an inadequate level of investment in efficient production capacity that had not kept pace with rivals in other regions.

Reduction in the average cost of mining was thus an immediate commercial priority, but this would have to be followed by a more considered program of rationalization of the company's available mining resources, leading progressively to a more efficient and economical long-term plan for development of its estate. These shorter- and longer-term commercial adjustments to meet the competition and the wants of the market will now be examined in greater depth.

COST OF MINING

During the early years of its existence the Coal and Iron was good enough to publish detailed information about its costs of mining, both on an annual and a monthly basis. These figures chart the aggregate effect of company policies on reducing costs and the rate at which these changes could be brought into effect. The annual data are shown in Table 15.1.

The figures in the table refer to the cost of delivering coal into cars at the breaker, but do not include any selling expenses. It would appear from the statement in the 1876 Annual Report that in the early years, at least, mining costs did not include any expenditure for new mine improvements, which would have been added to the capital account.[4] By the following year, however, such costs were charged to mining expenses, and thus would have been included in the cost of mining, although the purchase of existing improvements appears still to have been treated separately.[5]

At first sight the reduction of costs in the period 1873–1877 is particularly striking, as the figures were reduced from $2.50 a ton to just over $1, but there was also more general downward pressure on prices and wages during the depression of the period. The 1873 figure may appear extremely high, but it is interesting to use again the much later testimony of A.J. Davis, operator of the Warrior Run colliery near Wilkes-Barre, who declared that in 1875, his costs per ton were $2.50 and hence at broadly the same level.[6] As noted in previous chapters, average prices of anthracite held up after 1873 until the collapse of the combination in 1876. For Schuylkill white ash lump coal the cartel era prices had remained above $4, but they slumped to a low of $2.59 in 1877. Thus

TABLE 15.1
Average Cost of Mining for the Coal and Iron 1873–1887.[3]

Year	Average Cost of Mining ($/ton)	Year	Average Cost of Mining ($/ton)
1873	2.51	1881	1.50
1874	2.45	1882	1.47
1875	1.87	1883	1.50
1876	1.35	1884	1.48
1877	1.04	1885	1.44
1878	1.24	1886	1.54
1879	1.14	1887	1.45
1880	1.44		

cost reductions by the Coal and Iron during the cartel period were more significant than after it, because such changes increased the margin of profit, as opposed to simply following prices down, as happened in 1876–1877. Between 1873–1875 therefore, the 64-cent reduction in costs was a genuine improvement in profitability. Unfortunately, this benefit and those of other subsequent cost reductions were largely wiped out by the even more rapid fall of prices 1875–1877. Nevertheless, had these reductions not taken place, the financial consequences both for the Coal and Iron and the parent Reading Railroad would have been even more disastrous, and felt even more quickly, than they eventually were.

Examination of the more detailed monthly data shown in Figure 15.1 shows an important aspect of the problem of mining costs that cannot be discerned from the more aggregate annual figures. This relates to the fixed costs of pumping and ventilation over the winter months, which continued even if output was very small, raising costs per ton produced to potentially quite high levels. This can be seen clearly during the 1870s, when cartel output restrictions and the 1875 strike, in particular, caused 'spikes' in the graph. These spikes are far less apparent in the non-cartel years of 1877 and 1879, when production was unconstrained. They also effectively disappeared in later years, when monthly costs became relatively stable around the annual norms of $1.45–$1.50 a ton. These figures themselves were very close to that reported by Coxe in 1887, as noted in chapter 13. This convergence of costs for companies of different sizes, in widely varying geological circumstances, is itself noteworthy.

Analysis of the factors that contributed to cost reductions allow the impact of falling wages to be separated from that of other management strategies. It is clear from Annual Report statements that the urgent need to cut costs, led to the formulation of the Reading plan to reduce mining contract work by 20 percent and common labor by 10 percent at the beginning of 1875. These terms were more severe than those proposed in the Lehigh Region at the same time. Crucially, also, the minimum on the basis wages scale was to be abolished and the sliding scale continued downwards below $2.50. It is easy to read authorities such as Schlegel, to gain the impression that the purpose of this plan was mainly to destroy the Miners' Union. However, in the light of the experience of 1872 in Schuylkill and the figures on

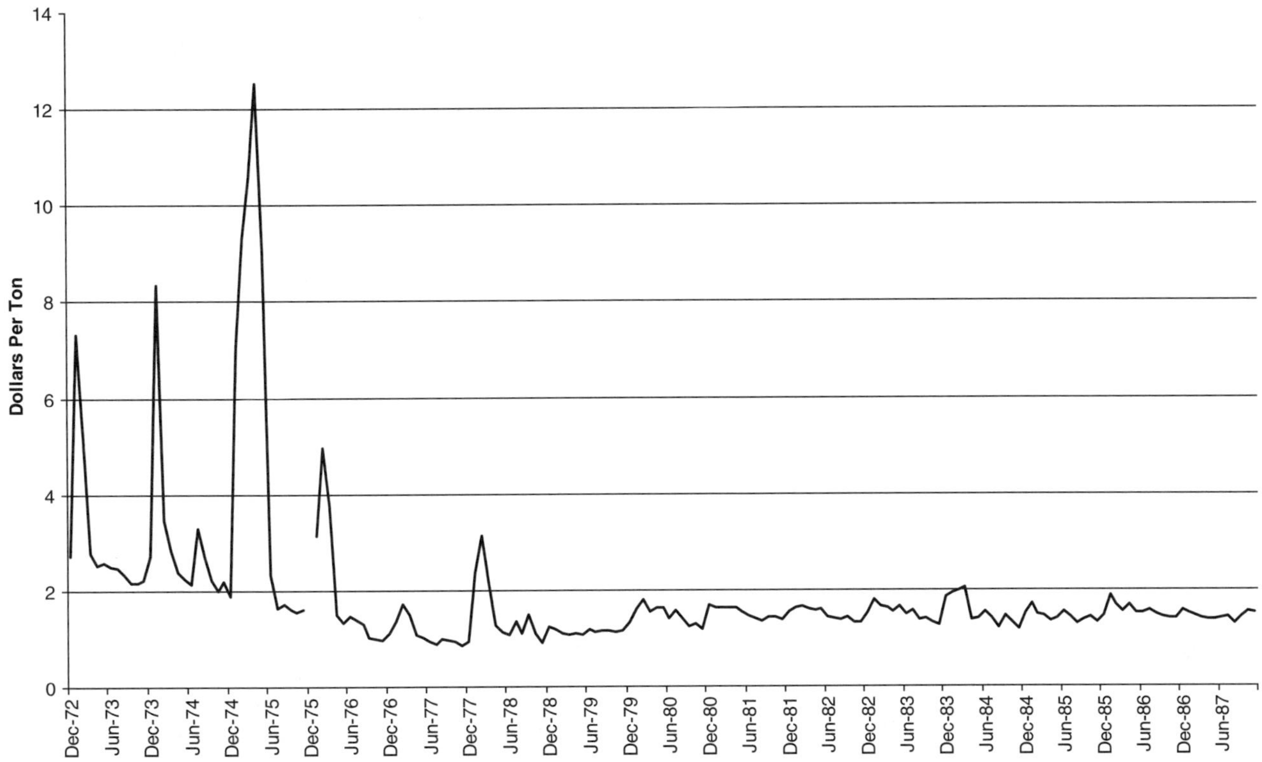

FIGURE 15.1
Philadelphia and Reading Coal and Iron Company: Average Monthly Cost of Mining 1872–1887.

mining costs, it would seem more reasonable to interpret the actions of President Gowen and his managers first and foremost in cold economic terms. Costs had to be reduced if the company was to survive. The Union was simply an impediment to this aim. If it continued to obstruct, by refusing to accept the removal of the basis minimum and a sliding scale with a more flexible downward ratchet, it would have to be confronted and/or swept aside. Not surprisingly, the miners refused to accept such swinging reductions in wages and the long strike of 1875 ensued. However, the men returned to work after five and a half months, on the company's terms, not their own. This effectively ended the power of the Union in the Schuylkill Region for many years to come. At the same time, it probably saved the Schuylkill coal trade, the Coal and Iron, and the Reading Railroad from imminent ruin. This was why the companies involved were willing to hold out effectively indefinitely, despite the short-term loss of trade and market share to the mining interests of the Northern Field.[7]

The combination of reductions, both in contract mining rates and hourly pay for company men, produced an approximate 15 percent saving in the average wage bill at the Coal and Iron in 1874–1875. However, actual cost reductions amounted to 24 percent, so other savings were clearly beginning to take effect. The benefit to the company, if not to the men, of removal of the minimum basis wage became apparent in late 1876, immediately after the collapse of the combination, when coal prices plummeted. Further wage reductions were made between September and November 1876, to bring the total percentage reduction in the wages bill since 1874 to 26 percent. Overall mining cost reductions now amounted to 46 percent, showing that other rationalization measures were increasingly contributing to improvements in the accounting bottom line. These measures were stated to include not only the renovation and improvement of older collieries, but also the successful development of new ones. Further cost reductions were forecast, as these measures continued to feed through into savings. The company also became able to operate its mines for at least ten months a year, thereby spreading the fixed costs more evenly over a larger tonnage.[8]

When the combination was re-established at the start of the 1880s, prices rose again from the ruinously low levels of 1877 and 1879, but, in consequence, there was also an increase in wages, which brought costs back up into the $1.45–$1.50 range. There is little sign, within the period covered by the graph, of any tendency for costs to continue rising with greater depth of mining, as might have been expected. In later years, reliable data are much harder to come by, but again some information from 1897 testimony by R. C. Luther, the Mining Superintendent of the Coal and Iron, provides a useful perspective. In his statement Luther claimed that costs per ton had risen to $1.96, excluding taxes, insurance and interest. The substance of his remarks suggested that the method of cost calculation was broadly comparable to the earlier period. On this basis, costs had clearly risen during the 1890s, to a greater extent than for the Lackawanna or the Lehigh Valley. The equivalent figure for the latter was $1.79 per ton.[9]

Part of the reason for rising costs may have been that the process of colliery refurbishment described by Henry Pleasants in 1876, was to prove much more protracted and costly than perhaps he realized at the time. An estate of the size and diversity that the company had acquired, and the large number of mines involved, meant that a seemingly endless amount of maintenance and rebuilding was required. The sharply worsening financial condition of the parent railroad after 1876 also meant that, in contrast to companies like the Lackawanna, available funds for investment in new mining plant were in perennially short supply, and only the most essential tasks could be undertaken at any given time. The company's strategies for management and development of its estate, in the light of the constraints placed upon it, will now be examined more closely.

OUTLINE OF RATIONALIZATION STRATEGIES ADOPTED

The early years of the operation of the Coal and Iron demonstrated a rapid evolution in the strategies adopted for development of the coal estate, in the light of their initial experience as first-time holders of coal lands. Important statements of the rationale for their actions can be found in the annual reports of the Reading Railroad for the period. These bear quotation at some length below, because of the number of different issues that are highlighted.

The first important point to note is that the railroad had embarked on its orgy of coal land acquisition with the aim of controlling long-term access to tonnage for the railroad, rather than from any particular desire to enter the mining business. However, it rapidly became apparent to the company's managers that the role of absentee landlord was not appropriate to safeguard the coal resources they had acquired. In the words of the 1874 Report:

> 'An experience of one or two years as landlords, showed how inadequate under existing circumstances, the individual tenants were, to develop and improve the estate. But few private persons had sufficient capital to open and conduct a colliery in any other manner than to make it profitable during a few years, after which it was thrown upon the hands of the landlords, who were forced to expend large sums upon it in order to protect it from destruction'.[10]

It was also noted that depression in the trade and repeated strikes in 1869–1871 had further weakened the surviving independent operators, reducing their capacity for future investment in existing and new improvements. The conclusion was obvious:

> 'If left to individual enterprise, the development of the coal region would be retarded for many years'.[11]

So the Coal and Iron, somewhat reluctantly embarked upon the same course as its competitors in the Northern Field, to become a miner of coal, as well as the landowner of its vast estate.

The first step in this was to begin securing by purchase, mining improvements made by others on lands it now owned. Since these were largely mines on coal leases, the Company would otherwise have had no control over the mining operations in question beyond the terms of the leases assigned to it as the new landowner. This process had begun by 1874. Two years later a further statement was made about the progress of these activities and the initial process of rationalization of operations:

'The Company has been engaged in renovating and remodelling such of the old collieries as were likely to produce coal at reasonable cost, and in opening new mines which are about completed.'

Also:

'The mining division has been employed in mining coal from such collieries as the Company became possessed, the least profitable ones being suspended and abandoned as fast as they could be spared'.[12]

With hindsight this looks like an expensive method of implementing cost-cutting measures. Existing mines were being purchased, so some could be closed down, effectively voiding the capital investment involved. At the same time, new collieries were being developed, the most prominent of which was the Pottsville mine, with its twin shafts more than 1,500 feet deep, which was necessarily a very expensive enterprise. At the time these policies were initiated, however, the cartel was in full operation; Gowen's expansionary zeal held sway with investors and the Reading's credit was still good. Raising $10 million via an issue of 6 percent gold coupon bonds to fund the Coal and Iron and other developments was therefore not perceived as problematic at the time.[13] Fortunately, as the above quotations indicate, much of this preliminary investment was completed before the collapse of the cartel in late 1876. While interest payments from these and other borrowings were eventually to bring the Railroad down, at least the Coal and Iron was not left with expensive and half-constructed improvements, which would have added insult to injury.

Thus, the Coal and Iron became involved in a complex, multi-faceted rationalization strategy designed to rationalize existing productive facilities, close uneconomic mines, and begin the development of its property in a more orderly fashion, by means of judicious investment in new mining improvements. Over time, additional aspects of the strategy began to emerge. These included the management of existing and new leases on its properties, takeover of improvements after the expiration of long-term leases, further selective abandonment of mines unsuited to deeper working, capacity management and idling of mines to meet the constraints of cartel operation. The size of the estate and the large number of mines involved compared to the Northern Field companies, gave the strategy considerable potential flexibility. At the same time, as a management undertaking it was very substantial, given the range of mining and geological risks that they faced. These were ever-present, but highly unpredictable in their temporal and geographical occurrence.

Unfortunately no regular commentary on the strategic decision-making of the Coal and Iron has survived, equivalent to the invaluable correspondence between Storrs and Sloan, although some of the engineering letter books are still extant for the period.[14] Had such correspondence survived, it would also have been very difficult to deal with, because of the large number of mines and the details involved. It is perhaps appropriate therefore to adopt a more aggregate, statistical approach to evaluation of the strategies of the Coal and Iron. This will utilize data from the Mine Inspectors' Reports and other sources, in an attempt to chart the major outlines of the Company's activities, as they unfolded across both the Southern and the Western Middle Fields.

BACKGROUND AND OVERVIEW OF MAJOR DEVELOPMENTS 1872–1901

Initially, it is appropriate to examine briefly the characteristics of the Schuylkill anthracite industry, in the period immediately prior to the Reading's coal land purchases. This will then serve as the basis for examination of the consolidation and retrenchment that took place in the region over the ensuing decades, largely, but not exclusively, as a result of the policies and strategies of the Coal and Iron. No attempt will be made to reiterate any detail of the earlier history of mining development in the Region, as this has been examined both quantitatively and qualitatively by other authors, and hence does not need to be repeated here.[15]

In 1860, Schuylkill County was the undisputed leader in the coal trade, boasting, according to the census, 95 collieries, capitalized at $5.87million, which produced approximately 3.5 million tons of coal. Four other counties, namely, Carbon, Columbia, Dauphin, and Northumberland, also overlapped portions of either the Southern or the Western Middle Field, but could only muster 31 collieries between them, with a total capitalization of $2.38 million, and an output of about 1.44 million tons. By way of comparison, the figures for Luzerne County, which had yet to be divided with Lackawanna County, were about 50 mines, capital invested of approximately $5.9 million and tonnage of 2.2 million, although it is likely that the capital investment figure is an underestimate.[16]

At the time, the industry was still heavily concentrated in the Southern Field, with more limited development of the Western Middle Field, except in its more westerly areas around Shamokin and Mount Carmel. During the Civil War years, there was substantial additional development in the Western Middle Field, on the Girard Estate Coal lands and around the growing settlements of Shenandoah and Mahanoy City in the eastern part of the Field. The various concentrations of mining activity can be seen in Figure 15.2, which is a version of Daddow and Bannan's famous map containing 1865 data, re-drawn onto a standard base map to facilitate comparison with later years. The main concentrations of mines in the Southern Field also stand out clearly. These centred on Pottsville and extended westwards to Tremont and eastwards to Tamaqua. Yearley's chronicling of the industry during the period has emphasized the relatively rapid turnover of mining operations, so care should be taken in assuming the locations of mining activity were set in stone. The point should also be borne in mind in interpreting Figure 15.3, which presents a different view of the status of the Schuylkill coal industry, just as the Reading's coal land acquisition phase swung into high gear. The graph is constructed from analysis of data on years in operation of Schuylkill mines for 131 of the mines in the region in 1872, in both Southern and Western Middle fields, but excluding most of the Shamokin district. Interestingly, this shows what might be termed 'echoes' of periods of speculative expansion in the industry, during earlier business cycle upswings in the 1830s, 1840s and 1850s. Most of these early mines were in the Southern Field. The peaks in the frequency of new mine starts during the Civil War and at the end of the 1860s are very noticeable. Many of this later group of mines were in the Western Middle Field.

Hence, at the beginning of the 1870s the Schuylkill industry could almost be characterized as having two distinct components. The first was a group of older 'surviving' mines in the Southern Field. The second group of relatively new enterprises was concentrated in the eastern part of the Western Middle Field. It is unfortunate that the data are not available for the Shamokin district at the other end of the same field, as this would probably have presented a more mixed group of both newer developments and older established enterprises.

In addition to having older and potentially less efficient mines, the beginnings of other related problems were already becoming apparent in the Southern Field, as early as 1860. Simple arithmetic on the summary census figures presented above indicates immediately that the average capitalization of Schuylkill County mines, in the days before the Western Middle Field was well-developed, was significantly below that of competitors in the Northern Field, and slightly less than in the other neighboring counties.

In an attempt to pursue the question of the scale of mining activity in different regions and relative changes over time, a comparison has been made between the Northern Field or Wyoming Region and the Schuylkill Region between 1875 and 1901. The results are given in Table 15.2. Average values are computed by ignoring all small operations producing less than 10,000 tons per annum, since the number of these was highly variable and many were producing coal for local consumption only.

The slightly uneven time coverage is necessitated by the absence of good data for the Northern Field before 1875, although it should be noted that the Schuylkill data for 1875 is itself rather partial, since tonnages were not recorded for all mines. The year 1901 was chosen rather than 1900, since the output figures for the latter year were greatly affected by the coal strike at that time.

The figures in the table tend to confirm the suggestion that Schuylkill mines were significantly smaller in the early period than those in the Wyoming Region, although the gap was beginning to close by 1880. A decade later, the tables were turned, because Schuylkill mines had doubled in average size, while Wyoming capacity had only increased by a limited amount. In the following few years it was time for the latter region to expand capacity and regain its superiority by a narrow margin at the turn of the century. These figures point to the 1880s as a period when great strides were made in competitiveness within the Schuylkill Region and consequently this period will merit closer investigation below.

Two further questions arise from examination of the aggregate figures in Table 15.2. The first of these is whether the developments in the Schuylkill Region were equally spread across both the Southern and the Western Middle Fields. The second is whether the expansion in average colliery size was

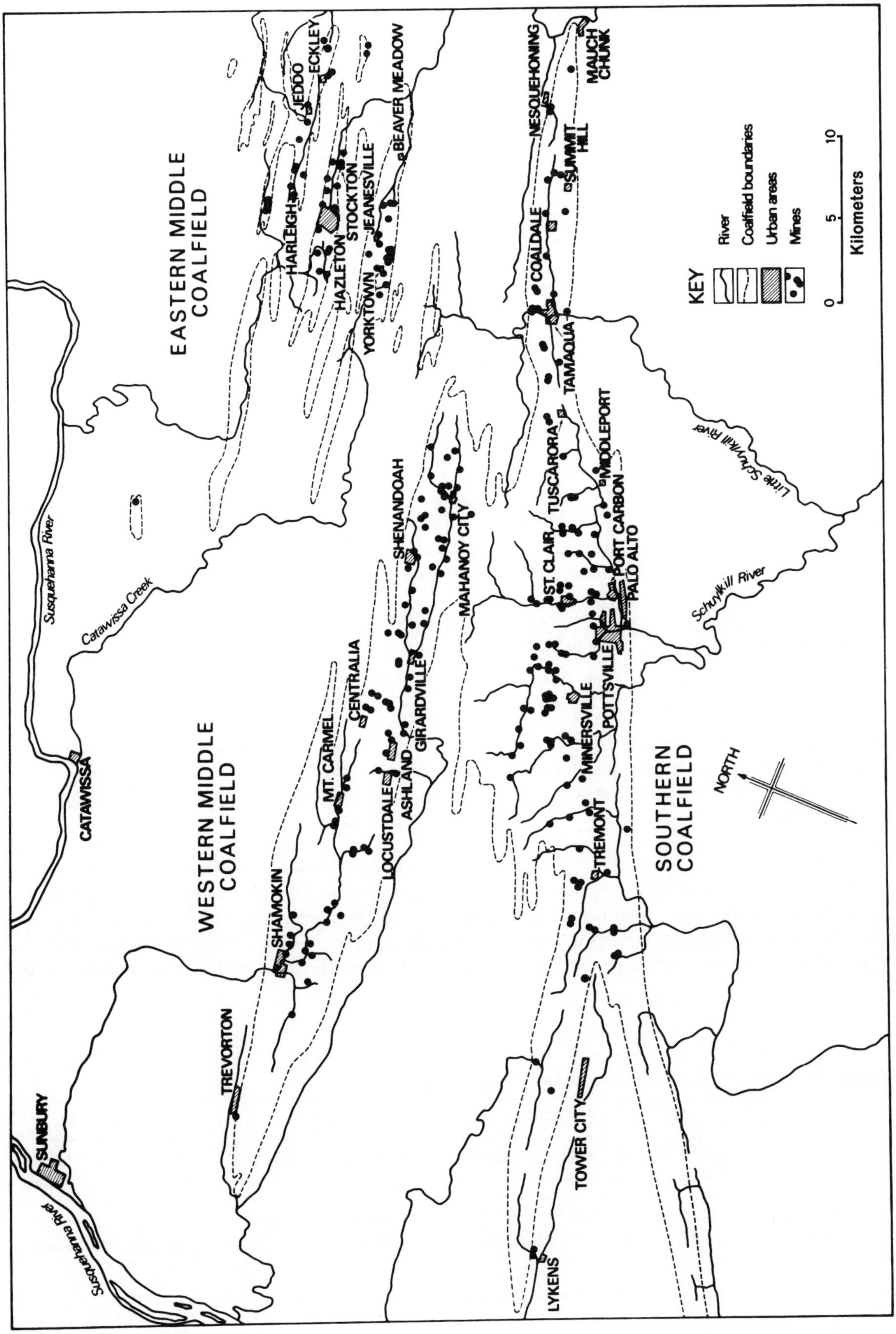

FIGURE 15.2
Map of the Distribution of Mining Activity in the Schuylkill and Lehigh Regions 1865.

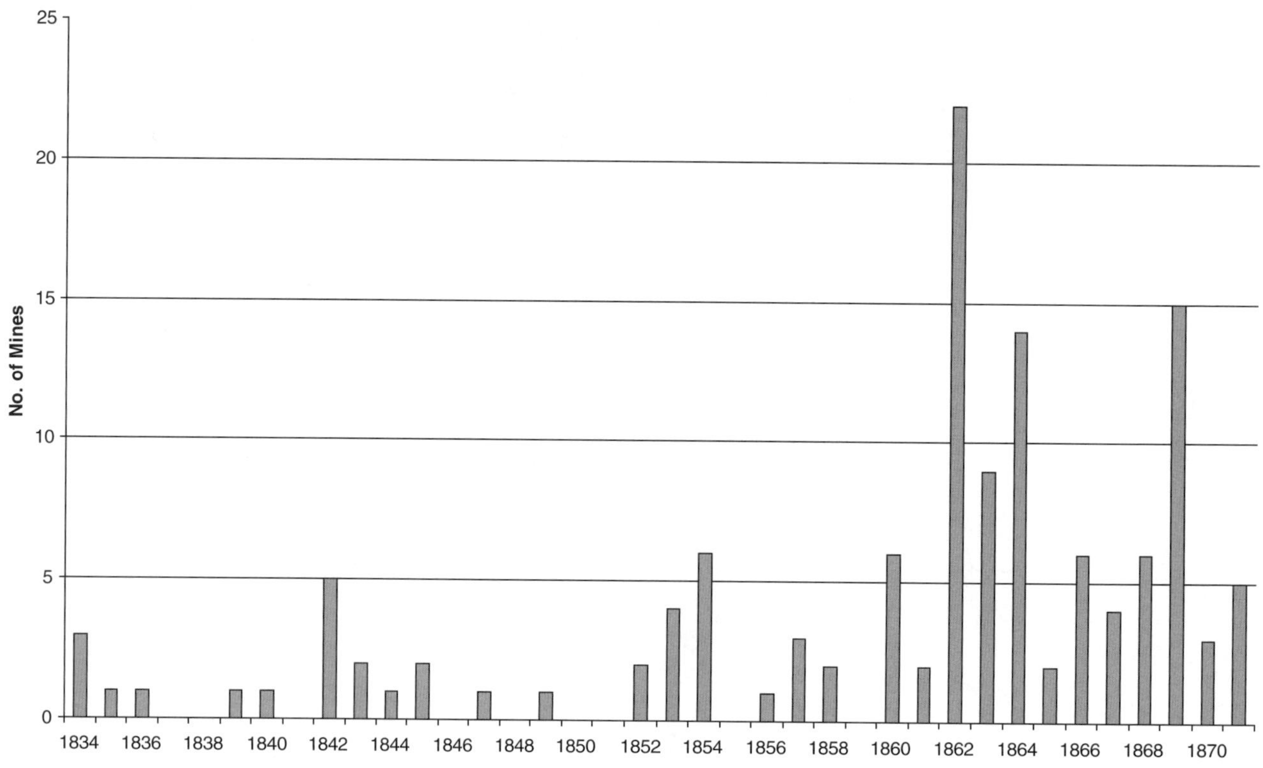

FIGURE 15.3
Start Dates of Early Schuylkill Mines Surviving to 1872.

attributable to the actions of the Coal and Iron, or to other companies still involved in the Schuylkill coal trade. Evidence that bears on these questions can be found in Tables 15.3 and 15.4.

The figures in Table 15.3 are not directly comparable to those in the previous table, because they refer to tons shipped rather than mined, but the 10,000-ton threshold has been applied as before. Again the picture is immediately clear. Expansion of capacity was concentrated in the Western Middle Field until the 1890s, before the Southern Field experienced a catch-up phase. It should be noted, however, that 61 mines make up the Western Middle Field average in 1901, but only 23 in the Southern Field. This indicates that a substantial thinning out of Southern Field operations had taken place, to concentrate on a few large efficient mines that could rival those of the Western Middle Field. Confirmation that the process had been underway for

TABLE 15.2
Average Production Per Mine, Wyoming and Schuylkill Regions 1875–1901.[17]

	1875		1880		1890		1901	
Region	Average Tonnage	No. of Mines	Average Tonnage	No. of Mines	Average Tonnage	No. of Mines	Average Tonnage	No. of Mines
Wyoming	88,354	127	93, 669	125	128,308	163	201,245	165
Schuylkill	57,470	60	76,311*	106	149,027	72	186,086	109

*All tonnages are tons mined rather than shipped, except the asterisked 1880 figure. Tons mined usually exceeded tons shipped by at least 5–10 percent.

TABLE 15.3
Estimated Average Shipments by Mine in the Southern and Western Middle Fields 1870–1901.[18]

Field	1870	1880	1890	1901
Southern	39,266	66,530	97,504	188,718
Western Middle	45,358	79,185	135,506	189,751

TABLE 15.4
Estimated Average Shipments by Mine for the Coal and Iron (PRCI) and Other Operators from the Schuylkill Region 1880–1901.[19]

	1880	1890	1901
PRCI	74,124	140,561	195,984
Non-PRCI	77,923	110,332	145,205

a number of years can be seen from Figure 15.4. This shows the distribution of mines in 1886 on the same base map as used for the 1865 data. The reduction in the level of activity in the Southern Field is already clearly apparent from a comparison with the earlier Figure 15.2.

Although the Coal and Iron was very much the dominant player in the Schuylkill Region after 1875, it was not in control of all mining activity. In the Western Middle Field, in particular, a number of other companies continued to operate in the Shamokin District, including Pennsylvania Railroad subsidiaries such as the Mineral Railroad and Mining Company. The Lehigh Valley also controlled sizeable blocks of territory and made large shipments from the Packer collieries on the Girard Estate, from the 1880s onwards. It is therefore appropriate to examine whether the Coal and Iron was largely responsible for the increasing size and concentration of mining activity, or whether the other companies played an equally large part. Table 15.4 shows a breakdown of shipments from the region based on whether or not the Coal and Iron was operating the mine, although a number of the mines not operated by them were leased properties on their lands.

Again, the nature of the changes can be seen very clearly in the figures. In 1880 the Coal and Iron was operating a set of mines that were below the average size, even for the Schuylkill Region. Given the rationalization and development that had already taken place by this date, it would seem that the company had acquired many of the smallest and probably least efficient mines in the entire coal regions between 1871 and 1875. This meant it had a very substantial task to improve profitability in the years that followed. During the 1880s, this rationalization activity finally began to pay dividends, driven by developments concentrated in the Western Middle Field. Additional investment in the Southern Field after 1890 allowed the trend to continue, albeit at a slower rate.

The different rationalization strategies, whose effects varied over time and by mining district, will now be examined in more detail. Initially, a more aggregate approach will be adopted, before moving down to the level of individual mines.

RATIONALIZATION STRATEGIES: MINE SUSPENSION AND CLOSURE

A good indication of the progressive effects of the Coal and Iron's mine closure policy can be seen in Table 15.5. Blank entries in the table indicate absence of reported data. In the initial stages of its coal land acquisition policy there were about 100 mines on its lands, nearly ¾ of which were leased to independent operators. The first step was to secure ownership of a greater proportion of these leased operations. The company more than doubled the number of mines under its direct control between 1872 and 1876, before the process of thinning out the actual number of mines themselves got underway. It should be noted that the total number of mines on company lands quoted in the table was always stated to exclude smaller operations. Unfortunately the basis for the exclusion was never given, although inspection of the Annual Report tonnage tables suggests the cut-off was at approximately 25,000 tons of output. However, it also appears to have varied somewhat between years. Hence the actual number of operating mines of all sizes on the company's lands was generally 10–20 percent more than the figures given in the table.[21]

Initially mines were suspended or abandoned temporarily, and the number of these can be seen to vary directly with business cycle conditions. Some 23 mines were placed in this category in 1878, a year both of depression and cartel constraint, but the number had dropped to less than ten at the

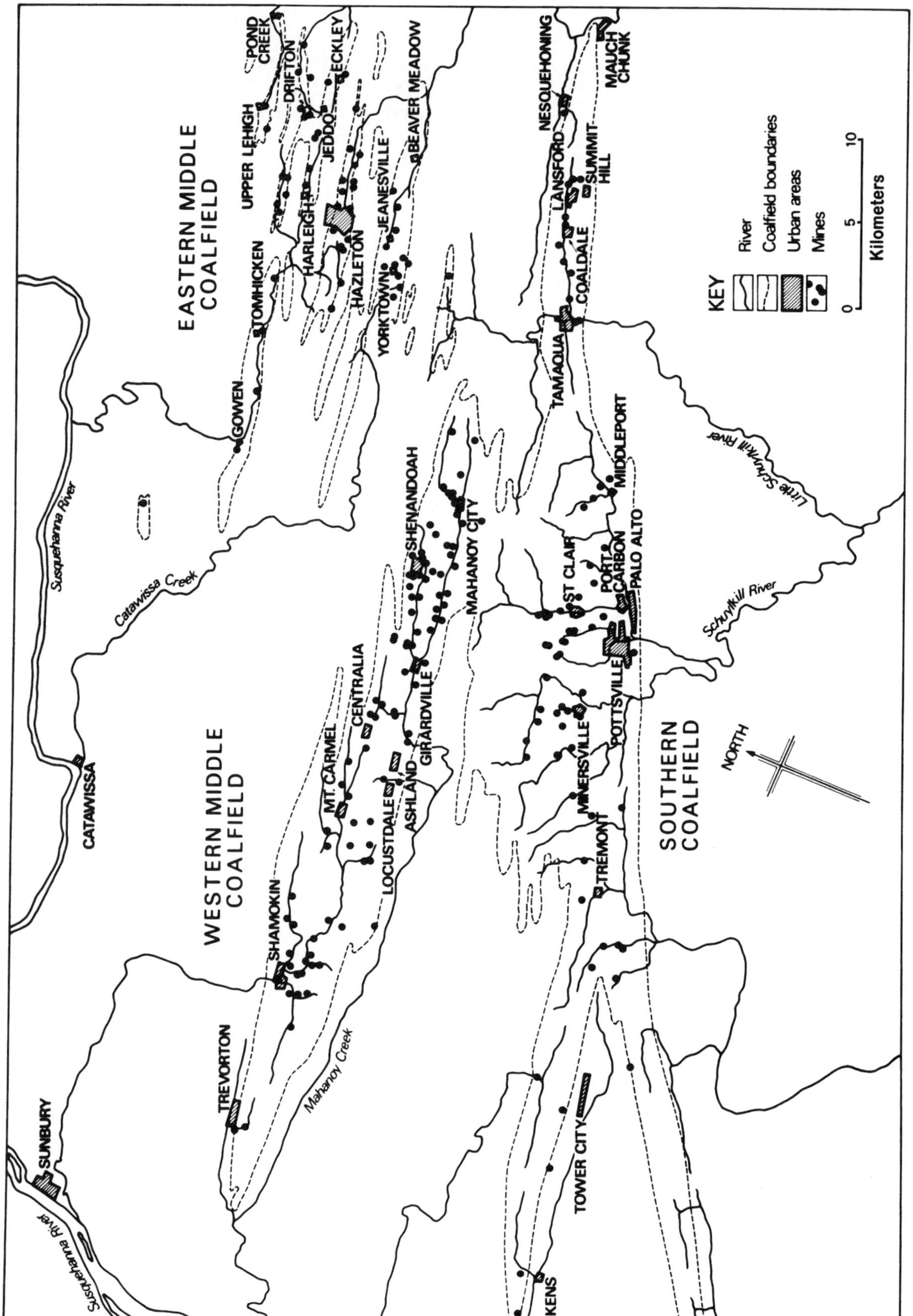

FIGURE 15.4
Map of the Distribution of Mining Activity in the Schuylkill and Lehigh Regions 1886.

TABLE 15.5

Mine Ownership and Leasing Statistics for the Coal and Iron Company 1872–1891.[20]

	Total No. of Mines on Company's Lands	Total Owned by Company	No. Owned and Under Development	No. Owned in Operation	No. Temporarily Suspended/ Abandoned	Leased by Company on Lands of Others	Tot. Leased to Others
1872	98	27					71
1873							
1874				37			45
1875							
1876	101	58	3	39	17		43
1877				48		7	38
1878	80			30	23	7	27
1879	79			41	13	8	25
1880	78			41	15	9	22
1881	78			41	16	9	21
1882	74			44	9	8	21
1883	74			47	8	12	19
1884	69			48	12	14	9
1885	66			35	21	11	10
1886	65			37	15	11	13
1887				44			
1888							
1889	56			36	8	13	14
1890	57			36	8	12	14
1891	63			40	10	11	13

peak of the business cycle in 1882–1883, before increasing again to 21 in 1885. It is interesting to note that throughout the period when data are available, the number of sizeable mines idle never dropped below 14–15 percent of the total capacity, and in 1885 it increased to more than 30 percent.

After an initial steep fall, the number of mines on the company's lands operated by lessees continued to decline during the 1870s, but stabilized in the early 1880s. When a number of leases expired at the beginning of 1884, the company took over the mines in question, which included the Kalmia, Lincoln, Suffolk, and Rausch Creek breakers.[22] One perhaps surprising feature of the data in the table is the number of mines leased by the company on lands that were not part of its estate. These included breakers such the Girard and the Hammond. During the 1880s, the number of mines in this category actually increased. It is difficult to understand why the company should have done this, when its own estate was in such urgent need of rationalization and development, to meet the interest charges upon its borrowed purchase money. Although the specific motive is not documented, it can only have been the customary reason, to prevent competitors securing the mines and diverting the coal traffic over their own lines.

Given the variation in output between different mines, simple counts of the numbers of breakers in operation only give a partial indication of the changes that were taking place under the company's management. Other aspects of these changes can be seen more clearly in Figures 15.5 to 15.7. In the first of these, Figure 15.5, the rapid fall in the proportion of total shipments attributable to tenant operators is immediately apparent and the sharp decline in 1884 matches with the statistics in Table 15.5. The effect of leasing breakers on lands owned by others can be seen in Figure 15.6. Although the overall proportion of tonnage derived from non-company lands declined progressively, as shipments from the company's own lands increased, the amount of coal obtained from this source remained at a significant level from the early 1880s until the turn of the century. Although Storrs in the Northern Field may have been justified in trying to let individual operators exhaust the coal on their leases, while the railroad retained the coal in its fee lands for future use, the same argument could scarcely hold for the Coal and Iron, since it

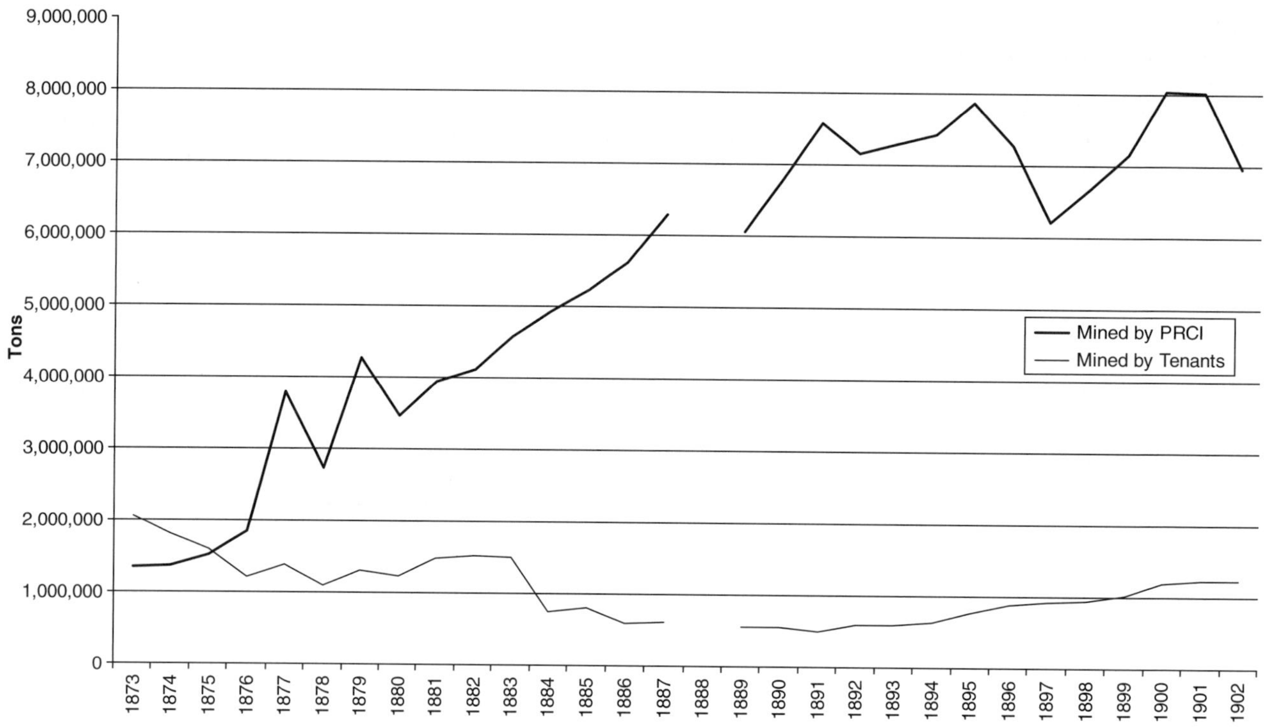

FIGURE 15.5
Tonnage Mined by Coal and Iron Company and by its Tenants 1873–1902.

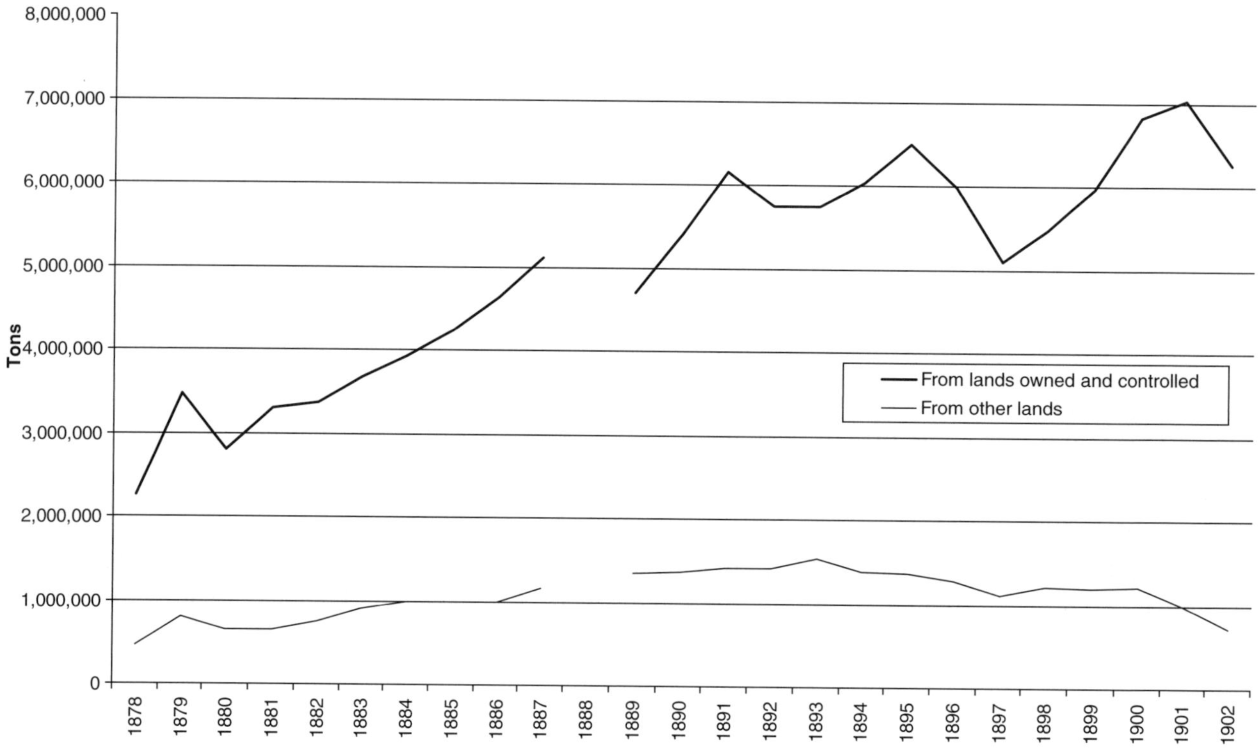

FIGURE 15.6
Coal and Iron Company: Tonnage Mined from Company's Lands and Other Lands 1878–1902.

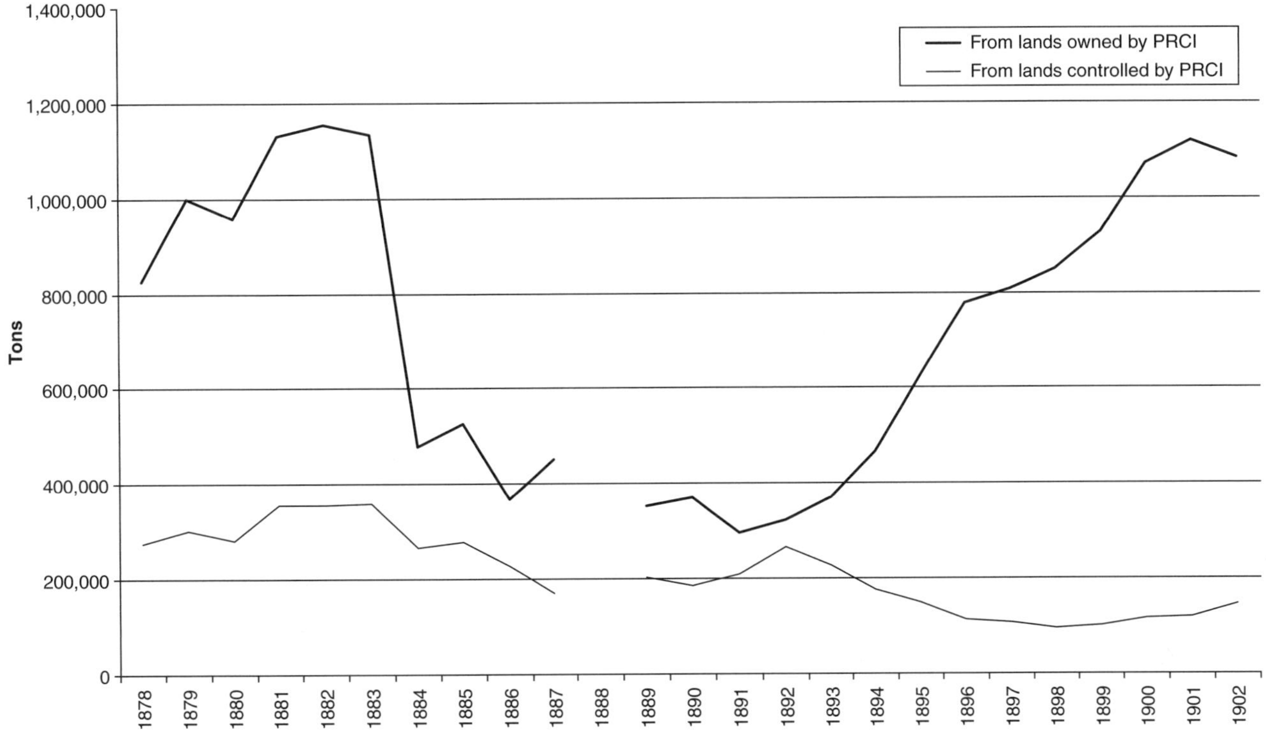

FIGURE 15.7
Tonnage Mined by Tenants from Coal and Iron Company's Lands 1878–1902.

controlled one-third of all remaining anthracite in the Coal Regions!

Conversely, it seems clear from Figure 15.7 that the Coal and Iron made several policy changes in relation to its own tenant operators. Although tenant operators had been reduced by nearly two-thirds in the early 1870s, the larger operators of this class that remained were rebuilding their contribution to the overall output from the Coal and Iron properties, until their numbers were halved again by lease expiry in 1884, as already mentioned. After the Railroad's third receivership in 1893, however, their tonnage contribution by mining from company lands returned progressively to levels last seen in the early 1880s. The output was sufficient to counterbalance the tonnage being mined by the Coal and Iron from lands not its own. Again, although the annual reports are silent on the matter, it is likely that in a time of financial stringency and relatively flat demand for anthracite, the Coal and Iron took the view that a modest expansion of output again by tenant operators was a low-cost, low-risk means of keeping up tonnages. The tenants paid royalty to the company on all the coal mined from its lands, they paid for the upkeep of their improvements, and they reduced the need for the company to make further investment in new improvements of its own, in the short and medium term.

Thus, in the first two decades of its existence up to 1891, the company had managed to reduce the number of mines on its lands by nearly 40 percent, while more than doubling the output from its lands and mines from 3.4 million tons in 1873 to 8 million tons in 1891. By the latter date it owned 51 mines, which included those on leased lands. This count had fallen still further by 1901 to only 39 mines in operation, indicating that the rationalization process continued until the end of the period. Figure 15.8 shows the location of the company's mines just before this time. In addition, several washeries reprocessing fine coal from culm banks were beginning to make a contribution to tonnage. By way of comparison, in that year 8 of the 13 mines in the Schuylkill Region shipping over 300,000 tons belonged to the Coal and Iron, but they faced stiff competition from the Northern Field which had 30 breakers shipping at this level.[23]

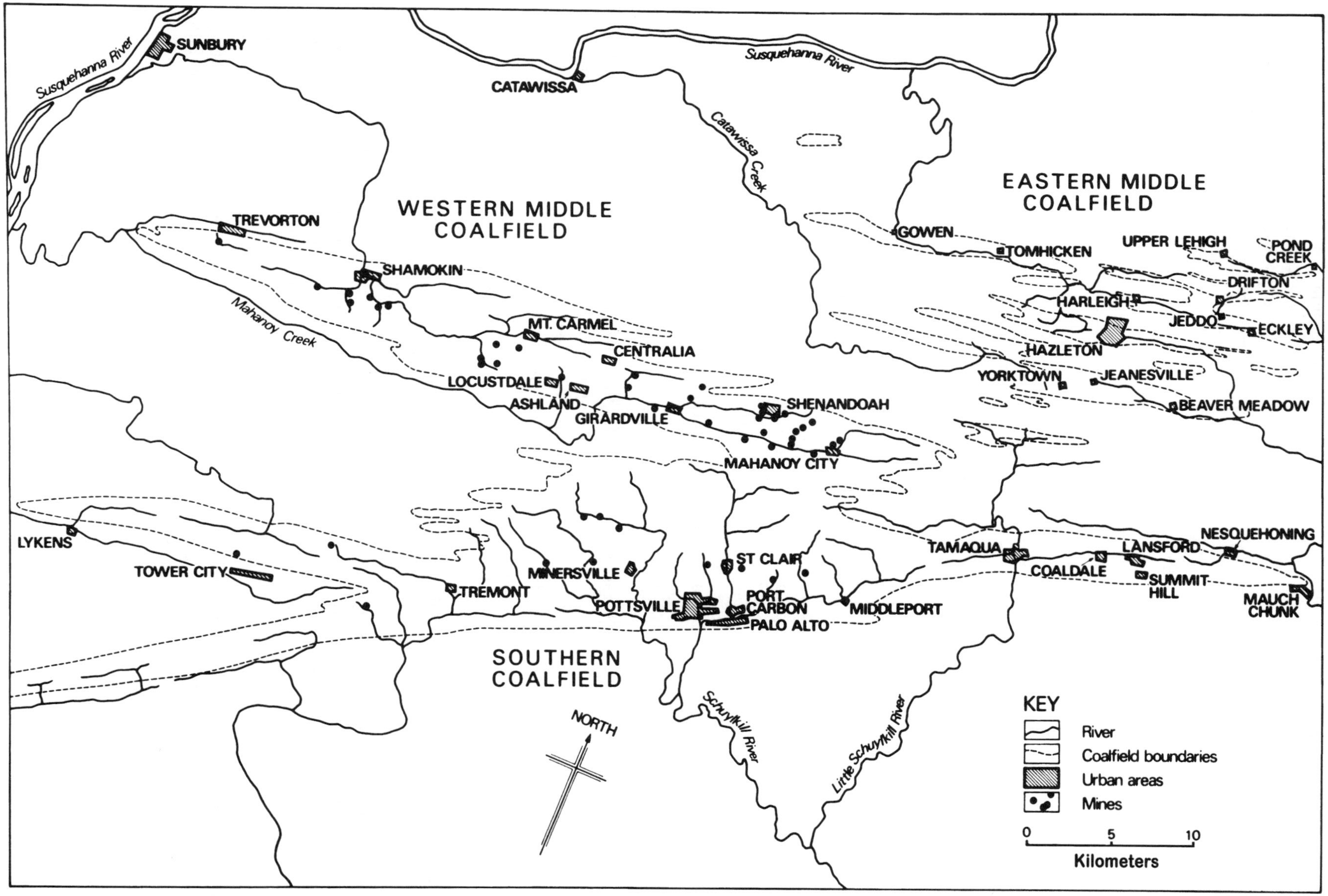

FIGURE 15.8
Location Map of Mines Operated by the Coal and Iron 1898.

RATIONALIZATION STRATEGIES: CHOICE OF INDIVIDUAL MINES FOR CLOSURE

The combination of abandonment, temporary abandonment, suspension of working and transfer of ownership or operation from tenants to the Coal and Iron Company, makes it a complex task to determine the exact operating status of many mines in any given year. This was especially true in the early period of 'shake-out' of uneconomic workings during the 1870s, when many of the changes were concentrated. Specific reasons for closure can rarely be documented, so recourse must be had to inference from statistical data or circumstantial evidence about the status of many operations prior to cessation of work.

As a first means of trying to identify the characteristics of mines that were acquired by the Coal and Iron, but rapidly shut down, data from the mine inspectors' reports for several different years were analyzed. More specifically, mines were examined that met several criteria. These were as follows: they were present in 1870 and were listed as being operated by the Coal and Iron in 1875, but had disappeared from the record by 1880. Further details about the mines that met these criteria were subsequently sought in the descriptive text of the reports during the 1870s, to provide additional contextual information, where it was available.

Several points emerge from this analysis. The first is that the mines subject to closure were all at the smaller end of the production spectrum, with outputs of about 30,000 tons or less. Secondly, a number of the breakers, such as the Manchester, the New Kirk and the Forestville were old, dating from the early 1840s, and had formerly been worked by tenants. However, the age criterion did not apply in all cases. Thirdly, as might be expected, in a number of cases problems relating to the operation of the mine had already been identified by the mine inspectors, before any closure decision by the Coal and Iron had been made. The types of problems identified included both adverse geological conditions and inadequate ventilation. For example, at the Kentucky mine near Tuscarora, the operators had suffered the serious misfortune of both gangways running into faulty rock dykes. Since they were not willing to make the necessary expenditures to drive through these faults, they were reduced to removing loose coal and robbing pillars in existing workings, at the time the Coal and Iron took over. So it is hardly surprising that this particular mine was shut down. The East Pine Knot colliery near St. Clair is perhaps a more interesting case. It dated from 1845 and the ventilation was badly arranged, such that the air currents circulated through old workings and accumulated noxious gases by the time they reached the miners, despite the deployment of a 25 horsepower fan. As a result, by 1872 the Mine Inspector was insisting that a number of improvements should be made. Doubtless to avoid such an expense at an old operation and fearing explosions from the gases being evolved in the mine, the entire operation was closed, despite a property valuation of $300,000.[24]

A second approach to the mine closure question is through examination of the Annual Reports of the Reading Railroad, which tabulate the outputs, both of the mines operated by the Coal and Iron and of those operated by tenants, from 1873 onwards. Analysis of these tabulations indicates that the brunt of closure was borne by the tenant operators, especially in the years 1873–1875 and 1877–1878. That said, the year 1874 also saw a number of new small tenant operations established, especially on the Valley Furnace tract. However, this short burst of unwarranted business optimism, in the aftermath of the 1873 panic, was not repeated in any subsequent year. Closure of mines that were in the hands of the Coal and Iron Company itself ran at only 2–3 a year in most years of the decade, but these closures were offset by a greater number of either new mine openings or takeover of operating mines formerly worked by tenants. After a more settled period of operation during the boom years of the early 1880s, a second 'shake-out' of tenants took place in 1883–1885. This involved mines such as the Staffordshire and the Hartford, which subsequently disappeared from the record, as well as others such as Lincoln and Suffolk, which were taken over by the Coal and Iron and continued in operation. These findings amplify the more aggregate statistics presented in Table 15.5.[25]

Further insight into the decision-making that lay behind some of these closures can be obtained from occasional references in other sources. These indicate both the complex issues facing the Coal and Iron Company under different mining and business conditions, and the need for a flexible response to changing circumstances. For example, a surviving letter written by then Associate Engineer J.S. Harris to his boss General Pleasants in December 1875, reveals the matters to be considered in relation to a possible acquisition of the Colket, Eureka, and West End breakers, located near Donaldson in the Southern Field. These small collieries were operated by tenants Owen, Eckel, Colket and Company on the Coal and Iron Company's lands. Harris asserted 'there is no prospect that the collieries can be made to pay without considerable expenditure of money—the openings are not in good condition and the work is badly designed for cheap production. I think the property of the lessees is worth actually nothing to us'.[26] Unfortunately, however, immediate economy of operation was not the only factor that had to be taken into account on this occasion. It emerged that there was a fire in the east workings (Mammoth Vein) at Colket and Harris was fearful that the lessees would not deal with it effectively, if they remained in possession of the property. Were this to be the case, a great deal of good coal could be consumed by the fire and the resulting underground damage could prevent the Coal and Iron from making further developments in the area for a number of years.[27] Unfortunately, further details of the subsequent negotiations between the lessees and the Coal and Iron Company have not been located. However, later information in the mine inspectors' reports and the annual reports of the Reading Railroad indicate that the concern over damage to the underground assets from the fire must have outweighed short-term profitability considerations, since the Coal and Iron duly took over the mines in question. Eureka was closed almost immediately, while West End was operated for a short period in 1877, after standing effectively idle in 1876, and was closed thereafter. Only Colket continued a modest level of production until 1884, when the last shipments from the colliery were made.[28]

A second item of correspondence from the Reading Railroad's papers reveals that significant concerns about the management of the estate of the Coal and Iron Company continued into the 1880s. Three specific charges were made. Firstly, that the interest cost on the $60 million borrowings used to finance the original acquisition of the lands, was not being factored into the reported costs of coal, when it was actually equivalent to 50 cents per ton produced. Secondly, as a result, the Company was underestimating its production costs and allowing sales margins to be reduced to the point of unprofitability. Thirdly, the Coal and Iron Company had been compelled to keep high-cost collieries in operation for the sake of tonnage for the railroad. The writer was so exercised about this state of affairs that he called for a:

> 'ruthless cleaning out of the whole nest of officials that have supinely allowed this magnificent Estate to sink deeper and deeper into Bankruptcy without an effort to stay the constant sagging of prices at all points'.[29]

From the perspective of this study, we would perhaps wish to be more forgiving of hard pressed company officials, bending to the dictates of the market during the depression of 1884–1885. Nevertheless, the above writer has materially assisted us by supporting his arguments with detailed figures on the production costs at the most expensive group of mines operated by the company on its own or leased lands. These figures are reproduced in Table 15.6. It is

TABLE 15.6
Production Costs at the Most Expensive Mines Operated by the Coal and Iron Company or Located on Its Lands, 1884.[30]

Mine	Tonnage	Cost of Mining (Royalty Added) ($/ton)
Lower Rausch Creek	3,751	9.323
Buck Ridge	6,740	9.206
Pottsville	16,075	4.712
Peerless	19,282	3.898
Greenback	93	3.496
Pine Forest	34,599	2.732
West Bear Ridge	45,995	2.647
Otto	53,700	2.539
Wadesville	30,098	2.463
Potts	16,340	2.456
Girard	59,199	2.331
Colket	33,685	2.306
Burnside	62,891	2.191
Glendower	41,269	2.145
Gilberton	63,716	2.105
Merriam	84,505	2.103
Middle Creek	57,064	2.103
Mine Hill Gap	67,456	1.934
Eagle	33,325	1.92
Richardson	68,390	1.902
Keystone	59,369	1.897
Forestville	36,804	1.865
Hammond	95,593	1.802
Phoenix Park No. 3	36,379	1.744
Eagle Hill	80,707	1.733
West Shenandoah	107,213	1.73
St. Nicholas	71,246	1.728
Locust Gap	67,210	1.706
Lincoln	62,529	1.701

apparent from the table that the extreme values (>$3 per ton) were very much a function of small output and the problem of not being able to spread fixed operating costs over a large product. However, from the remainder of the table there were clearly a substantial number of mines, producing significant quantities of very expensive coal. These high cost mines accounted for about 30 percent of output in 1884, offsetting much of the benefit of cheaper production at the company's other collieries.

Since the mines are individually identified, it is possible to determine whether these properties were subsequently targeted for closure or whether their poor economic performance was allowed to persist. In a number of cases, there is evidence that the former course of action was taken, at least temporarily. The case of Colket has already been mentioned above, and the Buck Ridge and Potts mines also ceased production by the end of 1884.[31] Similarly, the trade press reported at the same time that both Bear Ridge collieries, the Stanton at Mahanoy Plane and the Plank Ridge and Indian Ridge mines had all been identified as unprofitable and were to be shut down indefinitely. In the event, most of these mines were subsequently brought back into production at a later date. Lower Rausch Creek was closed in 1887.[32]

Clearly it was not possible to close all uneconomic mines at a stroke, since the company would then have had a shortage of output capacity. This in turn would have reduced its percentage market share in the cartel. Allowance also had to be made for unexpected accidents that could reduce output at individual collieries, sometimes for extended periods. Good examples of this were the fires at the Brookside and Mount Carmel mines in 1885 and 1886. In the first case, the Brookside breaker burned, but was rebuilt with sufficient speed, that after three months low level shipments could begin again.[33] A similar incident in June 1886 at the Mount Carmel breaker was particularly galling, because not only did the rebuilding cost have to be met, but in the process, the output of 120,000 tons of 'cheap coal' was lost to the company.[34] However, in both cases the opportunity was taken to rebuild the breakers at larger capacity, restrictions on throughput having been a particular problem at Mount Carmel before the fire.[35]

Thus, a number of expensive mines had to be kept in operation for an extended period. In some cases, little improvement in operating performance was seen over a number of years. For example, the Otto mine in the Southern Field was still reported as being unprofitable in 1890. In the same year, a similar assessment was made of the Bear Ridge collieries, which had been restarted in the latter part of the 1880s. Although the improvements were in good condition and the coal first class, the high royalty payments, in excess of 35 cents per ton, made it extremely difficult to generate any mining profit from these operations. At Middle Creek, uneconomic operation was a function of geological conditions, since nearly all the gangways had encountered faults of various kinds. Nevertheless, the company persisted at this location, because of the substantial remaining body of good Mammoth Vein coal.[36] The assessment of the Reliance colliery was different again. It was described as 'a profitable colliery when in operation, and an expensive one when idle'. This was all the more problematic, since it had apparently suffered from the prevailing dullness in the coal trade and had only managed 24¼ days work in the first half of 1890.[37]

In summary, it is apparent that the Coal and Iron enjoyed only partial success in closing down uneconomic mines under its own control, during the first two decades of the company's existence. The sharp reduction in the number of its tenants during the 1870s depression reduced its ability to increase shipments rapidly. This coupled with the need to keep up output in the cartel during the boom years of the early 1880s, meant that existing mining improvements had to be utilized on a continuing basis. Building new and more efficient works required a good deal of time and even more capital. The latter was in particularly short supply during the successive railroad receiverships of the 1880s. As the 1888 Reading Annual Report observed:

> 'During the long period of the Receivership, the Receivers were necessarily limited in their expenditure of moneys (sic), under the order of the court, to such sums as were absolutely required to preserve the property in the condition in which they found it. Not only during this period, but for many years previous, the property had not received the expenditures necessary to keep it in a condition to transact the business in the most economical manner, and to compete on equal terms with its competitors'.[38]

In consequence, the Coal and Iron had to make do for many years with essential maintenance and repairs, rebuilding breakers or re-organizing production only where a clear case for development or rationalization of operations could be made. Plans to open new mines in the manner that had characterized the first few expansive years of the Coal and Iron's operation, before the collapse of the railroad's finances, were not revived until 1888 after the second railroad receivership had ended. At this time a decision was taken to add four new collieries over a two-year period, three in the Southern and one in the Western Middle Field. These new mines would eventually provide a total increment of 900,000 tons of annual productive capacity. In the interim, the existing policy of developing collieries then in operation to their fullest extent, was to be 'rigidly adhered to.'[39]

Given the magnitude of the task involved in keeping up such a large coal estate, when funds were in such short supply for so long, it is appropriate now to assess the extent to which the Coal and Iron was able to succeed in deploying rationalization strategies of a less drastic kind than wholesale mine closure.

RATIONALIZATION STRATEGIES: PRODUCTION CONSOLIDATION AND BREAKER ENLARGEMENT

The idea of taking output from several mine openings and shipping it to a single breaker for processing was not a new one in the 1880s, since it had long been practiced by companies such as the Pennsylvania Coal Company and the Susquehanna Coal Company. However, as a rationalization and cost-cutting measure, abandonment of smaller breakers and transfer of output to new and larger breakers, was not perhaps as widely adopted as a management strategy in the latter part of the 19th century, as might have been expected. Part of the reason for this was related to the operation of the cartel. In the early years, the methods employed, which included suspensions for longer or shorter intervals, had not facilitated rationalization. By 1885, however, the adoption of a percentage allotment system enabled the Coal and Iron to decrease the number of working collieries and increase the working time at those kept in operation, thereby reducing fixed costs of pumping and ventilation.[40]

The best example of this rationalization process can be seen in the Shamokin Region at the western end of the Western Middle Field. The Coal and Iron had owned and controlled lands in this area since the early 1870s, but several of the mines on its lands, such as Henry Clay, Big Mountain, and Sterling, continued to be operated by lessees, such as Langdon and Company at Henry Clay, until the 1880s. After the latter mine was acquired by the Coal and Iron in 1885, it became the focus for an attempt to consolidatle production at a new large capacity breaker, taking coal not only from its own workings, but others in addition.[41] The first of these was the Peerless mine and breaker, situated on leased lands, which had been acquired in 1883.[42] At the same time as the new Henry Clay breaker was being built, the Sterling and Big Mountain collieries were also acquired in 1886. This meant that the new breaker could prepare the coal from four established operations. The impact of these changes on the output of the Henry Clay colliery can be seen clearly in Figure 15.9.

Under the control of lessees, Henry Clay had been able to produce a moderate output, except in the cartel year of 1878. Although shipments at the beginning of the 1880s were approximately double those of a decade previously, they were not sustained, and output fell away sharply in the final years of the lessees' period of operation. This compares with a much more stable pattern of output at the Luke Fidler breaker. Once the new Henry Clay breaker was completed and the coal from the four different mines began to reach it, production soared to nearly 260,000 tons in 1887, before peaking at 460,000 tons two years later. Unfortunately this level of output was only maintained for a limited span of years before a progressive decline in shipments set in from the mid 1890s onwards.

The managers of the Coal and Iron set out their intentions for the new Henry Clay breaker very clearly in the 1886 Annual Report:

> 'It is anticipated that from this concentration of outside work and expenses there will result a better and more uniform preparation of the coal and a decrease in the cost per ton'.[43]

It is interesting to note from this quotation that quality of product, which would enhance selling price, was just as much a consideration as cost reduction. The latter was certainly a factor for the output from the Peerless mine, which had ranked as one of the most expensive per ton, in the list given in Table 15.6. Unfortunately a detailed breakdown of costs per ton at each mine is not available for a sequence of years, to allow the effects of this rationalization to be charted closely in dollar terms. However, a second measure of overall productivity, tons shipped per man per day of breaker operation, can be calculated for most years after 1878, based on

FIGURE 15.9
Coal and Iron Company: Shipments from Henry Clay Mines 1870–1902.

data given in the mine inspectors' reports. This productivity measure is shown in Figure 15.10. The graph presents a rather erratic picture in productivity terms. It is apparent that productivity declined during the final years of the leasehold operation in the early 1880s, mirroring the decline in shipments. Similarly, it required several years for the efficiency benefits of the new 1886 breaker to become apparent and even these could only be partially sustained. It should be remembered that the rather crude measure employed encompasses both inside and outside workers. Thus an improvement in breaker productivity, which only affected outside labor, would only have a partial effect on overall productivity.

Examination of a single mine is obviously insufficient to gain a clear picture of the effects of rationalization policies across the Coal and Iron Estate and further investigation is needed. From the Annual Reports, it is apparent that there was a first period of breaker refurbishment in 1882, constrained by the tight financial controls imposed by the receivers. This was followed by a second, similar phase in 1886–1887, which developed into a major breaker-rebuilding program after the second receivership was ended. This program ran in tandem with further investment in new mine construction. A number of collieries benefited from the first phase, such as Bear Valley, Middle Creek, Brookside, and Glendower. Others, such as the Girard Mammoth and other breakers received investment in both the first and second phases. In 1888 a decision was made to rebuild the Glendower breaker to increase its capacity from 50 to 90 cars a day. Rebuilding work also took place at the North Franklin No. 1, Girard, and St. Nicholas breakers.[44]

The Glendower colliery was one of very few Southern Field mines to be in operation throughout the period. Its pattern of shipments and productivity changes are shown in Figures 15.11 and 15.12, together with comparative data for the Burnside colliery in the Shamokin Region, which had a somewhat similar pattern of productivity change. From Figure 15.11, it is apparent that both mines operated at a modest level in the 1870s, although the 1875 shipments from Burnside suggest this colliery was running well below capacity for an extended period until the mid-1880s. This may have been because it was a high cost operation, according to the list in Table 15.6. Between 1889 and 1891, output was doubled, despite no mention of major rebuilding work. This was largely as a result of high capacity utilization. The 1890 Annual Report declared the colliery to be

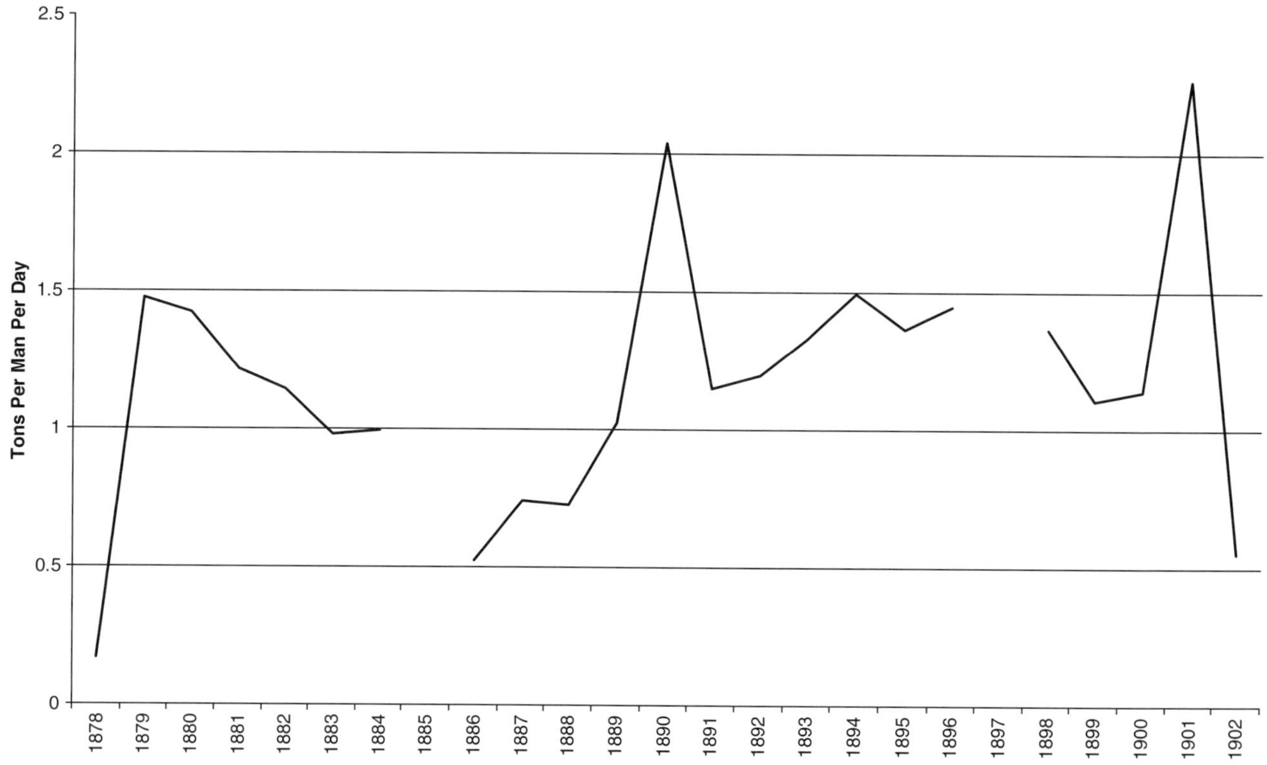

FIGURE 15.10
Coal and Iron Company: Productivity Changes Henry Clay Mines 1878–1902.

old, in bad repair and lacking modern improvements for the preparation of coal.[45] It was identified as a candidate for renewal/ refurbishment in the near term and judging by the limited number of days worked in 1892, the necessary remodelling took place at that time. By 1897 the breaker was able to produce almost the same amount of coal in 153 days that it had produced in 240 days in 1891.[46]

Repairs and rebuilding work at Glendower produced somewhat more modest results. Effective closure in 1882 for repairs allowed some increase in output to be obtained, although it required several years to achieve this. The same point applies to the 1888 breaker rebuilding. Sustained output in excess of 100,000 tons was not achieved until 1891 and shipments began to falter again in the late 1890s. Turning to the productivity changes associated with these shipment variations (Figure 15.12), several points emerge. The first of these is that both Burnside and Glendower suffered from low productivity of around one ton per man per day, until the late 1880s. Indeed, in 1884 they had exactly the same productivity level, while also having very similar and high production costs (Table 15.6). This suggests the crude productivity measure may be quite a good surrogate for more precise cost figures.

Secondly, the relationship between output and productivity is a complex one. In the case of Glendower, the 1882 refurbishment may have halted a productivity decline, but it did not produce any sustained productivity increase at prevailing production levels. At Burnside, although production doubled between 1881 and 1888, there was likewise no productivity improvement. Thereafter, up to 1891 a rapid further doubling of output resulted in a 33 percent productivity increase, before major breaker refurbishment took place. When subsequent breaker rebuilding is taken into account at both mines, it is apparent that the increased capacity at Glendower allowed shipments to double between 1889 and 1891, but productivity gains of about 50 percent were not fully realized until 1894, with some further improvement thereafter. At Burnside, the breaker improvements, post 1892, resulted in more rapid productivity gains than at Glendower, while shipment increases were much more limited. Taking a more aggregate and longer-term view, it is

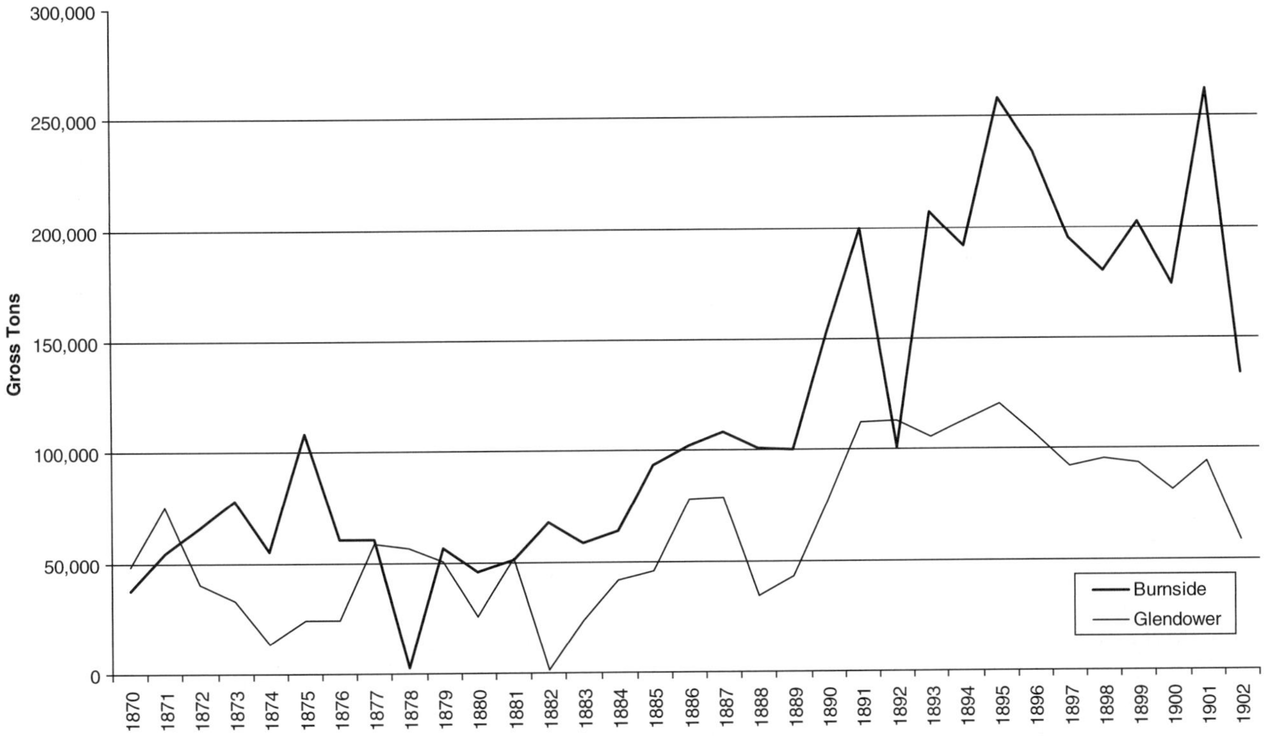

FIGURE 15.11
Coal and Iron Company: Mine Shipments 1870–1902 (Burnside and Glendower).

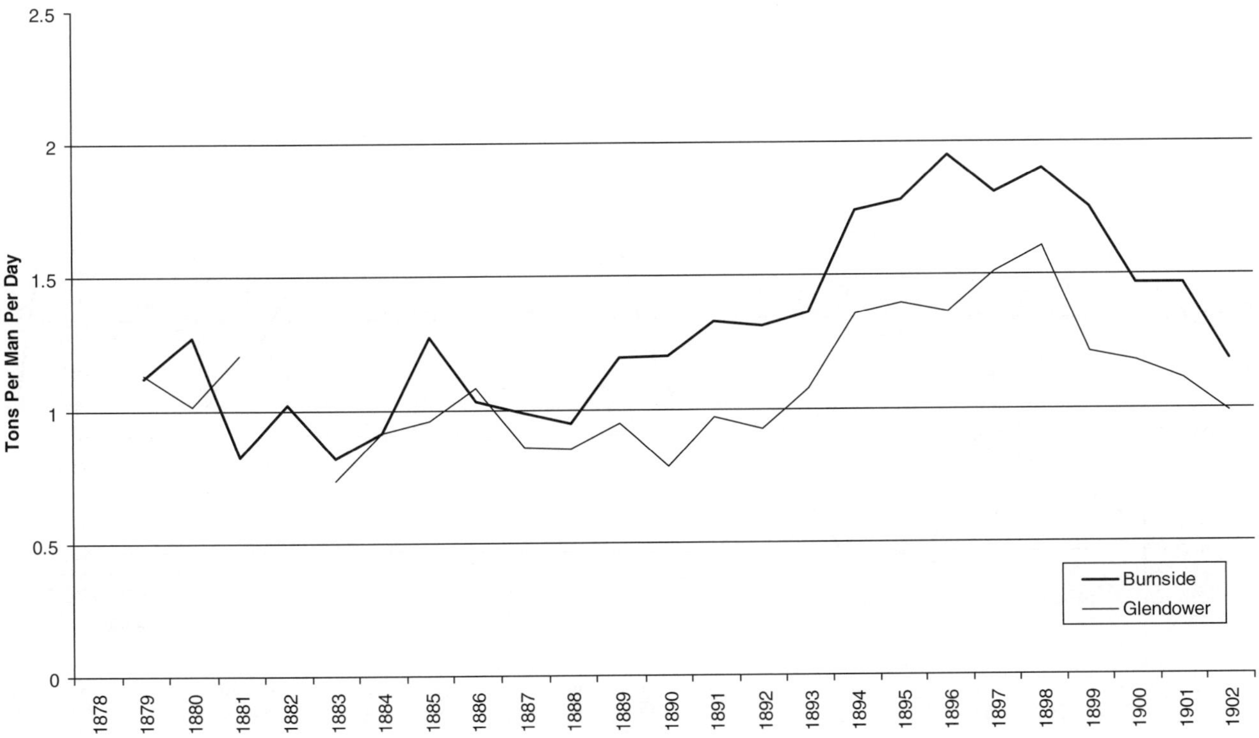

FIGURE 15.12
Coal and Iron Company: Mine Productivity Changes 1878–1902 (Burnside and Glendower).

interesting to note that, comparing the mid 1880s and 1890s, a doubling of shipments produced a 50 percent increase in productivity, while a 4–5 fold increase in shipments produced a near doubling of productivity. This suggests a possible linear relationship between the two variables. However, this is somewhat misleading by itself, because the overall effects are specifically attributable to major breaker-rebuilding episodes. Minor investment in repairs or breaker improvements was found to have much less of an effect or none at all.

In an attempt to determine whether the patterns of productivity change over time were replicated at other mines, the data for a further fifteen breakers were examined. Several of these were from other companies to provide a further element of comparison. While space precludes inclusion of all the resulting graphs, they indicated that several other mines did indeed have broadly similar patterns of productivity increase to Burnside and Glendower in the 1890s, albeit with a greater fluctuation about average levels at any given time. These included the Turkey Run and Tunnel Ridge breakers. The Suffolk breaker near Mahanoy City was also similar in the later period, but productivity had actually declined during the 1880s from more than 1.5 to just below 1 ton per man per day in 1889, before increasing steadily again to exceed 2 in the late 1890s.

A second group of mines showed no obvious productivity trend over time, but rather oscillations of varying productivity around a mean value of about 1.5 tons per man per day. The Mahanoy City and Shenandoah City breakers, whose shipments and productivity figures are shown in Figures 15.13 and 15.14, exemplify this group. In shipment terms, the two breakers contrast quite strongly. At the Mahanoy City, the increase in shipment levels between 1872 and 1898 was very limited, whereas at Shenandoah City there was a five-fold increase in shipments during the 1880s, although output levels fell back thereafter. Neither breaker was rebuilt prior to 1891, although Mahanoy City was refurbished in 1885.[47] Shenandoah City was originally a large triple breaker, but one side was not used for a number of years prior to 1884. At that time it was refurbished and put back into operation with a powerful new engine to drive all the machinery in the entire breaker.[48]

FIGURE 15.13
Coal and Iron Company: Shipments from Western Middle Field Mines 1870–1902 (Mahanoy City and Shenandoah City).

FIGURE 15.14
Coal and Iron Company: Productivity of Western Middle Field Mines (Mahanoy City and Shenandoah City).

In productivity terms (Figure 15.14), both mines support the earlier finding that refurbishment short of rebuilding had little impact on productivity. Further to this, in the case of the Mahanoy City colliery, fluctuations in output and productivity occurred at corresponding times. This implies that the mine was efficient and economical when running full, but rather more expensive to operate at lower output levels. The Shenandoah City productivity profile was doubtless something of a disappointment to managers at the time. It might have been reasonable to expect some economies of scale as output rose five-fold during the 1880s, but this appears not to have happened, other than temporarily and to a very limited degree. A further implication that might be drawn is that the 1870s design of triple-breaker conveyed no substantial advantage in productivity, as opposed to output terms, over the smaller breakers that were prevalent in other parts of the coalfield.

A third group of mines are represented by the Ellangowan breaker, whose shipments and productivity changes are shown in Figures 15.15 and 15.16, with the Gilberton breaker also shown by way of comparison. Ellangowan was developed by the Coal and Iron to be a major coal producer, on a quite different scale than the smaller operations the company had inherited. Although the first shipments were in 1871, the mine improvements were not completed until 1879. When fully operational the mine could initially ship 1000 tons per day.[49] Thereafter shipments grew rapidly over 2–3 years, in the early part of the 1880s, to reach an annual level of 300,000 tons. Output was pushed still higher in the latter part of the decade to exceed 400,000 tons, before falling back during the 1890s. The benefits of a new mine development whose breaker and inside workings were set up for a high level of shipments from the outset, can be seen in Figure 15.16. Ellangowan quickly reached a higher productivity level than most other breakers during the early 1880s and sustained it for longer, although there was a period of less productive operation when shipments declined in the early 1890s.

In comparison, the Gilberton breaker did not reach a productivity level of 2 tons per man per day until the end of the 1890s. This was a much smaller breaker than Ellangowan. It was situated on leased

FIGURE 15.15
Coal and Iron Company: Shipments from Western Middle Field Mines 1870–1902 (Ellangowan and Gilberton).

FIGURE 15.16
Coal and Iron Company: Productivity of Western Middle Field Mines 1880–1902 (Ellangowan and Gilberton).

lands and was not acquired by the Coal and Iron until 1879.[50] As Figure 15.16 shows, after an initial short period of quite effective operation, productivity slipped back to unsatisfactory levels until 1889. Thus, it is not surprising that the colliery also appears on the list of expensive mines in Table 15.6. 'Important improvements' were made at the colliery in 1887, although the Annual Report is not specific about exactly what these were.[51] They were clearly major, however, since shipments doubled in the period 1888–90, although productivity improvements were again somewhat short-lived. Relatively costly production may explain why shipments were restricted again during the depression of the 1890s.

ASSESSMENT OF COMPARATIVE PERFORMANCE OF THE COAL AND IRON COMPANY

From these examples and the other mines that they represent, it is clear from the outset that the Coal and Iron made considerable effort to improve the cost effectiveness of its mining operations. However, it was severely constrained by funding limitations after 1876, and especially during the receiverships of the 1880s. In the early years, the system of temporary suspensions of work employed by the cartel also limited its ability to rationalize inefficient operations. As might be expected, the evidence confirms that mines developed from the outset for large-scale operation were more productive than small collieries whose operations were subsequently expanded in an incremental manner. Similarly, minor breaker repairs and modifications allowed shipments to continue, but had limited, if any impact on productivity. Only substantial investment in major breaker overhauls or complete rebuilding to enlarge capacity had a substantial effect on productivity, and even this was not always sustained in the medium term. Hence the ultimate irony was that financial constraints greatly restricted the company from making just the kinds of major investments in new mining capacity that would have contributed to lower production costs and greater returns. Had the Estate been acquired over a longer period, using retained profits, rather than excessive borrowing, it might have been possible to have used spare borrowing capacity to finance a more substantial re-structuring of mining in the Schuylkill Region, and at an earlier date than was actually the case.

That said, the company was able to increase the productivity of many mines by the 1890s, or at least arrest any drift to lower productivity, while increasing shipment levels, closing smaller operations and concentrating output at cheaper collieries. It is also interesting to compare the experience of the Coal and Iron with that of other companies that continued operations during the period. Several examples are given in Figures 15.17 and and 15.18. The first two mines, the Cameron and the Luke Fidler, were long established operations in the Shamokin Region that had been taken over by the Mineral Railroad and Mining Company during the early 1870s. The latter company was a subsidiary of the Pennsylvania Railroad. As a major producer, the Cameron colliery provides a useful comparison with the larger mines of the Coal and Iron Company, while Luke Fidler was more comparable to the smaller breakers. At Cameron, a first phase of production expansion took place in the mid 1870s and a second phase at the beginning of the 1890s. It is interesting to note that production was sharply reduced for a time in the late 1880s, after the Pennsylvania Railroad joined the cartel with suitable encouragement from J.P. Morgan. Some expansion of operations took place at Luke Fidler by the early 1890s and this included sinking of a deep shaft, commenced in 1893.[52] Unfortunately a major fire broke out in the mine on October 8, 1894, that required more than 1 billion gallons of water to extinguish. This kept the mine out of production for an extended period.[53] The Excelsior breaker of the Excelsior Coal Company was an independent operation.

A striking feature of the productivity graphs (Figure 15.18) for Cameron and Luke Fidler is the marked reduction in productivity in the early 1890s. At Cameron this then stabilized at a low level, implying a high cost of production, while at Luke Fidler after a short-term improvement in 1886–1887, the position continued to deteriorate until the fire caused a temporary halt to operations. Examination of the Annual Reports of the Mineral Railroad and Mining Company confirms the adverse consequences of low productivity over an extended period. In 1884 and 1885, the company's net loss per ton was 13 cents and 11 cents per ton respectively. By 1894 this loss had widened to 95 cents per ton, although the figure included an extraordinary expenditure of $71,000 for pumping water out of Luke Fidler after the fire. A limited profit on operation was not restored until 1897.[54]

In contrast, the Excelsior Coal Company, despite an early downturn in productivity, was able to stabilize at a more reasonable level of about 1.5 tons per man per day until the early 1890s.

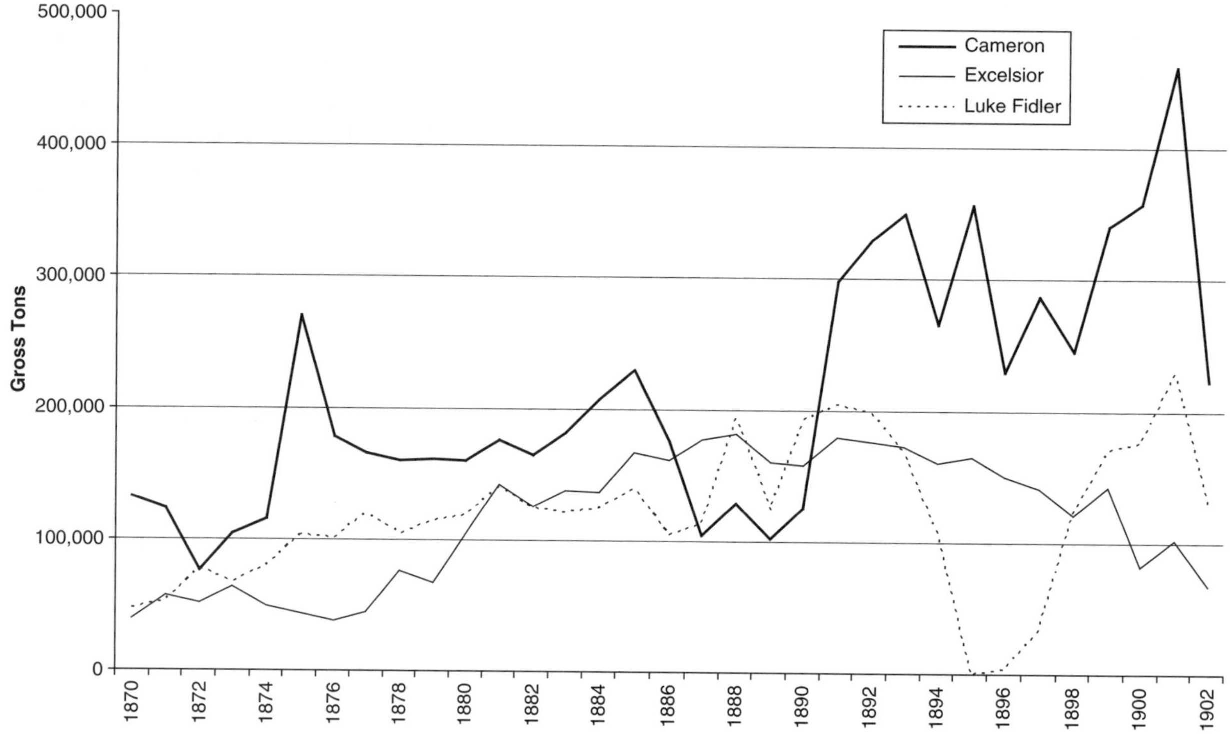

FIGURE 15.17
Non Coal and Iron Company Mines: Shipments from Shamokin Region 1870–1902.

Thereafter, it managed to achieve a series of productivity improvements, despite a general decline in shipments over the latter part of the decade. The findings both for this company and the Mineral Railroad and Mining Company, suggest two conclusions that can be drawn. Firstly, the Coal and Iron was not lagging behind all other railroad-backed companies in implementing productivity improvements. Indeed, compared to Luke Fidler and Cameron, its overall performance seems commendable by the 1890s. Secondly, despite this, it is also apparent that well-managed independents such as the Excelsior Coal Company were able to match the improvements achieved by larger companies and to do so at more modest shipment levels.

As a check on these findings, the productivity of all mines operating in the Southern and Western Middle Fields, excluding washeries and those classified as part of the Lehigh Region, was calculated and a rank ordering of the results was produced. These were divided into a series of productivity bands, both for the coalfields taken together, and for only those mines operated by the Coal and Iron. The results of this analysis are shown in Table 15.7. They enable the specific conclusions in the previous paragraph to be placed in a broader context.

Several key points emerge from the data in the table. The first is that only a limited proportion of the overall tonnage (15–16 percent) was produced in a very cost-effective way at a rate of more than two tons per man per day. For the Coal and Iron by itself, the figure was even lower, at 10 percent. Perhaps surprisingly, high productivity figures, in this year at least, are not primarily associated with high levels of average shipments. However, some care needs to be taken in interpreting the figures, since in the case of the Coal and Iron, the truly impressive productivity figure at Locust Gap of 4.6 tons per man per day, on shipments of 370,000 tons, is offset by low tonnages at the Hammond and West Shenandoah breakers, which also achieved high productivity levels.

Secondly, overall the dominant productivity category in tonnage terms was between 1.75 and 1.99 tons per man per day, although a substantial number of mines also fell into the two categories below this. By contrast, both numerically and in tonnage terms, the Coal and Iron mines were very

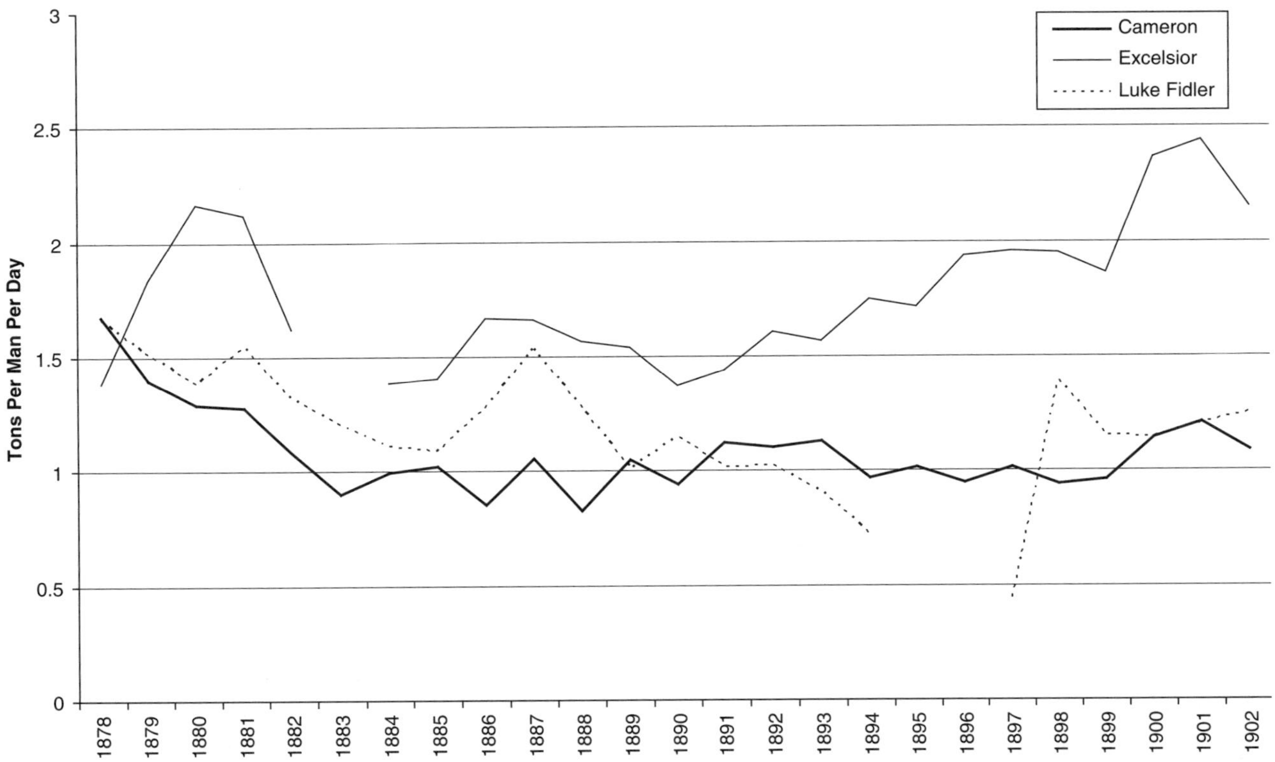

FIGURE 15.18
Non Coal and Iron Company Mines: Shamokin Region Productivity Changes 1878–1902.

much concentrated in these two lower categories, down to a figure of 1.25 tons per man per day. This was the case, despite their mines being substantially larger on average in these categories than for this segment of the industry as a whole. This contrast would be even sharper if only the non Coal and Iron Company mines were the basis for comparison, as opposed to the totals from both coalfields.

Thirdly, despite a preponderance of tonnage in the lower half of the productivity spectrum, the Coal and Iron had at least managed to minimize the proportion of its output that fell into the lowest two productivity bands, which had the highest production costs. Smaller mines also tended to congregate at these levels. This was most apparent across the coalfields as a whole in the lowest band,

TABLE 15.7
Comparative Mine Productivity in the Southern and Western Middle Fields 1901 (excluding Washeries).[55]

Productivity Category (tons/man/day)	Total No. of Mines	Total Tonnage	Average Tonnage	% Tonnage in Category	No. of Mines Operated by Coal & Iron	Total Coal & Iron Tonnage	Average Coal & Iron Tonnage	% Coal & Iron Tonnage in Category
>2.5	5	738,553	147,711	4.36	3	429,194	143,065	5.24
2.0–2.499	9	1,945,056	216,117	11.49	2	398,705	199,353	4.87
1.75–1.999	20	4,285,903	214,295	25.31	5	1,614,829	322,966	19.70
1.5–1.749	20	3,241,670	162,084	19.14	11	2,368,383	215,308	28.90
1.25–1.499	23	3,854,233	167,575	22.76	14	2,864,001	204,572	34.95
1.0–1.249	16	2,525,011	157,813	14.91	4	520,035	130,009	6.35
<1.0	5	342,016	68,403	2.02	0	–	–	–
Grand Total Tonnage		16,932,442				8,195,147		

and in the next-to-lowest for the Coal and Iron. However, it is also striking to note that across the productivity range of three bands from 1.0 to 1.75 tons per man per day, there does not appear to be any overall relationship with breaker size, at the level of the broader coalfield totals.

Fourthly, the bands also allow the examples of the Excelsior, Luke Fidler, and Cameron breakers to be set more clearly in context. There is no doubt that Excelsior was a high-flying performer, in productivity terms. Equally, the Cameron and Luke Fidler breakers fell into the next to lowest band in 1901. It is also interesting to note that the Williamstown and Short Mountain breakers, operated by another Pennsylvania Railroad subsidiary, were even lower down the rankings, although their high production costs tended to be offset by the premium prices which their coals usually attracted.

Finally, it must be concluded that although the Coal and Iron was not the worst performer in productivity terms by 1901, its cost-constrained efforts to rationalize its mining operations and the development of its estate had still not been sufficient to allow it to keep pace with a mixture of more nimble independent and railroad-backed operators in the Schuylkill Region. Since it shipped nearly half the region's tonnage this also meant that it continued to act as a drag on the overall cost structure of this part of the industry, even thirty years after its over-ambitious and debt-encumbered beginnings.

The Reading Railroad had not heeded the recommendations of the 1880 Harris Report to reduce the size of its estate,[56] choosing instead to soldier on through multiple financial reorganizations. By 1886, President Corbin was reported in the trade press to be advocating the leasing of the company's collieries to individual operators, on the grounds that they could produce coal more cheaply and the guaranteed royalty streams would fund interest payments on the Coal and Iron Company's bonds.[57] However, contrary arguments reiterated the old Schuylkill refrain that lessees developed properties for short-term gain to the long-term detriment of the larger estate. The proposal was put forward that large breakers were required to spread overhead costs across a large output, and that these should replace smaller ones as the latter become worn out.[58] Unfortunately, by the time the necessary capital was available to pursue such a policy, the profitable expansionary years of the 1880s had past and newer, more efficient methods of mining and preparing coal were needed, just to remain economically viable in the faltering markets that prevailed during the depression of the 1890s. It is also interesting to note that in 1893, after the panic and the onset of the Reading's third receivership, there were apparently strident calls from powerful interests within the company to revive Corbin's plan of leasing the mines to individual operators, but again they fell on deaf ears and the Coal and Iron clung on to its estate and mining operations.[59]

In his testimony before the Industrial Commission in 1901, Joseph S. Harris laid the blame for the relative retardation in the growth of the Schuylkill Region, not on poor management by the Reading, but on the difficulties inherent in mining in this part of the coalfields, as compared with the Wyoming Region. He attributed these difficulties firstly, to the greater depth of the seams and secondly, to the larger amount of waste and higher costs of mine drainage. More specifically, he asserted that when he had charge of the mines of the Lehigh Coal and Navigation at the eastern end of the Southern Field, three tons of coal and an estimated ten tons of water had to be raised from an average of 600 feet, for every ton actually shipped to market.

His final comment was as follows:

> 'The cheap mining in the Wyoming region and by the Lehigh Co. has been mostly done, and that is the reason why the Reading Co. can hold its own better than it could formerly'.[61]

While his optimism was no doubt well-founded, looking forward to the medium term, the actual situation in 1901 was less rosy than might be imagined. Examination of Table 15.8 and comparison with the earlier figures presented in Table 15.7, indicates that the competitive position of the Wyoming Region, from the productivity viewpoint, remained extremely strong. Both tables use the same productivity categories. It is immediately apparent that almost 80 percent of the tonnage shipped from the Wyoming Region had been produced at a rate of 1.75 tons per man per day or more. The equivalent figure for the Schuylkill Region as a whole was only 41 percent, while for the Coal and Iron itself, it was just short of 30 percent.

Such a productivity differential between major parts of the Coal Regions could hardly be expected

TABLE 15.8
Mine Productivity in the Northern Field 1901 (excluding Washeries).[60]

Productivity Category (tons/man/day)	Total No. of Mines	Total Tonnage	Average Tonnage	% Tonnage in Category
>2.5	19	4,268,755	224,671	15.28
2.0–2.499	45	9,477,100	210,602	33.93
1.75–1.999	39	8,578,621	219,965	30.71
1.5–1.749	17	2,584,336	152,020	9.25
1.25–1.499	11	2,071,533	188,321	7.42
1.0–1.249	3	373,511	124,504	1.34
<1.0	10	577,412	57,741	2.07
Grand Total Tonnage		27,931,268		

to disappear overnight, even if geological conditions were becoming less favorable as mining progressed in the Wyoming Region. Translated into increased profitability, such a productivity gap also gave the Northern Field companies a continuing ability to re-invest a greater proportion of earnings in improved equipment and breaker technology, that would enable the differential to be perpetuated for some time to come.

Given the extent to which the financial, geological and structural cards were so stacked against them for so long, it is perhaps appropriate to finish by recognizing the extraordinary skill, energy and persistence of the managers, engineers, and miners of the Coal and Iron. Despite numerous problems and setbacks, they had averted the collapse of their segment of the industry and rationalized production within the limitations of the constraints placed upon them. They had even restored a measure of competitiveness by the end of the period at their newer and more efficient breakers, though much more still remained to be done.

Chapter 16

REGIONAL DYNAMICS, DISEQUILIBRIUM TENDENCIES AND REGIONAL ECONOMIC DEVELOPMENT

INTRODUCTION

Previous chapters have examined in detail different facets of the evolving structure and operation of the anthracite coal industry, in the context of both the producing region and the railroad-mediated distribution region. The explanatory framework can perhaps be summarized in terms of a simple 'formula' of the following kind:

Regional Dynamics = Starting Conditions + National Economic Effects + Inter-Regional Competitive Effects + Company Performance + Interactions Between Companies.

Use of the plus sign is simply a representational convenience. There are obviously interactions between all the terms, not just between companies, and these effects are not merely additive in nature. The first and second terms are beyond the control of business decision-makers at any point in time, while the fourth measures the effectiveness of companies in utilizing the resources available to them and in responding appropriately to changing economic circumstances. The fifth, as we have seen, is of particular interest in relation to the anthracite industry, because of the early development of oligopolistic cartel structures to gain increasing control over anthracite coal supply. That control did not extend, however, to competing sources of bituminous coal production, whose influence was felt to a progressively greater degree in the last two decades of the 19th century. Effects of this kind, together with competition between individual Anthracite coal regions are encompassed by the third term in the formula.

All these effects and their associated adjustment processes were also very much time dependent. For example, as we have seen, it would be wrong to imagine the cartel as a fixed and unchanging organisation. High freight rates on relatively short haul anthracite shipments, as compared to long-haul bituminous shipments,[1] tempted occasional new entrants such as the New York, Susquehanna and Western and the New York, Ontario and Western. These companies injected vexing new elements of competition into the market for new tonnage contracts with individual operators. In the process, this led to a progressive reduction in the proportion of the final selling price for coal that found its way into the coffers of the shipping railroads. However, barriers to entry for new mining companies were much lower than for new railroads, at least until coal lands available for exploitation became very scarce. This scarcity developed in the Schuylkill Region during the late 1870s, but not until a decade later in the Northern Field. Under such circumstances, and in the light of the geographical constraints imposed by control of land tracts and means of access to shipping railroads, several unanticipated consequences, both of oligopolistic and competitive behavior began to emerge. These related particularly to the management of existing mining capacity and investment in new capacity. Superimposed on these in turn were speculative behavior in the market for coal lands and leases, linked to business-cycle changes, and the lengthy time lags inherent in bringing new productive capacity on stream, especially as shafts became deeper and breaker structures ever larger and more complex in construction.

Taken together, this wide range of interacting decision-making, investment, and productive processes constituted the complex driving mechanism for the economic development of the

Anthracite coal regions, although the levels of financial constraints governing the actions of major railroad players were also a function of operating success (or lack of it) in the wider anthracite distribution region.

To examine the resultant effects of these different processes, and the interaction between the anticipated and unanticipated consequences of actions taken by companies of differing sizes, it is necessary to change the scale of the analysis from that of the firm to that of entire coalfields or coal regions. Two key issues can then be examined. The first of these is the presence of trends or cycles in investment, output, employment, and capacity utilization at the regional level. The second is whether the system as a whole tended toward an equilibrium condition between market forces of supply and demand, or whether interactions within the industry tended toward disequilibrium conditions and the amplification of external fluctuations driven by changes in the economy at the national level. Analysis of these issues will shed new light on the long-standing questions from the literature on the Anthracite regions about the impact of overproduction and over-investment on the development of the industry. Finally, in the light of these findings, it will be appropriate to conclude with a brief assessment of the relative appropriateness of two different explanatory paradigms for the economic development of industrial regions like that of the Anthracite. The first of these is equilibrium-based, derived from neo-classical economic theory and generally favored in contemporary studies in American economic history. The other, based on adjustment theory and emphasizing disequilibrating tendencies, is less well-known, and, as noted in the introduction, prior to this study has not been deployed in historical or historical-geographical investigations of the present kind.

TRENDS AND CYCLES IN OUTPUT, EMPLOYMENT AND CAPACITY UTILIZATION

From a cursory examination of overall output trends by region, as already shown in Figure 4.3 the reader could be forgiven for assuming that cyclical behavior in the anthracite industry was of very limited importance, in contrast to long-term trends towards increasing output. At the same time, disparities in growth rates between the different coal regions became apparent during the period, the slowest growth being found in the Lehigh Region. Despite the long-term financial difficulties of the Coal and Iron Company, the Schuylkill Region was still able to more than double its output from about 7.5 million tons in 1880 to 16 million tons by 1901. However, it was out-paced by accelerating development in the Wyoming Region, or Northern Field, where output increased almost three-fold during the same period, from 11.4 million to about 30 million tons. As can be seen in Figure 16.1, employment growth in terms of head count, largely mirrored increases in output, rising from about 33,000 in 1880 to about 78,000 in 1901. This suggests, in a very broad sense that productivity improvements were very limited during the period. On closer inspection, between 1880 and 1893, both output and employment almost exactly doubled, suggesting no productivity gain at all during the interval. The corollary of this was that productivity improvements by 1901 were therefore concentrated in the period of the depression of the 1890s and its aftermath, when tight cost controls became the order of the day.

To consider mining productivity in this way is still unsatisfactory, however, because it neglects the all-important third variable of capacity utilization, as measured by the number of days worked by the coal breakers during the year. Put another way, an output of 100,000 tons from a mine and breaker working 200 days with a given workforce, reflected a lower productivity than the same output made by a workforce of the same size, at a breaker which only worked 150 days. Re-examining the observations just made on productivity in the Northern Field in the light of breaker days worked reveals two interesting points. Firstly, the conclusion about absence of productivity gain, 1880–1893, stands because the average number of breaker days worked by mines in the two years was fortuitously exactly the same at 185.6 days! Secondly, the conclusion about productivity improvements 1893–1901 is strengthened, because the average number of days worked by Northern Field mines in 1901 had fallen to only 170.

Incorporation of breaker days worked into the analysis suggests a further way of looking at overall trends, as can be seen in Figure 16.2. The graph shows the result of multiplying total employment at each mine by the number of days worked by the breaker in each case. Clearly this can only be regarded as an overall estimate, because employment levels could vary at each mine during the year, but only a single

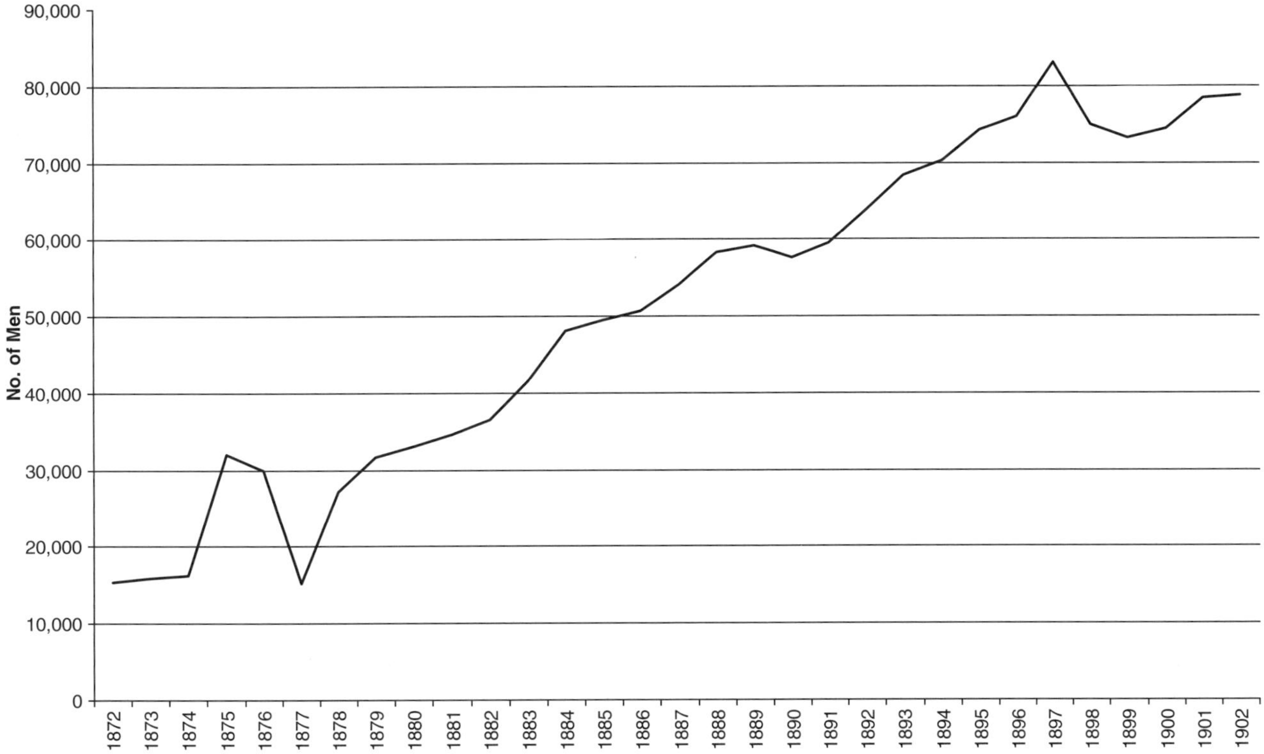

FIGURE 16.1
Northern Field: Estimated Employment in Coal Production 1872–1902.

employment figure is provided for each mine in the relevant mine inspector's report. The interesting aspect of this graph is that it reveals a somewhat different picture than the output figures referred to earlier. The adverse impact of the 1870s depression is very noticeable. This is followed by a decade-long trend of growth in labor input during the 1880s, punctuated only by a slight reverse in 1880 and a period of stasis during the depression of the mid 1880s. The benefits accruing to the workers of the Northern Field during the labor troubles of 1887–1888 in the other fields also stand out, but among the most significant aspects of the graph is the long standstill in utilized labor input during the depression of the Nineties. This continued until the surge in activity in 1901, which received some impetus from the strike of the previous year, and was followed by the great strike of 1902. Since wages accounted for 85 percent of the cost of coal production,[2] the implications of this standstill for regional wage receipts over an extended period were very significant.

More detailed examination of the contribution of changes in breaker days worked to questions of productivity change and capacity utilization can be undertaken with reference to Figure 16.3. Viewed from this perspective, the trend-related components of the development of the industry are less apparent and the cyclical components are more prominently displayed. The sharp dip in 1902 is obviously an artefact of the strike at that time. The first point to note is that there appears to be a long cycle, corresponding to a Kuznets-type periodicity, where days worked fell to only about 150 days a year during both the latter part of the 1870s depression and the equivalent period in the 1890s recession. This level represents only a 50 percent capacity utilization, based on the potential maximum of a six-day working week. A sustained period of 75 percent utilization, or about 225 days a year, was only approached during the early part of the recovery from the 1870s depression, in 1881–1883. Thereafter, several cyclical oscillations of smaller magnitude were experienced, beginning with the 1885 depression. Arguably, the peak in 1887–1888 was an artefact of the Labor Troubles at the time in the other regions and was followed by an expectable reduction in work time, after the other mines resumed work. Although the phase before the 1893 panic failed to produce any sharp increase in

FIGURE 16.2
Northern Field: Estimated Annual Manpower Input for Coal Production 1876–1902.

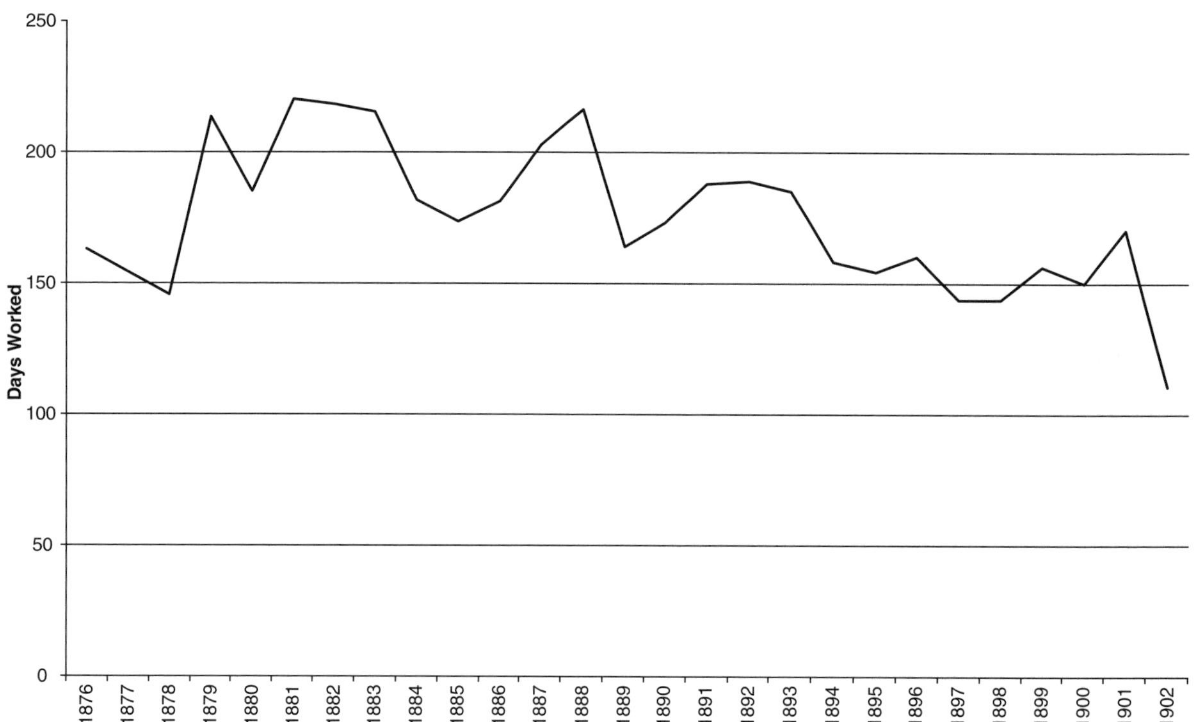

FIGURE 16.3
Northern Field: Estimated Average Breaker Days Worked 1876–1902.

TABLE 16.1
Average Breaker Days Worked in the Wyoming and Schuylkill Region 1880–1901.[3]

	All Mines				Mines Producing between 100,000 and 199,999 tons				Mines Producing between 200,000 and 299,999 tons				Mines Shipping ≥300,000 tons			
	Wyoming		Schuylkill		Wyoming		Schuylkill		Wyoming		Schuylkill		Wyoming		Schuylkill	
Year	ABD	NM	ABD	NM	ABD	NM	ABD	NM	ABD	NM	ABD	NM	ABD	NM	ABD	NM
1880	185	126	165	130	183*	37	203**	25	185*	6						
1890	173	171	189	111	194	73	222	42	204	16	240	12				
1901	170	182	199	113	172	40	203	23	200	42	227	31	219	30	240	13

*Only tons mined data available.
**Tons shipped.
ABD = Average breaker days worked.
NM = No. of mines reporting.

capacity utilization, the aftermath of the financial distress of the period certainly resulted in a sharp fall in utilization thereafter. These changes also need to be read in conjunction with other information on new additions to productive capacity over time, a topic that is addressed in a subsequent section.

In addition to the cyclical aspects of capacity utilization conveyed by the aggregate breaker days measure, further insight into the deployment of mining plant in relation to demand variations can be obtained, by examination of capacity utilization at mines of different sizes. A comparison of the Wyoming and Schuylkill Regions is given in Table 16.1. This shows the average breaker days worked by mines shipping at different output levels in the different years. Several interesting points emerge from the table. The first is that in both coal regions, there is a marked tendency for capacity utilization to increase, as annual output increases. In a partial sense this is a truism, since working more days will necessarily increase output, but the use of wide bands for the shipment classification means that this does not explain more than a small proportion of the variation observed. Secondly, however, the variation was not uniform between regions or between time periods. In 1880, in the Wyoming Region, there appears to have been remarkable concert of action, since mines of all sizes operated for very similar number of days. At this time, the Schuylkill Region was seriously under-performing, in terms of capacity utilization, by a figure of 20 days, but overall underperformance concealed the beginnings of a trend towards concentration of production activity at a small number of larger mines. These worked for nearly 40 days more than the average. A decade later, the trend had continued and the higher capacity utilization of more than 50 days above the average had shifted into the elite group of mines producing more than 200,000 tons. By the end of the period, the progressive disappearance of smaller operators had improved the overall average, but the greatest capacity utilization had again shifted into a higher output bracket. A similar broad trend was apparent in the Wyoming Region in 1890 and 1891, although the disparity between the average performance and the largest mines continued to widen by 1901. Thirdly, both the average figures and the disaggregated classification of the data by mine size shed further light on the rationalization strategies employed in the Schuylkill Region. When these were discussed in the previous chapter, the emphasis was placed on measures of labor productivity rather than capacity utilization. Although the average capacity utilization in the Schuylkill Region was relatively poor in 1880, the larger mines were already out-performing their Wyoming counterparts, in terms of days worked. By 1890, average days worked in Schuylkill had pulled ahead of the Wyoming Region and the gap had widened further by 1901. In the group of moderate shippers (>100,000 tons), Schuylkill's lead increased between 1880 and 1890 and was maintained thereafter, at about 30 days. The region's advantage in the next higher grouping was greatest in 1890, but the Wyoming Region had reduced the disparity to some degree by 1901.

Maintenance of this advantage in capacity utilization was doubtless made possible by the greater degree of monopolistic control of the industry in the Schuylkill Region by the Reading Coal and Iron Company. It also indicates a further element of success in the rationalization strategies adopted by the company's management. At the same time, it was clearly an essential aspect of retaining and enhancing the competitive position of the Schuylkill Region within the industry, given the productivity disadvantages under which it continued to labor throughout the period.

INVESTMENT CYCLES

From the discussion in earlier chapters, it is apparent that there were two main types of investment that characterized the mining industry in the second half of the 19th century. The first of these was investment in coal lands and existing mining improvements; the second was the construction and development of new mines. In the case of the Coal and Iron Company, the former dwarfed the latter, at least after the capitalized interests and commissions had added a further $12 million to the original $26 million expended [Table 10.3]. That said, the total expenditure on subsequent mining improvements was not inconsiderable. In the period of greatest financial stringency from 1877 to 1884, the Annual Reports record a total expenditure of at least $3.8 million, while in a second period when the figures are reported, running from 1888 to 1895, a further $5.75 million was expended on this account.[4] The second largest property owner, the Lehigh Valley, managed to expend some $15.8 million on lands and existing improvements 1868–1896, accompanied by an estimated $6 million on new mining developments 1871–1897. Of the latter sum, just over half was spent on 'new plant', $638,000 specifically on new coal breakers and breaker modifications, $285,000 on slopes, planes and tunnels, and just over $1 million on general maintenance of mine workings and equipment, especially when they were idle.[5] The $9 million invested in lands by the Lackawanna between 1868 and 1896 appears almost miserly in comparison, but it should be recalled that this company had begun acquiring lands in the 1850s and had precipitated the overall scramble for lands, as a result of its activities in the latter part of the Civil War period. It is therefore interesting to note that instead of storing up large reserves of coal lands for the indefinite future, the company concentrated on vigorous exploitation of its existing property, expending over $7 million on new mines and coal breakers during the same period, 1868–1896.[6] In general, the Lackawanna did not invest in the purchase of existing improvements in the same way as the Reading, the Lehigh Valley and the Delaware and Hudson, but it did acquire several failed individual operations at knock-down prices during the 1885 depression, as described in an earlier chapter.

Taking all the companies together, including the expenditure of more than $4.5 million on land and existing improvements by the Delaware and Hudson 1867–1896, results in a grand total of more than $50 million expended on coal estates, assuming the original cost figure of $26 million is used for the Reading. The equivalent total figure for new mine construction and new breakers was $22.5 for the Reading, Lackawanna and Lehigh Valley taken together. Had the relevant accounting data for the Delaware and Hudson been available, this total would doubtless have exceeded $25 million. Given the relative balance of expenditure by the Reading, it is not surprising that in the mid 1890s the company would bemoan the insistence of the cartel on using productive capacity to allocate shares of the trade, when according to its figures, ten times as much was invested in lands and railroads, as was invested in mines.[7]

Setting aside the sour grapes from Gowen's presidency, which set on edge the teeth of future generations of railroad and mine managers, the point remains that the large one-off investments in land bought a large measure of control of the industry for the major cartel players, but it did not actively develop the industry. Indeed, in parts of the Schuylkill Region it stultified it, or caused progressive abandonment of mining activity. In contrast, the money invested in new mine construction created new employment opportunities, caused the expansion of existing settlements and the creation of new ones and fundamentally altered the characteristics of the economic and physical landscape. This was especially true in the Wyoming Region, where development was most rapid and most far-reaching during the period. It should also be remembered, that while the major companies themselves were investing in new mine capacity, individual operators continued to do the same, particularly in the Wyoming Region, although to a lesser extent in Schuylkill, even into the 1890s. It would not be unreasonable, therefore to assume that the

$25 million figure mentioned above could be approximately doubled to account for all the other development activity that also took place. Sums of this magnitude, focussed on such a comparatively small area of geographical space, could only result in a radical transformation of the regional economy. In the process, the bucolic rural landscapes of areas such as the Wyoming and Lackawanna Valleys in the 1840s and 1850s were converted into the bustling industrial and urban landscapes of the 1890s. In the case of the major companies, much of the capital required was derived from earnings of the transportation business across the wider anthracite distribution region. In the case of the individual operators, much of the capital required was loaned by the transportation companies from the selfsame earnings. This highlights the close interdependence of the fortunes of the two regions, bound so closely together by the business of mining and shipping coal to market.

Having established the overall magnitude of investment in lands and new mine construction, the next step is to examine the aggregate patterns of inward regional investment over time. Although this is not entirely possible for coal land investment, owing to the enormous number of transactions involved, some indication of timing can be given by totalling information presented for several different companies in a previous chapter. The results of this, comprising data for the Lackawanna, Lehigh Valley, and Delaware and Hudson, are shown in Figure 16.4. The overall pattern confirms the earlier findings. Early stirrings are apparent from the activities of the Lackawanna during the Civil War, followed by a brief surge in activity in 1867–1868. Investment was apparently curtailed in 1869, possibly on account of strikes in the industry affecting business confidence. It resumed for a second phase in the early 1870s prior to the panic, before dwindling away to nothing during the economic depression, in the now familiar manner. Confidence was relatively slow to recover, following the upturn of 1879, but a period of sustained actively followed in the years 1882–1885, before the downturn again caused it to cease. By this time very little good land was left on the market in either the Wyoming or Lehigh Regions. The short

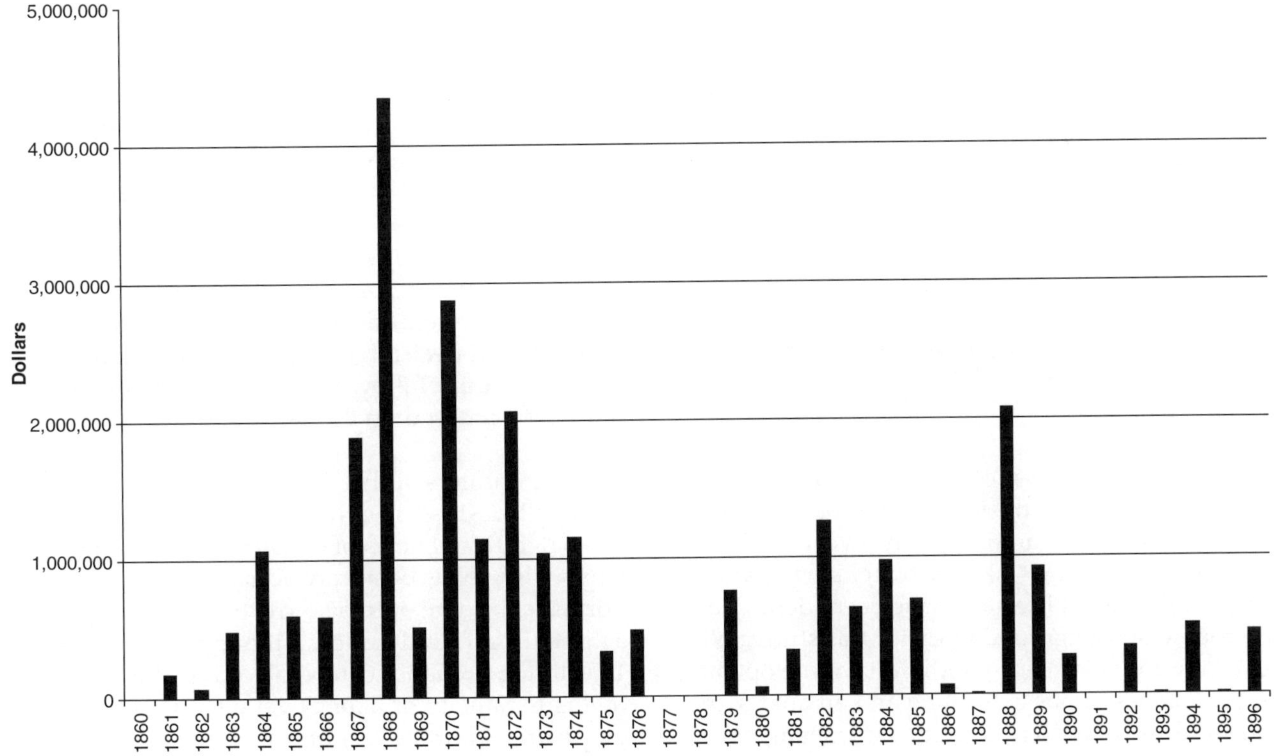

FIGURE 16.4
Combined Acquisition of Lands, Leaseholds, and Existing Improvements: Lackawanna, Lehigh Valley, and Delaware and Hudson 1860–1896.

burst of investment in 1888–1889 is perhaps surprising, as it was very largely undertaken by the Lehigh Valley. No very strong case for investment was offered by prevailing business conditions and the Railroad/Coal Company had been badly affected by the Labor Troubles of 1887–1888. Since the investment was mainly in the Schuylkill Region, it is possible that the adverse circumstances of the strike in the preceding period had thrown coal property on the market at what was considered a low price, so the company decided to take advantage of it. The coal land purchases of the Coal and Iron Company have not been included in the figures, since, as already discussed, they would dwarf the other data values. Their investment was, however, concentrated in the short period, 1872–1875. This confirms the focus of this type of investment on two periods, an initial start-up phase, 1863–1868, and the second main phase, 1870–1875, which had a huge peak in 1872, because of the $26 million, inclusive of commissions, much of which was spent by the Reading in that year.[8]

Although the large investment in coal lands had no direct effect on new mine capacity investment, it is worth noting that the timing of the latter investment was partly conditioned by the former, in that large companies could not start building new mine improvements until they controlled the lands they wished to improve. Prior to 1868, the Lehigh Valley had only acquired small amounts of coal land, by way of take-over of smaller companies, which enjoyed both railroad and mining privileges. Prior to 1871, the Reading officially held no lands at all. Conversely, companies like the Lackawanna, the Delaware and Hudson and the Lehigh Coal and Navigation Company had long held charter rights to coal land ownership, so were able to start new mine construction at an earlier stage, if they wished to do so.

Tracing the process of investment in new mine construction at the level of the coalfield or producing region is unfortunately a very difficult and time-consuming task. The better part of a thousand different breakers have been identified in the course of background work for this study. When further account is taken of individual mine openings such as slopes, tunnels, hoisting and air shafts, the total number of individual mine improvements where specific investment decisions were required, climbs to many thousands. In each case, there is also the question of when construction work started and when it was completed. Some of this information is tucked away in the descriptive text or tabulations of the Mine Inspectors' Reports from 1869–1870 onwards. Other data for earlier years are scattered through county histories, newspaper reports and company records, but these are sometimes inconsistent, or it is unclear whether dates refer to the start of construction or the start of production. A comprehensive chronology would therefore be a Herculean, if not impossible undertaking.

In an attempt to render the task feasible, while still shedding light on overall mine-construction activity, both by companies and individual entrepreneurs, it was decided to focus on the Northern Field, since a greater proportion of the development took place during the period when the records are better. Further than this, no attempt was made to enumerate all individual mine openings. Instead, two items of data were sought for each major mining operation, which might compose several openings and a breaker. These data items were the start of construction activity at the first opening and the date of the start of regular shipments from the breaker.[9] The cases of the Delaware and Hudson and Pennsylvania Coal Company also present difficulties, because in the early period especially, they operated a system of major mine openings, linked by railroad with a small number of large breakers, to which coal was brought from several locations. For these companies the rules have been slightly relaxed and details of both the main openings and the breakers have been included, in an attempt to reflect the real pattern of investment decisions as closely as possible. A final point is that when breakers were rebuilt specifically to increase output, the production start dates have also been included, since these also reflected a management decision to expand production. Breakers rebuilt following fires or other calamities have been ignored on the same basis.

The findings of this analysis are presented in Figures 16.5–16.7. The first graph shows the estimated frequencies of construction start dates. Unusually, it has been carried back before the normal start date of this study, to encompass activity in earlier decades, although prior to about 1854 limited reliance can be placed on the estimates, owing to paucity of data. Several features of the graph are of particular interest, especially in comparison with the previous graph of land investment cycles. Significant peaks are apparent in 1855–1856, 1864–1865, 1870, 1872, 1881–1883 and 1888–1889. The first of these reflects the speculative

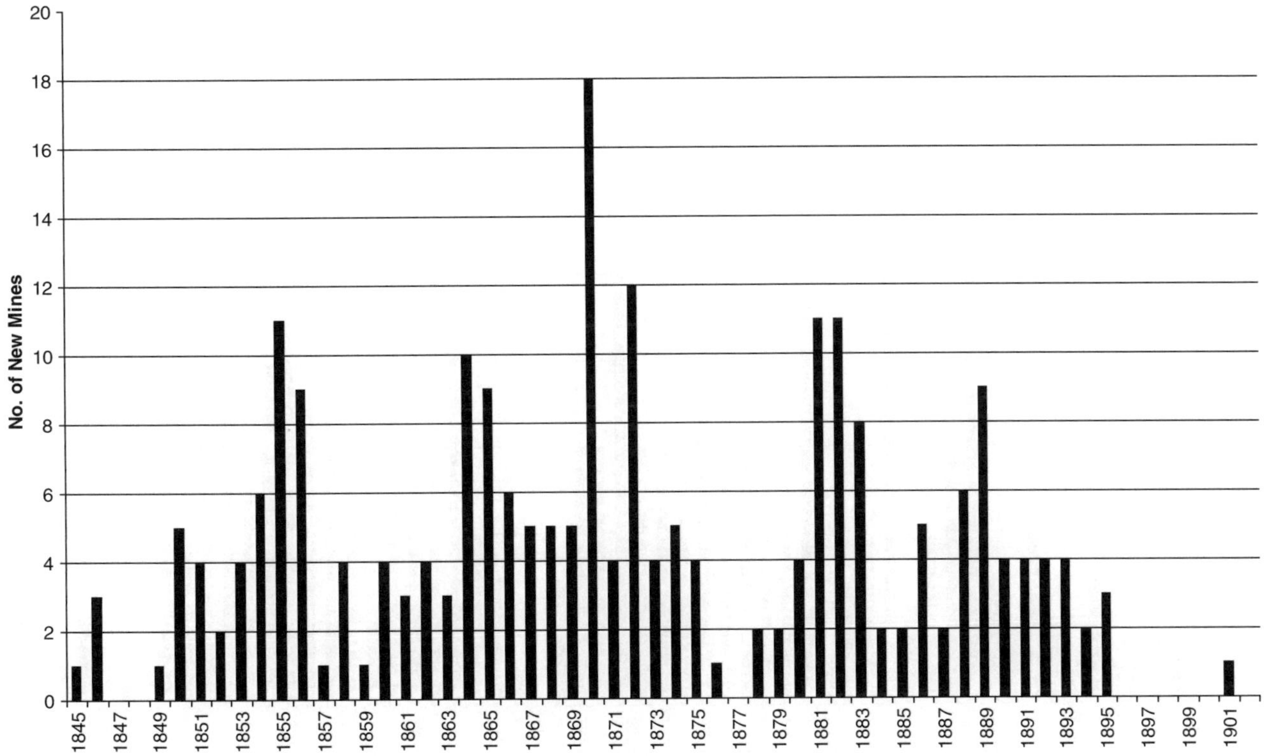

FIGURE 16.5
Northern Field: Estimated Start Dates of New Mine-Construction Activity 1845–1901.

'bubble' in mining activity just before the 1857 panic, while the second corresponds to the favorable financial circumstances of the Civil War. It is important to note that this surge in mining investment predated the main scramble for coal lands and represented an attempt to 'cash-in' on the windfall profits of the time. The 1870 and 1872 peaks correspond to a generally favorable investment outlook in the run up to 1873 panic and doubtless also an element of lagged response to some of the recently acquired coal lands. The sharp fall in 1871, when conditions were briefly depressed, however, suggests that the main determant of the investment climate at the time was short-term perception of prevailing business conditions, rather than long-term strategic planning. The surge in the early 1880s is in accordance with our expectations, based on earlier findings. Given the strength of recovery at that time, it is perhaps surprising that there were not even more mine starts. Expansion of developments in the period 1888–1889 is not attributable to the Lehigh Valley Coal Company, since its land purchases at the time were not in this Field. Instead it probably reflects the windfall profits of 1887–1888 obtained by most Northern Field operators, as a result of the Labor Troubles elsewhere. It also corresponds to the peak of the Juglar cycle [Table 2.1]. Finally, based on the experience of the speculative phases prior to each of the two previous panics in 1857 and 1873, it might have been anticipated that 1892–1893 would cause a repeat of this type of behavior. While a number of mines were indeed commenced at the beginning of the 1890s, such as the Bliss and Auchincloss mines by the Lackawanna, the shortage of available land for purchase or lease by this time would have necessarily dampened the enthusiasm of budding entrepreneurs.

To turn attention to the dips in the series, rather than the peaks, there can be no doubt of the ability of depressed conditions to dampen investor enthusiasm! After the 1857 panic and the depression that accompanied the onset of the Civil War, construction starts remained modest until inflationary profits became too large to ignore. The 1870s and 1890s depressions had their expected impact, while the 1885 depression, which corresponded to a levelling off, rather than a collapse in overall production, only left its mark for a short period.

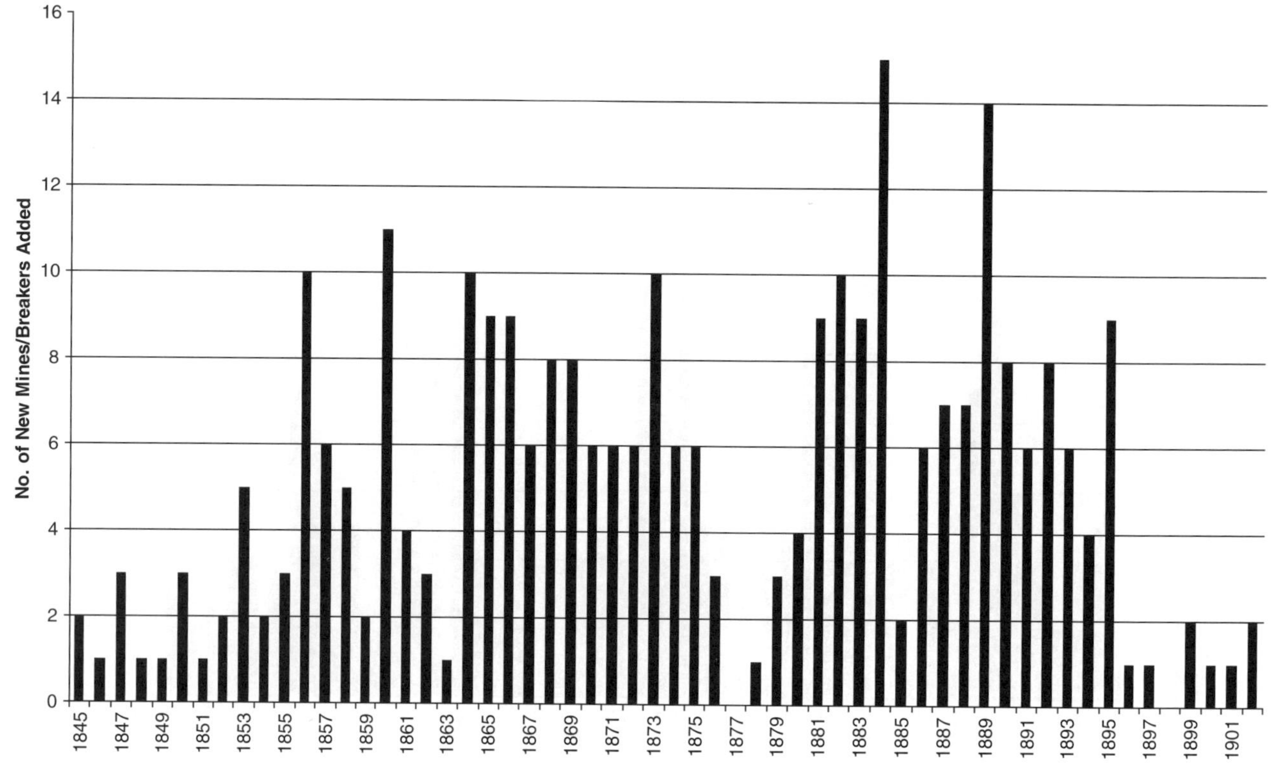

FIGURE 16.6
Northern Field: Estimated Addition of New Breakers/Mines by Date of First Shipment 1845–1902.

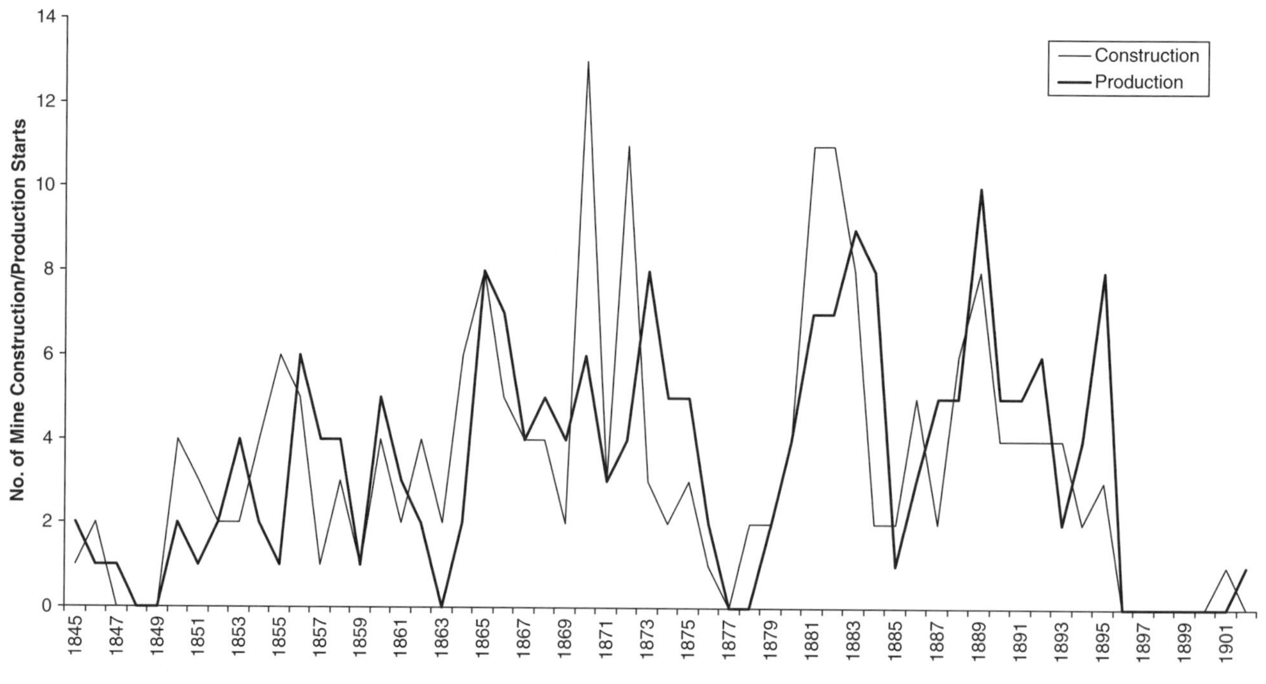

FIGURE 16.7
Northern Field: Start of Construction and Production at Mines where Both Dates are Known 1845–1902.

Although information on the construction start dates of 240 mines has been located, the equivalent data on production starts is more comprehensive, with some 297 mines/breakers represented. These dates are shown in Figure 16.6. As they cover more than 95 percent of all breakers built during the period, good reliance can be placed on the patterns observed, which compensate for several shortcomings in the more limited data shown in previous graphs. The first point to note is that several of the peaks already identified in the 1850s, 1864–1865, the early 1880s, and 1889 are clearly repeated in this graph, but with one very important difference. Namely, that a substantial portion of activity, if not the peak itself, is displaced 1–2 years to the right. In the case of the 1856 and 1884 surges, this has the effect of pushing a significant number of new production starts just to the point, or just beyond the point, when the next panic occurs, as in 1857 and in a smaller way 1884. This was extremely damaging to the financial fortunes of smaller companies and individual entrepreneurs, since the boom years had literally been spent sinking their capital in the ground. Yet just when they needed vigorous productive output to maintain cash flow and restore their depleted capital reserves, demand dropped off and prices declined, often driving the hapless firms, and sometimes their investors, to the wall.

Several other points also emerge. Firstly, there seems to be a significant peak in new productive capacity in 1860, taking off into the early Civil War years. This is not apparent from the more limited construction data. Secondly, the rightward shift of construction impulses, when they are converted into working mines, has resulted in a more than decade-long procession of regular capacity increments between 1864 and 1875. These resulted from the separate actions of a large number of investment decision-makers. As with the other cycles in 1857 and 1884, the same effect of lagged onset of production 1–2 years after the 1873 panic is also observed. Thirdly, with more comprehensive data, it appears that the period 1886–1895 constituted a further marked phase of new additions to capacity, that is not so obvious from the construction data. This phase certainly peaks with the Juglar cycle in 1889, but once again there is carry-over beyond the 1893 panic. The false dawn of 1895, which brought some temporary relief from the rigors of the 1890s depression, also seems to have called forth a new, if inappropriate burst of enthusiasm for new capacity, which quickly waned.

The importance of the two construction phases that were initiated in the 1880s needs to be emphasized, because the graph is simply reporting numbers of mine starts, not the productive-capacity investment associated with the individual mines. In the 1860s–1870s most new mines started out with a capacity of 600–800 tons/day. By the early 1880s this had increased to 1,000–1,200 tons per day and by the end of the decade to 1,800–2,000 tons/day or more, so available capacity was growing multiplicatively rather than arithmetically. It should also be remembered that a number of large mines, commenced in the early 1880s, such as the Woodward and Storrs, discussed in Chapter 14, were deliberately held back from entering production at an early date, because of the 1885 depression. Breaker construction did not start until later in the 1880s, and first production was in 1888 and 1889 respectively. Hence some investment decisions originally made in an earlier cycle, contributed new capacity at the peak of the next cycle. However, smaller concerns, such as the Amity Coal Company, lacked the flexibility to make such postponement decisions. Hence they joined the flood of new capacity coming on stream in 1884. Some subsequently collapsed, only to be salvaged by the Lackawanna by takeover or at the sheriff's sale.

Since the construction data are less complete than the production data, a final check was made by plotting only the data for mines where both construction and production dates are available, to ensure that the inference about production lags and business cycle timing was not simply an artifact of data limitations. This reduced the number of mines/breakers covered to 206, which rather thinned out the data series, as can be seen in Figure 16.7. Nevertheless, it is clear, especially in the 1850s, 1874–1875 and 1883–1884, that the lag between construction and production of 1–2 years was a real effect, which had major consequences for investment success, or lack of it.

While this temporal analysis of mine-construction activity has revealed a number of important phases in the aggregate development of the industry in the Northern Field, there is another distinct dimension to the problem, in terms of the evolving geographical distribution of mining and resultant employment opportunities across the coalfield. Stages in the development of the spatial pattern of mines can be followed by examination of Figures 16.8–16.10. These show

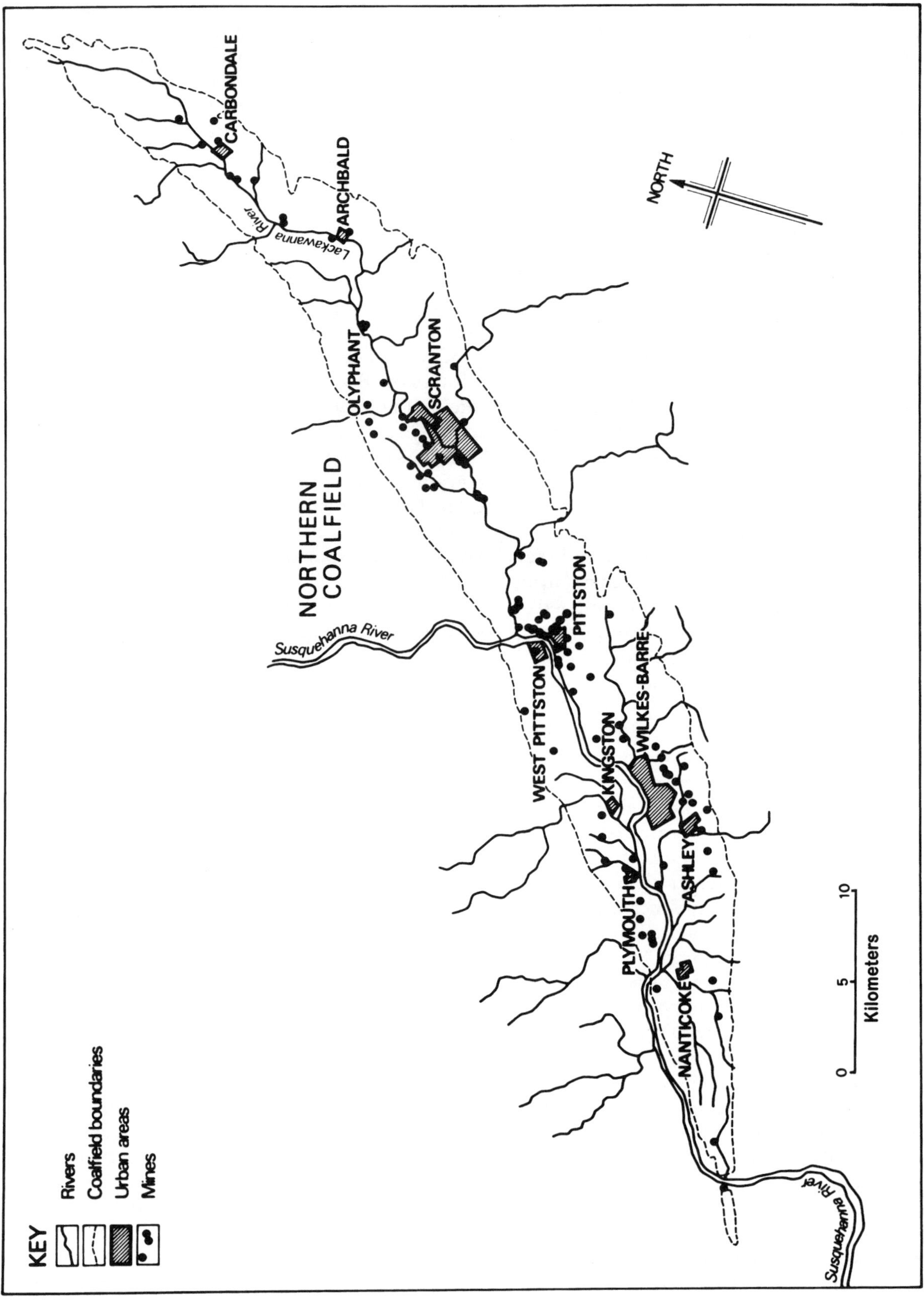

FIGURE 16.8
Map of the Distribution of Mining Activity in the Northern Field 1865.

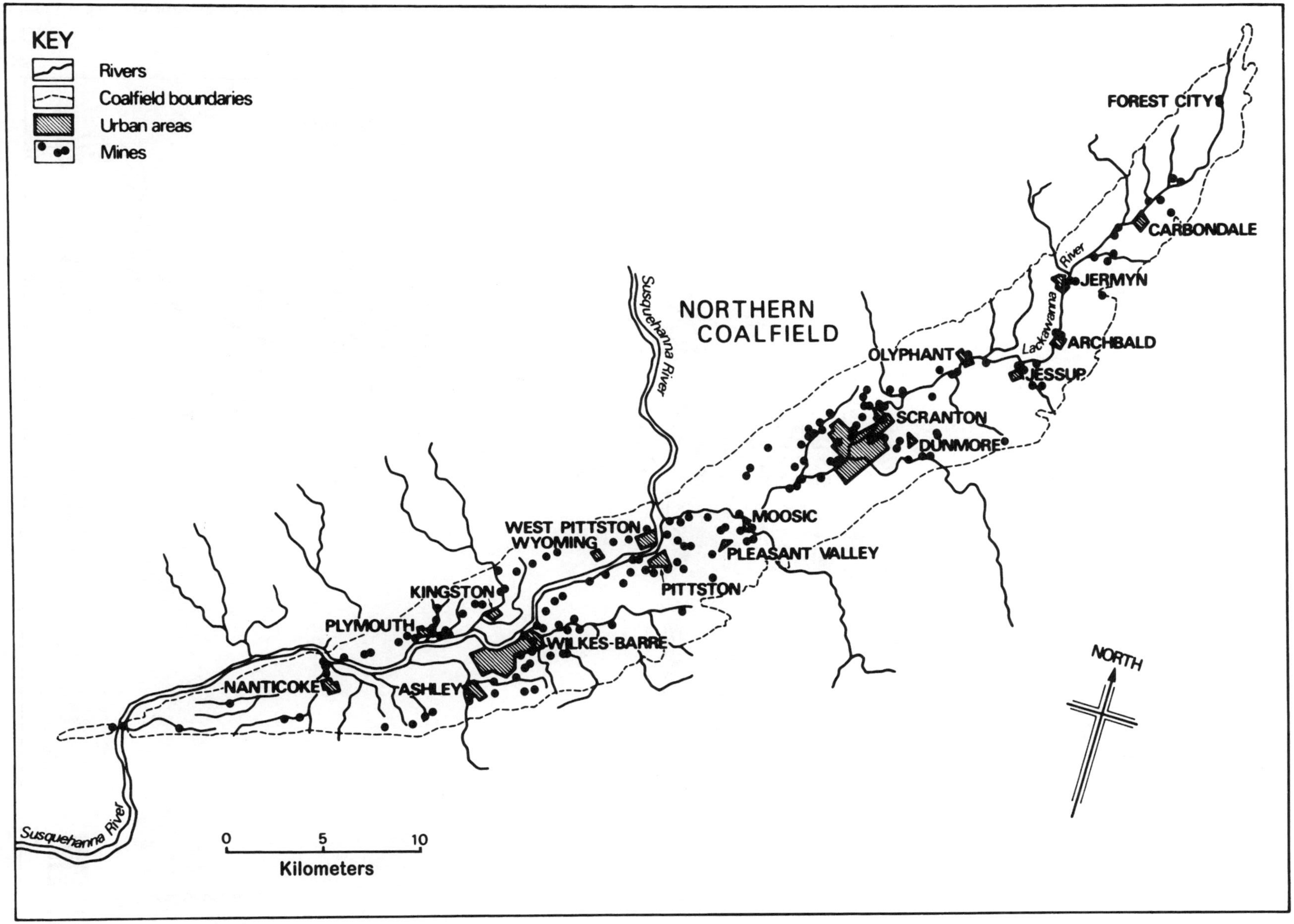

FIGURE 16.9
Map of the Distribution of Mining Activity in the Northern Field 1886.

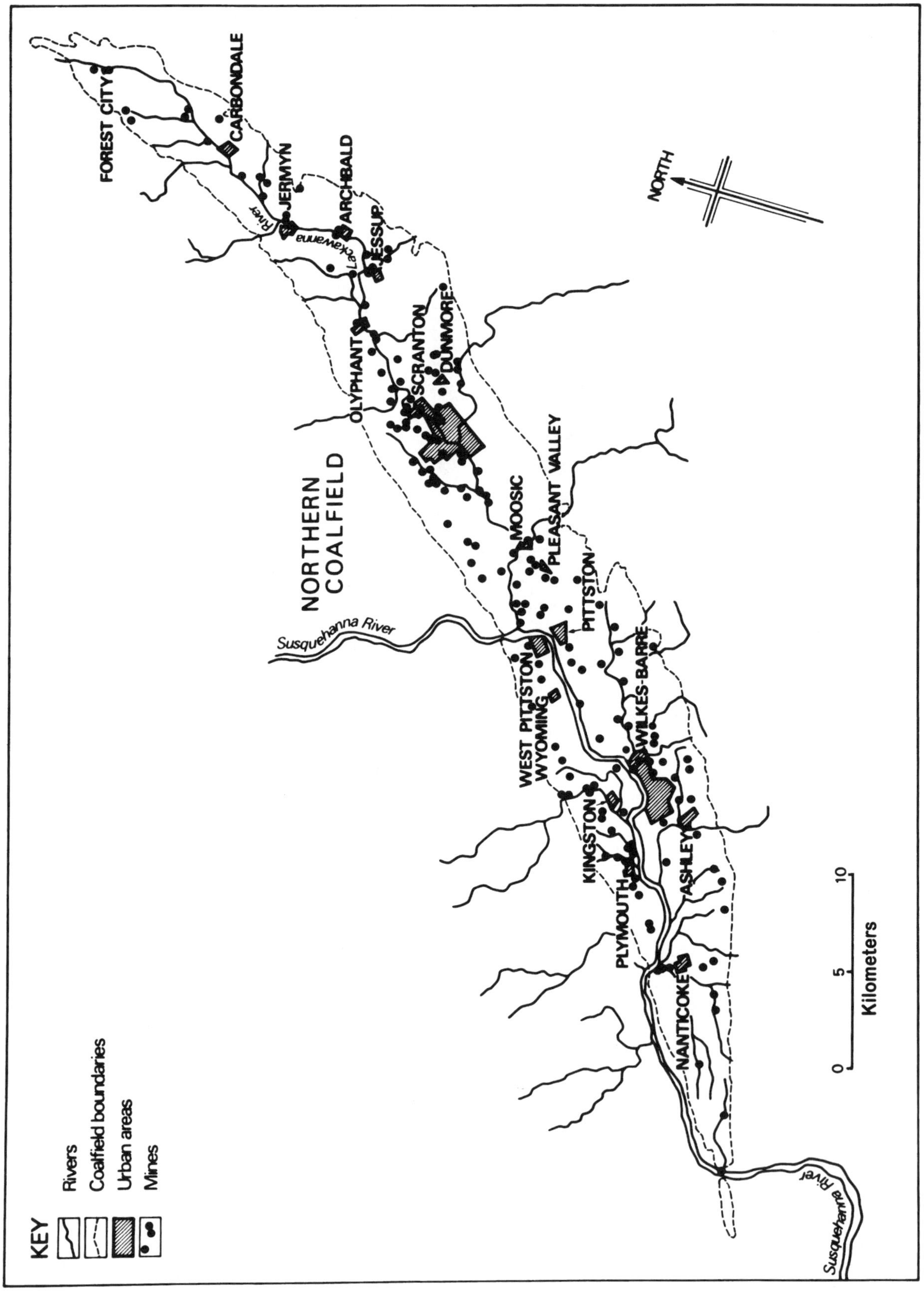

FIGURE 16.10
Map of the Distribution of Mining Activity in the Northern Field 1898.

the locations of the breakers in each of the years 1865, 1886 and 1898. Since very few breakers in the Field were abandoned during this period, each time-slice very largely builds on the pattern observable in the previous one. It is apparent that in 1865 mining activity was predominantly clustered around the major settlements, such as Carbondale, Scranton, Pittston, and Wilkes-Barre. By 1886, a process of in-filling had taken place between these settlements, as new developments, both by the major companies and individual operators, took place on remaining good quality coal lands, many of which were worked under lease. The final map, for 1898, shows the pattern of mining that was very largely to persist into the early decades of the 20th century. Development had now spread to more marginal coal lands, although there were also some new breakers on lands previously held in reserve by the major companies. Good examples of the latter were the Clinton breaker (Plate 19) of the Delaware and Hudson at Vandling in the northern part of the coalfield and the Auchincloss breaker (Plate 20) on the Hanover lands of the Lackawanna below Wilkes-Barre, at the other end of the coalfield. The Auchincloss post-dates the 1898 map since it did not come into full production until 1902, as the first breaker to be driven by electricity rather than steam power.

OVERPRODUCTION OR OVERINVESTMENT?

The magnitude and frequency of cycles of investment in both land and new mine capacity indicate that the path of economic development in the coal regions was anything but smooth. This finding necessitates a re-evaluation of one of the long-standing themes in the literature on the 19th century anthracite industry, where frequent reference is made to the problem of overproduction. Yearley, for example, referred to 'the overproduction characteristic of the whole anthracite industry'.[10] Miller and Sharpless put it slightly differently, 'the frenzy of the operators to raise production and the intense competition among them led to periodic cycles of overproduction and depression'.[11] In the light of the evidence already presented, however, this claim that cycles in output were endogenous, or internal to the anthracite industry, cannot be sustained, as it fails to take account of economic fluctuations at the national level. Yet identification of the problem of overproduction did not begin with recent authors such as these. As early as 1868, the Reading Railroad lamented the effect of high Civil War prices causing 'the erection of collieries capable of a production largely in excess of the consumption'.[12] The Scranton Republican took a slightly more personalized view of the same time period—'the unprecedented success turned myriads of covetous eyes to the coal business'.[13] A decade later, the Miner's Journal felt it necessary to defend the industry against charges of overproduction levelled by Iron Age, another trade-orientated newspaper.[14] Among the most perspicacious of assessments of the problem was made by the Wilkes-Barre Record in 1884, reflecting on the consequences of cartel operations, which had originally, of course, been justified as an attempt to prevent the ruinous effects of overproduction.[15] The Record railed against the practice of the companies in simultaneously restricting total output, while competing against one another for market share, which resulted in them constantly opening new mines and extending old ones. It regarded such behavior as injurious both to the men and to the economic health of the settlements where they lived, for the following reasons:

> 'It requires a constantly increasing army of mine workers to meet the demand while the mines are working steadily, and results in a proportionate increase of the idle and helpless during the periods of restriction'.[16]

A better summary of the conflicting trends in employment and days worked, presented in the previous section of this chapter, would be hard to find! The article concluded by proposing restriction on the development of new mines, rather than on tonnage output. Unfortunately, it was reasonably easy to identify the potential remedy in 1884, when much of the mining construction commenced in the earlier stages of the business cycle upturn was coming to fruition. It would have been much harder to deter the enthusiasm of entrepreneurs and companies alike, in the expansive days of 1881–1882. The Anthracite Coal Strike Commission of 1902 made a similar assessment about excessive opening of new mines, especially in the period 1893–1899. It is perhaps to them that the last contemporary word should be given:

> 'The endeavour to control this natural monopoly, has led many to miscalculate the anticipated increase in value of anthracite coal, due to said

increase being more gradual than they had thought, causing in the past undue competition, unwarranted development of the mines, and financial disaster to many of the interests participating in this competition, so that the business has been subject to violent fluctuations'.[17]

Returning to more recent commentary, Aurand, in his study of unionization in the Anthracite fields, observed that the Workingmen's Benevolent Association in the 1870s 'committed suicide', because it diagnosed the industry's ills as stemming from overproduction rather than overinvestment,[18] a finding that chimes with the observations of the Wilkes-Barre Record. Overinvestment in coal lands can largely be discounted as a factor, because of its very indirect association with actual production on a month-by-month basis. Yet even over-investment in mining capacity is a more problematic concept than it first appears, viewed in the light of the analysis of earlier chapters. Total investment at a certain point in time was based on the future expectations of the likely development of the trade, as identified by a multitude of different entrepreneurs and company managers, each with differing levels of financial constraint. Unfortunately, the time lag between an investment decision and the actual availability of new output capacity was at least 1–2 years on average, longer for large projects and deep mines. Such timescales exceeded the ability of even the sagest of operators to forecast future trading conditions accurately. Hence investment only became overinvestment if the business grew more slowly than anticipated or too many operators took too optimistic a view of future prospects, and overinvestment only became overproduction if operators were induced to continue shipping in excess of the wants of the market. In the days of the cartel, there were also potentially serious penalties for *underestimating* the rate of development of the industry, since other members would snap up market share from those who were unable to meet their quota of market demand. This aspect of the decision-making process has received far less attention in previous studies of the industry.

It would appear, therefore, that simple diagnoses of the industry's ills, whether in terms of over-production or over-investment, are insufficient to explain the complexities of tactical and strategic decision-making, coupled with lengthy and delay-prone construction processes that characterized the development of the Anthracite regions. Instead, it is more appropriate to summarize briefly the interacting factors involved, under two headings. The first of these is factors that exacerbated a tendency to expand productive capacity to a level that exceeded the wants of the market. The second is methods that could be used to mitigate adverse impacts of past decisions, made either by individual operators, groups of companies or as a result of the aggregate effects of actions across the industry. These adverse impacts might result from errors and miscalculations, unexpected consequences of intended actions or simply from circumstances beyond individual control, such as national scale economic fluctuations. Relative success, like relative failure, by entrepreneurs, companies, or entire coal regions, was attributable to temporally, and often geographically averaged performance, based on two business indicators. The first was ability to maximise the financial benefits of good decisions and strategies. The second was ability to minimize most effectively the costs attributable to errors, misjudgement, over-optimism or plain misfortune. As a useful by-product, this analysis of exacerbating and mitigating factors will also summarize and draw together many of the individual threads followed in detail in the earlier chapters of this study.

EXACERBATING FACTORS

Exacerbating factors can be divided into several groups. These relate to the nature of the mining business itself, to coal-lease provisions, to the actions of the cartel or individual railroads, and to expectations of the future of the coal business. In some cases the specific factors had direct effects on the resultant outcomes. In other cases, they produced unintended consequences, when their original direct effects were multiplied across a number of companies, or a whole coal region.

Several factors can be regarded as intrinsic to the development of mining itself. The first of these was that early mining above water level from drifts required little capital to enter the business, and there were limited cost penalties for maintenance of the opening, if the mine was idle. Hence output could be matched closely to demand and the mines could be operated as required. Once slopes and shafts descended below water level, start-up costs increased dramatically, as did maintenance and pumping costs, so long periods of idleness could be financially ruinous for operators.[19] An extension of this point, expressed in more standard economic terms, was that deep mining had high fixed costs of plant maintenance, pumping, and ventilation to prevent the accumulation of explosive or poisonous gases, regardless of whether any coal was actually being mined. So there was always a substantial incentive to increase output as much as possible, since this meant that the fixed costs were spread over a larger tonnage, thereby reducing their contribution to overall costs per ton.[20]

The bulky nature of coal as a commodity meant that it was costly and space intensive to stock it and customers were therefore unwilling to do so, once railroad transport made shipments on demand feasible. Likewise, stockpiling by the shipping companies, although possible to a limited extent, added 20 cents to the cost of coal in the 1880s, because it had to be dumped and re-loaded.[21] This reduced the extent to which demand variations were absorbed by changes in inventory, except in locations such as the Great Lakes' ports, where winter ice placed constraints on the navigation season. As a result, a marked seasonality in output characterized the industry throughout the period. The difficulty with this was that mines had to maintain a productive capacity large enough to meet the flood of orders in the busiest season in the Fall, rather than the average level of demand throughout the year. Therefore, mines had excess capacity part of the time and there was temptation to utilize it whenever possible, not only for cash flow reasons, but also because the workforce tended to drift away, if offered only limited or irregular employment.[22] A final problem that affected calculation of required production capacity was the contingency margin required to guard against unpredictable output losses. These might result from machinery breaks and boiler explosions, as well as the more familiar ravages of fire, flood, and occasional mine caves. While some individual operators might be fortunate to escape such problems, larger companies, with multiple mines in different parts of the coalfield, had to assume that such afflictions would be visited on different parts of their estate from time to time. For shipping railroads, which both owned mines and provided transportation for individual operators, fire, production problems, and abandonments could quickly add up to a serious capacity loss, as the Lehigh Valley discovered in 1897 (see Chapter 14). Nor could such problems be quickly overcome by working certain mines to the limit, since the later coal purchase contracts contained clauses 'providing for the maintenance of an equality as regards time worked at all the collieries tributary to the Railroad'.[23]

Most of these factors were unavoidable aspects of the mining business, but others related to the development of the cartel had to be added to them. It was the avowed intent of such a form of industrial organization to control tonnages and prices, the latter, either directly or indirectly, via output restrictions. However, it is apparent that insufficient consideration, especially in the early days, was given to the consequences that this would have for capacity management, when each major shipper looked to its own interests, while trying not to break the cartel rules in too blatant a fashion. Thus, the unintended consequence of cartel restriction, by means of periodic suspensions, was that mines had to be kept open, to maximize output on working days. This prevented rationalization of output into the smallest number of larger and efficient producers. Failure to maximize shipments on working days meant loss of share of tonnage. When this was factored in to the annual round of negotiations, it would result in progressive loss of percentage quota of overall tonnage for ensuing years. This kind of negative reinforcement of underperformance had to be avoided at all costs. Improvements in the method of quota allocation during the 1880s partially addressed this problem. The Reading, at least, took the opportunity to idle mines that 'could stand most cheaply and were least remunerative in working'.[24] Harris claimed that this action and similar moves by other companies, reduced the overall capacity of the anthracite industry from 38.13 million tons in 1884 to 36.48 million tons in 1885, thereby reducing the surplus of capacity over demand from 19.4 percent in 1884 to 13.3 percent the following year.[25]

However, the mines owned and operated by the coal subsidiaries of the major transportation companies were only part of the picture of tonnage share and capacity tributary to each railroad. The other part was associated with the individual and independent operators. Evidence presented in this and earlier chapters has confirmed the role of wider economic upturns in stimulating entrepreneurial enthusiasm and speculative activity. Unfortunately, this same enthusiasm relied very heavily on loans from the railroads to enable dreams to be converted into reality. No control over this aspect of the mining economy was exercised by the cartel, nor does it appear to have featured in any of the recorded negotiations pertaining to cartel agreements. Despite this, Storrs of the Lackawanna was in no doubt as to its effect on the industry in over-stimulating competition and inflating the value of coal properties. He also forecast in 1884 that depression and low prices would come and squeeze the operators, who had been enticed into the business by easily available loan finance, a prediction that was amply fulfilled before the year was out.[26] The difficulty was that no railroad could act alone, to curb this potential means of stimulating excessive new capacity construction by entrepreneurs, since these individuals would simply turn their attention to a different railroad, in their apparently insatiable appetite for other people's money![27] It is also worth noting, in this regard, that as the density of railroad tracks belonging to different companies increased over time, especially in the Wyoming Region, the ability of individual operators to play one railroad off against another also progressively increased.

A further railroad-related factor was that share of tonnage was much too precious a commodity to be left to market forces. So cash-constrained and largely bankrupt operators had to be kept going by the seemingly endless provision of rotating short-term loans, again from the railroads, paid back by deductions from the earnings on shipments. The Lackawanna, of course, could stand this kind of behavior better than most, because of the substantial operating reserves accumulated by its transportation division, especially during the 1880s.

Interactions between coal operators, large and small, and individuals or companies wishing to lease, rather than sell mineral rights, introduced yet other factors that encouraged rapid development of coal properties, thereby potentially contributing to excess capacity. Timing of the acquisition of coal leases by large companies was usually driven by speculative forces, rather than actual need for new territory. The fear was that if leases were not snapped up when offered, individuals such as Charles Parrish would acquire them, to offer back to the railroads at inflated royalty rates, at some point in the future. Unfortunately, this speculative timing could be reinforced by lease provisions, such as the one requiring development to begin within a specified limited time, or the lease would be forfeit. Over the longer term, if royalty rates were low, or very quickly, if they were high, the problems of advance royalties could create an absolute necessity to develop specific lands. This was to prevent large sums being paid to landowners, for which the return might be years, or even decades away. There is no doubt that this was an unintended consequence of a seemingly innocuous approach to leasing mineral rights. It arose from the lack of connection between the timing of lease acquisition and the commercial need for the coal it represented. Some companies, such as the Lackawanna, were eventually able to gain control of the advance royalty problem by bringing larger new mines into operation. Others, like the Lehigh Valley Coal Company, were not, at least within the time period of interest. For example, as late as 1898, the superintendent of the latter company lamented the fact that advanced royalties were still accruing at a rate of $250–300,000 a year. Even if all the company's Wyoming Region collieries were to be run flat out six days a week, he argued, they would only just keep up with the required royalty minima, never mind making in-roads into the accrued backlog of advanced royalties. Unfortunately, such full-time working was also out of the question, because it would necessitate working the company's collieries far more than those of the parent railroad's tributary individual operators. This would immediately leave the railroad open to charges of major discrimination, which was contrary to the terms of the purchase contracts, as noted above.[28]

A further group of exacerbating factors were related to the ways in which managers and entrepreneurs utilized past experience and perceived future possibilities in the coal trade. An underlying point to emphasize is that these individuals showed no evidence of understanding present-day concepts of business cycles. That said, the more sagacious or conservative of managers certainly entertained the possibility of future episodes of depressed business conditions, even if entrepreneurs as a group seemed largely unable to comprehend such a future possibility! Had they been better versed in the French language, they might have been able to relate Clement Juglar's writings on business fluctuations to Harris's observation about the old days in Schuylkill, that 'the whole crop of operators were cleared out about once in ten years by bankruptcy',[29] but apparently they did not.

For managers, the problem was that business fluctuations, and speculation in particular, brought long-term strategies into conflict with short-term expediency. Coal lands or leases that were not acquired when landowners offered them (usually on a cyclical upswing), regardless of actual immediate need, would be swallowed up by competitors, taking the properties off the market indefinitely. This would reduce the acreage tributary to the manager's own parent railroad in the long run. Such competition for coal property further encouraged speculation and price hikes by those fortunate enough still to hold valuable lands.

For entrepreneurs, the problem was one of persistent over-optimism, coupled with limited financial resources. As Storrs asserted, again in 1884, while low prices would drive the individual operators out of business, as soon as there was a margin again, they would try to come back in.[30] Shorter-term cyclical fluctuations over a 3–4 year period would almost be guaranteed to cause problems for this group, as the evidence from earlier in the chapter has shown. Their pent-up enthusiasm would tend to lag the upturn, peaking with the business cycle, when their mine construction would usually begin. Unfortunately, construction delays and cost overruns would bring them to the onset of production, with financial reserves exhausted and railroad-derived credit stretched to the limit, just as the market no longer required their new product.

A final factor that unfortunately encouraged some entrepreneurs to believe that the golden age of handsome profits was just around the corner was the occurrence of two widely spaced and brief periods when such profits had indeed been realized. The first of these was naturally the final two years of the Civil War, when profits were huge, even after taking account of price inflation. The second, which has been identified for the first time in this study, was during the Labor Troubles of 1887–1888, but on this occasion the substantial benefits accrued largely to the mines in the Northern Field. Such events were too widely spread and too unpredictable for any business plan to take account of them, and for every Florence Coal Company that could line its company coffers and the pockets of its shareholders with the proceeds, there was a Samuel Stetler, bankrupted and missing out on the windfall entirely. Nevertheless, such episodes fuelled the irrepressible optimism of those who could envisage only potential profits, not potential losses.

MITIGATING FACTORS

Mitigating factors can be viewed as the management actions or adjustment processes available in the attempt to match coal supply to fluctuating market demand. While the previous emphasis on problems of overproduction would suggest that downward adjustments predominated in the short term, account must also be taken of upwards adjustments, since the overall trend in output remained quite strongly positive during much of the period. Downward adjustments are already very familiar to the reader, so only need to be briefly summarized here. They included taking idle days or suspending operations for a period of a few weeks, as very short-term measures. Provided pumping and ventilation were not stopped, operations could be resumed quickly. Progressively more substantial measures would include closing a wing of the breaker, if it was a large structure, shutting off pumps and ventilation, and allowing the mine to fill with water. For more extended stoppages, temporary abandonment might take place, where in addition, some or all of the machinery might be removed to another site. Clearly, the more substantial the adjustment, the longer it would take to reverse it, in the event of a sharp, or unexpected upturn in demand. Thus, if the mine had filled with water, for example, pumping it out or hoisting out the water in tanks was a time-consuming and expensive process. Also, once the workings were pumped out they would have to be thoroughly checked for minor roof-falls and collapse of timbers, and the necessary repairs made, before normal work could re-commence. This also assumes that the workforce was effectively standing around waiting to be re-engaged when work re-started, but this was often not the case. Many men went to other areas seeking better prospects at such times and did not necessarily return. Good examples of this kind of process can be seen from the mines of the Lackawanna Coal Department during the latter part of the 1870s. A number of these mines had been temporarily suspended during the depressions years, including Continental, Dodge, Hampton and Hyde Park. Storrs indicated in his Annual Report for 1878 that these mines could be put back in service at short notice, but would require 'new equipment of mine cars, iron and mules'.[31] In the event, the sharp upturn of shipment in 1879 and the lack of cartel restraint resulted in very high capacity utilization at the mines that were still working. For example, the Central Breaker reached the very high figure of 286½ days work.[32] The company hurried to bring other capacity back on stream and managed to get Continental to work 153 days in that year, but the figures for Hampton and Dodge were much lower, at 49¼ and 35 days respectively.[33] The relatively slow pace of resumption was doubtless extended further by the reported shortages of both common and skilled labor in the second half of 1879, since all the operators would have been competing for the services of the same limited pool of un- or under-employed men.[34] Hyde Park did not re-start until the following year. The process was more prolonged at Archbald and Oxford, however. Storrs described these two mines as 'dismantled', since not only had the mine cars and T and strap iron (i.e. mine rails) been removed, as at the other breakers, but the pumps had also been removed and the workings allowed to fill with water. Several months would be required at a minimum, to bring such mines back into operation.[35] Consequently, it was 1881 before Archbald resumed shipments after a five year gap, and even longer at Oxford.

The final type of downward adjustment was, of course, permanent mine closure and breaker abandonment. This occurred relatively rarely in the Wyoming Region during the period, although much more frequently in the Schuylkill Region. Also, in the Wyoming Region, even if a mine was apparently abandoned, it was often the case that the shaft or workings were re-opened by another company at a much later date. This happened, for example, at the Buttonwood shaft below Wilkes-Barre. This was originally opened in 1855, but closed in 1866 because of problems with gas. It was re-opened and enlarged by the Parrish Coal Company in 1892.[36]

It might be thought that the adjustment processes required to increase output, at least in the short term, would be straightforward, since days worked could simply be increased up to the feasible maximum of around 300 days per year, at which point new sources of capacity would have to be found. While reasonable in theory, in practice this very rarely happened. One reason for this has already been given, namely labor shortages at times of peak output, since nearly every operator would be in the same position of wanting to expand production at the same time. The second was shortages of cars and even motive power for shipment, which is a reminder of the key role of railroad transportation and the absolute dependence of mining operations upon it. The point was well made by S.B. Whiting, General Superintendent of the Coal and Iron, while responding carefully to questioning during the inquiry into the 1887–1888 Labor Troubles. The dialogue ran as follows:

Q: How do you now ascertain the state of the markets: in other words, when you shall work the collieries or close them?
A: We would feel the want of coal by its accumulating in the cars and the cars not being returned to be filled.

Q: Do I understand you to say that you, as superintendent of that coal company, regulate the operation of that immense property simply by the fact that there are empty cars sent to you to be filled by the railroad company?
A: That is the method of computation now.[37]

Although Whiting was intending to avoid any suggestions that the Coal and Iron might be restricting shipments and therefore acting in restraint of trade, his point about the requirement for empty cars applied as much to times of peak output, as very low output. In the latter case, as during the 1870s, trains of coal cars might build up, unintentionally or otherwise, at the shipping points or on way sidings, preventing any further coal being mined until they were unloaded. At peak times, so many cars were in transit in coal trains, or detained while unloading at numerous destinations, that it could be difficult to ensure a full supply of empties would reach the mines on a regular basis.

A further, slightly unexpected, point in relation to full-time working is an assertion made by Charles Parrish during the 1875 Investigation of the Reading, that mines working ¾ time were more economical than mines working full time.[38] Unfortunately the claim was not elaborated, nor was it repeated elsewhere, but Parrish was an experienced operator and his comment deserves some credence. It is possible that productivity was better with slightly shorter hours or that a larger force of men was required to ensure reliable operation of machinery on a full-time basis, than under ¾ time working. In the latter circumstances, some maintenance could presumably be scheduled for idle days and be carried out more efficiently in this way. Be that as it may, the vicissitudes of operating complex, heavy-duty equipment in adverse environmental conditions, tended to ensure that there were sufficient breaks and day-to-day problems that full work could not be achieved, even if it was desired.

Beyond a reduction in idle days, several adjustments of progressively greater magnitude could be made to increase breaker throughput. However, these would only be effective if sufficient chambers were open underground to supply more coal, and there was spare hoisting capacity to raise it to the surface. These adjustments included building or re-opening an additional breaker wing, and incorporating labor-saving devices such as jigs and separators to process the finer coals. Alternatively, the decision could be taken to rebuild the breaker entirely, to incorporate the most modern design features and productivity improvements. While examples of this abounded, a good one was the breaker at the Scranton Coal Company's Slope, whose output prior to 1871 averaged 5–6,000 net tons a month at best. The idle days of the 1871 strike and its aftermath were utilized to rebuild the breaker, which commenced operations in September 1871 and by May 1872 had increased monthly output to 11,613 net tons.[39]

In general, the timescales involved for this kind of adjustment varied from a number of months, at a minimum, to a year or more, before the benefits of the investment were fully realized. Beyond this, new mine capacity was required and if it could not be bought, it would have to be constructed. If the opening was a slope or the depth of shaft required was not great, new capacity could begin production within a 1–2 year span. Examples of deeper shafts in previous chapters indicate that this timescale could quickly stretch to 3–4 years, assuming no interruptions to the construction process. If delays were introduced as a result of financial problems, or deliberate management decisions, the period of investment, without financial return, could extend to 8–9 years or even longer in the most extreme cases. In addition to this, it appears that the larger the mine and the more complex the breaker infrastructure, the longer it took for new mines to reach their full operating capacity. Output at the Storrs breaker climbed for a decade after first production and, at Woodward, tonnage was still growing beyond the end of the period, nearly two decades after it began shipping.

If account is also taken of the timing of decisions to acquire the relevant lands and leases associated with such mines, the time horizons for decision-making and its implications for future returns stretched into astonishingly long time frames of 30 years or more. This was far beyond the possibilities of any kind of forecasts of future demand conditions or their degree of variability. Hence, it is not surprising to find an 1898 example where the unwinding of decisions made many years previously, coupled with the depressed conditions of the trade, resulted in a proposal from the Lackawanna Coal Dept. to President Sloan that echoed a strategy employed twenty years earlier during the 1870s. This strategy was to idle a group of mines in order to provide for fuller work at the remainder. The proposal is of particular interest, not only because the correspondence makes explicit reference to the circumstances of the 1870s, but also because of the way it demonstrates the modifying influences of past decisions on appropriate management strategies for the time in question.

The proposal noted that by February 28, 1898, the company had paid advance royalty on 7.4 million tons of coal from 31 different leases and the situation was deteriorating year by year. At the same time, almost all the coal being mined from the Archbald, Taylor, Dodge, Pyne, and Bliss mines was from company lands in fee, so it seemed prudent to consider making some of the mines idle, to put more work in the direction of the other mines, so the advance royalties could be reduced. Archbald and Taylor were particular candidates, because their mining costs were higher than at Pyne. However, even if mines were idled, there was no plan to remove the pumps, because the cost of re-opening, if the mines were allowed to fill with water, would greatly exceed the cost of pumping, even if continued at the idled mines for several years. Concern was also expressed about the disturbing effect on the community of throwing 500 men and boys out of work at each mine affected, but the plan was also to re-employ married men whenever possible at the remaining mines doing fuller work, leaving the single men unfortunately to fend for themselves. Equally, however, it was noted that the present situation where no one was working more than 15 hours a week was maximizing shared misery for all.
A disturbing postscript to the implementation of this proposal was that three weeks later, on April 15, 1898, the Oxford breaker burned down. Many of the men believed the fire was incendiary in origin (i.e. arson), but W.H. Storrs initially attributed the cause to overheating in the fan machinery. However, he later revised his opinion to claim that Oxford had been set on fire so work could resume at Archbald.[40] Clearly the risks attendant on certain management strategies were more than just the geological or economic.

A further interesting aspect of this example is that despite the experience gained in the period between the 1870s and the 1890s, the degree of management control over events affecting the industry did not appear to have materially increased. If anything, the situation was worse, because the burden of advance royalties weighed much more heavily on the company's accounts by the end of the century that it had two decades previously. Nor can the fault be laid at the door of a new generation of management, who lacked the necessary knowledge and expertise. President Sloan was still in office in 1898, although his extraordinary tenure of more than thirty years was drawing to a close and W.R. Storrs was still active in an advisory capacity despite failing health, assisting his son W.H. Storrs, who dealt with the day-to-day responsibilities of the Coal Department.

INDIVIDUAL OVERCONFIDENCE AND CORPORATE UNDERPERFORMANCE

Given the range of factors that encouraged an imbalance between supply and the fluctuating demand for coal, and the variety of management strategies available to combat them, it can become difficult to discern which forces were most important in shaping the overall trajectory of the industry during the period.

A first step is to identify the motivation for individual investors and entrepreneurs to enter the coal business. Franklin B. Gowen, himself a failed coal operator turned railroad president, was perhaps as well placed as anyone to pursue this issue. During the 1875 Reading Investigation he quizzed James Neill of Neill and McCreary, who had been in the mining business since 1847. Neill had just observed that between 1870 and 1873, the prices received for coal at Port Carbon near Pottsville had not covered the running expenses of the colliery:

Q: (Gowen) 'Is it not a grave question at these prices whether or not a man would be justified in investing $2–300,000 in a colliery?
A: (Neill) Mining is [a] good deal like gambling, when a man gets into it he is likely to stick to it.

Q: Is it a delusive business?
A: Very.

Q: Is it always going to be better?
A: Yes.

Q: How many persons have you known shipwrecked in the business of mining since you commenced in 1847?
A: Their name is legion, almost'[41]

In the same vein, at a later stage in the investigation, Gowen was moved to observe that 'there is hardly any business in which the risk to capital is so great as in the coal trade'.[42]

Under cross examination at the 1887–1888 Labor Troubles Investigation more than a decade later, a similar view was expressed by J.S. Harris, speaking of the Schuylkill Region:

> 'It had the fascination that all mining has, people being willing to undergo ruin, which was not only the ruin of themselves but also of their creditors, for the sake of the possibly very large profits. Large profits during the war demoralized the trade for many years afterwards. People were always looking for the large profits and expecting a time would come again when the profits would be large'.[43]

Thus speculative gain, driven by the lure of potentially large returns, was a prime motivating force for individual investors. In a market where the long-run trend was clearly upwards, the downside of an industry, where possible high reward came at very high risk, was too often ignored. In the ceaseless strife between fear and greed, those emotions that drive the investment business, greed was an all too easy winner. Speculation in lands and leases can be traced back at least to the mid-1850s in the Wyoming Region and much earlier in Schuylkill, and it continued until there were no good lands left to acquire. Investment in speculative mine construction continued as long as there was property to be developed and a reasonable quantity of good coal to be won.

Speculative gains and losses were also a prominent feature of gambling on the success of new railroad enterprises. However, by the end of the Civil War the major anthracite railroads were established companies with institutional shareholders and the backing of leading capitalists in New York, Philadelphia, and London. For such organizations with millions of dollars of investment in railroad and terminal infrastructure, total capital loss could not be countenanced and future bond interest payments and shareholders dividends were to be safeguarded wherever possible. For financial interests on this scale, mere market participation was not sufficient to manage risk, owing to the level of dependence on a fluctuating and unreliable mining sector. A substantial element of market power and control over supply was the inevitable result, as expressed in the various cartels. In this context, it is interesting to note that during the 1875 Investigation, Gowen cast the Coal and Iron in the role of savior of the Schuylkill Region, since without it James Neill argued that as a result of the 1873 panic, the majority of Schuylkill collieries would have ended up in bankruptcy, to be sold at the Sheriff's sale for 10 cents on the dollar.[44] On this basis, restraint of trade and avoidance of capital destruction were two sides of the same coin.

However, the cartel was not able to withstand adverse market forces indefinitely, crumbling as it did in 1876, with a partial revival in 1878. When prosperous times and expanding production returned at the beginning of the 1880s, the cartel organization became easier to maintain, in the absence of serious downward pricing pressure. Unfortunately, improving profitability tempted the individual operators back into the industry. Such activity fuelled speculation in coal properties and the competition between railroads to secure tonnage and retain market share further encouraged new entrants and gave them access to railroad loans. This positive feedback process led to significant overcapacity problems, after construction time lags had worked their way through the system and interacted in unexpected ways with the depressed conditions of the mid-1880s. This process was more marked in the Wyoming than the Schuylkill Region. It was partly a result of the more favorable mining conditions in the former region, where lower costs increased the likelihood of profitable working. However, the smaller landholdings of the railroads in the Wyoming Region, as compared to the Coal and Iron in Schuylkill, meant that there was greater competition between cartel players and a lesser degree of spatial monopoly, owing to the proximity of tracks belonging to the different companies. This made the Wyoming Region more fertile ground for new entrants to establish a foothold.

Unfortunately for the Wyoming Region, the lesser degree of control that could be exercised by any one cartel member, as a result of this more competitive environment, reduced the effectiveness of adjustment processes operated by management in the different companies. Burgeoning advance royalties, that unwelcome by-product of innocuous contractual conditions and speculative lease acquisition ahead of business requirements, and the seemingly inexorable decline in days worked per annum, were the most obvious manifestations of this long-term failure to match production capacity to demand. Limited capacity utilization also created a favorable environment for the spread of unionization, prior to the strikes of 1900 and 1902. It is interesting to compare this situation with the Schuylkill Region, which managed to employ its mining capacity somewhat more effectively in the later years of the 19th century. Here the degree of control exerted by the Coal and Iron was much greater than its Wyoming Region colleagues could achieve, but it was still far from comprehensive. In 1890 the company mined about 55 percent of the Schuylkill Region total, but its relative share had declined by 1901 to about 47 percent.

We are then left with a curious irony. It appears that the more prudent levels of investment in mining lands by the Wyoming Region companies, coupled with the web of contractual relationships these railroads developed with their tributary operators, may have produced slightly less efficient overall development of mining operations than the excessive investments of the Coal and Iron in its large estates, which saddled its parent railroad with 30 years of debt, and denied shareholders a reasonable return on their capital. Or perhaps, with the benefit of hindsight, we are being too harsh on those who devoted their energy and financial resources to developing a major industry in the face of uncertain and fluctuating economic conditions. Perhaps J. S. Harris was correct to assert that 'under no system that could be devised . . . would it be possible to have just enough mines open to supply the demand and keep the men steadily employed'.[45]

Of course, mining efficiency and productivity was only part of the equation. The Lackawanna was able to develop its own mines rapidly, while making loans to entrepreneurs and keeping cash-strapped individual operators afloat. It could do this because of its large reserves and its masterly timing, in the construction of its connection to Buffalo, to benefit from the expansion of the western trade. These reserves also carried it through the revenue doldrums of the 1890s. The Lehigh Valley suffered from correspondingly poor railroad investment timing in later years. It did not profit from the 1880s expansion to the same extent as the Lackawanna and its Buffalo connection, completed in 1892, was made just as the growth of the previous decade evaporated. This, coupled with costly mining investments and failure to control the advance royalty problem had brought the company almost to its knees by the end of the 19th century. The Coal and Iron, and its parent the Reading Railroad, struggled on through multiple financial crises and loss of investor confidence both at home and overseas. Available funds for developing its property were too limited for too long to maximize the potential of its huge estate within the time period of interest, but the dogged persistence of its engineers, managers, and men finally brought some financial stability as the new century dawned.

There can be few sagas of regional economic development where such rapid growth and such high levels of investment were concentrated into such restricted areas of geographic space. Hundreds of thousands of men and their families were to participate in the industry and its mining communities. The product would be shipped across a large portion of the United States and even abroad, competing with declining success in industrial markets against other coals and cokes, but remaining pre-eminent as a domestic fuel of choice, wherever it was distributed. And in the process, there developed one of the world's most unique and abidingly fascinating regional industrial landscapes.

POSTSCRIPT ON THE NATURE OF EXPLANATION IN THE ANALYSIS OF REGIONAL ECONOMIC DEVELOPMENT

To recapitulate from Chapter 1, it has not been the purpose of this study to burden the reader with abstract theorizing or complex mathematical equations. Instead, the preferred approach has been to let the richness of economic life in a regional context speak, at least in some measure, for itself, through the many voices of the major actors in the drama. We have examined the basis for their decisions and attempted to trace the unintended consequences of their actions. Such consequences arose both from events over which they had no control, and from the outcomes of the particular mix of competitive and co-operative behavior exhibited by the industry, over the time period of interest.

Several points are not in doubt, based on the evidence presented. The first is that major deviations from a settled equilibrium between supply and demand for anthracite, and inefficient use of mining plant, characterized the hard coal industry. Although improvements in average capacity utilization were observed over time in Schuylkill, the reverse was the case in the Wyoming Region. Secondly, the levels of uncertainty over future market requirements for coal, the timing and amplitude of economic fluctuations and possible occurrences of mining accidents or calamities were very great. As a result, managers spent much of their time reacting to circumstances, some of which were the product of past decisions, whether good or bad, while others were not. Larger companies had more adjustment options available to them, because of better financing and the ability to shift production between mines. Despite this, both large and small companies had regular encounters either with the Receivers or the Sheriff's sale. The other difficulty faced by managers was that investments in coal estates and deep mines had implications that lasted for decades, while the business fluctuations across which they had to maintain cash flow, were of much shorter duration. That said, it is equally clear that the better managers in the coal industry pursued remarkably long-term strategic policies, to acquire, safeguard and develop their coal estates and tributary tonnage, not least because of their railroad president's wished to safeguard even larger investments in railroad infrastructure spanning several states.

On this basis, it is appropriate to concur with Clark, Gertler, and Whiteman, in their insistence that the major assumptions of neo-classical economic theory about the importance of economic equilibrium, based on actors with perfect information about the present state of the system and the effects of their actions on future system states, cannot be sustained.[46] As a by-product, such theorizing ignores temporal contingency, spatial constraints and the processes of adjustment to changing economic circumstances. In so doing it eliminates the history, geography, and complex dynamics of regional economic systems, taking with it, in the view of the present author, almost everything that is of real interest in regional economic analysis. This would be less of a problem were it not for the fact that neo-classical theory is the prevailing explanatory paradigm in economics and for a good deal of work in economic history that draws its inspiration from contemporary economic thought.[47] For studies of the present kind, neo-classical approaches are largely unproductive, because they have minimal explanatory power. Indeed, there is justification for claiming that they are counterproductive, because they focus attention on hypothesized processes that have no real world counterparts. On this basis it can be argued, using suitable mining terminology, that they should be consigned to the culm bank of dead ends in the history of ideas, where they rightfully belong, as far as the analysis of historical processes of regional economic development is concerned!

This study has aimed instead to demonstrate that the decisions and adjustments made by investors, entrepreneurs, managers, and labor, in the light of geographical and other constraints, and in response to a highly uncertain and fluctuating economic climate, were not incidental to the business of regional economic development, they were its very substance. The study has also shown that, despite the continual buffeting of adverse economic circumstances, genuine long-term vision (as opposed to 'visionary' schemes!) and strategic thinking also had an important part to play. In this regard, it is necessary to take issue with the assertion of Clark et al., that long-run patterns of regional growth, income and employment are simply the accumulation of short-run production decisions.[48] While this may be true in certain historical or contemporary case studies, it was not the case in the anthracite industry, at least as long as managers of the calibre of General Henry Pleasants, Thomas Dickson, William R. Storrs, and the indomitable Samuel Sloan had a say in how it was run.

NOTES FOR CHAPTERS 1–16

NOTES FOR CHAPTER 1

1. G.L. Clark, M.S. Gertler and J.E.M. Whiteman, *Regional Dynamics: Studies in Adjustment Theory* (London; Boston, Mass.: Allen and Unwin, 1986).
2. E. Jones, *The Anthracite Coal Combination in the United States* (Cambridge, Mass.: Harvard University Press, 1914).
3. M. Hollis and E.J. Nell, *Rational Economic Man: a Philosophical Critique of Neo-Classical Economics* (London: Cambridge University Press, 1975).
4. S. Dow, "Mainstream Economic Methodology," *Cambridge Journal of Economics* 21, 1 (1997): 73–93.
5. A. Giddens, *The Constitution of Society: Outline of the Theory of Structuration* (Cambridge: Polity Press, 1984).
6. N. Gregson, "Structuration Theory: Some Thoughts on the Possibilities for Empirical Research," *Environment and Planning D: Society and Space* 5 (1987): 73–91; N. Gregson, "On the (Ir)relevance of Structuration Theory to Empirical Research," in *Social Theory of Modern Societies: Anthony Giddens and his Critics.* Edited by D. Held and J.B. Thompson, (Cambridge: Cambridge University Press, 1989).
7. M.P. Conzen, "The American Urban System in the Nineteenth Century," in *Geography and the Urban Environment: Progress in Research and Applications,* ed. D.T. Herbert and R.J. Johnston, (New York: John Wiley, 1981); M.P. Conzen, "The Progress of American Urbanism, 1860–1930," in *North America: The Historical Geography of a Changing Continent.* Edited by R.D. Mitchell and P.A. Groves, (London: Hutchinson, 1987); D.R. Meyer, "A Dynamic Model of the Integration of Frontier Urban Places into the United States System of Cities," *Economic Geography* 56 (1980): 120–140; E.K. Muller, "Regional Urbanization and the Selective Growth of Towns in North American Regions," *Journal of Historical Geography* 3 (1977): 21–39; A.R. Pred, *The Spatial Dynamics of U.S. Urban-Industrial Growth, 1800–1914: Interpretive and Theoretical Essays* (Cambridge, Mass.: MIT Press, 1966); G.R. Taylor and I.D. Neu, *The American Railroad Network 1861–1890* (Cambridge, Mass.: Harvard University Press, 1956); J.E. Vance Jr., *The North American Railroad: Its Origin, Evolution and Geography* (Baltimore: Johns Hopkins University Press, 1995); D. Ward, *Cities and Immigrants: A Geography of Change in Nineteenth-Century America* (New York and London: Oxford University Press, 1971); Bureau of Census, *A Century of Population Growth in The United States 1790–1900* (Washington, DC: Bureau of Census, 1909).
8. D.R. Meyer, "Emergence of the American Manufacturing Belt: An Interpretation," *Journal of Historical Geography* 9 (1983): 145–174; D.R. Meyer, "Midwestern Industrialization and the American Manufacturing Belt in the Nineteenth Century," *Journal of Economic History* 49 (1989): 921–937.
9. C.P. DiCiccio, *Coal and Coke in Pennsylvania* (Harrisburg: Pennsylvania Historical and Museum Commission, 1996).
10. *Historical Statistics of the United States: Colonial Times to 1970* (Washington: Bureau of the Census, 1976), series M93–106, M123–137.
11. K. Warren, *Wealth, Waste and Alienation: Growth and Decline in the Connellsville Coke Industry* (Pittsburgh: University of Pittsburgh Press, 2001).
12. C.L. Bartholomew and L.E. Metz, *The Anthracite Iron Industry of the Lehigh Valley* (Easton, PA.: Center for Canal History and Technology, 1988).
13. H.M. Eavenson, *The First Century and a Quarter of the American Coal Industry* (Pittsburgh: Privately Printed, 1942), 252.
14. C.K. Yearley, *Enterprise and Anthracite: Economics and Democracy in Schuylkill County, 1820–1875* (Baltimore: Johns Hopkins Press, 1961).
15. J.I. Bogen, *The Anthracite Railroads: A Study in American Railroad Enterprise* (New York: Ronald Press, 1927).
16. Jones, *Anthracite Coal Combination,* 40.
17. P.W. MacAvoy, *The Economic Effects of Regulation: The Trunk-Line Railroad Cartels and the Interstate Commerce Commission before 1900* (Cambridge, Mass.: MIT Press, 1965).
18. Jones, *Anthracite Coal Combination.*
19. Ibid., 40; Yearley, *Enterprise and Anthracite,* 215; G.O. Virtue, "The Anthracite Combinations." *Quarterly Journal of Economics* 10 (April 1896): 296–323.
20. A. Marshall, *Principles of Economics* (London: Macmillan, 1920), 12.
21. B. Warf, "Regional Transformation, Everyday Life, and Pacific Northwest Lumber Production," *Annals, Association of American Geographers* 78, 2 (1988): 326–346.
22. H.W. Aurand, *From the Molly Maguires to the United Mine Workers: The Social Ecology of an Industrial Union, 1869–1897* (Philadelphia: Temple University Press, 1971); A.F.C. Wallace, *St. Clair: A Nineteenth-Century Coal Town's Experience with a Disaster-Prone Industry* (Ithaca: Cornell University Press, 1987).
23. U. S. Anthracite Coal Strike Commission, *Report to the President on the Anthracite Coal Strike of May–October, 1902* (Washington: Government Printing Office, 1903).

Hereafter referred to as *Report of the Anthracite Coal Strike Commission.*

24. P.K. Blatz, *Democratic Miners: Work and Labor Relations in the Anthracite Coal Industry, 1875–1925* (Albany: State University of New York Press, 1994).
25. Quoted in Blatz, *Democratic Miners*, 37.
26. G. Palladino, *Another Civil War: Labor, Capital, and the State in the Anthracite Regions of Pennsylvania, 1840–1868* (Urbana, IL: University of Illinois Press, 1990); D.L. Salay (Ed.), *Hard Coal, Hard Times: Ethnicity and Labor in the Anthracite Region* (Scranton, PA: Anthracite Museum Press, 1984); V.R. Greene, *The Slavic Community on Strike: Immigrant Labor in Pennsylvania Anthracite* (Notre Dame: University of Notre Dame Press, 1968); W.D. Jones, *Wales in America: Scranton and the Welsh 1860–1920* (Cardiff: University of Wales Press, 1993); D.L. Miller, and R.E. Sharpless, *The Kingdom of Coal: Work, Enterprise and Ethnic Communities in the Mine Fields* (Philadelphia: University of Pennsylvania Press, 1985).
27. P. Roberts, *The Anthracite Coal Industry* (New York: The Macmillan Company, 1901); P. Roberts, *Anthracite Coal Communities* (New York: The Macmillan Company, 1904).
28. B.W. Folsom, *Urban Capitalists: Entrepreneurs and City Growth in Pennsylvania's Lackawanna and Lehigh Regions, 1800–1920* (Baltimore: Johns Hopkins University Press, 1981); E.J. Davies, *The Anthracite Aristocracy: Leadership and Social Change in the Hard-Coal Regions of Northeastern Pennsylvania 1800–1930* (DeKalb, IL: Northern Illinois University Press, 1985).
29. W.G. Roy, "Interlocking Directorates and the Corporate Revolution," *Social Science History* 7, 2 (1983): 143–164.
30. A.D. Chandler, *The Visible Hand: the Managerial Revolution in American Business* (Cambridge, Mass.: Harvard University Press, 1977).
31. J. Allen, D. Massey, and A. Cochrane, with J. Charlesworth, G. Court, N. Henry, and P. Sarre, *Rethinking the Region* (London: Routledge, 1998), 56.

NOTES FOR CHAPTER 2

1. B. Thomas, *Migration and Economic Growth* (Cambridge, England Cambridge University Press, 1973); G.W. Van Vleck, *The Panic of 1857: An Analytical Study* (New York: Columbia University Press, 1943); D. Glasner (Ed.), *Business Cycles and Depressions: An Encyclopaedia* (New York: Garland, 1997).
2. C. Juglar, *A Brief History of Panics and Their Periodical Occurrence in the United States* (New York: Augustus M. Kelley, 1966) (Reprint of 1916 3rd English Edition of French Original Text published in Paris in 1862).
3. A. Tylecote, *The Long Wave in the World Economy: The Present Crisis in Historical Perspective* (London and New York: Routledge, 1991), 10.
4. J. S. Gibbons, *The Banks of New York: Their Dealers, the Clearing House, and the Panic of 1857* (New York: Appleton, 1858).
5. J. Kitchin, "Cycles and Trends in Economic Factors," *Review of Economics and Statistics* 5 (1923): 10–16.
6. S.S. Kuznets, *Secular Movements in Production and Prices: Their Nature and Their Bearing upon Cyclical Fluctuations* (Cambridge, Mass.: Riverside Press, 1930); S.S. Kuznets, "Long Swings in the Growth of Population and in Related Economic Variables," *Proceedings of the American Philosophical Society* 102 (1958): 25–52.
7. N.D. Kondratieff, "The Long Waves in Economic Life," *Review of Economics and Statistics* 17 (1935): 105–115.
8. J.J. van Duijn, *The Long Wave in Economic Life* (London: George Allen and Unwin, 1983), 6.
9. Kitchin, *Cycles and Trends.*
10. Tylecote, *Long Wave*, 7.
11. van Duijn, *Long Wave*, 12–13.
12. C. Hoffman, *The Depression of the Nineties: An Economic History* (Westport, Conn.: Greenwood Press, 1970), 237.
13. Sources of data for table columns: **Kitchin Cycles**: A.F. Burns and W.C. Mitchell, *Measuring Business Cycles* (New York: National Bureau of Economic Research (November, 1946), 78; **Juglar Cycles (peaks only)**: S. Solomou, *Phases of Economic Growth 1850–1973* (Cambridge: Cambridge University Press, 1987), 28; **Kuznets Cycles**: van Duijn, *Long Wave*, 16; **Kondratieff Cycles**: Tylecote, *Long Wave*, 12; **Currie's Rainfall Cycle**: Tylecote, *Long Wave*, 8.
14. Kondratieff, *Long Waves.*
15. R. Easterlin, *Population, Labor Force and Long Swings in Economic Growth: The American Experience* (New York: Columbia University Press for NBER, 1968), 6.
16. Kuznets, *Long Swings.*
17. M. Abramowitz, Long Swings in American Economic Growth, in *New Views on American Economic Development: A Selective Anthology of Recent Work.* Edited by R.L. Andreano (Cambridge, Mass.: Schenkman Publishing Company, 1965).
18. van Duijn, *Long Wave*, 15.
19. Thomas, *Migration and Economic Growth.*
20. Solomou, *Phases of Economic Growth*, 11.
21. Kondratieff, *Long Waves*, 111.
22. For additional discussion of this issue, see J.S. Goldstein, "Kondratieff Waves as War Cycles," *International Studies Quarterly* 29 (1985): 411–444.
23. van Duijn, *Long Wave.*
24. Tylecote, *Long Wave.*
25. van Duijn, *Long Wave*, 83.
26. Tylecote, *Long Wave*, Chapter 2.
27. C. Goodrich, *Canals and American Economic Development* (New York: Columbia University Press, 1961), 6–7.
28. Tylecote, *Long Wave*, 70.
29. Solomou, *Phases of Economic Growth*, 100.
30. Ibid., 61, 80–87.

31. Tylecote, *Long Wave*, 8.
32. J.W. Forrester, "Changing Economic Patterns," *Technology Review* 80, 8 (1978): 3–9; J.W. Forrester, "Innovation and Economic Change," *Futures* 13 (1981): 323–331; Mass, *Economic Cycles: An Analysis of Underlying Causes* (Cambridge, Mass.: Wright-Allen Press, 1975).
33. J.P. Warren, *The 50-Year Boom-Bust Cycle: The Case for Kondratieff's Long Wave Theory* (Godalming: Warren, Cameron and Company, 1982), 6–7.
34. Solomou, *Phases of Economic Growth*, 9.
35. A.W. Niemi, *State and Regional Patterns in American Manufacturing, 1860–1900* (Westport, Conn.: Greenwood Press, 1974).
36. G.H. Borts, "Regional Cycles of Manufacturing Employment in the United States, 1914–1953," *Journal of the American Statistical Association* 55 (March 1960): 151–211.
37. W. Isard, *Methods of Regional Analysis: An Introduction to Regional Science* (Cambridge, Mass.: MIT Press, 1960).
38. M. Marshall, *Long Waves of Regional Development* (Basingstoke: Macmillan, 1987).
39. While it is very difficult to find any literature that bears directly on this issue, useful background information on financial aspects of the problem can be found in L.E. Davis, "The Investment Market, 1870–1914: The Evolution of a National Market," *Journal of Economic History* 25 (1965): 355–399; other early papers that raise pertinent points about the study of regional business cycles include: P. Neff, "Interregional Cyclical Differentials: Causes, Measurement, and Significance," *American Economic Review* 39 (1949): 105–119; and R. Vining, "Location of Industry and Regional Patterns of Business-Cycle Behavior," *Econometrica* 14 (1946): 37–68; in contrast, F. Bateman and T. Weiss, "Comparative Regional Development in Antebellum Manufacturing," *Journal of Economic History* 35 (1975): 182–208, makes no mention of business-cycle-related regional dynamics.
40. van Duijn, *Long Wave*, 6.
41. W.W. Price, *We Have Recovered Before! A Comparison of the Present Depression with the Major Depressions of the Past Century: 1837, 1857, 1873, 1893* (New York: Harper and Brothers, 1933).
42. R.J. Cornell, *The Anthracite Coal Strike of 1902* (Washington: Catholic University of America Press, 1957).
43. Van Vleck, *Panic of 1857*, 38.
44. See J.L. Huston, *The Panic of 1857 and the Coming of the Civil War* (Baton Rouge: Louisiana State University Press, 1987), 34 footnote.
45. Ibid., 6; for a convenient summary chronology of the Crimean War, see *http://british-forces.com/fkac/conflicts/1800–1914/crimea.html*
46. Van Vleck, *Panic of 1857*, 48–50.
47. E.D. Fite, *Social and Industrial Conditions in the North During the Civil War* (Williamstown, Mass.: Corner House, 1976), 14–15.
48. Huston, *Panic of 1857*, 30–34.
49. D.C. North, *The Economic Growth of the United States, 1790–1860* (Englewood Cliffs, NJ: Prentice-Hall, 1961), 256; C.W. Calomiris and L. Schweikart, "The Panic of 1857: Origins, Transmission, and Containment," *Journal of Economic History* 51 (1991): 807–834. The latter authors dispute Huston's emphasis on the importance of changes in the transatlantic grain trade as a factor leading to the panic, but the price data they use as evidence do not allow the effects of local supply/demand conditions in the two countries, related to different harvest conditions, to be distinguished from the effects of changes in balancing levels of American grain exports.
50. Van Vleck, *Panic of 1857*, 27.
51. Huston, *Panic of 1857*, 21.
52. Gibbons, *Banks of New York*, 365.
53. Huston, *Panic of 1857*, 13.
54. Van Vleck, *Panic of 1857*, 70–74.
55. Huston, *Panic of 1857*, 29, 34.
56. Van Vleck, *Panic of 1857*, 91.
57. Huston, *Panic of 1857*, 212.
58. J.M. Swank, *History of the Manufacture of Iron in All Ages and Particularly in the United States from Colonial Times to 1891* (Philadelphia: American Iron and Steel Association, 1891), 376.
59. Huston, *Panic of 1857*, 62, 217; Van Vleck, *Panic of 1857*, 88–89.
60. R.L. Andreano (Ed.), *The Economic Impact of the Civil War* (Cambridge, Mass.: Schenkman, 1962).
61. D.T. Gilchrist and W.D. Lewis (Eds.), *Economic Change in the Civil War Era* (Greenville, Del.: Eleutherian Mills-Hagley Foundation, 1965).
62. P. O'Brien, *The Economic Effects of the American Civil War* (Basingstoke: Macmillan, 1988).
63. Ibid., 9–10.
64. T.S. Cochran, "Did the Civil War Retard Industrialisation?" in *The Economic Impact of the Civil War*, ed. R.L. Andreano (Cambridge, Mass.: Schenkman, 1962).
65. S. Salsbury, "The Effect of the Civil War on American Industrial Development," in *The Economic Impact of the Civil War*, ed. R.L. Andreano (Cambridge, Mass.: Schenkman, 1962).
66. W.W. Swanson, "The Crisis of 1860 and the First Issue of Clearing-House Certificates: II," *Journal of Political Economy* 16 (1908): 212–226.
67. Fite, *Social and Industrial Conditions*, 106–108.
68. Ibid., 141.
69. *Miner's Journal*, April 15, 1865, August 5, 1865.
70. Swank, *History of the Manufacture of Iron*, 376.
71. Fite, *Social and Industrial Conditions*, 184.
72. R.P. Sharkey, *Money, Class, and Party: An Economic Study of Civil War and Reconstruction* (Baltimore: Johns Hopkins Press, 1959), 181.
73. S.H. Daddow and B. Bannan, *Coal, Iron and Oil* (Pottsville, PA: B. Bannan, 1866), 781.
74. O'Brien, *Economic Effects*, 55.

75. Aurand, *Molly Maguires*; W.C. Broehl Jr., *The Molly Maguires* (New York: Vintage/Chelsea House, 1964).
76. American Iron and Steel Association (AISA), *Statistics of the American and Foreign Iron Trades for 1882* (Philadelphia: AISA, 1883), 57.
77. Thomas, *Migration and Economic Growth*, 402.
78. R. Fels, *American Business Cycles, 1865–1897* (Chapel Hill: University of North Carolina Press, 1959), 97–106.
79. Ibid., 110.
80. Jones, *Anthracite Coal Combination.*
81. S. Rezneck, "Distress, Relief, and Discontent in the United States during the Depression of 1873–1878," *Journal of Political Economy*, 58 (1950): 494–512; E.W. Martin [James Dabney McCabe], *The History of the Great Riots: Being a Full and Authentic Account of the Strikes and Riots on the Various Railroads of the United States and in the Mining Regions … Together with a Full History of the Mollie Maguires* (Philadelphia: J.C. McCurdy and Co., 1877); Bogen, *Anthracite Railroads.*
82. Warren, *The 50-Year Boom-Bust Cycle*; Fels, *American Business Cycles*, 168–172.
83. Fels, *American Business Cycles*, 173.
84. Hoffman, *Depression of the Nineties*, 54.
85. Fels, *American Business Cycles*, 185.
86. Hoffman, *Depression of the Nineties*, 57–62.
87. Ibid., 79–85.
88. Ibid., 63, 68, 86, 259.
89. Fels, *American Business Cycles*, 211–213.
90. C.P. Kindleberger, *Manias, Panics, and Crashes: A History of Financial Crises* (London: Macmillan, 1978), 126.
91. L.E. Davis and R.J. Cull, *International Capital Markets and American Economic Growth 1820–1914* (Cambridge: Cambridge University Press, 1994).
92. D.R. Adler, *British Investment in American Railways 1834–1898* (Charlottesville: University of Virginia Press, 1970).
93. Robert Fleming to W.H. Tillinghast, January 21, 1881, Special Letterbook, W.H. Tillinghast, September 1879–June 1885, Box 37, Lehigh and Wilkes-Barre Coal Company Records, 1869–1928, Luzerne County Historical Society, hereafter abbreviated to LWBCCR.
94. For further details of this important event and the developments leading up to it, see R.L. Thompson, *The History of the Telegraph Industry in the United States, 1832–1866* (Princeton: Princeton University Press, 1947).
95. Tylecote, *Long Wave*, 8.
96. Mr. Dillingham, *Reports of the Immigration Commission, Statistical Review of Immigration 1820–1910; Distribution of Immigrants 1850–1900, 61st Cong., 3rd Sess. Senate Doc. 756* (Washington, DC: Government Printing Office, 1911), 29.
97. O. MacDonough, "The Irish Famine Emigration to the United States," Perspectives in American History 10 (1976): 357–446.
98. H. Jerome, *Migration and Business Cycles* (New York: NBER, 1926), 187.
99. Thomas, *Migration and Economic Growth*, 116–117.
100. Jerome, *Migration and Business Cycles*, 206–207.
101. Ibid., 208, 241.
102. Thomas, *Migration and Economic Growth*, 245.
103. R.T. Berthoff, "The Social Order of the Anthracite Regions," *Pennsylvania Magazine of History and Biography* 89 (1965), 261–291; W.A. Gudelunas and W.G. Shade, *Before the Mollie Maguires: The Emergence of the Ethno-Religious Factor in the Politics of the Lower Anthracite Region, 1844–1872* (New York: Arno Press, 1976), 23.
104. Jones, *Wales in America.*
105. Dillingham, *Statistical Review of Immigration 1820–1910*, 13.
106. A.K. Knowles, *Calvinists Incorporated: Welsh Immigrants on Ohio's Industrial Frontier* (Chicago: The University of Chicago Press, 1997).
107. R. Lewis, West Virginia University, Pers. Comm., 2001.
108. Nationality Census of Miners, 1857, Miscellaneous Coal Department Records, Delaware, Lackawanna and Western Railroad Coal Department Records, Lackawanna Historical Society, hereafter referred to as LCDR.
109. *Reports of the Immigration Commission, Parts 17–20, (Part 19, Anthracite Coal Mining), 61st Cong. 2nd Sess. Senate Doc. 78* (Washington, DC: Government Printing Office, 1911), 591–593.
110. W.R. Storrs to S. Sloan, Jan. 25, 1881, Letterbook (LB) 32A (LCDR).
111. F.E. Saward, *The Coal Trade 1882* (data for 1881), 2. These are annual publications containing coal-trade statistics and market reports. The latest data are for the year preceding that of the publication date. To distinguish this series of publications clearly from the Coal Trade Journal, which was a trade newspaper, it will hereafter be referred to as Coal Trade Annual, followed by the publication date and the date of latest statistics in brackets, e.g. Coal Trade Annual 1882 (1881), 2.
112. c.f. *Report of the Inspectors of Mines of the Anthracite Coal Regions of Pennsylvania for the Year 1879* (Harrisburg: State Printer, 1880), 274–275. Hereafter referred to as AIM Report for the specified year.
113. *Reports of the Immigration Commission, Part 19, Anthracite Coal Mining*, 593.

DATA SOURCES FOR FIGURES

2.1 Both data series from van Duijn, *Long Wave*, 75–76.
2.2 Figures a and b, data from tabulation in Price, *We Have Recovered Before!*
2.3 Data for 1860–1903 from *Statistics of the American and Foreign Iron Trades for 1903* (Philadelphia: AISA, 1904), 85; data for 1842–1859 from *Industrial Statistics, 1873–1874, Annual Report of the Secretary of Internal Affairs of the Commonwealth of Pennsylvania* (Harrisburg, PA: State Printer), 290.
2.4 A. Fishlow, *American Railroads and the Transformation of the Ante-Bellum Economy* (Cambridge, Mass.: Harvard University Press, 1965), 397, table 53.

2.5 Ibid.
2.6 *Statistics of the American and Foreign Iron Trades for 1882* (Philadelphia: AISA, 1883), 56.
2.7 Series U1–25 (1900a and b figures averaged), *Historical Statistics of the United States.*
2.8 Dillingham, *Statistical Review of Immigration 1820–1910*, Table 1.
2.9 Ibid., Table 9.
2.10 Ibid.
2.11 Ibid.
2.12 Ibid.

NOTES FOR CHAPTER 3

1. F.M. Binder, *Coal Age Empire: Pennsylvania Coal and its Utilization to 1860* (Harrisburg: Pennsylvania Historical and Museum Commission, 1974), 64–67.
2. J.P. Lesley, E.V. d'Invilliers and A.D.W. Smith, *A Summary Description of the Geology of Pennsylvania, Vol. 3, Part I* (Harrisburg: Board of Commissioners of the Geological Survey of Pennsylvania, 1895), 1920–1925, 2151.
3. H.M. Chance, *Report on the Mining Methods and Appliances used in the Anthracite Coal Fields* (Harrisburg: Board of Commissioners of the Geological Survey of Pennsylvania, 1883), 8.
4. S.W. Campbell, *Businessmen and Anthracite: Aspects of Change in the Late 19th-Century Anthracite Industry* (Ph.D. Diss., University of Delaware, 1978), 5.
5. Chance, *Mining Methods*, 10–17.
6. Daddow and Bannan, *Coal, Iron and Oil*, 255–257; Yearley, *Enterprise and Anthracite*, 155; Binder, *Coal Age Empire*, 18.
7. Daddow and Bannan, *Coal, Iron and Oil*, 249.
8. *Pittston Gazette*, August 17, 1855.
9. See C.A. Ashburner, *First Report of Progress in the Anthracite Coal Region: The Geology of the Panther Creek Basin or Eastern End of the Southern Coalfield* (Harrisburg: Board of Commissioners of the Second Geological Survey, 1883), 180.
10. Chance, *Mining Methods*, 457–474.
11. Daddow and Bannan, *Coal, Iron and Oil*, 112–115; M. Knies, *Coal on the Lehigh 1790–1827: Beginnings and Growth of the Anthracite Industry in Carbon County, Pennsylvania (Easton: Canal History and Technology Press, 2001), 44.*
12. Chance, *Mining Methods*, photograph facing p. 77.
13. Daddow and Bannan, *Coal, Iron and Oil*, 411–412.
14. Yearley, *Enterprise and Anthracite*, 110.
15. T. Phillips, *History of Wilkes-Barre and Luzerne County* (Wilkes-Barre: Typescript, n.d.), 129, 275.
16. *Annual Report of the Lehigh Coal and Navigation Company for 1868*, 27.
17. Chance, *Mining Methods*, 57.
18. Yearley, *Enterprise and Anthracite*, 130.
19. Chance, *Mining Methods*, 58.
20. C.C. Ciesla, *1694–1994 History of Hanover Township: Including the Boroughs of Ashley, Sugar Notch, and Warrior Run and Portions of Nanticoke City* (Slocum township: C. Ciesla, 1994), 85.
21. Chance, *Mining Methods*, 130.
22. Chance, *Mining Methods*, 130–137.
23. Daddow and Bannan, *Coal, Iron and Oil*, 431.
24. Yearley, *Enterprise and Anthracite*, 122.
25. Chance, *Mining Methods*, 118, 144.
26. Daddow and Bannan, *Coal, Iron and Oil*, 432.
27. Chance, *Mining Methods*, 128.
28. *AIM Report for 1879*, 284; Wallace, *St. Clair*, 42–49.
29. Yearley, *Enterprise and Anthracite*, 114.
30. Blatz, *Democratic Miners*, 148–152.
31. Chance, *Mining Methods*, 443–445.
32. Ibid., 472.
33. Miller and Sharpless, *Kingdom of Coal*, 123.
34. J.L. Peresie, "Crusader with a Camera: Lewis Hine and his Battle Against Child 'Slavery'," *Pennsylvania Heritage* 23, 3 (1997): 4–13.
35. *Pittston Gazette*, Dec. 2, 1869.
36. J.H. Jones, *Industrial Espionage Report on the Pennsylvania Coal Company*, 1889, Reading Company Papers, Hagley Library, 83.
37. Data from Chance, *Mining Methods*, 455.
38. J.H. Jones, *Industrial Espionage Report on the Susquehanna Coal Company*, 1889, Reading Company Papers, 6.
39. R. Pumpelly, *Report on the Mining Industries of the United States (Exclusive of the Precious Metals): with Special Investigations into the Iron Resources of the Republic and into the Cretaceous Coal of the Northwest* (Washington: Government Printing Office, 1886), 632.
40. U.S. House of Representatives, *Labor Troubles in the Anthracite Regions of Pennsylvania 1887–1888*, 50th Cong. 2nd Sess., House Rept. No. 4147, Vol. 4 (Washington: Government Printing Office, 1889), hereafter referred to as *Labor Troubles*.
41. Daddow and Bannan, *Coal, Iron and Oil*, 483.
42. Chance, *Mining Methods*, 471.
43. W. Ferguson, *America by River and Rail* (London: James Nisbet and Company, 1856), 184–197.
44. *Scranton Republican*, September 11, 1869; Wallace, *St. Clair*, 293–313.
45. This photograph appears to be marked with the year 1888. Supporting evidence is provided by a check on the details of the locomotive shown in Plate 3. From photographs and engine rosters in T.T. III Taber, *The Delaware, Lackawanna and Western Railroad in the Nineteenth Century 1828–1899* (Muncy, Pennsylvania: The Author, 1977), it would appear that the engine is of a type not constructed until the 1880s, making a date of 1888 entirely reasonable. It should be noted that there is an earlier photograph of Avondale, believed to be derived from a Beckwith stereograph, which shows a different configuration of the breaker. Some have thought this might be the original pre-disaster breaker. Unfortunately, no original copy of this photograph is known to be in public hands at the time of this writing and its present whereabouts is unknown. However, according to information provided to the author by

C.F. Petrillo, there are no known examples of Beckwith's work prior to the early part of the 1870s. Hence the stereograph is perhaps more likely to represent the configuration of the breaker shortly after rebuilding. This would further imply that changes to the rebuilt breaker were made between the early 1870s and the time of the 1888 photograph reproduced in Plate 3.

46. W.R. Storrs to S. Sloan, September 13, 1869, LB21 (LCDR).
47. *AIM Report for 1898*, 97.
48. W.R. Storrs to S. Sloan, September 21, 1869, LB21 (LCDR).
49. The tone of individual mine reports by Mine Inspectors in 1872 as compared with 1870 is indicative of substantial progress towards compliance with the 1870 legislation, although the quality of ventilation was still giving cause for concern in many locations.
50. *AIM Report for 1886*, 7–12.
51. Ibid.
52. J. Shaughnessy, *Delaware and Hudson* (Berkeley: Howell-North Books, 1967), 36, 40, 48; *AIM Report for 1877*, frontispiece map. The Johnson Photograph Collection from which the plate is taken is sometimes dated to the beginning of the 1860s, but evidence from this and other photographs suggests many of them were in fact taken in the mid to late 1860s.
53. M.T. Connolly, *The Gravity: History of the Pennsylvania Coal Company Railroad, 1850–1885* (Scranton: Langan, 1972), 21–22.
54. Connolly, *The Gravity*, 37; Shaughnessy, *Delaware and Hudson*, 42.

SOURCES FOR FIGURES

3.1 Cartography by the author. The extent of the railroad network in 1867 was reconstructed from a variety of sources, including: Schaefer, P.W. *A Map of all the Anthracite Collieries and the Pottsville, Lehigh, Mahanoy and Shamokin Coal Basins*. Pottsville, 1859; Daddow and Bannan, *Coal, Iron and Oil*, frontispiece map; *Map of the Lehigh and Susquehanna Railroad*. (Philadelphia: published by the company, 1867); *AIM Report for 1872*, frontispiece map; *Map of the Philadelphia and Reading Railroad and Branches* (Philadelphia: published by the company, 1871*); Annual Reports of the Philadelphia and Reading Railroad for 1867–1872; Annual Reports of the Lehigh Coal and Navigation Company for 1867, 1870.*

3.2 Lesley et al., *Summary Description*, Plate 305, facing p. 1928; also printed on Atlas Plate XXIII to accompany Chance, *Mining Methods.*

3.3 Atlas Plate XVI to accompany Chance, *Mining Methods.*

3.4 Ibid.

NOTES FOR CHAPTER 4

1. See for example G.R. Taylor, *The Transportation Revolution, 1815–1860*, Vol. 4 of The Economic History of the United States (New York: Rinehart, 1951).
2. Eavenson, *First Century*, 139.
3. S. Pearce, *Annals of Luzerne County: A Record of Interesting Events, Traditions, and Anecdotes from the First Settlement at Wyoming to 1860* (Philadelphia: J.B. Lippincott and Company, 1860), 365–366.
4. Daddow and Bannan, *Coal, Iron and Oil*, 110; Pearce, *Annals*, 366–367.
5. Eavenson, *First Century*, 140.
6. Daddow and Bannan, *Coal, Iron and Oil*, 113.
7. Daddow and Bannan, *Coal, Iron and Oil*, 111; Pearce, *Annals*, 368.
8. *AIM Report for 1891*, 80; Eavenson, *First Century*, 141.
9. Phillips, *History of Luzerne County*, 127.
10. Eavenson, *First Century*, 143.
11. H.B. Powell, *Philadelphia's First Fuel Crisis: Jacob Cist and the Developing Market for Pennsylvania Anthracite* (State College: Pennsylvania State University Press, 1978), 19.
12. O.J. Harvey, *A History of Wilkes-Barre, Luzerne County from its First Beginnings to the Present Time* (6 vols.) (Wilkes-Barre: Reader Press, 1909–1930), Vol. 4, 1824–1825.
13. *AIM Report for 1891*, 81–83.
14. Phillips, *History of Luzerne County*, 266.
15. Eavenson, *First Century*, 144–145.
16. Daddow and Bannan, *Coal, Iron and Oil*, 116, 120–121.
17. Powell, *Philadelphia's First Fuel Crisis*, 24.
18. Ibid., 67–68.
19. C.L. Jones, *Economic History of the Anthracite Tide-Water Canals* (Philadelphia: J.C. Winston, 1908), 9.
20. Powell, *Philadelphia's First Fuel Crisis*, 82–85.
21. Jones, *Anthracite Tide-Water Canals*, 126–128.
22. Hudson Coal Company, *The Story of Anthracite* (New York: Hudson Coal Company, 1932), 34.
23. Jones, *Anthracite Tide-Water Canals*, 17.
24. J. Macfarlane, *The Coal Regions of America: Their Topography, Geology and Development* (New York: D. Appleton and Company, 1873), table facing p. 54.
25. Hudson Coal Company, *Story of Anthracite*, 40–42, 61.
26. Jones, *Anthracite Tide-Water Canals*, 76–78.
27. M.B. Wakefield, *Coal Boats to Tidewater: The Story of the Delaware and Hudson Canal* (Grahamsville, NY: Wakefair Press, 1965), 9.
28. Hudson Coal Company, *Story of Anthracite*, 87–90.
29. F.C. Petrillo, *Anthracite and Slackwater: The North Branch Canal 1828–1901* (Easton, PA: Center for Canal History and Technology, 1986), 39–49, 60–71.
30. Pearce, *Annals*, Appendix W.
31. Powell, *Philadelphia's First Fuel Crisis*, 97, 137.
32. Bogen, *Anthracite Railroads*, 11–13.
33. Hudson Coal Company, *Story of Anthracite*, 92–93, 99.
34. Philadelphia and Reading Railroad Company, *Report of the Engineers of the Philadelphia and Reading Railroad Company, with Accompanying Documents* (Philadelphia: John C. Clark, 1839).
35. Bogen, *Anthracite Railroads*, 22–27.
36. Jones, *Anthracite Tide-Water Canals*, 60–67, 104–111.
37. Macfarlane, *Coal Regions of America*, table facing p. 54.

38. R.F. Archer, *The History of the Lehigh Valley Railroad: The Route of the Black Diamond* (Berkeley: Howell-North Books, 1977), 31.
39. Bogen, *Anthracite Railroads*, 146–147.
40. Jones, *Anthracite Tide-Water Canals*, 139.
41. Ibid., 37.
42. *Pittston Gazette*, March 2, 1855.
43. Ibid., June 6, 1856.
44. Ibid., May 31, 1860; June 21, 1860.
45. Board of Managers Minutes, Pennsylvania Railroad, May 11, 1864; June 1, 1864, MG-286, Penn Central Collection, Pennsylvania State Archives, Harrisburg.
46. *An Act to Incorporate the Pennsylvania Coal Company, By an Act Incorporating the Washington Coal Company, passed 11th April, 1838 with the Several Supplements Thereto* (New York: Bourne and Company, 1869); Board of Managers Minutes, Pennsylvania Coal Company, May 16, 1849. According to M. Knies, "The D&H Coal Company: New Insights from the James Archbald Papers," Canal History and Technology Proceedings 20 (2001): 53–80; L. Lowenthal, From the Coal Fields to the Hudson: A History of the Delaware and Hudson Canal (Fleischmanns, N.Y.: Purple Mountain Press, 1997) states that the Wyoming Coal Association acquired the Washington Coal Company. However, the records of the latter contain no reference to such a take-over. In fact, the Wyoming Coal Association was absorbed by the Pennsylvania Coal Company, by an agreement dated May 30, 1851 (Board of Managers Minutes, Pennsylvania Coal Company, June 3, 1851, Pennsylvania Coal Company Records MG-282, Pennsylvania State Archives, Harrisburg).
47. Board of Directors Minutes, Washington Coal Company, Nov. 15, 1847, June 19, 1848, MG-300, Erie Railway Collection, Pennsylvania State Archives, Harrisburg.
48. Board of Directors Minutes, Washington Coal Company, Nov. 30, 1847.
49. P. Ruth, *Of Pulleys and Ropes and Gear: The Gravity Railroads of the Delaware and Hudson Canal Company and the Pennsylvania Coal Company* (Honesdale, PA: Wayne County Historical Society, 1997), 38; Board of Managers Minutes, Pennsylvania Coal Company, September 9, 1847.
50. Ruth, *Of Pulleys and Ropes and Gear*, 38.
51. S. Patton, "Delaware and Hudson Company vs. Pennsylvania Coal Company during the 1850s," *Canal History and Technology Proceedings* 9 (1992), 6–24, see p. 11.
52. Board of Directors Minutes, Washington Coal Company, January 7, 1848, May 16, 1849.
53. Board of Managers Minutes, Pennsylvania Coal Company, October 29, 1849.
54. Ibid., November 3, 1849.
55. Board of Managers Minutes, Delaware and Hudson Canal Company, July 15, 1847, National Canal Museum, Easton, PA.
56. Board of Managers Minutes, Delaware and Hudson Canal Company, August 2, 1847; Patton, "Delaware and Hudson", 11.
57. These figures derive from President Wurts' communication to the Board of Managers on December 23, 1846 and indicate that the figures given in Patton "Delaware and Hudson", p. 10 are incorrect.
58. Board of Managers Minutes, Delaware and Hudson Canal Company, August 2, 1847.
59. Ibid., December 23, 1846.
60. Knies, "D&H".
61. Board of Managers Minutes, Delaware and Hudson Canal Company, November 17, 1847; March 28, 1848.
62. Ibid., November 7, 1850.
63. Patton, "Delaware and Hudson", 12.
64. Ibid.
65. B.H. Throop, *A Half Century in Scranton* (Scranton: Scranton Republican Press, 1895), 227–228.
66. Hudson Coal Company, *Story of Anthracite*, 101.
67. Board of Managers Minutes, Pennsylvania Coal Company, June 3, 1851, August 6, 1851; Board of Managers Minutes, Delaware and Hudson Canal Company, July 29, 1851.
68. Board of Managers Minutes, Pennsylvania Coal Company, May 10, 1854.
69. Patton, "Delaware and Hudson".
70. *Pittston Gazette*, November 19, 1863.
71. Hudson Coal Company, *Story of Anthracite*, 103; Bogen, *Anthracite Railroads*, 77–79.
72. W.D. Lewis, "The Early History of the Lackawanna Iron and Coal Company: A Study in Technological Adaptation," *Pennsylvania Magazine of History and Biography* 96, 4 (1972): 424–468, see p. 438.
73. Lewis, "Lackawanna Iron and Coal Company," 465; Bogen, *Anthracite Railroads*, 79–80.
74. Bogen, *Anthracite Railroads*, 81–83; Hudson Coal Company, *Story of Anthracite*, 104.
75. R.J. Casey and W.A.S. Douglas, *The Lackawanna Story*. (New York: McGraw-Hill, 1951), 80.
76. *Pittston Gazette*, December 6, 1850.
77. Campbell, *Businessmen and Anthracite*, 15.
78. Bogen, *Anthracite Railroads*, 83.
79. Pennsylvania Coal Company, Land Record 1848–1894, Pennsylvania Coal Company Records, MG-282, Pennsylvania State Archives, Harrisburg; *Act to Incorporate the Pennsylvania Coal Company*.
80. Jones, *Anthracite Tide-Water Canals*, 7, 20.
81. Bogen, *Anthracite Railroads*, 120.
82. Yearley, *Enterprise and Anthracite*, 15; S.P. Adams, *Old Dominion, Industrial Commonwealth: Coal, Politics and Economy in Antebellum America* (Baltimore: Johns Hopkins Press, 2004), 196. Adams' study also contains an excellent discussion of the use of special and general incorporation acts by the Pennsylvania State Legislature to foster economic development in different parts of the state (see pp. 176–177). In the present writer's view, however, much more evidence would be required to make a convincing case that

political rather than economic forces dominated processes of 19th-century regional development in Pennsylvania. Although focused on the post-1860 period, the argument of the present work leans decidedly in the opposite direction.

83. *Coal Trade Annual 1888 (1887)*, 17.
84. Philadelphia and Reading Railroad, Coal Shipment and Earnings Data (1843–1898) V-TJ, Transportation Dept. Statement Book 1844–1882 in Accession 1520, The Reading Company, Hagley Library (hereafter referred to as Reading Company Papers).
85. *Pittston Gazette*, February 15, 1856.
86. Ibid., November 14, 1856, November 28, 1856; see Petrillo, *Anthracite and Slackwater*, Chapter 6 for a more detailed discussion.
87. Petrillo, *Anthracite and Slackwater*, 192–197.
88. Andreano, *New Views*, 24–25.
89. Wallace, *St. Clair.*
90. G. Jones, "Map of the Wyoming and Lackawanna Valleys, Principally from a Map Constructed by Col. John L. Butler of Wilkesbarre," *American Journal of Science* 18 (1830): facing p. 309.
91. Phillips, *History of Wilkes-Barre*, 128.
92. Powell, *Philadelphia's First Fuel Crisis*, 88.
93. Phillips, *History of Wilkes-Barre*, 129.
94. Ibid., 187; Ciesla, *History of Hanover Township*, 88.
95. Map of the Wyoming Coal Company's Lands in Luzerne County, n.p., n.d., Luzerne County Historical Society, Wilkes-Barre.
96. Phillips, *History of Wilkes-Barre.*, 130.
97. *Pittston Gazette*, Jan. 7, 1853.
98. Petrillo, *Anthracite and Slackwater*, 146.
99. *Pittston Gazette*, April 22, 1853.
100. Petrillo, *Anthracite and Slackwater*, 141–142; *The Charter and By-Laws of the Maryland Anthracite Coal Company of the Wyoming Coal Region with Reports on their Geology and Mining Resources of their Coal Lands, Together with Supplementary Remarks and Extracts* (Baltimore: John W. Woods, 1856). The author is indebted to Dr. Greg Elmes, University of West Virginia for obtaining a copy of this report.
101. *Pittston Gazette*, April 22, 1853.
102. U.S. Census of Manufactures, *Products of Industry, Luzerne County, Pennsylvania*, 1850.
103. Yearley, *Enterprise and Anthracite*, 111.
104. *Pittston Gazette*, December 17, 1852.
105. Ibid., May 20, 1853.
106. Ibid., January 27, 1854.
107. Ibid., December 5, 1856; Macfarlane, *Coal Regions of America*, table facing p. 54.
108. *Pittston Gazette*, January 18, 1856.
109. Ibid., July 20, 1855.
110. Harvey, *History of Wilkes-Barre*, vol. 4, 1994; *Pittston Gazette*, January 4, 1856.
111. *Pittston Gazette*, December 12, 1851.
112. F.C. Von Bergen, *Photographs of the Delaware, Lackawanna and Western Railroad Coal Department Collieries* (Unpublished bound volume of photographs, Lackawanna Historical Society, circa 1908).
113. *Pittston Gazette*, July 27, 1855, January 11, 1856; it should also be noted that the Luzerne Coal Company referred to here is not the same as the Luzerne Anthracite Company mentioned by Adams, *Old Dominion*, 175.
114. Book 303, Union Iron and Coal Company Records, Sub Series 4.k., Moses Taylor Collection, New York Public Library.
115. Board of Directors Minutes, Union Iron and Coal Company, entries for December 29, 1857, March 13, 1858, January 5, 1859. The direct quotation in the text is from the December 29 minutes.
116. Board of Managers Minutes, Delaware, Lackawanna and Western Railroad, in Delaware, Lackawanna and Western Railroad Company Records, George Arents Research Library, Syracuse University, September 26, 1866, January 25, 1867. The price quotation is found in the first minute, the second minute corrects the first which erroneously states that the lands were bought by the railroad's president, rather than the Steuben Coal Company; see also *Luzerne Union*, January 29, 1868; T.F. Murphy, *Jubilee History, Commemorative of the Fiftieth Anniversary of the Creation of Lackawanna County, Pennsylvania* (Topeka-Indianapolis: Historical Publishing Company, 1928), vol. 1, 149.
117. Shaughnessy, *Delaware and Hudson*, 39–41; Miner's Journal, January 15, 1859; Board of Managers Minutes, Delaware and Hudson Canal Company, December 3, 1859.
118. *Pittston Gazette*, July 13, 1855.
119. Board of Managers Minutes, Delaware and Hudson Canal Company, February 15, 1856, December 19, 1856.
120. *Miner's Journal*, January 15, 1859.
121. Phillips, *History of Wilkes-Barre*, 180.
122. *AIM Report for 1875*, 219.
123. H.D. Rogers, *The Geology of Pennsylvania* (*2 vols.*) (Philadelphia: J.B. Lippincott, 1858), vol. 2, 413–417; Yearley, *Enterprise and Anthracite*, 111.
124. Yearley, *Enterprise and Anthracite*, 115.
125. Phillips, *History of Wilkes-Barre*, 130, 150.
126. *Pittston Gazette*, April 22, 1853, January 27, 1854.
127. F.L. Hitchcock, *History of Scranton and its People* (New York: Lewis Historical Publishing Co., 1914), 64.
128. J. Wurts to J. Archbald, June 28, 1844, Archbald Papers, Lackawanna Historical Society.
129. Knies, 2001, 67.
130. Irad Hawley to J. Archbald, April 8, 1851, Archbald Papers.
131. Munsell and Co. *History of Luzerne, Lackawanna and Wyoming Counties* (New York: Munsell and Co., 1880), 447.
132. Hitchcock, *History of Scranton*, 76.
133. *Pittston Gazette*, January 11, 1856.
134. Ibid., March 21, 1861.
135. Phillips, *History of Wilkes-Barre*, 132.

136. *Pittston Gazette*, April 19, 1860.
137. Ibid., March 21, 1861.
138. Binder, *Coal Age Empire.*
139. Ibid., 13–19.
140. Ibid., 70–74, 166 Table H. The figures in Binder's table do not exactly match those given by the *Miner's Journal*, January 19, 1839.
141. E.T. Freedley, *Philadelphia and Its Manufactures: A Handbook Exhibiting the Development, Variety and Statistics of the Manufacturing Industry of Philadelphia in 1857* (Philadelphia: Edward Young, 1859); J.L. Bishop, *A History of American Manufactures from 1608 to 1860*, 3 Vols. (Philadelphia: Edward Young, 1868).
142. Swank, *History of the Manufacture of Iron*, 355–359.
143. *Pittston Gazette*, May 16, 1856.
144. J.B. Pearse, *A Concise History of the Iron Manufacture of the American Colonies up to the Revolution and of Pennsylvania until the Present Time* (Reprint edition) (New York: Burt Franklin, originally published 1876), 236–237.
145. *Pittston Gazette*, January 7, 1853.
146. Daddow and Bannan, *Coal, Iron and Oil*, 753.
147. R.C. Taylor, *Statistics of Coal* (Philadelphia: J.W. Moore, 1855), 423–424; Binder, *Coal Age Empire*, 69–70.
148. Lewis, *Lackawanna Iron and Coal Company*, 438.
149. Binder, *Coal Age Empire*, 68; Pearse, *Concise History of the Iron Manufacture of the American Colonies*, 238.
150. Thomas Iron Co., Weekly Blast Furnace Reports 1858, 1859, Moses Taylor Collection.
151. Daddow and Bannan, *Coal, Iron and Oil*, 715; P. Temin, *Iron and Steel in Nineteenth-Century North America: An Economic Inquiry* (Cambridge, Mass.: MIT Press, 1964), 65.
152. Iron production data from *Official Statistics—The Coal Trade for 1868* (Pottsville: B. Bannan, 1869), 22.
153. Iron production data from W.D. Walsh, *The Diffusion of Technological Change in the Pennsylvania Pig Iron Industry, 1850–1870* (New York: Arno Press, 1975), 208.
154. J. P. Lesley, ed. *The Iron Manufacturer's Guide to the Furnaces, Forges, and Rolling Mills of the United States* (New York: J. Wiley, 1859); A.K. Knowles, R.G. Healey, T.R. Stamp, D. Kidd, and M. Neirick, *"Mapping the Ante-Bellum Iron Industry"*, Social Science and History Conference, November, 1999, Fort Worth, Texas.
155. F.M. Binder, "Pennsylvania Coal and the Beginnings of American Steam Navigation," *Pennsylvania Magazine of History and Biography* 83, 4 (1959): 420–445.
156. Ibid., 439.
157. *Pittston Gazette*, December 19, 1851.
158. Binder, "Pennsylvania Coal," 432.
159. Binder, *Coal Age Empire*, 122–126, 164–165, tables E and F.
160. *Annual Report of the Delaware, Lackawanna and Western Railroad for 1860*, 8.
161. D. Lindstrom, *Economic Development in the Philadelphia Region 1810–1850* (New York: Columbia University Press, 1978); J.L. Livingood, *The Philadelphia-Baltimore Trade Rivalry 1780–1860* (Harrisburg: Pennsylvania Historical and Museum Commission, 1947).
162. Binder, *Coal Age Empire*, 140.
163. Jones, *Anthracite Tidewater Canals*, 58–59.
164. *Coal Trade Annual 1879(1878)*, 4.
165. *Anthracite Coal Trade for 1867* (Pottsville: Miner's Journal, 1868), 7. N.B. The data for 1866 and 1867 are reversed in the relevant tabulation.
166. *Annual Report of the Philadelphia and Reading Railroad for 1868)*, 47.
167. Ibid.
168. *Anthracite Coal Trade for 1867*, 12.
169. *Coal Statistical Register, 1870* (Pottsville: Miner's Journal, 1871), 16.
170. Livingood, *Philadelphia-Baltimore Trade Rivalry*, 99.
171. Binder, *Coal Age Empire*, 14.
172. *Official Statistics—The Coal Trade for 1868*, 12.
173. Binder, *Coal Age Empire*, 141.
174. Ibid., 144.
175. *Miner's Journal*, December 29, 1855, February 12, 1870, January 28, 1871, January 20, 1872.
176. R.G. Albion, *The Rise of New York Port 1815–1860* (New York: Charles Scribner's Sons, 1939).
177. Board of Managers Minutes, Delaware and Hudson Canal Company, Statement of Business for the Year 1860.
178. **Boston** data: see Miner's Journal, January 21, 1865, **New York** data: see calculations in Table 4.2c, **Philadelphia** data: see calculations in Table 4.2b.
179. **Philadelphia and Reading Railroad** and **Schuylkill Navigation Company**: *Coal Statistical Register, 1870*, 15.
Delaware Division Canal: Column for 'Reached Bristol' in Jones, *Anthracite Tidewater Canals*, 58–59. Judging by Taylor, (*Statistics of Coal*, 404), it is possible that some of this coal which reached Bristol was then shipped to Eastern ports other than Philadelphia, but the amounts and destinations are not known.
North Penn Railroad: Prior to 1858, there are no shipment data for the North Penn Railroad. In 1860, it shipped 91,327 tons (*Anthracite Coal Trade for 1867*, 12). Of this, 21.7 percent is assumed to have been used on the line, based on the first year (1869) for which this breakdown is given (*Coal Statistical Register, 1870*, 17).
180. **Morris Canal**: Jones, *Anthracite Tidewater Canals*, 122, but figures reduced by 10 percent based on an 1865 estimate of way tonnage (*Miner's' Journal*, January 20, 1866).
Delaware and Raritan: Jones, *Anthracite Tidewater Canals*, 58–59, no information on way tonnage is available, so a 10 percent reduction to allow for this has been made.
Central Railroad of New Jersey: *Anthracite Coal Trade for 1867*, 11. The figure is based on total Lehigh and Lackawanna shipments less a correction for line sales.

The correction factors are derived from the nearest available date (1866) for which precise figures on line sales are available. The correction factors are 21 percent for Lehigh and 1.6 percent for Lackawanna coal.

Belvidere Delaware Railroad: Actual 1860 tonnage less line tonnage proportion estimated from the first year when the line tonnage is known (1864). The proportion used is 8.1 percent. See *Anthracite Coal Trade for 1867*, 10. According to Binder, (*Coal Age Empire*, 144) coal on this railroad went to Trenton then was transferred into Delaware and Raritan canal boats for New York.

Port Richmond: For 1855, the figures are derived from the *Miner's Journal*, December 29, 1855. For 1860, the 1855 proportion of shipments going to New York (24.73 percent) is applied to the figure for total shipments from Port Richmond in that year from *Anthracite Coal Trade for 1867*, 8.

Delaware and Hudson: *Annual Report for year ending 1st March 1866*, shipments 'eastwards to tidewater' (reduced by 20 percent as estimated allowance for shipments north from Rondout).

Pennsylvania Coal Company: *Annual Report for 1863*. The same 20 percent reduction has been applied as for the Delaware and Hudson.

181. Binder, *Coal Age Empire*, 144–145.
182. cf. W.J.L. Parker, *The Great Coal Schooners of New England* (Mystic, CT: Marine Historical Association, 1948).
183. *Miner's Journal*, January 14, 1860.
184. *Coal Trade Annual 1882(1881)*, 38.
185. Jones, *Anthracite Tidewater Canals*, 58; *Anthracite Coal Trade for 1867*, 8.
186. Daddow and Bannan, *Coal, Iron and Oil*, 715.
187. *Official Statistics—The Coal Trade for 1868*, 13–19.
188. *Coal Trade Annual 1875(1874)*, 37.
189. W. Cronon, *Nature's Metropolis: Chicago and the Great West* (New York: W.W. Norton, 1991).
190. *Coal Trade Annual 1876(1875)*, 25.
191. Binder, *Coal Age Empire*, 145.
192. Board of Managers Minutes, Pennsylvania Coal Company, December 23, 1850.
193. Ibid., August 9, 1851, September 24, 1851, January 22, 1852.
194. Ibid., October 23, 1849.
195. Ibid., November 30, 1849.
196. Minutes of Special Committees (Coal Committee Minutes), Delaware, Lackawanna and Western Railroad, in Delaware, Lackawanna and Western Railroad Company Records, Hagley Library, Greenville, DE., October 2, 1856.
197. Board of Managers Minutes, Delaware, Lackawanna and Western Railroad, September 20, 1858; Coal Committee Minutes, Delaware, Lackawanna and Western Railroad, November 4, 1858.
198. Coal Committee Minutes, Delaware, Lackawanna and Western Railroad, June 14, 1861.
199. Coal Committee Minutes, Delaware, Lackawanna and Western Railroad, February 11, 1859; Board of Managers Minutes, Pennsylvania Coal Company, March 16, 1852; Coal Committee Minutes, Delaware, Lackawanna and Western Railroad, September 19, 1861.
200. Board of Managers Minutes, Pennsylvania Coal Company, April 18, 1856; Board of Managers Minutes, Delaware and Hudson, February 28, 1859.
201. Coal Committee Minutes, Delaware, Lackawanna and Western Railroad, July 7, 1856.
202. Annual Report of the Delaware, Lackawanna and Western Railroad for 1859, 6; Coal Committee Minutes, Delaware, Lackawanna and Western Railroad, October 22, 1858.
203. Coal Committee Minutes, Delaware, Lackawanna and Western Railroad, August 22, 1856.
204. Freedley, *Philadelphia and Its Manufactures*, 116–117.
205. Board of Managers Minutes, Delaware, Lackawanna and Western Railroad, May 27, 1856; Coal Committee Minutes, Delaware, Lackawanna and Western Railroad, June 9, 1856.
206. Binder, *Coal Age Empire*, 146.
207. Coal Committee Minutes, Delaware, Lackawanna and Western Railroad, December 29, 1858, January 27, 1859.
208. Ibid., January 6, 1860.
209. Coal Committee Minutes, Delaware, Lackawanna and Western Railroad, February 11, 1857.
210. Commonwealth of Pennsylvania Legislature, "Testimony Before the Committee to Investigate the Philadelphia and Reading Railroad Company and the Philadelphia and Reading Coal and Iron Company", *Legislative Documents* 4 (1876): 495–1,133, see pp. 651–660; Hereafter referred to as *Reading Railroad Investigation 1876.*
211. See Correspondence in Letterbook 12, Box 6, Sharpe Papers, MS-38, Luzerne County Historical Society, Wilkes-Barre.
212. Committee on the Judiciary, General, *Report of the Committee on the Judiciary, General, of the State of Pennsylvania in Relation to the Anthracite Coal Difficulties, with the Accompanying Testimony* (Harrisburg: B. Singerly, State Printer, 1871), 152 (hereafter referred to as *Committee on the Judiciary*); R.G. Healey, "Sharpe, Weiss and Company and the Lehigh Coal Trade, 1867–1874," *Proceedings of the Wyoming Historical and Geological Society* 24 (1984): 24–43.

DATA SOURCES FOR FIGURES

4.1 Macfarlane, *Coal Regions of America*, table facing p. 54.
4.2 Ibid. for North Branch tonnages except 1841; 1841 tonnage from Pearce, *Annals*, Appendix W; Delaware and Hudson data from *Annual Report of the Delaware and Hudson Canal Company for 1899.*

4.3 *Atlas to Accompany Part III of the Annual Report of the Geological Survey of Pennsylvania for 1886* (Harrisburg: Board of Commissioners for the Geological Survey, 1887).
4.4 Ibid.
4.5 Same as for note 154 above, this chapter.

NOTES FOR CHAPTER 5

1. R.L. Gunnarsson, *The Story of the Northern Central Railway: From Baltimore to Lake Ontario* (Sykesville, MD: Greenberg, 1991); H.W. Schotter, *The Growth and Development of the Pennsylvania Railroad Company* (Philadelphia: Allen, Lane and Scott, 1927), 85.
2. Shaughnessy, *Delaware and Hudson*, 61.
3. Board of Managers Minutes, Delaware and Hudson Canal Company, September 10, 1868.
4. Board of Managers Minutes, Delaware, Lackawanna and Western Railroad, March 16, 1872.
5. Bogen, *Anthracite Railroads*, 148–149.
6. Annual Report of the Lehigh Coal and Navigation Company for 1867, 16–18, 30.
7. T.T. Taber III, *The Delaware, Lackawanna and Western Railroad in the Nineteenth Century 1828–1899* (Muncy, PA: the author, 1977), 223.
8. Board of Managers Minutes, Delaware, Lackawanna and Western Railroad, April 29, 1870.
9. H.H. Pierce, *Railroads of New York: a Study of Government Aid, 1826–1875* (Cambridge, MA: Harvard University Press, 1953).
10. Board of Managers Minutes, Delaware, Lackawanna and Western Railroad, April 29, July 29, 1870.
11. Taber, *Delaware, Lackawanna and Western*, 245.
12. Shaughessy, *Delaware and Hudson*, 137–140.
13. Board of Managers Minutes, Delaware and Hudson Canal Company, May 19, September 13, 1871.
14. Shaughnessy, *Delaware and Hudson*, 142.
15. Delaware and Hudson Company, *A Century of Progress, History of the Delaware and Hudson Company, 1823–1923* (Albany: J.B. Lyon, 1925), 245–247.
16. Shaughnessy, *Delaware and Hudson*, 151; *Annual Report of the Delaware and Hudson Canal Company for 1876*.
17. Chandler, *Visible Hand*, 174.
18. J.D. Dilts, *The Great Road: the Building of the Baltimore and Ohio, the Nation's First Railroad, 1828–1853* (Stanford: Stanford University Press, 1993).
19. Schotter, *Pennsylvania Railroad*.
20. Chandler, *Visible Hand*, 101.
21. Quoted in Bogen, *Anthracite Railroads*, 43.
22. *Annual Report of the MH&SHRR for 1861*, 5.
23. Philadelphia and Reading Railroad, *Leases and Traffic Contracts 1860–1883*, Hagley Library.
24. J.V. Hare, *History of the Reading* (Philadelphia: John H. Stock, 1966), 133.
25. Bogen, *Anthracite Railroads*, 42.
26. Ibid., 44.
27. *Annual Report of the MH&SHRR for 1861*, 5.
28. *Annual Report of the MH&SHRR for 1862*, 3–4.
29. Bogen, *Anthracite Railroads*, 44; Hare, *History of the Reading*, 123, 136.
30. Philadelphia and Reading Railroad, *Leases and Traffic Contracts 1860–1883*. Term of lease of the Little Schuylkill Railroad in *Annual Report of the Auditor General of the State of Pennsylvania and of the Tabulations and Deductions from the Reports of the Railroad, Canal and Telegraph Companies for 1874* (Harrisburg: B.F Meyers), 281.
31. Mine Hill and Schuylkill Haven Railroad, Journal A and Ledger A, 1865–1870, Hagley Library.
32. *Annual Report of the Philadelphia and Reading RR for 1871*, 12, 19.
33. Jones, *Anthracite Tidewater Canals*, 146–147.
34. *Annual Report of the Philadelphia and Reading Railroad for 1875, 1879*, various tabulations.
35. Calculated from table in Jones, *Anthracite Tidewater Canals*, 158.
36. Archer, *Lehigh Valley Railroad*, 23.
37. Board of Managers Minutes, Lehigh Coal and Navigation Company, March 16, May 5, 1863.
38. Archer, *Lehigh Valley Railroad*, 39–40; Bogen, *Anthracite Railroads*, 116.
39. Jones, *Anthracite Tidewater Canals*, 38–39.
40. Board of Managers Minutes, Lehigh Coal and Navigation Company, May 2, 1865.
41. Both quotations from Board of Managers Minutes, Beaver Meadow Railroad and Coal Company, June 13, 1864.
42. Ibid.
43. Ibid.
44. Archer, *Lehigh Valley Railroad*, 40; Bogen, *Anthracite Railroads*, 117.
45. Board of Managers Minutes, Beaver Meadow Railroad and Coal Company, January 20, 1862; January 19, 1863; January 14, 1864.
46. Board of Managers Minutes, Lehigh Coal and Navigation Company, May 2, 1865.
47. C. Goodrich, *Canals and American Economic Development*.
48. Special Committee Minutes, Delaware, Lackawanna and Western Railroad, Aug. 29, 1856.
49. Archer, *Lehigh Valley Railroad*, 52.
50. *Annual Report of the Lehigh Valley Railroad for 1870*, 23.
51. *Annual Report of the Lehigh Valley Railroad for 1869*, 21; Archer, *Lehigh Valley Railroad*, 117.
52. Board of Managers Minutes, Lehigh Valley Railroad, December 13, 1870.
53. Archer, *Lehigh Valley Railroad*, 115.
54. Board of Managers Minutes, Lehigh Valley Railroad, November 14, 1871, April 9, 1872, February 11, 1873, January 18, 1876.
55. Ibid., September 18, 1880.
56. Ibid., December 13, 1881.
57. Archer, *Lehigh Valley Railroad*, 123.
58. Board of Managers Minutes, Lehigh Valley Railroad, December 13, 1881.
59. Ibid.

60. *Annual Report of the Lehigh Valley Railroad for 1871*, 25.
61. *Annual Report of the Lehigh Valley Railroad for 1881*, 26.
62. *Annual Reports of the Lehigh Valley Railroad for 1871–1885.*
63. Archer, *Lehigh Valley Railroad*, 123.
64. Board of Managers Minutes, Lehigh Valley Railroad, monthly reports 1887–1891.
65. Taber, *Delaware, Lackawanna and Western*, 152, 214–215.
66. *Annual Report of the Delaware, Lackawanna and Western Railroad for 1857.*
67. Board of Managers Minutes, Delaware, Lackawanna and Western Railroad, April 30, 1858.
68. Taber, *Delaware, Lackawanna and Western*, 218.
69. Ibid., 220–221.
70. *Annual Report of the Delaware, Lackawanna and Western Railroad for 1857*, 12.
71. Taber, *Delaware, Lackawanna and Western*, 221.
72. Board of Managers Minutes, Delaware, Lackawanna and Western Railroad, May 29, 1861.
73. Taber, *Delaware, Lackawanna and Western*, 221–223; Board of Managers Minutes, Delaware, Lackawanna and Western Railroad, December 31, 1869.
74. Taber, *Delaware, Lackawanna and Western*, 235.
75. Board of Managers Minutes, Delaware, Lackawanna and Western Railroad, February 24, 1865, August 26, 1881; Taber, *Delaware, Lackawanna and Western*, 172.
76. Taber, *Delaware, Lackawanna and Western*, 245.
77. Board of Managers Minutes, Delaware, Lackawanna and Western Railroad, April 29, July 29, 1870; Taber, *Delaware, Lackawanna and Western*, 245–246.
78. Taber, *Delaware, Lackawanna and Western*, 246.
79. Bogen, *Anthracite Railroads*, 97.
80. Executive Committee Minutes, Delaware, Lackawanna and Western Railroad March 25, 1870.
81. Board of Managers Minutes, Delaware and Hudson Canal Company, November 26, December 20, 1872; Taber, *Delaware, Lackawanna and Western*, 253.
82. Board of Managers Minutes, Delaware, Lackawanna and Western Railroad, December 30, 1875.
83. Board of Managers Minutes, Delaware and Hudson Canal Company, January 18, 1876; Taber, *Delaware, Lackawanna and Western*, 252–253. This discussion points up the apparent errors in the treatment of this agreement by Taber on the pages just specified. Firstly, he states that the Delaware and Hudson loaned the Oswego Midland $200,000 in 1873 to acquire the two leased roads, but this does not correspond with the entries in the Minute Books of the former company. Secondly, he states that there is no mention of the operation of the two lines in the Lackawanna's records. One of the references to the previous paragraph in the current study makes clear that this is not the case.
84. Taber, *Delaware, Lackawanna and Western*, 234.
85. Delaware, Lackawanna and Western Railroad, Journal K, 674, 678–679.
86. *Annual Report of the Delaware, Lackawanna and Western Railroad for 1900*, 22.
87. Annual Reports of General Coal Agent, DLW Coal Department Letter books (LCDR). N.B. Prior to 1879, the DLW reported its tonnages in *net tons*. The reported figures for 1860–1876 have been recalculated to gross tons to match the later figures. The total shipments given by the Lackawanna in its own reports do not always exactly match these in the 'official' tabulations of the anthracite trade produced by J.H. Jones.
88. Coal Committee Minutes, Delaware, Lackawanna and Western Railroad, January 24, February 8, 1861.
89. Board of Managers Minutes, Delaware, Lackawanna and Western Railroad, volume for 1888–1906, 69.
90. Bogen, *Anthracite Railroads*, 124.
91. *Annual Report of the Lehigh Valley Railroad for 1868*, 23.
92. Board of Managers Minutes, Lehigh Valley Railroad, March 14, 1876.
93. *Annual Reports of the Lehigh Valley Railroad for 1869*, 22; *for 1870*, 28; *for 1871*, 29.
94. *Annual Reports of the Lehigh Valley Railroad for 1872*, 27, 30; *for 1873*, 28.
95. Bogen, *Anthracite Railroads*, 125.
96. Board of Managers Minutes, Lehigh Valley Railroad, March 14, 1876.
97. *Annual Report of the Lehigh Valley Railroad for 1876*, 31.
98. Board of Managers Minutes, Lehigh Valley Railroad, December 13, 1881.
99. Tabulations in *Annual Reports of the Lehigh Valley Railroad for 1876–1885.*
100. Board of Managers Minutes, Lehigh Valley Railroad, June 13, 1882.
101. *Annual Report of the Lehigh Valley Railroad for 1879*, 26.
102. Board of Managers Minutes, Lehigh Valley Railroad, September 13, 1881, October 11, 1881.
103. *Annual Reports of the Lehigh Valley Railroad for 1881*, 25–26; *for 1882*, 24.
104. Board of Managers Minutes, Lehigh Valley Railroad, November 8, 1881; Archer, *Lehigh Valley Railroad*, 127.
105. *Annual Reports of the Lehigh Valley Railroad for 1883*, 7, 30; *for 1884*, 7, 26.
106. Bogen, *Anthracite Railroads*, 128.
107. Board of Managers Minutes, Lehigh Valley Railroad, April 9, December 10, 1889, May 13, 1890; Archer, *Lehigh Valley Railroad*, 129.
108. Board of Managers Minutes, Lehigh Valley Railroad, September 8, October 13, November 10, 1891.
109. Archer, *Lehigh Valley Railroad*, 135.
110. Letterbook 29-F, 236, 672 (LCDR).
111. Bogen, *Anthracite Railroads*, 98–99.
112. Taber, *Delaware, Lackawanna and Western*, 258.
113. Bogen, *Anthracite Railroads*, 99–100.
114. Board of Managers Minutes, Delaware, Lackawanna and Western Railroad, September 29, 1882; June 27, 1884.
115. Both quotations from Board of Managers Minutes, Delaware, Lackawanna and Western Railroad, February 26, 1884.
116. Bogen, *Anthracite Railroads*, 100.
117. Taber, *Delaware, Lackawanna and Western*, 258, 266.

118. Board of Directors Minutes, Delaware, Lackawanna and Western Railroad, July 31, 1885; Minutes 1888–1906, 71.
119. Letterbook 71, reports and tabulations (LCDR).
120. Board of Managers Minutes, Delaware and Hudson Canal Company, May 13, 1868.
121. Ibid., December 5, 1867.
122. Ibid., November 17, 1863, December 15, 1863, February 16, 1864, September 10, 1868.
123. Hudson Coal Company, *Story of Anthracite*, 90–91; Board of Managers Minutes, Delaware and Hudson Canal Company, May 15, 1867.
124. Pierce, *Railroads of New York*, 17; Shaughnessy, *Delaware and Hudson*, 74–75.
125. Board of Managers Minutes, Delaware and Hudson Canal Company, February 27, May 15, 1866, March 5, 1868.
126. Board of Managers Minutes, Delaware and Hudson Canal Company, April 23, September 10, 1868; Shaughnessy, *Delaware and Hudson*, 61–62, 75.
127. Shaughnessy, *Delaware and Hudson*, 76–83.
128. Board of Managers Minutes, Delaware and Hudson Canal Company, March 11, 1869, January 27, September 29, 1870, May 19, 1871; *Corporate History of the Delaware and Hudson Company and Subsidiary Companies* (n.p., 1906), vol. 1, 270.
129. *Century of Progress*, 243.
130. *Report to the Stockholders of the Delaware and Hudson Canal Company, by the Committee Appointed at the Annual Meeting Held on May 8th, 1877* (New York: P. Barnes, Steam Book and Job Printer, 1877), 7, 44, 48–49, 78; *Corporate History of the Delaware and Hudson* (1907), vol. 3, 477.
131. *Annual Reports of the Delaware and Hudson Canal Company for 1870–1899.*
132. *Century of Progress*, 267–270.
133. Bogen, *Anthracite Railroads*, 196.
134. Board of Managers Minutes, Delaware and Hudson Canal Company, n.d., following entry of March 22, 1877.
135. Bogen, *Anthracite Railroads*, 196.
136. *Report to the Stockholders of the Delaware and Hudson*, 50–52.
137. Engineering and Mining Journal, *Editorial Review of the Delaware and Hudson Canal Company's Report*, articles of September 15, 22, 29, October 6, 1877.

DATA SOURCES FOR FIGURES

5.1 Cartography by the author. Digital cartographic data for the relevant portions of the railroad network from the NE USA Historical GIS (see http://www.geog.port.ac.uk/usa/project/ for more details). Additional data from *Annual Reports of the Delaware and Hudson Canal Company* and historical maps reproduced in Shaughnessy, *Delaware and Hudson.*

5.2 Cartography by the author. Sources include: P.W. Schaefer, *A Map of all the Anthracite Collieries and the Pottsville, Lehigh, Mahanoy and Shamokin Coal Basins*; Daddow and Bannan, *Coal, Iron and Oil*, frontispiece map; *AIM Report for 1872*, frontispiece map; *Map of the Philadelphia and Reading Railroad and Branches* (Philadelphia: published by the company, 1871); Plates 1 and 2 in Atlas to accompany Part III of the Annual Report of the Geological Survey of Pennsylvania for 1886; C.T. Baer, *Canals and Railroads of the Mid-Atlantic States, 1800–1860* (Greenville, DE: Eleutherian Mills-Hagley Foundation, 1981), Maps A and B of the Anthracite Coalfields, 1860.

5.3 Cartography by the author. Digital cartographic data for the relevant portions of the railroad network from the NE USA Historical GIS. Additional data from *Annual Reports of the Lehigh Valley Railroad*, and from figures and historical maps reproduced in Archer, *Lehigh Valley Railroad* and Taber, *Delaware, Lackawanna and Western.*

5.4 *Annual Reports of the Lehigh Valley Railroad for 1872–1885.*

5.5 *Annual Reports of the Pennsylvania and New York Canal and Railroad Company appended to Annual Reports of the Lehigh Valley Railroad for 1869–1885.*

5.6 Ibid. *for 1871–1885.*

5.7 Ibid. *for 1873–1885;* Tabulations in *Schotter, Pennsylvania Railroad.*

5.8 *Annual Reports of the Pennsylvania and New York Canal and Railroad Company appended to Annual Reports of the Lehigh Valley Railroad for 1872–1885.*

5.9 Sources as for Figure 5.3.

5.10 In the early years, the data are derived from the General Coal Agent's manuscript annual reports in the Company Letter books (LCDR), but in the 1870s the information becomes very hard to find. Some of the 1870s data were pieced together from tonnage reports in various newspapers. In several cases they are slightly approximate, as the available reports do not correspond exactly to the end of the calendar year, but the accuracy is sufficient for the general observations being made here. Prior to 1879 the original tonnages are quoted in net tons. These have been recalculated to gross tons to make the data comparable across the period.

5.11 *Annual Reports of the Delaware and Hudson Canal Company for 1870–1899.*

5.12 Ibid. *for 1883–1899.*

NOTES FOR CHAPTER 6

1. Taber, *Delaware, Lackawanna and Western*, 66–80.
2. Board of Managers Minutes, Delaware, Lackawanna and Western Railroad March 26, 1869, April 30 1869.
3. Board of Managers Minutes, Delaware, Lackawanna and Western Railroad July 31, 1868.
4. Taber, *Delaware, Lackawanna and Western*, 79–80.
5. Board of Managers Minutes, Delaware, Lackawanna and Western Railroad, June 24, 1870, February 24, 1871; Taber, *Delaware, Lackawanna and Western*, 81–82.

6. Board of Managers Minutes, Delaware, Lackawanna and Western Railroad, September 29, 1871.
7. Taber, *Delaware, Lackawanna and Western*, 86.
8. Board of Managers Minutes, Delaware, Lackawanna and Western Railroad, May 25, 1877.
9. Taber, *Delaware, Lackawanna and Western*, 85.
10. See for example the vouchers for June and July 1870 in Coal Department Construction Voucher Records, Delaware, Lackawanna and Western Railroad Company Records, George Arents Research Library, Syracuse University.
11. Bogen, *Anthracite Railroads*, 96.
12. *Annual Report of the Central Railroad of New Jersey for 1870*, 12.
13. Taber, *Delaware, Lackawanna and Western*, 175.
14. *Annual Report of the Central Railroad of New Jersey for 1871*, 20–21; Board of Managers Minutes, Delaware, Lackawanna and Western Railroad, March 16, 1872.
15. S. Sloan to W.R. Storrs, December 23, 1873, LB Sloan to Storrs, 1870–1885 (LCDR).
16. Board of Managers Minutes, Delaware, Lackawanna and Western Railroad, December 20, 1872.
17. *Annual Report of the Central Railroad of New Jersey for 1872*, 16.
18. Executive Committee Minutes and Board of Managers Minutes of the Delaware, Lackawanna and Western Railroad, May 9, May 31, 1872 respectively.
19. Jones, *Anthracite Coal Combination*, 228.
20. *Annual Report of the Central Railroad of New Jersey for 1872*, 9.
21. Board of Managers Minutes, Delaware, Lackawanna and Western Railroad, December 11, 1872.
22. *Annual Report of the Central Railroad of New Jersey for 1872*, 16–17.
23. Bogen, *Anthracite Railroads*, 84.
24. Board of Managers Minutes, Delaware, Lackawanna and Western Railroad, December 11, 1872.
25. Executive Committee Minutes, Delaware, Lackawanna and Western Railroad, September 25, 1872.
26. *Annual Report of the Central Railroad of New Jersey for 1872*, 15; Board of Managers Minutes, Delaware, Lackawanna and Western Railroad, October 25, 1872.
27. Executive Committee Minutes, Delaware, Lackawanna and Western Railroad, December 16, 1872.
28. Bogen, *Anthracite Railroads*, 96.
29. Board of Managers Minutes, Delaware, Lackawanna and Western Railroad, February 28, 1873, *Annual Report of the Central Railroad of New Jersey for 1872*, 18–19.
30. Board of Managers Minutes, Delaware, Lackawanna and Western Railroad, May 21, 1873.
31. Board of Managers Minutes, Delaware, Lackawanna and Western Railroad, October 27, 1873.
32. Bogen, *Anthracite Railroads*, 96–97.
33. S. Sloan to W.R. Storrs, May 9, 1873, LB Sloan to Storrs, 1870–1885 (LCDR).
34. Board of Managers Minutes, Delaware, Lackawanna and Western Railroad, January 30, 1880, January 27, 1888.
35. Board of Managers Minutes, Delaware, Lackawanna and Western Railroad, June 24, 1881.
36. Journal K, 673–674, Delaware, Lackawanna and Western Railroad Records, Syracuse.
37. Ibid., 679.
38. *Annual Report of the Central Railroad of New Jersey for 1871*, 22.
39. Board of Managers Minutes, Januuary 30, 1891; Annual Report for 1900, Delaware, Lackawanna and Western Railroad, 15.
40. *Annual Report of the Lehigh Valley Railroad for 1871*, 6.
41. Bogen, *Anthracite Railroads*, 122.
42. Data tabulations in *Annual Reports of the Lehigh Valley Railroad*.
43. *Annual Report of the Lehigh Valley Railroad for 1871*, 6–7; Board of Directors Minutes, Lehigh Valley Railroad, April 28, 1871.
44. Board of Directors Minutes, Lehigh Valley Railroad, July 11, November 14, 1871.
45. Board of Directors Minutes, Lehigh Valley Railroad, June 13, 1871.
46. Taber, *Delaware, Lackawanna and Western*, 134–139.
47. Bogen, *Anthracite Railroads*, 123.
48. Archer, *Lehigh Valley Railroad*, 104.
49. Taber, *Delaware, Lackawanna and Western*, 139.
50. Board of Managers Minutes, Lehigh Valley Railroad, March 12, April 9, 1872, September 9, 1873.
51. *Annual Report of the Lehigh Valley Railroad for 1872*, 19; *for 1873*, 7, 15.
52. *Annual Report of the Lehigh Valley Railroad for 1874*, 6, 10; *for 1875*, 16, 19; *for 1877*, 8.
53. *Annual Report of the Lehigh Valley Railroad for 1873*, 16; *for 1876*, 18.
54. *Annual Report of the Lehigh Valley Railroad for 1876*, 7.
55. *Annual Report of the Lehigh Valley Railroad for 1877*, 7; Bogen, *Anthracite Railroads*, 125.
56. *Annual Report of the Lehigh Valley Railroad for 1877*, 7.
57. J.L. Holton, *The Reading Railroad: History of a Coal Age Empire, 2 vols.* (Laury's Station, PA: Garrigues House,1989–1992), 276–277.
58. Schotter, *Pennsylvania Railroad*, 96.
59. S. Sloan to W.R. Storrs, December 11, 1873, LB Sloan to Storrs, 1870–1885 (LCDR).
60. Jones, *Anthracite Tidewater Canals*, 157.
61. Archer, *Lehigh Valley Railroad*, 209.
62. Board of Managers Minutes, Lehigh Valley Railroad, June 14, 1887, March 11, October 14, 1890.
63. Archer, *Lehigh Valley Railroad*, 110–112.
64. *Annual Report of the Central Railroad of New Jersey for 1875*, 8–9.
65. *Annual Report of the Lehigh Coal and Navigation Company for 1873*, 8–9.
66. Bogen, *Anthracite Railroads*, 154–155; *Statement of Board of Directors to the Stockholders of Central Railroad of New Jersey, September 9th, 1876* (New York: G.F. Nesbitt, 1876), 7.
67. Bogen, *Anthracite Railroads*, 169.
68. *Annual Report of the Philadelphia and Reading Railroad for 1879*, 24–25; Holton, *Reading Railroad*, 258–259, 274.

69. Board of Managers Minutes, Delaware, Lackawanna and Western Railroad, November 28, 1879; *Annual Report of the Philadelphia and Reading Railroad for 1879*, 25.
70. Bogen, *Anthracite Railroads*, 173–174.
71. Board of Directors Minutes, Pennsylvania Railroad, April 26, May 15, 1871.
72. *Annual Reports of the Philadelphia and Reading Railroad for 1867–1875.*
73. *Annual Reports of the Philadelphia and Reading Railroad for 1872–1889.*
74. Bogen, *Anthracite Railroads*, ch. 9.
75. Cornell, *Anthracite Coal Strike.*
76. Holton, *Reading Railroad*, 252.
77. Tabulations in *Annual Reports of the Philadelphia and Reading Railroad.*
78. *Annual Report of the Philadelphia and Reading Railroad for 1882*, 14–15.
79. Philadelphia and Reading Railroad, Weekly Tonnage Reports, August 16, 23, 1884, Reading Company Papers.
80. *Annual Report of the Philadelphia and Reading Railroad for 1883*, 8, 86; *for 1884*, 85.
81. Bogen, *Anthracite Railroads*, 73.
82. *Annual Report of the Reading Company for 1899*, 7–8; *for 1900*, 10.
83. Tabulations in *Annual Report of the Philadelphia and Reading Railroad for 1892; Annual Report of the Reading Company for 1901*. It should be noted that on reorganization into the Reading Company in 1896, the financial year end was changed from November 30th to June 30th in each year and therefore data for 1897 partially overlaps with the data for 1896 reported on the old basis. A further effect of this is that the 1902 figures do not reflect the impact of the 1902 coal strike, whereas other figures, such as the Jones statistical series for the coal trade being based on calendar years, do reflect the loss of tonnage as a result of the strike.
84. Bogen, *Anthracite Railroads*, 69, 73.
85. Ibid., 99–101.
86. December journal entries, Delaware, Lackawanna and Western Railroad accounts, Syracuse; Annual Report Summaries in Board of Managers Minutes, various years.
87. In Bogen's brief discussion of this period, he presents values for the Lackawanna operating ratio for several years without giving a source. The values in the present figure are calculated from the transportation earnings and expenses as recorded in the accounts of the company. On this basis the ratios for 1881 and 1883 are 49.81 percent and 58.03 percent, not 53.23 percent and 63.94 percent as Bogen claims. His other figures match those derived from the accounts directly (Bogen, *Anthracite Railroads*, 101.)
88. See, for example, M.W. Schlegel, *Ruler of the Reading: The Life of Franklin B. Gowen 1836–1889* (Harrisburg: Archives Publishing Company, 1947), 56–57.
89. *Annual Reports of the Lehigh Valley Railroad for 1880–1883.*
90. See, for example, *Verbatim Report of the General Meeting of the Share and Bondholders of the Philadelphia and Reading Railroad Company and of the Bondholders of the Perkiomen Railroad Company Held at the City Terminus Hotel, Canon Street, London, June 6, 1877 (Thomas Wilde Powell Esq. in the Chair)* (London: Charles Skipper and East, Printers, 1877).
91. *Annual Report of the Philadelphia and Reading Railroad for 1887*, 14–15, 20.
92. *Plan and Agreement for the Reorganization of the Philadelphia and Reading Railroad Company and the Philadelphia and Reading Coal and Iron Company*, December 14, 1895 (n.p).
93. *Annual Report of the Reading Company for 1901*, 8.
94. *Annual Reports of the Philadelphia and Reading and Lehigh Valley Railroads for 1879–1881.*
95. *Annual Reports of the Lehigh Valley Railroad for 1881–1882*; Bogen, *Anthracite Railroads*, 131.
96. *Annual Report of the Lehigh Valley Railroad for 1895*, 13.
97. Bogen, *Anthracite Railroads*, 132–135.
98. See E.W. Martin [James Dabney McCabe], *History of the Great Riots.*
99. December journal entries 1877–1881, Delaware, Lackawanna and Western Railroad Accounts, Syracuse.
100. Board of Managers Minutes, Delaware, Lackawanna and Western Railroad, Annual Financial Summaries, 1888–1898.
101. *Annual Reports of the Philadelphia and Reading Railroad for 1887, 1889, 1892.*

DATA SOURCES FOR FIGURES

6.1 *Annual Report of the Central Railroad of New Jersey for 1874.*
6.2 Cartography by the author. Digital cartographic data for the relevant portions of the railroad network from the NE USA Historical GIS. Additional data from *Map of the Philadelphia and Reading Railroad and Branches*, 1871; figures and historical maps reproduced in Archer, *Lehigh Valley Railroad*; Holton, *Reading Railroad;* and Taber, *Delaware, Lackawanna and Western.*
6.3 1866–1868 Ledger C, 92; 1869–1872 Ledger D, 125; 1873–1878 Ledger E 125, Delaware, Lackawanna and Western Railroad Accounts, Syracuse.
6.4 *Annual Reports of the Lehigh Valley Railroad for 1866–1885.*
6.5 Ibid.
6.6 Delaware, Lackawanna and Western Railroad Accounts; Board of Managers Minutes, Lehigh Valley Railroad; V-77 Transportation Dept. Statement Book 1844–1882, Records of the Philadelphia and Reading Railroad, Reading Company Papers.
6.7 Lackawanna tonnages from 1851–1867 from a variety of sources including company records and newspaper reports; Delaware and Hudson tonnages for the same period from annual reports; tonnages for both companies for 1868–1869 from A.C. Bining Papers, Historical Society of Pennsylvania; tonnages for 1870–1886 from *Atlas to Accompany Part III of the Annual Report of the Geological Survey of Pennsylvania for 1886*; tonnages from 1887 onwards from annual tabulations in the *Coal Trade Annual*. It should be noted that after 1878

the total tonnages given by the exceed those of the standard reports by J.H. Jones, the anthracite tonnage accountant, which are the data series used in the figure for 1868–1902.

This is because the standard series were specifically created to eliminate double counting of the same tonnage transported for part of its journey by different railroads, whereas the Delaware and Hudson's own figures are inflated by this coal transported for others.

6.8 Tonnage figures from 1868 onwards derived from the same sources as the equivalent data in the previous figure.

6.9 Tabulations in *Annual Reports of the Philadelphia and Reading Railroad.*

6.10 Tabulations in *Annual Reports of the Lehigh Valley Railroad.*

6.11 Delaware, Lackawanna and Western Railroad Accounts, various years.

6.12 V-77 Transportation Dept. Statement Book 1844–1882, Records of the Philadelphia and Reading Railroad, Reading Company Papers.

6.13 Tabulations in *Annual Reports of the Philadelphia and Reading Railroad.*

6.14 Calculated from data in Delaware, Lackawanna and Western Railroad Accounts and *Annual Reports of the Delaware and Hudson Canal Company.*

6.15 Calculated from data in *Annual Reports of the Lehigh Valley Railroad and Philadelphia and Reading Railroad.*

6.16 Calculated from data in *Annual Reports of the Lehigh Valley Railroad.*

6.17 Calculated from data in *Annual Reports of the Philadelphia and Reading Railroad.*

6.18 For 1863–1891 calculated from data in Delaware, Lackawanna and Western Railroad Accounts; for 1892–1898 calculated from data in published *Annual Reports.*

6.19 Calculated from data in *Annual Reports of the Lehigh Valley Railroad.*

NOTES FOR CHAPTER 7

1. Campbell, *Businessmen and Anthracite*, 39, 57.
2. Ibid., ch. 3.
3. A comprehensive treatment of the Connellsville Coke Industry can be found in Warren, *Wealth, Waste and Alienation.*
4. Hoffman, *Depression of the Nineties.*
5. *Annual Report of the Secretary of the American Iron and Steel Association to December 31, 1874* (Philadelphia: American Iron and Steel Association, 1875); hereafter referred to as *AISA Annual Reports.*
6. *Coal Trade Journal*, March 5, 1879.
7. *Coal Trade Journal*, December 28, 1892.
8. *Coal Trade Journal*, June 27, 1877.
9. Temin, *Iron and Steel*, 157.
10. Ibid., 158–163; 200–205.
11. *Coal Trade Annual 1888* (1887), 90.
12. *Coal Trade Journal*, March 5, 1879, copy of article from *Iron Age.*
13. Ibid.
14. Daddow and Bannan, *Coal, Iron and Oil*, 698.
15. *Coal Trade Journal*, June 27, 1877.
16. *Miner's Journal*, April 6, 1877.
17. *Coal Trade Journal*, February 15, 1882, copy of table from 1880 census.
18. *Coal Trade Journal*, February 15, 1882, January 20, 1886.
19. *Iron Age*, April 4, 1889, quoted in K. Warren, *The American Steel Industry 1850–1950: A Geographical Interpretation*, (Oxford: Clarendon Press, 1973), 147.
20. *Coal Trade Annual 1887* (1886), 16, quoted in Campbell, *Businessmen and Anthracite*, 46; *Coal Trade Journal*, January 20, 1886.
21. *Coal Trade Journal*, March 5, 1879, copy of article from *Iron Age*; *Coal Trade Annual 1888* (1887), 90.
22. Data from a variety of sources, including Daddow and Bannan, *Coal, Iron and Oil;* Bartholomew and Metz, *Anthracite Iron Industry; AISA Annual Reports.*
23. *Miner's Journal*, January 30, 1869.
24. *Miner's Journal*, March 20, 1874.
25. Tabulations in *Annual Reports of the Philadelphia and Reading Railroad* for 1872, 1876.
26. *The Ironwork of the United States: A Directory of the Furnaces, Rolling Mills, Steel Works, Forges and Bloomaries in every State*, (Philadelphia: American Iron and Steel Association, 1876), 19, 79.
27. Percentage calculated from figures in Tables 1 and 2 in Healey, "Sharpe, Weiss and Company," 25–43, 29.
28. Healey, "Sharpe, Weiss and Company," Table 1, 29.
29. Ibid., 32.
30. Ibid., 36–37.
31. *Coal Trade Journal*, September 5, 1877.
32. F. Platt, *Special Report to the Legislature upon the Causes, Kinds and Amount of Waste in Mining Anthracite*, (Harrisburg: Board of Commissioners of the 2nd Geological Survey, 1881).
33. *Annual Report of the Philadelphia and Reading Railroad for 1883*, 68.
34. *Coal Trade Journal*, March 5, 1884.
35. *Annual Report of the Philadelphia and Reading Railroad for 1868*, 47.
36. *Annual Report of the Philadelphia and Reading Railroad for 1887*, 20.
37. *Annual Report of the Philadelphia and Reading Railroad for 1896*, 27.
38. *Annual Report of the Philadelphia and Reading Railroad for 1886*, 40.
39. *Annual Report of the Philadelphia and Reading Railroad for 1888*, 35.
40. *Coal Trade Journal*, January 11, 1893; tabulations in *Annual Reports of the Lehigh Valley Railroad for 1896–1901.*
41. *Annual Report of the Delaware, Lackawanna and Western Railroad for 1901*, 16.
42. *Annual Report of the Delaware, Lackawanna and Western Railroad for 1901*, 21; *Coal Trade Journal*, January 24, 1900.

43. R.C. Reed, *The New York Elevated*, (New York: A.S. Barnes and Co., 1978), 58, 134.
44. *Coal Trade Journal*, April 23, 1884; *Coal Trade Journal*, November 22, 1893; Campbell, *Businessmen and Anthracite*, 55.
45. J. Wilson to Col. Fred Martin, November 16, 1891, Letterbook James Wilson (Gen. Sales Agt), August 5, 1891–December 1, 1891, Box 65, Lehigh and Wilkes-Barre Coal Company Records; hereafter called Wilson LB.
46. *Pittston Gazette*, December 2, 1869.
47. *Annual Report of the Philadelphia and Reading Railroad for 1872*, 20–21.
48. Coal Committee Minutes, April 23, 1861, Delaware, Lackawanna and Western Railroad Records, Hagley Library.
49. Healey, "Sharpe, Weiss and Company," 29.
50. *Committee on the Judiciary*, 119.
51. Notebook, General Sales Agent, L.A. Powelson, Box 43, Lehigh and Wilkes-Barre Coal Company Records; hereafter called Powelson Notebook.
52. James Wilson to W.F. Herbert, August 31, 1891, Wilson LB.
53. *Coal Trade Annual 1887(1886)*, 16, quoted in Campbell, *Businessmen and Anthracite*, 46.
54. 1889 contracts, Powelson Notebook; J. Wilson to E.H. Lawall, October 28, 1891, Wilson LB.
55. J. Wilson to E.H. Lawall, October 6, 1891, Wilson LB.
56. J. Wilson to Columbia Oil Co., August 13, 1891, Wilson LB.
57. *Coal Trade Journal*, November 1, 1893.
58. Coal Committee Minutes, Delaware, Lackawanna and Western Railroad, February 17, 1859.
59. Fite, *Social and Industrial Conditions*, 26.
60. *Official Statistics—The Coal Trade for 1868*, 14.
61. W.R. Storrs (WRS) to Leicester Salt Works, January 28, 1892; WRS to LeRoy Salt Co., February 1, 1892; WRS to Royal Salt Co. February 12, 1892, LB51; WRS to Crystal Salt Co., September 12, 1893, LB69 (LCDR).
62. 1888, 1889 contracts, Powelson Notebook.
63. *Annual Report of the Lehigh Valley Railroad for 1890*, 13.
64. *Coal Trade Journal*, April 23, 1884.
65. 1888, 1889 contracts, Powelson Notebook; Campbell *Businessmen and Anthracite*, 55.
66. *Coal Trade Annual 1893* (1892), 62; *Coal Trade Annual 1898 (1897)*, 89.
67. *Coal Trade Journal*, April 23, 1884.
68. *Coal Trade Journal*, November 17, 1886.
69. *Coal Trade Journal*, April 6, 1892.
70. *Coal Trade Annual 1887* (1886), 16.
71. *Annual Report of the Philadelphia and Reading Railroad for 1872*, 20; *for 1874*, 18.
72. *Miner's Journal*, April 6, 1877.
73. *Coal Trade Annual 1893* (1892), 62.
74. *Coal Trade Journal*, November 15, 1893.
75. *Coal Trade Annual 1898* (1897), 63.
76. *Coal Trade Journal*, January 24, 1877.
77. *Coal Trade Journal*, April 23, 1879.

DATA SOURCES FOR FIGURES

7.1 Data for 1860–1875 from 1876 AISA Annual Report, given in W.D. Walsh, *The Diffusion of Technological Change in the Pennsylvania Pig Iron Industry, 1850–1870* (New York: Arno Press, 1975), 208, thereafter from *Coal Trade Annual* tabulations.

7.2 *Coal Trade Annual* tabulations to 1902 apart from 1887 from Temin, *Iron and Steel*, 267 (converted back to net tons). N.B. After 1882, as Temin's tabulation makes clear, the amount of iron made with anthracite alone, as opposed to an anthracite/coke mixture, declined rapidly. By 1890, pure anthracite pig iron was only 11.4 percent of the total anthracite/coke total and the proportion declined still further thereafter.

7.3 Anthracite Pig Iron 1854–1869 from *Annual Report of the Philadelphia Board of Trade* (Philadelphia: J.B. Chandler, 1870), 70; 1870 data from T. Dunlap, *Statistical Report of the National Association of Iron Manufacturers for 1872* (Philadelphia: J.A. Wagenseller, 1873), appendix; 1871 data from *Miner's Journal*, April 3, 1874; remaining data from *AISA Annual Reports*; raw bituminous/coke iron from *Miner's Journal*, April 3, 1874, thereafter from *AISA Annual Report for 1877*, 30–31; *for 1880*, 46–47.

7.4 Data for 1874 from *AISA Annual Report for 1874*, 33–34; subsequent years from reports in *Miner's Journal, Coal Trade Annual;* 1899, 1901–1902 data from *AISA Annual Report for 1902*, 41. Data apply to end of previous year/ start of new year.

7.5 Data for 1854–1869, 1871–1872 from *Miner's Journal*, April 3, 1874; 1870 data from T. Dunlap, *Statistical Report of the National Association of Iron Manufacturers for 1872*, appendix; remaining data from *AISA Annual Reports*.

7.6 Tabulations in *Annual Reports of the Philadelphia and Reading Railroad*.

7.7 Data derived from the Sharpe Papers and from the Coxe Estate Coal Rent Returns, formerly in the Coxe Family Mining Papers, Historical Society of Pennsylvania. Since these data were collected in 1979, the original coal rent returns have been destroyed. A copy of the spreadsheet data derived from them has therefore been deposited for archival purposes at the Anthracite Heritage Museum, Scranton, by the author.

7.8 Tabulations in *Annual Reports of the Philadelphia and Reading Railroad*.

NOTES FOR CHAPTER 8

1. *Annual Report of the Philadelphia and Reading Railroad for 1870*, 22–23.
2. Schotter, *Pennsylvania Railroad Company*, 214. The line was opened from Philadelphia to Pottsville in that year. In 1884 it had been opened as far as Reading (p. 205).
3. *Committee on the Judiciary*, 152.

4. Board of Managers Minutes, Delaware and Hudson Canal Company, January 18, 1873.
5. *Century of Progress*, 271.
6. *Delaware and Hudson Stockholders Report 1877*, 94.
7. Summary shipment reports in *Miner's Journal*, February 12, March 20, 1874; calculations based on the figures and J.H. Jones series of shipments reported each year in the *Coal Trade Annual*.
8. *Miner's Journal*, February 17, 1872 and for dates given in Note 7.
9. Sources as for note 8. Tidewater shipments for 1873 can only be (closely) estimated because inspection of the tabulation in the *Miner's Journal* indicates that the figures for the Pennsylvania Coal Company have been omitted in that year.
10. Annual figures reported in the *Coal Trade Annual*.
11. *Coal Trade Annual 1895* (1894), 21.
12. *Coal Trade Annual 1890* (1889), 107.
13. Schotter, *Pennsylvania Railroad*, 97.
14. *Coal Trade Annual 1887* (1886), 68.
15. Given and calculated from data in *Miner's Journal*, January 19, February 12, 1870, March 13, March 20, 1874. N.B. A further 20,020 tons was shipped to Manayunk in 1869 on the Schuylkill Navigation. Arguably this should be included with the Philadelphia figures to make them consistent with the railroad calculations.
16. *Annual Report of the Philadelphia Board of Trade*, January 25, 1875, 95.
17. Tabulations in the *Coal Trade Annual and Annual Reports of the Philadelphia and Reading Railroad*, various years.
18. *Coal Trade Annual 1890* (1889), 46.
19. *Coal Trade Annual 1897* (1896), 132.
20. *Coal Trade Annual 1893* (1892), 55; *1895* (1894), 58; *1897* (1896), 132.
21. *Coal Trade Annual 1901* (1900), 60–61.
22. *Coal Trade Annual 1897* (1896), 132; *1901* (1900), 60–61; *1902* (1901), 60.
23. Annual Report figures in *Miner's Journal*, various years.
24. *Miner's Journal*, February 12, 1870; January 22, 1870 (Schuylkill Navigation data).
25. Ibid., March 20, 1874.
26. *Coal Trade Annual 1882* (1881), 44.
27. *Coal Trade Annual 1892* (1891), 55; *Coal Trade Annual 1890* (1889), 65.
28. *Coal Trade Annual 1902* (1901), 62.
29. Ibid. Table figures do not sum to overall totals given in the text because no miscellaneous category is included.
30. *Coal Trade Annual* 1889 (1888), 90, quoted in Campbell, *Businessmen and Anthracite*, 183.
31. *Coal Trade Annual 1903* (1902), 61.
32. Ibid., 63.
33. *Coal Trade Annual 1888* (1887), 80.
34. Cornell, *Anthracite Coal Strike*.
35. *Coal Trade Annual 1890* (1889), 71.
36. *Coal Trade Annual 1893* (1892), 58.
37. *Coal Trade Annual 1900* (1899), 68.
38. *Coal Trade Annual 1901* (1900), 65.
39. *Coal Trade Annual 1903* (1902), 57.
40. Ibid., 56–57.
41. Ibid., 110.
42. *Coal Trade Annual 1893* (1892), 59.
43. Ibid., 59–60; *Coal Trade Annual 1903* (1902), 70.
44. *Coal Trade Annual 1883* (1882), 37.
45. *Coal Trade Annual 1893* (1892), 59; *Coal Trade Annual 1903* (1902), 70.
46. *Coal Trade Annual 1883* (1882), 36–37.
47. *Coal Trade Annual 1879* (1878), 27.
48. *Coal Trade Annual 1893* (1892), 59; *Coal Trade Annual 1903* (1902), 70.
49. *Coal Trade Annual 1879* (1878), 27.
50. W.H. Shank, *The Amazing Pennsylvania Canals* (York, PA: American Canal and Transportation Center, 1960), 72.
51. Livingood, *Philadelphia-Baltimore Trade Rivalry*, 78.
52. Ibid., 80.
53. *Annual Report of the Philadelphia and Reading Railroad for 1884*, 69.
54. Data from *Annual Reports of the Philadelphia and Reading Railroad*, various years; Shank, *Amazing Pennsylvania Canals*, 72.
55. Cronon, *Nature's Metropolis*.
56. *Labor Troubles*.
57. H. Perry Smith (Ed.), *History of Buffalo and Erie County*, vol. 2 (Syracuse: D. Mason and Co., 1884), 201.
58. *Delaware and Hudson Stockholders Report* 1877, 94–95.
59. Coal Committee Minutes, Delaware, Lackawanna and Western Railroad, February 8, 1861.
60. Perry Smith, *History of Buffalo*, 201. This source claims the Pittston and Elmira Coal Co. went out of business before 1870 and the claim is perpetuated by Campbell, *Businessmen and Anthracite*, 68. However, this is incorrect. Monthly shipment figures recorded in the letter-books of the Lackawanna Coal Dept. indicate that the company continued in business until 1874.
61. Ibid., 201.
62. Ibid., 209; *Annual Report of the Lehigh Valley Railroad for 1883*; Delaware, Lackawanna and Western Railroad Accounts, Year-end Journal Entries, 1883.
63. *Coal Trade Journal*, January 11, 1882.
64. Calculated from data in *Coal Trade Annual*, various years.
65. *Coal Trade Annual 1886* (1885), 35.
66. Ibid., *Coal Trade Annual 1885* (1884), 35.
67. *Coal Trade Journal*, November 4, 1885.
68. Schotter, *Pennsylvania Railroad*, 125.
69. *Coal Trade Journal*, November 14, 1885. This information contradicts Campbell's assertion that there was no direct evidence of price leadership by the Association in the 1880s (Campbell, *Businessmen and Anthracite*, 69).
70. Ibid.
71. *Labor Troubles*, 584–585.
72. H.C. Bradsby (Ed.) *History of Luzerne County*, vol. 1 (Chicago: S.B. Nelson, 1893), 290, quoted in Campbell, *Businessmen and Anthracite*, 67.

73. *Annual Report of the Philadelphia and Reading Railroad for 1884*, 85.
74. *Coal Trade Journal*, November 4, 1885.
75. Ibid. Calculations based on table of May–September allotments, shipments, and excesses over allotments.
76. *Coal Trade Journal*, November 12, 1884; Perry Smith, *History of Buffalo*, 206.
77. *Coal Trade Journal*, November 12, 1884.
78. *Coal Trade Annual 1886* (1885), 35.
79. *Coal Trade Annual 1902* (1901), 67.
80. *Coal Trade Annual*, Buffalo Trade Reports, various years.
81. *Coal Trade Annual 1901* (1900), 79.
82. Perry Smith, *History of Buffalo*, 202; Campbell, *Businessmen and Anthracite*, 68–69 (this author incorrectly quotes Perry Smith to state that Langdon 'owned' as opposed to 'handled' the product of half a dozen mines in the Shamokin region); *AIM Report for 1872*, 133; *for 1878*, 88. By 1865 they had been succeeded by H. S. Mercur and Co., according to Daddow and Bannan, *Coal, Iron and Oil*, 757. The Sweatland shaft was later to become the Delaware and Hudson No. 4.
83. *AIM Report for 1872*, 255, 257.
84. Perry Smith, *History of Buffalo*, 204–206 (this authority incorrectly states that Grassy Island was in the Pittston area); *Coal Trade Journal*, January 11, 1882; *AIM Report for 1884*, 174.
85. A.T. Andreas, *History of Chicago from the Earliest Period to the Present Time* (Chicago: A.T. Andreas Company, 1886), Vol. 3, 387.
86. Andreas, *History of Chicago*, vol. 2, 673.
87. *Annual Report of the Lehigh Valley Railroad for 1872*, 34; Perry Smith, *History of Buffalo*, 203, 205.
88. Board of Managers Minutes, June 26, 1868, December 28, 1883; Journal 1869, 88; Executive Committee Minutes, July 29, 1870, Delaware, Lackawanna and Western Railroad.
89. Andreas, *History of Chicago*, Vol. 3 (1886), 387.
90. *Coal Trade Annual*, various years; Macfarlane, *Coal Regions of America*, 673.
91. *Coal Trade Annual*, various years.
92. Ibid.
93. *Coal Trade Journal*, April 2, 1884.
94. Ibid., March 19, 1884.
95. Ibid., February 24, 1892.
96. Ibid.
97. *Coal Trade Annual 1893* (1892), 67.
98. *Coal Trade Journal*, March 16, 1892.
99. Ibid., April 27, 1892.
100. *Coal Trade Annual 1893* (1892), 66–67; *Coal Trade Journal, April 27*, 1892.
101. *Coal Trade Journal*, September 14, 1892.
102. *Coal Trade Annual 1893* (1892), 66.
103. *Coal Trade Journal*, January 15, 1879; *Coal Trade Annual 1902* (1901), 66.
104. *Coal Trade Annual*, tabulations, various years.
105. Ibid., Campbell, *Businessmen and Anthracite*, 76, 82.
106. *Coal Trade Journal*, January 20, 1892.
107. Ibid.
108. *Coal Trade Journal*, February 6, 1892; *Coal Trade Annual*, tabulations, various years.
109. *Coal Trade Journal*, September 20, 1893.
110. *Coal Trade Annual 1903* (1902), 97.
111. *Coal Trade Journal*, September 20, 1893.
112. Ibid.
113. Ibid.
114. *Labor Troubles*, 584–585.
115. *Labor Troubles*, 584.
116. *Coal Trade Journal*, September 20, 1893.
117. Ibid.
118. *Coal Trade Journal*, September 20, 1893; *Coal Trade Journal*, November 15, 1893.
119. *Coal Trade Annual 1887* (1886), 72; *Coal Trade Journal*, December 28, 1892 (tabulations of shipments to Superior by major companies).
120. *Coal Trade Annual 1893* (1892), 70.
121. *Coal Trade Journal*, October 10, 1900.
122. Ibid., December 28, 1892.
123. Ibid., December 13, 1899.
124. Ibid., September 20, 1893.
125. *Coal Trade Annual*, tabulations for various years.
126. *Coal Trade Annual 1895* (1894), 70–71.
127. Ibid., 70.

DATA SOURCES FOR FIGURES

8.1 Data derived from the Sharpe Papers and from the Coxe Estate Coal Rent Returns, formerly in the Coxe Family Mining Papers, Historical Society of Pennsylvania. See also note to figure 7.7, previous chapter.
8.2 Ibid.
8.3 Data from tabulations in *Miner's Journal, and Coal Trade Journal, and Coal Trade Annual.*
8.4 *Annual Reports of the Philadelphia and Reading Railroad.*
8.5 Tabulations in *Coal Trade Annual*, various years.
8.6–8.18 Ibid.

NOTES FOR CHAPTER 9

1. *Coal Trade Annual*, tabulations, various years. N.B. The reported tonnages shipped to PA, NY, and NJ in 1896 do not correspond to the reported percentage of shipments. Since the reported percentage gives a tonnage that more closely matches the overall shipment figures recorded elsewhere, the reported percentage has been used in the graph and the reported tonnage figure ignored (cf. *Coal Trade Annual* 1898 (1897), 22–23).
2. Journals, various years, Delaware, Lackawanna and Western Railroad Accounts.
3. Macfarlane, *Coal Regions of America*, 674.
4. E. Jones, *Anthracite Coal Combination*, 67.
5. Cornell, *Anthracite Coal Strike*; R. Janosov (Ed.), *Perspectives on the Great Strike* (Easton: Canal History and Technology Press, 2002).

6. See note on sources for Figure 9.8.
7. Monthly price reports 1891–1892, 1895–1902 in Coal Averages Books (LWBCCR).
8. Calculated from monthly report data from the Anthracite Coal Operators' Association for 1897 given in *Coal Trade Annual 1898* (1897), 23.
9. *Coal Trade Journal*, January 30, 1877.
10. *Labor Troubles*, 302.
11. *Annual Report of the Philadelphia and Reading Railroad for 1885*, 14–15.
12. *Labor Troubles*, 606, 664.
13. Campbell, *Businessmen and Anthracite*, 52.
14. *Annual Report of the Summit Branch Railroad for 1867*, 27.
15. *Miner's Journal*, January 30, 1869.
16. Board of Managers Minutes, Delaware, Lackawanna and Western Railroad 1858–1872, 399; *Miner's Journal*, January 30, 1869.
17. Healey, "Sharpe, Weiss and Company"; G.B. Newton to Sharpe, Weiss and Co., May 2, 1872, LB13, Box 7, Sharpe Papers.
18. *Coal Trade Annual 1890* (1889), 101.
19. W.R. Storrs (WRS) to 'My dear Mac' January 7, 1893, LB51 (LCDR).
20. **DL&W**: 1879 from General Coal Agent's Annual Report, Board of Managers Minute Book 1873–1888, 218f. 1887 from ibid., 465.
    1898 calculated from monthly tonnage reports in W.H. Storrs letter-books, (LCDR).
    L&WB: Calculated for all years from monthly data in Coal Averages books (LWBCCR).
    P&RC&I Co.:1887 from *Coal Trade Annual 1890* (1889), 101.
    Susquehanna Coal Co.: All years from *Annual Reports* for the years in question.
21. WRS to C.D. Simpson April 18, 1892, LB51 (LCDR).
22. Board of Managers Minutes 1858–1872, Delaware, Lackawanna and Western Railroad, 399; Ibid. 1873–1888, 465; Letterbook, W.H. Storrs, January 17, 1896–May 20, 1899, 321 (LCDR).
23. J.N. Hoffman, *Girard Estate Coal Lands in Pennsylvania 1801–1884*, Smithsonian Studies in History and Technology, no. 15 (Washington: Smithsonian Institution Press, 1972).
24. H.S. Thompson, *Notes on Waste in Mining and Preparing Coal and on the Percentage of Coal Won at Representative Collieries on the Girard Estate*, from the Report to the Directors of City Trusts, October 3, 1892, 7, 13.
25. *Coal Trade Annual for 1901* (1900), 26.
26. Ibid.
27. *Coal Trade Journal*, May 31, 1893. Other data in the table from which this was taken correspond to figures given in the Coal Trade Annual in earlier years for the 'Coal and Iron'.
28. *Report of the Anthracite Coal Strike Commission*, 25. *Coal Trade Journal*, May 31, 1893.
29. For general discussion of this type of problem in relation to Sharpe, Weiss and Company in the 1860s–1870s, see Healey, "Sharpe, Weiss and Company"; see also C. Miesse, *Points on Coal and the Coal Business* (Myerstown: Feese and Uhrich, 1887), 177.
30. *Coal Trade Annual 1898* (1897), 27–28 (reprint of paper by A.C. Leisenring on 'Small Sizes of Anthracite Coal').
31. *Annual Report of the Lehigh Coal and Navigation Company for 1877*, 14.
32. Ibid., 15.
33. Ibid., 16.
34. *Coal Trade Journal*, April 6, 1892.
35. *Report of the Commission Appointed to Investigate the Waste of Coal Mining, with a View to Utilizing the Waste* (Philadelphia: Allen, Lane and Scott, 1893); Coal Trade Journal, May 31, 1893; Platt, *Waste in Mining Anthracite*.
36. *Coal Trade Journal*, April 6, 1892.
37. *Report of the Anthracite Coal Strike Commission*, 24.
38. Blatz, *Democratic Miners*, Chapter 5.

DATA SOURCES FOR FIGURES

9.1 1882–1886 data from Plate 7, *Atlas to Accompany Part III of the Annual Report of the Geological Survey of Pennsylvania for 1886*; remaining data from tabulations in *Coal Trade Annual.*
9.2 Tabulations in *Coal Trade Annual.*
9.3 Ibid.
9.4 Series E123–134, *Historical Statistics of the United States.*
9.5 *AISA Annual Report for 1882*, 56.
9.6 Data from a wide variety of sources including monthly reports in *Miner's Journal, Pittston Gazette, New York Times, Coal and Iron Record, Scranton Republican, Coal Trade Annual,* Moses Taylor Collection, Storrs Letterbooks (LCDR).
9.7 Jones, *Anthracite Coal Combination*, 228.
9.8 Coal Averages Books (LWBCCR).
9.9 *AISA Annual Report for 1882*, 56; Cumberland Prices from *AISA Annual Report for* 1877, 46.
9.10 Sources same as for note to Figure 9.6.
9.11 Sources same as for notes to Figures 9.7 and 9.8.
9.12 Coal Averages Books (LWBCCR).
9.13 Ibid.
9.14 Tabulations in *Coal Trade Annual.*
9.15 Ibid.
9.16 Ibid.
9.17 Ibid.
9.18 Ibid.
9.19 Calculations based on data derived from the Sharpe Papers and from the Coxe Estate Coal Rent Returns, formerly in the Coxe Family Mining Papers. See also note to Figure 7.7.
9.20 Ibid.
9.21 Tabulations in *Annual Reports of the Lehigh Coal and Navigation Company.*
9.22 Calculations based on data in Coal Averages Books (LWBCCR).
9.23 Ibid.
9.24 Ibid.

9.25 Ibid. for Lehigh and Wilkes-Barre Coal Company; LB W.H. Storrs January 17, 1896–May 20, 1899 (for January 1896–April 1899) (LCDR); remaining 1899 data from President's Working Papers (S. Sloan) 1899–1904, Box 1, Lackawanna Railroad Records, Syracuse University.

9.26 Data sources same as for previous note.

NOTES FOR CHAPTER 10

1. Jones, *Anthracite Coal Combination*, 22; Ibid., quoting *Hazards Register* 13, 218.
2. S.P. Adams, "Different Charters, Different Paths: Corporations and Coal in Antebellum Pennsylvania and Virginia," *Business and Economic History* 27, 1 (1998): 78–90; Adams, *Old Dominion.*
3. Adams, *Different Charters*, 88.
4. See Jones, *Anthracite Coal Combination,* and Bogen, *Anthracite Railroads.*
5. Ibid.
6. Schlegel, *Ruler of the Reading;* Campbell, *Businessmen and Anthracite,* chapter 6.
7. *Virtue,* "Anthracite Combinations," 297, 299; Jones, *Anthracite Coal Combination,* 40.
8. Coal Committee Minutes, Delaware, Lackawanna and Western Railroad, December 8, 1856.
9. Ibid., January 14, February 23, March 24, 1860.
10. J.S. Magee to J.J. Albright, December 19, 1863, Correspondence Received by J.J. Albright 1858–1865 (LCDR).
11. Ibid., Perry Smith, *History of Buffalo*, 201.
12. Daddow and Bannan, *Coal, Iron and Oil*, 753; E.N. Frisbie to W.R. Storrs (WRS) June 12, 1867, Incoming Correspondence, WRS, General Coal Agent, Book 3, March–June 1867 (Microfilm), (LCDR).
13. Board of Managers Minutes, Delaware and Hudson Canal Company, February 20, 1862.
14. Magee to Albright, December 19, 1863 [see note 10].
15. See incoming letters from R. Law to WRS, Storrs Incoming Correspondence Book 3, e.g. May 7, 1867.
16. E.N. Frisbie to WRS, April 8, 1867, Storrs Incoming Correspondence Book 3.
17. R. Law to WRS, April 3, 1867; E.N. Frisbie to WRS, April 11, 1867, both in Storrs Incoming Correspondence Book 3.
18. R. Law to WRS, April 9, 1867, Storrs Incoming Correspondence Book 3.
19. J.D.F. Slee to WRS, April 25, 1867, Storrs Incoming Correspondence Book 3.
20. H. Rasbach to McKinney and Phelps, Agts., Binghamton, March 8, 1867, Storrs Incoming Correspondence Book 3.
21. A.G. Cook (Oswego Agt.) to WRS, April 10, 1867, Storrs Incoming Correspondence Book 3.
22. J.D.F. Slee to WRS, March 9, 1867.
23. *Luzerne Union*, August 15, 1866; monthly tonnage reports in Storrs Outgoing Letter Books (LCDR).
24. *Luzerne Union*, July 15, 1867.
25. Ibid.
26. *Luzerne Union*, October 30, 1867.
27. WRS to Samuel Sloan (SS), August 15, 1868, LB17 (LCDR).
28. WRS to SS, January 8, 1868, LB16; *Pittston Gazette*, Dec.12, 1867; Monthly shipment tabulations in letterbooks; WRS to SS, September 16, 1869, LB21; WRS to SS, April 16, 1868, LB16; Pittston and Elmira Account with Delaware, Lackawanna and Western Railroad 1868, following correspondence of September 3, 1869, LB20 (LCDR).
29. WRS to SS, April 16, 1868, LB16; WRS to SS, November 4, 1868, LB18 (LCDR).
30. WRS to SS, July 6, 1868, LB17; WRS to SS, December 16, 1868, LB18 (LCDR).
31. WRS to A. Cook, Oswego, February 18, 1869, LB19; R. Law to WRS, October 20, 1868, Storrs Incoming Correspondence Book 8.
32. Board of Managers Minutes, Delaware and Hudson Canal Company, January 27, 1870.
33. Executive Committee Minutes, Delaware, Lackawanna and Western Railroad, November 9, 1870.
34. WRS to SS, April 2, 1870, LB22 (LCDR).
35. WRS to SS, October 19, 1869, LB21; WRS to SS, January 29, 1872, LB25 (LCDR).
36. WRS to SS, September 24, 1873, LB26; SS to WRS, July 3, October 13, October 16, 1874, LB SS to WRS, 1870–1885 (LCDR).
37. Detailed recourse to county courthouse deed books was required to trace even this outline of the demise of the Pittston and Elmira Coal Company and a number of aspects of what happened still remain obscure. See for example the following deed book entries, DB181, p. 366; DB202, p. 379 f., DB218, p. 332, Luzerne County Courthouse; *AIM Report for 1875*, 202.
38. SS to WRS, October 16, 1874, LB SS to WRS 1870–1885; monthly shipment records in the Storrs letter books indicate the cessation of shipments over the Lackawanna at this time.
39. Board of Managers Minutes, Delaware and Hudson Canal Company, December 1, 1870; January 30, 1871.
40. Board of Managers Minutes, Delaware, Lackawanna and Western Railroad, March 26, 1869.
41. *Annual Report of the Philadelphia and Reading Railroad for 1869*, 20.
42. cf. Wallace, *St. Clair*. The present author is, however, uncomfortable with Wallace's thesis that propensity to disaster was one of the key defining features of the anthracite industry.
43. Aurand, *Molly Maguires*, 67, 73.
44. Ibid., 68.
45. *Miners Journal*, December 21, 1872.
46. Ibid.
47. Scranton auction prices were set at the end of the month to apply to orders for the following month (the month used in the text). Prices were assembled from a variety of sources, but mainly from newspaper reports

in the *Miner's Journal, the Pittston Gazette* and *the Scranton Republican.*

48. *Miner's Journal,* April 6, April 20, April 27, 1872.
49. Ibid., May 18, 1872.
50. Ibid., May 25, 1872.
51. Ibid., August 17, November 16, 1872.
52. Ibid., June 1, 1872.
53. Ibid., June 8, July 13, 1872.
54. Ibid., July 27, 1872; see also *Annual Report of the Philadelphia and Reading Railroad for 1872;* 18.
55. *Miner's Journal,* June 29, August 10, 1872.
56. Aurand, *Molly Maguires.* The summary details in the ensuing paragraphs are largely based on information derived from this study.
57. In July 1868, a number of mines in the Wyoming Region, such as the Baltimore Coal Company and those of the Pittston and Elmira Company, did come out on strike; see *Pittston Gazette,* July 23, 1868.
58. Wallace, *St. Clair,* 275.
59. Aurand, *Molly Maguires,* 68–70.
60. *Pittston Gazette,* July 23, 1868.
61. WRS to SS, n.d., (but position in letter book is between July 14 and 16, 1868), LB17 (LCDR).
62. WRS to SS, July 17, 1868, LB17 (LCDR).
63. WRS to SS, April 10, 1869, LB19 (LCDR).
64. WRS to SS, May 7, 1869, LB20 (LCDR).
65. WRS to SS, June 17, 1869, LB20 (LCDR).
66. WRS to SS, n.d., 845, LB20; WRS to Sharpe, Weiss and Co., Eckley, PA, August 26, 1869, LB20 (LCDR).
67. Aurand, *Molly Maguires,* 79.
68. *Annual Report of the Philadelphia and Reading Railroad for 1875,* 16–17.
69. WRS to SS, February 2, 1872, LB25 (LCDR).
70. WRS to SS, February 3, 1872, LB25 (LCDR).
71. Janosov, *Perspectives on the Great Strike.*
72. Executive Committee Minutes, Delaware, Lackawanna and Western Railroad, May 11, 1869.
73. *Miner's Journal,* December 19, 1868, January 30, 1869.
74. SS to WRS, December 23, 1873, LB SS to WRS, 1870–85.
75. *Committee on the Judiciary,* 242.
76. Bogen, *Anthracite Railroads,* 187.
77. DB101, p. 118, Luzerne County Courthouse; for details of the business and personal relationships between Pyne and Taylor, see D. Hodas, *The Business Career of Moses Taylor: Merchant, Finance Capitalist and Industrialist* (New York: New York University Press, 1976).
78. Ibid., DB101, p. 172, Luzerne County Courthouse.
79. Hodas, *Moses Taylor,* 261.
80. J. Brisbin to T.F. Hunt, February 6, 1867, President's LB (J. Brisbin), September 1866–March, 1868 (LCDR); J. Brisbin to P.R. Pyne, July 16, 1868, LB Counsel and General Land Agent (J. Brisbin) (LCDR);
81. Statement for New York shareholders, October 16, 1866, LB Nanticoke Coal and Iron Company (LCDR); J. Brisbin to Moses Taylor, December 5, 1866, President's LB (J. Brisbin) (LCDR).
82. *Luzerne Union,* August 28, 1867.
83. See, for example the letters of February 21 and 24, 1868, J.C. Phelps to A.J. Odell (Treasurer of the Lackawanna) LB Nanticoke Coal and Iron Company (LCDR).
84. Board of Managers Minutes, Delaware, Lackawanna and Western Railroad, 1869, p. 400.
85. J. Brisbin to A.T. McClintock, March 23, 1868, LB Counsel and General Land Agent (J. Brisbin) (LCDR).
86. J. Brisbin to Hon. L.D. Shoemaker, March 30, 1868, LB Counsel and General Land Agent (J. Brisbin) (LCDR).
87. See, for example, Board of Managers Minutes, April 29, 1870, Delaware, Lackawanna and Western Railroad.
88. Ibid., June 24, 1870; Lackawanna Coal Department Voucher Records for August 1870, Lackawanna Railroad Records, Syracuse.
89. Bogen, *Anthracite Railroads,* 94. Brisbin claimed the D&H could hold 13,000 acres but the amended 1859 charter of the latter only made provision for 8,000 acres, see Bogen, *Anthracite Railroads,* 186.
90. John Brisbin to Hon. George Landon, April 8, 1867, President's LB (John Brisbin) (LCDR).
91. Bogen, *Anthracite Railroads,* 94–95.
92. Board of Managers Minutes, Delaware and Hudson Canal Company, December 5, 1867.
93. Ibid., both quotations.
94. Board of Managers Minutes, Delaware and Hudson Canal Company, June 3, June 22, July 19, 1859. This transaction does not appear in the later company list of coal lands owned, however, (see Note 95) which raises a question as to whether it was ever recorded.
95. Exhibit A, Cost of Coal Lands, Box 61, Records of Delaware and Hudson and Affiliated Companies, New York State Archives, Albany; Board of Managers Minutes, Delaware and Hudson Canal Company, May 2, 1865; *History of Luzerne, Lackawanna and Wyoming Counties,* 470.
96. Board of Managers Minutes, Delaware and Hudson Canal Company, December 5, 1867.
97. Bogen, *Anthracite Railroads,* 188.
98. Board of Managers Minutes, Delaware and Hudson Canal Company, December 3, 1859; *Luzerne Union,* June 17, 1863, quoting *Lackawanna Herald.* A cursory reading of Shaughnessy, *Delaware and Hudson,* 39, might give the impression that this road was in operation in 1860, but that was not the case.
99. Bogen, *Anthracite Railroads,* 188.
100. *Corporate History of the Northern Coal and Iron Company, the Hudson Coal Company, etc.* (n.p., 1907), 1–6.
101. Daddow and Bannan, *Coal, Iron and Oil,* 754–755; *Scranton Republican,* January 23, 1868.
102. *Corporate History of the Northern Coal and Iron Company,* 124.
103. Ibid., 73f.
104. Board of Managers Minutes, Delaware and Hudson Canal Company, June 28, 1866, August 21, 1866.
105. *Corporate History of the Northern Coal and Iron Company,* 101, 111.
106. Ibid., 111f.

107. *Scranton Republican*, February 8, 1867.
108. *Scranton Republican*, August 10, September 21, November 2, 1866; *Pittston Gazette*, April 18, June 13, 1867; *Luzerne Union*, May 15, June 26, 1867.
109. Board of Managers Minutes, Delaware and Hudson Canal Company, October 15, 1867.
110. Ibid.
111. *Scranton Republican*, April 5, 1867.
112. *Annual Report of the Lehigh Coal and Navigation Company for 1867*, 14; for 1868, 14.
113. *Corporate History of the Northern Coal and Iron Company*, 132–133; Board of Managers Minutes, Delaware and Hudson Canal Company, October 15, 1867.
114. *Corporate History of the Northern Coal and Iron Company*, 173f., 191. Both the earlier pages of the Corporate History and Dickson's letter to the Managers referenced in the previous note indicate that a variety of funding mechanisms for the deal were postulated before the agreement was finally signed. However, the statement in Bogen, *Anthracite Railroads*, 187, that the Delaware and Hudson gave $1.3 million to the Union to buy the Baltimore, corresponds to one of the proposed mechanisms, not the final basis on which the transaction actually took place.
115. Board of Managers Minutes, Delaware and Hudson Canal Company, October 15, 1867; *Corporate History of the Northern Coal and Iron Company*, 163, 195.
116. *Scranton Republican*, November 30, 1867; January 16, 1868; January 23, 1868; *Pittston Gazette*, February 20, 1868; *Corporate History of the Northern Coal and Iron Company*, 208f.
117. *Scranton Republican*, January 23, 1868.
118. Board of Managers Minutes, Delaware and Hudson Canal Company, March 5, 1868; *Scranton Republican*, March 6, 1868; DB79, p. 577, Boston Coal Co. Agreement, April 18, 1860, Luzerne County Courthouse; Delaware and Hudson Company and affiliated companies, Cost of Coal Lands, Box 61, Exhibit A, New York State Archives, Albany.
119. *Corporate History of the Northern Coal and Iron Company*, 9, 31, 251, 258.
120. *Annual Report of the Lehigh Coal and Navigation Company for 1867*, 13.
121. Ibid.
122. Ibid., 14–15.
123. Ibid. for 1870, 37.
124. Board of Managers Minutes, Delaware, Lackawanna and Western Railroad, March 31, 1871; WRS to SS, December 12, 1870, LB24; WRS to SS, September 7, 1871, LB25; WRS to SS, February 23, 1872, LB25; WRS to SS, n.d., (April 1872), 889–890, LB25 (LCDR).
125. *Annual Report of the Central Railroad of New Jersey for 1871*, 15.
126. Ibid.
127. *Annual Report of the Lehigh Coal and Navigation Company for 1873*, 8–9; for 1876, 9; *Annual Report of the Central Railroad of New Jersey for 1873*, 15; Bogen, *Anthracite Railroads*, 161. The latter authority incorrectly states that the Honey Brook Coal Company owned 8,000 acres of lands in the Lehigh Region. The correct figures are given in the table in the text.
128. *Annual Report of the Central Railroad of New Jersey for 1873*, 16.
129. Ibid.
130. *Statement of the Board of Directors to the Stockholders of the Central Railroad of New Jersey*, 8–9.
131. *Annual Report of the Lehigh Valley Railroad for 1867*, 19.
132. *Annual Report of the Lehigh Valley Railroad for 1869*, 7.
133. D.G. Baird, A Statement of some of the Important Matters that have entered into the History of the Lehigh Valley Coal Company from its Inception to and including the Fiscal Year ending June 30, 1906, 13–15. Unpublished Typescript, 1907. Lehigh Valley Railroad Company Records b) Subsidiary Companies 2) Non-Transportation Companies MG-274, Pennsylvania State Archives, Harrisburg. (Hereafter called LV Coal Co. History).
134. Sources same as in previous note.
135. Report on Trevorton Lands. Harris Bros., 1868, Land Development File, 1849–1962, Box 4, Lehigh Valley Railroad Company Records.
136. Such calculations are difficult to perform accurately for several reasons. If a company leased land directly from a landowner, there was not usually any upfront payment, as there would have been if the land had been purchased. On this basis, the acreages of the Lehigh Valley lands leased in the Wyoming Region in Table 10.2 and the Lehigh Navigation and Greenwood lands in Table 10.1 have been excluded from the calculation. However, if one company purchased another company, which itself held leases, part of the cost involved would be for the acquisition of the right to operate those leases. Hence, under these circumstances, the acreage is included in the calculation. The costs attributed to the acquisition of stock of companies in Table 10.2 should also be viewed with circumspection. Although all the figures are taken from the sources given in Note 134, it is apparent that, in some cases, stock was progressively acquired over a number of years, resulting in a progressive increase in the book cost of control over the company in question. An attempt has been made to identify the figures that most closely represent the costs after initial acquisition, but these are subject to a margin of error because of data limitations.
137. Baird, *LV Coal Co. History*, tabulations and lease details spread throughout the work; tabulations, Israel Morris Papers (Lehigh Valley Coal Co. Land Agent), Luke Morris Section, Morris Family Papers, Historical Society of Pennsylvania, Philadelphia; *Annual Report of the Lehigh Valley Railroad for 1873*, 8; Bogen, *Anthracite Railroads*, 121.

138. *Report to the Stockholders of the Delaware and Hudson*, 67.
139. WRS to J. Brisbin, February 19, 1872, LB25 (LCDR).
140. Bogen, *Anthracite Railroads*, 95.
141. Land Record, 1848–1894, Pennsylvania Coal Company Records, MG-282, Pennsylvania State Archives, Harrisburg.
142. *Pittston Gazette*, November 4, 1869; Baird, LV Coal Co. History, 37; Bogen, *Anthracite Railroads*, 164.
143. Board of Managers Minutes, Lehigh Valley Railroad, July 11, 1871; *AIM Report 1872*, 168.
144. Baird, LV Coal Co. History, 15.
145. Ibid., 81.
146. Ibid., 11–13.
147. Ibid., 17; Mortgage of Lehigh Valley Coal Co to the Girard Life Insurance, Annuity and Trust Company of Philadelphia, December 9, 1884, Israel Morris Papers.
148. Schotter, *Pennsylvania Railroad*, 85, 105, 109.
149. See Note 136.
150. Board of Managers Minutes, Susquehanna Coal Company, vol. 1, 2–7. Pennsylvania Railroad Records, Pennsylvania State Archives; the North Branch was merged into the Pennsylvania Canal Company in 1869, see Petrillo, *Anthracite and Slackwater*, 202–205.
151. Board of Managers Minutes, Susquehanna Coal Company, vol. 1, 2–7, 10–13.
152. Ibid., 10–13.
153. Board File 225/108, Pennsylvania Railroad Company Records, Pennsylvania State Archives.
154. Board of Managers Minutes, Susquehanna Coal Company, vol. 1, November 21, 1870; Board of Directors Minutes, Pennsylvania Railroad, November 16, 1870.
155. Board File 225/108, Pennsylvania Railroad Company Records.
156. *Report of the Investigating Committee of the Pennsylvania Railroad Company*. (Philadelphia: Allen, Lane and Scott, 1874), 132–133; Board File 225, Hazleton Lands of the Pennsylvania Railroad Co., Final Report of their Exploration and Proving by Irving A. Stearns, June 25, 1884.
157. *Miner's Journal*, February 20, 1869; Wm. B. Fowle to J. Imbrie Miller, August 14, 1876, Board Files, Box 8, Pennsylvania Railroad Records; *Report of the Investigating Committee of the Pennsylvania Railroad*, 129–130; Board of Directors Minutes, Pennsylvania Railroad, November 26, 1873 (a slightly lower valuation of the Summit Branch stock is given for December 31, 1873 by Schotter, *Pennsylvania Railroad*, 128–129).
158. Unsigned letter to J. Edgar Thomson, February 6, 1874, Board Files, Box 2, Pennsylvania Railroad Records. *Annual Report of the Summit Branch Railroad for 1873*, in Board Files, Box 8, Pennsylvania Railroad Records.
159. *Miner's Journal*, February 20, 1869.
160. Legal history of the Shamokin Coal Company, Board File 58, Box 8, Pennsylvania Railroad Records; *Report of the Investigating Committee of the Pennsylvania Railroad*, 131–132.
161. *Report of the Investigating Committee of the Pennsylvania Railroad*, 140.
162. Ibid., 137.
163. *Annual Report of the Erie Railway for 1869*, 8–10.
164. Ibid., 9.
165. Ibid., 9–10; Board of Managers Minutes, Delaware and Hudson Canal Company, May 13, 1869.
166. *Annual Report of the Erie Railway for 1872*, 13.
167. WRS to SS, May 14, 1872, LB25 (LCDR).
168. Industrial Espionage Report on the Hillside Coal and Iron Company 1889, by J.H. Jones, Reading Company Papers, Hagley Library.
169. Board of Managers Minutes, Delaware and Hudson Canal Company, December 1, 1870. Although this entry makes reference to the 1869 agreement, unusually the agreement itself does not appear in the minutes, nor has any copy been located to date. The Whitmore tract purchase is referenced in the later agreement of November 2, 1870, recorded in the minutes on the date above.
170. This assertion is based on deduction from the available evidence. The Glenwood Coal Company only ever controlled three mines in this area, the Carbon Hill, the Oak Hill, and the Spring Brook. The latter two were served by the Delaware and Hudson Railroad, while Carbon Hill shipped over the Lackawanna. See *AIM Report for 1872*, 263–265; shipment tables in Lackawanna Coal Dept. Letterbooks; *Map of Wyoming Anthracite Field (Surveys to September 1870)* by R.P. Rothwell, Earth Conservancy, Ashley, PA. (I am indebted to Dr. R. Janosov for bringing the existence of this important early map to my attention and to the Director of Earth Conservancy for giving access to it); WRS to SS, (n.d. but must be May 1875), list of leases, p. 555, LB 28, (LCDR).
171. *AIM Report for 1872*, 297 indicates that the Erie was the only colliery, in the area in question, operated by the Glenwood Coal Co.
172. *Miner's Journal*, November 30, 1872.
173. WRS to SS, July 15, 1870, LB23 (LCDR).
174. Annual Report of the Lackawanna Coal Department for 1870, LB24, 656, (LCDR); *Rothwell Map of 1870* (see note 170); WRS to SS, July 15, 1870, LB23 (LCDR). It would appear that the Coray contract is the one referred to incorrectly as the 'Lehigh and Susquehanna Coal Company' in the Erie's Annual Report for 1869, published in 1870. The company of the latter name operated the Hartford (later Ashley No. 6) breaker at the foot of the Ashley planes below Wilkes-Barre and it was not capable of producing the large tonnages required by the contract as indicated in the Erie's Annual Report.
175. WRS to SS, December 19, 1870, LB24; WRS to SS, List of Leases, 555, LB28 (LCDR); *Rothwell Map of 1870*;

there is an element of guesswork in this interpretation of the chronology, based on the fact that the 1870 map shows Carbon Hill as operated by Malvern and Scott and the 1870 Storrs letter states that Carbon Hill had changed hands twice during the year, the second time being when it was bought out by the 'Erie ring'.

176. WRS to SS, December 19, 1870, LB24; August 11, 1870, LB23; Sept. 5, 1870, LB23 (LCDR).
177. WRS to SS, list of leases, p. 555, LB28 (LCDR).
178. *Miner's Journal,* November 30, 1872; *Saward's Coal Trade Circular,* June 4, 1873, quoting the *Scranton Times.*
179. *Pittston Gazette,* July 10, 1873; Baird, LV Coal Company History, 9. Board of Directors Minutes, Hillside Coal and Iron Company, 18, MG-300, Erie Railroad Co., Subsidiary Companies, Pennsylvania State Archives.
180. *AIM Report for 1876,* 171–172; J.H. Jones Industrial Espionage Report on the Hillside Coal and Iron Company 1889.
181. WRS to SS, August 31, 1871, WRS to SS, September 16, 1871; WRS to SS, September 30, 1871, LB25 (LCDR).
182. *Pittston Gazette,* June 5, 1873.
183. Board of Managers Minutes, Delaware and Hudson Canal Company, May 12, May 26, 1875. *Coal Trade Journal* 1893, 774–775.
184. *Pittston Gazette,* June 15, 1873, quoting the *Scranton Republican.*
185. Gowen F.B., in *Verbatim Report of the General Meeting … Held at the City Hotel, Cannon Street, London, June 6, 1877.*
186. *Miner's Journal,* April 24, 1869; Schlegel, *Ruler of the Reading,* 31–32. *Testimony before the Committee to Investigate the Philadelphia and Reading Railroad (1876),* 1036–1037.
187. *Miner's Journal,* March 25, 1871; February 3, 1872.
188. Schlegel, *Ruler of the Reading,* 32f.
189. Acreages from the listing in the *Miner's Journal,* February 3, 1872 and the *Annual Report of the Philadelphia and Reading Railroad for 1871,* 17. The expenditure figure is from the latter source.
190. Bogen, *Anthracite Railroads,* 53. This authority quotes as his source the Annual Report for 1871, but the present writer cannot locate his figures in the printed text of that publication.
191. *Miner's Journal,* March 27, 1874; J.S. Harris, *Report of Joseph S. Harris on the Coal Lands of the Philadelphia and Reading Coal and Iron Company.* (Philadelphia: Jackson Bros, 1880), 4.
192. Harris, *Report on Coal Lands,* 4.
193. *Miner's Journal,* February 3, 1872, March 27, 1874. The acreages for specific tracts are not always fully consistent between these two lists.
194. Compilation of the Capital Accounts, Tonnage Reports, Maps of Lands, etc. of the Philadelphia and Reading Coal and Iron Company, 735, Reading Anthracite Company, Pottsville, PA.
195. Ibid.
196. Schlegel, *Ruler of the Reading,* 196.
197. Bogen, *Anthracite Railroads,* 53. Bogen gives an average cost per acre of $293.
198. Statement showing original cost of Coal Estate of Philadelphia and Reading Coal and Iron Company 1871–1894, Reading Anthracite Company.
199. Compilation of the Capital Accounts, 724, Reading Anthracite Company.
200. Yearley, *Enterprise and Anthracite,* see the discussion of this period on 208–209.
201. *Annual Report of the Philadelphia and Reading Railroad for 1873,* 26.
202. *Annual Report of the Philadelphia and Reading Railroad for 1876,* 36.
203. Harris, *Report on Coal Lands,* 34, 78.
204. F.B. Gowen, *Statement of the Present Condition of the Philadelphia and Reading Railroad and the Philadelphia and Reading Coal and Iron Company with a Plan for their Financial Reorganization* (Philadelphia: Jackson Bros., 1880), 26.
205. See the interesting comments on these points by Schlegel, *Ruler of the Reading,* 210f.
206. *Annual Report of the Philadelphia and Reading Railroad for 1888,* 58; paper insert in back of Compilation of the Capital Accounts, Reading Anthracite Company.
207. *Annual Report of the Philadelphia and Reading Railroad for 1876,* 25; *Miner's Journal,* February 3, 1872.
208. *Miner's Journal,* March 27, 1874; Hoffman, *Girard Estate Coal Lands.*
209. For example, *Annual Report of the Lehigh Valley Railroad for 1886.* At that time Ario Pardee, G.B. Markle, and Charles Skeer of Linderman and Skeer were all listed as directors.

DATA SOURCES FOR FIGURES

10.1 Data from tabulation in Price, *We Have Recovered Before!*

10.2. P&RRR data from Accession 1520, Reading Company Papers, Records of P&RRR Co. V-77 Transportation Dept. Statement Book 1844–1882; LVRR data from Board of Managers Minutes, Lehigh Valley Railroad.

10.3. Monthly tabulations in the Storrs Letter Books (LCDR).

10.4. Digital data for land tract boundaries from the coal field land data layer of the NE USA Historical GIS, originally digitized and polygonized from the Mine Sheets of the *Atlases of the 2nd Pennsylvania Geological Survey*; additional data from *Rothwell Map of 1870.*

NOTES FOR CHAPTER 11

1. Virtue, "Anthracite Combinations"; Jones, *Anthracite Coal Combination.* The rest of the section draws heavily on these sources, hence, for brevity's sake, general points will not be individually referenced.

2. Jones, *Anthracite Coal Combination*, 41.
3. Ibid., 42.
4. Ibid., 46.
5. Ibid., Ch. 8. This chapter contains a wealth of information on the legal status of the combination and it describes the failure of the ICC to take firm action against the combination in the years following the 1902 Coal Strike, the impact of the Hepburn Act of 1906, and the prosecution of certain of the anthracite railroads in 1907 under the Sherman Anti-Trust Act.
6. Cf. chapter 10, note 70.
7. *Annual Report of the Philadelphia and Reading Railroad for 1871*, 17; Jones, *Anthracite Coal Combination*, 57.
8. *Annual Report of the Lehigh Valley Railroad for 1876*, 8.
9. Letter of Thomas Dickson, Board of Managers Minutes, Delaware and Hudson Canal Co., January 18, 1873.
10. Ibid.
11. Schlegel, *Ruler of the Reading*, 38.
12. Ibid., 44.
13. Ibid.
14. Statement L, Lehigh Valley Railroad Papers relating to the attempt to limit coal production and transportation 1877–1878, A.C. Bining Collection, Historical Society of Pennsylvania; Schlegel, *Ruler of the Reading*, 44f.
15. Schlegel, *Ruler of the Reading*, 44.
16. Virtue, "Anthracite Combinations," 302–303.
17. Copy of *Philadelphia Ledger* article in the *Pittston Gazette*, July 31, 1873.
18. Virtue, "Anthracite Combinations," 302.
19. Copy of *New York Tribune* article in the *Pittston Gazette*, January 15, 1874.
20. Ibid.
21. WRS to E. Frisbie, December 12, 1874, LB28 (LCDR).
22. Schlegel, *Ruler of the Reading*, 31.
23. J.J. Albright to J. Bowen, June 13, 1863, LB10; J.J. Albright to J. Brisbin, March 21, 1866, LB11; *Testimony before Committee to investigate the Philadelphia and Reading Railroad (1876)*, 638; J.J. Albright to Hallstead, May 9, 1866, LB11 (LCDR).
24. *Testimony before Committee to investigate the Philadelphia and Reading Railroad (1876)*, 617.
25. Ibid., 836; Schlegel, *Ruler of the Reading*, 44f.
26. *Testimony before Committee to investigate the Philadelphia and Reading Railroad (1876)*, 540, 617, 1023; Schlegel, *Ruler of the Reading*, 46.
27. WRS to SS, January 4, January 12, 1875, LB28 (LCDR).
28. *Testimony before Committee to investigate the Philadelphia and Reading Railroad (1876)*, 631, 761.
29. WRS to S.H. Daddow, June 27, 1874, LB27 (LCDR).
30. Aurand, *Molly Maguires*, 87f.
31. Schlegel, *Ruler of the Reading*, 73.
32. *Saward's Journal* quoted in *Wilkes-Barre Record of the Times* (*WBRT*), December 5, 1874.
33. *Testimony before Committee to investigate the Philadelphia and Reading Railroad (1876)*, 1075–1076.
34. Letter to *New York Tribune* quoted in the *WBRT*, January 5, 1875; *WBRT*, January 15, 1875.
35. *Saward's Journal* quoted in *WBRT*, November 28, 1874; *WBRT*, December 23, 1874.
36. *WBRT*, January 15, 1875; Aurand, *Molly Maguires*, 89.
37. *WBRT*, January 6, 1875.
38. WRS to E. Frisbie, December 31, 1874, LB28 (LCDR).
39. WRS to SS, December 31, 1874, LB28 (LCDR).
40. *WBRT*, January 13, 1875; Aurand, *Molly Maguires*, 89f.
41. *Scranton Republican*, March 9, 1875.
42. *Testimony before Committee to investigate the Philadelphia and Reading Railroad (1876)*, 621.
43. WRS to SS, February 3, 1875, LB28 (LCDR); monthly shipment returns in the same LB.
44. WRS to SS, letter referenced in previous note.
45. *Scranton Republican*, March 10, April 7, 1875.
46. WRS to SS, Feb.1, 1875, LB28 (LCDR); Statement K, Lehigh Valley Railroad Papers, A.C. Bining Collection.
47. *Coal Trade Annual 1880* (1879), 26.
48. *Scranton Republican*, May 19, 1875.
49. *Coal Trade Journal*, March 5, 1879.
50. *Coal Trade Journal*, 1876, 55; *Annual Report of the Philadelphia and Reading Railroad for 1876*, 19. The figure for the 1874 Delaware and Hudson shipments to interior points in the *Coal Trade Journal* table seems to have had 1 million tons added to it by mistake, making the arithmetic incorrect.
51. *Testimony before Committee to investigate the Philadelphia and Reading Railroad (1876)*, 621.
52. *WBRT*, August 25, 1876.
53. Virtue, "Anthracite Combinations," 303.
54. Schlegel, *Ruler of the Reading*, 154.
55. *Coal Trade Journal*, June 7, 1876, quoting the *New York Daily Bulletin*.
56. Schlegel, *Ruler of the Reading*, 155.
57. *WBRT*, August 25, 1876.
58. Virtue, "Anthracite Combinations," 305.
59. Quoted in the *Coal Trade Journal*, June 7, 1876.
60. *Annual Reports of the Philadelphia and Reading Railroad, Lehigh Valley Railroad and Delaware and Hudson Canal Company*; Main Ledgers and Journals of the Lackawanna, Railroad, Delaware, Lackawanna and Western Railroad Records, Syracuse.
61. *Annual Report of the Philadelphia and Reading Railroad for 1876*, 29–30.
62. *Testimony before Committee to Investigate the Philadelphia and Reading Railroad (1876)*, 1112–1113.
63. Board of Managers Minutes, Susquehanna Coal Company, January 27, 1876.
64. Newspapers used include the *Scranton Republican*, *Miner's Journal*, *WBRT*, and *Coal Trade Journal*.
65. Schlegel, *Ruler of the Reading*, 155f.
66. *Miner's Journal*, March 30, 1877.
67. Ibid., April 6, 1877.
68. Ibid.
69. *Miner's Journal*, June 8, 1877.
70. *Annual Reports of the Philadelphia and Reading Railroad for 1876, 1877*; *Miner's Journal*, September 28, 1877.
71. *Miner's Journal*, June 8, 1877.

72. *Miner's Journal*, March 23, 1877; *Bill in Equity, The Central Railroad Co. of New Jersey vs. The Lehigh and Wilkes-Barre Coal Co.*, May Term 1877, No. 62, Special Report of Joseph A. Clay Master, Circuit Court of the United States, Western District of Pennsylvania; Bogen, *Anthracite Railroads*, 166f.
73. *Annual Report of the Lehigh Coal and Navigation Company for 1877*, 7–9.
74. Board of Managers Minutes, Delaware and Hudson Canal Company, February 19, 1877.
75. Ibid.
76. Ibid., March 22, 1877, see also note 134, Chapter 5.
77. *Annual Report of the Delaware and Hudson Canal Company for 1877*, 4; *Report to the Stockholders of the Delaware and Hudson.*
78. *Engineering and Mining Journal*, articles of September 18, 22, 29, and October 6, 1877, reproduced as typewritten pamphlet, Pliny Fisk Library, Princeton University.
79. *Miner's Journal*, November 2, 1877; Bogen, *Anthracite Railroads*, 197.
80. A.C. Bining Collection, Table of Percentage Shipments by members of the Combination 1877; Schlegel, *Ruler of the Reading*, 167f.
81. Scranton Auctions Price Tabulations in *Coal Trade Annual 1877* (1876), 9; Jones, *Anthracite Coal Combination*, 228.
82. *Miner's Journal*, December 14, 1877.
83. *Philadelphia Ledger*, December 10, 1877, clipping in A.C. Bining Collection.
84. Copies of the tabulations used at the meeting survive in the A.C. Bining Collection. The record of the votes is contained in an undated letter in the collection addressed to the coal operators shipping by the Lehigh.
85. Schlegel, *Ruler of the Reading*, 169.
86. Virtue, "Anthracite Combinations," 307; *Miner's Journal*, January 25, 1878.
87. *Miner's Journal*, February 22, 1878.
88. *Miner's Journal*, February 1, 1878; circular from Lehigh Valley Railroad, February 6, 1878, cards circulated by Lehigh Valley Railroad, March 1, 1878, A.C. Bining Collection.
89. *Miner's Journal*, February 15, March 1, 1878.
90. Schlegel, *Ruler of the Reading*, 172.
91. Ibid., 173–174.
92. *Miner's Journal*, January 2, February 28, 1879.
93. *Coal Trade Journal*, 1880, 3; *Miner's Journal*, June 13, 1879.
94. Official figures are provided in the *Coal Trade Annual* for the relevant years.
95. *Miner's Journal*, June 20, 1879.
96. Reply by Asa Packer on back of letter of Thomas Dickson, December 13, 1877, A.C. Bining Collection.
97. See Annual Tonnage figures in the *Coal Trade Annual*, various years.
98. Ibid.
99. *Miner's Journal*, February 8, 1878.
100. *Miner's Journal*, October 3 and 10, November 14, December 5 and 12, 1879.
101. SS to WRS, March 6, 1879, LB Sloan to Storrs 1870–1885 (LCDR).
102. SS to WRS, May 17, July 31, 1879, LB Sloan to Storrs 1870–1885 (LCDR).
103. SS to WRS, December 4, 1879, LB Sloan to Storrs 1870–1885 (LCDR).
104. SS to WRS, December 11, 1879, LB Sloan to Storrs 1870–1885 (LCDR).
105. SS to WRS, December 4, 1879, LB Sloan to Storrs 1870–1885 (LCDR).
106. SS to WRS, December 15, 1879, LB Sloan to Storrs 1870–1885 (LCDR).
107. Virtue, "Anthracite Combinations," 310.
108. Letter copy, January 4, 1881, LB Sloan to Storrs 1870–1885 (LCDR); for other evidence of signature attempts, see for example W.H. Tillinghast to F.B. Gowen, March 27, 1883, Special Letterbook, W.H. Tillinghast, September 1879–June 1885, Box 37 (LWBCCR), hereafter referred to as Tillinghast Special Letterbook.
109. W.H. Tillinghast to T. Dickson, March 3, 1880, Tillinghast Special Letterbook.
110. W.H. Tillinghast to R.M. Olyphant, March 19, 1882; the quotation is from a later letter from W.H. Tillinghast to E.B. Leisenring, May 23, 1882, LB W.H. Tillinghast, January–November 1882, Box 1 (LWBCCR).
111. Virtue, "Anthracite Combinations," 310; SS to WRS, Dec.11, 1879; January 25, 1882, L.B. Sloan to Storrs 1870–1885 (LCDR).
112. W.H. Tillinghast to T. Dickson, June 23, 1880, Tillinghast Special Letterbook; SS to WRS, December 27, 1883, Sloan to Storrs LB 1870–1885 (LCDR).
113. W.H. Tillinghast to F.S. Lathrop, November 30, 1879, Tillinghast Special Letterbook; WRS to SS, November 30, December 17, 1883, LB37A (LCDR).
114. WRS to SS, May 9, July 30, 1881; January 4, 1882, LB32A (LCDR).
115. WRS to SS, April 26, 1882, LB32A (LCDR).
116. SS to WRS, November 30, 1883, LB Storrs to Sloan 1870–1885 (LCDR).
117. WRS to SS, April 12, November 29, 1884, LB37A (LCDR).
118. Board of Managers Minutes, Susquehanna Coal Company, February 28, 1879.
119. Ibid., May 24, 1879.
120. Bogen, *Anthracite Railroads*, 61.
121. SS to WRS, November 1, 1883, LB Sloan to Storrs, 1870–1885; WRS to SS, September 20, 1884, LB37A (LCDR).
122. WRS to SS, April 14, 1884, LB37A (LCDR).
123. WRS to SS, January 14, 1885, LB40A (LCDR).
124. Shipment figures from *Coal Trade Annual*, various years.
125. Virtue, "Anthracite Combinations," 311 (official suspension days); Annual Reports of WRS, Board of Managers Minutes, Delaware, Lackawanna and Western Railroad. The 1882 figure for the Lackawanna excludes the Manville Breaker, worked

one half of the time by the company, and the other half by the Delaware and Hudson.

126. Virtue, "Anthracite Combinations," 313.
127. *Labor Troubles*, 640.
128. Ibid., 641.
129. WRS to SS, December 31, 1884; January 2, January 5, 1885, LB40A (LCDR).
130. Campbell, *Businessmen and Anthracite*, 160–168.
131. *Labor Troubles*, 649.
132. Ibid., 649–650.
133. Campbell, *Businessmen and Anthracite*, 69–171; *Labor Troubles*, 649.
134. Virtue, "Anthracite Combinations," 314; WRS to SS, February 17, 1890, LB August 1887–April 1891 (LCDR).
135. Campbell, *Businessmen and Anthracite*, 174.
136. WRS to SS, n.d. (January 1890), LB August 1887–April 1891 (LCDR).
137. *Coal Trade Journal*, 1892, 56.
138. Bogen, *Anthracite Railroads*, 173–174.
139. Ibid., 62–71.
140. *Coal Trade Journal*, 1893, 678.
141. Virtue, "Anthracite Combinations," 315.
142. *Coal Trade Journal*, 1892, 105.
143. *Coal Trade Journal*, 1892, 143–144, 212, 240.
144. *Coal Trade Journal*, 1892, 473.
145. Ibid., 145.
146. Ibid., 473.
147. *Annual Report of the Philadelphia and Reading Railroad for 1895*, 14–16.
148. Pamphlet 'To The Members of the Anthracite Coal Operators' Association' (ACOA), Box 29, Folder 8, Reading Company Papers.
149. *Coal Trade Journal*, 1892, 73.
150. Pamphlet "To the Members of the ACOA", Reading Company Papers.
151. Ibid.
152. Page proofs of document describing plans for dividing up anthracite tonnage, n.d., Box 29, Folder 8, Reading Company Papers. Hereafter referred to as Tonnage Plans; *Annual Report of the Philadelphia and Reading Railroad for 1895*, 16–17.
153. Report by C.E. Henderson, Chairman, Aug. 15, 1894, Box 29, Folder 8, Reading Company Papers.
154. Typescript history of attempts to reach agreement for the regulation of the anthracite coal trade, n.d., Box 29, Folder 9, Reading Company Papers. Hereafter referred to as History of Regulation Attempts.
155. Tonnage Plans, 2.
156. Ibid.
157. Ibid.
158. *Annual Report of the Philadelphia and Reading Railroad for 1895*, 20.
159. Ibid., 21; History of Regulation Attempts, 3–4; Minutes of Meeting of Presidents of the Anthracite Roads, April 5, 1895; notes of a meeting between Messrs. Olyphant, Baker, Thomas et al. and the Reading Receivers, April 22, 1895, Box 29, Folder 8, Reading Company Papers.
160. *Annual Report of the Philadelphia and Reading Railroad for 1895*, 22.
161. Memorandum and Suggestions for monthly output of Anthracite Tonnage for balance of year commencing June 1, 1895, Box 29, Folder 9, Reading Company Papers; History of Regulation Attempts, 5–6.
162. History of Regulation Attempts, 5; *Annual Report of the Philadelphia and Reading Railroad for 1895*, 24. It should be noted that the final corrected tonnage statistics for 1895 actually gave the Reading a percentage of 21.47 percent, which was to prove its best showing during the decade of the 1890s.
163. Typescript paper on the Colliery Capacity Plan, February 20, 1895, Box 29, Folder 8; typescript paper on distribution of tonnage, April 4, 1895, Box 29, Folder 8, Reading Company Papers.
164. Press cutting, January 23, 1896 (*New York Times*), Box 29, Folder 9; Memorandum, January 7, 1898, Box 29, Folder 9, Reading Company Papers, hereafter called Memo 1898.
165. Memorandum, January 31, 1896, Box 29, Folder 9, Reading Company Papers.
166. Ibid.
167. Jones, *Anthracite Coal Combination*, 56.
168. J.S. Harris to E.B. Thomas (president, Erie Railway), October 11, 1897, Box 29, Folder 9, Reading Company Papers.
169. Ibid.
170. Memo 1898, Reading Company Papers.
171. Tables of bituminous tidewater tonnage 1897–1899, Box 29, Folder 9, Reading Company Papers.
172. Jones, *Anthracite Coal Combination*, see especially 67f., upon which the remainder of this section is based; detailed reference will not be made to individual points.
173. Ibid., 73f.

DATA SOURCES FOR FIGURES

11.1 Percentages to 1886 from *Atlas to Accompany Part III of the Pennsylvania Geological Survey Annual Report for 1886*; remaining figures calculated from data in *Coal Trade Annual*, various years.
11.2 Ibid.
11.3 Ibid.
11.4 Ibid.
11.5 Weekly tonnage reports from *Scranton Republican.*
11.6 Calculations based on tonnage reports from a number of sources but predominantly *Miner's Journal* and *Scranton Republican.*
11.7 Weekly tonnage reports from *Coal Trade Journal.*
11.8 Ibid.
11.9 Ibid.
11.10 Week-ending quotations nearest to month end from *New York Times.*
11.11 Weekly tonnage reports from *Coal Trade Journal.*
11.12 Weekly tonnage reports from *Miner's Journal.*
11.13 Ibid.
11.14 Table in Jones, *Anthracite Coal Combination*, 156–157.

NOTES FOR CHAPTER 12

1. Jones, *Anthracite Coal Combination*, 73f.
2. WRS to SS, January 20, 1883, LB37A (LCDR).
3. WRS to SS, February 26, 1891, LB August 1887–April 1891 (LCDR).
4. WRS to SS, February 3, 1885, LB40A (LCDR).
5. See, for example, October 22, December 31, 1880; February 25, 1881, Board of Directors Minutes, Susquehanna Coal Company; WRS to SS, May 9, 1891, LB50A (LCDR).
6. H.M. Chance, "Valuation of Coal Land," *Trans. AIME* 47 (1914): 111–146; quotation is from p. 130.
7. Chance, *Mining Methods*, 115.
8. W.H. Storrs to SS, January 4, 1898, LB 1897–1899 (LCDR). In the second stage of sinking, the shafts were lowered to the Dunmore Vein. The Sloan breaker, through which the output of both shafts was processed following the fire at the Central in 1890, was idle for 15 months from the beginning of 1898.
9. Lesley et al., *Summary Description*.
10. Baird, LV Coal Co. History, 190.
11. Chance, *Mining Methods*, 54.
12. Ibid. 137–138.
13. WRS to SS, December 27, 1886, LB40A (LCDR).
14. Chance, "Valuation of Coal Land," (discussion), 137f.
15. *Annual Report of the Philadelphia and Reading Railroad for 1876*, 27–28.
16. Harris, *Report on Coal Lands*, 41, 80–81.
17. Ibid., 31.
18. Chance, "Valuation of Coal Land," 135.
19. For an example of specific reference to the role of interest calculations in the valuation process, see WRS to SS, August 28, 1880, LB32A (LCDR).
20. Executive Committee Minutes, Delaware, Lackawanna and Western Railroad, October 17, 1869.
21. WRS to SS, August 28, 1880, LB32A (LCDR).
22. WRS to SS, April 13, 1881, LB32A (LCDR).
23. WRS to SS, April 14, 1881, LB32A (LCDR).
24. SS to WRS, April 16, April 18, April 20, 1881, Letterbook SS to WRS 1870–1885 (LCDR).
25. WRS to SS, March 19, 1881, LB32A (LCDR).
26. WRS to SS, April 15, 1893, LB50A (LCDR).
27. WRS to SS, March 21, 1899, LB1897–1899 (LCDR).
28. Yearley, *Enterprise and Anthracite*, 51.
29. Baird, LV Coal Company History, 125.
30. Ibid. 125–126.
31. DB81, p. 183f; DB127 p. 174f, Luzerne County Courthouse.
32. DB130, p. 28, Luzerne County Courthouse.
33. Report on Coal Lands, December 8, 1891, Israel Morris Papers.
34. Ibid.
35. Ibid.
36. See Yearley, *Enterprise and Anthracite*, 52, for a good example of this.
37. Journal T, p. 192, Lackawanna Railroad Records, Syracuse University.
38. *Annual Report of the Delaware and Hudson Canal Company for 1883*; Statement of the Coal Company's Property (typescript, 1884); *Report on Coal Lands*, December 8, 1891, Israel Morris Papers.
39. G.E. Stevenson, *Reflections of an Anthracite Engineer* (Scranton: published by the author, 1931).
40. Baird, LV Coal Company. History, 127–128.
41. Ibid., 52–54; W. Brewster, *History of Certified Township of Kingston, PA. 1769–1929* (Kingston, PA: Kingston Borough School District, 1929), 499.
42. W.H. Storrs to SS, January 25, January 27, 1894, LB50A (LCDR).
43. W.H. Storrs to SS, November 7, 1895, LB50A; May 28, 1897, Undated (June 1897), February 5, February 7, February 8, 1898, LB 1897–1899 (LCDR).
44. Lease No. 6506, Box 11, Contracts 2002–6320, Pennsylvania Coal Company Records, Pennsylvania State Archives.
45. Lease No. 801, B.T. Reed to the Delaware and Hudson Canal Company, May 16, 1863, Box 12, Contracts 6357–8453, Pennsylvania Coal Company Records, Pennsylvania State Archives.
46. Yearley, *Enterprise and Anthracite*, 53.
47. Ibid., 52.
48. Board of Managers Minutes, Pennsylvania Coal Company, November 17, 1851.
49. Board of Managers Minutes, Delaware and Hudson Canal Company, February 15, October 9, December 19, 1856.
50. Agreement, T.H. Oram and the Delaware, Lackawanna and Western Railroad, July 15, 1861, DB85, p. 419f, Luzerne County Courthouse.
51. Ibid.
52. The agreements can be found in the following deedbooks: DB80 p. 344f; DB80, p. 349f; DB80, p. 468f; DB79, p. 305(572)f; DB84, p. 80f; DB84, p. 85f; DB85, p. 419f; DB92, p. 478f; and DB92, p. 472f.
53. Agreement Wm. Connell et al. and the Delaware, Lackawanna and Western Railroad, September 1, 1870, DB144, p. 131f. (quote from p. 133), Luzerne County Courthouse.
54. Agreement Mount Pleasant Coal Co. and the Delaware, Lackawanna and Western Railroad, August 15, 1871, DB151, p. 551f; Agreement Charles Hutchison and the Delaware, Lackawanna and Western Railroad, July 10, 1872, DB160, p. 366f, Luzerne County Courthouse.
55. Agreement Lackawanna and Susquehanna Coal and Iron Company and the Delaware, Lackawanna and Western Railroad, July 24, 1872, DB160, p. 462f, Luzerne County Courthouse.
56. Monthly shipment records in letterbooks 1872–1875 (LCDR); *AIM Report for 1873*, 231; *Scranton Republican*, January 3, 1877.
57. The Lackawanna shipment records show that shipments from the Sibley began in mid-1873; shipments on the contract by the Coal and Iron

Company (which had been renamed the Pennsylvania Anthracite Coal Company) to the end of 1878 are found in the Minutes of the Board of Managers, Delaware, Lackawanna and Western Railroad, Jan. 31, 1879. The total given almost exactly matches the tonnages shipped by Sibley and Phinney 1873–1878, according to the Lackawanna records. Sibley appears in the 1878 list of Lackawanna shippers in the Board of Managers Minutes, for February 21, 1878, as being operated by the Pennsylvania Anthracite Coal Company, not Elliott, Koerner and Company, indicating that the business relationship had become much closer since 1872–1873, although the precise details are not known. The proposal by Elliott, Koerner and Company in 1872, for a contract with the Lackawanna, can be found in WRS to SS, March 23, 1872; WRS to F. Koerner, March 28, 1872, LB25 (LCDR). For details of the Carbon Hill arrangement, see the discussion in Chapter 10 on the Erie coal lands.

58. *Coal Trade Journal*, March 19, 1879.
59. WRS to SS, May 8, 1882, LB32A (LCDR).
60. William Connell and Company to WRS, October 30, 1874, copy in SS to WRS LB 1870–1885 (LCDR).
61. Agreement William Connell et al. and the Delaware, Lackawanna and Western Railroad, Sept. 1, 1870, DB144, p. 133, Luzerne County Courthouse.
62. Agreement William H. Meeker et al., and the Delaware, Lackawanna and Western Railroad, September 1, 1873, DB173, p. 496f, Luzerne County Courthouse.
63. See for example the Report transcribed into the Board of Directors Minutes, Delaware, Lackawanna and Western Railroad, January 31, 1879.
64. The Roaring Brook Coal Company (Nay Aug Colliery) contract had a clause of this kind, for example. See Agreement T.F. Hunt et al., and the Delaware, Lackawanna and Western Railroad, November 25, 1863, DB92, p. 472f, Luzerne County Courthouse.
65. W.H. Meeker agreement (see note 62).
66. WRS to SS, January 4, 1875, LB28 (LCDR).
67. Agreement John Jermyn and the Delaware, and Hudson Canal Company and the Delaware, Lackawanna and Western Railroad, April 1, 1876, DB195, p. 456, Luzerne County Courthouse.
68. Annual Report by WRS for 1882, Board of Managers Minutes, 1873–1888, Delaware, Lackawanna and Western Railroad, 324f.
69. J. Brisbin to T. Dickson, November 12, 1866, President's Letterbook, (J. Brisbin) September 1866–March 1868 (LCDR).
70. WRS to SS, June 15, 1886, LB40A (LCDR).
71. Jones, *Anthracite Coal Combination*, 88.
72. For general treatments of the history of these railroads see W.A. Lucas, *The History of the New York, Susquehanna and Western Railroad* (New York: Railroadians of America, 1939); W.F. Helmer, *O & W: The Long Life and Slow Death of the New York, Ontario and Western Railway* (Berkeley: Howell-North Books, 1959); J. Krause and E. Crist, *Susquehanna: The New York, Susquehanna and Western Railroad* (Newton, NJ: Carstens Publications Inc., 1980); H.S. Fredericks, *The Wilkes-Barre and Eastern Railroad* (Livingston, NJ: Railroadians of America, 1986); and R.E. Mohowski, *The New York, Susquehanna and Western Railroad* (Baltimore: Johns Hopkins University Press, 2003).
73. *Annual Report of the New York, Susquehanna and Western Railroad for 1882*; Board of Managers Minutes, Delaware, Lackawanna and Western Railroad, December 30, 1881.
74. WRS to SS, August 1, August 4, 1881, LB32A (LCDR).
75. *Coal Trade Annual 1897* (1896), 143.
76. WRS to SS, August 1, 1881, LB32A (LCDR).
77. WRS to SS, June 18, 1881, LB32A (LCDR).
78. WRS to SS, May 29, June 29, July 10, 1882, LB32A (LCDR).
79. WRS to SS, October 26, 1883, LB37A (LCDR).
80. WRS to SS, June 18, 1881, LB32A (LCDR).
81. WRS to SS, April 20, 1881, LB32A (LCDR).
82. WRS to SS, November 25, 1881, LB32A (LCDR).
83. Agreement, J.C. Haddock and C.F. Steel and the Delaware, Lackawanna and Western Railroad, February 24, 1882, DB226, p. 285f. Luzerne County Courthouse; WRS to SS, November 25, 1881, LB32A (LCDR).
84. WRS to SS, November 25, 1881, LB32A (LCDR).
85. WRS to SS, February 2, 1882, LB32A (LCDR).
86. Baird, LV Coal Company History, 190–198; Agreement, C.S. Maltby and the Delaware, Lackawanna and Western Railroad, January 13, 1879, DB232, p. 40f, Luzerne County Courthouse; WRS to SS, June 17, 1882, LB32A (LCDR).
87. *AIM Report for 1881*, 199; *for 1884*, 140; *for 1885*, 44–45.
88. WRS to SS, March 21, 1885, LB40A (LCDR).
89. Agreements; J.H. Swoyer and the Delaware, Lackawanna and Western Railroad, September 1, 1880, DB229, p. 396f, (Wyoming Breaker); and January 15, 1880, DB231, p. 305f, (Forty Fort Breaker), Luzerne County Courthouse.
90. WRS to SS, June 17, 1881, July 26, 1882, LB32A; September 5, 1882, LB37A (LCDR).
91. Agreement Wyoming Valley Coal Company and the Delaware, Lackawanna and Western Coal Company, July 19, 1883, DB238, p. 280f., Luzerne County Courthouse.
92. WRS to SS, April 10, April 11, 1884, LB37A (LCDR).
93. WRS to SS, September 24, 1884; October 15 and 16, 1884, LB37A (LCDR); March 28, 1885, LB40A (LCDR); Lease, Wyoming Valley Coal Company to the Lehigh Valley Coal Company, February 27, 1885; President's Report to the Directors, April 9,1885; President's Report to Stockholders, January 29, 1885; President's Report to Stockholders, January 28, 1886, all in Wyoming Valley Coal Company Papers 1878–1917 (1 box). LVRR Records, B) Subsidiary Companies 2. Non-transportation companies, MG-274, Pennsylvania State Archives, Harrisburg; hereafter referred to as

WVCCo. Papers; Tonnage figures from *AIM Report for 1884*, 140; *for 1885*, 44–45.

94. J.M. Crane to I.W. Morris, June 29, 1900, WVCCo. Papers.
95. WRS to SS, especially, June 27, August 2, September 24, September 26, 1887, see also comments in intervening correspondence, LB40A and LB, 1887–1891 (LCDR).
96. WRS to SS, December 27, 1888, January 1, 1889, LB 1887–1891 (LCDR); *Coal Trade Annual 1902* (1901), 30; *Coal Trade Journal*, December 6, 1893; Helmer, *O & W*, 60.
97. WRS to SS, January 16, April 23, May 28, 1889, LB 1887–1891 (LCDR).
98. Jones, *Anthracite Coal Combination*, 88; WRS to SS, February 28, 1891, LB 1887–1891 (LCDR).
99. *Coal Trade Journal*, January 20, 1892, November 8, December 6, 1893.
100. WRS to SS, January 28, 1891, LB 1887–1891 (LCDR).
101. Fredericks, *Wilkes-Barre and Eastern Railroad*, 10; WRS to SS, n.d. (June 1892), LB50A, 198; WRS to SS, January 2, March 23, 1893, LB50A (LCDR).
102. WRS to SS, April 2, 1892, LB50A (LCDR); *Coal Trade Journal*, October 25, 1893 (source of quotation).
103. Agreement Babylon Coal Company and the Philadelphia and Reading Coal and Iron Company, March 7, 1892, copy in LB70, 369–378 (LCDR).
104. WRS to SS, August 10, September 16, 1893, LB50A (LCDR).
105. WRS to SS, April 2, May 12, May 26, June 20, July 18, 1892, LB50A (LCDR).
106. C. Pardee Foulke and W.G. Foulke, *Calvin Pardee 1841–1923: His Family and his Enterprises* (Philadelphia: Pardee Company, 1979), 135–137; Interstate Commerce Commission, no. 150, *Coxe Brothers and Co. vs. The Lehigh Valley Railroad Co., Argument of Mr. Franklin B. Gowen for Complainants*, March 19 and 20, 1889 (Philadelphia: Allen Lane and Scott, 1889); E.B. Coxe to G.B. Roberts, November 14, 1887, Box 458, Folder 6, Coxe Family Mining Papers, 1767–1968, HSP.
107. *Coal Trade Journal*, October 25, 1893.
108. Fredericks, *Wilkes-Barre and Eastern Railroad*, 20–24; Krause and Crist, *New York, Susquehanna and Western Railroad*, 6.
109. Jones, *Anthracite Coal Combination*, 60–61; Krause and Crist, *New York, Susquehanna and Western Railroad*, 6; Minute Book 1887–1909, New York, Susquehanna and Western Coal Company, July 13, 1894, Pennsylvania State Archives.
110. *Coal Trade Journal*, 1892, 630; *Coal Trade Journal*, 1893, 729; W.H. Storrs to SS, January 12, 1899, LB 1897–1899 (LCDR); Jones, *Anthracite Coal Combination*, 74–76.
111. Jones, *Anthracite Coal Combination*, 76–82.
112. Temple Iron Company, Stock Ownership List, Carton 4, Folder 2, Corporation Book 1904, Lehigh Valley Railroad Company Records, Pennsylvania State Archives.
113. W.H. Storrs to SS, February 6, 1899, n.d., p. 257, LB 1897–1899 (LCDR); Jones, *Anthracite Coal Combination*, 78.
114. Jones, *Anthracite Coal Combination*, 83–84.
115. Ibid., 89–97. The section in question contains the full text of a standard 65 percent contract. The table referred to is on p. 96.
116. Bogen, *Anthracite Railroads*, 139–140.
117. Calculated from data in *Coal Trade Annual 1903* (1902), 26.
118. Ibid., 21.
119. U.S. Industrial Commission, *Report of the Industrial Commission on the Relations and Conditions of Capital and Labor Employed in the Mining Industry*, vol. 12. (Washington, DC: Government Printing Office, 1901), 701. For further discussion of these changes see R.G. Healey, "Disturbance of the Peace: the Operators' View of the 1902 Anthracite Coal Strike" in Janosov, *Perspectives on the Great Strike.*
120. *Coal Trade Journal*, 1900, 735.
121. SS to WRS, July 16, 1879, Letterbook SS to WRS, 1870–1885 (LCDR).
122. SS to WRS, Oct.16, 1885, copy of letter enclosed in LB40A (LCDR).
123. Annual Report of WRS to SS, for 1877, Board of Managers Minutes 1873–1888, Delaware, Lackawanna and Western Railroad, 169f.
124. *Annual Report of the Lehigh Valley Railroad for 1890*, 12.
125. WRS to SS, January 23, 1889, LB August 1887–April 1891 (LCDR). This letter explains the details of the $350,000 upfront payment made to Abram Nesbitt to secure the lease of his prime lands near Plymouth. An example of a payment to take over existing leases is given by the joint takeover of John Jermyn's No. 3 colliery in 1881 by the Delaware and Hudson and the Delaware, Lackawanna and Western. Each company paid $100,000 to gain control of the mine and its lease. Thereafter the two companies worked the breaker in alternate months, as indicated in the main text, see WRS to SS, September 8, October 27, 1881, LB32A (LCDR).
126. Von Bergen, *Photographs of the Delaware, Lackawanna and Western Railroad Coal Department Collieries.*
127. See note 125.
128. See, for example, Baird, LV Coal Co. History, 113, 330 for details of the acquisition and development of the Blackwood and York Farm properties in 1888–1889.
129. *Corporate History of the Northern Coal and Iron Company*, 1, 63; also see ch. 10 above.
130. WRS to SS, April 13, 1887, LB40A (LCDR).
131. Baird, LV Coal Co. History, 11–17.
132. See Schlegel, *Ruler of the Reading*, 32f.
133. Baird, LV Coal Co. History, 262.
134. WRS to SS, August 23, 1884, LB37A (LCDR).

DATA SOURCES FOR FIGURES

12.1 WRS to SS, Feb. 27, 1879, LB29F (LCDR).

12.2 Calculated from data derived from general coal agent's annual reports in letterbooks and minute books; newspaper reports; some 1888–1900 data from coal dept. expenses ledgers, Lackawanna Railroad Records (Syracuse).

12.3 Data as for previous figure.

12.4 Data as for Figure 12.2.

12.5 Data from various sources, including reports in the *Miner's Journal* and the *Pittston Gazette*; Daddow and Bannan, *Coal, Iron and Oil*, 751.

12.6 Calculated from data in *Annual Reports of the Lehigh Valley Railroad.*

12.7 Owing to the extent of 'charter restriction avoidance' that was indulged in by the Lackawanna Railroad, distinguishing the timing of transactions that genuinely represented strategic decision-making and investment, rather than attempts to obfuscate an audit trail, is a difficult undertaking. The major task was to distribute lands accumulated by the subsidiary, the Nanticoke Coal and Iron Company, back to the years when control of the lands in question actually passed to individuals or corporate entities acting in the interest of the railroad. Otherwise, a false impression would be created that the railroad only acquired control of the lands at the time of the merger of the Nanticoke Coal and Iron Company in 1870. Some of the earliest transactions involved purchases, by Lackawanna directors J.J. Phelps and Moses Taylor, during the Civil War. Some of these lands were subsequently signed over to subsidiaries, such as the Granby Coal Company later merged into the Nanticoke Coal and Iron Company. Others remained in trust until the deaths of the directors in question, after which they were assigned to the Railroad for the princely sum of $1! An attempt has therefore been made to use the earliest of the series of transaction dates associated with a land parcel where a connection to parties acting in the interest of the railroad could be established. This introduces a limited element of estimation at times, since some individuals who were not themselves railroad directors could have acted on behalf of the railroad, and there is also some evidence that not all transactions were recorded. However, comparison of the deed records with other company accounts suggests that lands falling into these categories did not constitute a significant proportion of the company's portfolio. The only exception to this is the lands of the ill-fated Union Iron and Coal Company, originally discussed in Chapter 4. Although Moses Taylor was a major player in this company, it was not a subsidiary of the railroad in the same way as the Nanticoke or Granby Coal Companies. Hence, control of its estate by the railroad is not dated back to the 1850s, but to 1861 when Percy Pyne bought it in on behalf of the Lackawanna at a sheriff's sale.

The information used to construct the graph has been derived from two main sources. The first is the deed book records in the Luzerne and Lackawanna county courthouses and the second is the coal department construction record and associated original accounting vouchers in the Lackawanna Railroad Records (Syracuse). It should be noted that the graph does not take into consideration three accounting issues related to land transactions. The first of these is that many of the land purchases were paid by mortgages, to spread the cost over a number of years. Secondly, it does not allow for depreciation charges, which progressively reduced the book value of coal land assets, and thirdly, it does not take account of sales of surface lots to individuals. Such sales reduced the book cost of the lands. The level of these sales became substantial in the late 1890s, especially in the city of Scranton.

12.8 Delaware and Hudson Company and Affiliated Companies, Cost of Coal Lands, Box 61, Delaware and Hudson Company Records, New York State Archives.

12.9 Data throughout text in Baird, LV Coal Co. History; tabulations and lists in Israel Morris Papers.

12.10 Original lease agreements, Deed Books, Luzerne and Lackawanna County courthouses.

12.11 Sources same as for Figure 12.9.

ADDITIONAL NOTES ON LEHIGH VALLEY COAL COMPANY LAND DIAGRAMS (FIGURES 12.9 AND 12.11)

As indicated above, the data required to assemble these time series were pieced together from information in Baird, LV Coal Co. History and the earlier tabulations in the Israel Morris Papers in the Historical Society of Pennsylvania. It should be noted that the recorded total acreages of tracts and the acreage on tracts lying within the coal measures often differ between these two sources, sometimes by a significant amount. Under these circumstances the later figures were used, as they would have reflected the results of additional surveys and prospecting activity.

Development of the time series for leases was complicated by several factors. Firstly, some leases were perpetual, i.e. they remained in force until all the coal was exhausted, while others, such as the Girard Estate leases, were fixed-term and had to be renewed. Secondly, control over a number of leases was acquired through acquisition of stock in the companies that were the original lessees, although neither the Lehigh Valley Railroad nor its coal company were actually named on the leases themselves. Thirdly, although in general the execution of leases did not entail payments at the point of execution, on occasion, bonus payments were made to landholders as part of the agreement. Payments were also made to purchase the 'right-to-mine' encapsulated in leases, often as part of the take-over of another mining concern. Finally, in the case of the Girard leases, the area lying within the coal measures on each of the different tracts has not been located, but the total acreage of each lease and the total acreage within the coal measures on all the tracts taken together, is known. From these data, estimates were made of the acreage of coal lands in each case, on a pro rata basis. Based on these points, several rules were adopted. Firstly, where leases had to be renewed, their acreage is included again at the time of lease renewal. Secondly, where control over leases was acquired by take-over or majority-stock interest, the acreage is reported at the acquisition date. Thirdly, all one-off payments for leases that were not royalty-related are included in the time series for the second graph on the purchase costs of fee lands, leaseholds, and mining improvements.

This second graph of purchase costs covers the three types of expense, either singly or in combination. In the earlier years, lands were often bought in fee without any mining improvements. Subsequently, they were quite frequently bought together or improvements might be purchased on expiry of leases or bankruptcy of lessees. In an analogous manner to leases, control over fee lands might be obtained through purchase of a majority of stock in the company owning the lands. Under these circumstances, the cost of acquiring the stock, rather than the original cost of the land, is used in the time series. In most cases the timing of stock acquisition is clear, but in others it was spread over a period, without the corresponding financial breakdown being available. Under these circumstances, the purchase is attributed to the earliest known date of stock acquisition. This may slightly exaggerate the level of land acquisition activity at the end of the 1860s.

Owing to the considerable complexity of land and lease transactions over long periods of time, the individual items of data have been checked and cross-checked against other summary tabulations and sources where it has been possible to do this. Time unfortunately precluded a detailed check against all the original county courthouse deed records and even this would not have guaranteed completeness, since there is evidence from the company history (p. 167) that in at least one case a lease was being operated, although no paperwork record had ever been executed. In the case of one or two small tracts, either the costs are not available, or the area within the coal measures has had to be assumed to be the same as the total area, for want of information to the contrary, based on the known location of the tracts within the coalfield. However, sufficient information is available for all the major tracts, leases, and transactions to give a very close estimate of the pattern of investment activity over the time period in question.

NOTES FOR CHAPTER 13

1. Board of Managers Minutes, Susquehanna Coal Company, November 20, 1870.
2. Baird, LV Coal Company History, 378.
3. Ibid., 415.
4. *Charter and By-Laws of the Hartford Coal Company* (Philadelphia: Oliver P. Glessner, Book and Job Printer, 1855). Author's Collection.
5. *Charter and By-Laws of the Honey Brook Coal Company, incorporated by Special Act of the Legislation of the State of Pennsylvania April 23rd, 1864* (Philadelphia: Bryson and Sons, Printers, 1864). F.C. Petrillo Private Collection.
6. *Scranton Republican*, January 3, 1879; WRS to SS, January 2, 1879, LB31 (LCDR).
7. *Scranton Republican*, January 21, 1879.
8. *Annual Report of the Secretary of Internal Affairs of the State of Pennsylvania, Part III, Industrial Statistics 1876–1877* (Harrisburg: Lane S. Hart, State Printer, 1878), 61–62; WRS to SS, February 1, 1879, LB31 (LCDR).
9. Journals 1 and 2, 1869–1908, Susquehanna Coal Company; *Annual Reports of the Susquehanna Coal Company*; Records and Board Files, Pennsylvania Railroad Records.
10. *Annual Reports of the Philadelphia and Reading Railroad for 1879–1902*. The floating debt appeared on the opposite side of the balance sheet from the profit and loss account, which fell under the income heading, as would be expected. See, for example the *Annual Report for 1886*, 60–61. The floating debt peaked at $2.8 million in 1885. Consistency of reporting across the period is affected by the change of fiscal year end in 1898 to June 30 instead of November 30, as had prevailed in earlier years.
11. Baird, LV Coal Co. History, 266.
12. Ibid.
13. Ibid., 416–417.
14. The figures are derived from summation of annual profits and losses shown for the years in question on the respective graphs.
15. In 1885, the *Annual Report of the Susquehanna Coal Company* records that the company's four main breakers worked 300, 307, 294 and 302 days. The equivalent average for the Lackawanna's mines was only 153 days, see Board of Managers Minutes, Delaware, Lackawanna and Western Railroad, January 26, 1886.
16. *Testimony before the Committee to Investigate the Philadelphia and Reading Railroad* (1876),1045.
17. Profits are derived from the main ledgers and journals, Delaware, Lackawanna and Western Railroad Records, Syracuse.
18. For example, the annual report data indicate that Breaker No. 6, which worked for almost 300 days in 1886, was only working 214 or 215 days by 1889–1890, cf. note 15.
19. *Miner's Journal*, August 19, 1865; see also Wallace, *St. Clair*, 75–76.
20. Daddow and Bannan, *Coal, Iron and Oil*, 768.
21. *Fourth Annual Report of the Susquehanna and Wyoming Valley Railroad and Coal Company, January 1, 1864.*
22. Daddow and Bannan, *Coal, Iron and Oil*, 768.

23. *Fourth Annual Report of the Susquehanna and Wyoming Valley Railroad and Coal Company.*
24. *AIM Report for 1882*, 182.
25. Board of Managers Minutes 1883–1903, Florence Coal Company, MG-300, Erie Lackawanna Railway Company Deposit, Erie Railroad Company, Subsidiary Companies; page 1, list of corporators; minutes for September 7, September 22, November 11, 1884, May 12, 1885; the start date is given in the *AIM Report for 1885*, 42, and the combined tonnage figures for 1883, 1884 are derived from the tabulations in the *AIM Reports* for these years.
26. Board of Managers Minutes, Florence Coal Company, February 13, April 17, 1884, May 11, 1885.
27. Ibid., June 3, 1885.
28. Ibid., July 1, July 16, 1885.
29. Ibid., October 27, 1887, February 2, 1893.
30. Ibid., December 16, 1887, January 27, February 23, August 30, 1888, July 28, 1892.
31. *Coal Trade Journal*, 1893, 729.
32. Wyoming Valley Coal Company, Profit and Loss Account, Israel Morris Papers.
33. *Charter and By-Laws of the Wyoming Valley Coal Company* (Wilkes-Barre: Record Book and Job Printing Dept., 1883). F.C. Petrillo Collection.
34. I.W. Morris to R.H. Sayre, July 26, 1890, Wyoming Valley Coal Company Papers.
35. Board of Managers Minutes, Florence Coal Company, August 20. 1885; *Annual Reports of the Susquehanna Coal Company*; Delaware, Lackawanna and Western Railroad, entries in main journals.
36. *Annual Reports of the Susquehanna Coal Company*; Coal Department Construction Record Book, Delaware, Lackawanna and Western Railroad Records, Syracuse.
37. I.W. Morris to A. Walter (president), February 21, 1898, Israel Morris Papers.
38. Campbell, *Businessmen and Anthracite*, 125.
39. Statement of A.J. Davis in "Extracts from Testimony Before Legislative Investigating Committee," typescript attached to letter of I.W. Morris to A. Walter [see n. 37], p. 1. The 1875 figure for the Susquehanna Coal Company is obtained by subtracting improvement costs per ton from mining costs per ton as given in the annual report data. However, it appears the 1875 figure was incorrectly printed as $1.39 instead of $1.59 based on calculations from other available accounting figures in *Annual Report of the Susquehanna Coal Company for 1876*, 10. The corrected figure of $1.59 has been adopted here as it is more consistent with the other data.
40. *Annual Report of the Susquehanna Coal Company for 1885*, 11.
41. *Annual Reports of the Susquehanna Coal Company for 1887–1893*, tabulations of expenses.
42. *Annual Reports of the Mineral Railroad and Mining Company and Summit Branch Railroad Company for 1885*, tabulations of expenses.
43. Operating Costs, William Penn Colliery, Pennsylvania Railroad Records.
44. Extracts from Testimony before Investigating Committee, 3 [see note 39].
45. Coal Expenses Record Account Books, Coal Department, Delaware, Lackawanna and Western Railroad Records, Syracuse.
46. The relevant tabulations are scattered through the faded letter books of W.H. Storrs (LCDR).
47. WRS to SS, August 22, 1885, LB40A (LCDR).
48. Extracts from Testimony before Investigating Committee, 1–6.
49. Reports on individual mines and companies, Industrial Espionage Reports, Reading Company Papers, Hagley Library.
50. Ibid., report on the Red Ash Coal Company.
51. *Labor Troubles*, 627.
52. Ibid., 618, 626.
53. Ibid., 593, 623.
54. Ibid., 624.
55. Ibid., 623.
56. Ibid., 624.
57. Journal 1869–1884, p. 26, Susquehanna Coal Company Records; *Annual Report of the Susquehanna Coal Company for 1879*, 8–9.
58. Baird, LV Coal Co. History, 365.
59. Board of Managers Minutes, Delaware and Hudson Coal Company, May 28, 1860.
60. Board of Managers Minutes, Lehigh Valley Railroad, September 9, October 11, 1879; William Montelius, a Western Middle Field operator requested and was granted a loan of $15,000.
61. It was reported in 1881 that the Jersey Central had furnished part of the money for improvements on a lease made by T. Waddell and E.A. Packer, WRS to SS May 11, 1881, LB32A (LCDR).
62. These credit ledgers are held at the Baker Library, Harvard School of Business.
63. WRS to SS, November 23, 1880, LB32A (LCDR).
64. WRS to SS, March 5 (?), 1880, LB32A (LCDR).
65. WRS to SS, March 4, 1880, LB32A (LCDR).
66. WRS to SS, April 20, 1881, LB32A (LCDR).
67. SS to WRS, December 13, 1880, LB SS to WRS 1870–85 (LCDR).
68. SS to WRS, April 21, 1883, LB SS to WRS 1870–85 (LCDR).
69. SS to WRS, August 16, 1883, LB SS to WRS 1870–85 (LCDR).
70. WRS to SS, November 25, 1881, LB32A (LCDR).
71. WRS to SS, August 21, 1881, LB32A (LCDR).
72. WRS to SS, February 23, 1883, LB37A (LCDR).
73. WRS to SS, February 11, 1884, LB37A (LCDR).
74. WRS to SS, October 26, 1883, LB37A (LCDR).
75. Entry 2/150, August 25, 1881, Pennsylvania Ledgers, R.G. Dun and Company Collecton (RGDR); WRS to SS, April 30, 1881, LB32A (LCDR).
76. WRS to SS, April 20, 1881, LB32A (LCDR).
77. Entry 2/150, August 25, 1881 (RGDR).
78. WRS to SS, March 5, March 8, 1880, LB32A (LCDR). The quotation is from the latter letter.

79. WRS to SS, July 14, November 2, 1880, LB32A (LCDR).
80. Entry 2/150, August 25, 1881 (RGDR).
81. There is extensive correspondence relating to this matter, see, for example, WRS and SS, April 2, April 20, April 30, May 11, June 1, June 3, 1881 (LCDR): the original labor basis contract can be found in DB220, p. 476f, Luzerne County Courthouse.
82. *AIM Report for 1881*, 221.
83. WRS to SS, May 12, 13, 30, 1882, LB32A (LCDR).
84. WRS to SS, July 29, 1882, LB37A (LCDR).
85. WRS to SS, August 29, 30, 1882, LB37A (LCDR); Entry 2/150, August 15, 1881, January 31, 1883 (RGDR).
86. *AIM Report for 1882*, 205.
87. Entry 2/269, October 19, 1882 (RGDR).
88. WRS to SS, September 27, 1881, LB32A (LCDR).
89. WRS to SS, October 3, 1882, LB37A; S.N. Stetler to SS, October 19, 1882, LB SS to WRS 1870–1885 (LCDR).
90. WRS to SS, October 23, 1882, LB37A; WRS to SS, October 24, 1882, LB SS to WRS 1870–1885 (LCDR).
91. Entry 2/269, December 26, 1882 (RGDR); DB232, p. 496f, Luzerne County Courthouse.
92. Entry 2/269, October 18, 1882 (RGDR); WRS to SS, September 29, 1882, LB37A (LCDR).
93. WRS to SS, August 7, August 14, 1883, LB37A (LCDR).
94. WRS to SS, August 14, 1883, LB37A (LCDR).
95. WRS to SS, August 14, December 1, December 3, 1883, LB37A (LCDR), SS to WRS, December 4, 1883, LB SS to WRS 1870–1885 (LCDR).
96. DB22, p. 394f, p. 402f, Lackawanna County Courthouse; WRS to SS, February 29, March 5, 1884, LB37A (LCDR).
97. WRS to SS, December 11, Dec. 12, 1883, LB37A (LCDR).
98. Entry 2/269, April 25, 1884 (RGDR); WRS to SS, April 24, 1884, LB37A (LCDR).
99. WRS to SS, May 22, May 23, June 6, August 14, 1884, LB37A (LCDR); Entry 2/269, August 12, 1884 (RGDR).
100. WRS to SS, August 14, 1884, LB37A (LCDR).
101. WRS to SS, November 8, 1884, LB37A (LCDR).
102. Ibid.
103. DB 252, p. 333f, Luzerne County Courthouse.
104. WRS to SS, November 5, November 29, December 3, 1884, LB37A (LCDR).
105. WRS to SS, December 22, 1884, January 9, 1885, LB40A (LCDR).
106. Entry 2/278, February 20, April 21, 1885 (RGDR).
107. WRS to SS, March 17, June 4, 1885, LB40A (LCDR).
108. WRS to SS, August 17, August 22, 1885, LB40A (LCDR).
109. WRS to SS, August 18, 1885, LB40A (LCDR).
110. Entry 2/278, May 8, 1883 (RGDR).
111. Calculated from tabulations in *AIM Reports* 1882–1902.
112. Entry 2/278, August 18, 1885 (RGDR).
113. WRS to SS, April 23, April 26, 1892, LB50A (LCDR).
114. WRS to SS, May 29, 1899, LB50A (LCDR).
115. See, for examples, the letters of WRS to SS, October 19, October 22, November 11, November 18, November 21 and November 25, 1881, LB32A (LCDR).
116. WRS to SS, May 21, December 9, 1891; March 26, 1892, LB50A (LCDR).
117. Both quotations from WRS to SS, March 30, 1892, LB50A (LCDR).
118. WRS to F.H. Gibbens, January 29, 1884, LB37A (LCDR).

DATA SOURCES FOR FIGURES

13.1 *Annual Reports of the Susquehanna Coal Company*; Main Account Books, Susquehanna Coal Company Records.
13.2 *Annual Reports of the Susquehanna Coal Company.*
13.3 *Annual Reports of the Philadelphia and Reading Railroad.*
13.4 Baird, LV Coal Co. History, 376 f; *Annual Reports of the Lehigh Valley Railroad.*
13.5 Main Account Books, Lackawanna Railroad Records, Syracuse; *Annual Reports of the Delaware, Lackawanna and Western Railroad* for data from 1889 onwards.
13.6 Board of Managers Minutes, Florence Coal Company.
13.7 Ibid.
13.8 Calculated from data in Board of Managers Minutes, Florence Coal Company.
13.9 Main Account Books, Susquehanna Coal Company Records.
13.10 Wyoming Valley Coal Company Papers.
13.11 Tabulations in *Annual Reports of the Susquehanna Coal Company.*
13.12 Ibid.
13.13 Ibid.
13.14 Coal Department Expenses Account Books 1888–1897 and 1898–1900, Lackawanna Railroad Records, Syracuse.

NOTES FOR CHAPTER 14

1. In the case of the Susquehanna Coal Company, the 'construction account' was officially closed during the mid 1870s, but the company continued to report the costs of new improvement work as a separate item, both in its Annual Reports until 1893 and in its monthly journal entries until 1897.
2. The apparent peak in the graph in 1887 is an artefact of the charging of the costs of acquiring the Fuller and Amity improvements to the construction accounts in that year, instead of 1885 when the purchases actually took place. The detailed reasons why this was done have not come down to us. See 1887 entries in the Coal Department Construction Record, Lackawanna Railroad Records, Syracuse.

3. For details of the costs of acquiring the individual breakers, see Tabulations of the Improvement Accounts, 1897, Israel Morris Papers; costs of mine fires and unspecified "maintenance" have been excluded from the graph as they do not qualify as capital expenditure. Se also Hoffman, *Girard Estate Coal Lands*.
4. For details of the start of construction at the Dorrance, see *AIM Report for 1881*, 141.
5. Improvement Accounts 1897, Israel Morris Papers.
6. *AIM Report for 1872*, 176–177; *Annual Report of the Susquehanna Coal Company for 1876*.
7. Susquehanna Coal Company section, p. 7–11, Industrial Espionage Reports, Reading Company Papers.
8. *Annual Report of the Susquehanna Coal Company for 1892*.
9. *AIM Reports for 1879–1881* and Annual Reports of WRS 1879–1881, Board of Managers Minutes, Delaware, Lackawanna and Western Railroad.
10. SS to WRS, January 29, 1881, LB SS to WRS 1870–1885 (LCDR).
11. Von Bergen, *Photographs*.
12. DB 80, p. 277f, DB130, p. 28f, Luzerne County Courthouse; see also DB19, p. 443f, Lackawanna County Courthouse for details of the 1864 lease and its eventual assignment to the Lackawanna.
13. DB 162, p. 54f and DB 226, p. 65f, Luzerne County Courthouse.
14. Annual Report of WRS to SS for 1869, LB22, 210f (LCDR).
15. WRS to SS, November 8, 1870, LB24 (LCDR).
16. Annual Report of WRS to SS for 1870, LB24, 656f (LCDR).
17. Annual Report of WRS to SS for 1871, LB25, 674f (LCDR).
18. WRS to SS, January 29, 1872, LB25 (LCDR).
19. Ibid.
20. Annual Report of WRS to SS for 1872, LB26, 504f (LCDR).
21. Ibid.
22. WRS to SS, March 24, 1881, LB 32A (LCDR); Letter from SS to all Superintendents and Master Mechanics on the Road, September 26, 1873, recorded in the Board of Managers Minutes, Delaware, Lackawanna and Western Railroad.
23. Annual Report of WRS to SS for 1873, LB27, 281f (LCDR).
24. WRS to SS , July 17, 1879, LB 31 (LCDR).
25. Annual Report of WRS for 1879, Board of Managers Minutes, Delaware, Lackawanna and Western Railroad.
26. Executive Committee Minutes, December 29, 1879, Delaware, Lackawanna and Western Railroad, Hagley Library.
27. See WRS to SS, November 16, 1880; March 21, 23, 24, 1881; April 2, 4, 20, 1881, LB 32A (LCDR).
28. Ibid. for 1878.
29. WRS to SS, February 15, 1883, LB 37A (LCDR).
30. See for instance the proposition by J. Mitchell to lease 156 acres near Woodward to the Company described in WRS to SS, August 17, 1881, LB 32A (LCDR).
31. WRS to SS, January 8, 1889, LB August 1887–April 1891 (LCDR).
32. WRS to SS, March 2, June 25, 1883, LB 37A (LCDR).
33. WRS to SS , September 12, 1882, LB 37A (LCDR).
34. WRS to SS, December 31, 1888, January 23, 1889, LB August 1887–April 1891 (LCDR).
35. WRS to SS, February 6, March 7, 1889, LB August 1887–April 1891 (LCDR); Grantee Index to Deeds, Luzerne County, no. 1, Luzerne County Courthouse.
36. *AIM Report for 1892*, 142.
37. DB 226, p. 164f, Luzerne County Courthouse.
38. Journal T, December 31, 1890, p. 197 and section describing coal leases, Lackawanna Railroad Records, Syracuse.
39. WRS to SS, November 25, 1880; January 7, June 23, 1881, LB 32A (LCDR).
40. WRS to SS, June 23, 1881, LB 32A (LCDR).
41. WRS to SS, June 30, 1881, LB 32A (LCDR).
42. WRS to SS, November 25, 1881, LB 32A (LCDR); *AIM Reports for 1880, 1881*.
43. *AIM Report for 1890*, 90.
44. Von Bergen, *Photographs*.
45. Baird, LV Coal Co. History, 264.
46. Ibid., 435–436. The company history stresses that individual operators mined for immediate return rather than maximizing the total output of properties and that if the coal company subsequently took them over, large amounts had to be extended to place the improvements in good working order.
47. See the summary of Pennsylvania Coal Mining Legislation in the *AIM Report for 1898*, p lxxxv f.
48. See, for example, the tabulation of contractors bids for sinking the Continental shaft, B. Hughes to WRS, December 26, 1867, Incoming LB no. 8 (LCDR).
49. Annual Report of WRS to SS, LB26, 504f (LCDR).
50. *AIM Report for 1876*, 168–170.
51. Exhaustive detail of the methodology used to analyse these voucher records can be found in the author's PhD thesis, Chapter 7 (R.G. Healey, "Object-oriented Simulation of Evolving Regional Economic Systems: A Case Study of Mining Development in the Anthracite Coalfields of Pennsylvania During the Post-Bellum Period." Ph.D. diss, University of Cambridge, 1993). The case study reported here is based on extracts from this chapter. The mine payrolls are located in the Lackawanna Historical Society, Scranton and the voucher records in the Lackawanna Railroad Records, Syracuse. If not otherwise specified, dates, costings and other accounting figures in this section are derived from these records or calculations based upon them.
52. See, for example, the voucher dated November 29, 1873 to Wm. Youle. Coal Department Construction Voucher Records, Lackawanna Railroad Records, Syracuse.

53. WRS to SS, June 23, 1881, LB 32A (LCDR); DB 226 p. 164, Luzerne County Courthouse.
54. WRS to SS, March 30, June 29, 1882, LB 32A (LCDR).
55. Totals calculated from entries in the Coal Department Construction Record 1853–1896 and Cash Book entries, Lackawanna Railroad Records, Syracuse.
56. Annual Report of WRS to SS for 1882, p. 327 in Board of Managers Minutes, Delaware, Lackawanna and Western Railroad.
57. *AIM Report for 1888*, 102.
58. *AIM Report for 1880*, 175.
59. *AIM Report for 1889*, 4.
60. The figures for Woodward are based on a count back of the known or estimated depths sunk in specific years, reported in Table 7.1 in Healey, *Object-oriented Simulation*; the figure for Storrs is given in the *AIM Report for 1889*, 4.
61. *AIM Report for 1880*, 175; *for 1882*, 124; for *1883*, 117; *for 1884*, 107; *for 1888*,102; for *1889*, 3–4, 70; Coal Department Construction Voucher Record 1880–88, Cash Books, 1880–1888, Lackawanna Railroad Records, Syracuse. The comment about reduction of dimensions at Woodward was made by B. Hughes in a letter to the mine inspector, G.M. Williams, March 9, 1889, LB B. Hughes, October 1885–June 1889 (LCDR).
62. Annual Reports by WRS to SS 1882–1886, Board of Managers Minutes, Delaware, Lackawanna and Western Railroad.
63. Ibid. for 1883.
64. WRS to SS, January 11, 1884, LB 37A (LCDR).
65. Ibid.
66. *AIM Report for 1884*, 102.
67. Correspondence of January 13, 1890, LB B. Hughes, October 1885–June 1899 (LCDR).
68. WRS to SS, November 22, 1884, LB37A (LCDR).
69. WRS to SS, September 1, 1884, LB37A (LCDR).
70. WRS to SS, August 29, 1885, LB40A (LCDR).
71. *Wilkes-Barre Telephone*, July 11, 1885.
72. *Wilkes-Barre Telephone*, April 21, 1886.
73. *Coal Trade Journal*, November 19, 1886.
74. Ibid.
75. The payroll entries for Lackawanna Coal Department contract miners listed the types and amounts of work done, rather than the days worked. While the mine was still being developed, the contract miners payroll sheets would be headed 'construction'. After production began, the heading changed to 'expenses' to indicate whether the payrolls were being charged to the construction accounts or the equivalent for operating expenses. The Payrolls are in the Lackawanna Coal Dept. Records (LCDR).
76. The payrolls are missing for two months during this period, so these are minimum estimates; the actual totals would have been somewhat larger.
77. All the detailed calculations in this section are taken from Healey, *Object-Oriented Simulation*, Chapter 7, where additional cost breakdowns can be found. For evidence of problems with quicksand at Pettebone, see *Wilkes-Barre Telephone*, August 14, 1886.
78. Figures from *Coal Trade Annual* quoted in Campbell, *Businessmen and Anthracite*, 116.
79. Calculated from tonnage figures in *AIM Reports* for the years 1876–1902.
80. Ibid., for the years 1888–1902.
81. W.A. Lathrop and I.W. Morris to E.P. Wilbur, May 26, 1897 and enclosed letter of December 16, 1897 to the same addressee, Israel Morris Papers.

DATA SOURCES FOR FIGURES

14.1 Coal Department Construction Record, Lackawanna Railroad Records, Syracuse.
14.2 *AIM Report for 1872*, frontispiece map; General Map of the Anthracite Coal Fields Showing the Position of Each Colliery, in *Atlas to Accompany Part III of the Annual Report of the Geological Survey of Pennsylvania for 1886;* Maps Showing Location of Coal Mines in the 1st–4th Anthracite Inspection Districts of Pennsylvania 1898, maps accompanying *AIM Report for 1898.*
14.3 Tabulation of Lehigh Valley Coal Co. Improvement Accounts 1897, Israel Morris Papers. Maintenance and mine fire accounts have been removed from final totals.
14.4 Journals, Susquehanna Coal Company; *Annual Reports of the Susquehanna Coal Company.*
14.5 Calculated from data in sources for Figures 14.1, 14.3 and 14.4.
14.6 Digital data for land tract boundaries from the coalfield land data layer of the NE USA Historical GIS, originally digitized and polygonized from the Mine Sheets of the *Atlases of the 2nd Pennsylvania Geological Survey.* Lease data derived from the Deed Books, Luzerne County Courthouse.
14.7 Calculated from Construction Payroll Records (LCDR).
14.8 Calculated from Coal Department Construction Voucher Records, Lackawanna Railroad Records, Syracuse; Construction Payroll Records (LCDR).
14.9 Ibid.
14.10 Ibid.
14.11 Ibid.

NOTES FOR CHAPTER 15

1. *Annual Report of the Philadelphia and Reading Railroad for 1876*, 105.
2. See Yearley, *Enterprise and Anthracite*, 185 for comment on this point.
3. Tabulations in *Annual Reports of the Philadelphia and Reading Railroad for 1876–1887.*
4. *Annual Report of the Philadelphia and Reading Railroad for 1876*, 106.
5. *Annual Report of the Philadelphia and Reading Railroad for 1877*, 103.

6. A.J. Davis Testimony, Extracts from Testimony before Legislative Investigating Committee, typescript attached to letter of I.W. Morris to A. Walter, February 21, 1898, Israel Morris Papers.
7. *Annual Report of the Philadelphia and Reading Railroad for 1876*, 105; Schlegel, *Ruler of the Reading*, 64f.
8. *Annual Report of the Philadelphia and Reading Railroad for 1876*, 106.
9. Testimony of R.C. Luther, W.A. Lathrop, Extracts from Testimony, Israel Morris Papers.
10. *Annual Report of the Philadelphia and Reading Railroad for 1874*, 26.
11. Ibid.
12. *Annual Report of the Philadelphia and Reading Railroad for 1876*, 105.
13. *Annual Report of the Philadelphia and Reading Railroad for 1874*, 27–28.
14. See the collection of Coal and Iron letterbooks in the archives of the Smithsonian Institution.
15. See Yearley, *Enterprise and Anthracite* for the qualitative approach and D.F. Schaefer, "A Quantitative Description and Analysis of the Growth of the Pennsylvania Anthracite Coal Industry, 1820–1865." (Ph.D. diss., Chapel Hill: University of North Carolina, 1967).
16. Schaefer, *Quantitative Description*, 59–60, quoting figures variously derived from the 1860 manuscript census returns and published census tables, and discussion of the accuracy of the estimates in the following pages.
17. Average tonnages are calculated from the tabulations in the *AIM Reports* for the respective years. Washeries are included, although their numbers are small.
18. As for note 17. A small number of mines are excluded from the calculations owing to uncertainty as to their exact location, because of lack of information in the reports. Washeries have also been excluded.
19. As for note 18. Small operations shipping less than 10,000 tons are again excluded.
20. Derived from data in *Annual Reports of the Philadelphia and Reading Railroad for 1872–1891.*
21. E.g., *Annual Report of the Philadelphia and Reading Railroad for 1886*, 65–67. In these tables there are 80 mines recorded, with tonnages ranging from 386,522 down to 44 tons, plus a few with a notional output of less than a car of coal!
22. *Annual Report of the Philadelphia and Reading Railroad for 1883*, 94f.
23. Calculations from tabulations in the *AIM Report for 1901.*
24. Analysis based on the tabulations in the *AIM Reports for 1870, 1875, and 1880.* Specific details of the East Pine Knot and Kentucky mines are given in the 1872 report, pages 78 and 94 respectively.
25. Analysis based on *Annual Reports of the Philadelphia and Reading Railroad for 1873–1885.*
26. J.S. Harris to General Pleasants, December 16, 1875, Letterbook of J.S. Harris, no. 4, 1875–1876, Pottsville, Letterbooks of the Philadelphia and Reading Coal and Iron Company, Smithsonian Institution.
27. Ibid.
28. *Annual Reports of the Philadelphia and Reading Railroad for 1876–1878, AIM Reports for 1876–1885.*
29. G.W. Dithridge to Mr. Sully, November 1, 1886, Box 19, Folder 1, Reading Company Papers.
30. Ibid.; the original table also contains two different rows of data both labelled 'Preston No. 2', but these have been omitted as it is not clear which row contains the correct information for this mine.
31. Tabulations in the *AIM Reports for 1885.*
32. *Coal Trade Journal*, December 31, 1884, February 9, 1887.
33. *AIM Report for 1885*, 237.
34. *Annual Report of the Philadelphia and Reading Railroad for 1886*, 63.
35. Ibid.
36. *Annual Report of the Philadelphia and Reading Railroad for 1890*, 51, 55, 56.
37. Ibid., 57.
38. *Annual Report of the Philadelphia and Reading Railroad for 1888*, 7.
39. Ibid., 61.
40. *Annual Report of the Philadelphia and Reading Railroad for 1885*, 73.
41. Ibid., 72.
42. *Annual Report of the Philadelphia and Reading Railroad for 1883*, 94f.
43. *Annual Report of the Philadelphia and Reading Railroad for 1886*, 63.
44. *Annual Report of the Philadelphia and Reading Railroad for 1882*, 79; *for 1886*, 63; *for 1887*, 73; *for 1888*, 62.
45. *Annual Report of the Philadelphia and Reading Railroad for 1890*, 51.
46. Tabulations in *AIM Reports for 1891, 1897.*
47. *AIM Report for 1885*, 183.
48. *AIM Report for 1884*, 37; *for 1885*, 187.
49. *Annual Report of the Philadelphia and Reading Railroad for 1879*, 107.
50. Ibid., 106.
51. *Annual Report of the Philadelphia and Reading Railroad for 1887*, 73.
52. *AIM Report for 1893*, 223.
53. *AIM Report for 1894*, 262–263.
54. *Annual Report of the Mineral Railroad and Mining Company for 1885*, 35; *for 1895*, 19–21; *for 1898*, 7. The 1897 report p. (21) also noted that it cost $148,210 simply to maintain Luke Fidler in an idle state during that year, and all its costs had to be borne by the Cameron colliery, which itself had high fixed charges (Annual Report for 1896, 21). It should be noted that in 1884 and 1885, the company was also operating the Hickory Ridge Colliery, so it made a contribution to costs and earnings in addition to Luke Fidler and Cameron.
55. Calculated from data in tabulations in *AIM Report for 1901.*
56. J.S. Harris, *Report on Coal Lands*, 78.
57. *Coal Trade Journal*, November 3, 1886.

58. *Coal Trade Journal*, November 17, 1886.
59. *Coal Trade Journal*, October 25, 1893.
60. Calculated from data in tabulations in *AIM Report for 1901*.
61. Testimony and quotation reported in the *Coal Trade Annual 1902 (1901)*, 82.

DATA SOURCES FOR FIGURES

15.1 Tabulations in *Annual Reports of the Philadelphia and Reading Railroad*.
15.2 The data for 1865 were derived or deduced from the following sources: Daddow and Bannan, *Coal, Iron and Oil*, frontispiece map, statistics of coal and mine descriptions, 702–775; *Anthracite Coal Trade for 1867* (Pottsville: Miner's Journal, 1868); *Official Statistics—The Coal Trade for 1868*. (Pottsville: Miner's Journal, 1869); *Coal Statistical Register, 1870* (Pottsville: Miner's Journal, 1871); G.B. Strauch and A.B. Cochran, Map of the Anthracite Coalfields, 1871 (Pottsville, 1871); Baer, *Canals and Railroads of the Mid-Atlantic States*, Map B of the Anthracite Coalfields, 1860; mine sheets of the *Atlas of the Anthracite Coalfields*; Report of the Inspector of Mines of Schuylkill County for 1869 (Harrisburg: Singerly, 1870).
15.3 Calculated from data in tabulations in *AIM Reports for 1870–1872*.
15.4 The data for 1886 were obtained from Plates 1 and 2 in *Atlas to accompany Part III of the Annual Report of the Geological Survey of Pennsylvania for 1886*. Changes in the names of mines 1865–1886 were traced using the *AIM Reports for 1870–1885*, the *Atlas of the Anthracite Coalfields* mine sheets, and G.B. Strauch and A.B. Cochran, Maps of the Anthracite Coalfields of Pennsylvania for 1878, 1879, 1880 (Pottsville, 1878–1880).
15.5 Tabulations in *Annual Reports of the Philadelphia and Reading Railroad*.
15.6 Ibid.
15.7 Ibid.
15.8 Maps Showing Location of Coal Mines in the 5th–8th Anthracite Inspection Districts of Pennsylvania 1898, maps accompanying *AIM Report for 1898*.
15.9 Tabulations in *AIM Reports*.
15.10 Calculated from tabulations in *AIM Reports*.
15.11 Tabulations in *AIM Reports*.
15.12 Calculated from tabulations in *AIM Reports*.
15.13 Tabulations in *AIM Reports*.
15.14 Calculated from tabulations in *AIM Reports*.
15.15 Tabulations in *AIM Reports*.
15.16 Calculated from tabulations in *AIM Reports*.
15.17 Tabulations in *AIM Reports*.
15.18 Calculated from tabulations in *AIM Reports*.

NOTES FOR CHAPTER 16

1. *Labor Troubles*, 669.
2. Ibid., 667.
3. Calculated from tabulations in the *AIM Reports* for the specified years. Gaps in the tabulation reflect the fact that there are either no mines in the category or a very small number (three or less) such that an average may not be very indicative.
4. Account item 'costs of improvements charged to expenses', *Annual Reports of the Philadelphia and Reading Coal and Iron Company*, appended to the reports of the Railroad, various years.
5. The first figure for lands and improvement costs is derived from details given in Baird, LV Coal Co. History and other tabulations in the Israel Morris Papers. The specific account headings for new improvements are given in improvement account tabulations in the Israel Morris Papers, dating from 1897. Owing to the different ways in which these tabulations were drawn up and the complexity of the calculations involved, there may be a small element of double counting involved.
6. Totals based on figures in the Coal Dept. Construction Record, Lackawanna Railroad Records, Syracuse.
7. Undated typescript 1895–1896 (probably authored by J. S. Harris), Box 29, Folder 9, Reading Company Papers.
8. Compilation of the Capital Accounts, Tonnage Reports, etc. of the Philadelphia and Reading Coal and Iron Company, 735, Reading Anthracite Company, Pottsville.
9. Unfortunately these measures themselves are open to interpretation. Sometimes there were old, shallow workings on a property before it was taken over by another company. These have been disregarded in identifying dates of first construction. Also, breakers were sometimes finished before the mine itself was ready for production, and coal derived from gangway driving might be passed through the breaker in small quantities. These small shipments have likewise been disregarded.
10. Yearley, *Enterprise and Anthracite*, 215.
11. Miller and Sharpless, *Kingdom of Coal*, 80.
12. *Annual Report of the Philadelphia and Reading Railroad for 1867*, 13, (published 1868).
13. *Scranton Republican*, December 21, 1866.
14. *Miner's Journal*, April 6, 1877.
15. *Annual Report of the Philadelphia and Reading Railroad for 1876*, 29.
16. *Coal Trade Journal*, quoting the *Wilkes-Barre Record*, April 16, 1884.
17. *Report of the Anthracite Coal Strike Commission*, 23.
18. Aurand, *Molly Maguires*, 93.
19. Virtue, "Anthracite Combinations," 319.
20. J.S. Harris gave a very clear exposition of this point, in 1886, in an affidavit to the Dauphin County Court of Common Pleas. It was reported in F.A. Hull, *Annual Report of the Geological Survey of Pennsylvania, Part III, Report on the Anthracite Region*. (Harrisburg: Edwin K. Meyers, State Printer, 1886), 1051–1052. Hereafter referred to as Harris, Anthracite Region Report.

21. Ibid., 1056.
22. Ibid., 1053; Virtue, *Anthracite Combinations*, 319.
23. W.A. Lathrop to A. Walter, February 15, 1898, Israel Morris Papers.
24. *Labor Troubles*, 662.
25. Harris, *Anthracite Region Report*, 1055.
26. WRS to SS, January 28, 1884, LB37A (LCDR), see the end of Chapter 13 for a verbatim quotation.
27. This problem was recognized by Samuel Sloan at an early stage, see S. Sloan to W.R. Storrs, Dec. 13, 1880, Sloan to Storrs LB, 1870–1885.
28. W.A. Lathrop to A.Walter, February 15, 1898, Israel Morris Papers.
29. *Labor Troubles*, 671.
30. WRS to SS, September 17, 1884, LB37A (LCDR).
31. Annual Report for 1878, WRS to SS, Board of Managers Minutes, Delaware, Lackawanna and Western Railroad, 197.
32. *AIM Report for 1879*, 274.
33. Ibid.
34. Annual Report for 1879, WRS to SS, Board of Managers Minutes, Delaware, Lackawanna and Western Railroad, 218f.
35. Annual Report for 1878, WRS to SS, Board of Managers Minutes, Delaware, Lackawanna and Western Railroad, 196–197.
36. Ciesla, *History of Hanover Township*, 85; *AIM Report for 1879*, 131.
37. *Labor Troubles*, 151.
38. *Testimony before the Committee to Investigate the Philadelphia and Reading Railroad Company* (1876), 1001.
39. Monthly tonnage returns, Storrs Letter Books (LCDR).
40. WRS to SS, March 22, April 15, 1898, February 8, 1899; LB 1897–1899 (LCDR); *Scranton Republican*, April 16, 1898.
41. *Testimony before the Committee to Investigate the Philadelphia and Reading Railroad Company* (1876), 725.
42. Ibid., 1075–1076.
43. *Labor Troubles*, 671.
44. *Testimony before the Committee to Investigate the Philadelphia and Reading Railroad Company* (1876), 725.
45. Harris, *Anthracite Region Report*, 1054.
46. Clark, et al. *Regional Dynamics*, see especially the early chapters.
47. See R.L. Martin, "Critical Survey. The New 'Geographical Turn' in Economics: Some Critical Reflections," *Cambridge Journal of Economics* 23 (1999): 65–91; see also, for example, R.W. Fogel and S.L. Engermen, (eds.), *The Re-Interpretation of American Economic History* (New York: Harper and Row, 1971); an incisive overview of current trends in economic thought can be found in the following: Daniel M. Hausman, "Philosophy of Economics," *The Stanford Encyclopedia of Philosophy (Winter 2003 Edition)*, Edward N. Zalta (ed.), URL = <http://plato.stanford.edu/archives/win2003/entries/economics/>.
48. Clark, et al. *Regional Dynamics*, 42.

DATA SOURCES FOR FIGURES

16.1 Calculated from mine data tabulations in *AIM Reports*.
16.2 Ibid.
16.3 Ibid.
16.4 Calculated from data in sources for Figures 12.7, 12.8, and 12.9.
16.5 The individual sources are too numerous to list in detail here, since so many mines are involved. Major sources include *AIM Reports*, the Industrial Espionage Reports in the Reading Company Papers, County Histories, reports in Northern Field Newspapers and correspondence in the Storrs Letter Books (LCDR).
16.6 As for Figure 16.5.
16.7 As for Figure 16.5.
16.8 The data for 1865 were derived or deduced from the following sources: Daddow and Bannan, *Coal, Iron and Oil*, frontispiece map, statistics of coal, and mine descriptions, 702–775; Baer, *Canals and Railroads of the Mid-Atlantic States,* Map A of the Anthracite coalfields, 1860; mine sheets of the *Atlas of the Anthracite Coalfields;* Coal Dept. Construction Record, Lackawanna Railroad Records, Syracuse.
16.9 As for Figure 15.4.
16.10 Maps Showing Location of Coal Mines in the 1st–4th Anthracite Inspection Districts of Pennsylvania 1898; maps accompanying *AIM Report for 1898.*

Preface to Bibliography: A Note on Empirical Sources

As mentioned in the Introduction, one of the explanations for the lack of adoption of adjustment theory in empirical studies to date, is that the data requirements to confirm the nature and time scales of decision-making and adjustment processes are very great. In addition, both quantitative and qualitative data are required. A potential weakness of the original regional dynamics study by Clark et al. is its heavy reliance on statistical inference from quantitative data, at the expense of qualitative sources of information. Conversely, most previous historical investigations of the anthracite industry are seriously lacking in quantitative data, and especially data that would shed light on spatio-temporal dynamics.

This study has been designed not only to demonstrate the application of adjustment theory in terms of the spatio-temporal dynamics of the anthracite industry but also to utilize both quantitative and qualitative data as fully as possible, and in complementary ways. For example, company correspondence of senior corporate managers contains extensive evidence of criteria used in decision-making strategies, but is often very limited in showing how decisions resulted in the timing of business expenditure over periods of months or years. This can be undertaken much more effectively by quantitative analysis of accounting records. The result is that each type of information amplifies the other, to give a clearer picture of the initiation and operation of adjustment processes over time and space. It should also be noted that the analysis referred to here is largely reconstitution of cost series from accounting voucher data that are effectively complete, so the question of statistical inference does not arise, as it would for, say, a sample of census data.

Fortunately, the available sources relating to many of the major corporations involved in the anthracite industry in the 19th century, both railroads and associated coal companies, are voluminous and wide-ranging. Of particular importance are the records of the Delaware, Lackawanna and Western Railroad and its Coal Department. In terms of quantitative data, these include the major railroad account books, as well as Coal Department mine payrolls and construction accounting vouchers. These are complemented by a remarkable series of letter books, which contain an almost daily record of correspondence between the General Coal Agent in Scranton and the Railroad President in New York, from the Civil War period through to the end of the century. Other important railroad collections include those for the Lehigh Valley Railroad, the Philadelphia and Reading, and, of course, the mighty Pennsylvania. In addition, significant collections of records for several major coal companies and other railroad coal mining subsidiaries have survived and are very useful for comparative purposes. These include the Pennsylvania Coal Company, the Hudson Coal Company, the Lehigh Valley and Lehigh and Wilkes-Barre Coal Companies, and the Susquehanna Coal Company. While there is a relative dearth of detailed records for small companies during the period, some materials have survived, and the extent of their interactions with the railroads in relation to shipping contracts has meant that, in several cases, quite detailed files pertaining to their activities have come down to us. In addition, therefore, it is most fortunate that the credit ledgers of R. G. Dun and Company are accessible to scholars in the Baker Library, Harvard Business School. While their coverage is variable up to the 1860s, in a large number of cases they contain significant financial information relating to small mining companies and entrepreneurs about whom we would otherwise know almost nothing.

Company records are complemented by a number of other types of manuscript and published archival sources. The schedules of the manuscript industrial censuses of Pennsylvania for the decade points 1850–1870 are particularly important. In the early period they serve as snapshots of mining activity when few other records are available, and in the later years they give an indication of the development of ancillary mining related industries. However, as indicators of adjustment dynamics, on their own they are of very limited value. A further key manuscript resource is the County Courthouse Deed Book records. Importantly, these contain both details of changing land ownership and copies of the contracts for all coal land leases. Since the anthracite coalfields had the most fragmented non-urban land ownership in the United States at the time, systematic analysis of land ownership data is so complex it is almost impossible to undertake without the full-scale deployment of Geographical Information Systems (GIS) technology. It comes as no surprise therefore, that researchers have not previously attempted such analysis. Fortunately, the published map sheets of the 2nd Pennsylvania Geological Survey, at a scale of 800 ft to the inch, which date from the 1880s, contain detailed land parcel boundaries. A large GIS dataset of land parcels in the Northern Field has been developed from these, which has made it possible to utilize the land-deed records effectively for specific companies. These ownership and lease records are extremely important in understanding decision-making and adjustment processes concerned with the management of local coal estates and their development by means of new mine construction. It is also worth noting that a comprehensive,—as opposed to selective—analysis of land-ownership change for all the coal fields is regrettably such a large undertaking, it would require the efforts of a number of investigators over many years. Therefore it has not been possible to pursue it as part of this study.

Major published sources include the Anthracite Mine Inspectors' Reports from 1870 onwards, the Pennsylvania Bureau of Industrial Statistics Reports and newspapers of the period. The Mine Inspectors' Reports contain detailed qualitative and quantitative information on all mines, in operation and under construction, together with some data on abandoned mines. The Geological Survey maps also give the locations of all existing and abandoned workings as of the 1880s. It has proved possible to use these, together with earlier printed and manuscript maps to match mine locations in the GIS to their associated data from the Mine Inspectors' Reports and other sources. The resulting GIS dataset is essential for the analysis of spatio-temporal cycles in mining investment and it effectively covers the totality of anthracite mines in the Northern Field during the period, not a sample of firms.

Several regional newspapers are extremely important both for their weekly reports on the coal trade and the more detailed tables of coal shipments, prices, and freight rates they contain for extended periods of time. Prominent among these are the Pottsville Miner's Journal and the Scranton Republican, which have, for example, enabled a reconstruction to be made of the coal prices achieved at the monthly Scranton Coal Auctions between 1864 and 1879. In a similar way the New York Times has been used to build up monthly time series of Wall Street stock prices of the major railroads. Finally, Saward's Coal Trade Annual, a publication that began in the 1870s, has proved of great value in charting the changing markets for anthracite across the distribution region up to 1902.

Bibliography

ABBREVIATIONS FOR LIBRARY LOCATIONS

EMHL Hagley Library, Greenville, Delaware

GARL Special Collections Research Center, Syracuse University Library

HSP Historical Society of Pennsylvania, Philadelphia

LHS Lackawanna Historical Society, Scranton

LCHS Luzerne County Historical Society, Wilkes-Barre

NCM National Canal Museum, Easton

NYSA New York State Archives, Albany

NYPL New York Public Library

PSA Pennsylvania State Archives, Harrisburg

PSL Pennsylvania State Library, Harrisburg

MANUSCRIPT COLLECTIONS

Archbald Papers, LHS.

Beaver Meadow Railroad and Coal Company Records, in Lehigh Valley Railroad Records, PSA (see below).

A.C. Bining Collection, HSP.

Coxe Family Mining Papers, 1767–1968, HSP.

Delaware, Lackawanna and Western Railroad, Minute Book of Special Committees (Coal Committee Minutes), in Delaware, Lackawanna and Western Railroad Company Records, EMHL.

Delaware, Lackawanna and Western Railroad, Board of Managers Minute Books, in Delaware, Lackawanna and Western Railroad Company Records, Corporate Records of the Lackawanna Division of the Erie-Lackawanna Railroad Company, GARL. (Hereafter referred to as Lackawanna Railroad Records, Syracuse.)

Delaware, Lackawanna and Western Railroad, Main Account Books (Ledgers, Journals, Cashbooks), Lackawanna Railroad Records, Syracuse.

Delaware, Lackawanna and Western Railroad, Coal Department Construction Accounting Records, Lackawanna Railroad Records, Syracuse.

Delaware, Lackawanna and Western Railroad, Coal Department Records, LHS.

Delaware and Hudson Canal Company, Board of Managers Minute Books, NCM.

Delaware and Hudson Canal Company Records, NYSA.

R. G. Dun and Company Collection, Baker Library, Harvard Business School.

Edward Carey Gardiner Collection, HSP.

Erie, Lackawanna Railway Company Deposit, MG-300, PSA.

Florence Coal Company Records, Erie Lackawanna Railway Company Deposit, Erie Railroad Company, Subsidiary Companies, PSA.

Glen Alden Coal Company Records, Earth Conservancy, Ashley, PA.

Israel Morris Papers, Luke Morris Section, Morris Family Papers, HSP.

Lehigh Coal and Navigation Company Records, MG-311, PSA.

Lehigh Valley Railroad Records, MG-274, PSA.

Lehigh and Wilkes-Barre Coal Company Records, LCHS.

Mine Hill and Schuylkill Haven Railroad Records, Reading Company Archive, EMHL.

New York, Susquehanna and Western Coal Company, Minute Book, Erie, Lackawanna Railway Company Deposit, Erie Railroad Company, Subsidiary Companies, PSA.

Pennsylvania Coal Company Records, MG-282, PSA.

Pennsylvania Coal Company Records, NCM.

Pennsylvania Railroad, Board of Directors Minute Books and Board Files, MG-286, Penn Central Collection, PSA.

Philadelphia and Reading Coal and Iron Company, Letter Book Collection, Smithsonian Institution.

Philadelphia and Reading Railroad Records, The Reading Company Archive, EMHL.

Sharpe Papers, LCHS.

Susquehanna Coal Company Records, Pennsylvania Railroad, Subsidiary Companies, Penn Central Collection, PSA.

Thomas Iron Company, Weekly Blast Furnace Reports 1858–1861, Vols. 800–802, Moses Taylor Collection, NYPL.

Union Iron and Coal Company Records, Sub Series 4.k., Moses Taylor Collection, NYPL.

U.S. Census of Manufactures, Products of Industry, Lackawanna, Luzerne, and Schuylkill Counties, Pennsylvania, 1850–1880, Microfilm of Manuscript Census Schedules, PSL.

Washington Coal Company, Board of Directors Minute Book 1847–1849, Erie Lackawanna Railway Company Deposit, Erie Railroad Company, Subsidiary Companies, PSA.

Wyoming Valley Coal Company Papers 1878–1917, Lehigh Valley Railroad Records, Subsidiary Companies, PSA.

ANNUAL REPORTS

Printed and/or manuscript Annual Reports of the following companies and organizations:

American Iron and Steel Association (AISA)

Central Railroad of New Jersey

Delaware and Hudson Canal Company

Delaware, Lackawanna and Western Railroad

Erie Railway

Lehigh Valley Railroad

Lykens Valley Coal Company

Mine Hill and Schuylkill Haven Railroad

Mineral Railroad and Mining Company

Northern Central Railway

Pennsylvania Coal Company

Pennsylvania Railroad

Philadelphia and Reading Railroad

Philadelphia Board of Trade

Summit Branch Railroad

Susquehanna Coal Company

Susquehanna and Wyoming Valley Railroad and Coal Company

Wyoming Valley Coal Company

Also the following series of Annual Reports issued by the State of Pennsylvania:

Annual Reports of the Auditor General of the State of Pennsylvania and of the Tabulations and Deductions from the Reports of the Railroad, Canal and Telegraph Companies.

Annual Reports of the Secretary of Internal Affairs of the State of Pennsylvania, Part III, Industrial Statistics.

MAPS

Atlas of the Anthracite Coal Fields of Pennsylvania. Harrisburg: Board of Commissioners of the Second Geological Survey of Pennsylvania, 1882–1891.

Daddow, S.H. *Map of the Anthracite Coalfields of Pennsylvania.* Pottsville: B. Bannan, 1866.

General Map of the Anthracite Coal Fields Showing the Position of Each Colliery, in *Atlas to Accompany Part III of the Annual Report of the Geological Survey of Pennsylvania for 1886.* Harrisburg: Board of Commissioners of the Second Geological Survey of Pennsylvania, 1887.

Jones, G. "Map of the Wyoming and Lackawanna Valleys, Principally from a Map Constructed by Col. John L. Butler of Wilkes-Barre." *American Journal of Science* 18 (1830): facing p. 309.

Map of the Lehigh and Susquehanna Railroad. Philadelphia: Published by the Company, 1867.

Map of the Philadelphia and Reading Railroad and Branches. Philadelphia: Published by the Company, 1871.

Maps Showing Location of Coal Mines in the Anthracite Inspection Districts of Pennsylvania 1898, maps accompanying *AIM Report for 1898*.

Map of the Wyoming Coal Company's Lands in Luzerne County, n.p., n.d. at LCHS, Wilkes-Barre.

R.P. Rothwell. *Map of Wyoming Anthracite Field (Surveys to September 1870*). Earth Conservancy, Ashley, PA.

Schaefer, P.W. *A Map of all the Anthracite Collieries and the Pottsville, Lehigh, Mahanoy and Shamokin Coal Basins.* Pottsville, 1859.

Strauch, G. B., and Cochran, A. B. *Map of the First and Second Anthracite Coalfields of Pennsylvania, 1871*. Pottsville, 1871.

Strauch, G.B. and Cochran, A.B. *Maps of the Anthracite Coalfields of Pennsylvania for 1878, 1879, 1880.* Pottsville, 1878–1880.

NEWSPAPERS/TRADE JOURNALS

Carbondale Advance

Coal and Iron Record

Coal Trade Journal

Engineering and Mining Journal

Iron Age

Lackawanna Herald

Luzerne Union

New York Times

Pittston Gazette

Pottsville Miner's Journal

Scranton Republican

Wilkes-Barre Record of the Times

Wilkes-Barre Telephone

JOURNAL ARTICLES AND PAPERS

Adams, S.P. "Different Charters, Different Paths: Corporations and Coal in Antebellum Pennsylvania and Virginia." *Business and Economic History* 27, 1 (1998): 78–90.

Bateman, F. and T. Weiss. "Comparative Regional Development in Antebellum Manufacturing." *Journal of Economic History* 35 (1975): 182–208.

Berthoff, R.T. "The Social Order of the Anthracite Regions." *Pennsylvania Magazine of History and Biography* 89 (1965): 261–291.

Binder, F.M. "Pennsylvania Coal and the Beginnings of American Steam Navigation." *Pennsylvania Magazine of History and Biography* 83, 4 (1959): 420–445.

Borts, G.H. "Regional Cycles of Manufacturing Employment in the United States, 1914–1953." *Journal of the American Statistical Association* 55 (March 1960): 151–211.

Calomiris, C.W. and L. Schweikart. "The Panic of 1857: Origins, Transmission, and Containment." *Journal of Economic History* 51 (1991): 807–834.

Chance, H.M. "Valuation of Coal Land." *Trans. AIME* 47 (1914): 111–146.

Commonwealth of Pennsylvania Legislature. "Testimony Before the Committee to Investigate the Philadelphia and Reading Railroad Company and the Philadelphia and Reading Coal and Iron Company." *Legislative Documents*, 4 (1876): 495–1133.

Davis, L.E. "The Investment Market, 1870–1914: The Evolution of a National Market." *Journal of Economic History* 25 (1965): 355–399.

Dow, S. "Mainstream Economic Methodology." *Cambridge Journal of Economics* 21, 1 (1997): 73–93.

Forrester, J.W. "Changing Economic Patterns." *Technology Review* 80, 8 (1978): 30–9.

Forrester, J.W. "Innovation and Economic Change." *Futures* 13 (1981): 323–331.

Goldstein, J.S. "Kondratieff Waves as War Cycles." *International Studies Quarterly* 29 (1985): 411–444.

Gregson, N. "Structuration Theory: Some Thoughts on the Possibilities for Empirical Research." *Environment and Planning D: Society and Space* 5 (1987): 73–91.

Healey, R.G. "Sharpe, Weiss and Company and the Lehigh Coal Trade, 1867–74." *Proceedings of the Wyoming Historical and Geological Society* 24 (1984): 24–43.

Kitchin, J. "Cycles and Trends in Economic Factors." *Review of Economics and Statistics* 5 (1923): 10–16.

Knies, M. "The D&H Coal Company: New Insights from the James Archbald Papers." *Canal History and Technology Proceedings* 20 (2001): 53–80.

Knowles, A.K., R.G. Healey, T.R. Stamp, D. Kidd, and M. Neirick. *Mapping the AnteBellum Iron Industry.* Social Science and History Conference, November 1999, Fort Worth, Texas.

Kondratieff, N.D. "The Long Waves in Economic Life." *Review of Economics and Statistics* 17 (1935): 105–115.

Kuznets, S.S. "Long Swings in the Growth of Population and in Related Economic Variables." *Proceedings of the American Philosophical Society* 102 (1958): 25–52.

Lewis, W.D. "The Early History of the Lackawanna Iron and Coal Company: A Study in Technological Adaptation." *Pennsylvania Magazine of History and Biography* 96, 4 (1972): 424–468.

MacDonough, O. "The Irish Famine Emigration to the United States." *Perspectives in American History* 10 (1976): 357–446.

Martin, R.L. "Critical Survey. The New 'Geographical Turn' in Economics: Some Critical Reflections." *Cambridge Journal of Economics* 23 (1999): 65–91.

Meyer, D.R. "A Dynamic Model of the Integration of Frontier Urban Places into the United States System of Cities." *Economic Geography* 56 (1980): 120–140.

Meyer, D.R. "Emergence of the American Manufacturing Belt: An Interpretation." *Journal of Historical Geography* 9 (1983): 145–174.

Meyer, D.R. "Midwestern Industrialization and the American Manufacturing Belt in the Nineteenth Century." *Journal of Economic History* 49 (1989): 921–937.

Muller, E.K. "Regional Urbanization and the Selective Growth of Towns in North American Regions." *Journal of Historical Geography* 3 (1977): 21–39.

Neff, P. "Interregional Cyclical Differentials: Causes, Measurement, and Significance." *American Economic Review* 39 (1949): 105–119.

Patton, S. "Delaware and Hudson Company vs. Pennsylvania Coal Company During the 1850s." *Canal History and Technology Proceedings* 9 (1992): 6–24.

Peresie, J.L. "Crusader with a Camera: Lewis Hine and his Battle Against Child 'Slavery'." *Pennsylvania Heritage* 23, 3 (1997): 4–13.

Rezneck, S. "Distress, Relief, and Discontent in the United States during the Depression of 1873–1878." *Journal of Political Economy* 58 (1950): 494–512.

Roy, W.G. "Interlocking Directorates and the Corporate Revolution." *Social Science History* 7, 2 (1983): 143–164.

Swanson, W.W. "The Crisis of 1860 and the First Issue of Clearing-House Certificates: II." *Journal of Political Economy* 16 (1908): 212–226.

Vining, R. "Location of Industry and Regional Patterns of Business Cycle Behavior." *Econometrica* 14 (1946): 37–68.

Virtue, G.O. " The Anthracite Combinations." *Quarterly Journal of Economics* 10 (April 1896): 296–323.

Warf, B. "Regional Transformation, Everyday Life, and Pacific Northwest Lumber Production." *Annals, Association of American Geographers* 78, 2 (1988): 326–346.

BOOKS AND PAMPHLETS

Abramowitz, M. "Long Swings in American Economic Growth." In *New Views on American Economic Development: A Selective Anthology of Recent Work*. Edited by R.L. Andreano. Cambridge, Massachusetts: Schenkman Publishing Company, 1965.

Allen, J., D. Massey, and A. Cochrane, with J. Charlesworth, G. Court, N. Henry, and P. Sarre. *Rethinking the Region.* London: Routledge, 1998.

An Act to Incorporate the Pennsylvania Coal Company, By an Act Incorporating the Washington Coal Company, passed April 11, 1838 *with the Several Supplements Thereto.* New York: Bourne and Company, 1869.

Adams, S.P. *Old Dominion, Industrial Commonwealth: Coal, Politics and Economy in Antebellum America.* Baltimore: Johns Hopkins Press, 2004.

Adler, D.R. *British Investment in American Railways 1834–1898.* Charlottesville: University of Virginia Press, 1970.

American Iron and Steel Association (AISA). *Statistics of the American and Foreign Iron Trades for 1882.* Philadelphia: AISA, 1883.

Albion, R.G. *The Rise of New York Port 1815–1860.* New York: Charles Scribner's Sons, 1939.

Andreano, R.L., ed. *The Economic Impact of the Civil War.* Cambridge, Massachusetts: Schenkman Publishing Company, 1962.

Andreano, R.L., ed. *New Views on American Economic Development.* Cambridge, Massachusetts: Schenkman Publishing Company, 1965.

Andreas, A.T. *History of Chicago from the Earliest Period to the Present Time, 3 Vols.* Chicago: A.T. Andreas Company, 1884–1886.

Anthracite Coal Trade for 1867. Pottsville: Miner's Journal, 1868.

Archer, R.F. *The History of the Lehigh Valley Railroad: The Route of the Black Diamond.* Berkeley: Howell-North Books, 1977.

Ashburner, C.A. *First Report of Progress in the Anthracite Coal Region: The Geology of the Panther Creek Basin or Eastern End of the Southern Coalfield.* Harrisburg: Board of Commissioners of the Second Geological Survey, 1883.

Aurand, H.W. *From the Molly Maguires to the United Mine Workers: The Social Ecology of an Industrial Union, 1869–1897.* Philadelphia: Temple University Press, 1971.

Baer, C.T., *Canals and Railroads of the Mid-Atlantic States, 1800–1860.* Greenville, DE: Eleutherian Mills-Hagley Foundation, 1981.

Bartholomew, C.L., and L.E. Metz. *The Anthracite Iron Industry of the Lehigh Valley.* Easton: Center for Canal History and Technology, 1988.

Bill in Equity, The Central Railroad Co. of New Jersey vs. The Lehigh and Wilkes-Barre Coal Co., May Term 1877, No. 62, Special Report of Joseph A. Clay Master, Circuit Court of the United States, Western District of Pennsylvania.

Binder, F.M. *Coal Age Empire: Pennsylvania Coal and its Utilization to 1860.* Harrisburg, PA: Pennsylvania Historical and Museum Commission, 1974.

Bishop, J.L. *A History of American Manufactures from 1608 to 1860.* Philadelphia: E. Young, 1868.

Blatz, P.K. *Democratic Miners: Work and Labor Relations in the Anthracite Coal Industry, 1875–1925.* Albany: State University of New York Press, 1994.

Bogen, J. *The Anthracite Railroads.* New York: Ronald Press, 1927.

Bradsby, H.C., ed. *History of Luzerne County*. Chicago: S.B. Nelson, 1893.

Brewster, W. *History of Certified Township of Kingston, PA. 1769–1929*. Kingston, PA: Kingston Borough School District, 1929.

Broehl, W.C., Jr. The *Molly Maguires.* New York: Vintage/Chelsea House, 1964.

Bureau of the Census. *A Century of Population Growth in The United States 1790–1900.* Washington, DC: Bureau of the Census, 1909.

Burns, A.F., and W.C. Mitchell. *Measuring Business Cycles.* New York: National Bureau of Economic Research (NBER), 1946.

Campbell, S.W., *Businessmen and Anthracite: Aspects of Change in the Late Nineteenth Century Anthracite Industry.* Ph.D. Diss. University of Delaware, 1978.

Casey, R.J. and W.A.S Douglas. *The Lackawanna Story.* New York: McGraw-Hill, 1951.

Chance, H.M. *Report on the Mining Methods and Appliances Used in the Anthracite Coal Fields.* Harrisburg: Board of Commissioners of the Geological Survey of Pennsylvania, 1883.

Chandler, A.D. *The Visible Hand: the Managerial Revolution in American Business.* Cambridge, MA: Harvard University Press, 1977.

Charter and By-Laws of the Hartford Coal Company. Philadelphia: Oliver P. Glessner, Book and Job Printer, 1855.

Charter and By-Laws of the Honey Brook Coal Company, incorporated by Special Act of the Legislation of the State of Pennsylvania April 23, 1864. Philadelphia: Bryson and Sons, Printers, 1864.

Charter and By-Laws of the Maryland Anthracite Coal Company of the Wyoming Coal Region with Reports on the Geology and Mining Resources of their Coal Lands, Together with Supplementary Remarks and Extracts. Baltimore: John W. Woods, 1856.

Charter and By-Laws of the Wyoming Valley Coal Company. Wilkes-Barre: Record Book and Job Printing Dept., 1883.

Ciesla, C.C., *1694–1994 History of Hanover Township: Including the Boroughs of Ashley, Sugar Notch, and Warrior Run and Portions of Nanticoke City.* Slocum, PA: Slocum Township, 1994.

Clark, G.L., M.S. Gertler, and J.E.M. Whiteman. *Regional Dynamics: Studies in Adjustment Theory.* London: Allen and Unwin, 1986.

Coal Statistical Register, 1870. Pottsville: Miner's Journal, 1871.

Cochran, T.S. "Did the Civil War Retard Industrialization?" In *The Economic Impact of the Civil War.* Edited by R.L. Andreano. Cambridge, MA: Schenkman Publishing Company, 1962.

Committee on the Judiciary, General, *Report of the Committee on the Judiciary, General, of the State of Pennsylvania in Relation to the Anthracite Coal Difficulties, with the Accompanying Testimony.* Harrisburg: B. Singerly, State Printer, 1871.

Connolly, M.T. *'The Gravity': History of the Pennsylvania Coal Company Railroad, 1850–1885.* Scranton: Langan, 1972.

Conzen, M.P. "The American Urban System in the Nineteenth Century." In *Geography and the Urban Environment: Progress in Research and Applications.* Edited by D.T. Herbert and R.J. Johnston. New York: John Wiley, 1981.

Conzen, M.P. "The Progress of American Urbanism, 1860–1930." in *North America: The Historical Geography of a Changing Continent.* Edited by R.D. Mitchell and P.A. Groves. London: Hutchinson, 1987.

Cornell, R.J. T*he Anthracite Coal Strike of 1902.* Washington: Catholic University of America Press, 1957.

Corporate History of the Delaware and Hudson Company and Subsidiary Companies. n.p., 1906–1907, 3 vols.

Corporate History of the Northern Coal and Iron Company, The Hudson Coal Company, etc. n.p., 1907.

Cronon, W. *Nature's Metropolis: Chicago and the Great West.* New York: W.W. Norton, 1991.

Daddow, S.H., and B. Bannan. *Coal, Iron and Oil.* Pottsville, PA: B. Bannan, 1866.

Davies, E.J. *The Anthracite Aristocracy: Leadership and Social Change in the Hard Coal Regions of Northeastern Pennsylvania 1800–1930.* DeKalb, IL: Northern Illinois University Press, 1985.

Davis, L.E., and R.J. Cull. *International Capital Markets and American Economic Growth 1820–1914.* Cambridge: Cambridge University Press, 1994.

Delaware and Hudson Company. *A Century of Progress, History of the Delaware and Hudson Company, 1823–1923.* Albany: J.B. Lyon Company, Printers, 1925.

DiCiccio, C.P. *Coal and Coke in Pennsylvania.* Harrisburg: Pennsylvania Historical and Museum Commission, 1996.

Dillingham. *Reports of the Immigration Commission, Statistical Review of Immigration 1820–1910; Distribution of Immigrants 1850–1900, 61st Cong., 3rd Sess., Senate Doc. 756.* Washington: Government Printing Office, 1911.

Dillingham. *Reports of the Immigration Commission, Parts 17–20, (Part 19, Anthracite Coal Mining), 61st Cong. 2nd Sess., Senate Doc. 78.* Washington: Government Printing Office, 1911.

Dilts, J.D. *The Great Road: the Building of the Baltimore and Ohio, the Nation's First Railroad, 1828–1853.* Stanford: Stanford University Press, 1993.

Dunlap, T. *Statistical Report of the National Association of Iron Manufacturers for 1872.* Philadelphia: J.A. Wagenseller, 1873.

Easterlin, R. *Population, Labor Force, and Long Swings in Economic Growth: The American Experience.* New York: Columbia University Press for NBER, 1968.

Eavenson, H.M. *The First Century and a Quarter of the American Coal Industry.* Pittsburgh: Privately Printed, 1942.

Engineering and Mining Journal, *Editorial Review of the Delaware and Hudson Canal Company's Report,* articles of September 15, 22, 29, and October 6, 1877.

Fels, R. *American Business Cycles, 1865–1897.* Chapel Hill: University of North Carolina Press, 1959.

Ferguson, W. *America by River and Rail.* London: James Nisbet and Company, 1856.

Fishlow, A. *American Railroads and the Transformation of the Ante-Bellum Economy.* Cambridge, MA: Harvard University Press, 1965.

Fite, E.D. *Social and Industrial Conditions in the North during the Civil War*. New York: The Macmillan Company, 1910.

Fogel, R.W., and S.L. Engermen, eds. *The Re-Interpretation of American Economic History*. New York: Harper and Row, 1971.

Folsom, B.W. *Urban Capitalists: Entrepreneurs and City Growth in Pennsylvania's Lackawanna and Lehigh Regions, 1800–1920.* Baltimore: Johns Hopkins University Press, 1981.

Fredericks, H.S. *The Wilkes-Barre and Eastern Railroad*. Livingston, NJ: Railroadians of America, 1986.

Freedley, E. *Philadelphia and Its Manufactures: A Handbook Exhibiting the Development, Variety and Statistics of the Manufacturing Industry of Philadelphia in 1857.* Philadelphia: Edward Young, 1859.

Gibbons, J.S. *The Banks of New York: Their Dealers, the Clearing House, and the Panic of 1857*. New York: Appleton, 1858.

Giddens, A. *The Constitution of Society: Outline of the Theory of Structuration*. Cambridge: Polity Press, 1984.

Gilchrist, D.T., and W.D. Lewis, eds. *Economic Change in the Civil War Era.* Greenville, DE: Eleutherian Mills-Hagley Foundation, 1965.

Glasner, D., ed. *Business Cycles and Depressions: An Encyclopaedia*. New York: Garland, 1997.

Goodrich, C., ed. *Canals and American Economic Development*. New York: Columbia University Press, 1961.

Gowen, F.B. *Statement of the Present Condition of the Philadelphia and Reading Railroad and the Philadelphia and Reading Coal and Iron Company with a Plan for Their Financial Reorganization*. Philadelphia: Jackson Bros., 1880.

Greene, V.R. *The Slavic Community on Strike: Immigrant Labor in Pennsylvania Anthracite.* Notre Dame: University of Notre Dame Press, 1968.

Gregson, N. "On the (Ir)relevance of Structuration Theory to Empirical Research." In *Social Theory of Modern Societies: Anthony Giddens and his Critics*. Edited by D. Held and J.B. Thompson. Cambridge: Cambridge University Press, 1989.

Gudelunas, W.A., and W.G. Shade. *Before the Mollie Maguires: The Emergence of the Ethno-Religious Factor in the Politics of the Lower Anthracite Region, 1844–1872.* New York: Arno Press, 1976.

Gunnarsson, R.L. *The Story of the Northern Central Railway: From Baltimore to Lake Ontario*. Sykesville, MD: Greenberg Publishing Company, 1991.

Hare, J.V. *History of the Reading*. Philadelphia: John H. Stock, 1966.

Harris, J.S. *Report of Joseph S. Harris on the Coal Lands of the Philadelphia and Reading Coal and Iron Company*. Philadelphia: Jackson Bros., 1880.

Harvey, O.J. *A History of Wilkes-Barre, Luzerne County from Its First Beginnings to the Present Time. (6 vols).* Wilkes-Barre, PA: Reader Press, 1909–1930.

Hausman, D. M. "Philosophy of Economics." In *The Stanford Encyclopedia of Philosophy (Winter 2003 Edition).* Edited by E. N. Zalta , URL = http://plato.stanford.edu/archives/win2003/entries/economics/

Healey, R.G. *Object-Oriented Simulation of Evolving Regional Economic Systems: A Case Study of Mining Development in the Anthracite Coalfields of Pennsylvania During the Post-Bellum Period.* Ph.D. Diss. University of Cambridge, 1993.

Healey, R.G. "Disturbance of the Peace: the Operators' View of the 1902 Anthracite Coal Strike." In *Perspectives on the Great Strike.* Edited by R. Janosov. Easton: Canal History and Technology Press, 2002.

Helmer, W.F. *O & W: The Long Life and Slow Death of the New York, Ontario and Western Railway.* Berkeley: Howell-North Books, 1959.

Historical Statistics of the United States: Colonial Times to 1970. Washington: Bureau of the Census, 1976.

History of Luzerne, Lackawanna, and Wyoming Counties. New York: Munsell and Co., 1880.

Hitchcock, F.L. *History of Scranton and its People.* New York: Lewis Historical Publishing Co., 1914.

Hodas, D. *The Business Career of Moses Taylor: Merchant, Finance Capitalist and Industrialist.* New York: New York University Press, 1976.

Hoffman, C. *The Depression of the Nineties: An Economic History.* Westport, CT: Greenwood Press, 1970.

Hoffman, J.N. *Girard Estate Coal Lands in Pennsylvania 1801–1884,* Smithsonian Studies in History and Technology, No. 15. Washington: Smithsonian Institution Press, 1972.

Hollis, M., and E.J. Nell. *Rational Economic Man: a Philosophical Critique of Neo-Classical Economics.* London: Cambridge University Press, 1975.

Holton, J.L. *The Reading Railroad: History of a Coal Age Empire, 2 vols.* Laury's Station, PA: Garrigues House, 1989–1992.

Hudson Coal Company. *The Story of Anthracite.* New York: Hudson Coal Company, 1932.

Hull, F.A. *Annual Report of the Geological Survey of Pennsylvania, Part III, Report on the Anthracite Region.* Harrisburg: Edwin K. Meyers, State Printer, 1886.

Huston, J.L. *The Panic of 1857 and the Coming of the Civil War.* Baton Rouge: Louisiana State University Press, 1987.

Inspectors of Mines, *Reports of the Inspectors of Mines of the Anthracite Coal Regions, 1870–.* Harrisburg: State Printer, 1870–.

Interstate Commerce Commission. *No. 150: Coxe Brothers and Co. vs. The Lehigh Valley Railroad Co., Argument of Mr. Franklin B. Gowen for Complainants, March 19 and 20, 1889.* Philadelphia: Allen Lane and Scott, 1889.

Isard, W. *Methods of Regional Analysis: An Introduction to Regional Science.* Cambridge, MA: MIT Press, 1960.

Janosov, R., ed. *Perspectives on the Great Strike.* Easton: Canal History and Technology Press, 2002.

Jerome, H. *Migration and Business Cycles.* New York: NBER, 1926.

Jones, C.L. *Economic History of the Anthracite Tide-Water Canals.* Philadelphia: J.C. Winston, 1908.

Jones, E. *The Anthracite Coal Combination in the United States.* Cambridge, MA: Harvard University Press, 1914.

Jones, W.D. *Wales in America: Scranton and the Welsh 1860–1920.* Cardiff: University of Wales Press, 1993.

Juglar, C. *A Brief History of Panics and Their Periodical Occurrence in the United States.* New York: Augustus M. Kelley, 1966. (Reprint of 1916 3rd English Edition of French Original Text published in Paris in 1862).

Knies, M. *Coal on the Lehigh 1790–1827: Beginnings and Growth of the Anthracite Industry in Carbon County, Pennsylvania.* Easton: Canal History and Technology Press, 2001.

Knowles, A.K. *Calvinists Incorporated: Welsh Immigrants on Ohio's Industrial Frontier.* Chicago: The University of Chicago Press, 1997.

Kindleberger, C.P. *Manias, Panics, and Crashes: A History of Financial Crises.* London: Macmillan, 1978.

Krause, J., and E. Crist. *Susquehanna: The New York, Susquehanna and Western Railroad.* Newton, New Jersey: Carstens Publications, Inc., 1980.

Kuznets, S.S. *Secular Movements in Production and Prices: Their Nature and Their Bearing upon Cyclical Fluctuations.* Cambridge, MA: Riverside Press, 1930.

Lesley, J.P., ed. *The Iron Manufacturer's Guide to the Furnaces, Forges, and Rolling Mills of the United States.* New York: J. Wiley, 1859.

Lesley, J.P., E.V. d'Invilliers, and A.DW. Smith. *A Summary Description of the Geology of Pennsylvania, Vol. 3, Part I.* Harrisburg: Board of Commissioners of the Geological Survey of Pennsylvania, 1895.

Lindstrom, D. *Economic Development in the Philadelphia Region, 1810–1850.* New York: Columbia University Press, 1978.

Livingood, J.L. *The Philadelphia-Baltimore Trade Rivalry, 1780–1860.* Harrisburg: Pennsylvania Historical and Museum Commission, 1947.

Lowenthal, L. *From the Coal Fields to the Hudson: A History of the Delaware and Hudson Canal.* Fleischmanns: Purple Mountain Press, 1997.

Lucas, W.A. *The History of the New York, Susquehanna and Western Railroad.* New York: Railroadians of America, 1939.

MacAvoy, P.W. *The Economic Effects of Regulation: The Trunk-Line Railroad Cartels and the Interstate Commerce Commission before 1900.* Cambridge, MA: MIT Press, 1965.

Macfarlane, J. *The Coal Regions of America: Their Topography, Geology and Development.* New York: D. Appleton and Company, 1873.

Marshall, A. *Principles of Economics.* London: Macmillan, 1920.

Marshall, M. *Long Waves of Regional Development.* Basingstoke: Macmillan, 1987.

Martin E.W., [James Dabney McCabe]. *The History of the Great Riots: Being a Full and Authentic Account of the Strikes and Riots on the Various Railroads of the United States and in the Mining Regions … Together with a Full History of the Mollie Maguires.* Philadelphia: J.C. McCurdy and Co., 1877.

Mass, N., *Economic Cycles: An Analysis of Underlying Causes.* Cambridge, MA: Wright-Allen Press, 1975.

Miesse, C. *Points on Coal and the Coal Business.* Myerstown: Feese and Uhrich, 1887.

Miller, D.L. and R.E. Sharpless. *The Kingdom of Coal: Work, Enterprise and Ethnic Communities in the Mine Fields.* Philadelphia: University of Pennsylvania Press, 1985.

Mohowski, R.E. *The New York, Susquehanna and Western Railroad.* Baltimore: Johns Hopkins University Press, 2003.

Murphy, T.F. *Jubilee History, Commemorative of the Fiftieth Anniversary of the Creation of Lackawanna County, Pennsylvania.* (2 Vols.) Topeka-Indianapolis: Historical Publishing Company, 1928.

Niemi, A.W. *State and Regional Patterns in American Manufacturing, 1860–1900.* Westport, CT: Greenwood Press, 1974.

North, D.C. *The Economic Growth of the United States, 1790–1860.* Englewood Cliffs, NJ: Prentice-Hall, 1961.

O'Brien, P., *The Economic Effects of the American Civil War.* Basingstoke: Macmillan, 1988.

Official Statistics—The Coal Trade for 1868. Pottsville: Miner's Journal, 1869.

Palladino, G. *Another Civil War: Labor, Capital, and the State in the Anthracite Regions of Pennsylvania, 1840–1868.* Urbana: University of Illinois Press, 1990.

Pardee Foulke, C. and W.G. Foulke. *Calvin Pardee 1841–1923: His Family and his Enterprises.* Philadelphia: Pardee Company, 1979.

Parker, W.J. Lewis. *The Great Coal Schooners of New England.* Mystic, CT: Marine Historical Association, 1948.

Pearce, S. *Annals of Luzerne County: A Record of Interesting Events, Traditions, and Anecdotes from the First Settlement at Wyoming to 1860.* Philadelphia: J.B. Lippincott and Company, 1860.

Pearse, J.B. *A Concise History of the Iron Manufacture of the American Colonies up to the Revolution and of Pennsylvania until the Present Time.* (Reprint of 1876 edition). New York: Burt Franklin, 1970.

Perry Smith, H., ed. *History of Buffalo and Erie County.* (2 Vols.) Syracuse: D. Mason and Company, 1884.

Pred, A.R. *The Spatial Dynamics of U.S. Urban-Industrial Growth, 1800–1914: Interpretive and Theoretical Essays.* Cambridge, MA: MIT Press, 1966.

Petrillo, F.C. *Anthracite and Slackwater: the North Branch Canal 1828–1901.* Easton: Center for Canal History and Technology, 1986.

Philadelphia and Reading Railroad Company. *Report of the Engineers of the Philadelphia and Reading Railroad Company, with Accompanying Documents.* Philadelphia: John C. Clark, 1839.

Phillips' History of Wilkes-Barre and Luzerne County. Wilkes-Barre: Unpublished Typescript, Luzerne County Historical Society, n.d.

Pierce, H.H. *Railroads of New York: a Study of Government Aid, 1826–1875.* Cambridge, MA: Harvard University Press, 1953.

Plan and Agreement for the Reorganization of the Philadelphia and Reading Railroad Company and the Philadelphia and Reading Coal and Iron Company. December 14, 1895 (n.p.).

Platt, F. *Special Report to the Legislature upon the Causes, Kinds and Amount of Waste in Mining Anthracite.* Harrisburg: Board of Commissioners of the 2nd Geological Survey, 1881.

Powell, H.B. *Philadelphia's First Fuel Crisis: Jacob Cist and the Developing Market for Pennsylvania Anthracite.* State College: Pennsylvania State University Press, 1978.

Price, W.W. *We Have Recovered Before! A Comparison of the Present Depression with the Major Depressions of the Past Century: 1837, 1857, 1873, and 1893.* New York: Harper and Brothers, 1933.

Pumpelly, R. *Report on the Mining Industries of the United States (Exclusive of the Precious Metals): with Special Investigations into the Iron Resources of the Republic and into the Cretaceous Coal of the Northwest.* Washington: Government Printing Office, 1886.

Reed, R.C. *The New York Elevated.* New York: A.S. Barnes and Co., 1978.

Report of the Commission Appointed to Investigate the Waste of Coal Mining, with a View to Utilizing the Waste. Philadelphia: Allen, Lane and Scott, 1893.

Report of the Inspector of Mines of Schuylkill County for 1869. Harrisburg: Singerly, 1870.

Report of the Investigating Committee of the Pennsylvania Railroad Company. Philadelphia: Allen, Lane and Scott, 1874.

Report to the Stockholders of the Delaware and Hudson Canal Company, by the Committee Appointed at the Annual Meeting Held on May 8th, 1877. New York: P. Barnes, Steam Book and Job Printer, 1877.

Roberts, P. *The Anthracite Coal Industry.* New York: Macmillan, 1901.

Roberts, P. *Anthracite Coal Communities.* New York: Macmillan, 1904.

Rogers, H.D. *The Geology of Pennsylvania.* (2 vols.) Philadelphia: J.B. Lippincott, 1858.

Ruth, P. *Of Pulleys and Ropes and Gear: The Gravity Railroads of the Delaware and Hudson Canal Company and the Pennsylvania Coal Company.* Honesdale, PA: Wayne County Historical Society, 1997.

Salay, D.L., ed. *Hard Coal, Hard Times: Ethnicity and Labor in the Anthracite Region.* Scranton, PA: Anthracite Museum Press, 1984.

Salsbury, S. "The Effect of the Civil War on American Industrial Development." In *The Economic Impact of the Civil War.* Edited by R.L. Andreano. Cambridge, MA: Schenkman Publishing Company, 1962.

Saward, F.E. *The Coal Trade.* New York: F.E. Saward, 1875–.

Schaefer, D. F. *A Quantitative Description and Analysis of the Growth of the Pennsylvania Anthracite Coal Industry, 1820–1865.* Ph.D. Diss. University of North Carolina, 1967.

Schlegel, M.W. *Ruler of the Reading: The Life of Franklin B. Gowen 1836–1889.* Harrisburg: Archives Publishing Company, 1947.

Schotter, H.W. *The Growth and Development of the Pennsylvania Railroad Company.* Philadelphia: Allen, Lane and Scott, 1927.

Shank, W.H. *The Amazing Pennsylvania Canals.* York, PA: American Canal and Transportation Center, 1960.

Sharkey, R.P. *Money, Class, and Party: An Economic Study of Civil War and Reconstruction.* Baltimore: Johns Hopkins Press, 1959.

Shaughnessy, J. *Delaware and Hudson.* Berkeley, CA: Howell-North Books, 1967.

Solomou, S. *Phases of Economic Growth 1850–1973.* Cambridge, England: Cambridge University Press, 1987.

Statement of Board of Directors to the Stockholders of the Central Railroad of New Jersey, September 9, 1876. New York: G.F. Nesbitt, 1876.

Statistics of the American and Foreign Iron Trades for 1882. Philadelphia: American Iron and Steel Association, 1883.

Statistics of the American and Foreign Iron Trades for 1903. Philadelphia: American Iron and Steel Association, 1904.

Stevenson, G.E. *Reflections of an Anthracite Engineer.* Scranton: Published by the author, 1931.

Swank, J.M. *History of the Manufacture of Iron in All Ages and Particularly in the United States from Colonial Times to 1891.* Philadelphia: American Iron and Steel Association, 1892.

Taber, T.T., III. *The Delaware, Lackawanna and Western Railroad in the Nineteenth Century 1828–1899.* Muncy, PA: Published by the author, 1977.

Taylor, G.R. *The Transportation Revolution, 1815–1860.* (Vol. 4 of *The Economic History of the United States*). New York: Rinehart, 1951.

Taylor, G.R. and I.D. Neu. *The American Railroad Network 1861–1890.* Cambridge, MA: Harvard University Press, 1956.

Taylor, R.C. *Statistics of Coal.* Philadelphia: J.W. Moore, 1855.

Temin, P. *Iron and Steel in Nineteenth-Century North America: An Economic Inquiry.* Cambridge, MA: MIT Press, 1964.

The Ironworks of the United States: A Directory of the Furnaces, Rolling Mills, Steel Works, Forges and Bloomaries in Every State. Philadelphia: American Iron and Steel Association, 1876.

Thomas, B. *Migration and Economic Growth.* Cambridge: Cambridge University Press, 1973.

Thompson, H.S. *Notes on Waste in Mining and Preparing Coal and on the Percentage of Coal Won at Representative Collieries on the Girard Estate, from the Report to the Directors of City Trusts.* Philadelphia: n.p. October 3, 1892.

Thompson, R.L. *The History of the Telegraph Industry in the United States, 1832–1866.* Princeton: Princeton University Press, 1947.

Throop, B.H. *A Half-Century in Scranton.* Scranton, PA: Scranton Republican Press, 1895.

Tylecote, A. *The Long Wave in the World Economy: The Present Crisis in Historical Perspective.* New York: Routledge, 1991.

U.S. Anthracite Coal Strike Commission. *Report to the President on the Anthracite Coal Strike of May-October, 1902.* Washington: Government Printing Office, 1903.

U.S. House of Representatives. *Labor Troubles in the Anthracite Regions of Pennsylvania 1887–1888*, 50th Cong. 2nd Sess., House Rept. No. 4147, Vol. 4. Washington: Government Printing Office, 1889.

U.S. Industrial Commission. *Report of the Industrial Commission on the Relations and Conditions of Capital and Labor Employed in the Mining Industry*, Vol. 12. Washington: Government Printing Office, 1901.

van Duijn, J.J. *The Long Wave in Economic Life.* London: George Allen and Unwin, 1983.

Van Vleck, G.W. *The Panic of 1857: An Analytical Study.* New York: Columbia University Press, 1943.

Vance, J.E., Jr. *The North American Railroad: Its Origin, Evolution, and Geography.* Baltimore: Johns Hopkins University Press, 1995.

Verbatim Report of the General Meeting of the Share and Bondholders of the Philadelphia and Reading Railroad Company and of the Bondholders of the Perkiomen Railroad Company, Held at the City Terminus Hotel, Cannon Street, London, June 6, 1877. (Thomas Wilde Powell Esq. in the Chair). London: Charles Skipper and East, printers, 1877.

Von Bergen, F.C. *Photographs of the Delaware, Lackawanna and Western Railroad Coal Department Collieries.* Lackawanna Historical Society: Unpublished bound volume of photographs, circa 1908.

Wakefield, M.B. *Coal Boats to Tidewater: the Story of the Delaware and Hudson Canal.* Grahamsville, NY: Wakefair Press, 1965.

Wallace, A.F.C. *St. Clair: A Nineteenth-Century Coal Town's Experience with a Disaster-Prone Industry.* Ithaca: Cornell University Press, 1987.

Walsh, W.D. *The Diffusion of Technological Change in the Pennsylvania Pig Iron Industry, 1850 – 1870.* New York: Arno Press, 1975.

Ward, D. *Cities and Immigrants: A Geography of Change in Nineteenth-Century America.* London: Oxford University Press, 1971.

Warren, J.P. *The 50-Year Boom-Bust Cycle: The Case for Kondratieff's Long Wave Theory.* Godalming: Warren, Cameron and Co., 1982.

Warren, K. *The American Steel Industry, 1850–1970: A Geographical Interpretation.* Oxford: Clarendon Press, 1973.

Warren, K. *Wealth, Waste, and Alienation: Growth and Decline in the Connellsville Coke Industry.* Pittsburgh: University of Pittsburgh Press, 2001.

Yearley, C.K. *Enterprise and Anthracite: Economics and Democracy in Schuylkill County, 1820–1875.* Baltimore: Johns Hopkins Press, 1961.

INDEX